DIE GRUNDLEHREN DER

MATHEMATISCHEN WISSENSCHAFTEN

IN EINZELDARSTELLUNGEN MIT BESONDERER
BERÜCKSICHTIGUNG DER ANWENDUNGSGEBIETE

HERAUSGEGEBEN VON

R. GRAMMEL · F. HIRZEBRUCH · E. HOPF
H. HOPF · W. MAAK · W. MAGNUS · F. K. SCHMIDT
K. STEIN · B. L. VAN DER WAERDEN

BAND 60

THE NUMERICAL TREATMENT OF DIFFERENTIAL EQUATIONS

BY

LOTHAR COLLATZ

THIRD EDITION

Springer-Verlag Berlin Heidelberg GmbH

1960

THE NUMERICAL TREATMENT
OF DIFFERENTIAL EQUATIONS

BY

DR. LOTHAR COLLATZ
O. PROFESSOR IN THE UNIVERSITY OF HAMBURG

THIRD EDITION

TRANSLATED FROM A SUPPLEMENTED VERSION OF THE
SECOND GERMAN EDITION
BY P. G. WILLIAMS, B. SC.
MATHEMATICS DIVISION, NATIONAL PHYSICAL LABORATORY,
TEDDINGTON, ENGLAND

WITH 118 DIAGRAMS

AND 1 PORTRAIT

Springer-Verlag Berlin Heidelberg GmbH

1960

ISBN 978-3-662-22986-6   ISBN 978-3-662-24934-5 (eBook)
DOI 10.1007/978-3-662-24934-5

By Springer-Verlag Berlin Heidelberg 1960
Ursprünglich erschienen bei Springer-Verlag OHG. Berlin. Göttigen. Heidelberg 1960.
Softcover reprint of the hardcover 3rd edition 1960

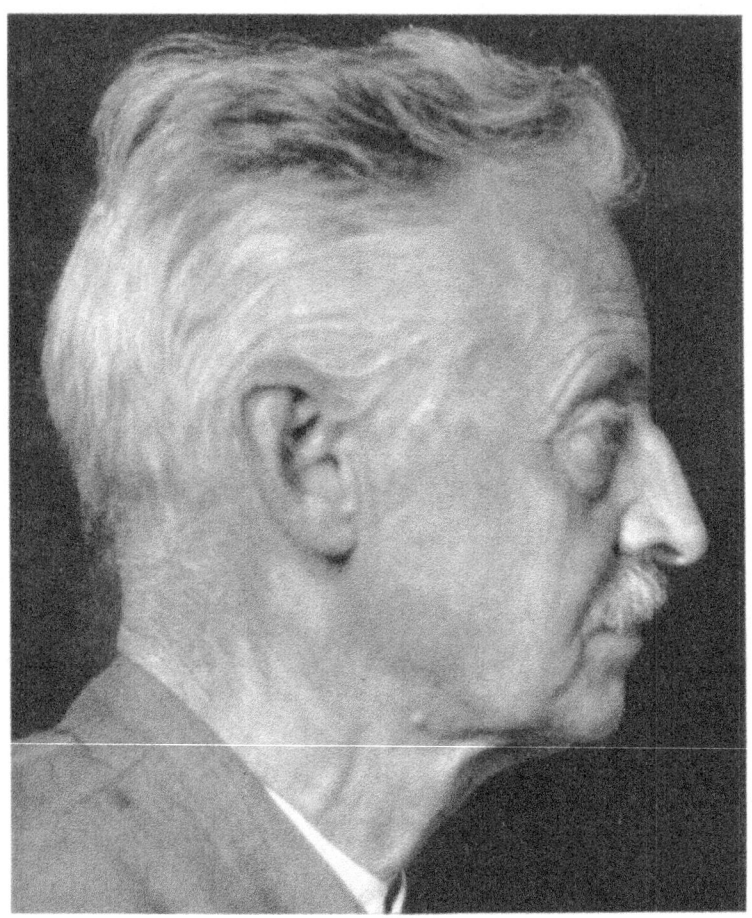

From the preface to the first edition

This book constitutes an attempt to present in a connected fashion some of the most important numerical methods for the solution of ordinary and partial differential equations. The field to be covered is extremely wide, and it is clear that the present treatment cannot be remotely exhaustive; in particular, for partial differential equations it has only been possible to present the basic ideas, and many of the methods developed extensively by workers in applied fields — hydrodynamics, aerodynamics, etc. —, most of which have been developed for specific problems, have had to be dismissed with little more than a reference to the literature.

However, the aim of the book is not so much to reproduce these special methods, their corresponding computing schemes, etc., as to acquaint a wide circle of engineers, physicists and mathematicians with the general methods, and to show with the aid of numerous worked examples that an idea of the quantitative behaviour of the solution of a differential equation problem can be obtained by numerical means with nothing like the trouble and labour that widespread prejudice would suggest. This prejudice may be partly due to the kind of mathematical instruction given in technical colleges and universities, in which, although the theory of differential equations is dealt with in detail, numerical methods are gone into only briefly. I have always observed that graduate mathematicians and physicists are very well acquainted with theoretical results, but have no knowledge of the simplest approximate methods. If approximate methods were more well known, perhaps many problems would be solved with their aid which hitherto have simply not been tackled, despite the fact that interest in their solution has existed throughout. Especially with partial differential equations it has been the practice in many applied fields to restrict attention to the simplest cases — sometimes even to the cases for which the solution can be obtained in closed form —, while advancing technology demands the treatment of ever more complex problems. Further, considerable effort has often been put into the linearization of problems, because of a diffidence in tackling non-linear problems directly; many approximate methods are, however, immediately applicable also to non-linear problems, though clearly heavier computation is only to be expected; nevertheless, it is my belief that there will be a great increase in the importance of non-linear problems in the future.

As yet, the numerical treatment of differential equations has been investigated far too little, both in theoretical and practical respects, and approximate methods need to be tried out to a far greater extent than hitherto; this is especially true of partial differential equations and non-linear problems. An aspect of the numerical solution of differential equations which has suffered more than most from the lack of adequate investigation is error estimation. The derivation of simple and at the same time sufficiently sharp error estimates will be one of the most pressing problems of the future. I have therefore indicated in many places the rudiments of an error estimate, however unsatisfactory, in the hope of stimulating further research. Indeed, in this respect the book can only be regarded as an introduction.

Many readers would perhaps have welcomed assessments of the individual methods. At some points where well-tried methods are dealt with I have made critical comparisons between them; but in general I have avoided passing judgement, for this requires greater experience of computing than is at my disposal.

Hannover, December 1950

LOTHAR COLLATZ

From the preface to the second edition

In this new edition I have incorporated, so far as they have been accessible to me, the advances which have been made since the publication of the first edition. With the intense active interest which is now being taken in the numerical solution of differential equations the world over, new results are being obtained in a gratifyingly large number of topics. I always welcome especially the derivation of new error estimates, and in the present edition I have in fact been able to include error estimates with a large number of examples for which not even the rudiments of an error estimate were given in the first edition. May such further progress be made that in the future an error estimate will be included as a matter of course in the numerical treatment of any differential equation of a reasonable degree of complexity.

In spite of the fact that the book has been allowed to expand, considerable care has been necessary in choosing what extra material should be taken in; with such a large field to cover it is hardly possible to achieve completeness. It is certain that many readers will notice the omission of something which in their opinion ought to have been included. In such cases I am always grateful for criticism and interesting suggestions.

My very especial thanks are due to Dr. JOHANN SCHRÖDER, Dr. JULIUS ALBRECHT and Dr. HELMUT BARTSCH, who have inspected the proof-sheets with great care and in doing so have made numerous valuable suggestions for improvement.

Hamburg, Summer 1954

LOTHAR COLLATZ

Preface to the third edition

This English edition was translated from the second German edition by Mr. P. G. WILLIAMS, B. Sc., Mathematics Division, National Physical Laboratory, Teddington, England. It differs in detail from the second edition in that throughout the book a large number of minor improvements, alterations and additions have been made and numerous further references to the literature included; also new worked examples have been incorporated. Mr. WILLIAMS has made a series of suggestions for improving the presentation, which I gratefully acknowledge. My especial thanks are due to him, to his wife, Mrs. MARION WILLIAMS, and to my assistant, Dr. PETER KOCH, for the proof-reading, and also to Springer-Verlag for their continued ready compliance with all my wishes.

Hamburg, Summer 1959

LOTHAR COLLATZ

Contents

Chapter I

Mathematical preliminaries and some general principles

Chapter II

Initial-value problems in ordinary differential equations

Chapter III

Boundary-value problems in ordinary differential equations

Chapter IV

Initial- and initial-/boundary-value problems
in partial differential equations

Chapter V

Boundary-value problems in partial differential equations

Chapter VI

Integral and functional equations

Appendix

Chapter I

Mathematical preliminaries and some general principles

In this chapter we collect together some mathematical results which will be needed later and state some general approximation principles which are applicable in all the following chapters.

Some notes on the numerical examples

1. The numerical examples will be used solely for illustrating the methods. Consequently it is sufficient in many cases to exhibit only the early stages of the computation and it is permissible to simplify the calculations by using a rather large finite-difference interval or by retaining only a few terms in a Ritz approximation, etc. The results obtained with such simplifications are often quite crude, but it should be borne in mind that their accuracy can always be improved by using a smaller interval or taking in more terms as the case may be (to save space this will not be stated explicitly in each individual case). The explanatory treatment of the examples in the text should enable the reader to effect such improvement in accuracy without difficulty, though this may not be necessary in many cases; in technical applications, for instance, quite a low accuracy, permitting an error of several per cent maybe, is often quite sufficient.

2. To study accuracy I have computed the results of many of the examples to more significant figures than are given in the tables and elsewhere. These are rounded values and hence anyone who works through an example with the number of significant figures given is liable to arrive at a slightly different result.

3. *One should be wary of drawing general conclusions as to the merit or demerit of a method on the basis of individual examples, for a great deal of experience is needed before a sound assessment can be made.* Furthermore, the efficacy of a method is strongly dependent on the computing technique of the individual, the degree to which he is accustomed to the method, the resources at his disposal, and many other factors.

4. Several important checks are mentioned in the text. However, the numerous checks which were, in fact, always applied during the calculation of the examples are generally not reproduced in the interests

of economy of space. *In carrying out a computation, whether for a new problem or as a check on results already calculated, one should apply as many current checks as possible;* the beginner usually regards checks as superfluous until he makes a deep-rooted mistake whose location and correction takes longer than a proper computation carried out with current checks. One should beware the hasty calculation and heed the well-known proverb: "More haste, less speed."

Of course, a study of possible checking techniques is a necessary preliminary. For any calculation it is important to consider how one can control effectively which sources of error shall be present and how suitable checks can be kept on the errors arising from these sources. Only when sufficient checks have been satisfied to inspire confidence in the accuracy of the calculation up to the current point should one proceed further. Even though checks have been satisfied, it is still possible that the calculation may have gone astray; the experienced computer knows that he can never be too suspicious of the calculation in hand. Whenever possible the results should be confirmed by a second calculation based on a different method or by a recalculation performed by another person.

5. Elementary calculations such as the solution of algebraic equations or systems of linear equations and the evaluation of elementary integrals have been omitted. Hence the actual work involved in a calculation is often considerably greater than a glance at the printed reproduction would suggest and the reader is warned not to be misled by this.

§ 1. Introduction to problems involving differential equations

1.1. Initial-value and boundary-value problems in ordinary differential equations

The general solution of an n-th order differential equation

$$F\left(x, y(x), y'(x), y''(x), \ldots, y^{(n)}(x)\right) = 0 \qquad (1.1)$$

for a real function $y(x)$ normally depends on n parameters $c_1, \ldots, c_n$. In an initial-value problem these parameters are determined by prescribing the values

$$y_0^{(\nu)} = y^{(\nu)}(x_0) \qquad (\nu = 0, 1, 2, \ldots, n - 1) \qquad (1.2)$$

at a fixed point $x = x_0$. If the conditions are based on more than one point x, then the problem is called a boundary-value problem. The "boundary conditions" can have the form

$$V_\nu\left(y_{x_1}, y'_{x_1}, \ldots, y_{x_1}^{(n-1)}, y_{x_2}, y'_{x_2}, \ldots, y_{x_2}^{(n-1)}, \ldots, y_{x_k}, y'_{x_k}, \ldots, y_{x_k}^{(n-1)}\right) = 0$$
$$(\nu = 0, \ldots, n - 1), \qquad\qquad (1.3)$$

where the x_s ($x_1 < x_2 < \cdots < x_k$, say) are prescribed points which may include $\pm \infty$ and the $y_{x_s}^{(\varrho)}$ denote the values of the ϱ-th derivative of $y(x)$ at the points $x = x_s$:

$$y_{x_s}^{(\varrho)} = \left(\frac{d^\varrho y}{d x^\varrho} \right)_{x=x_s}.$$

F and V_ν are given functions which, in general, will be non-linear. To solve the "boundary-value problem" (1.1), (1.3) is to find that function $y(x)$ which satisfies equations (1.1) and (1.3). The boundary conditions need not be restricted to the form (1.3); for instance, we can impose a condition such as

$$\int_{x_1}^{x_2} V\big(x, y(x), \ldots, y^{(n-1)}(x)\big)\, dx = 0, \tag{1.4}$$

where V is a prescribed integrable function.

Depending on the functions F and V_ν, such a boundary-value problem may have no solutions, one solution, several solutions or even infinitely many; for example, the problem

$$y'' + y = 0; \qquad y(0) = y(\pi) = 1$$

has no solution and the problem

$$y'' + y = 0; \qquad y(0) = 1; \qquad y(\pi) = -1$$

has infinitely many solutions.

Further, as an example in which the length of the range of integration is an additional unknown, we mention the gunnery problem in which, for a fixed muzzle velocity v_0, the angle of elevation ϑ_0 is to be determined so as to hit a given target. If the projectile has unit mass and for simplicity the problem is assumed to be two-dimensional, then the co-ordinates of the projectile after a time t satisfy the fourth-order system of differential equations

$$\ddot{x} = -w \cos \vartheta, \qquad \ddot{y} = -w \sin \vartheta - g(y).$$

Here, dots denote differentiation with respect to time, the acceleration g due to gravity is a given function of the height y, the air resistance w is a given function of height and projectile velocity, and the angle between the trajectory and the horizontal is denoted by $\vartheta = \tan^{-1}(dy/dx)$. With the firing point as the origin of the co-ordinates and (x_1, y_1) as the co-ordinates of the target we have (corresponding to the four constants of integration and the unknown time of flight t_1) the five boundary conditions

$$t = 0: \quad x = 0, \qquad y = 0, \qquad \dot{x}^2 + \dot{y}^2 = v_0^2;$$
$$t = t_1: \quad x = x_1, \qquad y = y_1.$$

Note that in the condition $\dot{x}^2 + \dot{y}^2 = v_0^2$ we have a non-linear boundary condition.

1.2. Linear boundary-value problems

We shall deal in greater detail with the special class of boundary-value problems *for which the differential equation and the boundary conditions are linear (these are called linear boundary-value problems)* and

1*

for which the boundary conditions are based on only two points, say $x_1 = a$ and $x_2 = b$, where $a < b$. The differential equation may be written in the form

$$L[y] = r(x), \qquad (1.5)$$

where

$$L[y] = \sum_{r=0}^{n} f_r(x) y^{(r)} = f_0(x) y + f_1(x) y' + f_2(x) y'' + \cdots + f_n(x) y^{(n)}, \quad (1.6)$$

and similarly the boundary conditions (assumed to be linearly independent) may be written

$$U_\nu[y] = \gamma_\nu \qquad (\nu = 1, 2, \ldots, n), \qquad (1.7)$$

where

$$U_\nu[y] = \sum_{k=0}^{n-1} \left(\alpha_{\nu,k} \, y^{(k)}(a) + \beta_{\nu,k} \, y^{(k)}(b) \right). \qquad (1.8)$$

The $f_\nu(x)$ and $r(x)$ are given functions which we normally assume to be continuous and the γ_ν, $\alpha_{\nu,k}$, $\beta_{\nu,k}$ are given constants.

The differential equation is called homogeneous when $r(x) \equiv 0$ in the interval $a \leq x \leq b$ and otherwise inhomogeneous; similarly a boundary condition is called homogeneous when the γ_ν associated with it are zero and otherwise inhomogeneous. The boundary-value problem is called homogeneous when the differential equation and all the boundary conditions are homogeneous.

When the differential equation is of even order $n = 2m$, the boundary conditions can be divided into "essential" and "suppressible"[1,2] *(the senses in which they are essential or suppressible, respectively, will be explained in Ch. III §§ 5, 6 when we deal with* Ritz's *method). As many as possible linearly independent linear combinations of the $2m$ given boundary conditions (1.7) are formed such that derivatives of the m-th and higher orders are removed. By this process, k (say) linearly independent boundary conditions are obtained which contain only derivatives of up to the $(m-1)$-th order. These are called "essential" boundary conditions. A further $2m - k$*

---

[1] Following E. Kamke: Math. Z. **48**, 67—100 (1942), who called the two types of boundary condition "wesentlich" and "restlich". Biezeno-Grammel: Technische Dynamik, Vol. I, 2nd ed., p. 136, Berlin 1953, uses a different terminology; from mechanical ideas he derives the terms "geometrische" and "dynamische" for boundary conditions in second- and fourth-order problems corresponding to our "essential" and "suppressible", respectively.

[2] For special problems the following distinctions are also found in the literature: For differential equations of the second order ($m = 1$) one calls the conditions

$y(a) = \alpha$,	$y(b) = \beta$	boundary conditions of the first kind,
$y'(a) = \alpha$,	$y'(b) = \beta$	boundary conditions of the second kind,
$c_1 y(a) + c_2 y'(a) = \alpha$,	$d_1 y(b) + d_2 y'(b) = \beta$	boundary conditions of the third kind or
(where $|c_1| + |c_2| > 0$ and $|d_1| + |d_2| > 0$)		sometimes Sturm's boundary conditions.

boundary conditions, which may be termed "suppressible", can be specified such that no more "essential" boundary conditions can be obtained from them by linear combination and such that the combined "essential" and "suppressible" boundary conditions are equivalent to the original system of 2m boundary conditions.

For example, if the boundary conditions $y(0) + y'(1) = 1$, $y(1) + 2y'(1) = 3$ are given with a second-order differential equation $(m = 1)$, they can be combined linearly to give the new boundary condition $2y(0) - y(1) = -1$ in which the first derivative no longer appears. No more linear combinations of these boundary conditions can be found which contain no derivatives and hence we have one "essential" boundary condition $2y(0) - y(1) = -1$ and one "suppressible" boundary condition, say $y(0) + y'(1) = 1$.

A homogeneous boundary-value problem (sometimes called completely homogeneous) will usually possess only the trivial solution $y(x) \equiv 0$. One therefore considers those problems in which a parameter λ occurs, either in the differential equation or in the boundary conditions, and investigates the values of λ, the so-called "eigenvalues", for which the boundary-value problem has a "non-trivial" solution, i.e. a solution which does not vanish identically. Such a solution is called an eigenfunction and the whole problem is called an "eigenvalue problem" (see Example III in § 1.2 of Ch. III and also § 8 of that chapter).

1.3. Problems in partial differential equations

Examples of initial-value and boundary-value problems in partial differential equations which will acquaint the reader with the types of problems that arise will be found in Ch. IV, § 1 and Ch. V, §1. In this section we go directly to the general formulation and pose the problem of determining a function $u(x_1, x_2, \ldots, x_n)$ of n independent variables which satisfies a partial differential equation

$$F(x_1, \ldots, x_n, u, u_1, \ldots, u_n, u_{11}, \ldots, u_{nn}, \ldots) = 0 \quad \text{in } B \qquad (1.9)$$

and certain boundary conditions

$$V_\mu(x_1, \ldots, x_n, u, u_1, \ldots, u_n, u_{11}, \ldots, u_{nn}, \ldots) = 0 \quad \text{on } \Gamma_\mu. \qquad (1.10)$$

The subscripts on u denote partial differentiation with respect to the x_i; for example,

$$u_j = \frac{\partial u}{\partial x_j}, \qquad u_{jk} = \frac{\partial^2 u}{\partial x_j \partial x_k}. \qquad (1.11)$$

B is a given region of the $(x_1, \ldots, x_n)$ space, the Γ_μ are $(n-1)$-dimensional "hyper-surfaces" of this space and F and the V_μ are given functions which will be assumed to be continuous.

Here also the linear problems play a special rôle. A partial differential equation is said to be linear when it is linear in u and the derivatives of u, i.e. when it has the form

$$L[u] \equiv \sum_{\alpha_1 + \cdots + \alpha_n \leq m} A_{\alpha_1, \alpha_2, \ldots, \alpha_n} \frac{\partial^{\alpha_1 + \cdots + \alpha_n} u}{\partial x_1^{\alpha_1} \ldots \partial x_n^{\alpha_n}} = r(x_1, \ldots, x_n), \quad (1.12)$$

where the $A_{\alpha_1, \alpha_2, \ldots, \alpha_n}$ and r are given functions of $x_1, \ldots, x_n$ and m is the order of the differential equation (in so far as a derivative of the m-th order actually appears). As with ordinary differential equations the terms homogeneous and inhomogeneous are applied to the equations with $r \equiv 0$ and $r \not\equiv 0$, respectively.

A partial differential equation is called quasi-linear when it is linear in the highest order derivatives occurring (say of the m-th order), i.e. when it has the form

$$\sum_{\alpha_1 + \cdots + \alpha_n = m} A_{\alpha_1, \ldots, \alpha_n} \frac{\partial^{\alpha_1 + \cdots + \alpha_n} u}{\partial x_1^{\alpha_1} \ldots \partial x_n^{\alpha_n}} = r,$$

where r and the $A_{\alpha_1, \ldots, \alpha_n}$ are now given functions of $x_1, \ldots, x_n$, u and partial derivatives of u of up to and including the $(m-1)$-th order.

If the boundary conditions are linear, i.e. linear in u and its partial derivatives, then we write them in the form

$$\left. \begin{array}{c} U_\mu(x_1, \ldots, x_n, \ u, u_1, \ldots, u_n, \ u_{11}, \ldots, u_{nn}, \ldots) = \gamma_\mu \\ \text{on } \Gamma_\mu \quad (\mu = 1, \ldots, k), \end{array} \right\} \quad (1.13)$$

where the γ_μ are given functions of position on the Γ_μ and U_μ is linear and homogeneous in u and its derivatives. Again we say the boundary conditions are homogeneous when $\gamma_\mu \equiv 0$ and inhomogeneous when $\gamma_\mu \not\equiv 0$.

A boundary-value problem is termed linear when the differential equation and all the boundary conditions are linear. Any linear boundary-value problem can be reduced to one in which either the differential equation or the boundary conditions (which should not be self-contradictory, of course) are homogeneous. This is achieved by introducing a new function $u^* = u - u_0$, where u_0 is a function satisfying either the inhomogeneous differential equation or the inhomogeneous boundary conditions, respectively.

§2. Finite differences and interpolation formulae

We assume here that the reader is familiar with the foundations of the calculus of finite differences so that we may be brief.

2.1. Difference operators and interpolation formulae

Suppose that we know the values $f_\nu = f(x_\nu)$ of a function $f(x)$ at the $(N+1)$ equidistant points $x_\nu = x_0 + \nu h$, where $\nu = 0, 1, 2, \ldots, N$ (sometimes ν may be non-integral). h is variously called the interval of tabulation, pivotal interval, step size, or just the interval and is always taken to be positive. For any function $F(x)$ we can define the difference operators Δ, ∇, δ for the interval h as follows:

$$\Delta F(x) = F(x+h) - F(x),$$
$$\nabla F(x) = F(x) - F(x-h),$$
$$\delta F(x) = F(x+\tfrac{1}{2}h) - F(x-\tfrac{1}{2}h);$$

we call

$$\Delta f_k = f_{k+1} - f_k \qquad \text{forward differences,}$$
$$\nabla f_k = f_k - f_{k-1} \qquad \text{backward differences,}$$

and

$$\delta f_k = f_{k+\frac{1}{2}} - f_{k-\frac{1}{2}} \quad \text{central differences.}$$

Each of these may be extended to higher differences; thus, for example,

$$\Delta^2 f_k = \Delta(\Delta f_k) = \Delta(f_{k+1} - f_k) = f_{k+2} - 2f_{k+1} + f_k,$$

and in general

$$\Delta^p f_k = \Delta(\Delta^{p-1} f_k), \quad \nabla^p f_k = \nabla(\nabla^{p-1} f_k), \quad \delta^p f_k = \delta(\delta^{p-1} f_k) \quad (p=1,2,3,\ldots).$$

To extend these definitions to $p=0$ we write

$$\Delta^0 f_k = \nabla^0 f_k = \delta^0 f_k = f_k.$$

For the general non-central differences we have

$$\Delta^p f_k = \sum_{\varrho=0}^{p} (-1)^\varrho \binom{p}{\varrho} f_{k+p-\varrho}, \qquad \nabla^p f_k = \sum_{\varrho=0}^{p} (-1)^\varrho \binom{p}{\varrho} f_{k-\varrho}.$$

The "first" $(p=1)$ and "higher" $(p>1)$ differences are often conveniently written down in the "difference table" of the function $f(x)$ in which the difference of two entries appears opposite the gap between them as in Table I/1.

Table I/1. *The "sloping" difference table of a function $f(x)$*

x	f	First ∇f	Second $\nabla^2 f$	Third $\nabla^3 f$	differences $\ldots$
x_{-2}	f_{-2}		$\nabla^2 f_{-1}$		$\ldots$
		∇f_{-1}		$\nabla^3 f_0$	
x_{-1}	f_{-1}		$\nabla^2 f_0$		$\ldots$
		∇f_0		$\nabla^3 f_1$	
x_0	f_0		$\nabla^2 f_1$		$\ldots$
		∇f_1		$\nabla^3 f_2$	
x_1	f_1		$\nabla^2 f_2$		$\ldots$
		∇f_2		$\nabla^3 f_3$	
x_2	f_2		$\nabla^2 f_3$		$\ldots$

Occasionally, however, it is more convenient to write the differences $\nabla^p f_k$ $(p = 0, 1, 2, 3, \ldots)$ with the same k all on the same line as in Table I/2.

Table I/2. *The "horizontal" difference table of $f(x)$*

x	f	First ∇f	Second $\nabla^2 f$	differences $\ldots$
x_{-1}	f_{-1}	∇f_{-1}	$\nabla^2 f_{-1}$	$\ldots$
x_0	f_0	∇f_0	$\nabla^2 f_0$	$\ldots$
x_1	f_1	∇f_1	$\nabla^2 f_1$	$\ldots$

In the theory of interpolation, polynomials are derived having the property that they coincide with the function $f(x)$ at certain points x_ν; we reproduce here only those polynomials which will be used later. With the abbreviation $u = \dfrac{x - x_0}{h}$ "NEWTON's interpolation polynomial with backward differences" can be written

$$
\left. \begin{aligned}
N_p(x) = f_0 &+ \frac{u}{1!} \nabla f_0 + \frac{u(u+1)}{2!} \nabla^2 f_0 + \\
&+ \frac{u(u+1)(u+2)}{3!} \nabla^3 f_0 + \cdots + \frac{u(u+1)\ldots(u+p-1)}{p!} \nabla^p f_0;
\end{aligned} \right\} \quad (2.1)
$$

it takes the values of $f(x)$ at the points $x_0, x_{-1}, \ldots, x_{-p}$. There is a point $x = \xi$ inside the smallest interval of the x axis containing the points $x, x_0, x_{-1}, \ldots, x_{-p}$ such that the remainder term at non-tabular points is given by[1]

$$
R_{p+1}(x) = f(x) - N_p(x) = \frac{u(u+1)\ldots(u+p)}{(p+1)!} h^{p+1} f^{(p+1)}(\xi) . \quad (2.2)
$$

The corresponding Newtonian interpolation formula with forward differences is

$$
\left. \begin{aligned}
N_p^*(x) = f_0 &+ \frac{u}{1!} \Delta f_0 + \frac{u(u-1)}{2!} \Delta^2 f_0 + \cdots + \binom{u}{p} \Delta^p f_0 + \\
&+ \binom{u}{p+1} h^{p+1} f^{(p+1)}(\xi) .
\end{aligned} \right\} \quad (2.3)
$$

STIRLING's interpolation polynomial, which we write down only for even p, is

$$
\left. \begin{aligned}
St_p(x) = f_0 &+ u \frac{\nabla f_0 + \nabla f_1}{2} + \frac{u^2}{2!} \nabla^2 f_1 + \frac{u(u^2-1)}{3!} \frac{\nabla^3 f_1 + \nabla^3 f_2}{2} + \\
&+ \frac{u^2(u^2-1)}{4!} \nabla^4 f_2 + \cdots + \frac{u^2(u^2-1)\ldots\left(u^2 - \left(\frac{p}{2}-1\right)^2\right)}{p!} \nabla^p f_{p/2};
\end{aligned} \right\} \quad (2.4)
$$

[1] For the whole of this section see, for example, G. SCHULZ: Formelsammlung zur praktischen Mathematik. Sammlung Göschen, Vol. 1110. Leipzig and Berlin 1945; and FR. A. WILLERS: Methoden der praktischen Analysis, 2nd ed. Berlin 1950, particularly p. 76 et seq.

it coincides with $f(x)$ at the points $x_{-p/2}, \ldots, x_{-1}, x_0, x_1, \ldots, x_{p/2}$ and the remainder term at other points is given by

$$R_{p+1}(x) = f(x) - St_p(x) = \binom{u + \dfrac{p}{2}}{p+1} h^{p+1} f^{(p+1)}(\xi).$$

From STIRLING's polynomial we can obtain a symmetric formula[1] whose remainder term is of higher order:

$$\left. \begin{aligned} \frac{f(x_0 + u\,h) + f(x_0 - u\,h)}{2} &= f_0 + \frac{u^2}{2!}\, \nabla^2 f_1 + \frac{u^2(u^2 - 1)}{4!}\, \nabla^4 f_2 + \cdots + \\ &\quad + \frac{u^2(u^2 - 1) \ldots \left(u^2 - \left(\dfrac{p}{2} - 1\right)^2\right)}{p!}\, \nabla^p f_{p/2} + R_p^{**}, \end{aligned} \right\} \quad (2.5)$$

where

$$R_p^{**} = \frac{u^2(u^2 - 1) \ldots \left(u^2 - \left(\dfrac{p}{2}\right)^2\right)}{(p+2)!}\, h^{p+2}\, f^{(p+2)}(\xi).$$

2.2. Some integration formulae which will be needed later

From the polynomials of the preceding section we can obtain formulae for the approximate integration of $f(x)$ over specified intervals by replacing $f(x)$ by one of these polynomials in the required interval. Thus by integrating NEWTON's interpolation formula (2.1), (2.2) with respect to x over the interval x_0 to $x_0 + h$ we obtain

$$\left. \begin{aligned} \int_{x_0}^{x_0+h} f(x)\, dx &= h\left[f_0 + \frac{1}{2}\nabla f_0 + \frac{5}{12}\nabla^2 f_0 + \frac{3}{8}\nabla^3 f_0 + \cdots\right] + \\ &\quad + \int_{x_0}^{x_0+h} R_{p+1}(x)\, dx = h\sum_{\varrho=0}^{p} \beta_\varrho \nabla^\varrho f_0 + S_{p+1}, \end{aligned} \right\} \quad (2.6)$$

where

$$\beta_\varrho = \frac{1}{\varrho!} \int_0^1 u(u+1) \ldots (u + \varrho - 1)\, du \qquad (\varrho = 1, 2, \ldots). \quad (2.7)$$

The first few β_ϱ are

$$\beta_0 = 1, \ \beta_1 = \frac{1}{2}, \ \beta_2 = \frac{5}{12}, \ \beta_3 = \frac{3}{8}, \ \beta_4 = \frac{251}{720}, \ \beta_5 = \frac{95}{288}, \ \beta_6 = \frac{19087}{60480}.$$

For S_{p+1} we have the estimate

$$\left. \begin{aligned} |S_{p+1}| &= \left| \int_{x_0}^{x_0+h} \frac{u(u+1) \ldots (u+p)}{(p+1)!}\, h^{p+1} f^{(p+1)}(\xi)\, dx \right| \\ &\leq h^{p+2} \beta_{p+1} |f^{(p+1)}|_{\max}. \end{aligned} \right\} \quad (2.8)$$

---

[1] STEFFENSEN, J. F.: Interpolation, p. 29. Baltimore 1927.

If we replace $f(x)$ by the Newtonian polynomial based on the points $x_1, x_0, x_{-1}, \ldots, x_{-p+1}$, i.e. the polynomial

$$f_1 + \frac{u-1}{1!} \nabla f_1 + \frac{(u-1)\,u}{2!} \nabla^2 f_1 + \cdots + \frac{(u-1)\,u \ldots (u+p-2)}{p!} \nabla^p f_1,$$

then integration over the interval x_0 to x_0+h gives

$$
\left.
\begin{aligned}
\int_{x_0}^{x_0+h} f(x)\,dx &= h\left[f_1 - \frac{1}{2}\nabla f_1 - \frac{1}{12}\nabla^2 f_1 - \frac{1}{24}\nabla^3 f_1 - \cdots\right] + S_{p+1}^* \\
&= h\sum_{\varrho=0}^{p} \beta_\varrho^* \nabla^\varrho f_1 + S_{p+1}^*,
\end{aligned}
\right\} \tag{2.9}
$$

where

$$\beta_\varrho^* = \frac{1}{\varrho!} \int_0^1 (u-1)\,u\,(u+1) \ldots (u+\varrho-2)\,du.$$

The first few β_ϱ^* are

$$\beta_0^* = 1, \quad \beta_1^* = -\frac{1}{2}, \quad \beta_2^* = -\frac{1}{12}, \quad \beta_3^* = -\frac{1}{24}, \quad \beta_4^* = -\frac{19}{720},$$

$$\beta_5^* = -\frac{3}{160}, \quad \beta_6^* = -\frac{863}{60480}.$$

An estimate for S_{p+1}^* is given by

$$
\left.
\begin{aligned}
|S_{p+1}^*| &= \left| \int_{x_0}^{x_0+h} \frac{(u-1)\,u\,(u+1)\ldots(u+p-1)}{(p+1)!}\, h^{p+1} f^{(p+1)}(\xi)\,dx \right| \\
&\leq h^{p+2} |\beta_{p+1}^*|\,|f^{(p+1)}|_{\max}.
\end{aligned}
\right\} \tag{2.10}
$$

If we apply (2.9) to the function $\tilde{f}(x) = f(2x_0 + h - x)$, we obtain

$$
\left.
\begin{aligned}
\int_{x_0}^{x_0+h} \tilde{f}(x)\,dx &= \int_{x_0}^{x_0+h} f(2x_0 + h - x)\,dx = h\sum_{\varrho=0}^{p} \beta_\varrho^* (-1)^\varrho \nabla^\varrho f_0 + \tilde{S}_{p+1} \\
&= h\left(f_0 + \frac{1}{2}\nabla f_1 - \frac{1}{12}\nabla^2 f_2 + - \cdots\right) + \tilde{S}_{p+1}
\end{aligned}
\right\} \tag{2.11}
$$

with a corresponding form for the remainder term $\tilde{S}_{p+1}$.

The β_ϱ and β_ϱ^* are connected by the relation

$$\beta_\varrho + \beta_{\varrho+1}^* = \beta_{\varrho+1} \quad (\varrho = 0, 1, 2, \ldots), \tag{2.12}$$

for

$$\beta_\varrho + \beta_{\varrho+1}^* = \int_0^1 \left[\binom{u+\varrho-1}{\varrho} + \binom{u+\varrho-1}{\varrho+1}\right] du = \int_0^1 \binom{u+\varrho}{\varrho+1} du = \beta_{\varrho+1}.$$

By adding equations (2.12) for $\varrho = 0, 1, 2, \ldots, p-1$, together with $\beta_0 = \beta_0^* \ (=1)$, we obtain the relation

$$\sum_{\varrho=0}^{p} \beta_\varrho^* = \beta_p. \tag{2.13}$$

Integration of STIRLING's formula (2.5) over the interval $x_0 - h$ to $x_0 + h$ gives

$$\begin{aligned}
\int_{x_0-h}^{x_0+h} f(x)\,dx &= h\left[2f_0 + \frac{1}{3}\nabla^2 f_1 - \frac{1}{90}\nabla^4 f_2 + \frac{1}{756}\nabla^6 f_3 - \cdots\right] + S_p^{**} \\
&= h\sum_{\varrho=0}^{p/2} \beta_\varrho^{**}\nabla^{2\varrho} f_\varrho + S_p^{**},
\end{aligned} \tag{2.14}$$

where

$$\beta_\varrho^{**} = \frac{2}{(2\varrho)!}\int_0^1 u^2(u^2-1)(u^2-4)\ldots(u^2-(\varrho-1)^2)\,du. \tag{2.15}$$

An upper bound for the magnitude of the remainder term is

$$|S_p^{**}| \leq h^{p+3}\left|\beta_{\frac{p}{2}+1}^{**}\right|\,|f^{(p+2)}|_{\max}. \tag{2.16}$$

For $p=2$ (2.14) reduces to SIMPSON's rule

$$\int_{x_0-h}^{x_0+h} f(x)\,dx = h\left[2f_0 + \frac{1}{3}\nabla^2 f_1\right] + S_2^{**}, \tag{2.17}$$

where

$$|S_2^{**}| \leq \frac{h^5}{90}\,|f^{(IV)}|_{\max}. \tag{2.18}$$

2.3. Repeated integration

If we integrate NEWTON's interpolation formula (2.1), (2.2) over the interval x_0 to x $(x = x_0 + uh)$, we obtain the indefinite integral

$$\begin{aligned}
\int_{x_0}^{x} f(x)\,dx &= h\int_0^u N_p(x(u))\,du + \int_{x_0}^x R_{p+1}\,dx = {}'f(x) - {}'f(x_0) \\
&= h\left[uf_0 + \frac{u^2}{2}\nabla f_0 + \frac{3u^2+2u^3}{12}\nabla^2 f_0 + \cdots\right] + R_{1,\,p+1},
\end{aligned} \tag{2.19}$$

and a second integration gives

$$\begin{aligned}
\int_{x_0}^{x}\int_{x_0}^{x} f(x)\,(dx)^2 &= {}''f(x) - {}''f(x_0) - {}'f(x_0)(x-x_0) \\
&= h^2\left[\frac{u^2}{2}f_0 + \frac{u^3}{6}\nabla f_0 + \cdots\right] + R_{2,\,p+1},
\end{aligned} \tag{2.20}$$

where $'f$ is an indefinite integral of $f(x)$ $(=^{(0)}f)$, $''f$ is a indefinite second integral of $f(x)$ and in general $^{(\nu)}f$ denotes a ν-th repeated indefinite integral of $f(x)$ or, shortly, a ν-th integral of $f(x)$, so that

$$^{(\nu)}f = \int^x {}^{(\nu-1)}f\, dx \qquad (\nu = 1, 2, 3, \ldots).\tag{2.21}$$

Similarly we define (with $R_{0,\,p+1} = R_{p+1}$)

$$R_{\nu,\,p+1} = \int_{x_0}^x R_{\nu-1,\,p+1}\, dx \qquad (\nu = 1, 2, 3, \ldots).\tag{2.22}$$

By further integrations we obtain the general n-th integral

$$\left.\begin{aligned}
\int_{x_0}^x \cdots \int_{x_0}^x f\,(dx)^n &= {}^{(n)}f(x) - \sum_{\nu=0}^{n-1} {}^{(n-\nu)}f(x_0)\,\frac{(x-x_0)^\nu}{\nu!} \\
&= h^n \sum_{\varrho=0}^p P_{n,\,\varrho}(u)\,\nabla^\varrho f_0 + R_{n,\,p+1}(x),
\end{aligned}\right\}\tag{2.23}$$

where the $P_{n,\,\varrho}(u)$ are defined by the n-th integrals

$$\left.\begin{aligned}
P_{n,\,\varrho}(u) &= \int_0^u \cdots \int_0^u \frac{u(u+1)\cdots(u+\varrho-1)}{\varrho!}\,(du)^n \quad \text{for} \quad \varrho \geq 1 \\
P_{n,\,0}(u) &= \frac{1}{n!}\,u^n \qquad\qquad\qquad (n = 1, 2, \ldots).
\end{aligned}\right\}\tag{2.24}$$

For small n and ϱ the $P_{n,\,\varrho}$ have already been exhibited as polynomials in u in the formulae (2.19), (2.20) above; the polynomials for several higher values of n and ϱ are given in Table I/3.

Table I/3. *The polynomials* $P_{n,\,\varrho}(u)$

	$\varrho=0$	$\varrho=1$	$\varrho=2$	$\varrho=3$	$\varrho=4$
$n=1$	u	$\frac{1}{2}u^2$	$\frac{1}{12}(3u^2+2u^3)$	$\frac{1}{24}(4u^2+4u^3+u^4)$	$\frac{1}{720}(90u^2+110u^3+45u^4+6u^5)$
$n=2$	$\frac{1}{2}u^2$	$\frac{1}{6}u^3$	$\frac{1}{24}(2u^3+u^4)$	$\frac{1}{360}(20u^3+15u^4+3u^5)$	$\frac{1}{1440}(60u^3+55u^4+18u^5+2u^6)$
$n=3$	$\frac{1}{6}u^3$	$\frac{1}{24}u^4$	$\frac{1}{240}(5u^4+2u^5)$	$\frac{1}{720}(10u^4+6u^5+u^6)$	$\frac{1}{10080}(105u^4+77u^5+21u^6+2u^7)$
$n=4$	$\frac{1}{24}u^4$	$\frac{1}{120}u^5$	$\frac{1}{720}(3u^5+u^6)$	$\frac{1}{5040}(14u^5+7u^6+u^7)$	

The value of the n-th indefinite integral (2.23) when $x = x_1$ is

$$\left.\begin{aligned}
\int_{x_0}^{x_1}\int_{x_0}^x \cdots \int_{x_0}^x f\,(dx)^n &= {}^{(n)}f(x_1) - \sum_{\nu=0}^{n-1} {}^{(n-\nu)}f(x_0)\,\frac{h^\nu}{\nu!} \\
&= h^n \sum_{\varrho=0}^p \beta_{n,\,\varrho}\,\nabla^\varrho f_0 + R_{n,\,p+1}(x_1),
\end{aligned}\right\}\tag{2.25}$$

where the coefficients $\beta_{n,\varrho}$ denote the numbers

$$\beta_{n,\varrho} = \int_0^1 \int_0^u \cdots \int_0^u \frac{u(u+1)(u+2)\ldots(u+\varrho-1)}{\varrho!}\,(du)^n = P_{n,\varrho}(1). \quad (2.26)$$

These coefficients $\beta_{n,\varrho}$ include the numbers β_ϱ of (2.7) as the special case $n=1$, i.e. $\beta_\varrho = \beta_{1,\varrho}$. The first few $\beta_{n,\varrho}$ are given in Table I/4. From (2.2) and (2.26) we see that an estimate for the remainder term in (2.25) is given by

$$\left| R_{n,\,p+1}(x_1) \right| \leq \beta_{n,\,p+1} h^{p+n+1} \left| f^{(p+1)} \right|_{\max}. \quad (2.27)$$

Table I/4. *The numbers* $\beta_{n,\varrho}$

	$\varrho=0$	$\varrho=1$	$\varrho=2$	$\varrho=3$	$\varrho=4$	$\varrho=5$
$n=1$	1	$\dfrac{1}{2}$	$\dfrac{5}{12}$	$\dfrac{3}{8}$	$\dfrac{251}{720}$	$\dfrac{95}{288}$
$n=2$	$\dfrac{1}{2}$	$\dfrac{1}{6}$	$\dfrac{1}{8}$	$\dfrac{19}{180}$	$\dfrac{3}{32}$	$\dfrac{863}{10\,080}$
$n=3$	$\dfrac{1}{6}$	$\dfrac{1}{24}$	$\dfrac{7}{240}$	$\dfrac{17}{720}$	$\dfrac{41}{2016}$	$\dfrac{731}{40\,320}$
$n=4$	$\dfrac{1}{24}$	$\dfrac{1}{120}$	$\dfrac{1}{180}$	$\dfrac{11}{2520}$	$\dfrac{89}{24\,192}$	$\dfrac{5849}{181\,440}$

If we take the upper limit in (2.25) to be x_{-1} instead of x_1, we obtain

$$\left.\begin{aligned}
\int_{x_0}^{x_{-1}} \cdots \int_{x_0}^{x} f\,(dx)^n &= {}^{(n)}f(x_{-1}) - \sum_{\nu=0}^{n-1} {}^{(n-\nu)}f(x_0)\frac{(-h)^\nu}{\nu!} \\
&= (-1)^{n+\varrho} h^n \sum_{\varrho=0}^{p} \gamma_{n,\varrho}\,\nabla^\varrho f_0 + R_{n,\,p+1}(x_{-1}),
\end{aligned}\right\} \quad (2.28)$$

where

$$\left.\begin{aligned}
\gamma_{n,\varrho} &= (-1)^{n+\varrho} \int_0^{-1}\int_0^u \cdots \int_0^u \binom{u+\varrho-1}{\varrho}(du)^n \\
&= \int_0^1 \int_0^u \cdots \int_0^u \binom{u}{\varrho}(du)^n = (-1)^{n+\varrho} P_{n,\varrho}(-1).
\end{aligned}\right\} \quad (2.29)$$

The first few numbers $\gamma_{n,\varrho}$ are given in Table I/5.

Table I/5. *The numbers* $\gamma_{n,\varrho}$

	$\varrho=0$	$\varrho=1$	$\varrho=2$	$\varrho=3$	$\varrho=4$
$n=1$	1	$\dfrac{1}{2}$	$-\dfrac{1}{12}$	$\dfrac{1}{24}$	$-\dfrac{19}{720}$
$n=2$	$\dfrac{1}{2}$	$\dfrac{1}{6}$	$-\dfrac{1}{24}$	$\dfrac{1}{45}$	$-\dfrac{7}{480}$
$n=3$	$\dfrac{1}{6}$	$\dfrac{1}{24}$	$-\dfrac{1}{80}$	$\dfrac{1}{144}$	$-\dfrac{47}{10\,080}$
$n=4$	$\dfrac{1}{24}$	$\dfrac{1}{120}$	$-\dfrac{1}{360}$	$\dfrac{1}{630}$	

Clearly we can also obtain these numbers by integrating NEWTON's formula (2.3) with forward differences:

$$\int_{x_0}^{x_1}\int_{x_0}^{x}\cdots\int_{x_0}^{x} f\,(d\,x)^n = {}^{(n)}f(x_1) - \sum_{\nu=0}^{n-1} {}^{(n-\nu)}f(x_0)\,\frac{h^{\nu}}{\nu!}$$
$$= h^n \sum_{\varrho=0}^{p} \gamma_{n,\varrho}\,\Delta^{\varrho}f_0 + S. \tag{2.30}$$

An estimate for the remainder term S can easily be derived by integration of the remainder term in (2.3):

$$|S| \leq h^{p+n+1}|\gamma_{n,\,p+1}|\,|f^{(p+1)}|_{\max}. \tag{2.31}$$

In addition to the coefficients introduced already we shall need the quantities $\beta^*_{n,\varrho}$ which are obtained in the same way as the $\beta_{n,\varrho}$ except that the lower limit x_0 in (2.19) to (2.26) is replaced by x_{-1}. Performing the individual integrations afresh, we obtain

$$\int_{x_{-1}}^{x} f(x)\,dx = h\int_{-1}^{u} N_p\big(x\,(u)\big)\,du + \int_{x_{-1}}^{x} R_{p+1}\,dx = {}'f(x) - {}'f(x_{-1})$$
$$= h\left[(u+1)\,f_0 + \frac{u^2-1}{2}\,\nabla f_0 + \frac{2u^3+3u^2-1}{12}\,\nabla^2 f_0 + \cdots\right] + R^*_{1,\,p+1}, \tag{2.32}$$

$$\int_{x_{-1}}^{x}\int_{x_{-1}}^{x} f(d\,x)^2 = {}''f(x) - {}''f(x_{-1}) - {}'f(x_{-1})\,(x-x_{-1})$$
$$= h^2\left[\frac{(u+1)^2}{2}\,f_0 + \frac{u^3-3u-2}{6}\,\nabla f_0 + \cdots\right] + R^*_{2,\,p+1}, \tag{2.33}$$

and generally, for n integrations,

$$\int_{x_{-1}}^{x}\cdots\int_{x_{-1}}^{x} f(d\,x)^n = {}^{(n)}f(x) - \sum_{\nu=0}^{n-1} {}^{(n-\nu)}f(x_{-1})\,\frac{(x-x_{-1})^{\nu}}{\nu!}$$
$$= h^n \sum_{\varrho=0}^{p} P^*_{n,\varrho}(u)\,\nabla^{\varrho}f_0 + R^*_{n,\,p+1}(x), \tag{2.34}$$

where the polynomials $P^*_{n,\varrho}$ are given by the n-th indefinite integrals

$$P^*_{n,\varrho}(u) = \int_{-1}^{u}\cdots\int_{-1}^{u}\binom{u+\varrho-1}{\varrho}(d\,u)^n \quad \text{for } \varrho \geq 1$$
$$P^*_{n,0}(u) = \frac{1}{n!}\,(u+1)^n \quad (n=1,2,\ldots). \tag{2.35}$$

Table I/6 gives the polynomials $P^*_{n,\varrho}$ for a few values of n and ϱ. For $x=x_0$ (2.34) becomes

$$\int_{x_{-1}}^{x_0}\cdots\int_{x_{-1}}^{x} f(d\,x)^n = {}^{(n)}f(x_0) = \sum_{\nu=0}^{n-1} {}^{(n-\nu)}f(x_{-1})\,\frac{h^{\nu}}{\nu!}$$
$$= h^n \sum_{\varrho=0}^{p} \beta^*_{n,\varrho}\,\nabla^{\varrho}f_0 + R^*_{n,\,p+1}(x_0). \tag{2.36}$$

Table I/6. *The polynomials $P_{n,\varrho}^*(u)$*

	$\varrho=0$	$\varrho=1$	$\varrho=2$
$n=1$	$u+1$	$\frac{1}{2}(u^2-1)$	$\frac{1}{12}(2u^3+3u^2-1)$
$n=2$	$\frac{1}{2}(u+1)^2$	$\frac{1}{6}(u^3-3u-2)$	$\frac{1}{24}(u^4+2u^3-2u-1)$
$n=3$	$\frac{1}{6}(u+1)^3$	$\frac{1}{24}(u^4-6u^2-8u-3)$	$\frac{1}{240}(2u^5+5u^4-10u^2-10u-3)$
$n=4$	$\frac{1}{24}(u+1)^4$	$\frac{1}{120}(u^5-10u^3-20u^2-15u-4)$	$\frac{1}{720}(u^6+3u^5-10u^3-15u^2-9u-2)$

	$\varrho=3$	$\varrho=4$
$n=1$	$\frac{1}{24}(u^4+4u^3+4u^2-1)$	$\frac{1}{720}(6u^5+45u^4+110u^3+90u^2-19)$
$n=2$	$\frac{1}{360}(3u^5+15u^4+20u^3-15u-7)$	$\frac{1}{1440}(2u^6+18u^5+55u^4+60u^3-$
$n=3$	$\frac{1}{720}(u^6+6u^5+10u^4-15u^2-14u-4)$	$-38u-17)$
$n=4$	$\frac{1}{5040}(u^7+7u^6+14u^5-35u^3-49u^2-28u-6)$	

Corresponding to (2.22) and (2.26) we have

$$R_{\nu,\,p+1}^* = \int_{x_{-1}}^{x} R_{\nu-1,\,p+1}^*\,dx \qquad (\nu=1,2,3,\ldots) \tag{2.37}$$

(with $R_{0,\,p+1}^* = R_{p+1}^*$) and

$$
\left.
\begin{aligned}
\beta_{n,\,\varrho}^* &= \int_{-1}^{0}\int_{-1}^{u}\cdots\int_{-1}^{u}\binom{u+\varrho-1}{\varrho}(du)^n \qquad \text{for } \varrho\geqq 1, \\
\beta_{n,\,0}^* &= \frac{1}{n!}.
\end{aligned}
\right\} \tag{2.38}
$$

Here also the coefficients $\beta_{n,\,\varrho}^*$ contain the numbers β_ϱ^* of (2.9) as the special case $n=1$ since $\beta_\varrho^*=\beta_{1,\,\varrho}^*$. The first few $\beta_{n,\,\varrho}^*$ are given in Table I/7. Exactly as in (2.12) it can be shown that

$$\beta_{n,\,\varrho} + \beta_{n,\,\varrho+1}^* = \beta_{n,\,\varrho+1} \qquad (\varrho=0,1,2,\ldots)\ (n=1,2,3,\ldots). \tag{2.39}$$

By adding equations (2.39) for $\varrho=0,1,2,\ldots,p-1$, together with $\beta_{n,\,0}=\beta_{n,\,0}^*\left(=\frac{1}{n!}\right)$, we obtain a generalized form of (2.13), namely

$$\sum_{\varrho=0}^{p}\beta_{n,\,\varrho}^* = \beta_{n,\,p}. \tag{2.40}$$

Table I/7. *The numbers $\beta_{n,\,\varrho}^*$*

	$\varrho=0$	$\varrho=1$	$\varrho=2$	$\varrho=3$	$\varrho=4$
$n=1$	1	$-\frac{1}{2}$	$-\frac{1}{12}$	$-\frac{1}{24}$	$-\frac{19}{720}$
$n=2$	$\frac{1}{2}$	$-\frac{1}{3}$	$-\frac{1}{24}$	$-\frac{7}{360}$	$-\frac{17}{1440}$
$n=3$	$\frac{1}{6}$	$-\frac{1}{8}$	$-\frac{1}{80}$	$-\frac{1}{180}$	$-\frac{11}{3360}$
$n=4$	$\frac{1}{21}$	$-\frac{1}{30}$	$-\frac{1}{360}$	$-\frac{1}{840}$	$-\frac{83}{120960}$

From (2.2), (2.37), (2.38) we see that limits for the remainder term $R^*_{n,\,p+1}$ are given by

$$|R^*_{n,\,p+1}(x_0)| \leq h^{p+n+1}|\beta^*_{n,\,p+1}|\,|f^{(p+1)}|_{\max}. \tag{2.41}$$

Finally we derive some repeated integration formulae from STIRLING's interpolation formula (2.5). With

$$\Phi(x) = \tfrac{1}{2}\left[f(x_0+uh) + f(x_0-uh)\right] = \tfrac{1}{2}\left[f(x) + f(2x_0-x)\right] \tag{2.42}$$

we have

$$\begin{aligned}
\int_{x_0}^{x} \Phi\,dx &= \frac{1}{2}\left['f(x_0+uh) - 'f(x_0-uh)\right]\\
&= h\left[uf_0 + \frac{u^3}{6}\nabla^2 f_1 + \frac{-5u^3+3u^5}{360}\nabla^4 f_2 +\right.\\
&\quad \left.+ \frac{28u^3 - 21u^5 + 3u^7}{15\,120}\nabla^6 f_3 + \cdots\right] + R^{**}_{1,\,p}\,,
\end{aligned} \right\} \tag{2.43}$$

$$\begin{aligned}
\int_{x_0}^{x}\int_{x_0}^{x}\Phi(dx)^2 &= \frac{1}{2}\left[''f(x_0+uh) + ''f(x_0-uh)\right] - ''f(x_0)\\
&= h^2\left[\frac{u^2}{2}f_0 + \frac{u^4}{24}\nabla^2 f_1 + \frac{-5u^4+2u^6}{1440}\nabla^4 f_2 + \cdots\right] + R^{**}_{2,\,p}\,,
\end{aligned} \right\}$$

$$\begin{aligned}
\int_{x_0}^{x}\!\!\int\!\!\int\Phi(dx)^3 &= \frac{1}{2}\left['''f(x_0+uh) - '''f(x_0-uh)\right] - ''f(x_0)\,(x-x_0)\\
&= h^3\left[\frac{u^3}{6}f_0 + \frac{u^5}{120}\nabla^2 f_1 + \frac{-7u^5+2u^7}{10\,080}\nabla^4 f_2 + \cdots\right] + R^{**}_{3,\,p}\,,
\end{aligned} \right\} \tag{2.44}$$

$$\begin{aligned}
\int_{x_0}^{x}\!\!\int\!\!\int\!\!\int\Phi(dx)^4 &= \frac{1}{2}\left[''''f(x_0+uh) + ''''f(x_0-uh)\right] -\\
&\quad - ''''f(x_0) - ''f(x_0)\frac{(x-x_0)^2}{2}\\
&= h^4\left[\frac{u^4}{24}f_0 + \frac{u^6}{720}\nabla^2 f_1 + \frac{-14u^6+3u^8}{120\,960}\nabla^4 f_2 + \cdots\right] + R^{**}_{4,\,p}\,.
\end{aligned} \right\}$$

We write down the general n-th repeated integral of $\Phi(x)$ only for $x=x_1$:

$$\int_{x_0}^{x_1}\int_{x_0}^{x}\cdots\int_{x_0}^{x}\Phi(dx)^n = h^n\sum_{\varrho=0}^{p/2}\beta^{**}_{n,\,\varrho}\nabla^{2\varrho}f_\varrho + R^{**}_{n,\,p}(x_1)$$

$$=\begin{cases}
\dfrac{1}{2}\left[^{(n)}f(x_0+h) + {}^{(n)}f(x_0-h)\right] - \displaystyle\sum_{\varrho=0}^{(n/2)-1}\dfrac{h^{2\varrho}}{(2\varrho)!}\,{}^{(n-2\varrho)}f(x_0)\\
\hspace{6cm}\text{for } n \text{ even,}\\[2mm]
\dfrac{1}{2}\left[^{(n)}f(x_0+h) - {}^{(n)}f(x_0-h)\right] - \displaystyle\sum_{\varrho=0}^{(n-3)/2}\dfrac{h^{2\varrho+1}}{(2\varrho+1)!}\,{}^{(n-1-2\varrho)}f(x_0)\\
\hspace{6cm}\text{for } n \text{ odd,}
\end{cases} \tag{2.45}$$

where

$$R^{**}_{\nu,p} = \int_{x_0}^{x} R^{**}_{\nu-1,p}\,dx \qquad (2.46)$$

and

$$R^{**}_{0,p} = R^{**}_{p} \quad \text{from (2.5).}$$

The first few numbers $\beta^{**}_{n,\varrho}$ are given in Table I/8.

Table I/8. *Some values of the numbers $\beta^{**}_{n,\varrho}$*

	$\varrho=0$	$\varrho=1$	$\varrho=2$	$\varrho=3$
$n=1$	1	$\frac{1}{6}$	$-\frac{1}{180}$	$\frac{1}{1512}$
$n=2$	$\frac{1}{2}$	$\frac{1}{24}$	$-\frac{1}{480}$	$\frac{31}{120\,960}$
$n=3$	$\frac{1}{6}$	$\frac{1}{120}$	$-\frac{1}{2016}$	
$n=4$	$\frac{1}{24}$	$\frac{1}{720}$	$-\frac{11}{120\,960}$	

Table I/9. *The numbers $r^{**}_{n,p}$*

	$p=2$	$p=4$
$n=1$	$\frac{2}{15}$	$\frac{10}{21}$
$n=2$	$\frac{1}{20}$	$\frac{31}{168}$
$n=3$	$\frac{1}{84}$	
$n=4$	$\frac{11}{5040}$	

From (2.5) an estimate for the remainder term for even p is

$$|R^{**}_{n,p}| \leqq r^{**}_{n,p} \frac{h^{p+n+2}}{(p+2)!} \left| f^{(p+2)} \right|_{max}, \qquad (2.47)$$

where

$$r^{**}_{n,p} = \int_0^1 \int_0^u \cdots \int_0^u \left| u^2 (u^2-1)(u^2-4)\cdots\left(u^2-\left(\frac{p}{2}\right)^2\right)\right| (du)^n; \qquad (2.48)$$

the first few values of these numbers are given in Table I/9.

2.4. Calculation of higher derivatives

From a differential equation

$$y^{(n)} = f(x, y, y', \ldots, y^{(n-1)}) \qquad (2.49)$$

we can calculate the higher derivatives $y^{(n+1)}$, $y^{(n+2)}$, ... by repeated differentiation (assuming that f is differentiable a sufficient number of times with respect to each of its arguments). Thus for the first derivative we have

$$y^{(n+1)} = \frac{\partial f}{\partial x} + \frac{\partial f}{\partial y} y' + \frac{\partial f}{\partial y'} y'' + \cdots + \frac{\partial f}{\partial y^{(n-1)}} f.$$

We now introduce the notation

$$\left.\begin{array}{l} y = u_0, \quad y' = u_1, \quad y'' = u_2, \ldots, \quad y^{(n)} = u_n = f, \\[2mm] f_x = \frac{\partial f}{\partial x}, \quad f_\nu = \frac{\partial f}{\partial u_\nu}, \quad f_{\mu\nu} = \frac{\partial^2 f}{\partial u_\mu \partial u_\nu}, \quad f_{x\nu} = \frac{\partial^2 f}{\partial x\,\partial u_\nu}, \ldots \end{array}\right\} \quad (2.50)$$

and also the symbolic operator D whose operation on any function $\varphi(x, u_0, u_1, \ldots, u_{n-1})$ is defined by

$$D\varphi = \frac{\partial \varphi}{\partial x} + u_1 \frac{\partial \varphi}{\partial u_0} + u_2 \frac{\partial \varphi}{\partial u_1} + \cdots + u_{n-1} \frac{\partial \varphi}{\partial u_{n-2}} + f \frac{\partial \varphi}{\partial u_{n-1}}, \quad (2.51)$$

so that

$$y^{(n+1)} = Df. \quad (2.52)$$

For the higher derivatives it is convenient to introduce further operators E and F. By the operator $D^{(r)}$ we understand the operator which results from formal expansion of the power

$$\left(\frac{\partial}{\partial x} + u_1 \frac{\partial}{\partial u_0} + u_2 \frac{\partial}{\partial u_1} + \cdots + u_{n-1} \frac{\partial}{\partial u_{n-2}} + f \frac{\partial}{\partial u_{n-1}}\right)^r,$$

treating the u_ν and f as constant factors so that they can be collected together as coefficients in front of each differential operator; thus, for example,

$$D^{(2)} = \frac{\partial^2}{\partial x^2} + u_1^2 \frac{\partial^2}{\partial u_0^2} + \cdots + u_{n-1}^2 \frac{\partial^2}{\partial u_{n-2}^2} + f^2 \frac{\partial^2}{\partial u_{n-1}^2} + 2u_1 \frac{\partial^2}{\partial x \, \partial u_0} +$$

$$+ 2u_2 \frac{\partial^2}{\partial x \, \partial u_1} + \cdots + 2u_1 u_2 \frac{\partial^2}{\partial u_0 \, \partial u_1} + \cdots + 2f u_{n-1} \frac{\partial^2}{\partial u_{n-2} \, \partial u_{n-1}}.$$

If we now differentiate Df with respect to x, the differentiation of the derivatives $\frac{\partial f}{\partial x}, \frac{\partial f}{\partial u_\nu}$ occurring in the individual products gives precisely $D^{(2)} f$; in addition there are the terms arising from the differentiation of the factors u_ν and f. Thus

$$\left. \begin{array}{l} y^{(n+2)} = \dfrac{d}{dx} Df \\[2mm] = D^{(2)} f + u_2 f_0 + u_3 f_1 + \cdots + u_{n-1} f_{n-3} + f f_{n-2} + f_{n-1} Df = Ef, \end{array} \right\} (2.53)$$

say, where E is the operator defined by the preceding expression.

For a further differentiation it is expedient to derive first the formula

$$\frac{d}{dx} (D^{(2)} f) = D^{(3)} f +$$

$$+ 2 [u_2 Df_0 + u_3 Df_1 + u_4 Df_2 + \cdots + u_{n-1} Df_{n-3} + f Df_{n-2} + Df \cdot Df_{n-1}]$$

by writing down $D^{(2)} f$ in detail and applying the operator D. Then if we define another operator F by

$$\left. \begin{array}{l} Ff = D^{(3)} f + \\[2mm] + 3 [u_2 Df_0 + u_3 Df_1 + \cdots + u_{n-1} Df_{n-3} + f Df_{n-2} + Df \cdot Df_{n-1}] + \\[2mm] + [u_3 f_0 + u_4 f_1 + \cdots + u_{n-1} f_{n-4} + f f_{n-3}], \end{array} \right\} (2.54)$$

we can write

$$y^{(n+3)} = Ff + f_{n-1} Ef + f_{n-2} Df. \quad (2.55)$$

2.5. HERMITE'S generalization of TAYLOR'S formula

Equations (2.57), (2.58) below constitute a generalization[1] of TAYLOR's formula; they give a relation between the derivatives of a function $f(x)$ at two points a and b, where $f(x)$ has continuous derivatives of the $(k+m+1)$-th order in the interval $\langle a, b \rangle$ (k, m are non-negative integers).

If $g(x) = (x-a)^k (x-b)^m$, then, by integration by parts $k+m$ times, we have

$$
\begin{aligned}
\int_a^b f'(x) \, g^{(k+m)}(x) \, dx = [f'(x) \, g^{(k+m-1)}(x) &- f''(x) \, g^{(k+m-2)}(x) + \\
&+ \cdots + (-1)^{k+m-1} f^{(k+m)}(x) \, g(x)]_a^b + \\
&+ (-1)^{(k+m)} \int_a^b f^{(k+m+1)}(x) \, g(x) \, dx.
\end{aligned}
\tag{2.56}
$$

Since $g^{(k+m)}(x) = (k+m)!$, the value of the integral on the left-hand side is

$$
(k+m)! \left(f(b) - f(a) \right).
$$

It remains to calculate the values of the derivatives of g at the boundary points a, b. By LEIBNIZ's rule for the differentiation of a product we have

$$
g^{(r)}(x) = \sum_{\varrho=0}^r \binom{r}{\varrho} \binom{k}{\varrho} \varrho! \, (x-a)^{k-\varrho} \binom{m}{r-\varrho} (r-\varrho)! \, (x-b)^{m-r+\varrho}.
$$

Bearing in mind that the binomial coefficients $\binom{k}{\varrho}$ and $\binom{m}{r-\varrho}$ are zero for $\varrho > k$ and $\varrho < r - m$, respectively, we see that $\sum\limits_{\varrho=0}^r$ can be replaced by $\sum\limits_{\varrho=r-m}^k$ when $r \leq k+m$ and that the sum is zero when $r > k+m$. If we now put $x=a$, only the term with $\varrho = k$ remains and we have

$$
g^{(r)}(a) = \binom{r}{k} k! \binom{m}{r-k} (r-k)! \, (a-b)^{m-r+k} = r! \binom{m}{r-k} (a-b)^{m+k-r};
$$

similarly

$$
g^{(r)}(b) = r! \binom{k}{r-m} (b-a)^{m+k-r}.
$$

---

[1] Different proofs are given by S. HERMITE: Œuvres **3**, 438 (1912). — KOWALEWSKI, G.: Dtsch. Math. **6**, 349—351 (1942). — OBRESCHKOFF, N.: Abh. Preuss. Akad. Wiss., Math.-naturw. Kl. **1940**, Nr. 4, 1—20. — PFLANZ, E.: Z. Angew. Math. Mech. **28**, 167—172 (1948). — BECK, E.: Z. Angew. Math. Mech. **30**, 84—93 (1950).

With these values for the derivatives at the end-points and with h denoting the interval length $b-a$, (2.56) becomes

$$(k+m)!\ (f(b)-f(a)) + \left[\sum_{\nu=1}^{k+m}(-1)^{\nu} f^{(\nu)} g^{(k+m-\nu)}\right]_a^b$$

$$= \sum_{\nu=0}^{k+m} (k+m-\nu)!\ h^{\nu}\left\{(-1)^{\nu}\binom{k}{k-\nu}f^{(\nu)}(b) - \binom{m}{m-\nu}f^{(\nu)}(a)\right\}$$

$$= (-1)^{k+m}\int_a^b f^{(k+m+1)} g\, dx.$$

Still bearing in mind that $\binom{k}{k-\nu}=0$ when $\nu>k$ and $\binom{m}{m-\nu}=0$ when $\nu>m$, we see that, after division by $(k+m)!$,

$$\sum_{\nu=0}^{k}(-1)^{\nu} f^{(\nu)}(b)\, \frac{h^{\nu}}{\nu!}\, \frac{\binom{k}{\nu}}{\binom{k+m}{\nu}} = \sum_{\nu=0}^{m} f^{(\nu)}(a)\, \frac{h^{\nu}}{\nu!}\, \frac{\binom{m}{\nu}}{\binom{k+m}{\nu}} + R_{k,m} \qquad (2.57)$$

with the remainder term given by

$$R_{k,m} = \frac{(-1)^{k+m}}{(k+m)!}\int_a^b (x-a)^k (x-b)^m f^{(k+m+1)}(x)\, dx. \qquad (2.58)$$

This result includes TAYLOR's theorem as the special case $k=0$.

When $k=1$ and $m=2$, we have the formula

$$f(b) - \frac{h}{3} f'(b) = f(a) + \frac{2h}{3} f'(a) + \frac{h^2}{6} f''(a) + R_{1,2}, \qquad (2.59)$$

which will be used later in Ch. II, § 1.5.

Formula (2.57), like TAYLOR's formula, can be extended to more independent variables; in fact the extension can be derived using precisely the same device as is used for TAYLOR's formula. For simplicity we consider the extension only for a function $u(x, y)$ of two independent variables and for two points (x_0, y_0), (x_1, y_1). All we have to do is to apply formula (2.57) to the function

$$f(t) = u\big(x_0 + t(x_1 - x_0),\ y_0 + t(y_1 - y_0)\big)$$

in the interval $\langle a, b\rangle = \langle 0, 1\rangle$; then, assuming that all derivatives occurring are continuous, we have

$$\left.\begin{aligned}
\sum_{\nu=0}^{k}(-1)^{\nu}\left(h\frac{\partial}{\partial x} + l\frac{\partial}{\partial y}\right)^{\nu} u(x_1, y_1)\, \frac{1}{\nu!}\, \frac{\binom{k}{\nu}}{\binom{k+m}{\nu}} \\
= \sum_{\nu=0}^{m}\left(h\frac{\partial}{\partial x} + l\frac{\partial}{\partial y}\right)^{\nu} u(x_0, y_0)\, \frac{1}{\nu!}\, \frac{\binom{m}{\nu}}{\binom{k+m}{\nu}} + R,
\end{aligned}\right\} \qquad (2.60)$$

where $h = x_1 - x_0$, $l = y_1 - y_0$ and the symbolic notation $\left(h \dfrac{\partial}{\partial x} + l \dfrac{\partial}{\partial y} \right)^{\nu}$ is that generally used in TAYLOR's formula[1]. The form of the remainder term R may be found from (2.58).

§3. Further useful formulae from analysis

3.1. GAUSS'S and GREEN'S formulae for two independent variables

GAUSS's integral theorem in two dimensions: Let B be a closed, finite region of the (x, y) plane bounded by a piecewise-smooth curve Γ without double points. We denote the inward normal to this boundary curve by ν and the arc length measured anti-clockwise from a fixed point on the curve by s (see Fig. I/1). If $f(x, y)$ and $g(x, y)$ are continuous functions with continuous first partial derivatives in B, then[2]

$$\iint_B \frac{\partial f(x, y)}{\partial x} \, dx \, dy$$
$$= - \int_\Gamma f(x, y) \cos(\nu, x) \, ds, \quad \Bigg\} \quad (3.1)$$

$$\iint_B \frac{\partial g(x, y)}{\partial y} \, dx \, dy$$
$$= - \int_\Gamma g(x, y) \cos(\nu, y) \, ds. \quad \Bigg\} \quad (3.2)$$

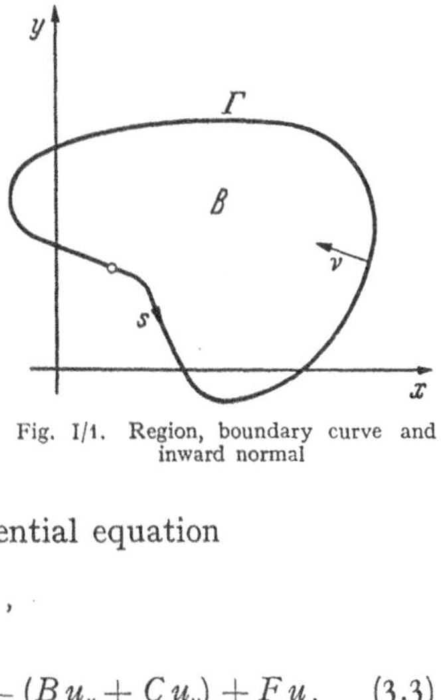

Fig. I/1. Region, boundary curve and inward normal

To apply these formulae to the differential equation

$$L[u] = r(x, y),$$

where

$$L[u] = - \frac{\partial}{\partial x} (A u_x + B u_y) - \frac{\partial}{\partial y} (B u_x + C u_y) + F u, \quad (3.3)$$

in which A, B, C, F are continuous functions with A, B, C also possessing continuous derivatives of at least the first order, we put

$$f = (A \psi_x + B \psi_y) \varphi$$

in (3.1) and

$$g = (B \psi_x + C \psi_y) \varphi$$

in (3.2) and add the two resulting equations (here subscripts denote partial differentiation, e.g. $u_x = \partial u / \partial x$). We then see that any two

---

[1] Cf., for example, H. v. MANGOLDT and K. KNOPP: Einführung in die Höhere Mathematik, Vol. II, 10th ed., p. 355. Stuttgart 1947.

[2] See, for instance, H. v. MANGOLDT and K. KNOPP: Einführung in die Höhere Mathematik, Vol. III, 10th ed., p. 346. Stuttgart 1948.

functions $\varphi(x, y)$ and $\psi(x, y)$ with continuous partial derivatives of up to the second order satisfy GREEN's formula

$$\left.\begin{array}{r} \iint\limits_{B} [A\,\varphi_x\psi_x + B\,(\varphi_x\psi_y + \varphi_y\psi_x) + C\,\varphi_y\psi_y + F\,\varphi\psi]\,dx\,dy - \\ - \iint\limits_{B} \varphi L\,[\psi]\,dx\,dy = -\int\limits_{\Gamma} \varphi L^*\,[\psi]\,ds, \end{array}\right\} \quad (3.4)$$

where L^* is defined by

$$L^*\,[\psi] = (A\,\psi_x + B\,\psi_y)\cos(\nu, x) + (B\,\psi_x + C\,\psi_y)\cos(\nu, y). \quad (3.5)$$

If φ and ψ are interchanged in (3.4) and the resulting equation is subtracted from (3.4), the first integrals cancel out and we obtain the equation

$$\iint\limits_{B} (\varphi L\,[\psi] - \psi L\,[\varphi])\,dx\,dy = \int\limits_{\Gamma} (\varphi L^*\,[\psi] - \psi L^*\,[\varphi])\,ds. \quad (3.6)$$

This is often known as GREEN's formula also. In the special case $A \equiv C \equiv 1$, $B \equiv F \equiv 0$, equations (3.4) and (3.6) become

$$\iint\limits_{B} (\varphi_x\psi_x + \varphi_y\psi_y)\,dx\,dy + \iint\limits_{B} \varphi\nabla^2\psi\,dx\,dy = -\int\limits_{\Gamma} \varphi\frac{\partial\psi}{\partial\nu}\,ds \quad (3.7)$$

and

$$\iint\limits_{B} (\varphi\nabla^2\psi - \psi\nabla^2\varphi)\,dx\,dy = -\int\limits_{\Gamma} \left(\varphi\frac{\partial\psi}{\partial\nu} - \psi\frac{\partial\varphi}{\partial\nu}\right)ds, \quad (3.8)$$

where ∇^2 is the usual symbol for LAPLACE's differential operator

$$\nabla^2\varphi = \frac{\partial^2\varphi}{\partial x^2} + \frac{\partial^2\varphi}{\partial y^2}$$

and $\partial/\partial\nu$ denotes differentiation along the inward normal:

$$\frac{\partial\psi}{\partial\nu} = \frac{\partial\psi}{\partial x}\cos(\nu, x) + \frac{\partial\psi}{\partial y}\cos(\nu, y). \quad (3.9)$$

3.2. Corresponding formulae for more than two independent variables

We also need GAUSS's integral theorem in space, say the space S_m of m dimensions. Let B be a closed, finite region in the $(x_1, x_2, \ldots, x_m)$ space bounded by a surface Γ [an $(m-1)$-dimensional hypersurface] which may consist of a finite number of "faces", each having continuous tangential hyperplanes. Let v be a given vector field in B with components $a_1, a_2, \ldots, a_m$, each being a continuous function of $x_1, x_2, \ldots, x_m$ with continuous partial derivatives of the first order. If $d\tau = dx_1\,dx_2 \ldots dx_m$ is the volume element in B, dS the surface element on Γ, ν the inward normal to Γ and $v_\nu = (v, \nu)$ the component of v along the inward normal (i.e. the scalar product of v with the unit vector ν directed inwards from and perpendicular to Γ), then GAUSS's integral theorem states that

$$\int_B \operatorname{div} v \, d\tau = -\int_\Gamma v_\nu \, dS \tag{3.10}$$

i.e.

$$\int_B \sum_{i=1}^m \frac{\partial a_i}{\partial x_i} \, d\tau = -\int_\Gamma \sum_{i=1}^m a_i \cos(\nu, x_i) \, dS. \tag{3.11}$$

From this we now derive a formula which will be useful when we deal with the differential equation

$$L[u] = -\sum_{i,k=1}^m \frac{\partial}{\partial x_i}\left(A_{ik}\frac{\partial u}{\partial x_k}\right) + q u = r, \tag{3.12}$$

where the coefficients q and r are continuous and the A_{ik} have continuous first derivatives. If, in (3.11), we put

$$a_i = \varphi \sum_{k=1}^m A_{ik}\frac{\partial \psi}{\partial x_k},$$

then, for two functions φ and ψ with continuous second derivatives, we have [in the notation of (3.12)]

$$\left.\begin{aligned}
\int_B \varphi L[\psi]\, d\tau &= \int_B \left(-\varphi\sum_{i,k=1}^m \frac{\partial}{\partial x_i}\left(A_{ik}\frac{\partial\psi}{\partial x_k}\right) + q\,\varphi\psi\right) d\tau \\
&= \int_B \left\{-\sum_{i=1}^m \frac{\partial}{\partial x_i}\left(\varphi\sum_{k=1}^m A_{ik}\frac{\partial\psi}{\partial x_k}\right) + \sum_{i,k=1}^m A_{ik}\frac{\partial\varphi}{\partial x_i}\frac{\partial\psi}{\partial x_k} + q\,\varphi\psi\right\} d\tau \\
&= J[\varphi,\psi] + \int_\Gamma \varphi L^*[\psi]\, dS,
\end{aligned}\right\} \tag{3.13}$$

where

$$J[\varphi,\psi] = \int_B \left(\sum_{i,k=1}^m A_{ik}\frac{\partial\varphi}{\partial x_i}\frac{\partial\psi}{\partial x_k} + q\,\varphi\psi\right) d\tau, \tag{3.14}$$

$$L^*[\psi] = \sum_{i,k=1}^m A_{ik}\frac{\partial\psi}{\partial x_k}\cos(\nu, x_i). \tag{3.15}$$

3.3. Co-normals and boundary-value problems in elliptic differential equations

If we introduce a positive number A and a direction σ so that

$$\sum_{i=1}^m A_{ik}\cos(\nu, x_i) = A_k^* = A\cos(\sigma, x_k),$$

then (3.15) can be written

$$L^*[\psi] = A\frac{\partial\psi}{\partial\sigma}.$$

The half-line running from Γ in the direction σ is called the "co-normal"[1].

---

[1] See, for example, A. G. WEBSTER and G. SZEGÖ: Partielle Differential-gleichungen der mathematischen Physik, p. 311. Leipzig and Berlin 1930.

If the matrix A_{ik} is symmetric $(A_{ik}=A_{ki})$, then, by interchanging φ and ψ in (3.13) and subtracting the resulting equation from (3.13), we obtain GREEN's formula

$$\left.\begin{aligned}\int_B (\varphi L [\psi] - \psi L [\varphi])\, d\tau &= \int_\Gamma (\varphi L^* [\psi] - \psi L^* [\varphi])\, dS \\ &= \int_\Gamma A \left(\varphi \frac{\partial \psi}{\partial \sigma} - \psi \frac{\partial \varphi}{\partial \sigma}\right) dS.\end{aligned}\right\} \quad (3.16)$$

In the special case

$$A_{ik} = \delta_{ik} = \begin{cases} 0 & \text{for} \quad i \neq k \\ 1 & \text{for} \quad i = k, \end{cases}$$

we have $A = 1$ and the co-normal σ coincides with the inward normal ν. If, in addition, $q = 0$, then with

$$\nabla^2 = \sum_{i=1}^m \frac{\partial^2}{\partial x_i^2}$$

the formulae (3.13), (3.16) become

$$-\int_B \varphi \nabla^2 \psi\, d\tau = \int_B (\text{grad}\, \varphi, \text{grad}\, \psi)\, d\tau + \int_\Gamma \varphi \frac{\partial \psi}{\partial \nu}\, dS, \quad (3.17)$$

$$\int_B (\varphi \nabla^2 \psi - \psi \nabla^2 \varphi)\, d\tau = -\int_\Gamma \left(\varphi \frac{\partial \psi}{\partial \nu} - \psi \frac{\partial \varphi}{\partial \nu}\right) dS, \quad (3.18)$$

where $\partial/\partial \nu$ denotes, in the usual way, differentiation in the direction of the inward normal:

$$\frac{\partial \varphi}{\partial \nu} = \sum_{i=1}^m \frac{\partial \varphi}{\partial x_i} \cos(\nu, x_i). \quad (3.19)$$

If the differential equation is of elliptic type, i.e. the matrix A_{ik} is positive definite, so that for arbitrary real numbers $p_1, p_2, \ldots, p_m$, not all zero,

$$\sum_{i,k=1}^m A_{ik} p_i p_k > 0, \quad (3.20)$$

the co-normal points into the interior of the region B since the scalar product of the co-normal vector with the inward normal is then positive:

$$\sum_{k=1}^m A_k^* \cos(\nu, x_k) = \sum_{i,k=1}^m A_{ik} \cos(\nu, x_i) \cos(\nu, x_k) > 0. \quad (3.21)$$

Boundary conditions of the form

$$A_1 u + A_2 L^* [u] = A_1 u + A_2 A \frac{\partial u}{\partial \sigma} = A_3, \quad (3.22)$$

where the A_1, A_2, A_3 are given functions on the boundary Γ, are often associated with differential equations of the type (3.12). In the case

$A_2 = 0$, $A_1 \neq 0$, the problem of determining u is called the first boundary-value problem; when $A_1 = 0$, $A_2 \neq 0$, it is called the second boundary-value problem and when neither A_1 nor A_2 are zero it is called the third boundary-value problem. In the case $A_2 \neq 0$ it is usually assumed, as will be done in this book, that the boundary Γ is piecewise smooth, i.e. consists of a finite number of sections Γ_ϱ each of which is an $(m-1)$-dimensional closed hypersurface with an $(m-2)$-dimensional boundary Γ_ϱ^* such that the inward normal ν to Γ_ϱ is continuous and tends to a definite limiting direction as any specific boundary point P on Γ_ϱ^* is approached from a point inside Γ_ϱ.

3.4. GREEN'S functions

GREEN's functions are often of great value for theoretical investigations but they can also be used effectively for establishing formulae which are useful in numerical work. For ordinary differential equations one occasionally uses the GREEN's function directly, but for partial differential equations direct application is usually avoided because the associated GREEN's functions are usually either too complicated or even not specifiable explicitly at all.

Let us consider the boundary-value problem (1.12), (1.13). There are classes of such boundary-value problems for which it may be shown that there exists a GREEN's function $G(x_1, \ldots, x_n, \xi_1, \ldots, \xi_n)$, or shortly $G(x_i, \xi_i)$, with the following property:

For any continuous function $r(x_1, \ldots, x_n)$ the boundary-value problem (1.12), (1.13) with homogeneous boundary conditions, i.e.

$$L[u] = r, \qquad U_\mu[u] = 0, \tag{3.23}$$

is equivalent to

$$u(x_j) = \int_B G(x_j, \xi_j)\, r(\xi_j)\, d\xi_j. \tag{3.24}$$

Thus the boundary-value problem (3.23) may be solved with the aid of the GREEN's function by means of (3.24), or in other words, a function calculated from (3.24) with given r satisfies the boundary-value problem (3.23). If a GREEN's function exists, then the boundary-value problem (1.12), (1.13) with inhomogeneous boundary conditions is soluble, for it can be reduced to the case with homogeneous boundary conditions by the introduction of a new function u^* as in § 1.3. The detailed theory of GREEN's functions is covered in text books on differential equations[1].

We note here only two simple examples of GREEN's functions which will be used later[2].

---

[1] See, for example, R. COURANT and D. HILBERT: Methoden der math. Physik, Vol. I, 2nd ed., p. 302 et seq. Berlin 1931.

[2] A short collection of GREEN's functions can be found in L. COLLATZ: Eigenwertaufgaben mit technischen Anwendungen, p. 425/426. Leipzig 1949.

1. The GREEN's function for the boundary-value problem

$$L[u] = -u'' = r(x), \qquad u(0) = u(l) = 0 \tag{3.25}$$

is

$$G(x, \xi) = \begin{cases} \dfrac{x}{l}(l - \xi) & \text{for } x \leqq \xi, \\ \dfrac{\xi}{l}(l - x) & \text{for } x \geqq \xi. \end{cases} \tag{3.26}$$

2. If $\dfrac{nl}{2\pi}$ is not an integer, then

$$L[u] = -u'' - n^2 u = r(x), \qquad u(0) - u(l) = u'(0) - u'(l) = 0 \tag{3.27}$$

has the GREEN's function

$$G(x, \xi) = -\frac{1}{2n \sin \dfrac{nl}{2}} \times \begin{cases} \cos n\left(\dfrac{l}{2} - \xi + x\right) & \text{for } x \leqq \xi, \\ \cos n\left(\dfrac{l}{2} - x + \xi\right) & \text{for } x \geqq \xi. \end{cases} \tag{3.28}$$

3.5. Auxiliary formulae for the biharmonic operator

A formula which is used repeatedly in biharmonic problems, namely (3.38) below, can be derived easily from (3.8). Let u, v, w be three functions of x, y (all functions mentioned here should possess derivatives of orders as high as occur) such that

$$\nabla^4 u = \nabla^4 v \quad \text{in } B \tag{3.29}$$

and

$$u = w, \qquad \frac{\partial u}{\partial v} = \frac{\partial w}{\partial v} \quad \text{on } \Gamma, \tag{3.30}$$

where the region B, boundary Γ and inward normal v are as used in § 3.2 and ∇^4 is the biharmonic operator defined by

$$\nabla^4 u \equiv \frac{\partial^4 u}{\partial x^4} + 2\frac{\partial^4 u}{\partial x^2 \partial y^2} + \frac{\partial^4 u}{\partial y^4}. \tag{3.31}$$

If we demand further that u shall satisfy the boundary-value problem

$$\nabla^4 u = p(x, y) \quad \text{in } B \tag{3.32}$$

with

$$u = f(s), \qquad u_v = g(s) \quad \text{on } \Gamma \quad \left(u_v = \frac{\partial u}{\partial v}\right), \tag{3.33}$$

then v and w will be functions satisfying the differential equation and boundary conditions, respectively.

Let us define

$$D[\varphi, \psi] = \iint_B \nabla^2 \varphi \, \nabla^2 \psi \, dx \, dy, \qquad D[\varphi] = D[\varphi, \varphi], \tag{3.34}$$

for any two functions φ and ψ. Then

$$D[\varphi + \psi] = D[\varphi] + 2D[\varphi, \psi] + D[\psi]. \tag{3.35}$$

Further from SCHWARZ's inequality we have

$$(D[\varphi, \psi])^2 \leqq D[\varphi] D[\psi]. \tag{3.36}$$

If we replace φ by $V^2\varphi$ in (3.8), we obtain

$$\left.\begin{aligned} D[\varphi, \psi] &= \iint_B V^2\varphi \, V^2\psi \, dx \, dy \\ &= \iint_B \psi \, V^4\varphi \, dx \, dy + \int_\Gamma \left(\psi (V^2\varphi)_\nu - \psi_\nu V^2\varphi \right) ds. \end{aligned}\right\} \tag{3.37}$$

We now calculate

$$D[v - w] = D[v - u + u - w] = D[v - u] + D[u - w] + \\ + 2D[v - u, u - w].$$

If $\varphi = v - u$, $\psi = u - w$, then $V^4\varphi = 0$ from (3.29) and $\psi = \psi_\nu = 0$ on Γ from (3.30); putting these values in (3.37), we have

$$D[v - u, \ u - w] = 0.$$

Hence, if u, v, w satisfy (3.29) and (3.30),

$$D[v - w] = D[v - u] + D[u - w]. \tag{3.38}$$

All three terms in this equation are non-negative so that

$$D[v - u] \leqq D[v - w] \quad \text{and} \quad D[u - w] \leqq D[v - w]. \tag{3.39}$$

If the boundary-value problem (3.32), (3.33) possesses a solution $u(x, y)$, then from (3.37) we can obtain a formula for the value $u(x_0, y_0)$ at the point (x_0, y_0) by using the "fundamental" solution

$$\varrho(x, y, x_0, y_0) = r^2 \ln r, \quad \text{where} \quad r = +\sqrt{(x - x_0)^2 + (y - y_0)^2}. \tag{3.40}$$

We have

$$\varrho_r = \frac{\partial \varrho}{\partial r} = r(2 \ln r + 1), \quad V^2\varrho = \varrho_{rr} + \frac{1}{r}\varrho_r = 4(\ln r + 1), \quad (V^2\varrho)_r = \frac{4}{r}, \quad V^4\varrho = 0.$$

We now form $D[\varrho, u] - D[u, \varrho]$ using (3.37), integrate over the region which consists of that part of B outside of a small circle C with radius δ and centre (x_0, y_0) and then let δ tend to zero. Since the contribution from $\int_C u(V^2\varrho)_r \, ds$ becomes $8\pi u(x_0, y_0)$ and all other contributions from integrals around C tend to zero as $\delta \to 0$, we have

$$8\pi u(x_0, y_0) = \iint_B \varrho \, V^4 u \, dx \, dy + \int_\Gamma [u_\nu V^2\varrho - u(V^2\varrho)_\nu + \varrho(V^2 u)_\nu - \varrho_\nu V^2 u] \, ds. \tag{3.41}$$

§4. Some error distribution principles

We describe here some approximate methods which use various principles for distributing the error as uniformly as possible throughout the domain of the solution. These are generally applicable in all the following chapters, including Ch. VI, but for convenience they are described here with reference to partial differential equations.

4.1. General approximation. "Boundary" and "interior" methods

Problems in differential equations are often attacked by assuming as an approximation to the solution $u(x_1, \ldots, x_n)$ or $y(x)$ of an initial-value or boundary-value problem (1.9), (1.10) an expression of the form

$$u \approx w(x_1, \ldots, x_n, a_1, \ldots, a_p) \qquad (4.1)$$

which depends on a number of parameters $a_1, \ldots, a_p$ and is such that, for arbitrary values of the a_ϱ,

(1) the differential equation is already satisfied exactly ("boundary" method),

or (2) the boundary conditions are already satisfied exactly ("interior" method),

or (3) w satisfies neither the differential equation nor the boundary conditions, in which case we speak of a "mixed" method.

One then tries to determine the parameters a_ϱ so that w satisfies

in case (1) (boundary method) the boundary conditions,
in case (2) (interior method) the differential equation,
in case (3) (mixed method) the boundary conditions and the differential equations

as accurately as possible in some sense yet to be defined (this is done in § 4.2).

For ordinary differential equations interior methods are used mostly, for if we did know the general solution of the differential equation, the fitting of the parameters to the boundary conditions would still require the solution of a set of simultaneous (possibly non-linear) equations. For partial differential equations, on the other hand, both boundary and interior methods are used, but in general boundary methods are to be preferred since their use, in so far as integration is involved, requires the evaluation of integrals over the boundary rather than throughout the region. This applies also when collocation is used: the two types of method offer the alternatives of boundary collocation and collocation throughout the region, the former being the more acceptable and less prone to error (see I a of the following section). Mixed methods (case 3) are used when the differential equations and boundary conditions are rather complicated.

If, in case (1), we insert the approximation w in a boundary condition $V_\mu = 0$, we are left with an error function

$$\left.\begin{aligned}\varepsilon_\mu(x_1, \ldots, x_n, a_1, \ldots, a_p) \\ = V_\mu(x_1, \ldots, x_n, w, w_1, \ldots, w_n, w_{11}, \ldots, w_{n1}, \ldots)\end{aligned}\right\} \quad (4.2)$$

defined on Γ_μ (the subscripts to w denote precisely the same partial derivatives of w as were defined for u in (1.11), e.g. $w_{12} = \dfrac{\partial^2 w}{\partial x_1 \partial x_2}$).
Similarly, for case (2), we have the error function

$$\varepsilon(x_j, a_\varrho) = F(x_j, w, w_1, \ldots, w_{11}, \ldots) \quad (4.3)$$

defined in the region B and for case (3) we have two error functions ε_μ and ε. The parameters a_ϱ must then be determined so that these error functions ε_μ and ε approximate the zero function as nearly as possible on Γ_μ and in B, respectively. Various principles can be formulated for doing this, which we now describe; for brevity we will usually refer only to the error function ε for case (2), but what is said is naturally applicable also to ε_μ for case (1) and to both ε and ε_μ for case (3).

4.2. Collocation, least-squares, orthogonality method, partition method, relaxation

I a. Pure collocation. The error ε is made to vanish at p points $P_1, \ldots, P_p$ — the "collocation" points. One tries to distribute these p points fairly uniformly over the region B or boundary surfaces Γ_μ. If the co-ordinates of the P_ϱ are $x_{1\varrho}, \ldots, x_{n\varrho}$, then the equations for determining the a_σ read

$$\varepsilon(x_{1\varrho}, x_{2\varrho}, \ldots, x_{n\varrho}, a_1, \ldots, a_p) = 0 \quad (\varrho = 1, \ldots, p). \quad (4.4)$$

In general, collocation should be used as a boundary method wherever possible, for a reasonable uniformity in the distribution of the collocation points can be achieved more easily on the boundaries Γ_μ than in the region B. (It is, for example, easier to arrange p points uniformly around the circumference of a circle than throughout its interior.) Few investigations have been made into the suitable choice of collocation points.

I b. Collocation with derivatives. The calculation is sometimes simplified if ε is not equated to zero at p points but at q $(<p)$ points, the number of the equations for the a_σ then being made up to p by equating to zero derivatives of ε at several points which may or may not coincide with any of the first q points.

II a. Least-squares method. Here we require that the mean square error shall be as small as possible:

$$J = \int_B \varepsilon^2 \, d\tau + \sum_{\mu=1}^k \int_{\Gamma_\mu} \varepsilon_\mu^2 \, dS = \text{minimum.} \quad (4.5)$$

For a boundary method (case 1) the first integral, over the region B, would not appear and for an interior method (case 2) the second term, involving integrals over the boundary surfaces, would be absent. The requirement that J should be a minimum leads to the well-known necessary conditions

$$\frac{1}{2}\frac{\partial J}{\partial a_\varrho} = \int_B \varepsilon \frac{\partial \varepsilon}{\partial a_\varrho}\,d\tau + \sum_{\mu=1}^{k}\int_{\Gamma_\mu} \varepsilon_\mu \frac{\partial \varepsilon_\mu}{\partial a_\varrho}\,dS = 0 \qquad (\varrho = 1, \dots, p), \qquad (4.6)$$

which constitute p equations, non-linear in general, for the determination of the a_σ.

II b. Least-squares with weighting functions[1]. Let $P(x_1, \dots, x_n)$ and $P_\mu(x_1, \dots, x_n)$ be chosen as positive weighting functions in B and on Γ_μ, respectively. Then instead of (4.5) we require that

$$J = \int_B P\varepsilon^2\,d\tau + \sum_{\mu=1}^{k}\int_{\Gamma_\mu} P_\mu \varepsilon_\mu^2\,dS = \text{minimum.} \qquad (4.7)$$

III. Orthogonality method. For an interior method (and similarly for a boundary method) we choose p linearly independent functions $g_\varrho(x_1, \dots, x_n)$ and require that ε shall be orthogonal to these functions in the region B, i.e.

$$\int_B \varepsilon g_\varrho\,d\tau = 0 \qquad (\varrho = 1, \dots, p); \qquad (4.8)$$

the g_ϱ are often chosen to be the first p functions of a complete system of functions in B.

IV. Partition method. The region B (and similarly each bounding surface Γ_μ for a boundary method) is partitioned into p sub-regions $B_1, \dots, B_p$ and we require that the integral of ε over each sub-region be zero, i.e.

$$\int_{B_\varrho} \varepsilon\,d\tau = 0 \qquad (\varrho = 1, \dots, p). \qquad (4.9)$$

V. Relaxation. We choose a set of values of the a_ϱ and calculate the corresponding values of ε (or ε_μ) at a large number of points in B (or on Γ_μ); these values are called the "residuals" in relaxational parlance. We then note what changes are produced in these residuals by altering, or "relaxing", each a_ϱ in turn by an amount δa_ϱ say. It is frequently quite easy to see from this how much we must alter the individual a_ϱ by in order to make the magnitudes of the residuals as small as possible. This method allows one considerable latitude in the actual calculation; with practice it often leads to the required result more quickly than any other method.

---

[1] PICONE, M.: Analisi quantitativa ed esistenziale nei problemi di propagazione. Atti del 1° Congresso dell'Unione Matematica Italiana 1937.

This list of principles can easily be extended but those already mentioned are, in fact, the ones which have been most used hitherto. While on the subject of error distribution, we should also mention the principle of the smallest maximum error, which is of importance for boundary-value problems with elliptic differential equations (see Ch. V).

Combinations and variations of these principles may also be used; for example, one can choose $q \ (>p)$ points $P_1, \ldots, P_q$ in B and require that

$$ J = \sum_{\sigma=1}^{q} [\varepsilon(P_\sigma, a_1, \ldots, a_p)]^2 = \text{minimum}, $$

so that the a_ϱ are determined by the equations $\dfrac{\partial J}{\partial a_\varrho} = 0 \ (\varrho = 1, \ldots, p)$. Many further variants can be devised; see, for example, Note 3 in Ch. III, § 5.3.

4.3. The special case of linear boundary conditions

We now assume that the boundary conditions are linear and of the form (1.13), although the differential equation may still be non-linear. We may therefore take the approximate solution w to be a linear expression

$$ w = v_0(x_1, \ldots, x_n) + \sum_{\varrho=1}^{p} a_\varrho v_\varrho(x_1, \ldots, x_n) \tag{4.10} $$

in the parameters a_ϱ with v_0 satisfying the inhomogeneous boundary conditions and the v_ϱ the corresponding homogeneous conditions, i.e.

$$ U_\mu(v_0) = \gamma_\mu, \qquad \left(\mu = 1, \ldots, k \right) \tag{4.11} $$
$$ U_\mu(v_\varrho) = 0 \qquad \left(\varrho = 1, \ldots, p \right). \tag{4.12} $$

This ensures that the w given by (4.10) satisfies the prescribed boundary conditions (1.13). The v_ϱ are naturally assumed to be linearly independent.

All the principles described in § 4.2 can, of course, be applied to problems of this particular type, but we single out for mention here a noteworthy special case of the orthogonality method (principle III of § 4.2).

III a. GALERKIN'S method. This is the special case of the orthogonality method in which the functions v_ϱ are used for the g_ϱ; thus the equations (4.8), with ε replaced by its explicit formula (4.3), here read

$$ \int_B F(x_1, \ldots, x_n, w, w_1, \ldots, w_n, \ldots) v_\varrho(x_1, \ldots, x_n) \, d\tau = 0 $$
$$ (\varrho = 1, \ldots, p). \tag{4.13} $$

These are GALERKIN's equations and are easily remembered (cf. § 5.3, Ch. III).

If the differential equation is linear as well as the boundary conditions, then the equations determining the a_ϱ are also linear for any of the principles listed in § 4.2. Further, the least-squares method can be regarded as the special case of the orthogonality method in which $g_\varrho = \partial\varepsilon/\partial a_\varrho$.

As in (1.12), let the linear differential equation be written

$$L[u] = r, \tag{4.14}$$

where $L[u]$ is a linear expression in u and its partial derivatives and $r(x_1, \ldots, x_n)$ is a position function in B. The error function ε is now linear in the a_ϱ:

$$\varepsilon = L[w] - r = \sum_{\varrho=1}^{p} a_\varrho L[v_\varrho] + L[v_0] - r = \sum_{\varrho=1}^{p} a_\varrho V_\varrho + V_0 - r, \tag{4.15}$$

where we have put $V_\varrho(x_1, \ldots, x_n) = L[v_\varrho]$.

I. If collocation is used, p collocation points P_ϱ with co-ordinates $x_{1\varrho}, \ldots, x_{n\varrho}$ are chosen and the a_ϱ found from the p equations

$$\sum_{\nu=1}^{p} v_{\varrho\nu} a_\nu = T_\varrho, \tag{4.16}$$

where $\quad v_{\varrho\nu} = V_\nu(x_{1\varrho}, x_{2\varrho}, \ldots, x_{n\varrho})$ $\qquad (\varrho = 1, \ldots, p).$

and $\quad T_\varrho = r(x_{1\varrho}, \ldots, x_{n\varrho}) - V_0(x_{1\varrho}, \ldots, x_{n\varrho}) \tag{4.17}$

III. Similarly, if the orthogonality method is used, p functions $g_\varrho(x_1, \ldots, x_n)$ are chosen and the a_ϱ determined from the p equations

$$\sum_{\nu=1}^{p} w_{\varrho\nu} a_\nu = t_\varrho, \tag{4.18}$$

where $\quad w_{\varrho\nu} = \int_B g_\varrho V_\nu \, d\tau$ $\qquad (\varrho = 1, \ldots, p).$

and $\quad t_\varrho = \int_B g_\varrho(r - V_0) \, d\tau \tag{4.19}$

As mentioned above, these include as special cases GALERKIN'S[1] equations when $g_\varrho = v_\varrho$ (principle IIIa) and the equations for the least-squares method when $g_\varrho = L[v_\varrho] = V_\varrho$ (principle II).

[1] GALERKIN, B. G.: Reihenentwicklungen für einige Fälle des Gleichgewichts von Platten und Balken. Wjestnik Ingenerow Petrograd 1915, H. 10 [Russian]. — HENCKY, H.: Eine wichtige Vereinfachung der Methode von RITZ zur angenäherten Behandlung von Variationsaufgaben. Z. Angew. Math. Mech. 7, 80—81 (1927). — DUNCAN, W. J.: The principles of the Galerkin method. Rep. and Mem. No. 1848 (3694), Aero. Res. Comm. 1938, 24 pp.

IV. If the region B is divided into sub-regions B_ϱ, then for the determination of the a_ϱ by the partition method we have the p equations

$$\sum_{\nu=1}^{p} z_{\varrho\nu} a_\nu = \zeta_\varrho,$$ (4.20)

where

$$z_{\varrho\nu} = \int_{B_\varrho} V_\nu \, d\tau$$

$$\left. \right\} \quad (\varrho = 1, \dots, p).$$

and

$$\zeta_\varrho = \int_{B_\varrho} (r - V_0) \, d\tau$$ (4.21)

4.4. Combination of iteration and error distribution

Consider the problem (1.9), (1.10) with the differential equation in the form

$$M[u] = P[u],$$

where M and P are functions of the x_j, u and its derivatives. We define an iterative scheme in which a sequence of approximations u_n is generated from an arbitrary function u_0 by the equations

$$M[u_{n+1}] = P[u_n],$$
$$U_\mu[u_{n+1}] = \gamma_\mu \quad (\mu = 1, \dots, k) \left. \right\} \quad (n = 0, 1, \dots).$$

u_0 can still be chosen to depend on p parameters $a_1, \dots, a_p$, in which case u_1 will also be a function of these a_ϱ; we can then demand that u_0 and u_1 shall be as close as possible, i.e. that the difference

$$\zeta = \zeta(a_\varrho, x_j) = u_1 - u_0$$

shall be as small as possible. The error distribution principles[1] described in § 4.2 are at our disposal for satisfying this requirement.

If $P[u]$ is linear in u and contains no derivatives of u, i.e. the differential equation has the form

$$M[u] = p(x_j) + q(x_j) u,$$

then

$$q(x_j)\zeta = p(x_j) + q(x_j) u_1 - M[u_1]$$

and, apart from the factor $q(x_j)$ multiplying ζ, the method coincides with the ordinary error distribution methods of § 4.2 only now they are applied to the first iterate u_1 obtained from u_0. With a suitable choice of M and P we may expect better results from this method than

---

[1] NovožiLov, V. V.: On an approximate method of solution of boundary problems for ordinary differential equations. Akad. Nauk SSSR. Prikl. Mat. Meh. **16**, 305–318 (1952) [Russian]. [Reviewed in Math. Rev. **13**, 993 (1952); also in Zbl. Math. **46**, 343.] NovoziLov gives particular prominence to least-squares and collocation.

if we had applied the error distribution methods directly to the error function formed with u_0. An example using a combination of iteration and collocation will be found in Ch. III, § 4.8, I.

§5. Some useful results from functional analysis

In this section we first prove a fundamental general theorem on iteration which can be used effectively in several places in this book. In order to avoid placing needless restrictions on its range of application from the start, it is necessary to employ a general formulation. The representation of functional analysis can be used with advantage for this purpose, for, although it seems rather abstract at first, it proves to be very fruitful. Certain manifolds of functions (or other "elements") are considered and each "element" is regarded as a "point" of a "space". The introduction in § 5.1 is purposely kept broad.

5.1. Some basic concepts of functional analysis with examples

Let S be an abstract space containing elements denoted by $f, f_1, f_2, \ldots$ In this book these elements will be continuous functions but this fact is not used in the general theorem of § 5.2. The symbol $\in$ used in $f \in S$ signifies that f "is contained in" S. For any two elements f_1, f_2 of the space (or, shortly, for $f_1, f_2 \in S$) we define a "distance" as a real number $\|f_1 - f_2\|$ with the following properties:

1. Symmetry: $\|f_1 - f_2\| = \|f_2 - f_1\|$.

2. Positive definiteness: $\|f_1 - f_2\| \geq 0$ for any $f_1, f_2 \in S$ and $\|f_1 - f_2\| = 0$ if and only if $f_1 = f_2$. $\left.\right\}$ (5.1)

3. Triangular inequality: $\|f_1 - f_2\| \leq \|f_1 - f_3\| + \|f_3 - f_2\|$ for $f_1, f_2, f_3 \in S$. $\left.\right\}$ (5.2)

For the space S^0 of single-valued continuous functions $f(x)$ of the real variable x in the closed interval $\langle a, b \rangle$ we might define the "distance" between f_1 and f_2 as the maximum absolute value of the difference $f_1 - f_2$:

$$\|f_1 - f_2\| = \max_{a \leq x \leq b} |f_1(x) - f_2(x)|; \qquad (5.3)$$

it can be seen immediately that this definition possesses the three properties listed above. For the same space we could define a more general distance by

$$\|f_1 - f_2\| = \max_{a \leq x \leq b} \frac{|f_1(x) - f_2(x)|}{W(x)}, \qquad (5.4)$$

where $W(x)$ is a fixed, positive, continuous function in $\langle a, b \rangle$, e.g. e^x.

For the applications in this book there is usually a commutative addition of elements, and with respect to this addition S forms an additive group. S contains a zero element Θ such that, for $f_1 \in S$, $\Theta + f_1 = f_1$ and to each f_1 there corresponds an "inverse element" $-f_1$ with $f_1 + (-f_1) = \Theta$. (In the above example $S = S^0$, the function $f(x) \equiv 0$ is the zero element.) For such function spaces one can work with the "norm" $\|f\|$ of an element f instead of the "distance"; the norm is the distance from the zero element:

$$\|f\| = \|f - \Theta\|. \tag{5.5}$$

Further, we need the idea of the completeness of a space S or sub-space F of S. A sub-space F of S (which may be improper, i.e. coincident with S) is termed complete when to every sequence $f_1, f_2, \ldots$ of elements in F such that

$$\lim_{m, n \to \infty} \|f_m - f_n\| = 0 \tag{5.6}$$

there is a limit element f such that

$$\lim_{n \to \infty} \|f - f_n\| = 0 \tag{5.7}$$

which also belongs to F.

It is an important fact that the space S^0 of continuous functions in $\langle a, b \rangle$, considered as an example above, is complete[1] under the distance definition (5.4), as also is the sub-space of these functions for which $u_1(x) \leq f(x) \leq u_2(x)$, where $u_1(x)$ and $u_2(x)$ are two fixed, continuous functions in $\langle a, b \rangle$ with $u_1 \leq u_2$. This can be seen from (5.4) and (5.6), which imply that

$$|f_m - f_n| \leq \varrho \|f_m - f_n\|,$$

where $\varrho = \max_{a \leq x \leq b} W(x)$, and hence the sequence $f_n(x)$ is uniformly convergent; as is well known, the limit function of a uniformly convergent sequence of continuous functions in a closed interval is also continuous and hence belongs to the space considered, which is therefore complete.

We now define an "operator" (or "transformation") T which associates a unique element Tf of a sub-space F^* of S with each element f of a sub-space F (the inverse mapping of F^* onto F need not be unique, nor need F^* be different from F).

In the space S^0 we can construct a considerable variety of operators, for example,

$$Tf = \int_a^x f(\xi) \, d\xi, \quad Tf = f(x) \, W(x), \quad Tf = \sin(f(x));$$

on the other hand, $Tf = df/dx$ need not be an admissible operator, for, if the sub-space F is chosen appropriately, differentiation can lead to functions not belonging to S^0.

---

[1] BANACH, ST.: Théorie des opérations linéaires. Warsaw 1932. Recent impression New York 1949, p. 11.

An operator T is said to be Lipschitz-bounded in F with reference to the chosen distance definition if there is a "Lipschitz constant" K such that

$$\| T f_1 - T f_2 \| \leqq K \| f_1 - f_2 \| \tag{5.8}$$

for all $f_1, f_2 \in F$.

To give an example, let us consider again the space S^0 of continuous functions in $\langle a, b \rangle$ with the distance definition (5.4) and define an operator T by

$$T f(x) = \int_a^x G(x, \xi) f(\xi) \, d\xi,$$

where $G(x, \xi)$ is a given, continuous, bounded function with $|G(x, \xi)| \leqq C$ for $a \leqq x, \xi \leqq b$. This operator is Lipschitz-bounded under the distance definition (5.4) since

$$\| T f_1 - T f_2 \| = \max_{a \leqq x \leqq b} \frac{\left| \int_a^x G(x, \xi) [f_1(\xi) - f_2(\xi)] \, d\xi \right|}{W(x)}$$

$$\leqq \| f_1 - f_2 \| \max_{a \leqq x \leqq b} \frac{\int_a^x |G(x, \xi)| W(\xi) \, d\xi}{W(x)} .$$

If we put $\gamma = \max_{a \leqq x \leqq b} \dfrac{\int_a^x W(\xi) \, d\xi}{W(x)}$, then γC can be used as the Lipschitz constant K.

5.2. The general theorem on iterative processes

Suppose that we require the solutions of the equation

$$f = T f \tag{5.9}$$

(or the "fixed points" or fixed-point elements of the operator T). Let us set up the iterative process[1]

$$u_{n+1} = T u_n \qquad (n = 0, 1, 2, \ldots) \tag{5.10}$$

whereby we proceed from a function $u_0 \in F$ and form $u_1 = T u_0$, $u_2 = T u_1$ and so on up to u_{n+1} as long as u_n lies in F. Then, if F is complete and T Lipschitz-bounded in F, we have for $0 < n < m$

$$\left. \begin{array}{l} \| u_m - u_n \| = \| T u_{m-1} - T u_{n-1} \| \leqq K \| u_{m-1} - u_{n-1} \| \leqq \cdots \\ \qquad\qquad \leqq K^r \| u_{m-r} - u_{n-r} \| \qquad \text{(for } 0 \leqq r \leqq m, n), \end{array} \right\} \tag{5.11}$$

---

[1] KANTOROVICH, L.: The method of successive approximations for functional equations. Acta Math. **71**, 63—97 (1939). — WEISSINGER, J.: Zur Theorie und Anwendung des Iterationsverfahrens. Math. Nachr. **8**, 193—212 (1952).

and from (5.2), (5.11), it follows that

$$||u_m - u_n|| \leqq \sum_{s=n}^{m-1} ||u_{s+1} - u_s||$$

$$\leqq \sum_{t=1}^{m-n} K^t ||u_n - u_{n-1}|| \leqq K^{n-1} \sum_{t=1}^{m-n} K^t ||u_1 - u_0||.$$

We now make the crucial assumption that $K < 1$, so that

$$\sum_{t=1}^{m-n} K^t \leqq \frac{K}{1-K} \quad \text{and} \quad ||u_m - u_n|| \leqq \frac{K^n}{1-K} ||u_1 - u_0||. \quad (5.12)$$

We observe now that an important sub-space of S is the "sphere" Σ containing all the elements h in S with

$$||h - u_1|| \leqq \frac{K}{1-K} ||u_1 - u_0||, \quad (5.13)$$

for it has the property that, if it is contained in the sub-space F, then the iteration process can be carried on indefinitely, i.e. no u_n falls outside of F; this follows from (5.12) with $n = 1$:

$$||u_m - u_1|| \leqq \frac{K}{1-K} ||u_1 - u_0||, \quad (5.14)$$

for this implies that u_m lies in the sphere Σ (and hence also in F) for $m = 1, 2, 3, \ldots$. It follows further from (5.12) that

$$\lim_{m, n \to \infty} ||u_m - u_n|| = 0$$

since $K < 1$, and hence, on account of the completeness of F, there exists a limit element u in F such that

$$\lim_{n \to \infty} ||u - u_n|| = 0. \quad (5.15)$$

Since u lies in F, we can form Tu; then from (5.2), (5.8) we have

$$||Tu - u|| \leqq ||Tu - Tu_n|| + ||Tu_n - u|| \leqq K||u - u_n|| + ||u_{n+1} - u||$$

for all n. Now according to (5.15) the right-hand side tends to zero as $n \to \infty$; hence

$$||Tu - u|| = 0,$$

so that, from the "distance" property (5.1), $Tu = u$.

Thus the limit element u is a solution of (5.9), and the existence of a solution is demonstrated.

Suppose now that there exists another solution v of (5.9), so that $v = Tv$. Then from (5.8) we have

$$||u - v|| = ||Tu - Tv|| \leqq K||u - v||,$$

and since $K < 1$, we must have $\|u - v\| = 0$, i.e. $u = v$. Thus (5.9) can have only one solution in F, and the uniqueness of u in F is also demonstrated.

Finally, it follows from (5.2), (5.14) that

$$\|u - u_1\| \leqq \|u - u_m\| + \|u_m - u_1\| \leqq \|u - u_m\| + \frac{K}{1 - K} \|u_1 - u_0\|,$$

and since, according to (5.15), the term $\|u - u_m\| \to 0$ as $m \to \infty$, we have

$$\|u - u_1\| \leqq \frac{K}{1 - K} \|u_1 - u_0\|, \qquad (5.16)$$

i.e. u itself lies in the sphere Σ. This formula provides an error estimate.

Summary: *Let there be an equation $f = Tf$ (5.9) for an element f of a space S. If we can define a distance satisfying the three conditions of § 5.1 and can choose a sub-space F so that*

1. Tf is defined uniquely for all f in F,

2. F is complete under the chosen distance definition,

3. T satisfies a Lipschitz-condition in F with $K < 1$,

4. besides the chosen u_0, F contains the whole sphere Σ defined by (5.13),

then the iteration (5.10) may be continued indefinitely and the sequence u_n converges in the sense of (5.15) to an element u in Σ which is the unique solution of the given equation (5.9) in the sub-space F. At the same time, (5.16) gives an estimate for the error in the approximation u_1[1].

In many cases it is obvious that all u_n lie in F and the condition 4 can be omitted. Closer error estimates can be obtained in several cases by using pseudo-metric spaces[2]; the distance $\|f - g\|$ between two elements f, g of such a space is an element of a semi-ordered space and thus is not restricted to being a real number, cf. Ch. III, § 4.8.

5.3. The operator T applied to boundary-value problems

Consider the boundary-value problem with the differential equation

$$L[u] = \varphi(x_1, \ldots, x_n, u) \quad \text{in } B \qquad (5.17)$$

---

[1] Another general method based on functional analysis has been investigated by P. C. ROSENBLOOM: The method of steepest descent. Proc. Symp. Appl. Math. VI, New York-Toronto-London 1956, pp. 127−176.

[2] KUREPA, L.: Tableaux ramifiés d'ensembles, espaces pseudo-distanciés. C. R. **198**, 1563−1565 (1934). − SCHRÖDER, J.: Nichtlineare Majoranten beim Verfahren der schrittweisen Näherung. Arch. Math. **7**, 471−484 (1956). − Neue Fehlerabschätzungen für verschiedene Iterationsverfahren. Z. Angew. Math. Mech. **36**, 168−181 (1956). − Über das Newtonsche Verfahren. Arch. Rational Mechanics **1**, 154−180 (1957). − Das Iterationsverfahren bei allgemeinerem Abstandsbegriff. Math. Z. **66**, 111−116 (1956).

and the linear boundary conditions as in § 1.3

$$U_\mu[u] = \gamma_\mu \quad \text{on } \Gamma_\mu \qquad (\mu = 1, \ldots, k), \tag{5.18}$$

where $L[u]$ and $U_\mu[u]$ are linear homogeneous differential expressions in a function $u(x_1, \ldots, x_n)$ and φ is a given function with a continuous derivative with respect to u and which we will assume to be continuous with respect to its other arguments. We can proceed from an arbitrarily chosen function u_0 according to the iterative formulae

$$\left.\begin{aligned} L[u_{q+1}] &= \varphi(x_j, u_q) \quad \text{in } B \\ U_\mu[u_{q+1}] &= \gamma_\mu \quad \text{on } \Gamma_\mu \end{aligned}\right\} \qquad (q = 0, 1, 2, \ldots). \tag{5.19}$$

Thus u_{q+1} is determined from u_q by solution of a linear boundary-value problem.

Now let us assume that for any continuous functions r, γ_μ the problem

$$L[\psi] = r \quad \text{in } B, \qquad U_\mu[\psi] = \gamma_\mu \quad \text{on } \Gamma_\mu \tag{5.20}$$

always possesses a unique solution ψ; this will be the case, for example, when a GREEN's function $G(x_j, \xi_j)$ exists (see § 3.4). Then the iteration cycle can be repeated indefinitely provided that $\varphi(x_j, u_q)$ is defined for all $u = u_q$.

With the aid of a function $\tilde{u}$ satisfying the inhomogeneous boundary conditions, and also possessing continuous derivatives of as high an order as is necessary for the formation of $L[\tilde{u}]$, we can transform the problem into one with homogeneous boundary conditions (this transformation is used only in establishing an error estimate and is not needed in carrying out the computation). Corresponding to a function f we can define a function g by

$$L[g] = \varphi(x_j, \tilde{u} + f) - L[\tilde{u}] \quad \text{in } B, \qquad U_\mu[g] = 0 \quad \text{on } \Gamma_\mu; \tag{5.21}$$

we write this correspondence as $g = Tf$, thus defining an operator T. If now we put $u = \tilde{u} + v$ and $u_q = \tilde{u} + v_q$, then

$$v_{q+1} = T v_q \quad (q = 0, 1, \ldots) \quad \text{and} \quad v = T v, \tag{5.22}$$

and since

$$u_q - u = v_q - v, \tag{5.23}$$

the error at the q-th stage is unaltered by the transformation from u to v.

To determine the Lipschitz constant K of the transformation T, we form

$$\left.\begin{aligned} L[g_1 - g_2] &= \varphi(x_j, \tilde{u} + f_1) - \varphi(x_j, \tilde{u} + f_2), \\ U_\mu[g_1 - g_2] &= 0, \end{aligned}\right\} \tag{5.24}$$

where $g_1 = T f_1$ and $g_2 = T f_2$. We then choose a domain D of the $(x_1, \ldots, x_n, u)$ space which is convex with respect to u and contains a solution u and the iterates u_q to it and assume that φ satisfies the Lipschitz condition

$$|\varphi(x_j, z) - \varphi(x_j, z^*)| \leq N(x_j)|z - z^*| \qquad (5.25)$$

in D. Since φ possesses a continuous derivative with respect to u, we could put $N = \max\limits_{u} \left| \dfrac{\partial \varphi}{\partial u} \right|$. From (5.24) and (5.25) we have

$$|L[H]| \leq N|h| \quad \text{in } B, \qquad U_\mu[H] = 0 \quad \text{on } \Gamma_\mu, \qquad (5.26)$$

where $h = f_1 - f_2$ and $H = T f_1 - T f_2$. From here we can proceed in two ways.

1. Using the GREEN's function. Let $W(x_i)$ be a positive (or possibly non-negative) function in B and, as in (5.4), let the norm of f be defined by

$$\|f\| = \operatorname*{upper\ lim}_{\text{in } B} \left| \frac{f}{W} \right|. \qquad (5.27)$$

Now the inequality in (5.26) can be replaced by the equation

$$L[H] = \vartheta N|h|, \qquad \text{where} \qquad |\vartheta| \leq 1,$$

which can be solved for H by means of the GREEN's function, as in (3.24)

$$H = \int_B \vartheta(\xi_j) G(x_j, \xi_j) N(\xi_j) |h(\xi_j)| d\xi_j;$$

hence

$$|H| \leq \int_B |G(x_j, \xi_j) N(\xi_j) h(\xi_j)| d\xi_j \leq \|h\| \int_B |G(x_j, \xi_j)| N(\xi_j) W(\xi_j) d\xi_j, \qquad (5.28)$$

and we can use

$$K = \operatorname*{upper\ lim}_{\text{in } B} \frac{\int_B |G(x_j, \xi_j)| N(\xi_j) W(\xi_j) d\xi_j}{W(x_j)} \qquad (5.29)$$

as Lipschitz constant.

If the boundary-value problem is linear, so that the differential equation (5.17) can be put in the form

$$L[u] = p(x_j) u + q(x_j), \qquad (5.30)$$

then (5.25) holds with $N(x_j) = |p(x_j)|$.

A simple, if sometimes crude, error estimate can be given when a non-negative GREEN's function exists in B and the eigenvalue problem

$$L[z] = \lambda z \quad \text{in } B, \qquad U_\mu[z] = 0 \quad \text{on } \Gamma_\mu \qquad (5.31)$$

possesses a non-negative eigenfunction $z(x_j)$ in B corresponding to the eigenvalue $\lambda = \lambda_z$. We can then choose $W(x_j) = z(x_j)$, so that, since

$$\lambda_z \int_B G(x_j, \xi_j) z(\xi_j) d\xi_j = z(x_j) \tag{5.32}$$

from (3.24), (5.29) simplifies to

$$K = \frac{\max\limits_{\text{in } B} N(x_j)}{\lambda_z}. \tag{5.33}$$

2. Using a monotonic property. Here we make use of the concept of a boundary-value problem of monotonic type, thus limiting ourselves to real values (see §§ 5.4, 5.6). For our present purposes we modify the definition slightly and say that the problem (5.17), (5.18) has a monotonic character when

$$L[v] \leqq L[w] \quad \text{in } B, \quad U_\mu[v] = U_\mu[w] = 0 \quad \text{on } \Gamma_\mu \tag{5.34}$$

implies that $v \leqq w$, v and w being two functions with continuous partial derivatives of as high an order as is required to form L and U_μ.

Now let $z(x_j)$ be a function such that

$$L[z] = A(x_j) \geqq \alpha > 0 \quad \text{in } B, \quad U_\mu[z] = 0 \quad \text{on } \Gamma_\mu, \tag{5.35}$$

where α is constant. Then under our assumptions it follows from

$$|L[\psi]| \leqq D = \text{const} \quad \text{in } B, \quad U_\mu[\psi] = 0 \quad \text{on } \Gamma_\mu \tag{5.36}$$

that $|\psi|$ is majorized by zD/α:

$$|\psi| \leqq \frac{zD}{\alpha}; \tag{5.37}$$

for we have

$$\left.\begin{aligned}
L\left[\frac{-zD}{\alpha}\right] &\leqq -D \leqq L[\psi] \leqq D \leqq L\left[\frac{zD}{\alpha}\right], \\
U_\mu\left[\pm \frac{zD}{\alpha}\right] &= U_\mu[\psi] = 0,
\end{aligned}\right\} \tag{5.38}$$

and hence, from the monotonic property,

$$-\frac{zD}{\alpha} \leqq \psi \leqq \frac{zD}{\alpha}. \tag{5.39}$$

We now apply (5.36), (5.37) to (5.26); this gives

$$|H| \leqq \frac{|Nh|_{\max} z}{\alpha}. \tag{5.40}$$

Using the same norm as defined in (5.27) with a similar function $W(x_j)$, we obtain

$$\|H\| \leqq \frac{|Nz|_{\max}}{\alpha} \|h\|; \tag{5.41}$$

we can therefore use

$$K = \frac{|N z|_{\max}}{\alpha} \qquad (5.42)$$

as Lipschitz constant.

Detailed numerical examples will be found in Ch. III, § 4.8. The methods of § 5.2 are also applied to a non-linear integral equation in Example III of Ch. VI, § 1.5.

5.4. Problems of monotonic type[1]

Let R be a real space of elements $u, v, \ldots, f, g, \ldots$. These elements may be real numbers, vectors with real components or real-valued functions of real variables and the signs $\leq, <, >, \geq$ shall have their usual significance; applied to vectors, for example, the sign $\leq$ signifies that the inequality holds for all components. (In the terminology of the theory of abstract spaces, R is a semi-ordered space.)

Let there be given an operator T which associates with each element of a proper or improper sub-space M of R a unique element of an image space M^* which is also a sub-space of R. The operator T need not be linear. Now let f be a given element of R; then we ask for the solution u (an element of M) of the equation $Tu = f$. We say this problem is "of monotonic type" (or T is a monotonic operator) when $Tv \leq Tw$ implies that $v \leq w$ for arbitrary elements v, w in M. More accurately we should say that the problem is "written as a problem of monotonic type", for it can certainly happen that a particular problem of analysis can be formulated in different ways such that one formulation is of monotonic type while another is not.

If we assume the existence of a solution u, there is the possibility of "bracketing" it when the problem is of monotonic type: if v_1 and v_2 are two approximate solutions in M such that

$$T v_1 \leq f \leq T v_2, \qquad (5.43)$$

then

$$v_1 \leq u \leq v_2. \qquad (5.44)$$

There is also the possibility of deducing an error estimate for result obtained by the relaxation method. With the relaxation technique we define the "defect" or, in the customary terminology of relaxation, the "residual", $d = d(v) = Tv - f$ of an approximation v and try to make it approximate the zero element as closely as possible by making alterations in v, usually in the nature of small corrections. If we find that $d \geq 0$, we know that $v \geq u$ and similarly $v \leq u$ follows from $d \leq 0$

---

[1] COLLATZ, L.: Aufgaben monotoner Art. Arch. Math. **4**, 366–376 (1952).

The monotonic method requires that we know that a solution u exists; this knowledge can be acquired in several ways:

1. It is often possible to make sure that a solution exists by starting the iteration procedure of § 5.2, possibly with quite a crude first iterate, and appealing to the general theorem of § 5.2 (cf. Ch. III, § 4.8, Ex. II).

2. With the aid of topology and functional analysis, generalizations of BROUWER's fixed-point theorem can be used to establish results concerning the existence of a fixed point for a wide range of problems; for example, non-linear boundary-value problems for elliptic differential equations have been considered by SCHAUDER and LERAY[1] and non-linear integral equations by ROTHE[2], among others.

3. For certain classes of linear and non-linear boundary-value problems the existence of a solution is assured by special theorems.

5.5. Application to systems of linear equations of monotonic type

Consider a system of real linear equations, which may be written in matrix form

$$A x = r, \tag{5.45}$$

where $A = (a_{jk})$ is the matrix of coefficients, $x = (x_j)$ the column matrix (or vector) of the unknowns and $r = (r_j)$ the column matrix of the right-hand sides. A corresponds to the operator T. The problem of determining x is of monotonic type when $A z \geq 0$ implies that $z \geq 0$ for any real vector z. Such a matrix A is called "monotonic" or "of monotonic type", as also is the system of equations $A x = r$. By reversing the sign of z we see also that $A z \leq 0$ implies $z \leq 0$ if A is monotonic; therefore $A z = 0$ implies that $z = 0$, and the determinant of A cannot be zero. For a monotonic matrix A, $A y \leq r \leq A w$ implies that $y \leq x \leq w$, for this is equivalent to: $A(y - x) \leq 0 \leq A(w - x)$ implies that $y - x \leq 0 \leq w - x$; in particular, we can conclude from $|A x| \leq A y$ that $|x| \leq y$ (corresponding to the definition in § 5.4, $|x| \leq y$ signifies that $|x_j| \leq y_j$ for $j = 1, \dots, n$).

A necessary and sufficient condition for a matrix A to be monotonic is that all elements of the inverse matrix A^{-1} be non-negative[3]; for practical work, however, we need simpler criteria which do not involve

---

[1] SCHAUDER, J.: Zur Theorie stetiger Abbildungen in Funktionalräumen. Math. Z. **26**, 47−65 (1927) and notes thereto 417−431. − LERAY, J., and J. SCHAUDER: Topologie et équations fonctionelles. Ann. Sci. École norm. sup. **51**, 45−78 (1934).

[2] ROTHE, E.: Zur Theorie der topologischen Ordnung und der Vektorfelder in Banachschen Räumen. Comp. Math. **5**, 117−197 (1938).

[3] Another criterion which includes our Theorem 1 is given by A. OSTROWSKI: Über die Determinanten mit überwiegender Hauptdiagonale. Comm. Math. Helv. **10**, 69−96 (1937).

determinants and can be applied easily, even if they are only sufficient conditions. Such a criterion[1,2] is furnished by:

Theorem 1. *If an $n \times n$ real matrix A is such that*

1. $a_{jk} \leq 0$ for $j \neq k$,

2. A does not "decompose",

3. there exist non-zero vectors y and r such that $y > 0$, $r \geq 0$ and $A y = r$,

then A is a monotonic matrix.

We say that a matrix $A = (a_{jk})$ "decomposes" when for some integer m in $1 \leq m \leq n - 1$ we can separate the integers $1, \ldots, n$ into two groups $\varrho_1, \ldots, \varrho_m$; $\sigma_1, \ldots, \sigma_{n-m}$ in such a way that $a_{\varrho_\nu \sigma_\mu} = 0$ for $\nu = 1, \ldots, m$; $\mu = 1, \ldots, n - m$.

Proof. For $x \geq 0$ to follow from $A x \geq 0$ we need only show that the two assumptions

(a) z is a vector with at least one negative component, say $z_q < 0$,

(b) $A z \geq 0$

are contradictory.

From assumption (b) it follows that $A[(1 - \lambda) y + \lambda z] \geq 0$ if the real parameter λ lies in the interval $0 \leq \lambda \leq 1$, y being the vector of condition 3. Now under assumption (a) we can choose $\lambda = \Lambda$ in $0 < \lambda < 1$ so that the vector $w = (1 - \Lambda) y + \Lambda z$ has no negative components, but has at least one zero component and is not the zero vector. For since $y_q > 0$ and $z_q < 0$, there is a $\lambda = \lambda_q$ in $\langle 0, 1 \rangle$ such that $(1 - \lambda_q) y_q + \lambda_q z_q = 0$, and similarly for any other negative component of z, say z_r, there is a λ_r in $\langle 0, 1 \rangle$ such that $(1 - \lambda_r) y_r + \lambda_r z_r = 0$; we then let Λ be the smallest of these values λ_r (there are at most n of them); w is not the zero vector since with $0 < \Lambda < 1$, $A y \geq 0$ and $\dfrac{1 - \Lambda}{\Lambda} y = -z$ imply that $A z \leq 0$ and with assumption (b) this implies that $A z = A y = 0$, in contradiction to condition 3 that r is a non-zero vector.

We now use conditions 1 and 2 to contradict the deduction from (b) that $A w \geq 0$. Let $w_{\varrho_1}, \ldots, w_{\varrho_m}$ and $w_{\sigma_1}, \ldots, w_{\sigma_{n-m}}$ be the zero and non-zero (and therefore positive) components of w, respectively. Now

---

[1] A similar criterion is demonstrated geometrically by G. Schulz: Interpolationsverfahren zur numerischen Quadratur gewöhnlicher Differentialgleichungen. Z. Angew. Math. Mech. **12**, 53 (1932). The conditions which Schulz obtains are: $\det A \neq 0$, $a_{jj} > 0$, $a_{ij} < 0$ for $j \neq k$, and our condition 3. with $r > 0$.

[2] That the criterion is not necessary, even in the sense that at least one rearrangement (by row or column interchanges) of a monotonic matrix must satisfy it, is easily shown by simple examples such as the monotonic matrix

$$\begin{pmatrix} 1 & -2 & 8 \\ -2 & 1 & -2 \\ 8 & -2 & 1 \end{pmatrix}.$$

$1 \leq m \leq n-1$, so that, since A does not decompose (condition 2), there is at least one element $a_{\varrho_\nu \sigma_\mu}$ which is non-zero and therefore negative (condition 1). Then calculating the ϱ_ν-th component of Aw, we obtain

$$(Aw)_{\varrho_\nu} = \sum_{k=1}^{n} a_{\varrho_\nu k} w_k = \sum_{\mu=1}^{n-m} a_{\varrho_\nu \sigma_\mu} w_{\sigma_\mu} < 0, \qquad (5.46)$$

so that at least one component of Aw is negative.

If the vector y of condition 3 satisfies the stronger condition in which $r > 0$, condition 2 may be relaxed. For in this case $A\big((1-\lambda)y + \lambda z\big) > 0$ if $0 \leq \lambda < 1$, so that $Aw > 0$; ϱ_ν is therefore chosen arbitrarily this time and we cannot say that there is at least one negative $a_{\varrho_\nu \sigma_\mu}$; however, since $a_{\varrho_\nu \sigma_\mu} \leq 0$, (5.46) holds with $\leq$ instead of $<$ and we still have a contradiction, this time with $Aw > 0$.

As the special case $y_j = 1$, Theorem 1 includes a criterion which is used very frequently in applications: condition 3 becomes the weak "row-sum" criterion[1]

3*.
$$\sum_{k=1}^{n} a_{jk} \begin{cases} \geq 0 & \text{for all } j \\ > 0 & \text{for at least one } j = j_0; \end{cases}$$

consequently conditions 1, 2, 3* are sufficient for A to be monotonic. The case where $r > 0$ in condition 3 corresponds to the ordinary row-sum criterion as defined below. We may therefore state the less general

Theorem 2. *If the coefficients a_{jk} of an $n \times n$ matrix A satisfy the conditions*

1. sign distribution[2]: $a_{jj} > 0$, $a_{jk} \leq 0$ for $j \neq k$,

2a. the "weak row-sum criterion":

$$\sum_{k=1}^{n} a_{jk} \begin{cases} \geq 0 & \text{for } j = 1, 2, \ldots, n, \\ > 0 & \text{for at least one } j = j_0, \end{cases} \qquad (5.47)$$

and 2b. the non-decomposition of A,

or, instead of 2a. and 2b., the stronger condition

2c. the "ordinary row-sum criterion":

$$\sum_{k=1}^{n} a_{jk} > 0 \quad \text{for } j = 1, \ldots, n, \qquad (5.48)$$

then A is monotonic and in particular $\det A \neq 0$.

---

[1] COLLATZ, L.: Aufgaben monotoner Art. Arch. Math. **3**, 373 (1952).

[2] It is convenient to include $a_{jj} > 0$ as a condition of the theorem although it can be deduced from the remaining conditions and could therefore have been omitted.

The solution x of the system of real equations $Ax = r$ can then be bracketed by $v_1 \leqq x \leqq v_2$ provided that two approximations v_1 and v_2 can be found such that $A v_1 \leqq r \leqq A v_2$.

Systems of equations of monotonic type occur frequently in the solution of boundary-value problems by finite differences. As these systems of equations are often solved by iteration we conclude this section with mention of a theorem which will be used later in this context.

A series of iteration procedures can be defined by writing the given matrix A as the sum of two matrices B and C:

$$A = B + C;$$

then starting from an arbitrarily chosen vector x_0 we determine a sequence of vectors x_k from the iterative formula

$$B x_{k+1} + C x_k = r \qquad (k = 0, 1, \ldots). \tag{5.49}$$

We assume that $\det A \neq 0$ and $\det B \neq 0$. If $B = (b_{jk})$ is chosen as the diagonal matrix with

$$b_{jk} = \begin{cases} a_{jk} & \text{for } j = k \\ 0 & \text{for } j \neq k, \end{cases} \qquad c_{jk} = \begin{cases} 0 & \text{for } j = k \\ a_{jk} & \text{for } j \neq k, \end{cases} \tag{5.50}$$

the components of the current iterate are calculated directly from the preceding iterate; this iterative scheme is usually called the "total-step process". If B is chosen to be triangular with

$$b_{jk} = \begin{cases} a_{jk} & \text{for } j \geqq k \\ 0 & \text{for } j < k, \end{cases} \qquad c_{jk} = \begin{cases} 0 & \text{for } j \leqq k \\ a_{jk} & \text{for } j < k, \end{cases} \tag{5.51}$$

then the components of the current iterate are calculated successively; this is usually called the "single-step process".

The following theorem[1] gives sufficient conditions for the applicability of these processes.

Theorem 3. *If the matrix A of the system of equations (5.45) satisfies the ordinary row-sum criterion*

$$\sum_{\substack{k=1 \\ k \neq j}}^{n} |a_{jk}| < a_{jj} \qquad (j = 1, \ldots, n),$$

or if it does not decompose and satisfies the weak row-sum criterion

$$\sum_{\substack{k=1 \\ k \neq j}}^{n} |a_{jk}| \begin{cases} \leqq a_{jj} & \text{for } j = 1, \ldots, n \\ < a_{jj} & \text{for at least one } j = j_0, \end{cases}$$

then $\det A \neq 0$ and both the total-step and single-step iterative processes will converge when applied to the equations (5.45).

---

[1] Cf., for example, Math. Z. **53**, 149—161 (1950).

Several matrices occur in later chapters for which we show that the conditions of Theorem 2 are satisfied. We therefore observe here that those conditions are included in the conditions of Theorem 3 as the special case for which the sign distribution of condition 1. of Theorem 2 holds and that the above-mentioned iterative processes therefore converge for these matrices.

5.6. Non-linear boundary-value problems

As in (5.17), (5.18), consider the linear or non-linear differential equation

$$L[u] + F(x_j, u) = 0 \quad \text{in } B \tag{5.52}$$

for a function $u(x_1, \ldots, x_n)$ of n real variables $x_1, \ldots, x_n$ subject to the linear boundary conditions

$$U_\mu[u] = \gamma_\mu \quad \text{on } \Gamma_\mu \quad (\mu = 1, \ldots, k). \tag{5.53}$$

Here $F(x_1, \ldots, x_n, u)$ is a given function in B possessing a continuous partial derivative with respect to u. Let us define an operator T (not the same operator as in § 5.3) by

$$T v = L[v] + F(x_j, v) \tag{5.54}$$

with a domain of definition D restricted to the domain of functions v which satisfy the boundary conditions $U_\mu[v] = \gamma_\mu$ and possess partial derivatives of order sufficiently high for the formation of the differential expressions $U_\mu[v]$ and $L[v]$.

Then, for any two functions v and w in D, we have

$$T v - T w = L[\varepsilon] + \varepsilon A(x_j), \tag{5.55}$$

where $\varepsilon = v - w$ and $A(x_j)$ is obtained from TAYLOR's theorem

$$F(x_j, v) - F(x_j, w) = \varepsilon \left(\frac{\partial F}{\partial u}\right)_{(x_j, u = \tilde{w})} = \varepsilon A(x_j), \tag{5.56}$$

$\tilde{w}(x_j)$ being a point intermediate between v and w; for fixed v and w, $A(x_j)$ is a function only of position. Now let H be a domain of the $(x_1, \ldots, x_n, u)$ space which is convex with respect to u and contains v and w (and hence also $\tilde{w}$); the assumption that A is non negative in H can be fruitful in several applications. One will frequently choose for H a domain which is known to contain a solution u of (5.52), (5.53) and also an approximation to it, say v; then, with $w = u$, ε is the error $\zeta = v - u$ in the approximation v. If, on the basis of further special properties of the problem (5.52), (5.53), we can show that in H

$$L[\varepsilon] + \varepsilon A(x_j) \geqq 0 \quad \text{in } B, \tag{5.57}$$

$$U_\mu[\varepsilon] = 0 \quad \text{on } \Gamma_\mu, \tag{5.58}$$

with non-negative $A(x_j)$, implies that $\varepsilon \geqq 0$, then the problem (5.52), (5.53) is of monotonic type in H. The non-linear problem is accordingly reduced to the discussion of a linear problem with homogeneous boundary conditions, cf. Ch. III § 4.9 and Ch. V § 4.1.

Chapter II

Initial-value problems
in ordinary differential equations

§ 1. Introduction

First of all, in §§ 1.1—1.3, we discuss some quite general points mainly concerning accuracy. In §§ 1.4—1.7 some crude methods are described which would be used only if a rough idea of the solution over a fairly short range were wanted quickly.

1.1. The necessity for numerical methods

Even with quite simple differential equations it can happen that their solutions are not expressible in closed form and that a numerical approach is the most convenient way of dealing with the problem. For example, the differential equation

$$\frac{dy}{dx} = x + x^2 + y^2$$

does not possess a closed solution in terms of elementary functions, although its solution can be expressed in a complicated way in terms of little-tabulated Bessel functions of fractional order. Of course, when the coefficients appearing in the differential equation are empirical functions (such as the air resistance in external ballistics problems), some kind of approximate method, whether graphical, numerical or involving the use of an analogue machine, is the only way of obtaining a solution at all.

General literature for this chapter
(in chronological order)

Runge, C., and H. König: Numerisches Rechnen. Berlin 1924.

Lindow, M.: Numerische Infinitesimalrechnung. Berlin and Bonn 1928.

Scarborough, James B.: Numerical mathematical analysis. 416 pp., in particular Ch.s XI—XIII. Baltimore and London 1930.

Levy, H., and E. A. Baggot: Numerical studies in differential equations, Vol. 1. London 1934.

Kamke, E.: Differentialgleichungen, Lösungsmethode und Lösungen, Vol. 1, Ch.A § 8: Numerische, graphische und instrumentelle Integrationsverfahren, 3rd ed. 666 pp. Leipzig 1944.

SANDEN, H. v.: Praxis der Differentialgleichungen, 3rd ed. Berlin 1945.
SCHULZ, GÜNTHER: Formelsammlung zur praktischen Mathematik. Sammlung Göschen, Vol. 1110. Berlin and Leipzig 1945.
WILLERS, FR. A.: Methoden der praktischen Analysis, 2nd ed. 410 pp. Berlin 1950.
HARTREE, D. R.: Numerical analysis. 287 pp. Oxford 1952.
MINEUR, H.: Techniques de Calcul numérique. 605 pp. Paris et Liège 1952.
MILNE, W. E.: Numerical solution of differential equations. 275 pp. New York-London 1953.
KOPAL, ZDENĚK: Numerical analysis. 556 pp. London 1955.

1.2. Accuracy in the numerical solution of initial-value problems

In an initial-value problem we have to determine approximately in some interval $x_0 \leqq x \leqq \xi$ that solution $y(x)$ of an n-th order differential equation

$$y^{(n)} = f(x, y, y', \dots, y^{(n-1)}) \tag{1.1}$$

which has prescribed "initial" values

$$y^{(\nu)}(x_0) = y_0^{(\nu)} \qquad (\nu = 0, 1, 2, \dots, n-1) \tag{1.2}$$

at the "initial" point $x = x_0$. The existence and uniqueness of such a solution $y(x)$ in this interval will be assumed. Most approximate methods in current use yield approximations $y_1, \dots, y_k, \dots$ to the values $y(x_1), \dots, y(x_k), \dots$ of the exact solution at a number of discrete points $x_1, \dots, x_k, \dots$.

The choice of method from among the numerous approximate methods available and the whole arrangement of the calculation is governed decisively by the number of steps, i.e. the number of points x_k- and the accuracy desired. In initial-value problems conditions particularly unfavourable to accuracy are met with; not only is a lengthy calculation involved, in which inaccuracies at the beginning of the calculation influence all subsequent results (such is also the case, for example, when a large set of linear equations is solved by elimination), but inaccuracies in the individual $y_1, y_2, \dots$ cause additional increases in the error; it can happen that solutions of the differential equation which lie close together at $x = x_1$ diverge considerably from one another at a subsequent point $x = x_n$. This last remark, sounding as it does rather trite, is nevertheless of such consequence as to be worthy of amplification by an example. Consider the initial-value problem

$$\left. \begin{array}{l} y'' = 10 y' + 11 y \\ y = 1, \quad y' = -1 \quad \text{for} \quad x = 0, \end{array} \right\} \tag{1.3}$$

whose solution $y(x)$ is to be calculated approximately in the interval $0 \leqq x \leqq 3$. The exact solution is $y(x) = e^{-x}$; this has the value $y(3) \approx 0 \cdot 0498$ at $x = 3$, but the usual approximate methods will give completely false values for it. The general solution of the given differential equation is $y = c_1 e^{-x} + c_2 e^{11x}$, where c_1, c_2 are constants of integration. Now if we calculate successively the values $y_1, y_2, \ldots$ at the points $x_1, x_2, \ldots$ by any approximate method, the unavoidable rounding errors alone are bound to generate a component $c_2 e^{11x}$; even if we had an ideal method with no inherent error, this component, whose value at some point $x = \zeta$ is ε, say, would grow to $\varepsilon \times e^{22} \approx 3 \cdot 6 \times 10^9 \varepsilon$ by the time the point $x = \zeta + 2$ was reached, so that, if we had been computing to seven decimals, for example, the exact solution e^{-x} would already be completely swamped by it. Fortunately the conditions are usually not so unfavourable as in this "viscious" example, but it should always be borne in mind that carrying out an approximate step-by-step integration of an initial-value problem without some idea of the behaviour of the solution is like skating on ice of unknown thickness.

Accordingly we distinguish two cases:

Case 1. The solution is wanted only for a short interval and with moderate accuracy. Here we can use any of the crude methods of § 1.5; also analogue or graphical methods[1] could be used. In favourable cases with not many steps the calculation could even be performed with a slide-rule if need be.

Case 2. The solution is wanted for a long interval (or for a short interval with higher accuracy). Here, even if quite low accuracy is sufficient in the solution, the calculation must be very accurate and should certainly be based on one of the more accurate methods such as the finite-difference or Runge-Kutta methods. Moreover the accuracy to which we work must be substantially higher than that to which the results are required (for example, two or three additional decimals, the so-called "guarding figures", must be carried); consequently a slide-rule is no longer adequate and the calculation has to be carried out with the aid of a calculating machine. Also it is essential to exercise the utmost care at the beginning of the calculation to guard against the introduction of avoidable errors.

1.3. Some general observations on error estimation for initial-value problems

If approximate values y_ν of the solution $y(x)$ of the initial-value problem (1.1), (1.2) have been calculated at the points x_ν by some approximate method, then the question of the magnitude of the error

$$\varepsilon_\nu = y_\nu - y(x_\nu) \qquad (\nu = 1, 2, \ldots) \tag{1.4}$$

---

[1] WILLERS, FR. A.: Graphische Integration, pp. 45—104. Sammlung Göschen. Berlin and Leipzig 1920. — Methoden der praktischen Analysis, 2nd ed., p. 327 et seq. Berlin 1950. — SANDEN, H. v.: Praxis der Differentialgleichungen, 3rd ed., pp. 6—18. Berlin 1945. — KAMKE, E.: Differentialgleichungen, Lösungsmethoden und Lösungen, 3rd ed., Vol. I. Ch.A § 8. Leipzig 1944.

is of great importance. It might at least be possible to estimate the order of magnitude of this error ε_ν and find out how many decimals can be safely guaranteed in the results.

While we possess simple and useful error estimates for many problems in other spheres of practical analysis, we are still far from possessing such estimates for initial-value problems, at any rate as far as integration over anything but a very short interval (with only a few steps) is concerned. For many methods — the finite-difference method, for example — rigorous limits have been established for the error, but for a large number of steps these limits are usually so wide that they far exceed the actual error in magnitude and hence their practical value is illusory.

This difficulty is fundamental. It can even happen that approximate methods fail completely so rapidly do the errors grow, as the example (1.3) shows, and since a strict upper bound for the error must cover all cases, including the most unfavourable, we cannot expect it to give an accurate estimate of the magnitude of the error in a favourable case where the error remains small. Probably better limits will be established only if, instead of considering general differential equations such as $y'=f(x, y)$ or $y''=f(x, y, y')$, we treat each case on its merits so that we can use special properties of the function f. Nevertheless general error limits are derived in §§ 1, 4 and 5 in order to exhibit the underlying principles on which other error limits can be based.

Thus for integration over long intervals, we must usually be satisfied at present with "indications" of the actual error. Such indications are:

I. Comparison of two approximations with different lengths of step. If we perform the calculation with a step h, obtaining approximate values $y_1, y_2, \ldots$ at the points $x_p = x_0 + p h$, and then repeat the calculation with a step $\tilde{h}$, say $\tilde{h} = 2h$, obtaining values $\tilde{y}_1, \tilde{y}_2, \ldots, \tilde{y}_p$ corresponding to the previous values $y_2, y_4, \ldots, y_{2p}$, differences $\delta_p = y_{2p} - \tilde{y}_p$ will be revealed.

We now define a certain order k of an approximate method. Assuming that the values $y_1, y_2, \ldots, y_p$ are exact, i.e. $y_\nu = y(x_\nu)$ for $\nu = 0, 1, \ldots, p$, we calculate y_{p+1} by the method in question. If we now expand $y(x_{p+1})$ and y_{p+1} into Taylor series based on the point $x = x_p$ thus

$$y(x_{p+1}) = y(x_p) + \frac{h}{1!} y'(x_p) + \frac{h^2}{2!} y''(x) + \cdots,$$

$$y_{p+1} = y(x_p) + \frac{h}{1!} a_1 + \frac{h^2}{2!} a_2 + \cdots,$$

there will be a last term whose coefficient is the same in both series. The corresponding exponent of h is called the order k of the method; thus

$$a_1 = y'(x_p), \ldots, a_k = y^{(k)}(x_p), \quad \text{but} \quad a_{k+1} \neq y^{(k+1)}(x_p). \tag{1.5}$$

4*

It is customary to reason (by no means rigorously) as follows: in the first calculation, with a step length h, an error proportional to h^{k+1} is introduced at each step, so that at the point $\xi = x_0 + 2nh$ the error is $A \times 2nh^{k+1} = A(\xi - x_0) h^k$, where A is a constant of proportionality; and in the second calculation, with a step length $2h$, an error proportional to $(2h)^{k+1}$ is introduced at each of the n steps up to the same point ξ, so that the error is $A \times n(2h)^{k+1} = A(\xi - x_0) 2^k h^k$, where A is the same constant of proportionality; thus we assume that approximately

$$y(\xi) \approx y_{2n} - A(\xi - x_0) h^k \approx \tilde{y}_n - A(\xi - x_0) 2^k h^k. \tag{1.6}$$

Solving for an approximate value of A, we find the error in the calculation with the small interval to be one $(2^k - 1)$-th part of the difference between the results of the two calculations:

$$A(\xi - x_0) h^k \approx \frac{\tilde{y}_n - y_{2n}}{2^k - 1} \quad \text{and} \quad y(\xi) \approx y_{2n} - \frac{\tilde{y}_n - y_{2n}}{2^k - 1}. \tag{1.7}$$

II. The terminal check. Here we calculate the values

$$f_k = f(x_k, y_k, y'_k, \ldots, y_k^{(n-1)}) \tag{1.8}$$

from the approximate solution and use them in the approximate evaluation of the repeated integral in the formula

$$\begin{aligned} y(x) = y_0 + y'_0(x - x_0) + y''_0 \frac{(x - x_0)^2}{2!} + \cdots + y_0^{(n-1)} \frac{(x - x_0)^{n-1}}{(n-1)!} + \\ + \int_{x_0}^{x} \cdots \int_{x_0}^{\xi_3} \int_{x_0}^{\xi_2} f(\xi_1, y(\xi_1), y'(\xi_1), \ldots, y^{(n-1)}(\xi_1)) \, d\xi_1 \, d\xi_2 \ldots d\xi_n, \end{aligned} \tag{1.9}$$

which is derived by repeated integration from (1.1), (1.2) (it may be added that most of the values f_k would have been found during the original calculation, anyway). Provided that we use a sufficiently accurate quadrature formula, the value for $y(x)$ obtained from (1.9) will be more accurate than that obtained in the approximate solution; it is then usually assumed that these values agree with the exact values to the same accuracy as that to which they agree with each other, an assumption which is not proved of course. For practical application of this type of terminal check, see § 2.5.

1.4. Differential equations of the first order. Preliminaries

Consider the differential equation

$$y' = \frac{dy}{dx} = f(x, y). \tag{1.10}$$

Let $f(x, y)$ be a given continuous function in a domain D of the real (x, y) plane; for certain considerations we will assume that it is bounded in D, i.e.

$$|f(x, y)| \leqq M \quad \text{in } D, \tag{1.11}$$

and satisfies a Lipschitz condition there, i.e. there is a constant K such that for all pairs of points (x, y_1), (x, y_2) in the domain D we have

$$|f(x, y_1) - f(x, y_2)| \leqq K|y_1 - y_2|. \qquad (1.12)$$

We require that solution $y(x)$ of the differential equation (1.10) in the domain D which passes through a given initial point $x = x_0$, $y = y_0$ (Fig. II/1); in actual fact we require an approximation to this solution, i.e. we have to calculate approximations y_n to the values $y(x_n)$ of the exact solution at the points x_n $(n = 1, 2, \ldots)$. The difference $\varDelta x_n = x_{n+1} - x_n$ is called the step length or simply the step; it is usually denoted by h and is always taken to be positive. It often suffices to keep h constant (i.e. independent of n).

Many methods are based on the equation

$$y(x_{n+1}) = y(x_n) + \int_{x_n}^{x_{n+1}} f(x, y(x)) \, dx, \qquad (1.13)$$

Fig. II/1. Solution of a differential equation of the first order

which is obtained from (1.10) by integration, the integral on the right hand side being replaced by some approximate expression [1].

1.5. Some methods of integration

We discuss briefly a rather crude approximate method — a summation method — which would be used only when no great accuracy is required and only a small number of integration steps are involved. In this method, which we may call the "polygon" method, formula (1.13) is used in the crudest possible way: f is assumed to be constant at its initial value f_n over the step interval $\langle x_n, x_{n+1} \rangle$. Thus we calculate the next y_{n+1} from y_n by the formula

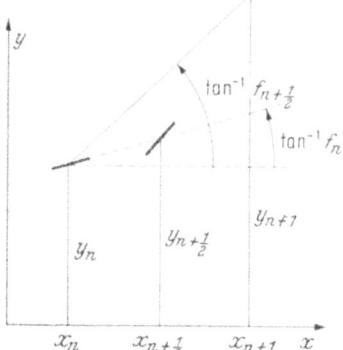

$$y_{n+1} = y_n + h f_n, \qquad (1.14)$$

where

$$f_n = f(x_n, y_n). \qquad (1.15)$$

Somewhat more accurate is the "improved polygon method" in which we use the slope at an intermediate point (see Fig. II/2): we find an

Fig. II/2. The improved polygon method

[1] The basic idea was used by LEONHARD EULER: Institutiones calculi integralis. Petersburg 1768—1770, published in Leonardi Euleri Opera omnia, series prima, Vol. 11, Leipzig and Berlin 1913, pp. 424—427; Vol. 12 (1914) pp. 271—274 (in Latin).

approximate intermediate point

$$x_{n+\frac{1}{2}} = x_n + \frac{h}{2}, \qquad y_{n+\frac{1}{2}} = y_n + \frac{1}{2}\,h f_n \qquad (1.16)$$

by using (1.14), (1.15) and then replace the f_n in (1.14) by

$$f_{n+\frac{1}{2}} = f(x_{n+\frac{1}{2}}, y_{n+\frac{1}{2}}), \qquad (1.17)$$

thus obtaining the improved approximation

$$y_{n+1} = y_n + h f_{n+\frac{1}{2}}. \qquad (1.18)$$

In the "improved Euler-Cauchy method" we first of all define rough values at x_{n+1} by

$$\tilde{y}_{n+1} = y_n + h f_n; \qquad \tilde{f}_{n+1} = f(x_{n+1}, \tilde{y}_{n+1}) \qquad (1.19)$$

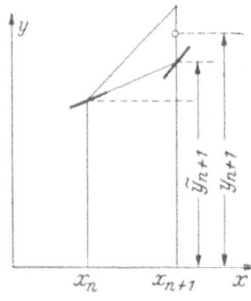

Fig. II/3. The improved Euler-Cauchy method

and then use the arithmetic mean of f_n and $\tilde{f}_{n+1}$ as an intermediate slope (see Fig. II/3). In this way we obtain the approximation

$$y_{n+1} = y_n + \frac{h}{2}\left(f_n + \tilde{f}_{n+1}\right). \qquad (1.20)$$

Duffing's method[1] is based on formula (2.59) of Ch. I. The interval $\langle a, b\rangle$ now becomes $\langle x_n, x_{n+1}\rangle$, the function $f(x)$ becomes $y(x)$, and $f'(x)$ becomes $y'(x) = f[x, y(x)]$. If we neglect the remainder term, we obtain the approximate equation

$$y_{n+1} = \frac{h}{3} f(x_{n+1}, y_{n+1}) + y_n + \frac{2h}{3} f(x_n, y_n) + \frac{h^2}{6}\left[\frac{d}{dx} f(x, y)\right]_{(x_n, y_n)}.$$

In general, y_n being already calculated, this is a non-linear equation for y_{n+1} and is usually solved iteratively: an estimated value of y_{n+1} is first inserted in the right-hand side, which then yields a new value of y_{n+1} (see also Ex. 3 in § 5.10).

Lotkin[2] goes one step further in this direction and proposes as an accurate method the use of the Euler-Maclaurin quadrature formula in

[1] Duffing, G.: Zur numerischen Integration gewöhnlicher Differentialgleichungen 1. and 2. Ordnung. Forsch.-Arb. Ing.-Wes. 224, 29–50 (1920). The derivations and improvements of Duffing's method given here are to be found in E. Pflanz: Bemerkungen über die Methode von G. Duffing zur Integration von Differentialgleichungen. Z. Angew. Math. Mech. 28, 167–172 (1948). The method is extended by E. Beck: Zwei Anwendungen der Obreschkoffschen Formel. Z. Angew. Math. Mech. 30, 84–93 (1950).

[2] Lotkin, M.: A new integrating procedure of high accuracy. J. Math. Phys. 31, 29–34 (1952).

the form (with $b - a = h$)

$$\int_a^b \varphi(t)\, dt = \frac{h}{2} \left(\varphi(a) + \varphi(b)\right) + \frac{h^2}{10} \left(\varphi'(a) - \varphi'(b)\right) +$$

$$+ \frac{h^3}{120} \left(\varphi''(a) + \varphi''(b)\right) + R;$$

here the remainder term R is of the seventh order. Starting from an approximate value

$$y_{r+1}^{[0]} = y_r + h f_r + \frac{h^2}{2} f_r' + \frac{h^3}{6} f_r'',$$

which determines the starting values $f_{r+1}^{[0]}, f_{r+1}'^{[0]}, f_{r+1}''^{[0]}$, we calculate successive iterates $y_{r+1}^{[\sigma]}$ from the formula

$$y_{r+1}^{[\sigma+1]} = y_r + \frac{h}{2} \left(f_r + f_{r+1}^{[\sigma]}\right) + \frac{h^2}{10} \left(f_r' - f_{r+1}'^{[\sigma]}\right) + \frac{h^3}{120} \left(f_r'' + f_{r+1}''^{[\sigma]}\right).$$

Implicit in the presentation of this method is the assumption that $f' = df/dx = f_x + f_y f$ and f'' may be calculated sufficiently easily.

Another method, which is sometimes effective also for differential equations of higher order, is to replace the differential equation by a simpler one solvable in closed form; for example, we can often obtain an approximate solution by replacing variable coefficients by piecewise-constant coefficients[1].

A simple example

The initial-value problem

$$y' = y - \frac{2x}{y}, \qquad y(0) = 1 \tag{1.21}$$

possesses an exact solution, namely

$$y = \sqrt{2x + 1}, \tag{1.22}$$

so that we can immediately assess the accuracy of the approximate methods for this particular case. Application of the approximate formulae (1.14) to (1.20) is so straightforward that the *modus operandi* is self-evident from the tables reproduced here.

Table II/1 gives the working for the improved polygon method with $h = 0.2$. As can be recommended for other calculations, we have indicated at the head of each column, numbered ①, ②, ③, ..., how its contents are obtained from previously calculated numbers in other columns; once this has been done, the calcula-

---

[1] In the case $y'' = f(y)$, for example, $f(y)$ can be replaced by linear functions over suitable intervals in y; then an approximation for y can be obtained by piecing together the solutions obtained in these intervals. Cf. K. KLOTTER: Technische Schwingungslehre, 2nd ed., Vol. I, p. 154. Berlin-Göttingen-Heidelberg 1951.

tion can proceed in an entirely systematic fashion without further reference to the formulae. This facilitates the actual computation considerably and renders it suitable for an assistant.

Table II/1. *Numerical integration of a first-order differential equation by the improved polygon method*

x_n	y_n	$-\dfrac{2x_n}{y_n}$	$\dfrac{h}{2} f_n$	$y_{n+\frac{1}{2}}$	$x_{n+\frac{1}{2}}$	$-4\dfrac{x_{n+\frac{1}{2}}}{y_{n+\frac{1}{2}}}$	$h f_{n+\frac{1}{2}}$	y_{n+1}
①	②	③	④	⑤	⑥	⑦	⑧	⑨
	$\left(\text{from } ⑨\right)$	$=-\dfrac{2\times①}{②}$	$=\dfrac{②+③}{10}$	$=②+④$	$=①+0.1$	$=-\dfrac{4\times⑥}{⑤}$	$=\dfrac{2\times⑤+⑦}{10}$	$=②+⑧$
0	1	0	0·1	1·1	0·1	−0·36364	0·18364	1·18364
0·2	1·18364	−0·33794	0·084570	1·26821	0·3	−0·94622	0·15902	1·34266
0·4	1·34266	−0·59583	0·074683	1·41734	0·5	−1·41109	0·14236	1·48502
0·6	1·48502	−0·80807	0·067695	1·55272	0·7	−1·80329	0·13021	1·61523
0·8	1·61523	−0·99057	0·062466	1·67770	0·9	−2·14580	0·12096	1·73619
1	1·73619	−1·15195	0·058424	1·79461	1·1	−2·45178	0·11374	1.84993
1·2	1·84993	−1·29735	0·055258	1·90519	1·3	−2·72939	0·10810	1·95803
1·4	1·9580	−1·4300	0·05280	2·0108	1·5	−2·9838	0·1038	2·0618
1·6	2·0618	−1·5520	0·05098	2·1128	1·7	−3·2185	0·1007	2·1625
1·8	2·1625	−1·6647	0·04978	2·2123	1·9	−3·4353	0·0989	2·2615
2	2·2615							

In order to compare the accuracy obtained by the various methods, we have carried more decimals than would otherwise be significant.

Table II/2 shows the corresponding calculations for the improved Euler-Cauchy method with the same step length $h = 0.2$; here the integration is taken only as far as the point $x = 1$.

Table II/2.
Numerical integration of a first-order equation by the improved Euler-Cauchy method

x_n	y_n	$-\dfrac{2x_n}{y_n}$	$\dfrac{h}{2} f_n$	$\tilde{y}_{n+1}$	x_{n+1}	$-\dfrac{2x_{n+1}}{\tilde{y}_{n+1}}$	$\tilde{f}_{n+1}$	$\dfrac{h}{2}(f_n+\tilde{f}_{n+1})$	y_{n+1}
①	②	③	④	⑤	⑥	⑦	⑧	⑨	⑩
	$\left(\text{from } ⑩\right)$	$=-\dfrac{2\times①}{②}$	$=\dfrac{②+③}{10}$	$=②+2\times④$		$=-\dfrac{2\times⑥}{⑤}$	$=⑤+⑦$	$=④+\dfrac{⑧}{10}$	$=②+⑨$
0	1	0	0·1	1·2	0·2	−0·33333	0·8667	0·18667	1·18667
0·2	1·18667	−0·33708	0·084959	1·35658	0·4	−0·58972	0·7669	0·16165	1·34832
0·4	1·34832	−0·59333	0·075499	1·49932	0·6	−0·80036	0·6990	0·14540	1·49372
0·6	1·49372	−0·80336	0·069036	1·63179	0·8	−0·98052	0·6513	0·13416	1·62788
0·8	1·6279	−0·9829	0·06450	1·7569	1	−1·1384	0·6185	0·12635	1·7543

In Table II/3 comparison is made with the exact solution (1.22) and also with results obtained by using the more accurate methods discussed in §§ 2 and 3; below each result the error is written in brackets (the error in an approximation ξ to a true value x is defined as $\xi - x$).

Table II/3. *Comparison of results obtained by various methods*

x	Exact solution $\sqrt{2x+1}$	Improved Euler-Cauchy | polygon method		Runge-Kutta method		Central-difference method
		$h=0\cdot2$	$h=0\cdot2$	$h=0\cdot4$	$h=0\cdot2$	$h=0\cdot1$
0·2	1·1832160	1·18667 (+0·00345)	1·18364 (+0·00042)		1·1832293 (+0·0000133)	1·1832215 (+0·0000055)
0·4	1·3416408	1·34832 (0·00668)	1·34266 (0·00102)	1·342066 (+0·000425)	1·3416669 (0·0000262)	1·3416491 (0·0000084)
0·6	1·4832397	1·49372 (0·0105)	1·48502 (0·00178)		1·4832815 (0·0000418)	1·4832508 (0·0000111)
0·8	1·6124515	1·62788 (0·0154)	1·61523 (0·00278)	1·613449 (0·000997)	1·6125140 (0·0000624)	1·6124665 (0·0000149)
1	1·7320508	1·7543 (0·02233)	1·73619 (0·00414)		1·7321419 (0·0000911)	1·7320713 (0·0000205)
1·2	1·8439089		1·84993 (0·00602)	1·84600 (0·00209)	1·8440401 (0·000131)	1·843937 (0·000028)
1·4	1·9493589		1·95803 (0·00867)		1·949547 (0·000188)	1·949398 (0·000039)
1·6	2·0493902		2·06181 (0·0124)	2·05367 (0·00428)	2·049660 (0·000270)	2·049445 (0·000055)
1·8	2·1447611		2·16252 (0·0178)		2·145148 (0·000387)	2·144839 (0·000078)
2	2·2360680		2·26145 (0·0254)	2·24486 (0·00789)	2·236624 (0·000556)	2·236179 (0·000111)

The Runge-Kutta method involves about twice the amount of work per step as the central-difference method so that it is reasonable to compare the Runge-Kutta results with $h = 0\cdot2$ with those of the central-difference method with $h = 0\cdot1$.

1.6. Error estimation

We assume here that the function $f(x, y)$ in the differential equation (1.10) is bounded as in (1.11) and satisfies the Lipschitz condition (1.12) in the domain D as in § 1.4 and also that it possesses bounded partial derivatives in so far as it satisfies the conditions

$$\left|\frac{df(x, y)}{dx}\right| = \left|\frac{\partial f}{\partial x} + f\frac{\partial f}{\partial y}\right| \le N_1; \quad \left|\frac{d^p f}{dx^p}\right| \le N_p \quad (p = 1, 2, \ldots). \quad (1.23)$$

For simplicity we will ignore rounding errors in obtaining the following error estimates, i.e. we estimate only the "inherent" or "truncation" errors. Rounding errors will, of course, be propagated throughout the calculation, but we take the view that, knowing an upper bound for the inherent errors from one of the error estimates, we can calculate with a number of decimals sufficient to ensure that the rounding errors remain within this bound. In principle it would not be difficult to derive similar estimates for the rounding errors.

We define the "error" in the approximate value y_n to be the quantity

$$\varepsilon_n = y_n - y(x_n); \tag{1.24}$$

from (1.13) the error increment $\Delta \varepsilon_n$ over one step is therefore

$$\Delta \varepsilon_n = \varepsilon_{n+1} - \varepsilon_n = \underbrace{y_{n+1} - y_n}_{\Delta y_n} - \int_{x_n}^{x_{n+1}} f(x, y(x)) \, dx. \tag{1.25}$$

The expression for the difference Δy_n is characteristic of each individual method, and further analysis in which we insert this expression into (1.25) must be referred to specific methods.

I. Polygon method. From (1.14) the expression for Δy_n is

$$\Delta y_n = h f_n;$$

therefore, with $F(x) = f[x, y(x)]$, (1.25) becomes

$$\Delta \varepsilon_n = h (f_n - F(x_n)) + J_n, \tag{1.26}$$

where

$$J_n = h F(x_n) - \int_{x_n}^{x_{n+1}} F(x) \, dx. \tag{1.27}$$

Since we have assumed that $f(x, y)$ satisfies the Lipschitz condition (1.12), limits for the first term on the right of (1.26) can be obtained from

$$|f_n - F(x_n)| = |f(x_n, y_n) - f(x_n, y(x_n))| \leq K |y_n - y(x_n)| = K |\varepsilon_n|. \tag{1.28}$$

For the second term J_n we first transform the integral by integrating by parts:

$$\int_{x_n}^{x_{n+1}} 1 \cdot F(x) \, dx = [(x - x_{n+1}) F]_{x_n}^{x_{n+1}} - \int_{x_n}^{x_{n+1}} (x - x_{n+1}) \frac{dF}{dx} \, dx;$$

it then follows that

$$\left.
\begin{aligned}
\left| h F(x_n) - \int_{x_n}^{x_{n+1}} F(x) \, dx \right| &= \left| \int_{x_n}^{x_{n+1}} (x - x_{n+1}) \frac{dF}{dx} \, dx \right| \\
&\leq N_1 \int_{x_n}^{x_{n+1}} (x_{n+1} - x) \, dx = \frac{1}{2} N_1 h^2,
\end{aligned}
\right\} \tag{1.29}$$

where N_1 is as defined in (1.23).

Taking absolute values in (1.26) and using (1.28), (1.29), we obtain

$$|\varDelta \varepsilon_n| = |\varepsilon_{n+1} - \varepsilon_n| \leqq hK|\varepsilon_n| + \tfrac{1}{2} N_1 h^2,$$

and since

$$|\varepsilon_{n+1} - \varepsilon_n| \geqq |\varepsilon_{n+1}| - |\varepsilon_n|,$$

we arrive at the "recursive error estimate"

$$|\varepsilon_{n+1}| \leqq (1 + hK)|\varepsilon_n| + \tfrac{1}{2} N_1 h^2. \tag{1.30}$$

Thus if an estimate for ε_n is known, ε_{n+1} can also be estimated.

We can now easily derive an "independent error estimate", i.e. one in which limits for $|\varepsilon_{n+1}|$ are determined without knowledge of limits for the preceding errors. The inequality (1.30) has the form

$$|\varepsilon_{n+1}| \leqq a|\varepsilon_n| + b, \text{ where } a \geqq 0, b \geqq 0, \varepsilon_0 = 0, \quad (n = 0, 1, 2, \ldots). \tag{1.31}$$

Consequently

$$|\varepsilon_1| \leqq b, \qquad |\varepsilon_2| \leqq a|\varepsilon_1| + b \leqq b(1 + a),$$

and in general, as can be proved immediately by induction,

$$|\varepsilon_n| \leqq b(1 + a + a^2 + \cdots + a^{n-1}) = \frac{b(a^n - 1)}{a - 1}. \tag{1.32}$$

We have assumed here that $a \neq 1$.

If we now insert the values $a = 1 + hK$, $b = \tfrac{1}{2} h^2 N_1$, corresponding to (1.30), we obtain, for $K > 0$, the independent error estimate

$$|\varepsilon_n| \leqq \frac{h N_1}{2K} [(1 + hK)^n - 1]. \tag{1.33}$$

Since $1 + u < e^u$ for $u > 0$, we can also use the estimate

$$|\varepsilon_n| \leqq \frac{h N_1}{2K} (e^{K(x_n - x_0)} - 1), \tag{1.34}$$

where nh has been replaced by $x_n - x_0$. From this it follows in particular that at a fixed point x_n the error limits tend to zero as $h \to 0$, i.e. the method is convergent.

1.7. Corresponding error estimates for the improved methods

II. Improved polygon method. We follow the same reasoning here so we may be brief. In this case the expression for $\varDelta y_n$ is, from (1.16) to (1.18),

$$\varDelta y_n = y_{n+1} - y_n = h f(x_{n+\frac{1}{2}}, y_n + \tfrac{1}{2} h f_n),$$

and hence (1.25) becomes

$$\Delta \varepsilon_n = h\,\Phi_n + J_n^*,\qquad(1.35)$$

where

$$\Phi_n = f\big(x_{n+\frac{1}{2}},\, y_n + \tfrac{1}{2}h f(x_n, y_n)\big) - f\big(x_{n+\frac{1}{2}},\, y(x_{n+\frac{1}{2}})\big)\qquad(1.36)$$

and

$$J_n^* = h F(x_{n+\frac{1}{2}}) - \int_{x_n}^{x_{n+1}} F(x)\,dx.\qquad(1.37)$$

For the bounds of Φ_n we employ the Lipschitz condition (1.12) again:

$$|\Phi_n| \leqq K\,|y_n + \tfrac{1}{2}h f(x_n, y_n) - y(x_{n+\frac{1}{2}})|$$
$$= K\,|y_n - y(x_n) + \tfrac{1}{2}h\,[f(x_n, y_n) - F(x_n)] - [y(x_{n+\frac{1}{2}}) - y(x_n) - \tfrac{1}{2}h F(x_n)]|;$$

formula (1.29) applied over an interval of length $\tfrac{1}{2}h$ yields

$$\left|\tfrac{1}{2}h F(x_n) - [y(x_{n+\frac{1}{2}}) - y(x_n)]\right| \leqq \tfrac{1}{2} N_1 \left(\frac{h}{2}\right)^2;$$

hence for Φ_n we have the estimate

$$|\Phi_n| \leqq K\{|\varepsilon_n| + \tfrac{1}{2}h K|\varepsilon_n| + \tfrac{1}{8}h^2 N_1\}.\qquad(1.38)$$

For the quadrature error J_n^* there are the known[1] bounds

$$|J_n^*| \leqq \frac{h^3}{24} N_2,\qquad(1.39)$$

where N_2 is as defined in (1.23). Consequently from (1.35), (1.38), (1.39) it follows that for $K>0$

$$\left.\begin{array}{l} |\varepsilon_{n+1}| - |\varepsilon_n| \leqq |\Delta \varepsilon_n| \leqq h K\{(1 + \tfrac{1}{2}h K)|\varepsilon_n| + \tfrac{1}{8}h^2 N_1\} + \tfrac{1}{24}h^3 N_2 \\[2mm] \text{i.e.}\quad |\varepsilon_{n+1}| \leqq |\varepsilon_n|\{1 + h K + \tfrac{1}{2}h^2 K^2\} + \tfrac{1}{8}h^3(K N_1 + \tfrac{1}{3}N_2). \end{array}\right\}\quad(1.40)$$

This recursive error estimate also has the form (1.31), and using (1.32) we immediately obtain the independent estimate

$$|\varepsilon_n| \leqq \frac{1}{8}h^2\left(N_1 + \frac{N_2}{3K}\right)\frac{(1 + h K + \tfrac{1}{2}h^2 K^2)^n - 1}{1 + \tfrac{1}{2}h K}.\qquad(1.41)$$

As with the polygon method the convergence of the method follows directly from this independent error estimate; here, however, the error at any fixed point x tends to zero like h^2 as $h \to 0$.

III. Improved Euler-Cauchy method. Proceeding exactly as in the previous case, we have from (1.19), (1.20)

$$\Delta \varepsilon_n = h\,\Psi_n + J_n^{**},$$

---

[1] See, for example, R. COURANT: Differential and Integral Calculus, 2nd ed., Vol. I, p. 347. Glasgow 1937.

where

$$\Psi_n = \tfrac{1}{2}\left[f(x_n, y_n) + f(x_{n+1}, y_n + h f_n) - F(x_n) - F(x_{n+1})\right]$$

and

$$J_n^{**} = \tfrac{1}{2} h\left[F(x_n) + F(x_{n+1})\right] - \int_{x_n}^{x_{n+1}} F(x)\, dx.$$

From the Lipschitz condition it follows that

$$|\Psi_n| \leqq \tfrac{1}{2} K\left(|\varepsilon_n| + |y(x_{n+1}) - y_n - h f(x_n, y_n)|\right).$$

Now[1] by TAYLOR's theorem there is a point $x = \xi_n$ in $\langle x_n, x_{n+1}\rangle$ such that

$$y(x_{n+1}) - y_n - h f(x_n, y_n) = y(x_n) + h y'(x_n) + \frac{h^2}{2} y''(\xi_n) -$$

$$- y_n - h f(x_n, y_n) = - \varepsilon_n + h\left[f(x_n, y(x_n)) - f(x_n, y_n)\right] + \frac{h^2}{2} y''(\xi_n),$$

and hence, since $|y''(\xi_n)| \leqq N_1$ from (1.23), we have

$$|y(x_{n+1}) - y_n - h f(x_n, y_n)| \leqq |\varepsilon_n| + h K|\varepsilon_n| + \frac{h^2}{2} N_1.$$

For the quadrature error J_n^{**} in the trapezium approximation used in this method there is the well-known estimate

$$|J_n^{**}| \leqq \frac{1}{12} h^3 N_2.$$

Altogether, we have

$$|\varepsilon_{n+1}| - |\varepsilon_n| \leqq |\Delta \varepsilon_n| \leqq h|\Psi_n| + |J_n^{**}| \leqq \frac{h K}{2}\left(|\varepsilon_n|(2 + h K) + \frac{h^2}{2} N_1\right) + \frac{h^3}{12} N_2,$$

i.e.

$$|\varepsilon_{n+1}| \leqq |\varepsilon_n|\left(1 + h K + \frac{h^2 K^2}{2}\right) + \frac{1}{4} h^3\left(K N_1 + \frac{1}{3} N_2\right).$$

Again the recursive error estimate arrived at has the form of the inequality (1.31) so that the independent estimate can be immediately written down from (1.32):

$$|\varepsilon_n| \leqq \frac{h^2}{4}\left(N_1 + \frac{N_2}{3 K}\right)\frac{(1 + h K + \tfrac{1}{2} h^2 K^2)^n - 1}{1 + \tfrac{1}{2} h K}. \tag{1.42}$$

These limits are precisely twice as wide as the limits (1.41) for the improved polygon method.

Once more we can infer the convergence of the method. In this case also, the error limits at a fixed point x tend to zero quadratically as $h \to 0$.

§ 2. The Runge-Kutta method for differential equations of the n-th order

The Runge-Kutta method is a much-used method for accurate integration. The general formulae derived in §§ 2.1 — 2.3 by some rather laborious calculations need

---

[1] The estimate given here constitutes a sharpening of the estimate given in the first German edition; I am indebted for it to a written communication from Dr. WILLY RICHTER, Neuchâtel.

not concern the reader whose interest lies chiefly in the numerical integration of specific problems; he will find the computational formulae for equations of up to the fourth order collected together in § 2.4.

2.1. A general formulation

Suppose that the n-th order differential equation

$$y^{(n)} = f(x, y, y', y'', \ldots, y^{(n-1)}) \tag{2.1}$$

is to be integrated subject to prescribed initial values

$$y_0^{(\nu)} = y^{(\nu)}(x_0) \qquad (\nu = 0, 1, 2, \ldots, n-1) \tag{2.2}$$

at the point $x = x_0$. Approximate values will be derived for y and its derivatives $y^{(\nu)}$ at the point $x_1 = x_0 + h$ one step ahead.

Rough approximations can be obtained by using the prescribed initial values in Taylor series truncated at the terms in $y_0^{(n-1)}$: for $y_1^{(\nu)}$ we have the value

$$y_0^{(\nu)} + \frac{h}{1!} y_0^{(\nu+1)} + \frac{h^2}{2!} y_0^{(\nu+2)} + \cdots + \frac{h^{n-\nu-1}}{(n-\nu-1)!} y_0^{(n-1)} = T_\nu(1),$$

the notation T_ν being defined, somewhat more generally, by

$$\left.\begin{aligned} T_\nu(\alpha) = y_0^{(\nu)} + \frac{\alpha h}{1!} y_0^{(\nu+1)} + \frac{(\alpha h)^2}{2!} y_0^{(\nu+2)} + \cdots + \\ + \frac{(\alpha h)^{n-\nu-1}}{(n-\nu-1)!} y_0^{(n-1)} \quad (\nu = 0, 1, 2, \ldots, n-1). \end{aligned}\right\} \tag{2.3}$$

In order to obtain better approximations we add corrections to these truncated Taylor series; we derive quantities $k, k', k'', \ldots, k^{(n-1)}$ such that the approximation

$$y_1^{(\nu)} = T_\nu(1) + \frac{\nu!}{h^\nu} k^{(\nu)} \qquad (\nu = 0, 1, \ldots, n-1) \tag{2.4}$$

to the ν-th derivative $y^{(\nu)}(x_1)$ is as accurate as possible (the factor $\nu! h^{-\nu}$ will be found convenient later).

First of all we describe a general formulation which includes many well-known numerical methods of integration as special cases.

The "corrections" $k^{(\nu)}$ are expressed as linear combinations

$$k^{(\nu)} = \sum_{\varrho=1}^{r} \gamma_{\nu\varrho} k_\varrho \qquad (\nu = 0, 1, \ldots, n-1) \tag{2.5}$$

of certain auxiliary quantities $k_1, k_2, \ldots, k_r$; these are defined in terms of the values of the function f at intermediate points in such a way that they can be calculated successively:

$$k_1 = \frac{h^n}{n!} f\left(x_0,\, T_0(0),\, T_1(0),\, \ldots,\, T_{n-1}(0)\right),$$

$$k_2 = \frac{h^n}{n!} f\left(x_0 + \alpha_1 h,\; T_0(\alpha_1) + a_{10} k_1,\; T_1(\alpha_1) + a_{11} k_1,\, \ldots,\right.$$
$$\left. T_{n-1}(\alpha_1) + a_{1,n-1} k_1\right),$$

$$k_3 = \frac{h^n}{n!} f\left(x_0 + \alpha_2 h,\; T_0(\alpha_2) + a_{20} k_1 + b_{20} k_2,\, \ldots,\right.$$
$$\left. T_{n-1}(\alpha_2) + a_{2,n-1} k_1 + b_{2,n-1} k_2\right).$$

$$(2.6)$$

From the definition of the $T_\nu(\alpha_\varrho)$ we know that the r "points"

$$x = x_0 + \alpha_\varrho h, \quad y = T_0(\alpha_\varrho), \quad y' = T_1(\alpha_\varrho), \ldots, y^{(n-1)} = T_{n-1}(\alpha_\varrho)$$

$(\varrho = 1, \ldots, r)$ of the $(x, y, y', \ldots, y^{(n-1)})$ space lie near the true solution. The constants α, $a_{\nu\varrho}$, $b_{\nu\varrho}, \ldots, \gamma_{\nu\varrho}$ are at our disposal for making the Taylor series for the $y_1^{(\nu)}$ coincide with the Taylor series for the $y^{(\nu)}(x_1)$ to as high an order as possible, i.e. for making the power of h in the last identical terms as high as possible. With the notation of (2.3) the Taylor series for $y^{(\nu)}(x_1)$ can be written

$$y^{(\nu)}(x_1) = T_\nu(1) + \frac{h^{n-\nu}}{(n-\nu)!} y_0^{(n)} + \frac{h^{n-\nu+1}}{(n-\nu+1)!} y_0^{(n+1)} + \cdots$$
$$= T_\nu(1) + \frac{\nu!}{h^\nu}\left[\binom{n}{\nu}\frac{h^n y_0^{(n)}}{n!} + \binom{n+1}{\nu}\frac{h^{n+1} y_0^{(n+1)}}{(n+1)!} + \cdots\right],$$

$$(2.7)$$

which becomes

$$y^{(\nu)}(x_1) = T_\nu(1) + \frac{\nu!}{h^\nu}\left[\binom{n}{\nu}\frac{h^n}{n!} f + \binom{n+1}{\nu}\frac{h^{n+1}}{(n+1)!} Df +\right.$$
$$\left. + \binom{n+2}{\nu}\frac{h^{n+2}}{(n+2)!} Ef + \binom{n+3}{\nu}\frac{h^{n+3}}{(n+3)!}(Ff + f_{n-1} Ef + f_{n-2} Df) + \cdots\right]$$

$$(2.8)$$

when the expressions (2.52), (2.53), (2.55) of Ch. I are inserted (when no particular argument values are specified for f, the values at the initial point $x = x_0$ are to be taken). We determine the $k^{(\nu)}$ as accurately as possible by choosing them so that when their Taylor expansions are inserted in the expressions (2.4), the latter coincide with (2.8) to as high an order as possible:

$$k^{(\nu)} = \frac{h^n}{n!}\left[\binom{n}{\nu} f + \binom{n+1}{\nu}\frac{h}{n+1} Df + \binom{n+2}{\nu}\frac{h^2}{(n+1)(n+2)} Ef +\right.$$
$$\left. + \binom{n+3}{\nu}\frac{h^3}{(n+1)(n+2)(n+3)}(Ff + f_{n-1} Ef + f_{n-2} Df) + \cdots\right].$$

$$(2.9)$$

2.2. The special Runge-Kutta formulation

Here we discuss the special case of the general formulation of § 2.1 which gives rise to the Runge[1]-Kutta[2] formulae. These have found favour on account of their simple coefficients and symmetrical form; they are exhibited for differential equations of up to the fourth order in the tables in § 2.4. Some other formulae of Runge-Kutta type are collected together in Table I of the appendix.

The basic idea due to C. Runge[3] was applied to first-order equations in more accurate form by Heun[4] and Kutta[5] and extended to second-

---

[1] Carl David Tolmé Runge, one of the most notable of German applied mathematicians, also known as a physicist through his investigations on spectral series, was one of the pioneers in the application of mathematical methods to the numerical treatment of technical problems. E. Trefftz writes [Z. Angew. Math. Mech. 6, 423—424 (1926)] of him (translated from the German): "If Runge succeeded in bridging the gap between mathematics and technology, his success was due to two characteristics, which mark a true applied mathematician: his profound mathematical knowledge ..., and his unflagging energy in perfecting his methods with particular regard to their practical usefulness."

Runge was born in Bremen on the 30th August, 1856 and spent his early childhood in Havanna, where his father was in charge of the administration of the Danish consulate. From 1876 to 1880 he studied first in Munich and then in Berlin, where he took his doctor's degree in 1880; he became an unsalaried lecturer at the University of Berlin in 1883 and was later, 1886, appointed as professor of mathematics in the Technische Hochschule, Hannover. From 1904 to 1924 he was professor of applied mathematics in Göttingen, where he died on the 3rd January, 1927.

His interests outside of scientific spheres were many and varied. He undertook several very lengthy journeys; in particular he accompanied Schwarzschild on an expedition to Algeria in connection with the solar eclipse of 1906. He also travelled to New York in 1909, where he was exchange professor in the Columbia University for the winter semester. Right into his old age he continued to engage in sports, gymnastics, swimming and rowing. Recalling some facets of his character L. Prandtl [Naturwiss. 15, 227—229 (1927)] writes (translated from the German): "He was of a kindly disposition, yet strongly independent in his opinions, even to severe condemnations of that which appeared to him unfair or narrow-minded. With regard to himself he was extremely modest."

Details can be found in the biography by his daughter Iris Runge: Carl Runge und sein wissenschaftliches Werk. Göttingen 1949.

[2] Martin Wilhelm Kutta, born on the 3rd November, 1867 in Pitschen (Upper Silesia), studied in Breslau from 1885 to 1890, then went to Munich, where he took his doctor's degree in 1901 and became an unsalaried lecturer in 1902. He spent 1898—1899 in Cambridge. In 1910 he was appointed to Aachen and in 1911 to Stuttgart as ordinary professor of mathematics (emeritus 1935). He died on the 25th December, 1944 in Fürstenfeldbruck (near Munich), where he was staying with his brother.

[3] Runge, C.: Math. Ann. 46, 167—178 (1895) and Nachr. Ges. Wiss. Göttingen, Math.-phys. Kl. 1905, 252—257.

[4] Heun, K.: Z. Math. Phys. 45, 23—38 (1900).

[5] Kutta, W.: Z. Math. Phys. 46, 435—453 (1901).

order equations by NYSTRÖM[1]. ZURMÜHL continued the extension, first to third-order equations[2] and then to equations of the n-th order[3].

In this particular formulation[4] we use just four auxiliary quantities k_1, k_2, k_3, k_4. We choose $\alpha_1 = \alpha_2 = \frac{1}{2}$, $\alpha_3 = 1$, so that k_2 and k_3 are based on the mid-point of the step interval and k_4 on the last [k_1 is based on the first, being defined always as in (2.6)]. The $a_{\nu\varrho}, b_{\nu\varrho}, \ldots$ are chosen in an obvious way for k_2: the additional terms in the arguments of f merely extend the truncated Taylor series. For k_4 these constants are chosen so that the additional terms are calculated in a similar fashion but from k_3 instead of k_1. The definition of k_3 differs from that of k_2 only in the last argument of f, where k_2 is used instead of k_1, so that $k_2 = k_3$ when f does not depend explicitly on $y^{(n-1)}$. Thus

$$
\left.
\begin{aligned}
k_2 &= \frac{h^n}{n!} f\left[x_0 + \frac{h}{2}, \quad T_0\left(\frac{1}{2}\right) + \frac{k_1}{2^n}, \quad T_1\left(\frac{1}{2}\right) + \binom{n}{1}\frac{1!\,k_1}{2^{n-1}h}, \right. \\
&\quad T_2\left(\frac{1}{2}\right) + \binom{n}{2}\frac{2!\,k_1}{2^{n-2}h^2}, \ldots, \\
&\quad \left. T_{n-2}\left(\frac{1}{2}\right) + \binom{n}{n-2}\frac{(n-2)!\,k_1}{4h^{n-2}}, \quad T_{n-1}\left(\frac{1}{2}\right) + \frac{n!\,k_1}{2h^{n-1}}\right], \\
k_3 &= \frac{h^n}{n!} f\left[x_0 + \frac{h}{2}, \quad T_0\left(\frac{1}{2}\right) + \frac{k_1}{2^n}, \quad T_1\left(\frac{1}{2}\right) + \binom{n}{1}\frac{1!\,k_1}{2^{n-1}h}, \right. \\
&\quad T_2\left(\frac{1}{2}\right) + \binom{n}{2}\frac{2!\,k_1}{2^{n-2}h^2}, \ldots, \\
&\quad \left. T_{n-2}\left(\frac{1}{2}\right) + \binom{n}{n-2}\frac{(n-2)!\,k_1}{4h^{n-2}}, \quad T_{n-1}\left(\frac{1}{2}\right) + \frac{n!\,k_2}{2h^{n-1}}\right], \\
k_4 &= \frac{h^n}{n!} f\left[x_0 + h, \quad T_0(1) + k_3, \quad T_1(1) + \binom{n}{1}\frac{1!\,k_3}{h}, \right. \\
&\quad T_2(1) + \binom{n}{2}\frac{2!\,k_3}{h^2}, \ldots, \\
&\quad \left. T_{n-2}(1) + \binom{n}{n-2}\frac{(n-2)!\,k_3}{h^{n-2}}, \quad T_{n-1}(1) + \frac{n!\,k_3}{h^{n-1}}\right].
\end{aligned}
\right\} \quad (2.10)
$$

(For the practical evaluation of these arguments of f see § 2.4.)

---

[1] NYSTRÖM, E. J.: Über die numerische Integration von Differentialgleichungen. Acta Soc. Sci. Fenn. 50, Nr. 13, 1—55 (1925).

[2] ZURMÜHL, R.: Zur numerischen Integration gewöhnlicher Differentialgleichungen zweiter und höherer Ordnung. Z. Angew. Math. Mech. 20, 104—116 (1940).

[3] ZURMÜHL, R.: Runge-Kutta-Verfahren zur numerischen Integration von Differentialgleichungen n-ter Ordnung. Z. Angew. Math. Mech. 28, 173—182 (1948).

[4] The general formulation of § 2.1 is pursued further by J. ALBRECHT: Beiträge zum Runge-Kutta-Verfahren. Z. Angew. Math. Mech. 35, 100—110 (1955).

2.3. Derivation of the Runge-Kutta formulae

We now expand k_1, k_2, k_3, k_4 into Taylor series based on the point $(x_0, y_0, y_0', \ldots, y_0^{(n-1)})$. This is trivial for k_1 since it is already in the required form, but for the other k_ν the expansion presents a rather laborious calculation, for, in general, all the $n+1$ arguments of f are different from their values at x_0, as for instance, in

$$\frac{n!}{h^n} k_2 = f\left(x_0 + \frac{h}{2}, \quad y_0 + y_0' \frac{h}{2} + y_0'' \frac{h^2}{4 \cdot 2!} + y_0''' \frac{h^3}{8 \cdot 3!} + \cdots + f \frac{h^n}{2^n \cdot n!},\right.$$

$$\left. y_0' + y_0'' \frac{h}{2} + \cdots, \cdots\right).$$

We shall continue the expansions of the required values of f only as far as the terms in h^3. It may help the reader if we first work through a simple case, say with $n=2$: with $y'=u$ and subscripts denoting derivatives of f we have

$$\frac{2}{h^2} k_2 = f\left(x_0 + \frac{h}{2}, \quad y_0 + \frac{h}{2} u + \frac{h^2}{8} f, \quad u + \frac{h}{2} f\right)$$

$$= f + \frac{h}{2} f_x + \left(\frac{h u}{2} + \frac{h^2}{8} f\right) f_y + \frac{h}{2} f f_u + \frac{h^2}{8} f_{xx} + \left(\frac{h^2 u}{4} + \frac{h^3 f}{16}\right) f_{xy} +$$

$$+ \frac{h^2}{4} f f_{xu} + \left(\frac{h^2 u^2}{8} + \frac{h^3 u f}{16}\right) f_{yy} + \left(\frac{h^2 u f}{4} + \frac{h^3 f^2}{16}\right) f_{yu} + \frac{h^2 f^2}{8} f_{uu} + \cdots.$$

(Again all function values are evaluated at $x=x_0$ unless otherwise specified.) In general, using the notation of (2.51), (2.53), (2.55) in Ch. I and neglecting powers of h greater than the third, we obtain the expressions

$$\left.\begin{aligned}
k_1 &= \frac{h^n}{n!} f, \\
k_2 &= \frac{h^n}{n!} \left[f + \frac{h}{2} Df + \frac{h^2}{8} (Ef - f_{n-1} Df) + \frac{h^3}{48} (Ff - 3 Df Df_{n-1})\right], \\
k_3 &= \frac{h^n}{n!} \left[f + \frac{h}{2} Df + \frac{h^2}{8} (Ef + f_{n-1} Df) + \right. \\
&\qquad \left. + \frac{h^3}{48} (Ff + 3 Df Df_{n-1} + 3 f_{n-1}(Ef - f_{n-1} Df))\right], \\
k_4 &= \frac{h^n}{n!} \left[f + h Df + \frac{h^2}{2} Ef + \right. \\
&\qquad \left. + \frac{h^3}{6} \left(Ff + \frac{3}{4} f_{n-1}(Ef + f_{n-1} Df) + \frac{3}{2} f_{n-2} Df\right)\right].
\end{aligned}\right\} \quad (2.11)$$

The corrections $k^{(0)}, k^{(1)}, \ldots, k^{(n-1)}$ from which the new values of y and its derivatives $y^{(\nu)}$ at $x = x_0 + h$ will be calculated are now formed by linearly combining these four expressions for the k_ϱ as in (2.5). The coefficients of the combination are then determined by equating factors of terms involving powers of h up to the $(n+2)$-th with the corresponding factors in (2.9); we find that

from $\dfrac{h^n}{n!} f$ $\qquad$ $\gamma_{\nu 1} + \gamma_{\nu 2} + \gamma_{\nu 3} + \gamma_{\nu 4} = \dbinom{n}{\nu}$,

from $\dfrac{h^{n+1}}{n!} Df$ $\qquad$ $\dfrac{1}{2}\gamma_{\nu 2} + \dfrac{1}{2}\gamma_{\nu 3} + \gamma_{\nu 4} = \dbinom{n+1}{\nu}\dfrac{1}{n+1}$,

from $\dfrac{h^{n+2}}{n!} Ef$ $\qquad$ $\dfrac{1}{8}\gamma_{\nu 2} + \dfrac{1}{8}\gamma_{\nu 3} + \dfrac{1}{2}\gamma_{\nu 4} = \dbinom{n+2}{\nu}\dfrac{1}{(n+1)(n+2)}$,

from $\dfrac{h^{n+2}}{n!} f_{n-1} Df$ $\qquad$ $-\dfrac{1}{8}\gamma_{\nu 2} + \dfrac{1}{8}\gamma_{\nu 3} = 0$.

These four equations for the $\gamma_{\nu 1}, \ldots, \gamma_{\nu 4}$ have the solution

$$\left.\begin{aligned}
\gamma_{\nu 1} &= \binom{n+2}{\nu}\frac{(n-\nu)^2}{(n+1)(n+2)}, \\
\gamma_{\nu 2} = \gamma_{\nu 3} &= \binom{n+2}{\nu}\frac{2(n-\nu)}{(n+1)(n+2)}, \\
\gamma_{\nu 4} &= \binom{n+2}{\nu}\frac{2-(n-\nu)}{(n+1)(n+2)}.
\end{aligned}\right\} \qquad (2.12)$$

Substituting back into (2.5), we obtain the final expressions for the $k^{(\nu)}$:

$$\left.\begin{aligned}
k^{(\nu)} = \binom{n+2}{\nu}&\frac{1}{(n+1)(n+2)} \times \\
&\times [(n-\nu)^2 k_1 + 2(n-\nu)(k_2+k_3) + (2-(n-\nu))k_4].
\end{aligned}\right\} (2.13)$$

With these expressions for the $k^{(\nu)}$ and with the k_1, k_2, k_3, k_4 given by (2.11), let us now compare the next terms in the Taylor series for $\dfrac{h^\nu}{\nu!} y_1^{(\nu)}$ and $\dfrac{h^\nu}{\nu!} y^{(\nu)}(x_1)$, i.e. the terms in h^{n+3}. The term from $\dfrac{h^\nu}{\nu!} y_1^{(\nu)}$ is the term in (2.13) after substitution from (2.11), i.e.

$$\frac{h^{n+3}}{n!}\binom{n+2}{\nu}\frac{1}{(n+1)(n+2)} \times$$
$$\times \left[\frac{4-q}{12}Ff + \frac{1}{4}f_{n-1}Ef + \frac{1-q}{4}f_{n-1}^2 Df + \frac{2-q}{4}f_{n-2}Df\right],$$

where $q = n-\nu$; the corresponding term from $\dfrac{h^\nu}{\nu!} y^{(\nu)}(x_1)$ is the last term in (2.9), which may be written

$$\frac{h^{n+3}}{n!}\binom{n+2}{\nu}\frac{1}{(n+1)(n+2)}\frac{1}{q+3}[Ff + f_{n-1}Ef + f_{n-2}Df].$$

These expressions are identical if, and only if, $q=1$, i.e. $\nu = n-1$. We have therefore obtained identity of the two Taylor series as far as the terms in h^{n+2} for $y, hy', \ldots, \dfrac{h^{n-2}}{(n-2)!} y^{(n-2)}$ and as far as the terms in h^{n+3} for $\dfrac{h^{n-1}}{(n-1)!} y^{(n-1)}$.

Thus, as a method for determining the ordinate y (cf. the remarks at the end of the next subsection), the Runge-Kutta method is of the $(n+2)$-th order (in the sense of § 1.3) for $n>1$ and of the fourth order for $n=1$. Actually this assertion about the case $n=1$ does not follow immediately from our formulae, for they show only that the method is of at least the fourth order when $n=1$, but it can in fact be shown (which we do not do here) that the terms in h^5 in the Taylor series for y_1 and $y(x_1)$ are not identical.

2.4. Hints for using the Runge-Kutta method

For practical calculation it is usually more convenient to work with the quantities

$$v_\nu = \frac{h^\nu}{\nu!}\, y^{(\nu)} \qquad (\nu = 0, 1, \ldots, n-1) \tag{2.14}$$

than with the derivatives $y^{(\nu)}$ direct and to transform the function f and the truncated Taylor series T_ν correspondingly. We denote the initial values of the v_ν by

$$v_{\nu,0} = \frac{h^\nu}{\nu!}\, y_0^{(\nu)} \tag{2.15}$$

and introduce the notation

$$f\left(x, y, \frac{1!}{h}v_1, \frac{2!}{h^2}v_2, \ldots, \frac{(n-1)!}{h^{n-1}}v_{n-1}\right) = \Phi(x, v_0, v_1, v_2, \ldots, v_{n-1}), \tag{2.16}$$

$$\left.\begin{aligned}
t_\nu(\alpha) = \frac{h^\nu}{\nu!}\, T_\nu(\alpha) = v_{\nu,0} + \alpha\binom{\nu+1}{\nu}v_{\nu+1,0} + \alpha^2\binom{\nu+2}{\nu}v_{\nu+2,0} + \cdots + \\
+ \alpha^{n-\nu-1}\binom{n-1}{\nu}v_{n-1,0} \qquad (\nu = 0, 1, \ldots, n-1).
\end{aligned}\right\} \tag{2.17}$$

With this new notation, (2.10) becomes

$$\left.\begin{aligned}
k_1 &= \frac{h^n}{n!}\, \Phi(x_0, v_{0,0}, v_{1,0}, v_{2,0}, \ldots, v_{n-1,0}), \\[2mm]
k_2 &= \frac{h^n}{n!}\, \Phi\left(x_0 + \frac{h}{2},\ t_0\left(\frac{1}{2}\right) + \frac{k_1}{2^n},\ t_1\left(\frac{1}{2}\right) + \frac{k_1}{2^{n-1}}\binom{n}{1}, \right. \\
&\qquad t_2\left(\frac{1}{2}\right) + \frac{k_1}{2^{n-2}}\binom{n}{2}, \ldots, t_{n-2}\left(\frac{1}{2}\right) + \frac{k_1}{4}\binom{n}{n-2},\ t_{n-1}\left(\frac{1}{2}\right) + \frac{k_1}{2}\binom{n}{n-1}\Bigg), \\[2mm]
k_3 &= \frac{h^n}{n!}\, \Phi\left(x_0 + \frac{h}{2},\ t_0\left(\frac{1}{2}\right) + \frac{k_1}{2^n},\ t_1\left(\frac{1}{2}\right) + \frac{k_1}{2^{n-1}}\binom{n}{1}, \right. \\
&\qquad t_2\left(\frac{1}{2}\right) + \frac{k_1}{2^{n-2}}\binom{n}{2}, \ldots, t_{n-2}\left(\frac{1}{2}\right) + \frac{k_1}{4}\binom{n}{n-2},\ t_{n-1}\left(\frac{1}{2}\right) + \frac{k_2}{2}\binom{n}{n-1}\Bigg), \\[2mm]
k_4 &= \frac{h^n}{n!}\, \Phi\left(x_0 + h,\ t_0(1) + k_3,\ t_1(1) + k_3\binom{n}{1},\ t_2(1) + k_3\binom{n}{2}, \ldots, \right. \\
&\qquad t_{n-2}(1) + k_3\binom{n}{n-2},\ t_{n-1}(1) + k_3\binom{n}{n-1}\Bigg),
\end{aligned}\right\} \tag{2.18}$$

while the formula (2.13) for the corrections $k^{(\nu)}$ remains unaltered. The new values

$$v_{\nu,1} = \frac{h^{\nu}}{\nu!}\, y_1^{(\nu)}$$

are then found from the transformed version of (2.4), namely

$$v_{\nu,1} = t_{\nu}(1) + k^{(\nu)}. \tag{2.19}$$

We discern the following

Computing procedure: First of all, the right-hand side of the given differential equation (2.1) and the initial values (2.2) are transformed by the introduction of the auxiliary variables v_{ν} of (2.14) into a function Φ and initial values $v_{\nu,0}$, respectively, as in (2.16), (2.15). The values of Φ needed in (2.18) for determing k_1, k_2, k_3, k_4 are calculated with the aid of the truncated Taylor series $t_{\nu}(\frac{1}{2})$, $t_{\nu}(1)$ defined in (2.17); then the corrections $k^{(\nu)}$ are found from the k_{ϱ} using (2.13); finally (2.19) gives the new values $v_{\nu,1}$. These serve as starting values for the next step in the continuation of the solution.

Tables II/4—II/7 exhibit the appropriate formulae for first-, second-, third- and fourth-order equations (coefficients for corresponding computing schemes[1] for n-th order differential equations where $5 \leq n \leq 10$

Table II/4. *Runge-Kutta scheme for differential equations of the first order.*
$$y' = f(x, y)$$

x	y	$k_{\nu} = h \cdot f(x, y)$	Correction
x_0	y_0	k_1	$k = \frac{1}{6}(k_1 + 2k_2 + 2k_3 + k_4)$
$x_0 + \frac{1}{2}h$	$y_0 + \frac{1}{2}k_1$	k_2	
$x_0 + \frac{1}{2}h$	$y_0 + \frac{1}{2}k_2$	k_3	
$x_0 + h$	$y_0 + k_3$	k_4	
$x_1 = x_0 + h$	$y_1 = y_0 + k$		

Table II/5. *Runge-Kutta scheme for differential equations of the second order.*
$$y'' = f(x, y, y')$$

x	y	$h y' = v_1$	$k_{\nu} = \frac{h^2}{2} f\left(x, y, \frac{v_1}{h}\right)$	Correction
x_0	y_0	v_{10}	k_1	$k = \frac{1}{3}(k_1 + k_2 + k_3)$
$x_0 + \frac{1}{2}h$	$y_0 + \frac{1}{2}v_{10} + \frac{1}{4}k_1$	$v_{10} + k_1$	k_2	
$x_0 + \frac{1}{2}h$	$y_0 + \frac{1}{2}v_{10} + \frac{1}{4}k_1$	$v_{10} + k_2$	k_3	$k' = \frac{1}{3}(k_1 + 2k_2 + 2k_3 + k_4)$
$x_0 + h$	$y_0 + v_{10} + k_3$	$v_{10} + 2k_3$	k_4	
$x_1 = x_0 + h$	$y_1 = y_0 + v_{10} + k$	$v_{11} = v_{10} + k'$		

---

[1] ZURMÜHL, R.: Runge-Kutta-Verfahren zur numerischen Integration von Differentialgleichungen n-ter Ordnung. Z. Angew. Math. Mech. **28**, 173—182 (1948).

have been calculated by ZURMÜHL[1]). The actual calculations are best set out in tables of similar form.

Table II/6. *Runge-Kutta scheme for differential equations of the third order.*

$$y''' = f(x, y, y', y'')$$

x	y	$hy' = v_1$	$\dfrac{h^2}{2}y'' = v_2$	$k_v = \dfrac{h^3}{6}f\left(x, y, \dfrac{v_1}{h}, \dfrac{2v_2}{h^2}\right)$	
x_0	y_0	v_{10}	v_{20}	k_1	k
$x_0 + \tfrac{1}{2}h$	$y_0 + \tfrac{1}{2}v_{10} + \tfrac{1}{4}v_{20} + \tfrac{1}{8}k_1$	$v_{10} + v_{20} + \tfrac{3}{4}k_1$	$v_{20} + \tfrac{3}{2}k_1$	k_2	k'
$x_0 + \tfrac{1}{2}h$	$y_0 + \tfrac{1}{2}v_{10} + \tfrac{1}{4}v_{20} + \tfrac{1}{8}k_1$	$v_{10} + v_{20} + \tfrac{3}{4}k_1$	$v_{20} + \tfrac{3}{2}k_2$	k_3	k''
$x_0 + h$	$y_0 + v_{10} + v_{20} + k_3$	$v_{10} + 2v_{20} + 3k_3$	$v_{20} + 3k_3$	k_4	
$x_1 = x_0 + h$	$y_1 = y_0 + v_{10} + v_{20} + k$	$v_{11} = v_{10} + 2v_{20} + k'$	$v_{21} = v_{20} + k''$		

where $k \;= \tfrac{1}{20}(9k_1 + 6k_2 + 6k_3 - k_4)$,
      $k' = k_1 + k_2 + k_3$,
      $k'' = \tfrac{1}{2}(k_1 + 2k_2 + 2k_3 + k_4)$.

Table II/7. *Runge-Kutta scheme for differential equations of the fourth order.*

$$y^{IV} = f(x, y, y', y'', y''')$$

x	y	$hy' = v_1$
x_0	y_0	v_{10}
$x_0 + \tfrac{1}{2}h$	$y_0 + \tfrac{1}{2}v_{10} + \tfrac{1}{4}v_{20} + \tfrac{1}{8}v_{30} + \tfrac{1}{16}k_1$	$v_{10} + v_{20} + \tfrac{3}{4}v_{30} + \tfrac{1}{2}k_1$
$x_0 + \tfrac{1}{2}h$	$y_0 + \tfrac{1}{2}v_{10} + \tfrac{1}{4}v_{20} + \tfrac{1}{8}v_{30} + \tfrac{1}{16}k_1$	$v_{10} + v_{20} + \tfrac{3}{4}v_{30} + \tfrac{1}{2}k_1$
$x_0 + h$	$y_0 + v_{10} + v_{20} + v_{30} + k_3$	$v_{10} + 2v_{20} + 3v_{30} + 4k_3$
$x_1 = x_0 + h$	$y_1 = y_0 + v_{10} + v_{20} + v_{30} + k$	$v_{11} = v_{10} + 2v_{20} + 3v_{30} + k'$

$\dfrac{h^2}{2}y'' = v_2$	$\dfrac{h^3}{6}y''' = v_3$	$k_v = \dfrac{h^4}{24}f\left(x, y, \dfrac{v_1}{h}, \dfrac{2v_2}{h^2}, \dfrac{6v_3}{h^3}\right)$	
v_{20}	v_{30}	k_1	k
$v_{20} + \tfrac{3}{2}v_{30} + \tfrac{3}{2}k_1$	$v_{30} + 2k_1$	k_2	k'
$v_{20} + \tfrac{3}{2}v_{30} + \tfrac{3}{2}k_1$	$v_{30} + 2k_2$	k_3	k''
$v_{20} + 3v_{30} + 6k_3$	$v_{30} + 4k_3$	k_4	k'''
$v_{21} = v_{20} + 3v_{30} + k''$	$v_{31} = v_{30} + k'''$		

where $k \;\;= \tfrac{1}{15}(8k_1 + 4k_2 + 4k_3 - k_4)$,
      $k' \;= \tfrac{1}{5}(9k_1 + 6k_2 + 6k_3 - k_4)$,
      $k'' = 2(k_1 + k_2 + k_3)$,
      $k''' = \tfrac{2}{3}(k_1 + 2k_2 + 2k_3 + k_4)$.

---

[1] Similar schemes can be found in E. BUKOVICS: Eine Verbesserung und Verallgemeinerung des Verfahrens von BLAESS zur numerischen Integration gewöhnlicher Differentialgleichungen. Öst. Ing.-Arch. **4**, 338—349 (1950).

Simplifications in particular cases. If the function $f(x, y, y', \ldots,$ $y^{(n-1)})$ in the differential equation (2.1) does not depend explicitly on $y^{(n-1)}$, then, as noted in § 2.2, $k_2 = k_3$. In such cases there is consequently one less row to compute in the Runge-Kutta scheme. If f does not depend on some derivative, $y^{(r)}$ say, then only the end value v_{r1} in the column corresponding to $y^{(r)}$ need be calculated since the three auxiliary values between v_{r0} and v_{r1} are superfluous.

Rule of thumb for the size of step length h. A rough guide which is often used for finding a reasonable length of step is that h should be such that k_2 and k_3 are approximately equal; to be more specific, the difference between k_2 and k_3 should not exceed in magnitude a few per cent of the difference between k_1 and k_2, otherwise a smaller step should be taken. The ratio $\dfrac{k_2-k_3}{k_1-k_2}$ is a measure of "sensitiveness" (cf. the "step index" S defined in § 3.4, which can also serve as a guide to the length of step for the Runge-Kutta method with $n = 1$. S is approximately twice as big as the above measure of sensitiveness).

Indications of the magnitude of the error. A satisfactory estimate for the error $|y_1^{(v)} - y^{(v)}(x_1)|$ does not yet exist for the general case (cf. the remarks in § 1.3) but a rough guide which is often employed is provided by a comparison between solutions calculated with steps of length h and $2h$, respectively, as discussed in § 1.3. For the case $n = 1$ the Taylor series for the approximation to y coincides with the Taylor series for the exact solution as far as the term in h^4, so that, from (1.7), the error in the calculation with steps of length h should be roughly $\frac{1}{15}$ of the difference between the results of this calculation and those of the calculation with steps of length $2h$. For the cases with $n > 1$ not all the quantities $y, y', \ldots, y^{(n-1)}$ are determined to the same order of accuracy at each step, but over a large number of steps it may be expected that the influence of the least accurately calculable derivative, namely $y^{(n-1)}$, will determine the overall accuracy; since the method is always of the fourth order (§ 1.3, I) for this derivative, we may still appropriately use the factor $\frac{1}{15}$ as long as no better error estimates exist.

For the first-order equation $y' = f(x, y)$ BIEBERBACH[1] has used a Taylor series expansion to establish the error estimate

$$|y_1 - y(x_1)| < \frac{6MN|x_1 - x_0|^5 |N^5 - 1|}{|N - 1|},$$

[1] BIEBERBACH, L.: Theorie der Differentialgleichungen, 3rd ed., p. 54. Berlin 1930. — On the remainder of the Runge-Kutta formula in the theory of ordinary differential equations. Z. Angew. Math. Phys. **2**, 233—248 (1951); this paper also contains error estimates for n-th order equations (2.1) and for systems of first-order equations. Other error estimates have been given by E. BUKOVICS: Beiträge zur numerischen Integration, II. Mh. Math. **57**, H. 4 (1953), and J. ALBRECHT: Beiträge zum Runge-Kutta-Verfahren. Z. Angew. Math. Mech. **35**, 100—110 (1955).

in which M and N are numbers such that

$$|f(x,y)| < M, \qquad \left|\frac{\partial^{j+k}f}{\partial x^j \partial y^k}\right| < \frac{N}{M^{k-1}} \quad \text{for} \quad j+k \leq 3, \; |x-x_0| N < 1$$

in a domain $|x-x_0| < a$, $|y-y_0| < b$ with $a \geq h$ and $b > Ma$.

2.5. Terminal checks and iteration methods

The incorporation of current checks[1] in the calculation is to be strongly recommended. A very good check is provided by a second calculation with a double-length step; it is best used as a current check (with the two calculations carried out concurrently) so that any errors which arise are noticed as soon as possible. In addition we can apply the following terminal check when the calculation has been taken as far as is required. First we construct the difference table of a convenient multiple of f, say $h^n f$ (this in itself may reveal some errors by a lack of smoothness in the higher differences); the differences can then be used in a finite-difference quadrature formula to evaluate the repeated integral occurring in (1.9). In general, differences of a sufficiently high order do not exist near the ends of the table, so that one cannot use the same formula throughout the table.

In the case of a differential equation of the first order, for example, one would use the following formulae:

at the beginning of the table

$$\int_{x_r}^{x_{r+1}} f \, dx \approx h \sum_{\varrho=0}^{p} (-1)^\varrho \beta_\varrho^* \, \nabla^\varrho f_{r+\varrho} = h \left[f_r + \frac{1}{2} \nabla f_{r+1} - \frac{1}{12} \nabla^2 f_{r+2} + \right.$$
$$\left. + \frac{1}{24} \nabla^3 f_{r+3} - \frac{19}{720} \nabla^4 f_{r+4} + \frac{3}{160} \nabla^5 f_{r+5} - \cdots \right], \tag{2.20}$$

in the middle of the table

$$\int_{x_{r-1}}^{x_{r+1}} f \, dx \approx h \left[2f_r + \frac{1}{3} \nabla^2 f_{r+1} - \frac{1}{90} \nabla^4 f_{r+2} + \frac{1}{756} \nabla^6 f_{r+3} - \cdots \right], \tag{2.21}$$

at the end of the table

$$\int_{x_{r-1}}^{x_r} f \, dx \approx h \left[f_r - \frac{1}{2} \nabla f_r - \frac{1}{12} \nabla^2 f_r - \frac{1}{24} \nabla^3 f_r - \cdots \right]. \tag{2.22}$$

These are respectively formulae (2.11), (2.14) and (2.9) of Ch. I.

---

[1] Cf. E. LINDELÖF: Remarques sur l'intégration .... Acta Soc. Sci. Fenn. A 2 1938, Nr. 13, 11. — SANDEN, H. v.: Praxis der Differentialgleichungen, 3rd ed., p. 29. Berlin 1945.

For differential equations of higher order, terminal checks can be applied in a variety of ways. For instance, instead of using formula (1.9), we could check all derivatives individually by using

$$y^{(\nu)}(x_{r+1}) = y^{(\nu)}(x_r) + \int_{x_r}^{x_{r+1}} y^{(\nu+1)}(x)\, dx \tag{2.23}$$

with $y^{(n)} = f$ and evaluating the integrals by the formulae (2.20) to (2.22) just quoted.

Another way is to evaluate the integral in formula (1.9) by the repeated-integration formulae (2.30), (2.45) of Ch. I; thus at the beginning of the table we use

$$y_{r+1} \approx y_r + h y_r' + \frac{h^2}{2} y_r'' + \cdots + \frac{h^{n-1}}{(n-1)!} y_r^{(n-1)} + h^n \sum_{\varrho=0}^{p} \gamma_{n,\varrho} \Delta^\varrho f_r, \tag{2.24}$$

and in the middle

$$y_{r+1} \approx \begin{cases} \begin{Bmatrix} y_{r-1} + 2 \sum_{\varrho=0}^{\frac{n-3}{2}} \frac{h^{2\varrho+1}}{(2\varrho+1)!} y_r^{(2\varrho+1)} \\[2mm] 2y_r - y_{r-1} + 2 \sum_{\varrho=1}^{\frac{n}{2}-1} \frac{h^{2\varrho}}{(2\varrho)!} y_r^{(2\varrho)} \end{Bmatrix} + \\[4mm] + 2h^n \sum_{\varrho=0}^{p/2} \beta_{n,\varrho}^{**} \nabla^{2\varrho} f_{r+\varrho} \end{cases} \begin{array}{l} \text{for } n \text{ odd,} \\[4mm] \text{for } n \text{ even.} \end{array} \tag{2.25}$$

If corresponding formulae are written down for each derivative occurring in the differential equation, they can be used as the basis of an iterative method of solution: if we have an approximate solution, say the ν-th approximation, denoted by

$$y_r^{[\nu]}, y_r'^{[\nu]}, \ldots, y_r^{(n-1)[\nu]}$$

we can, in general, obtain a better approximation, the $(\nu+1)$-th, by calculating the corresponding f values $f_r^{[\nu]}$ and using these in the right-hand sides of (2.24), (2.25) and the corresponding formulae for the derivatives which occur.

2.6. Examples

I. A differential equation of the first order. Consider again the example of § 1.5, i.e. the initial-value problem (1.21)

$$y' = y - \frac{2x}{y}; \qquad y(0) = 1.$$

We calculate a solution first with a step length $h = 0.2$. The calculation for the first few steps is reproduced in Table II/8, in which an extra column is kept for recording the values of the auxiliary quantity $2x/y$

needed in the calculation of $hf(x, y)$. As far as the point $x = 0.6$ the actual results of the calculation are distinguished by bold-face type; for steps beyond this point only the results are reproduced.

Table II/8. *A Runge-Kutta calculation for a first-order equation*

x	y	$\dfrac{2x}{y}$	$\dfrac{h}{2}f = 0{\cdot}1\left(y - \dfrac{2x}{y}\right)$	$3k$ and k
0	**1**	0	0·1	
0·1	**1·1**	0·181 818	0·091 8182	0·549 6877
0·1	**1·091 818**	0·183 181	0·090 8637	0·183 2292
0·2	**1·181 727**	0·338 488	0·084 3239	
0·2	**1·183 229 2**	0·338 058	0·084 5171	
0·3	**1·267 746**	0·473 281	0·079 4465	0·475 3128
0·3	**1·262 676**	0·475 181	0·078 7495	0·158 4376
0·4	**1·340 728**	0·596 691	0·074 4037	
0·4	**1·341 666 8**	0·596 274	0·074 5393	
0·5	**1·416 206**	0·706 112	0·071 0094	0·424 8538
0·5	**1·412 676**	0·707 831	0·070 4845	0·141 6179
0·6	**1·482 636**	0·809 369	0·067 3267	
0·6	**1·483 281 5**	0·809 016	0·067 4269	
0·8	1·612 5140		0·062 0282	
1	1·732 1419		0·057 7512	
1·2	1·844 0401		0·054 2565	
1·4	1·949 547		0·051 334	
1·6	2·049 660		0·048 846	
1·8	2·145 148		0·046 698	
2	2·236 624		0·044 828	

For checking, and also for assessing the accuracy, the calculation is repeated with a double-length step $h = 0.4$; the results are given in Table II/9.

Table II/9. *Corresponding calculation with a double-length step $h = 0.4$*

x	y	$\dfrac{2x}{y}$	$\dfrac{h}{2}f = 0{\cdot}2\left(y - \dfrac{2x}{y}\right)$	$3k$ and k
0	**1**	0	0·2	
0·2	**1·2**	0·333 333	0·173 3333	1·026 1974
0·2	**1·173 333**	0·340 909	0·166 4848	0·342 0658
0·4	**1·332 970**	0·600 164	0·146 5612	
0·4	**1·342 065 8**	0·596 096	0·149 1940	
0·6	**1·491 260**	0·804 689	0·137 3142	0·814 1486
0·6	**1·479 380**	0·811 151	0·133 6458	0·271 3829
0·8	**1·609 358**	0·994 185	0·123 0346	
0·8	**1·613 448 7**			
1·2	1·846 00			
1·6	2·053 67			
2	2·244 86			

We now collect together (in Table II/10) the results for the points $x = 0.4 \times k$ $(k = 1, 2, \ldots, 5)$ and calculate corrections as in (1.7). Comparison with the exact solution $y = \sqrt{2x + 1}$ yields the actual errors in the values $\tilde{y}$ obtained with $h = 0.2$ and in the "improved" values $[y] = \tilde{y} - \delta$; the latter are seen to be considerably smaller than the former.

Table II/10. *Comparison of the accuracies of the Runge-Kutta solution and the improved solution*

x	Results of Runge-Kutta calculation with		$\delta = \frac{1}{15}(\tilde{\tilde{y}} - \tilde{y})$	$[y] = \tilde{y} - \delta$	Error in $\tilde{y}$	Error in $[y]$
	$h = 0.2$ $\tilde{y}$	$h = 0.4$ $\tilde{\tilde{y}}$				
0·4	1·341 666 9	1·342 065 8	0·000 026 6	1·341 640 3	+0·000 026 2	−0·000 000 4
0·8	1·612 514 0	1·613 448 7	0·000 062 3	1·612 451 7	0·000 062 4	+0·000 000 1
1·2	1·844 040	1·845 99	0·000 130	1·843 910	0·000 131	+0·000 001 5
1·6	2·049 660	2·053 64	0·000 266	2·049 395	0·000 270	0·000 004 4
2	2·236 624	2·244 80	0·000 545	2·236 079	0·000 556	0·000 010 9

II. A System of first-order differential equations. Here we consider the Euler equations for the motion of an unsymmetrical top (principle moments of inertia A, B, C) subject to a frictional resistance proportional to the instantaneous angular velocity vector u_1, u_2, u_3. With differentiation in time denoted by a dot the equations run[1]

$$A \dot{u}_1 = (C - B) u_2 u_3 - \varepsilon u_1,$$

$$B \dot{u}_2 = (A - C) u_3 u_1 - \varepsilon u_2,$$

$$C \dot{u}_3 = (B - A) u_1 u_2 - \varepsilon u_3.$$

If $B = 2A$, $C = 3A$, $\varepsilon = 0.6$ A/sec., we obtain the system

$$\dot{u}_1 = u_2 u_3 - 0.6 u_1, \qquad \dot{u}_2 = -u_1 u_3 - 0.3 u_2, \qquad \dot{u}_3 = \tfrac{1}{3} u_1 u_2 - 0.2 u_3.$$

Let the initial angular velocity be given by $u_1 = 1$, $u_2 = 1$, $u_3 = 0$ at $t = 0$.

Application of the Runge-Kutta method occasions little difficulty. We have only to set up three schemes, one for each unknown, and it suffices to exhibit the first two steps with the step $h = 0.2$ and, for comparison, one step with the double length interval $h = 0.4$. This is done in Table II/11.

III. A differential equation of higher order. In the calculation of the laminar boundary layer on a flat plate parallel to a stream, the following boundary-value problem arises:

$$\frac{d^3\zeta}{d\xi^3} = -\zeta \frac{d^2\zeta}{d\xi^2}; \qquad \zeta(0) = \zeta'(0) = 0, \qquad \zeta' \to 2 \quad \text{as} \quad \xi \to \infty.$$

---

[1] See, for example, Handbuch der Physik, Vol. V, p. 405, Berlin 1927, or A. FÖPPL: Vorlesungen über technische Mechanik, Vol. 4 (Dynamik), 10th ed., p. 209. München und Berlin 1944.

Table II/11. *Runge-Kutta method applied to a system of three first-order differential equations*

t	u_1	u_2	u_3	u_2u_3	$\frac{h}{2}\dot{u}_1$	3k and k	$-u_1u_3$	$\frac{h}{2}\dot{u}_2$	3k and k	$\frac{1}{3}u_1u_2$	$\frac{h}{2}\dot{u}_3$	3k and k
0	1	1	**0**	0	−0·06		0	−0·03		0·33333	0·033333	
0·1	0·94	0·97	0·033333	0·03233	−0·053167	−0·322194	−0·03133	−0·032233	−0·191523	0·30393	0·029727	0·178546
0·1	0·946833	0·967767	0·029727	0·02877	−0·053933	−0·107398	−0·02815	−0·031848	−0·063841	0·30544	0·029549	0·059515
0·2	0·892134	0·936304	0·059098	0·05534	−0·047994		−0·05272	−0·033361		0·27843	0·026661	
0·2	**0·892602**	**0·936159**	**0·059515**	0·05572	−0·047985		−0·05312	−0·033397		0·27854	0·026664	
0·3	0·844617	0·902762	0·086179	0·07780	−0·042897	−0·259764	−0·07279	−0·034362	−0·205048	0·25416	0·023692	0·142967
0·3	0·849705	0·901797	0·083207	0·07504	−0·043478	−0·086588	−0·07070	−0·034124	−0·068349	0·25542	0·023878	0·047656
0·4	0·805646	0·867911	0·107271	0·09310	−0·039029		−0·08642	−0·034679		0·23308	0·021163	
0·4	**0·806014**	**0·867810**	**0·107171**	$2u_2u_3$			$-2u_1u_3$			$\frac{2}{3}u_1u_2$		
0	1	1	**0**	0	−0·12		0	−0·06		0·66667	0·066667	
0·2	0·88	0·94	0·066667	0·12533	−0·093067	−0·581606	−0·11733	−0·068133	−0·396675	0·55147	0·052480	0·322239
0·2	0·906933	0·931867	0·052480	0·09781	−0·099051	−0·193869	−0·09519	−0·065431	−0·132225	0·56343	0·054244	0·107413
0·4	0·801898	0·869138	0·108488	0·18858	−0·077370		−0·17399	−0·069547		0·46464	0·042124	
0·4	**0·806131**	**0·867755**	**0·107479**									

This may be reduced to the initial-value problem[1]

$$y''' = - y y'', \quad y(0) = y'(0) = 0, \quad y''(0) = 1.$$

The Runge-Kutta calculations for $h = 0·5$ are set out in Table II/12; for the steps from $x = 1$ to $x = 3$ only the results are given.

Table II/12. *Solution of a third-order differential equation by* RUNGE-KUTTA

x	y	$v_1 = \frac{1}{2} y'$	$v_2 = \frac{1}{8} y''$	$y v_2 = -6k$	k	$\frac{k_2+k_3}{2k''}$ $20k$	$\frac{k'}{k''}$ k
0	0	0	0·125	0	0		
0·25	0·03125		0·125	0·0039063	−0·0006510	−0·0012969	−0·0012969
0·25	0·03125		0·1240234	0·0038756	−0·0006459	−0·0051444	−0·0025722
0·5	0·1243541		0·1230622	0·0153033	−0·0025506	−0·0052308	−0·0002615
0·5	0·1247385	0·2487031	0·1224278	0·0152715	−0·0025453		
0·75	0·2793798		0·1186099	0·0331372	−0·0055229	−0·0108378	−0·0133831
0·75	0·2793798		0·1141435	0·0318894	−0·0053149	−0·0329269	−0·0164635
1	0·4905545		0·1064831	0·0522358	−0·0087060	−0·0792285	−0·0039614
1	0·4919080	0·4801756	0·1059643	0·0521247	−0·0086874		
1·5	1·0679173	0·6605306	0·0722886		−0·0128664		
2	1·787924	0·767123	0·035606		−0·010610		
2·5	2·581205	0·811538	0·012286		−0·005285		
3	3·400614	0·824154	0·003316		−0·001880		

Table II/13. *Corresponding calculation with double-length step*

x	y	$v_1 = y'$	$v_2 = \frac{1}{2} y''$	k
0	0	0	0·5	0
0·5	0·125		0·5	−0·020833
0·5	0·125		0·468750	−0·019531
1	0·480469		0·441406	−0·070694
1	0·491425	0·959636	0·424289	−0·069502
1·5	1·068627		0·320036	−0·114000
1·5	1·068627		0·253289	−0·090224
2	1·785126		0·153618	−0·091409
2	1·787377	1·534488	0·139609	−0·083178
2·5	2·579126		0·014842	−0·012760
2·5	2·579126		0·120469	−0·103568
3	3·357906		−0·171096	+0·191508
3	3·379570	1·614200	0·077446	

[1] MOHR, E.: Dtsch. Math. **4**, 485 (1939). — SCHLICHTING, H.: Grenzschicht-Theorie. 483 pp. Karlsruhe 1951. — Modern Developments in Fluid Dynamics (ed. S. GOLDSTEIN), Vol. I, p. 135. London: Oxford University Press 1938.

For comparison the calculation was repeated with the double-length step $h = 1$; this is exhibited in Table II/13, from which several auxiliary columns have been omitted.

For the last step, from $x = 2$ to $x = 3$, an alternation in sign appears in the v_2 and k columns, indicating that the interval is much too large and that we can no longer have any confidence in the values obtained; these values for the point $x = 3$ do, in fact, show a considerable deviation from those obtained by the calculation with the smaller interval.

§3. Finite-difference methods for differential equations of the first order

Among the approximate methods which exist at present, the finite-difference methods are the most accurate in general. Among these, the interpolation methods are superior to the extrapolation methods in that they give a considerable improvement in accuracy with only a moderate increase in the amount of labour involved. Consequently, interpolation methods are in far more extensive use nowadays than extrapolation methods. Of the interpolation methods, the central-difference method has several advantages over the Adams interpolation method: the computations are simpler, the convergence is more rapid and smaller error limits can be derived. Naturally only very general advice can be given to guide the computer in his choice of a suitable method. *Finite-difference methods are very suitable when the functions being dealt with are smooth and the differences decrease rapidly with increasing order; calculations with these methods are best carried out with a fairly small length of step. On the other hand, if the functions are not smooth, perhaps given by experimental results, or if we want to use a larger step, then the Runge-Kutta method is to be preferred; it is also advantageous to use this method when we have to change the length of step frequently, particularly when this change is a decrease. Clearly we should not choose too large a step even for the Runge-Kutta method.* (Cf.[1] the remarks in § 2.4.)

3.1. Introduction

Suppose that the differential equation (1.10)

$$y' = f(x, y)$$

is to be integrated numerically with the initial condition $y = y_0$ at $x = x_0$. As in § 1.4, let the range of integration be covered by the equally spaced points $x_0, x_1, \ldots, x_n$ with the constant difference $h = \Delta x_\nu = x_{\nu+1} - x_\nu$

---

[1] MILNE, W. E.: Note on the Runge-Kutta method. Research Paper RP 2101, J. Res. Nat. Bur. Stand. **44**, 549—550 (1950), gives examples for which the central-difference method yields substantially better results than the Runge-Kutta method.

(the step length) and let y_ν be an approximation to the value $y(x_\nu)$ of the exact solution at the point x_ν. The finite-difference methods are based on the integrated form (1.13)

$$y(x_{r+1}) = y(x_r) + \int_{x_r}^{x_{r+1}} f\left(x, y(x)\right) dx \qquad (3.1)$$

of equation (1.10), as were the crude methods of § 1.5, but here the integral is approximated more accurately. Suppose that the integration has already been carried as far as the point $x = x_r$ so that approximations $y_1, \ldots, y_{r-2}, y_{r-1}, y_r$, and hence also approximate values $f_r = f(x_r, y_r)$, are known. We now have to calculate y_{r+1}. In finite-difference methods, formulae for doing this are derived by replacing the integrand in (3.1) by a polynomial $P(x)$ which takes the values f_ν at a certain number of points x_ν and then integrating this polynomial over the interval x_r to x_{r+1}. This basic idea can be used in a variety of ways (cf. § 3.3)[1] but we always need to have a sequence of approximations f_ν before we can start the step-by-step procedures defined by the finite-difference formulae (see § 3.3); consequently the finite-difference methods have two distinct stages: the first is the calculation of "starting values", in which the first few approximations $y_1, y_2, \ldots$, the "starting values" (we reserve the term "initial values" for values at the initial point $x = x_0$), sufficiently many to calculate the values f_ν required for the first application of the finite-difference formula, are obtained by some other means; then follows the main calculation, in which the finite-difference formula is used to continue the solution step by step as far as required. The starting values should be calculated as accurately as possible (cf. § 1.2).

3.2. Calculation of starting values

The starting values needed for the main calculation can be obtained in a variety of ways. As has already been stressed, particular care must be exercised in the calculation of these starting values, for the whole calculation can be rendered useless by inaccuracies in them.

We now mention several possible ways of obtaining starting values; anyone with little experience might restrict himself to the first two only to begin with.

I. Using some other method of integration. Provided that it is sufficiently accurate, any method of integration which does not require starting values (as distinct from initial values) can be used. Bearing in mind the high accuracy desired, one would normally choose the

---

[1] The interpolation idea is used by W. QUADE: Numerische Integration von gewöhnlichen Differentialgleichungen nach HERMITE. Z. Angew. Math. Mech. 37, 161—169 (1957).

Runge-Kutta method; further, one would work preferably with a step of half the length to be used in the main calculation and with a greater number of decimals.

II. Using the Taylor series for $y(x)$. If the function $f(x, y)$ is of simple analytical form, we can determine the derivatives $y'(x_0)$, $y''(x_0)$, $y'''(x_0)$, ... by differentiation of the differential equation (1.10); starting values can then be calculated from the Taylor series

$$y(x_\nu) = y(x_0) + \frac{\nu h}{1!} y'(x_0) + \frac{(\nu h)^2}{2!} y''(x_0) + \cdots, \tag{3.2}$$

of which as many terms are taken as are necessary for the truncation not to affect the last decimal carried (always assuming that the series converges).

Several of the finite-difference methods need three starting values, and for these it suffices to use (3.2) for $\nu = \pm 1$; this usually possesses advantages over using (3.2) for $\nu = 1, 2$, particularly as regards convergence.

Example. For the example (1.21)

$$y' = y - \frac{2x}{y}, \qquad y(0) = 1$$

of § 1.5 we have

$$y'y - y^2 + 2x = 0,$$
$$y''y + y'^2 - 2y'y + 2 = 0,$$
$$y'''y + 3y''y' - 2y''y - 2y'^2 = 0,$$
$$\cdots \cdots \cdots \cdots \cdots \cdots \cdots$$

Putting $x = 0$ and $y = 1$ we obtain successively the initial values of the derivatives:

$$y'(0) = 1; \quad y''(0) = -1; \quad y'''(0) = 3; \quad y^{IV}(0) = -15; \quad y^V(0) = 105;$$

[generally, $y^{(n)}(0) = (-1)^{n-1} \times 1.3.5. \ldots (2n-3)$ for $n = 2, 3, \ldots$]. With $h = 0.1$, for example, we have the starting values

$$y(\pm 0.1) = 1 \pm 0.1 - \frac{1}{2!}(0.1)^2 \pm \frac{3}{3!}(0.1)^3 - \frac{15}{4!}(0.1)^4 \pm \frac{105}{5!}(0.1)^5 - \left.\begin{matrix} \\ \\ \end{matrix}\right\}$$
$$\left.- \frac{945}{6!}(0.1)^6 \pm \frac{10395}{7!}(0.1)^7 - \cdots, \right\} \tag{3.3}$$

which to seven decimals are

$$y(0.1) = 1.0954451,$$

$$y(-0.1) = 0.8944272.$$

IIa. MILNE'S starting procedure. W. E. MILNE[1] has given formulae which bring in the derivatives at the point x_1. They are obtained by eliminating $y^{IV}(x_0)$, $y^V(x_0)$, $y^{VI}(x_0)$ between the Taylor series for $y(x_1)$,

---

[1] MILNE, W. E.: A note on the numerical integration of differential equations. Research paper RP 2046, J. Res. Nat. Bur. Stand. **43**, 537—542 (1949).

$y'(x_1)$, $y''(x_1)$, $y'''(x_1)$ truncated after the term in $y^{VI}(x_0)$. The result is the formula

$$y(x_1) - y(x_0) = \frac{h}{2}\left[y'(x_1) + y'(x_0)\right] - \frac{h^2}{10}\left[y''(x_1) - y''(x_0)\right] +$$
$$+ \frac{h^3}{120}\left[y'''(x_1) + y'''(x_0)\right] + R_6,$$

where

$$R_6 = -\frac{h^7}{100\,800}\,y^{VII}(\xi).$$

The intermediate point ξ in this remainder term lies in the interval $x_0 \leqq \xi \leqq x_1$.

This formula is applicable when the higher derivatives may be expressed in simple analytical form in terms of x and y. For the same degree of accuracy, fewer derivatives need be calculated than with method II but since y_1 appears on the right-hand side, the formula is an equation, rather than an explicit expression, for y_1 and for non-linear differential equations y_1 will usually have to be determined by iteration (cf. § 3.3 method II). The formula can also be used for the main calculation. If we also expand $y(x_2)$ into a Taylor series and truncate after the term in $y^{VI}(x_0)$, we can derive in like manner the formula

$$y(x_2) - 2y(x_1) + y(x_0) = 7h\left[y'(x_1) - y'(x_0)\right] - 3h^2\left[y''(x_1) + y''(x_0)\right] +$$
$$+ \frac{h^3}{12}\left[11\,y'''(x_1) - 5\,y'''(x_0)\right] + \frac{1}{480}\,h^7\,y^{VII}(\xi).$$

III. Using quadrature formulae. Using formulae (2.11) and (2.14) of Ch. I, we build up rough values of y_ν, hf_ν, hVf_ν, ... for the first few values of ν and then improve them by an iterative process. Various schemes can be arranged for doing this; we give here a procedure[1] which is suitable for the construction of two (y_1, y_2) or three (y_1, y_2, y_3) starting values. The procedure is completely described by the following formulae (for two starting values, only the formulae framed in dots are needed and the B equations are to be used also for $\nu = 0$):

A. *Rough values.*

$$
\begin{aligned}
&1.\ \ \tilde{y}_1 = y_0 + hf_0, \quad \text{thence} \quad \tilde{f}_1 = f(x_1, \tilde{y}_1); \quad V\tilde{f}_1 = \tilde{f}_1 - f_0, \\
&2.\ \ y_1^{[0]} = y_0 + h\left(f_0 + \tfrac{1}{2}V\tilde{f}_1\right), \quad \text{thence} \quad f_1^{[0]}, Vf_1^{[0]}, \\
&3.\ \ y_2^{[0]} = y_0 + 2hf_1^{[0]}, \quad \text{thence} \quad f_2^{[0]}, Vf_2^{[0]}, V^2f_2^{[0]}, \\
&4.\ \ y_1^{[1]} = y_0 + h\left(f_0 + \tfrac{1}{2}Vf_1^{[0]} - \tfrac{1}{12}V^2f_2^{[0]}\right), \\
&\quad\ \ y_2^{[1]} = y_0 + h\left(2f_1^{[0]} + \tfrac{1}{3}V^2f_2^{[0]}\right), \text{ thence } f_1^{[1]}, f_2^{[1]}, Vf_1^{[1]}, Vf_2^{[1]}, V^2f_2^{[1]}, \\
&5.\ \ y_3^{[1]} = y_1^{[1]} + h\left(2f_2^{[1]} + \tfrac{1}{3}V^2f_2^{[1]}\right), \text{ thence } f_3^{[1]}, Vf_3^{[1]}, V^2f_3^{[1]}, V^3f_3^{[1]}.
\end{aligned}
\tag{3.4}
$$

---

[1] Other iterative schemes can be found in G. SCHULZ: Interpolationsverfahren zur numerischen Integration gewöhnlicher Differentialgleichungen. Z. Angew. Math. Mech. **12**, 44—59 (1932). — TOLLMIEN, W.: Über die Fehlerabschätzung beim Adamsschen Verfahren. Z. Angew. Math. Mech. **18**, 83—90 (1938), in particular p. 87.

B. *Iterative improvement for* $v = 1, 2, \ldots$ (or $v = 0, 1, \ldots$)

$$
\left.
\begin{aligned}
y_1^{[v+1]} &= y_0 + h\left(f_0 + \tfrac{1}{2}\nabla f_1^{[v]} - \tfrac{1}{12}\nabla^2 f_2^{[v]} + \tfrac{1}{24}\nabla^3 f_3^{[v]}\right), \\
y_2^{[v+1]} &= y_0 + h\left(2f_1^{[v]} + \tfrac{1}{3}\nabla^2 f_2^{[v]}\right), \\
y_3^{[v+1]} &= y_1^{[v+1]} + h\left(2f_2^{[v]} + \tfrac{1}{3}\nabla^2 f_3^{[v]}\right).
\end{aligned}
\right\} \tag{3.5}
$$

Thus we alternately improve the three y values and revise the function values $f_j^{[v]} = f(x_j, y_j^{[v]})$ and their differences. This starting process should be carried out with a sufficiently small step length (see § 3.4).

Table II/14. *Iterative calculation of starting values*

From		x	y	$hf = 0\cdot 1(x+y)$	$h\nabla f$	$h\nabla^2 f$	$h\nabla^3 f$
(3.4)	1.	0·1	0	0·01	0·01		
	2.	0·1	0·005	0·0105	0·0105		
	3.	0·2	0·021	0·0221	0·0116	0·0011	
(3.4)	4.	0·1	0·005158	0·0105158	0·0105158		
		0·2	0·021367	0·0221367	0·0116209	0·0011051	
	5.	0·3	0·049800	0·0349800	128433	12224	0·0001173
Iteration (3.5) $v = 2$		0·1	0·005170	0·0105170	0·0105170		
		0·2	0·021400	221400	116230	0·0011060	
		0·3	0·049851	349851	128451	12221	0·0001161
$v = 3$		0·1	0·005171	105171	105171		
		0·2	0·021403	221403	116232	11061	
		0·3	0·049858	349858	128455	12223	1162
$v = 4$		0·3	0·049859	349859	128456	12224	1163

Example. With $h = 0\cdot 1$ application of this starting procedure to the initial-value problem

$$y' = x + y, \qquad y(0) = 0$$

yields the numbers in Table II/14. The value of h is, in fact, rather large (cf. the remarks on the length of step interval in § 3.4); with $h = 0\cdot 05$ so many iterations would not have been necessary.

3.3. Formulae for the main calculation

We now describe how the next approximate value y_{r+1} can be obtained once the values $y_1, y_2, \ldots, y_r$ at the points $x_1, x_2, \ldots, x_r$ have been computed.

I. The Adams extrapolation method[1,2]. In the "extrapolation methods", which we consider first, the function $f(x, y(x))$ under the integral in equation (3.1) is replaced by the interpolation polynomial[3] $P(x)$ which takes the values $f_{r-p}, \ldots, f_{r-1}, f_r$ at the points $x_{r-p}, \ldots, x_{r-1}, x_r$ [where $f_\varrho = f(x_\varrho, y_\varrho)$]. In effect we evaluate the integral by means of the quadrature formula (2.6) of Ch. I; thus, with y_{r+1} and y_r replacing $y(x_{r+1})$ and $y(x_r)$, (3.1) becomes

$$\left.\begin{aligned}
y_{r+1} &= y_r + h \sum_{\varrho=0}^{p} \beta_\varrho \nabla^\varrho f_r \\
&= y_r + h\left(f_r + \frac{1}{2}\nabla f_r + \frac{5}{12}\nabla^2 f_r + \frac{3}{8}\nabla^3 f_r + \frac{251}{720}\nabla^4 f_r + \cdots\right),
\end{aligned}\right\} \quad (3.6)$$

in which the β_ϱ are given generally by formula (2.7) in Ch. I. The exact solution $y(x)$ satisfies the corresponding exact form

$$y(x_{r+1}) = y(x_r) + h \sum_{\varrho=0}^{p} \beta_\varrho \nabla^\varrho f(x_r, y(x_r)) + S_{p+1}. \qquad (3.7)$$

(2.8) of Ch. I gives an estimate for the remainder term S_{p+1}.

There are occasions when the extrapolation formula (3.6) is used in a somewhat different form (sometimes called the "Lagrangian" form) in which the differences $\nabla^\varrho f_r$ are expressed in terms of the function values f_s. If the terms involving each individual function value are collected together we obtain new coefficients $\alpha_{p\varrho}$ which depend on the number p of differences retained in (3.6):

$$y_{r+1} = y_r + h \sum_{\varrho=0}^{p} \alpha_{p\varrho} f_{r-\varrho}. \qquad (3.8)$$

---

[1] JOHN COUCH ADAMS, the English astronomer, born on the 5th June 1819 in Laneast, became a Fellow of St. Johns College, Cambridge, and tutor in mathematics there; then in 1849 he was appointed Director of the Observatory and in 1858 Lowndean Professor of Astronomy and Geometry in the University. He died on the 22nd January 1892 in London. He was one of the discoverers of the planet Neptune. As early as 1841 he tried to explain the perturbations in the motion of Uranus by the influence of an unknown planet and to calculate the orbit of this planet by first assuming it to be circular and then modifying it to elliptical form by deriving corrections from the perturbations of Uranus. In 1844 he communicated his results to Prof. CHALLIS and asked him if he would look out for the planet in the calculated position.

[2] BASHFORTH, F., and J. C. ADAMS: An attempt to test the theories of capillary action, p. 18. Cambridge 1883.

[3] Trigonometric interpolation polynomials are recommended by W. QUADE: Numerische Integration von Differentialgleichungen bei Approximation durch trigonometrische Ausdrücke. Z. Angew. Math. Mech. **31**, 237—238 (1951); exponential sums are used by P. BROCK and F. J. MURRAY: The use of exponential sums in step-by-step integration. M. T. A. C. **6**, 63—78 (1952).

The values of the first few $\alpha_{p\varrho}$ are given in Table II/15 (a check on the calculation of the $\alpha_{p\varrho}$ is given by the relation $\sum\limits_{\varrho=0}^{p}\alpha_{p\varrho}=1$).

Table II/15. *The numbers* $\alpha_{p\varrho}$

	$\varrho=$			
	0	1	2	3
$p=1$	$\dfrac{3}{2}$	$-\dfrac{1}{2}$		
$p=2$	$\dfrac{23}{12}$	$-\dfrac{16}{12}$	$\dfrac{5}{12}$	
$p=3$	$\dfrac{55}{24}$	$-\dfrac{59}{24}$	$\dfrac{37}{24}$	$-\dfrac{9}{24}$

In order to calculate y_{r+1} from (3.6) we need the values which are "boxed" in Table II/16. The values of $f, \nabla f, \ldots$ associated with the function f are conveniently tabulated with the factor h; the coefficients which we multiply them by in (3.6) are noted at the heads of the corresponding columns. When y_{r+1} is being calculated, all the numbers above the dotted line are known, and each further step in the calculation yields in turn another line of entries "parallel" to the dotted line. For convenience, the differences are often arranged as in Table I/2 so that this line of new entries is horizontal (cf. the example in § 3.5).

Table II/16. *The Adams extrapolation method*

		$\times 1$	$\times 1$	$\times\dfrac{1}{2}$	$\times\dfrac{5}{12}$	$\times\dfrac{3}{8}$	
x		y	hf	$h\nabla f$	$h\nabla^2 f$	$h\nabla^3 f$	
							$\boxed{h\nabla^3 f_r}$
x_{r-1}		y_{r-1}	hf_{r-1}		$\boxed{h\nabla^2 f_r}$		
				$\boxed{h\nabla f_r}$			
x_r		$\boxed{y_r}$	$\boxed{hf_r}$				
x_{r+1}		$\boxed{y_{r+1}}$					

If we use the equation

$$y(x_{r+1}) = y(x_{r-1}) + \int\limits_{x_{r-1}}^{x_{r+1}} f(x, y(x))\, dx \tag{3.9}$$

instead of (3.1) and, as above, replace the integrand by the polynomial $P(x)$ which takes the values f_ϱ at $x=x_\varrho$ ($\varrho=r-p,\ldots,r$), we obtain NYSTRÖM's extrapolation formula

$$\left.\begin{aligned}
y_{r+1} &= y_{r-1} + h\left[2f_r + \frac{1}{3}\nabla^2 f_r + \frac{1}{3}\nabla^3 f_r + \frac{29}{90}\nabla^4 f_r + \frac{14}{45}\nabla^5 f_r + \cdots\right] \\
&= y_{r-1} + h\left[2f_r + \frac{1}{3}(\nabla^2 f_r + \nabla^3 f_r + \nabla^4 f_r + \nabla^5 f_r) - \right. \\
&\qquad\qquad\left. - \frac{1}{90}(\nabla^4 f_r + 2\nabla^5 f_r) + \cdots\right],
\end{aligned}\right\} \tag{3.10}$$

which follows immediately from (2.32) of Ch. I with $u = 1$ (we have only to replace x_{-1}, x_0, x by x_{r-1}, x_r, x_{r+1}, respectively). If truncated after the term in $\nabla^3 f_r$, this formula has simpler coefficients than the Adams formula (3.6). The corresponding difference scheme is shown in Table II/17 with the values needed for the calculation of y_{r+1} boxed as before.

Table II/17. NYSTRÖM's *extrapolation method*

x	y	$\times 1$ hf	$\times 2$ $h\nabla f$	$\times \frac{1}{3}$ $h\nabla^2 f$	$\times \frac{1}{3}$ $h\nabla^3 f$
x_{r-1}	$\boxed{y_{r-1}}$	hf_{r-1}	$h\nabla f_{r-1}$	$\boxed{h\nabla^2 f_r}$	$\boxed{h\nabla^3 f_r}$
x_r	y_r	$\boxed{hf_r}$	$h\nabla f_r$		
x_{r+1}	$\boxed{y_{r+1}}$	hf_{r+1}			

II. The Adams interpolation method. Here the integrand $f(x, y(x))$ in the equation (3.1) is replaced by the polynomial $P^*(x)$ which takes the values $f_{r-p+1}, \ldots, f_{r-1}, f_r, f_{r+1}$ at the points $x_{r-p+1}, \ldots, x_{r-1}, x_r, x_{r+1}$. Then from the quadrature formula (2.9) of Ch. I it follows that

$$\left.\begin{aligned} y_{r+1} &= y_r + h \sum_{\varrho=0}^{p} \beta_\varrho^* \nabla^\varrho f_{r+1} \\ &= y_r + h\left(f_{r+1} - \frac{1}{2}\nabla f_{r+1} - \frac{1}{12}\nabla^2 f_{r+1} - \frac{1}{24}\nabla^3 f_{r+1} - \cdots\right), \end{aligned}\right\} \tag{3.11}$$

where the β_ϱ^* are given generally by formula (2.9) of Ch. I.

For the exact solution $y(x)$ we have the corresponding formula

$$y(x_{r+1}) = y(x_r) + h\sum_{\varrho=0}^{p}\beta_\varrho^* \nabla^\varrho f(x_{r+1}, y(x_{r+1})) + S_{p+1}^*$$

with remainder term S_{p+1}^*, for which an estimate is given by (2.10) of Ch. I.

Formula (3.11) is also used occasionally in Lagrangian form corresponding to (3.8):

$$y_{r+1} = y_r + h\sum_{\varrho=0}^{p}\alpha_{p\varrho}^* f_{r+1-\varrho}. \tag{3.12}$$

The first few values of the $\alpha_{p\varrho}^*$ are given in Table II/18.

As with the $\alpha_{p\varrho}$, here also we have the check $\sum_{\varrho=0}^{p}\alpha_{p\varrho}^* = 1$. From (2.7), (2.13) of Ch. I it follows that

$$\alpha_{p0}^* = \beta_p. \tag{3.13}$$

The quantities which appear in (3.11) are indicated in Table II/19, where the finite-difference scheme is set out in the same way as for the extrapolation method.

Table II/18. *The numbers* $\alpha^{*}_{p\varrho}$

	$\varrho=$				
	0	1	2	3	4
$p=1$	$\dfrac{1}{2}$	$\dfrac{1}{2}$			
$p=2$	$\dfrac{5}{12}$	$\dfrac{8}{12}$	$-\dfrac{1}{12}$		
$p=3$	$\dfrac{9}{24}$	$\dfrac{19}{24}$	$-\dfrac{5}{24}$	$\dfrac{1}{24}$	
$p=4$	$\dfrac{251}{720}$	$\dfrac{646}{720}$	$-\dfrac{264}{720}$	$\dfrac{106}{720}$	$-\dfrac{19}{720}$

In the application of (3.11) the difficulty arises that the quantities depending on $f_{r+1}=f(x_{r+1}, y_{r+1})$ which appear on the right-hand side are not yet known. Consequently the unknown y_{r+1} appears on both

Table II/19. *The Adams interpolation method*

x	$\times 1$ y	$\times 1$ hf	$\times\left(-\dfrac{1}{2}\right)$ $h\nabla f$	$\times\left(-\dfrac{1}{12}\right)$ $h\nabla^2 f$	$\times\left(-\dfrac{1}{24}\right)$ $h\nabla^3 f$
x_{r-1}	y_{r-1}	hf_{r-1}		$h\nabla^2 f_r$	
			$h\nabla f_r$		$h\nabla^3 f_{r+1}$
x_r	$\boxed{y_r}$	hf_r		$\boxed{h\nabla^2 f_{r+1}}$	
			$\boxed{h\nabla f_{r+1}}$		
x_{r+1}	$\boxed{y_{r+1}}$	$\boxed{hf_{r+1}}$			

sides of the equation and only in special cases is it possible to solve this equation exactly for y_{r+1}. Equation (3.11) is, however, very suitable for the iterative calculation of y_{r+1} provided that h is sufficiently small. We put an approximate value $y^{[\sigma]}_{r+1}$ in the right-hand side, forming $f^{[\sigma]}_{r+1}=f(x_{r+1}, y^{[\sigma]}_{r+1})$ and the differences $\nabla^\varrho f^{[\sigma]}_{r+1}=\nabla^{\varrho-1}f^{[\sigma]}_{r+1}-\nabla^{\varrho-1}f_r$, and then calculate

$$y^{[\sigma+1]}_{r+1}=y_r+h\sum_{\varrho=0}^{p}\beta^{*}_\varrho\nabla^\varrho f^{[\sigma]}_{r+1}\qquad(\sigma=0,1,2,\ldots)\qquad(3.14)$$

as the $(\sigma+1)$-th iterate.

III. Central-difference interpolation method. If we integrate both sides of the differential equation (1.10) over the interval x_r-h to x_r+h using STIRLING's interpolation formula, as in Ch. I (2.14), (2.15), we obtain (with p even)

$$y(x_{r+1}) - y(x_{r-1}) = h \sum_{\varrho=0}^{p/2} \beta_\varrho^{**} \, \nabla^{2\varrho} f\big(x_{r+\varrho}, y(x_{r+\varrho})\big) + S_p^{**}.$$

If the remainder term is neglected, we have an equation relating the approximations y_ϱ, namely

$$\left.\begin{aligned}
y_{r+1} - y_{r-1} &= h \sum_{\varrho=0}^{p/2} \beta_\varrho^{**} \, \nabla^{2\varrho} f_{r+\varrho} \\
&= h\left(2f_r + \frac{1}{3}\,\nabla^2 f_{r+1} - \frac{1}{90}\,\nabla^4 f_{r+2} + \frac{1}{756}\,\nabla^6 f_{r+3} - \cdots\right).
\end{aligned}\right\} \quad (3.15)$$

Usually this formula is truncated after the term in ∇^2, which gives SIMPSON's rule:

$$y_{r+1} = y_{r-1} + h\left(2f_r + \frac{1}{3}\,\nabla^2 f_{r+1}\right) = y_{r-1} + \frac{h}{3}\,(f_{r-1} + 4f_r + f_{r+1}). \quad (3.16)$$

An estimate for the remainder term S_2^{**} in the corresponding formula

$$y(x_{r+1}) = y(x_{r-1}) + h[2f(x_r, y(x_r)) + \tfrac{1}{3}\nabla^2 f(x_{r+1}, y(x_{r+1}))] + S_2^{**} \quad (3.17)$$

for the exact solution is given by (2.18) in Ch. I.

As with formula (3.11), (3.16) also includes the unknown y_{r+1} on both sides of the equation, so that here also one determines y_{r+1} iteratively in general. Thus the next approximation $y_{r+1}^{[\sigma+1]}$ is obtained from the current value $y_{r+1}^{[\sigma]}$ according to the formula

$$y_{r+1}^{[\sigma+1]} = y_{r-1} + h(2f_r + \tfrac{1}{3}\nabla^2 f_{r+1}^{[\sigma]}). \quad (3.18)$$

The form of the tabular scheme for this method can be seen in Table II/20.

Table II/20. *The central-difference method*

x	y ×1	hf ×2	$h\nabla f$	$h\nabla^2 f$ ×$\frac{1}{3}$
x_{r-1}	y_{r-1}	$h f_{r-1}$		$h\nabla^2 f_r$
			$h\nabla f_r$	
x_r	y_r	$h f_r$		$h\nabla^2 f_{r+1}$
			$h\nabla f_{r+1}$	
x_{r+1}	y_{r+1}	$h f_{r+1}$		

LINDELÖF[1] has suggested a method based on formula (3.15) in which the term in $\nabla \cdot f_{r+2}$ is taken into account as well. He rewrites the equation

---

[1] LINDELÖF, E.: Remarques sur l'intégration numérique des équations différentielles. Acta Soc. Sci. Fenn. A 2 **1938**, Nr. 13 (21 pp.). He also gives a further refinement of the method.

in the form

$$y_{r+1} = y_{r-1} + h\left[2f_r + \tfrac{1}{3}(\nabla^2 f_r + \nabla^3 f_r)\right] + \delta$$

with

$$\delta = \frac{h}{3}\left(\nabla^4 f_{r+1} - \frac{1}{30}\nabla^4 f_{r+2}\right),$$

and then, assuming that the solution has already been computed up to the point $x = x_r$, uses

$$y_{r+j+1}^{[0]} = y_{r+j-1} + h\left[2f_{r+j} + \tfrac{1}{3}(\nabla^2 f_{r+j} + \nabla^3 f_{r+j})\right]$$

for $j = 0$ and $j = 1$ to obtain tentative values for the next two points. These are then used to build up the difference table temporarily so that approximate values of $\nabla^4 f_{r+1}$ and $\nabla^4 f_{r+2}$ are available for the correction δ to y_{r+1}. The new value of y_{r+1} extends the final difference table up to the point $x = x_{r+1}$.

IV. Mixed extrapolation and interpolation methods. With the methods II and III, the f_{r+1} on the right-hand sides of equations (3.11) and (3.16), respectively, appears as an unknown as well as y_{r+1} and must be either estimated or calculated from an extrapolation formula. MILNE[1] recommends the latter procedure. A rough value y_{r+1}^* is calculated from an extrapolation formula, then $f_{r+1}^* = f(x_{r+1}, y_{r+1}^*)$ is formed and the difference table completed temporarily so that sufficient differences are available to determine y_{r+1} from an interpolation formula. If need be, this value of y_{r+1} can be still further improved by iteration using the interpolation formula[2]. In particular, MILNE gives the formulae

$$y_{r+1}^* = y_{r-3} + 4hf_{r-1} + \frac{8h}{3}\nabla^2 f_r,$$

$$y_{r+1} = y_{r-1} + 2hf_r + \frac{h}{3}\nabla^2 f_{r+1}^*,$$

the second of which is the formula (3.16) of the central-difference method.

3.4. Hints for the practical application of the finite-difference methods

I. Estimation of the highest difference occurring in an interpolation method. If the requisite starting values have been obtained by one of the starting procedures described in §3.2, then, in order to begin the iterations (3.14) or (3.18), we must estimate f_{r+1}; equivalently, we can estimate the highest difference occurring $\nabla^p f_{r+1}$ ($p = 2$ for the central-difference method), which is much easier. Once the calculation is

---

[1] MILNE, W. E.: Numerical solution of differential equations, p. 65. New York and London 1953.

[2] A further variant is mentioned by P. O. LÖWDIN: On the numerical integration of ordinary differential equations of the first order. Quart. Appl. Math. **10**, 97—111 (1952).

properly under way there exists a series of values of $\nabla^p f$ from which we can easily extrapolate for a good estimate of the next value by noting the trend of either the $\nabla^p f$ or $\nabla^{p+1} f$ values. The better the estimate, the less work there is involved in the iteration.

II. Length of step h. The step length h should be kept small enough for

(a) the iteration [(3.14) or (3.18)] to converge sufficiently rapidly — more explicitly, to settle to the required accuracy after one or two cycles — and for

(b) the first term neglected in the formula being used [(3.6), (3.11) or (3.15)] to have a negligible effect to the accuracy required.

In §§ 4.1, 4.3 a significant factor in predicting convergence and estimating the error is found to be the quantity (the "step index")

$$S = k\,h, \quad \text{where} \quad k = \left| \frac{\partial f}{\partial y} \right|.$$

For moderate accuracy S should be of the order 0·05 to 0·1, the smaller value being preferable for the starting calculations. In the example of § 3.2, III, the step was chosen so that $S = 0\cdot 1$ and consequently rather too many iterations were needed. If k varies considerably, it is advisable to adjust the length of step so that S remains approximately constant. We call $h = \text{const}/k$ the "natural step length", where the constant is chosen according to the accuracy required, say in the range 0·05 to 0·1 as mentioned above[1]. If the step used in the calculation is considerably longer than the natural step, then the differences do not decrease sufficiently rapidly with increasing order, the iteration converges too slowly (so that three cycles, or even more in certain circumstances, must be computed before the numbers settle) and the highest differences carried show such large fluctuations that confidence in their accuracy is no longer justified. On the other hand, if too small a step is used, the calculation runs extremely smoothly without any difficulty but, on account of the large number of steps required to cover the same range, more work is, in fact, involved.

III. Change of step length[2]. Doubling the step $(\bar{h} = 2h)$ merely requires the values $y_r, y_{r-2}, y_{r-4}, \ldots$ to be tabulated afresh and a new difference table of the corresponding values of $\bar{h} f$ to be constructed.

Halving the step, on the other hand, is more laborious and requires a new starting calculation. If the calculation has proceeded up to the point x_r, so that $f_r, \nabla f_r, \nabla^2 f_r, \nabla^3 f_r$ are known, then interpolation by formula (2.1) of Ch. I gives the intermediate values

$$\left. \begin{aligned} h f_{r-\frac{1}{2}} &= h\left(f_r - \frac{1}{2} \nabla f_r - \frac{1}{8} \nabla^2 f_r - \frac{1}{16} \nabla^3 f_r \right), \\ h f_{r+\frac{1}{2}} &= h\left(f_r + \frac{1}{2} \nabla f_r + \frac{3}{8} \nabla^2 f_r + \frac{5}{16} \nabla^3 f_r \right). \end{aligned} \right\} \tag{3.19}$$

---

[1] COLLATZ, L.: Natürliche Schrittweite … Z. Angew. Math. Mech. **22**, 216—225 (1942).

[2] Formulae for an arbitrary change in step length and for calculation with arbitrary non-equidistant abscissae (also for differential equations of higher order) are given by P. W. ZETTLER-SEIDEL: Improved Adams method of numerical integration of differential equations. Lecture at Internat. Math. Congr., Cambridge (U.S.A.), 1950.

Then the values $\tilde{h} f_j$ $(j = r - 1, r - \frac{1}{2}, r, r + \frac{1}{2}$, or $j = 0, 1, 2, 3$, if a new numbering is adopted) which correspond to the points at the smaller interval $\tilde{h} = \frac{1}{2} h$, together with their differences, serve as starting data for the refining iteration of (3.5).

IV. Simplification of the iterative procedure in the interpolation methods. The computation involved in the iterations (3.14) and (3.18) is simplified if only the changes in y_{r+1} through each cycle of the iteration are calculated[1]. These can be determined quite simply as follows. Consider, for example, the iteration (3.14); we have

$$y_{r+1}^{[\sigma+1]} - y_{r+1}^{[\sigma]} = h \sum_{\varrho=0}^{p} \beta_\varrho^* [\nabla^\varrho f_{r+1}^{[\sigma]} - \nabla^\varrho f_{r+1}^{[\sigma-1]}].$$

Now since the changes

$$\delta = \nabla^\varrho f_{r+1}^{[\sigma]} - \nabla^\varrho f_{r+1}^{[\sigma-1]} \tag{3.20}$$

are independent of ϱ, they can be taken outside of the summation; hence, using the result (2.13) of Ch. I, we obtain

$$y_{r+1}^{[\sigma+1]} = y_{r+1}^{[\sigma]} + h \beta_p \delta. \tag{3.21}$$

For the central-difference method the iteration (3.18) may be replaced by a similarly modified iteration:

$$y_{r+1}^{[\sigma+1]} = y_{r+1}^{[\sigma]} + \frac{1}{3} h \delta^*, \tag{3.22}$$

where

$$h \delta^* = h \nabla^2 f_{r+1}^{[\sigma]} - h \nabla^2 f_{r+1}^{[\sigma-1]}. \tag{3.23}$$

V. Development of unevenness[2] with the central-difference method. Formula (3.16) provides a direct relation between a value y_{r-1} and the next but one value y_{r+1}, but between consecutive values of y there exists only an indirect link through the differential equation. In the course of a calculation over a large number of steps this can cause (in consequence of irregularities in empirically defined functions, for example) the two approximate solutions represented by the interlaced sequences $y_{r-4}, y_{r-2}, y_r, \ldots$ and $y_{r-3}, y_{r-1}, y_{r+1}, \ldots$ separately to diverge slightly from one another. Thus an unevenness develops in the y values and since it also affects the f values through the differential equation, it is accentuated in the differences of f — in fact, the building up of these irregularities is first noticed as fluctuations in the values of $\nabla^2 f$. When these irregularities reach significant proportions, they can be removed by a smoothing process[3], coupled, possibly, with a new starting iteration, after which the calculation is continued as before.

---

[1] STOHLER, K.: Eine Vereinfachung bei der numerischen Integration gewöhnlicher Differentialgleichungen. Z. Angew. Math. Mech. **23**, 120—122 (1943).

[2] Compare with the theory in § 4.7.

[3] Cf., for example, L. COLLATZ and R. ZURMÜHL: Beiträge zu den Interpolationsverfahren der numerischen Integration von Differentialgleichungen erster und zweiter Ordnung. Z. Angew. Math. Mech. **22**, 42—55 (1942). They give another smoothing procedure which is more systematic.

The smoothing can often be accomplished very simply as follows. We make small corrections $\pm\varepsilon$ in the values of $h\nabla^3 f$, say, so that the values $h\nabla^3 f_{r-j}+(-1)^j\varepsilon$, i.e. $h\nabla^3 f_{r-3}-\varepsilon$, $h\nabla^3 f_{r-2}+\varepsilon$, $h\nabla^3 f_{r-1}-\varepsilon$, $h\nabla^3 f_r+\varepsilon$, run smoothly. For consistency we must make further corrections $\pm\frac{1}{2}\varepsilon$, $\pm\frac{1}{4}\varepsilon$, $\pm\frac{1}{8}\varepsilon$ in the second, first, and zero-th differences; thus we replace

$$h\nabla^k f_{r-j} \quad\text{by}\quad h\nabla^k f_{r-j}+(-1)^j\frac{\varepsilon}{2^{3-k}}. \tag{3.24}$$

Finally we have to modify the y values so as to obtain the required corrections $\pm\frac{1}{8}\varepsilon$ in the f values. This can be done by varying a y value, say y_r, by a small amount δ and noting what change ζ this produces in f_r; then the correction to be added to y_{r-j} is $(-1)^j\frac{\delta}{\zeta}\cdot\frac{\varepsilon}{8}$ (cf. the example in § 3.5).

3.5. Examples

I. Extrapolation method. If the magnetic characteristic of a coil wound on an iron core is assumed to be of cubic form, the sudden application of a periodic voltage

$$e = e_0 \sin\omega t$$

across the coil gives rise to the initial-value problem[1]

$$e = iR + \frac{d\psi}{dt}, \quad\text{where}\quad i = a\psi + b\psi^3,$$

with $\psi(0) = 0$ (notation: voltage e, current i, resistance R, magnetic flux ψ).

With reduced variables x, y defined by

$$e_0 y = aR\psi, \quad x = aRt$$

the initial-value problem becomes

$$y' = -y - 2y^3 + \sin 2x; \quad y(0) = 0$$

for the cases with $be_0^2 = 2a^3 R^2$, $2aR = \omega$.

This problem will be treated by the Adams extrapolation method. We obtain the necessary starting values from a power series solution found by the method of indetermined coefficients. This solution is

$$y = x^2 - \frac{1}{3}x^3 - \frac{1}{4}x^4 + \frac{1}{20}x^5 + \frac{13}{360}x^6 - \frac{733}{2520}x^7 + \frac{1903}{6720}x^8 + \cdots,$$

in which sufficient terms have been given to calculate $y(x)$ at $x = \pm 0\cdot 1$, $\pm 0\cdot 2$ to six decimals. These starting values, and also the corresponding values of hf with the differences required to proceed with the main calculation, are shown in Table II/21. Further values of y_{r+1} are calculated from (3.6) truncated after the term in $\nabla^4 f_r$.

[1] See, for example, K. KÜPFMÜLLER: Einführung in die theoretische Elektrotechnik, 4th ed., p. 401. Berlin-Göttingen-Heidelberg 1952.

Table II/21. *A non-linear problem treated by the Adams extrapolation method*

x	y	$2y^3$	$\sin 2x$	hf	$h\nabla f$	$h\nabla^2 f$	$h\nabla^3 f$	$h\nabla^4 f$
-0.2	0.042253	0.00015	-0.38942	-0.043182				
-0.1	0.010308	0.00000	-0.19867	-0.020898	22284			
0	0	0	0	0	20898	-1386		
0.1	0.009643	0.00000	0.19867	0.018903	18903	-1995	-609	
0.2	0.036951	0.00010	0.38942	0.035237	16334	-2569	-574	$+35$
0.3	0.079082	0.00099	0.56464	0.048457	13220	-3114	-545	$+29$
0.4	0.132657	0.00467	0.71736	0.058003	9546	-3674	-560	-15
0.5	0.193687	0.01453	0.84147	0.063325	5322	-4224	-550	$+10$
0.6	0.257710	0.03423	0.93204	0.064010				

II. Interpolation method. We consider again the problem (1.21)

$$y' = y - \frac{2x}{y}, \qquad y(0) = 1$$

and use it now to illustrate the central-difference method. Sufficient starting values for this method have already (§ 3.2) been calculated for $h = 0.1$ by the Taylor series method. Thus y_{-1}, y_0, y_1 are known and the corresponding function values f_{-1}, f_0, f_1, and their differences $\nabla f_0, \nabla f_1, \nabla^2 f_1$ can be calculated; this completes the first three rows of Table II/22 and we can now start the main calculation.

The iteration (3.18) for the first step of the main calculation is

$$y_2^{[\sigma+1]} = y_0 + h(2f_1 + \tfrac{1}{3}\nabla^2 f_2^{[\sigma]}); \qquad (3.25)$$

to start it we must first estimate the new second difference $\nabla^2 f_2$. If we have nothing to go on, the previous value $\nabla^2 f_1$ may be taken as a first approximation; but in the present case it is better to make use of the third difference $\nabla^3 f_1$, which can be estimated quite easily from the derivatives $y^{IV}(0) = -15$, $y^V(0) = 105$ already calculated for the starting values. Since $f = y'$, we have $h\nabla^3 f_1 = h^4 y^{IV}(0) + \cdots$, so that we may expect that $h\nabla^3 f_1 \approx (0.1)^4 \times y^{IV}(0) = -0.0015$; the next difference is of opposite sign [being approximately proportional to $y^V(0)$, which is positive], so $|h\nabla^3 f|$ will begin to decrease as the calculation progresses.

If we try $h\nabla^3 f_1 \approx -0.001$, our first estimated second difference (a separate column is provided in the table for these estimates) is

$$h\nabla^2 f_2^{[0]} = 0.003 - 0.001 = 0.002,$$

and from (3.25) the first iterate is

$$y_2^{[1]} = 1 + 2 \times 0.09128709 + \tfrac{1}{3} \times 0.002 = 1.1832408;$$

the row is completed by forming the corresponding $hf_2^{[1]}$, $h\nabla f_2^{[1]}$, $h\nabla^2 f_2^{[1]}$.

Table II/22. Application of the central-difference method to a non-linear equation of the first order

x	y	$+\dfrac{2x}{y}$	$hf = 0\cdot1\left(y-\dfrac{2x}{y}\right)$	$h\nabla f$	$h\nabla^2 f$	$h\nabla^4 f$	Estimated value of $h\nabla^2 f$	$h\delta^*$
−0·1	0·8944272	−0·2236068	0·11180340	−0·01180340	0·00309049			−0·00005556 0237
0	1	0	0·1	−0·00871291				+1
0·1	1·0954451	0·1825742	0·09128709	−0·00676847	0·00194444 94196	−0·00114853	0·002	−0·00017913 821
0·2	1·1832408 2223 2215	0·338 0546 0598 0601	0·084 51862 51625 51614	−0·00545008 77095	0·00132087		0·0015	37
0·3	1·2647774 9177 9150 9149	0·474 3168 3392 3402 3402	0·079 06606 5785 5748 5747	−0·004 52250	31228 93800	−0·00062968	0·0009	+0·00003617 174 9
0·4	1·3416364 6485 6491	0·596 2867 814 811	0·074 53497 3671 3680	−0·00383288	0·00093617 69679	−0·00037428	0·00072	2321 116
0·5	1·4142285 208 204	0·707 0993 1032 1034	0·070 71292 1176 1170	−0·00329159	69557 53351	−0·00024243	0·000504	+ 6
0·6	1·4832405 503 508	0·809 0394 341 338	0·067 42011 2162 2170	−0·00287032 34	53510 41968 66	16047	0·00042	2951 151
0·7	1·5492038 37	0·903 6900 01	0·064 55138 338	−0·00253206	33828	11544	0·00033	32
0·8	1·6124635 63 65	0·992 2705 688 687	0·062 01930 75 78	−0·00225584 86	33876 27574 72	8090	0·000276	+ 828 45
0·9	1·6733353 52	1·075 6959 60	0·059 76394 92	−0·0020255	0·00025304	−0·00006304	0·000231	−0·00000026
1	1·7320713	1·154 687	0·057 7384	−0·0018331	1924	−0·0000453	0·000192	−0·0000006
1·1	1·788876	1·229 823	0·055 9053	−0·0016679	1652	380	165	+ 4
1·2	1·843937	1·301 563	0·054 2374	−0·0015275	1404	272	138	+ 2
1·3	1·897397 398	1·370 298 298	0·052 7099 7100	−0·0014043 74	5	253		4
1·4	1·949398	1·436 341	0·051 5057	−0·0012980	0·0001231 1063	−0·0000174	0·000123	+0·0000001
1·5	2·000044	1·499 967	0·050 0077	−0·0012030	950	168	105	+ 13
1·6	2·049445	1·561 398	0·048 8047	−0·0011203	827	113	−0·000096	10
1·7	2·097681	1·620 837	0·047 6844	−0·0010453	750	123	84	− 13
1·8	2·144839	1·678 448	0·046 6391	−0·0009793	560	77	75	0
1·9	2·190981	1·734 383	0·045 6598	−0·0009185	0·0000508	90	66	0
2	2·236179	1·788 766	0·044 7413			−0·0000052	0·0000609	−0·0000001
1·8	2·144849	1·678440	0·0466409	−0·0009188	0·0000601	−0·0000065	0·0000546	−0·0000002
1·9	2·190982	1·788766	0·0447411	8644	544	57	495	0
2	2·236178	1·841716	0·438767	8149	495	49	453	0
2·1	2·280483	1·893330	0·430618	7696	453	42	417	1
2·2	2·323948	1·943699	0·422922	7280	416	37	0·0000384	+0·0000001
2·3	2·366621	1·992904	0·415642	0·0006895	0·0000385	−0·0000033		
2·4	2·408546	2·041015	0·0408747					
2·5	2·449762							

Another auxiliary column is provided for recording the differences of the successive iterates for $h\nabla^2 f$ as defined in (3.23). In the present case $h\nabla^2 f_2$ has changed by

$$h\,\delta^* = 0.001\,944\,44 - 0.002 = -0.000\,055\,56;$$

from (3.22) a third of this difference, i.e. -0.0000185, added to $y_2^{[1]}$ yields the new value $y_2^{[2]} = 1.183\,2223$.

Now we have only to work out the corresponding function values $f_2^{[2]}$ — the new differences need not be worked out since the change in the second difference required to form $h\,\delta^*$ is the same as the change in $f^{[\sigma]}$ itself [as in (3.20)] —, then $y_2^{[3]}$ can be calculated from (3.22). On forming $f_2^{[3]}$, we find that the change $h\,\delta^*$ is now only $-0.0000000\,11$, so that one third of it is smaller in magnitude than 0.5×10^{-7} and no longer affects the y values; $y_2^{[3]}$ is therefore taken as the final value $y_2 = y_2^{[3]}$ and the corresponding row of differences filled in. Further steps can now be dealt with in a similar manner.

At first the estimation of the new values of $h\nabla^2 f$ occasions a little difficulty, but after a number of steps a good idea of the run of the third differences can be formed and the calculation then proceeds quite happily; progress is extremely rapid and in fact the steps from $x = 1$ to $x = 2$ (with the exception of the point $x = 1.3$) are each accomplished in one row of computation. Gradually, however, the irregularities described in § 3.4 creep in, showing themselves in the third differences, which are alternately too large and too small; with the last decimal as unit the third differences for the points $x = 1.7$ to $x = 2.0$ run (with a factor h)

$$-123, \quad -77, \quad -90, \quad -52.$$

We try altering them alternately by $\pm\varepsilon$, say $-\varepsilon, \varepsilon, -\varepsilon, \varepsilon$, and find that $\varepsilon = -13 \times 10^{-7}$ gives the considerably smoother sequence

$$-110, \quad -90, \quad -77, \quad -65.$$

The remaining differences at $x = 2.0$ are now corrected correspondingly in accordance with (3.24); for example, the first difference $h\nabla f$ is altered by $\frac{1}{4}\varepsilon \approx -0.000\,000\,3$ to $-0.000\,9188$.

To find the corrections to the y values which will give the correct changes $\pm \varepsilon/8$ in the hf values, we alter y at $x = 1.8$ by $\delta = 0.000010$, say, and note that the new y value $2.144\,849$ produces a change $\zeta = 0.000\,0018$ in hf. To alter hf by $\varepsilon/8$ instead of ζ, we must therefore change the values y_{r-j} at $x = 2.0 - jh$ by $\delta' = \dfrac{\delta}{\zeta} \cdot \dfrac{\varepsilon}{8} \approx 0.000001$. When these changes are made at $x = 1.9$ and $x = 2.0$ the calculation proceeds smoothly again until the irregularities begin to creep in once more.

For comparison, the solution obtained here is tabulated in Table II/3 alongside the results obtained by other methods. Of particular interest is the comparison with the Runge-Kutta method with step interval $h = 0.2$, which, as is mentioned in § 1.5, corresponds roughly to the central-difference method with $h = 0.1$ in that it involves approximately the same amount of computation. In this example the central-difference method shows up to advantage but one should bear in mind the warning given on page 1 of the danger of making hasty general assessments of methods.

3.6. Differential equations in the complex plane

Let a function $w = w(z)$ of a complex variable z be defined by the differential equation

$$w' = F(z, w)$$

and the initial condition

$$w(z_0) = w_0.$$

Here F will be assumed to be an analytic function of z and w. We wish to calculate $w(z)$ numerically over some desired region of the z plane. There are various ways open to us. For instance, all quantities involved can be split into real and imaginary parts

$$z = x + iy, \quad w = u + iv, \quad F = U + iV$$

and the integration performed parallel to the real or imaginary axis; thus, integrating parallel to the real axis, we have

$$\frac{\partial u}{\partial x} = U, \quad \frac{\partial v}{\partial x} = V, \tag{3.26}$$

a pair of simultaneous first-order equations for the functions $u(x, y_0)$ and $v(x, y_0)$, which can be treated by the methods of §2 and §3.

Another way is to introduce a lattice of points in the z plane defined by $z = z_0 + jh + ikl$, where $j, k = 0, \pm 1, \pm 2, \ldots$ and h, l are the mesh widths, and derive formulae which use the values of w at a group of mesh points to give approximate values of w at neighbouring mesh points.

H. E. SALZER[1] gives such formulae for differential equations of the first and second order. He uses a square mesh ($h = l$) and obtains, for example, approximations for the values at the points $Q = (0, 2)$, $(1, 2)$, $(2, 1)$ and $(2, 0)$ from the values at the four points $P(j, k) = (0, 0)$, $(0, 1)$, $(1, 0)$ and $(1, 1)$.

Firstly some ordinary extrapolation formulae of such form which are exact for polynomials of the third (and less) degree are derived:

$$\left.\begin{aligned}
F_2 &= (2 - i) F_0 + (2 + 4i) F_1 - 2i F_i + (-3 - i) F_{1+i}, \\
F_{2+i} &= 2i F_0 + (-3 + i) F_1 + (2 + i) F_i + (2 - 4i) F_{1+i}, \\
F_{1+2i} &= -2i F_0 + (2 - i) F_1 + (-3 - i) F_i + (2 + 4i) F_{1+i}, \\
F_{2i} &= (2 + i) F_0 + 2i F_1 + (2 - 4i) F_i + (-3 + i) F_{1+i}.
\end{aligned}\right\} \tag{3.27}$$

For compactness, argument values are here denoted by subscripts.

Further, formulae for approximating

$$\Phi(z) = C + \int_0^z F(\zeta)\, d\zeta$$

---

[1] SALZER, H. E.: Formulas for numerical integration of first and second order differential equations in the complex plane. J. Math. Phys. **29**, 207—216 (1950).

are given:

$$\left.\begin{aligned}
24\,\Phi_0/h &= 0 + C,\\
24\,\Phi_1/h &= (9 + 5i)\,F_0 + (9 - 5i)\,F_1 + (3 + i)\,F_i + (3 - i)\,F_{1+i} + C,\\
24\,\Phi_i/h &= (5 + 9i)\,F_0 + (1 + 3i)\,F_1 + (-5 + 9i)\,F_i + (-1 + 3i)\,F_{1+i} + C,\\
24\,\Phi_{1+i}/h &= (8 + 8i)\,F_0 + (4 + 4i)\,F_1 + (4 + 4i)\,F_i + (8 + 8i)\,F_{1+i} + C.
\end{aligned}\right\} \quad (3.28)$$

These likewise are exact for polynomials of up to the third degree.

If we know approximate values of w, and hence of F, at the four points P (to start the calculation we must first calculate w at three points by some starting technique, say a series solution), we can obtain approximations to F at the points Q from the extrapolation formulae (3.27). Then, using the approximation to the differential equation obtained from (3.28), we calculate w at the points Q and, if necessary, improve these values of w by repeating the process with revised values of F [1].

3.7. Implicit differential equations of the first order

Occasionally it happens that when a differential equation is presented in the form

$$F(x, y, y') = 0 \tag{3.29}$$

it cannot be solved for y' in closed form. There are various ways of dealing with this situation:

1. In general, (3.29) can be transformed as follows into a pair of simultaneous explicit differential equations of the first order [2], which can then be integrated numerically by one of the methods of § 1, § 2 or § 3. We assume that the required solution is such that $y''(x) \neq 0$; then to $t = y'(x)$ there is an inverse function $x = x(t)$. Differentiation of (3.29) with respect to t yields

$$F_x \frac{dx}{dt} + F_y t \frac{dx}{dt} + F_t = 0,$$

---

[1] In addition to the "four-point formulae" quoted above Salzer (see previous footnote) gives formulae for 3 to 9 points and also formulae for repeated integration which can be used for second-order differential equations; thus for

$$\Psi(z) = A\,z + B + \int_0^z \int_0^\zeta F(s)\,ds\,d\zeta$$

he obtains the formulae

$$\begin{aligned}
120\,\Psi_0/h^2 &= 0 + A\,z_0 + B,\\
120\,\Psi_1/h^2 &= (33 + 13i)\,F_0 + (12 - 12i)\,F_1 + (8 + 2i)\,F_i + (7 - 3i)\,F_{1+i} + A\,z_1 + B,\\
120\,\Psi_i/h^2 &= (-33 + 13i)\,F_0 + (-8 + 2i)\,F_1 + (-12 - 12i)\,F_i + \\
&\qquad\qquad\qquad\qquad + (-7 - 3i)\,F_{1+i} + A\,z_i + B,\\
120\,\Psi_{1+i}/h^2 &= 56i\,F_0 + (4 + 20i)\,F_1 + (-4 + 20i)\,F_i + 24i\,F_{1+i} + A\,z_{1+i} + B.
\end{aligned}$$

[2] Kamke, E.: Differentialgleichungen reeller Funktionen, 2nd ed. (436 pp.), p. 112. Leipzig 1944.

so that, if we assume further that $F_x + F_y t \neq 0$, we obtain the system

$$\left.\begin{aligned}
\frac{dx}{dt} &= -\frac{F_t}{F_x + F_y t} = f(x, y, t), \\
\frac{dy}{dt} &= \frac{dy}{dx}\frac{dx}{dt} = -\frac{t F_t}{F_x + F_y t} = t f(x, y, t).
\end{aligned}\right\} \tag{3.30}$$

The initial conditions are $x = x_0$, $y = y_0$ at $t = y_0'$, where y_0' is determined from $F(x_0, y_0, y_0') = 0$. Solution of a set of two first-order equations by a finite-difference method normally requires two difference tables, but since the differences of tf can be expressed in terms of the differences of f [e.g. $\nabla^\varrho(t_k f_k) = t_k \nabla^\varrho f_k + \varrho h \nabla^{\varrho-1} f_{k-1}$, where h is the step length], the computation for the equations (3.30) can be arranged[1] so as to use only the difference table of f.

2. Differentiation of (3.29) with respect to x yields the second-order equation

$$y'' = -\frac{F_x + y' F_y}{F_{y'}}, \tag{3.31}$$

provided we assume that $F_{y'} \neq 0$. This is of explicit form and can be treated by the usual methods for second-order differential equations.

3. If we define

$$g(x, y, y') = y' - \frac{F(x, y, y')}{F_{y'}(x_0, y_0, y_0')}, \tag{3.32}$$

we can set up the iterative process

$$y_\nu(x) = y_0 + \int_{x_0}^{x} y_\nu'(x)\, dx; \quad y_{\nu+1}'(x) = g(x, y_\nu(x), y_\nu'(x)) \quad (\nu = 0, 1, 2, \ldots). \tag{3.33}$$

WEISSINGER[1] gives a convergence proof and error estimates for this method.

§ 4. Theory of the finite-difference methods

We start here by investigating the convergence of the iterations required in the interpolation methods of the preceding section, then describe how error estimates can be derived for all finite-difference methods. Formula (4.43) deserves particular mention as being a fundamental error formula for the central-difference method. § 4.7 deals with the danger of instability in finite-difference methods.

4.1. Convergence of the iterations in the main calculation

Here, and in the rest of § 4, it will be assumed that the function (x, y) appearing in the differential equation (1.10) satisfies a Lipschitz condition (1.12) with constant K. In practice K is usually taken to be the largest absolute value of the derivative $\partial f/\partial y$ within the domain under consideration, i.e.

$$K = k_{max} \quad \text{where} \quad k = \left|\frac{\partial f}{\partial y}\right|. \tag{4.1}$$

---

[1] WEISSINGER, J.: Numerische Integration impliziter Differentialgleichungen. Angew. Math. Mech. 33, 63—65 (1953).

With this assumption, quite a simple analysis suffices to examine the convergence of the iterations required in the interpolation methods. First of all we investigate the iterations (3.14) and (3.18) which occur in the main calculation.

For this we consider the equations in their Lagrangian form, in which the differences are expressed in terms of the function values, as in (3.12) for the interpolation formula (3.11). Using (3.13) we can write the iteration equations for the Adams interpolation method (3.14) in the form

$$y_{r+1}^{[\sigma+1]} = y_r + h \left[\beta_p f_{r+1}^{[\sigma]} + \sum_{\varrho=1}^{p} \alpha_{p\varrho}^* f_{r+1-\varrho} \right] \qquad (\sigma = 0, 1, 2, \ldots). \quad (4.2)$$

If the corresponding expression for $y_{r+1}^{[\sigma]}$ is subtracted from this equation, we see that the change

is given by
$$\left. \begin{array}{l} \delta^{[\sigma]} = y_{r+1}^{[\sigma+1]} - y_{r+1}^{[\sigma]} \\[6pt] \delta^{[\sigma]} = h\beta_p \left[f(x_{r+1}, y_{r+1}^{[\sigma]}) - f(x_{r+1}, y_{r+1}^{[\sigma-1]}) \right]. \end{array} \right\} \quad (4.3)$$

The Lipschitz condition (1.12) provides limits for the right-hand side:

and it follows that
$$\left. \begin{array}{l} |\delta^{[\sigma]}| \leq hK\beta_p |\delta^{[\sigma-1]}|, \\[6pt] |\delta^{[\sigma]}| \leq (hK\beta_p)^\sigma |\delta^{[0]}| \qquad (\sigma = 0, 1, 2, \ldots). \end{array} \right\} \quad (4.4)$$

We now suppose that h is so small that

$$Kh < \frac{1}{\beta_p}; \quad (4.5)$$

then the geometric series $|\delta^{[0]}| \sum\limits_{\sigma=0}^{\infty} (hK\beta_p)^\sigma$, which majorizes the series $\sum\limits_{\sigma=0}^{\infty} \delta^{[\sigma]}$, has a ratio $h\beta_p K$ which is less than one and therefore converges. Consequently the series

$$\lim_{\sigma \to \infty} y_{r+1}^{[\sigma]} = y_{r+1}^{[0]} + \delta^{[0]} + \delta^{[1]} + \delta^{[2]} + \cdots$$

converges (absolutely, in fact) to a value y_{r+1}, which provides a solution of (3.11). Thus the inequality (4.5) represents a sufficient condition for the convergence of the iteration. For the values of p normally used we have

$$\left. \begin{array}{llll} Kh < \dfrac{12}{5} &= 2 \cdot 4 & \text{for} & p = 2, \\[8pt] Kh < \dfrac{8}{3} &\approx 2 \cdot 67 & \text{for} & p = 3, \\[8pt] Kh < \dfrac{720}{251} &\approx 2 \cdot 87 & \text{for} & p = 4. \end{array} \right\} \quad (4.6)$$

For the central-difference method (3.15) the rearranged form of the iteration formula (3.18) is

$$y_{r+1}^{[\sigma+1]} = y_{r-1} + \frac{h}{3} (f_{r-1} + 4f_r + f_{r+1}^{[\sigma]}). \quad (4.7)$$

The same considerations as for the Adams interpolation method now yield

$$\delta^{[\sigma]} = \frac{h}{3}\{f(x_{r+1}, y_{r+1}^{[\sigma]}) - f(x_{r+1}, y_{r+1}^{[\sigma-1]})\}$$

for the change $\delta^{[\sigma]}$ of (4.3) and hence a sufficient condition for the convergence of the iteration in the central-difference method is

$$K\,h < 3.\qquad(4.8)$$

4.2. Convergence of the starting iteration

For investigating the convergence of the starting iteration (3.5) we make use of a theorem in matrix theory. This is concerned with a sequence of sets of quantities

$$x_1^{[\nu]}, x_2^{[\nu]}, \ldots, x_p^{[\nu]}\qquad(\nu = 0, 1, 2, \ldots),$$

which are bounded successively thus

$$\left|x_\varrho^{[\nu]}\right| \le \sum_{\sigma=1}^{p}\left|A_{\varrho\sigma}\right|\left|x_\sigma^{[\nu-1]}\right|\qquad(\nu = 1, 2, \ldots;\ \varrho = 1, 2, \ldots, p).$$

The theorem states[1]: For the convergence of the p series

$$\sum_{\nu=0}^{\infty}\left|x_\varrho^{[\nu]}\right|\qquad(\varrho = 1, 2, \ldots, p)$$

it is sufficient that the absolute values of all the characteristic roots $\varkappa$ of the matrix

$$A = \begin{pmatrix}\left|A_{11}\right| & \left|A_{12}\right| & \cdots & \left|A_{1p}\right| \\ \cdot & \cdot & \cdots & \cdot \\ \left|A_{p1}\right| & \left|A_{p2}\right| & \cdots & \left|A_{pp}\right|\end{pmatrix}\qquad(4.9)$$

be less than unity, i.e. $\left|\varkappa_\sigma\right| < 1$ for $\sigma = 1, \ldots, p$.

The characteristic roots $\varkappa_B$ of a matrix B with elements b_{jk} are defined as the roots $\varkappa$ of the "characteristic equation" of B:

$$\begin{vmatrix} b_{11} - \varkappa & b_{12} & \cdots b_{1p} \\ b_{21} & b_{22} - \varkappa \cdots b_{2p} \\ \cdot & \cdot \cdots \cdot \\ b_{p1} & b_{p2} & \cdots b_{pp} - \varkappa \end{vmatrix} = 0.\qquad(4.10)$$

We now turn to the starting iteration. The equation (3.5) which describes it becomes

$$y_\varrho^{[\nu+1]} = y_0 + h\left[a_{\varrho 0}f_0 + \sum_{\sigma=1}^{p} a_{\varrho\sigma}f_\sigma^{[\nu]}\right]\qquad(\varrho = 1, 2, \ldots, p;\ \nu = 1, 2, \ldots)\ (4.11)$$

---

[1] See, for example, L. COLLATZ: Eigenwertaufgaben, p. 311. Leipzig 1949.

when written in terms of the function values. The values of the $a_{\varrho\sigma}$ for $p=2$, i.e. using only the formulae of (3.5) within the dotted frame, are given in Table II/23 and for $p=3$, i.e. using the complete set of formulae (3.5), in Table II/24.

Table II/23. *The $a_{\varrho\sigma}$ for $p=2$*

	$\sigma=$		
	0	1	2
$\varrho=1$	$\dfrac{5}{12}$	$\dfrac{8}{12}$	$-\dfrac{1}{12}$
$\varrho=2$	$\dfrac{1}{3}$	$\dfrac{4}{3}$	$\dfrac{1}{3}$

Table II/24. *The $a_{\varrho\sigma}$ for $p=3$*

	$\sigma=$			
	0	1	2	3
$\varrho=1$	$\dfrac{9}{24}$	$\dfrac{19}{24}$	$-\dfrac{5}{24}$	$\dfrac{1}{24}$
$\varrho=2$	$\dfrac{1}{3}$	$\dfrac{4}{3}$	$\dfrac{1}{3}$	0
$\varrho=3$	$\dfrac{3}{8}$	$\dfrac{9}{8}$	$\dfrac{9}{8}$	$\dfrac{3}{8}$

From (4.11) we find that the changes

$$\delta_{\varrho}^{[\nu]} = y_{\varrho}^{[\nu+1]} - y_{\varrho}^{[\nu]}$$

are given by

$$\delta_{\varrho}^{[\nu]} = h \sum_{\sigma=1}^{p} a_{\varrho\sigma}\{f(x_{\sigma}, y_{\sigma}^{[\nu]}) - f(x_{\sigma}, y_{\sigma}^{[\nu-1]})\} \tag{4.12}$$

$$(\varrho = 1, 2, \ldots, p;\ \nu = 1, 2, \ldots);$$

using the Lipschitz condition (1.12) we obtain the inequalities

$$|\delta_{\varrho}^{[\nu]}| \leq K h \sum_{\sigma=1}^{p} |a_{\varrho\sigma}|\,|\delta_{\sigma}^{[\nu-1]}|. \tag{4.13}$$

The convergence of the series

$$\lim_{\nu\to\infty} y_{\varrho}^{[\nu]} = y_{\varrho}^{[0]} + \delta_{\varrho}^{[0]} + \delta_{\varrho}^{[1]} + \delta_{\varrho}^{[2]} + \cdots$$

is assured by the above-mentioned matrix theorem provided that the absolute values of all characteristic roots μ of the matrix A with elements $Kh|a_{\varrho\sigma}|$ are less than unity, i.e. provided that the absolute values of all characteristic numbers μ^* of the matrix

$$A^* = \begin{pmatrix} |a_{11}| & \cdots & |a_{1p}| \\ \cdot & \cdot & \cdot \\ |a_{p1}| & \cdots & |a_{pp}| \end{pmatrix}$$

are less than $(Kh)^{-1}$.

Given the matrix A^*, which we are, this condition provides an upper bound for Kh. Thus for $p=2$

$$A^* = \begin{pmatrix} \dfrac{8}{12} & \dfrac{1}{12} \\ \dfrac{4}{3} & \dfrac{1}{3} \end{pmatrix}$$

and the μ^* are the roots of

$$\begin{vmatrix} \dfrac{8}{12} - \mu^* & \dfrac{1}{12} \\[2ex] \dfrac{4}{3} & \dfrac{1}{3} - \mu^* \end{vmatrix} = 0, \qquad \text{namely} \qquad \mu^* = \frac{1}{2} \pm \frac{1}{6}\sqrt{5} \approx \begin{cases} 0\cdot873 \\ 0\cdot127. \end{cases}$$

Consequently convergence is ensured by choosing h such that

$$Kh < \frac{1}{0\cdot873}, \qquad \text{i.e.} \qquad Kh < 1\cdot14. \tag{4.14}$$

For $p = 3$ the corresponding equation is

$$\begin{vmatrix} 19 - \tau & 5 & 1 \\ 32 & 8 - \tau & 0 \\ 27 & 27 \cdot 9 - \tau \end{vmatrix} = 0,$$

where $\tau = 24\,\mu^*$, whose largest root is $\mu^* \approx 1\cdot24$, and the iteration certainly converges if

$$Kh < \frac{1}{1\cdot24}, \qquad \text{i.e.} \qquad Kh < 0\cdot8. \tag{4.15}$$

The limits (4.14), (4.15) on Kh for the starting iteration are more restrictive than the limits (4.5), (4.8) for the iterations in the main calculation. This fact accords with the particular sensitivity of the starting iteration.

4.3. Recursive error estimates

In this and the following section we describe[1] how error estimates for the finite-difference methods can be derived; we may note that

---

[1] For literature on error estimation see the following list: MISES, R. v.: Zur numerischen Integration von Differentialgleichungen. Z. Angew. Math. Mech. **10**, 81—92 (1930). — SCHULZ, G.: Interpolationsverfahren zur numerischen Integration gewöhnlicher Differentialgleichungen. Z. Angew. Math. Mech. **12**, 44—59 (1932). — Fehlerabschätzung für das Störmersche Integrationsverfahren. Z. Angew. Math. Mech. **14**, 224—234 (1934). — COLLATZ, L.: Natürliche Schrittweite bei numerischer Integration von Differentialgleichungssystemen. Z. Angew. Math. Mech. **22**, 216—225 (1942). — Differenzenschemaverfahren zur numerischen Integration von gewöhnlichen Differentialgleichungen n-ter Ordnung. Z. Angew. Math. Mech. **29**, 199—209 (1949). — HAMEL, G.: Zur Fehlerabschätzung bei gewöhnlichen Differentialgleichungen erster Ordnung. Z. Angew. Math. Mech. **29**, 337—341 (1949). — WEISSINGER, J.: Eine verschärfte Fehlerabschätzung zum Extrapolationsverfahren von ADAMS. Z. Angew. Math. Mech. **30**, 356—363 (1950). — Eine Fehlerabschätzung für die Verfahren von ADAMS und STÖRMER. Z. Angew. Math. Mech. **32**, 62—67 (1952). — TOLLMIEN, W.: Bemerkung zur Fehlerabschätzung beim Adamsschen Interpolationsverfahren. Z. Angew. Math. Mech. **33**, 151—155 (1953). — UHLMANN, W.: Fehlerabschätzungen bei Anfangswertaufgaben gewöhnlicher Differentialgleichungssysteme 1. Ordnung. Z. Angew. Math. Mech. **37**, 88—99 (1957). — Fehlerabschätzungen bei Anfangswertaufgaben mit einer gewöhnlichen Differentialgleichung höherer Ordnung. Z. Angew. Math. Mech. **37**, 99—111 (1957). — VIETORIS, L.: Der Richtungsfehler einer durch das Adamssche Interpolationsverfahren gewonnenen Näherungslösung einer Gleichung $y' = f(x, y)$. Öst. Akad. Wiss., Math.-naturw. Kl., S.-Ber. II a **162**, 157—167, 293—299 (1953).

the general remarks on error estimation made in § 1.3 apply here. The first rigorous error bounds were obtained by R. v. MISES[1].

To simplify matters we make the following assumptions:

1. *The number of decimals carried in the calculation is sufficient for rounding errors to be neglected.*

2. *The iterations* (3.5), (3.14), (3.18) *are always repeated until the numbers settle to the number of decimals carried so that it can be assumed that the values* y_r *are exact solutions of the corresponding equations without the bracketed superscripts.*

As will appear later, the following will be needed in the derivation of the error estimates:

1. A Lipschitz constant K [as in (4.1)].

---

[1] RICHARD EDLER VON MISES was one of the most eminent of applied mathematicians. The modern broad conception of applied mathematics is due to him, and his exceedingly numerous and diverse contributions to this comprehensive applied mathematics have had no little influence on its present-day importance; he published fundamental and pioneering work in almost every constituent subject: in practical analysis, geometry, probability theory, mathematical statistics, and in various branches of mechanics, from strength of materials and theory of machines to the mechanics of plastic media and hydro- and aerodynamics. Over and above these specific accomplishments he strove for philosophical understanding; he set out his unified "positivistic" view of the world, which was influenced by ERNST MACH, in a book "KLEINES LEHRBUCH DER POSITIVMUS", which embraced even religion, art and poetry. He had a great love of German literature and was particularly interested in the works of the Austrian poet RAINER MARIA RILKE; V. MISES was one of the greatest connoisseurs of RILKE and possessed one of the most notable collections of his works.

v. MISES was born in Lemberg, Austria, on the 19th April 1883. He obtained his doctorate at the Technische Hochschule in Vienna in 1908 and in the same year qualified as lecturer in Brünn. In 1909 at the age of 26 he took up a post as extraordinary professor at Strassburg. After the first world war, in which he saw flying service and also designed a large aeroplane of some 600 h.p. bearing his name, he went as professor first (1919) to Dresden and then (1920) to the University of Berlin, where he set up and directed the Institut für Angewandte Mathematik, which was to become famous later. Also in 1920 he founded the journal ZEITSCHRIFT FÜR ANGEWANDTE MATHEMATIK UND MECHANIK, the forerunner of many similar journals founded later; he was editor until 1934. During the years 1933—1939 he worked at the University of Istanbul, where, at the request of the Turkish government, he founded an institute for pure and applied mathematics. In 1939 he accepted an invitation to Harvard University, Cambridge, Mass., where he remained, becoming GORDON McKAY Professor of Aerodynamics and Applied Mathematics in 1943.

To the last years of his life he was exceedingly productive in the scientific field and retained his great versatility and nimbleness of mind. Highly honoured with the honorary degrees of numerous colleges, loved and respected by a vast number of friends and pupils from all parts of the world (the author of this book was one of his pupils), he died in Boston on the 14th July 1953, leaving a widow, HILDA GEIRINGER, his colleague for many decades.

2. Error limits Y_ϱ for the starting values y_ϱ. If, for example, the necessary starting values are calculated by method II of § 3.2 (Taylor series method), the error can be usually be estimated very easily; the maximum rounding error, i.e. $\frac{1}{2} \times 10^{-d}$ for a d decimal number, will often provide a suitable upper bound. If the iteration method (method III of § 3.2) is used to obtain the starting values, the error can be estimated by a special method[1] (cf. § 4.5).

3. An upper bound for the absolute value of a certain derivative $f^{(q)} = \dfrac{d^q f[x, y(x)]}{dx^q}$. When $f(x, y)$ has a simple analytic form, the estimation of $f^{(q)}$ from the explicit expression obtained by differentiating f is usually quite straightforward. Mostly, however, this method is very complicated and in fact sometimes may not be possible at all (for example, when empirical laws are involved). In such cases we must be content with an approximate value for $|f^{(q)}|_{\max}$ inferred from the difference table, using the fact that

$$f^{(q)}\left(x_k, y(x_k)\right) \approx \frac{1}{h^q} \nabla^q f_{k+\frac{q}{2}};$$

if the q-th differences run smoothly, we can get a fair idea of the maximum absolute value of $f^{(q)}$. Of course, a rigorous error estimate cannot be obtained by this method.

We investigate first the Adams interpolation method, which is based on the formula (3.11). In Lagrangian form this formula reads [as in (3.12)]

$$y_{r+1} = y_r + h \sum_{\varrho=0}^{p} \alpha_{p\varrho}^* f_{r+1-\varrho}. \tag{4.16}$$

A similar relation, but with a remainder term S_{p+1}^*, holds for the exact solution:

$$y(x_{r+1}) = y(x_r) + h \sum_{\varrho=0}^{p} \alpha_{p\varrho}^* f\left(x_{r+1-\varrho}, y(x_{r+1-\varrho})\right) + S_{p+1}^*. \tag{4.17}$$

For this remainder term, or "truncation error", there exists the estimate (2.11) of Ch. I:

$$|S_{p+1}^*| \leq C^*, \quad \text{where} \quad C^* = h^{p+2}|\beta_{p+1}^*| |f^{(p+1)}|_{\max}. \tag{4.18}$$

For the error

$$\varepsilon_r = y_r - y(x_r)$$

subtraction of (4.17) from (4.16) yields the relation

$$\varepsilon_{r+1} = \varepsilon_r + h \sum_{\varrho=0}^{p} \alpha_{p\varrho}^* \{f_{r+1-\varrho} - f\left(x_{r+1-\varrho}, y(x_{r+1-\varrho})\right)\} - S_{p+1}^*.$$

---

[1] See also p. 57—58 of G. Schulz: Interpolationsverfahren ... . Z. Angew. Math. Mech. **12**, 44—59 (1932), which has already been mentioned.

The differences of the f values which occur in the summation can be estimated[1] by means of the Lipschitz condition (1.12):

$$|\varepsilon_{r+1}| \leq |\varepsilon_r| + hK \sum_{\varrho=0}^{p} |\alpha^*_{p\varrho}|\,|\varepsilon_{r+1-\varrho}| + C^*. \qquad (4.19)$$

If the equality sign is written in place of "$\leq$", an equation results which determines recursively a sequence Y_r of upper limits for the errors ε_r once upper limits Y_s for the first p errors ε_s ($s = 0, 1, \ldots, p-1$) are known. In principle, therefore, an error estimate can be obtained from the equations

$$\left.\begin{aligned}(1 - hK\beta_p)\,Y_{r+1} = \left(1 + hK|\alpha^*_{p1}|\right) Y_r + hK \sum_{\varrho=2}^{p} |\alpha^*_{p\varrho}|\,Y_{r+1-\varrho} + C^* \\ (r = p-1, p, p+1, \ldots),\end{aligned}\right\} \quad (4.20)$$

where β_p has been substituted for α^*_{p0} in accordance with (3.13).

Here h is to be chosen so small that $Kh\beta_p < 1$. This is the same condition as that which ensures a convergent iteration in the main calculation, namely (4.5).

---

[1] WEISSINGER, J.: Z. Angew. Math. Mech. **30**, 356—363 (1950); **32**, 62—67 (1952), has succeeded in refining this estimate somewhat by performing additional manipulations before taking absolute values. He writes down the equation for $r, r-1, \ldots, r-l$, where $l \geq p$, and adds, obtaining (with the different truncation errors distinguished by the notation $s^*_{\sigma, p+1}$)

$$\varepsilon_{r+1} = \varepsilon_{r-l} + h\sum_{\sigma=0}^{p+l} b_\sigma\{f_{r+1-\sigma} - f(x_{r+1-\sigma}, y(x_{r+1-\sigma}))\} - \sum_{\sigma=0}^{l} s^*_{\sigma, p+1},$$

where

$$b_\sigma = \begin{cases} \displaystyle\sum_{\tau=0}^{\sigma} \alpha^*_{p\tau} & \text{for} \quad 0 \leq \sigma \leq p \\[2ex] \displaystyle\sum_{\tau=0}^{p} \alpha^*_{p\tau} = 1 & \text{for} \quad p \leq \sigma \leq l \\[2ex] \displaystyle\sum_{\tau=\sigma-l}^{p} \alpha^*_{p\tau} & \text{for} \quad l \leq \sigma \leq p+l. \end{cases}$$

Taking absolute values now yields

$$|\varepsilon_{r+1}| \leq |\varepsilon_{r-l}| + hK \sum_{\sigma=0}^{p+l} |b_\sigma|\,|\varepsilon_{r+1-\sigma}| + (l+1)\,C^*,$$

This is the equation which corresponds to (4.19) and the further considerations leading to an independent error estimate which are applied to (4.19) in § 4.4 apply here quite analogously: z is to be determined now from

$$z^{l+p} - z^{p-1} - Kh \sum_{\sigma=0}^{l+p} |b_\sigma|\,z^{l+p-\sigma} = 0$$

instead of from (4.28).

Precisely similar considerations applied to the Adams extrapolation method (3.6) lead to the equation

$$Y_{r+1} = Y_r + hK \sum_{\varrho=0}^{p} |\alpha_{p\varrho}| Y_{r-\varrho} + C \qquad (r = p,\ p+1, \ldots), \quad (4.21)$$

where

$$C = h^{p+2} \beta_{p+1} |f^{(p+1)}|_{\max}; \qquad (4.22)$$

applied to the central-difference method they give

$$(1 - \tfrac{1}{3}Kh) Y_{r+1} = \tfrac{4}{3} Kh Y_r + (1 + \tfrac{1}{3}Kh) Y_{r-1} + C^{**} \qquad (r = 1, 2, \ldots), \quad (4.23)$$

where

$$C^{**} = \tfrac{1}{90} h^5 |f^{IV}|_{\max}. \qquad (4.24)$$

Here again the condition on h, namely $Kh < 3$, is identical with the convergence condition (4.8) for the iteration in the main calculation.

4.4. Independent error estimates

Practical application of these recursive estimates is laborious so we now establish an upper bound for Y_s which does not entail the calculation of the previous Y_ϱ. We forfeit something thereby, as might be expected, for the error limits so obtained generally prove to be less precise than those calculated recursively.

Again we deal with the Adams interpolation method first. The error limits Y_r for this method satisfy the linear inhomogeneous difference equation (4.20) with constant positive coefficients. The solution of this equation which is determined by the starting values $Y_0, Y_1, \ldots, Y_{p-1}$ is majorized by any particular solution W_r such that

$$W_\varrho \geqq Y_\varrho \geqq 0 \quad \text{for} \quad \varrho = 0, 1, \ldots, p-1, \qquad (4.25)$$

for these inequalities remain valid for all positive ϱ on account of the positive coefficients in (4.20).

Now the general solution of a linear inhomogeneous difference equation with constant coefficients can be given in closed form. In precisely the same way as with the corresponding type of differential equation the solution can be written as the sum of a particular solution $W^{(1)}$ of the inhomogeneous equation and the general solution $W^{(2)}$ of the homogeneous equation. In the present case $W^{(1)}$ can be taken as a constant W^* and by substitution in (4.20) we find that

$$W^* = - \frac{C^*}{Kh \sum_{\varrho=0}^{p} |\alpha^*_{p\varrho}|}. \qquad (4.26)$$

The well-known method for solving the homogeneous equation is to assume a solution of the form

$$W_r = z^r; \tag{4.27}$$

from (4.20) we must have

$$(1 - Kh\beta_p) z^{r+1} = \left(1 + Kh|\alpha_{p1}^*|\right) z^r + Kh \sum_{\varrho=2}^{p} |\alpha_{p\varrho}^*| z^{r+1-\varrho},$$

which yields the "characteristic equation"

$$z^p - z^{p-1} = Kh \sum_{\varrho=0}^{p} |\alpha_{p\varrho}^*| z^{p-\varrho} \tag{4.28}$$

for z.

This equation always has a positive root z greater than unity, for the left-hand side is zero, and therefore smaller than the positive right-hand side, when $z = 1$, and, since the coefficient $Kh\beta_p$ of the z^p on the right-hand side is less than unity [assumption (4.5)], the left-hand side must be greater than the right-hand side for sufficiently large values of z. In the following we take z to be the smallest of the roots of (4.28) which are greater than unity. We can then determine a constant A so that the

$$W_\varrho = W^* + A z^\varrho \tag{4.29}$$

satisfy the inequalities (4.25) for $\varrho = 0, 1, \ldots, p-1$ and hence also for all positive ϱ. If all $|\varepsilon_\varrho|$ with $\varrho = 0, 1, \ldots, p-1$ are less than ε, we can put $A = \varepsilon - W^*$ (W^* is negative), thus obtaining the error limit

$$|\varepsilon_r| \leq \varepsilon z^r - W^*(z^r - 1); \tag{4.30}$$

with the values substituted from (4.18), (4.26) this becomes

$$|\varepsilon_r| \leq \varepsilon z^r + \gamma_p^* h^{p+1} \frac{1}{K} |f^{(p+1)}|_{\max} (z^r - 1) \qquad (r = 1, 2, \ldots), \tag{4.31}$$

where the γ_p^* are defined by

$$\gamma_p^* = \frac{|\beta_{p+1}^*|}{\sum\limits_{\varrho=0}^{p} |\alpha_{p\varrho}^*|} \tag{4.32}$$

and are given numerically for the first few values of p by

$$
\begin{aligned}
\gamma_1^* &= \frac{1}{12} \approx 0.0833, \\[4pt]
\gamma_2^* &= \frac{1}{28} \approx 0.0357, \\[4pt]
\gamma_3^* &= \frac{19}{1020} \approx 0.0186, \\[4pt]
\gamma_4^* &= \frac{27}{2572} \approx 0.0105.
\end{aligned}
\tag{4.33}
$$

This method of finding a solution W_r of the difference equation which majorizes the error limits Y_r can also be applied to the Adams extrapolation formula. The appropriate difference equation (4.21) now has the particular solution

$$W = -\frac{C}{Kh\sum\limits_{\varrho=0}^{p}|\alpha_{p\varrho}|}, \qquad (4.34)$$

and for the solution (4.27) of the corresponding homogeneous equation we determine z as the smallest positive root greater than unity of the equation

$$z^{p+1} - z^{p} = Kh\sum\limits_{\varrho=0}^{p}|\alpha_{p\varrho}|\,z^{p-\varrho}. \qquad (4.35)$$

If ε has the same significance as for the interpolation method above, we obtain here the limits

$$|\varepsilon_r| \leqq \varepsilon z^r - W(z^r - 1). \qquad (4.36)$$

With the values (4.22), (4.34) this reads

$$|\varepsilon_r| \leqq \varepsilon z^r + \gamma_p h^{p+1}\frac{1}{K}|f^{(p+1)}|_{\max}(z^r - 1) \qquad (r = 1, 2, \ldots), \quad (4.37)$$

where

$$\gamma_p = \frac{\beta_{p+1}}{\sum\limits_{\varrho=0}^{p}|\alpha_{p\varrho}|}. \qquad (4.38)$$

Numerical values for the first few γ_p are

$$\left.\begin{aligned}
\gamma_1 &= \frac{5}{24} \approx 0{\cdot}2033, \\
\gamma_2 &= \frac{9}{88} \approx 0{\cdot}1023, \\
\gamma_3 &= \frac{251}{4800} \approx 0{\cdot}0523.
\end{aligned}\right\} \qquad (4.39)$$

Finally we derive in a similar fashion an estimate for the error in the central-difference method. A particular solution of the pertinent difference equation (4.23) is

$$W^{**} = -\frac{C^{**}}{2Kh}, \qquad (4.40)$$

and for the solution of the corresponding homogeneous difference equation we again use the smallest positive root z greater than unity of its characteristic equation, which here reads

$$3(z^2 - 1) = Kh(z^2 + 4z + 1). \qquad (4.41)$$

For the determination of A in the particular solution

$$W_\varrho = W^{**} + A z^\varrho$$

we ought strictly to distinguish between the cases in which the limits $|\varepsilon_\varrho| \leqq \varepsilon$ are known for $\varrho = 0, 1, 2$ and $\varrho = 0, \pm 1$, respectively, but since the second case can be reduced to the first by re-numbering the ϱ values (displacing them by 1), we need only write the error estimate in the one form

$$|\varepsilon_r| \leqq \varepsilon z^r - W^{**}(z^r - 1) \tag{4.42}$$

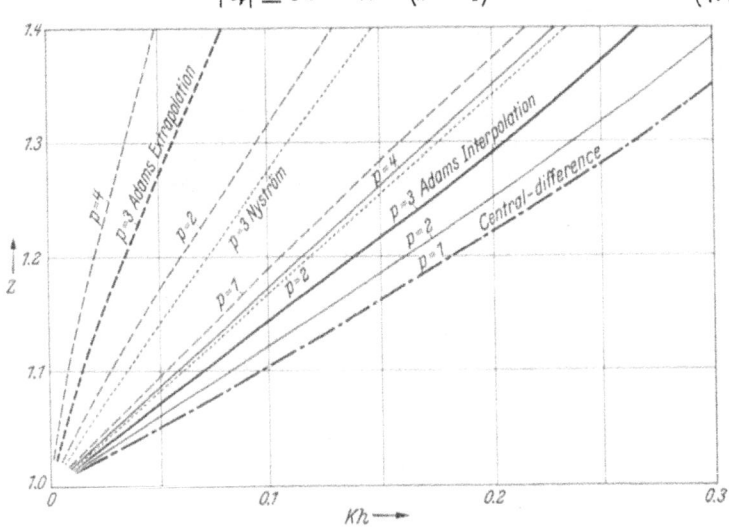

Fig. II/4. Curves of the quantity z, which determines the growth of the error limits, plotted against hK for various methods and orders of approximation. — — — ADAMS' extrapolation method (3.6), ······· NYSTRÖM's extrapolation method (3.10), ——— ADAMS' interpolation method (3.11), —·—·— Central-difference method (3.16)

corresponding to the first case; substituting from (4.24) and (4.40), we have

$$|\varepsilon_r| \leqq \varepsilon z^r + \frac{h^4}{180} \frac{1}{K} |f^{(4)}|_{\max} (z^r - 1). \tag{4.43}$$

For large values of r the growth of the error limits (4.31), (4.37), (4.43) is determined by the power z^r. Consequently the usefulness of the error estimates for the various methods may be compared by comparing the corresponding values of z for given Kh. This is done in Fig. II/4, where curves of z against Kh are drawn for small Kh for the three methods which we have been considering and also for NYSTRÖM's extrapolation method; except for the central-difference method, for which $p = 2$, the curves for several values of p are shown for each method[1].

---

[1] Although not exactly, the curve for the Adams interpolation method with $p = 1$ coincides to within the accuracy of Fig. II/4 with the curve for the central-difference method.

For ease of comparison the curves for $p=3$ are printed more heavily; we choose $p=3$ so that the truncation error is of the same order (h^4) as the central-difference method for all the methods compared. We see that the z values are smallest, and therefore the error limits most effective, for the central-difference method. It is unfortunate that the z values increase with p so that, although in general the calculation will be more accurate for larger p (so long as the differences do not start increasing or fluctuating), the error limits for large r become less precise as p increases.

4.5. Error estimates for the starting iteration (3.5)

We describe briefly how one can estimate the error in the starting iteration (3.5). For the exact solution we have

$$
\left.\begin{aligned}
y(x_1) &= y_0 + h\left(F_0 + \tfrac{1}{2}\nabla F_1 - \tfrac{1}{12}\nabla^2 F_2 + \tfrac{1}{24}\nabla^3 F_3\right) + \tilde{S}_1, \\
y(x_2) &= y_0 + h\left(2F_1 \qquad + \tfrac{1}{3}\nabla^2 F_2\right) \qquad\qquad + \tilde{S}_2, \\
y(x_3) &= y(x_1) + h\left(2F_2 \quad + \tfrac{1}{3}\nabla^2 F_3\right) \qquad\qquad + \tilde{\tilde{S}}_3,
\end{aligned}\right\}
\tag{4.44}
$$

in which we have introduced the notation $F_\nu = f(x_\nu, y(x_\nu))$ [the notation $\tilde{S}_p$ for the remainder terms is not the same as in Ch. I (2.11)].

If $N_p = \left|\dfrac{d^p f}{dx^p}\right|_{\text{max}}$ [cf. (1.23)], we have from Ch. I (2.10)

$$
|\tilde{S}_1| \leq h^5 |\beta_4^*|\, N_4 = \frac{19}{720}\, h^5 N_4
\tag{4.45}
$$

and from Ch. I (2.18)

$$
|\tilde{S}_2| \leq \frac{h^5}{90}\, N_4, \qquad |\tilde{\tilde{S}}_3| \leq \frac{h^5}{90}\, N_4.
\tag{4.46}
$$

Now we imagine the differences in (4.44) expressed in terms of the function values so that we can subtract it from (4.11), the corresponding form of (3.5); the coefficients $a_{\varrho\sigma}$ are given in Tables II/23, II/24. Using the Lipschitz condition to estimate the function differences which arise on subtraction, we find that the absolute values of the errors $|\varepsilon_j| = |y_j - y(x_j)|$ satisfy the inequalities

$$
\left| |\varepsilon_\varrho| - hK\sum_{\sigma=1}^{3} |a_{\varrho\sigma}|\,|\varepsilon_\sigma| \right| \leq |\tilde{S}_\varrho| \qquad (\varrho = 1, 2, 3),
\tag{4.47}
$$

where $\tilde{S}_3 = \tilde{S}_1 + \tilde{\tilde{S}}_3$.

We can replace the right-hand sides by the upper bounds given by (4.45), (4.46), and we then consider the corresponding set of equations; these determine quantities which we will denote by Y_ϱ. In matrix form we have

$$
Y - hKBY = S,
\tag{4.48}
$$

where

$$Y = \begin{pmatrix} Y_1 \\ Y_2 \\ Y_3 \end{pmatrix}, \qquad B = \begin{pmatrix} |a_{11}| & |a_{12}| & |a_{13}| \\ |a_{21}| & |a_{22}| & |a_{23}| \\ |a_{31}| & |a_{32}| & |a_{33}| \end{pmatrix}, \qquad S = \begin{pmatrix} S_1 \\ S_2 \\ S_3 \end{pmatrix}$$

and

$$|\tilde{S}_1| \leqq S_1 = \frac{19}{720} h^5 N_4, \qquad |\tilde{S}_2| \leqq S_2 = \frac{8}{720} h^5 N_4, \qquad |\tilde{S}_3| \leqq S_3 = \frac{27}{720} h^5 N_4.$$

If I denotes the unit matrix, the solution of the matrix equation (4.48) is given by

$$Y = (I + hKB + h^2 K^2 B^2 + \cdots) S \qquad (4.49)$$

provided that the matrix series converges. Now the condition ensuring the convergence of the initial iteration (§ 4.2), i.e. that the absolute values of all the characteristic roots of the matrix hKB shall be less than unity, also ensures the convergence of this matrix series[1], so (4.49) is the solution and the limit matrix is the inverse of the original matrix $I - hKB$. Clearly this limit matrix can have only non-negative coefficients so the system must be monotonic (see Ch. I, § 5.5) and it follows that the Y_ϱ are upper limits for the $|\varepsilon_\varrho|$.

These limits can be found by solving the equations (4.48) but it is quicker, though less precise, of course, to derive upper limits for the Y_ϱ from the solution (4.49). If b is the largest element of B, then any element of B^2 is at most $3b^2$ and in general the elements of B^q cannot exceed $\frac{1}{3}(3b)^q$; thus the matrix series in (4.49) is majorized by the series $I + \sum_{s=0}^{\infty} \frac{1}{3}(3bhK)^s J$, where $J = (j_{\varrho\sigma})$ is the matrix with $j_{\varrho\sigma} = 1$. If $3bhK < 1$, we can sum the geometric series to obtain the error limits

$$|\varepsilon_\varrho| \leqq Y_\varrho \leqq S_\varrho + \frac{hKb}{1 - 3hKb}(S_1 + S_2 + S_3) \qquad (\varrho = 1, 2, 3). \qquad (4.50)$$

4.6. Systems of differential equations

W. Richter[2] extends the above results to the system

$$y'_\nu(x) = f_\nu(x, y_1(x), \ldots, y_n(x)) \qquad (\nu = 1, \ldots, n)$$

with prescribed initial values $y_\nu(0)$.

Let the f_ν satisfy Lipschitz conditions of the form

$$|f_\nu(x, y_1, \ldots, y_n) - f_\nu(x, y_1^*, \ldots, y_n^*)| \leqq K \sum_{\mu=1}^{n} |y_\mu - y_\mu^*| \qquad (\nu = 1, \ldots, n)$$

[1] See, for example, L. Collatz: Eigenwertaufgaben mit technischen Anwendungen, p. 311. Leipzig 1949.

[2] Richter, Willy: Examination de l'erreur commise dans la méthode de M. W. E. Milne … (43 pages). Diss. Neuchâtel 1952.

in a certain domain D of the $(x, y_1, \ldots, y_n)$ space and let

$$R = \max_{\nu, D} \frac{h^5}{180} \left| \frac{d^4 f_\nu(x)}{d x^4} \right| ;$$

further, put $q = \dfrac{h K n}{3}$ and $Q = \dfrac{2(1+28q) R}{6q + 20q^2}$. For the approximations $Y_{\nu r}$ to $y_\nu(r h)$ obtained for $r = -1, +1, +2$ from a starting iteration (convergent for $q < 1$) and for $r > 2$ from MILNE's formulae (see § 3.3, IV) it is deduced that there exist error limits ω_r $(|Y_{\nu r} - y_\nu(r h)| \leqq \omega_r)$ with the upper bounds [corresponding to (4.30)]

$$\omega_r \leqq \omega z_1^{r-1} + Q(z_1^{r-1} - 1),$$

where $\omega = \max(\omega_{-1}, \omega_1, \omega_2)$ and z_1 is the positive root of

$$z^4 - 4q(1 + 2q) z^3 - (1 + q + 4q^2) z^2 - 8q^2 z - q = 0.$$

4.7. Instability in finite-difference methods

It can happen that an approximate solution calculated by a finite-difference method is unstable[1] even though the differential equation is inherently stable. This is particularly so when the difference equation used is of higher order than the differential equation, for it then has more independent solutions than the differential equation and among them there may be increasing solutions (which, on account of the ever-present rounding errors, finally determine the behaviour of the approximate solution) even when the differential equation possesses only decreasing solutions.

The following theory, developed by RUTISHAUSER[2], offers an explanation of the phenomenon. Several simplifying assumptions are made.

We can survey the situation in a rough way by assuming that f_y for the differential equation (1.10) can be treated as piecewise constant. We then consider the calculation of the y_r, say by the central-difference method (3.16), over an interval J of constant f_y. Let $y_r + \eta_r$ be another solution of the equation (3.16) for which the η_r are small, i.e. in the nature of perturbations from the solution y_r, in fact so small that quadratic terms in the η_r are negligible in comparison with the linear terms. Then these perturbations η_r satisfy the linearized "variation equation"

$$\eta_{r+1} = \eta_{r-1} + \frac{h}{3} f_y(\eta_{r-1} + 4\eta_r + \eta_{r+1}).$$

---

[1] The development of unevenness with the central-difference method which was mentioned in § 3.4, V [for which reference was made to L. COLLATZ and R. ZURMÜHL: Z. Angew. Math. Mech. 22, 46 (1942)] is also due to such a condition of instability. For further examples of unstable behaviour see J. TODD: Solution of differential equations by recurrence relations. Math. Tables and Other Aids to Computation 4, 39—44 (1950).

[2] RUTISHAUSER, H.: Über die Instabilität von Methoden zur Integration gewöhnlicher Differentialgleichungen. Z. Angew. Math. Phys. 3, 65—74 (1952). — See also W. LINIGER: Stabilität der Differenzenschemaverfahren. Diss. ETH Zürich 1956.

Here we have already used the assumption that f_y is constant in J. This linear homogeneous difference equation can be solved in the usual way by assuming a solution of the form $\eta_r = \lambda^r$; we find that λ must satisfy the quadratic equation

$$\lambda^2 \left(1 - \frac{H}{3}\right) - \frac{4H}{3} \lambda - \left(1 + \frac{H}{3}\right) = 0, \qquad (4.51)$$

where $H = h f_y$, whose roots are

$$\lambda_1 = 1 + H + \frac{H^2}{2} + \frac{H^3}{6} + \frac{H^4}{24} + \cdots \approx e^H$$

$$\lambda_2 = -1 + \frac{H}{3} - \frac{H^2}{18} + \cdots \approx -e^{-\frac{1}{3}H}.$$

We now treat similarly a perturbation of the exact solution $y(x)$ of the differential equation $y' = f(x, y)$. We select another solution $y(x) + \eta(x)$ and, on the assumption that $\eta(x)$ is small in the same sense as for η_r above, derive the differential equation

$$\eta'(x) = f_y \eta(x)$$

for $\eta(x)$, which must therefore be proportional to $e^{f_y x}$. Such a function changes by a factor e^H over an interval of length h. Now if $f_y > 0$, the component λ_2^r in the central-difference solution dies away exponentially and since $\lambda_1 \approx e^H$, a perturbation η_r grows at approximately the same rate as the same perturbation in the solution of the differential equation. If, on the other hand, $f_y < 0$, then the differential equation is stable, i.e. small perturbations die away, but, in general, small perturbations η_r in the solution of the difference equation increase exponentially; a component proportional to λ_2^r is always introduced by the inevitable rounding errors. Consequently the method will be described as unstable for the case $f_y < 0$. This does not necessarily mean that the method is unusable for this case but it is advisable to estimate the error which may arise through instability of the method as being roughly of the order $e^{-\frac{1}{3}h/f_y r} \times 10^{-q}$ at the r-th step of the integration, where q is the number of decimals carried.

In the work referred to, RUTISHAUSER also shows that for the Runge-Kutta method and the Adams extrapolation method no instability need be feared provided that h is chosen sufficiently small[1].

---

[1] In a note RUTISHAUSER shows that, with increasing h, instability first sets in much later for the interpolation method than for the extrapolation method. A similar result was found by A. R. MITCHELL and J. W. CRAGGS: Stability of difference relations in the solution of ordinary differential equations. Math. Tables and Other Aids to Computation 7, 127—129 (1953).

4.8. Improvement of error estimates by use of a weaker Lipschitz condition

The estimates which we have derived for the errors in finite-difference solutions of $y' = f(x, y)$ can lead to better results if the Lipschitz condition (1.12) on $f(x, y)$, which involves absolute values, is replaced by the weaker condition[1]

$$\frac{f(x, y_1) - f(x, y_2)}{y_1 - y_2} \leqq L.$$

This affords a more realistic and accurate discription of the actual situation than the condition which also specifies a lower limit for the quotient, particularly when a negative value can be chosen for L.

Consider the more general system

$$y_j' = f_j(x, y_1, y_2, \ldots, y_n) \qquad (j = 1, 2, \ldots, n) \qquad (4.52)$$

with the initial conditions $y_j(x_0) = y_{j0}$, where the f_j are given continuous (real) functions in a domain D of the $(x, y_1, y_2, \ldots, y_n)$ space which contains the initial point $(x_0, y_{10}, \ldots, y_{n0})$. For any two points $(x, \bar{y}_1, \ldots, \bar{y}_n)$, $(x, \underline{y}_1, \ldots, \underline{y}_n)$ in D with the same x let the function

$$\left.\begin{aligned} &L^*(x, \bar{y}_1, \ldots, \bar{y}_n, \underline{y}_1, \ldots, \underline{y}_n) \\ &= \frac{\sum\limits_{j=1}^{n} [f_j(x, \bar{y}_1, \ldots, \bar{y}_n) - f_j(x, \underline{y}_1, \ldots, \underline{y}_n)] (\bar{y}_j - \underline{y}_j)}{\sum\limits_{j=1}^{n} (\bar{y}_j - \underline{y}_j)^2} \end{aligned}\right\} \qquad (4.53)$$

satisfy the condition

$$L^*(x, \bar{y}_1, \ldots, \bar{y}_n, \underline{y}_1, \ldots, \underline{y}_n) \leqq L. \qquad (4.54)$$

If we take two sets of functions $\bar{y}_j(x)$ and $\underline{y}_j(x)$ which, for $x_0 \leqq x \leqq x_0 + a$, are differentiable and lie within D and insert them into the differential equations, we will be left with the "error functions"

$$\left.\begin{aligned} \bar{\varepsilon}_j(x) &= \bar{y}_j'(x) - f_j(x, \bar{y}_1, \ldots, \bar{y}_n) \\ \underline{\varepsilon}_j(x) &= \underline{y}_j'(x) - f_j(x, \underline{y}_1, \ldots, \underline{y}_n). \end{aligned}\right\} \qquad (4.55)$$

Now suppose that $\bar{y}_j(x) \geqq \underline{y}_j(x)$ for $j = 1, 2, \ldots, n$ and with

$$\bar{y}_j(x) - \underline{y}_j(x) = z_j(x) \geqq 0, \qquad \bar{\varepsilon}_j(x) - \underline{\varepsilon}_j(x) = \varepsilon_j(x)$$

define

$$+\sqrt{\sum_{j=1}^{n} z_j^2(x)} = z(x), \qquad +\sqrt{\sum_{j=1}^{n} \varepsilon_j^2(x)} = \varepsilon(x), \qquad \frac{\sum\limits_{j=1}^{n} z_j(x)\, \varepsilon_j(x)}{z(x)} = \tilde{\varepsilon}(x).$$

[1] See H. ELTERMANN: Fehlerabschätzung bei näherungsweiser Lösung von Systemen von Differentialgleichungen erster Ordnung. Math. Z. **62**, 469—501 (1955). Here we shall only sketch the basic idea of the error estimate. The assumptions actually made by ELTERMANN are slightly weaker still.

According to SCHWARZ's inequality, $\varepsilon(x) \geq |\tilde{\varepsilon}(x)|$; we will assume that $z(x) > 0$ for $x_0 > x \geq x_0 + a$. From (4.55) it follows that

$$\varepsilon_j(x) = z_j'(x) - f_j(x, \bar{y}_1(x), \ldots, \bar{y}_n(x)) + f_j(x, \underline{y}_1(x), \ldots, \underline{y}_n(x));$$

then multiplication by $\frac{z_j(x)}{z(x)}$ and summation over j from 1 to n yields

$$\tilde{\varepsilon}(x) = z'(x) - L^*(x) z(x), \tag{4.56}$$

where $L^*(x)$ is written for the function $L^*(x, \bar{y}_1(x), \ldots, \bar{y}_n(x), \underline{y}_1(x), \ldots, \underline{y}_n(x))$. This is a linear differential equation for $z(x)$ with the solution

$$z(x) = z(x_0) \exp\left(\int_{x_0}^{x} L^*(t)\, dt\right) + \int_{x_0}^{x} \tilde{\varepsilon}(s) \exp\left(\int_{s}^{x} L^*(t)\, dt\right) ds.$$

Finally, using (4.54) and the fact that $|\tilde{\varepsilon}(x)| \leq \varepsilon(x)$, we arrive at the estimate

$$z(x) \leq z(x_0)\, e^{L(x - x_0)} + \int_{x_0}^{x} \varepsilon(s)\, e^{L(x - s)}\, ds. \tag{4.57}$$

If $\bar{y}_j(x)$ and $\underline{y}_j(x)$ take the prescribed initial values, then $z(x_0) = 0$ and the first term on the right-hand side disappears. [The resulting upper limit for $z(x)$ implies the uniqueness of the solution of the initial-value problem, for if $\bar{y}_j(x)$ and $\underline{y}_j(x)$ are both solutions of the problem, $\varepsilon(s) \equiv 0$ and hence also $z(x) \equiv 0.$] If $\bar{y}_j$ and $\underline{y}_j$ are respectively approximate and exact values of the solution, then (4.57) provides an error estimate for $\bar{y}_j$.

4.9. Error estimation by means of the general theorem on iteration

Error estimates which depend only on knowledge of a Lipschitz function $L(x)$ defined by

$$|f(x, y) - f(x, y^*)| \leq L(x) |y - y^*| \tag{4.58}$$

and not on bounds for the higher derivatives $f^{(q)}$ [cf. (1.23)] may be established by means of the general theory of iterative processes discussed in Ch. I, § 5.2. We introduce an operator T such that

$$T \varphi(x) = y_0 + \int_{x_0}^{x} f(\xi, \varphi(\xi))\, d\xi \tag{4.59}$$

and define a norm in an interval $\langle x_0, z \rangle$ by

$$\|\varphi\| = \max_{\langle x_0, z \rangle} \frac{|\varphi(x)|}{W(x)}, \tag{4.60}$$

where $W(x)$ is a fixed positive function at our disposal. If f satisfies the condition (4.58), we can find a Lipschitz constant K for the operator

T as follows:

$$|T\varphi_1 - T\varphi_2| = \left| \int_{x_0}^{x} [f(\xi, \varphi_1(\xi)) - f(\xi, \varphi_2(\xi))]\, d\xi \right|$$

$$\leq \int_{x_0}^{x} L(\xi)\,|\varphi_1(\xi) - \varphi_2(\xi)|\, d\xi$$

$$\leq \int_{x_0}^{x} L(\xi)\,\|\varphi_1 - \varphi_2\|\, W(\xi)\, d\xi;$$

therefore

$$\|T\varphi_1 - T\varphi_2\| \leq K\,\|\varphi_1 - \varphi_2\|,$$

where

$$K = \max_{\langle x_0, z\rangle} \frac{\int_{x_0}^{x} L(\xi)\, W(\xi)\, d\xi}{W(x)}. \tag{4.61}$$

Provided that $K<1$, we can now apply the estimate (5.16) of Ch. I —
at least, in principle. By choosing $W(x) = e^{\lambda x}$ with a suitable value of λ
we can, in fact, ensure that the condition $K<1$ is satisfied; for numerical
purposes functions other than $e^{\lambda x}$ may be more suitable.

If $v(x)$ is any approximation to $y(x)$ with $v(x_0) = y_0$, we define the
corresponding "defect" $d(x)$ as the error in satisfying the differential
equation:

$$d(x) = v'(x) - f\big(x, v(x)\big).$$

We can express the integral $D(x)$ of the defect as

$$D(x) = \int_{x_0}^{x} d(\xi)\, d\xi = v(x) - v(x_0) - \int_{x_0}^{x} f\big(\xi, v(\xi)\big)\, d\xi = v - Tv = u_0 - u_1,$$

where

$$v(x) = u_0(x), \quad u_1 = Tv.$$

Then (5.16) of Ch. I gives an error estimate for the function $u_1(x) = v(x) - D(x)$.

Example. For the Example I of § 2.6, namely

$$y' = f(x, y) = y - \frac{2x}{y}, \quad y(0) = 1,$$

the approximation

$$u_0 = \frac{10x + 7}{3x + 7}$$

yields

$$u_1 = 1 + \frac{353}{150}\, x - \frac{3}{10}\, x^2 - \frac{49}{9}\, \ln\left(1 + \frac{3x}{7}\right) + \frac{343}{500}\, \ln\left(1 + \frac{10}{7}\, x\right).$$

Let the sub-space F (domain of definition of the operator T as used in
§§ 5.1, 5.2) consist of the continuous functions $u(x)$ in $0 \leq x \leq 0.4$ which

8*

satisfy

$$|u - u_0| \leq 0.001.$$

Then in F, $L \leq 1.447$ $\left(\text{this is the maximum value of } \left|\frac{\partial f}{\partial y}\right| = 1 + \frac{2x}{y^2} \text{ in } F\right)$ and with $W(x) = e^{2x}$

$$\frac{\int_0^x W(\xi)\, d\xi}{W(x)} = \frac{1 - e^{-2x}}{2} \leq 0.2753 = \varrho;$$

this choice of $W(x)$ also gives $\|u_1 - u_0\| = 0.0006$ and with $K = L\varrho = 0.398$, so that $\frac{K}{1-K} = 0.662$, (5.16) of Ch. I yields

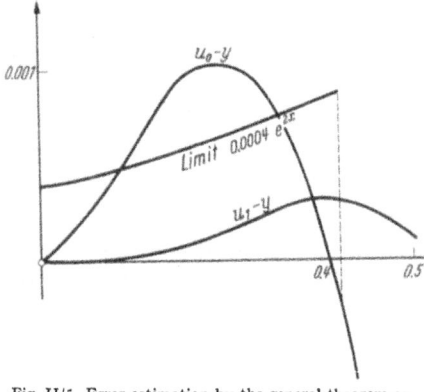

Fig. II/5. Error estimation by the general theorem on iteration

$$\|y - u_1\| \leq 0.662 \times 0.0006$$
$$= 0.0004.$$

Since the functions w with $\|w - u_1\| \leq 0.0004$ lie in F, all the assumptions are satisfied and the required error estimate in $0 \leq x \leq 0.4$ is

$$|y - u_1| \leq 0.0004\, e^{2x}. \quad (4.62)$$

The actual errors $u_0 - y$ and $u_1 - y$ are compared with this limit in Fig. II/5.

§5. Finite-difference methods for differential equations of higher order

5.1. Introduction

Firstly we remark that the general observations made at the beginning of §3 on the relative merits of the various approximate methods still hold when the methods are applied to differential equations of orders higher than the first.

Secondly we note that for the most part the extension to higher orders is achieved by straightforward modifications of the considerations for the first-order case (§§ 3, 4). We can therefore be brief.

An obvious way of dealing with an n-th-order differential equation (2.1) is to convert it into a system of n first-order equations

$$y_1' = y_2, \quad y_2' = y_3, \quad y_3' = y_4, \ldots, \quad y_{n-1}' = y_n, \quad y_n' = f(x, y_1, y_2, y_3, \ldots, y_n)$$

and treat these by one of the numerical methods already described for first-order equations. However, if we do this, not only is the amount of labour involved in the computation greatly increased (thus with the finite-difference methods, for example, n difference tables will be needed), but also the accuracy suffers; in the case of the finite-difference methods this loss in accuracy is due to the fact that n functions are replaced by polynomials instead of their full Taylor expansions in powers of h and consequently n truncation errors are commited[1]. Thus the treatment of an n-th-order differential equation as it stands offers fundamental advantages over the use of the equivalent system of first-order equations. For the same reasons a given system of differential equations of low order can be transformed with advantage into a system of fewer equations of correspondingly higher order, provided that no analytical difficulties attend the transformation and that the functions appearing in the transformed equations are not too complicated.

As in § 2, consider the n-th-order differential equation (2.1)

$$y^{(n)} = f(x, y, y', \ldots, y^{(n-1)})$$

with the initial values (2.2). We assume that the function f satisfies a Lipschitz condition of the form

$$|f(x, y, y', \ldots, y^{(n-1)}) - f(x, y^*, y^{*\prime}, \ldots, y^{*(n-1)})| \leq \sum_{\nu=0}^{n-1} K_\nu |y^{(\nu)} - y^{*(\nu)}| \quad (5.1)$$

for all pairs of points $(x, y, y', \ldots, y^{(n-1)})$, $(x, y^*, y^{*\prime}, \ldots, y^{*(n-1)})$ in a convex domain D of the $(x, y, y', \ldots, y^{(n-1)})$ space in which the solution and all approximations lie. In practice we often put

$$K_\nu = \left| \frac{\partial f}{\partial y^{(\nu)}} \right|_{\text{max in } D} \quad (\nu = 0, 1, \ldots, n - 1). \quad (5.2)$$

Sometimes it suffices to assume that f satisfies the simpler but cruder condition

$$|f(x, y, y', \ldots, y^{(n-1)}) - f(x, y^*, y^{*\prime}, \ldots, y^{*(n-1)})| \leq K \sum_{\nu=0}^{n-1} |y^{(\nu)} - y^{*(\nu)}|, \quad (5.3)$$

where

$$K = \max_\nu K_\nu.$$

As with first-order equations, the calculation consists of two quite separate steps:

1. Approximations $y_\nu, y'_\nu, \ldots, y_\nu^{(n-1)}$, from which values of f_ν can be calculated, are obtained by some other means for the first few points (calculation of "starting values").

2. The solution is continued step by step by the finite-difference formulae; these give the values of y and its derivatives at the point x_{r+1} once the values at $x_r, x_{r-1}, \ldots$ are known ("main calculation"). Again we give several such step-by-step procedures (cf. § 5.4 et seq.).

[1] COLLATZ, L., and R. ZURMÜHL: Zur Genauigkeit verschiedener Integrationsverfahren bei gewöhnlichen Differentialgleichungen. Ing.-Arch. **13**, 34—36 (1942).

5.2. Calculation of starting values

Substantially the same means are at our disposal here as for differential equations of the first order (§ 3.2):

I. Using a "self-starting" method of integration. As for first-order equations, the Runge-Kutta method is foremost in this category and, because of the particular importance of accuracy in the starting values, is best used with a step of half the length of the main step.

II. Using the Taylor series for $y(x)$ and its derivatives. (As in II and IIa of § 3.2.)

III. Using quadrature formulae. A set of starting values can also be determined by an iterative procedure based on the formulae mentioned in § 2.5 for use as a terminal check. For the first point $x = x_1$ we use formulae corresponding to (2.24); thus for $y_1, y_1', \ldots, y_1^{(n-1)}$ we have

$$\left. \begin{aligned} y_1^{(m)} &= y_0^{(m)} + h\, y_0^{(m+1)} + \frac{h^2}{2}\, y_0^{(m+2)} + \cdots + \frac{h^{n-m-1}}{(n-m-1)!}\, y_0^{(n-1)} + \\ &\quad + h^{n-m} \sum_{\varrho=0}^{p} \gamma_{n-m,\varrho}\, \varDelta^{\varrho} f_0. \end{aligned} \right\} \quad (5.4)$$

The main calculation will normally use second and perhaps also third differences, so we need further starting values at x_2 and perhaps also at x_3. For these we use the formulae corresponding to (2.25):

$$y_{r+1}^{(m)} = \left\{ \begin{aligned} & y_{r-1}^{(m)} + 2 \sum_{\varrho=0}^{\frac{n-m-3}{2}} \frac{h^{2\varrho+1}}{(2\varrho+1)!}\, y_r^{(m+2\varrho+1)} \\ & 2 y_r^{(m)} - y_{r-1}^{(m)} + 2 \sum_{\varrho=0}^{\frac{n-m-2}{2}} \frac{h^{2\varrho}}{(2\varrho)!}\, y_r^{(m+2\varrho)} \end{aligned} \right\} \left. \begin{aligned} & + 2h^{n-m}(\beta_{n,0}^{**} f_r + \\ & + \beta_{n,1}^{**} \nabla^2 f_{r+1}) \\ & \quad (r = 1, 2), \end{aligned} \right\} \quad (5.5)$$

in which the upper alternative applies when $n - m$ is odd and the lower when $n - m$ is even. These formulae are used iteratively in the usual manner: first, rough values of $y_1^{(m)}, y_2^{(m)}, y_3^{(m)}$ are obtained, say from their Taylor expansions truncated after the term in $y_0^{(n)}$, and the corresponding values f_1, f_2, f_3 and their differences evaluated; these values are then inserted in the right-hand sides of (5.4), (5.5) to give the next approximations, on which the process can be repeated. We now examine this method in detail for the special case of a second-order equation.

5.3. Iterative calculation of starting values for the second-order equation $y'' = f(x, y, y')$

As in § 3.2, we suggest a preliminary scheme suitable for constructing rough values at two or three points with which to start the iteration. If starting values are required at only two points x_1, x_2, the calculation represented by the equations framed in dots suffices.

A. Rough values

$$
\begin{aligned}
&\text{1. } \tilde{y}_1 = y_0 + h y_0' + \tfrac{1}{2} h^2 f_0, \qquad h \tilde{y}_1' = h y_0' + h^2 f_0; \\
&\quad \text{thence } \tilde{f}_1 = f(x_1, \tilde{y}_1, \tilde{y}_1'), \qquad \nabla \tilde{f}_1 = \tilde{f}_1 - f_0; \\
&\text{2. } y_1^{[0]} = y_0 + h y_0' + h^2 (\tfrac{1}{2} f_0 + \tfrac{1}{6} \nabla \tilde{f}_1), \\
&\quad h y_1'^{[0]} = h y_0' + h^2 (f_0 + \tfrac{1}{2} \nabla \tilde{f}_1), \quad \text{thence } f_1^{[0]}, \ \nabla f_1^{[0]}; \\
&\text{3. } y_2^{[0]} = 2 y_1^{[0]} - y_0 + h^2 f_1^{[0]}, \\
&\quad h y_2'^{[0]} = h y_0' + 2 h^2 f_1^{[0]}, \quad \text{thence } f_2^{[0]}, \ \nabla f_2^{[0]}, \ \nabla^2 f_2^{[0]}.
\end{aligned}
\tag{5.6}
$$

$$
\left.
\begin{aligned}
&\text{4. } y_3^{[0]} = 2 y_2^{[0]} - y_1^{[0]} + h^2 (f_2^{[0]} + \tfrac{1}{12} \nabla^2 f_2^{[0]}), \\
&\quad h y_3'^{[0]} = h y_1'^{[0]} + h^2 (2 f_2^{[0]} + \tfrac{1}{3} \nabla^2 f_2^{[0]}).
\end{aligned}
\right\}
\tag{5.7}
$$

B. Iterative improvement for $v = 1, 2, \ldots$

$$
\begin{aligned}
y_1^{[v+1]} &= y_0 + h y_0' + h^2 (\tfrac{1}{2} f_0 + \tfrac{1}{6} \nabla f_1^{[v]} - \tfrac{1}{24} \nabla^2 f_2^{[v]} && + \tfrac{1}{45} \nabla^3 f_3^{[v]}) \\
h y_1'^{[v+1]} &= h y_0' + h^2 \ (f_0 + \tfrac{1}{2} \nabla f_1^{[v]} - \tfrac{1}{12} \nabla^2 f_2^{[v]} && + \tfrac{1}{24} \nabla^3 f_3^{[v]}) \\
y_2^{[v+1]} &= 2 y_1^{[v+1]} - y_0 + h^2 (f_1^{[v]} + \tfrac{1}{12} \nabla^2 f_2^{[v]}) \\
h y_2'^{[v+1]} &= h y_0' + h^2 (2 f_1^{[v]} + \tfrac{1}{3} \ \nabla^2 f_2^{[v]})
\end{aligned}
\tag{5.8}
$$

$$
\left.
\begin{aligned}
y_3^{[v+1]} &= 2 y_2^{[v+1]} - y_1^{[v+1]} + h^2 (f_2^{[v]} + \tfrac{1}{12} \nabla^2 f_3^{[v]}) \\
h y_3'^{[v+1]} &= \phantom{2 y_2^{[v+1]}} h y_1'^{[v+1]} + h^2 (2 f_2^{[v]} + \tfrac{1}{3} \nabla^2 f_3^{[v]}).
\end{aligned}
\right\}
\tag{5.9}
$$

The improved values obtained at each stage are used to re-calculate the function values $f_j^{[v]} = f(x_j, y_j^{[v]}, y_j'^{[v]})$; the differences of these values are then formed for use in the next cycle. Very few cycles will be necessary if the step interval h is chosen sufficiently small; it should not be too large in any case, as has been stressed before (cf. § 3.4).

The scheme for calculating rough values which was given in § 3.2 incorporated an iteration on the first two points; this gave better values with which to calculate the rough value of y for the third point. Better values for starting the iteration B can be obtained here in a similar way: the rough values for the first two points are improved by an iteration using the formulae framed in B before step 4 is carried out; then the improved values are used in step 4 to give better rough values for the third point.

It may be noticed that the form of the equations in A and B is such that if y' does not occur explicitly in the differential equation, i.e.

it has the form $y'' = f(x, y)$, then the calculations to find rough and improved values of the y'_j are not needed at all.

We now investigate the convergence of this starting iteration, proceeding along lines quite analogous to those of § 4.2. We first express all the differences occurring in (5.8), (5.9) in terms of function values. The iterative scheme then reads

$$\left.\begin{aligned} y_j^{[\nu+1]} &= y_0 + j\,h\,y_0' + h^2 \sum_{k=0}^{3} \sigma_{jk} f_k^{[\nu]} \\ y_j'^{[\nu+1]} &= y_0' \qquad\quad + h \sum_{k=0}^{3} \tau_{jk} f_k^{[\nu]} \end{aligned}\right\} \qquad (j = 1, 2, 3), \qquad (5.10)$$

in which the σ_{jk} and τ_{jk} have the values given in Tables II/25 and II/26. $\left(\text{Checks on the calculation of these numbers are } \sum_{k=0}^{3} \sigma_{jk} = \tfrac{1}{2} j^2, \ \sum_{k=0}^{3} \tau_{jk} = j.\right)$

Table II/25. *The coefficients σ_{jk}*

σ_{jk}	$k=$			
	0	1	2	3
$j = 1$	$\dfrac{97}{360}$	$\dfrac{19}{60}$	$-\dfrac{13}{120}$	$\dfrac{1}{45}$
2	$\dfrac{28}{45}$	$\dfrac{22}{15}$	$-\dfrac{2}{15}$	$\dfrac{2}{45}$
3	$\dfrac{39}{40}$	$\dfrac{27}{10}$	$\dfrac{27}{40}$	$\dfrac{3}{20}$

Table II/26. *The coefficients τ_{jk}*

τ_{jk}	$k=$			
	0	1	2	3
$j = 1$	$\dfrac{3}{8}$	$\dfrac{19}{24}$	$-\dfrac{5}{24}$	$\dfrac{1}{24}$
2	$\dfrac{1}{3}$	$\dfrac{4}{3}$	$\dfrac{1}{3}$	0
3	$\dfrac{3}{8}$	$\dfrac{9}{8}$	$\dfrac{9}{8}$	$\dfrac{3}{8}$

If we subtract from equations (5.10) the corresponding equations with ν replaced by $\nu - 1$ and define $f_0^{[\nu]} = f_0$, we obtain

$$\delta_j^{[\nu]} = y_j^{[\nu+1]} - y_j^{[\nu]} = h^2 \sum_{k=1}^{3} \sigma_{jk} (f_k^{[\nu]} - f_k^{[\nu-1]}),$$

$$\delta_j'^{[\nu]} = y_j'^{[\nu+1]} - y_j'^{[\nu]} = h \sum_{k=1}^{3} \tau_{jk} (f_k^{[\nu]} - f_k^{[\nu-1]}).$$

Bounds for the function differences involved here are provided by the Lipschitz condition (5.1) and taking absolute values we have

$$\left.\begin{aligned} |\delta_j^{[\nu]}| &\le h^2 \sum_{k=1}^{3} |\sigma_{jk}| \left(K_0 |\delta_k^{[\nu-1]}| + K_1 |\delta_k'^{[\nu-1]}|\right) \\ |\delta_j'^{[\nu]}| &\le h \sum_{k=1}^{3} |\tau_{jk}| \left(K_0 |\delta_k^{[\nu-1]}| + K_1 |\delta_k'^{[\nu-1]}|\right). \end{aligned}\right\} \qquad (5.11)$$

Hence the quantities $v_j^{[\nu]}$ defined by

$$v_j^{[\nu]} = K_0 |\delta_j^{[\nu]}| + K_1 |\delta_j'^{[\nu]}|$$

satisfy the set of inequalities

$$v_j^{[\nu]} \le \sum_{k=1}^{3} \left(K_0 h^2 |\sigma_{jk}| + K_1 h |\tau_{jk}|\right) v_k^{[\nu-1]} \qquad (j = 1, 2, 3). \qquad (5.12)$$

If we regard the (positive) numbers

$$K_0 h^2 |\sigma_{jk}| + K_1 h |\tau_{jk}| = \alpha_{jk} \tag{5.13}$$

as the elements of a matrix A, we can apply the theorem quoted in § 4.2 exactly as we did there for the first-order case. This guarantees the convergence of the series

$$\sum_{\nu=0}^{\infty} v_j^{[\nu]}$$

and hence also the absolute convergence of the iterations for y_j and y_j', provided that all characteristic roots of the matrix A are less than 1 in absolute value.

For a numerical investigation of these characteristic roots we introduce the positive parameter p defined by

$$K_1 h = p K_0 h^2.$$

Then the characteristic roots of a second matrix

$$B = \frac{1}{K_0 h^2} A = (|\sigma_{jk}| + p |\tau_{jk}|)$$

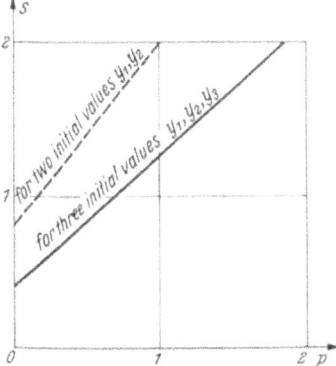

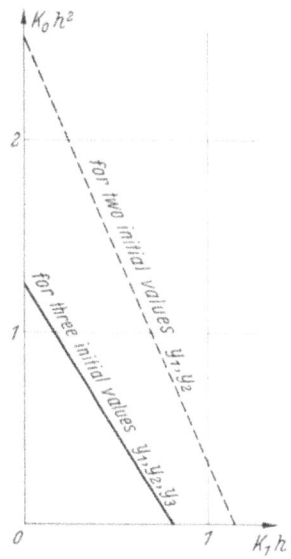

Fig. II/6. Plots of the auxiliary function $s(p)$

Fig. II/7. Boundaries corresponding to a sufficient condition for the starting iteration (5.8), (5.9)

will be functions of p only; let the greatest in absolute value be denoted by s (since B is a non-zero matrix with non-negative elements, s will in fact be positive[1]). The function $s(p)$ can be calculated point by point; the results are shown graphically in Fig. II/6 for the cases of two and three starting values.

Now the condition that all characteristic roots of A should be less than 1 in absolute value is equivalent to the condition

$$s K_0 h^2 < 1.$$

[1] See G. FROBENIUS: S.-B. Preuss. Akad. Wiss., Math.-phys. Kl. 1. HBd. 1912, pp. 456—477, in particular p. 457.

Hence the region of the $(K_1 h, K_0 h^2)$ plane for which convergence of the starting iteration is assured (Fig. II/7) is bounded by the curve with the parametric representation (parameter p)

$$K_0 h^2 = \frac{1}{s(p)}, \qquad K_1 h = \frac{p}{s(p)}. \tag{5.14}$$

Assuming that h is chosen so small that this convergence condition is satisfied, we can estimate the errors in the starting values

$$\varepsilon_j = y_j - y(x_j), \qquad \varepsilon_j' = y_j' - y'(x_j) \qquad (j = 1, 2, 3).$$

Again we proceed along the same lines as for the first-order case (§ 4.5) and can be brief.

For the exact solution we have

$$\left. \begin{aligned}
y(x_1) &= y(x_0) + hy'(x_0) + h^2(\tfrac{1}{2}F_0 + \tfrac{1}{6}\nabla F_1 - \tfrac{1}{24}\nabla^2 F_2 + \tfrac{1}{45}\nabla^3 F_3) + \widetilde{\widetilde{T}}_1, \\
y(x_2) &= 2y(x_1) - y(x_0) + h^2(F_1 + \tfrac{1}{12}\nabla^2 F_2) + \widetilde{\widetilde{T}}_2, \\
y(x_3) &= 2y(x_2) - y(x_1) + h^2(F_2 + \tfrac{1}{12}\nabla^2 F_3) + \widetilde{\widetilde{T}}_3, \\
y'(x_1) &= y'(x_0) + h(F_0 + \tfrac{1}{2}\nabla F_1 - \tfrac{1}{12}\nabla^2 F_2 + \tfrac{1}{24}\nabla^3 F_3) + \widetilde{S}_1, \\
y'(x_2) &= y'(x_0) + h(2F_1 + \tfrac{1}{3}\nabla^2 F_2) + \widetilde{S}_2, \\
y'(x_3) &= y'(x_1) + h(2F_2 + \tfrac{1}{3}\nabla^2 F_3) + \widetilde{\widetilde{S}}_3,
\end{aligned} \right\} \tag{5.15}$$

where $F_\nu = f(x_\nu, y(x_\nu), y'(x_\nu))$. With the notation

$$\left| \frac{d^p f}{dx^p} \right|_{\max} = N_p$$

the estimates of Ch. I (2.31), (2.46), (2.47) for the remainder terms occurring in (5.15) read

$$|\widetilde{\widetilde{T}}_1| \leq \frac{7}{480} h^6 N_4 = s_1, \quad |\widetilde{\widetilde{T}}_2| \leq \frac{1}{240} h^6 N_4 = \tilde{s}_2, \quad |\widetilde{\widetilde{T}}_3| \leq \frac{1}{240} h^6 N_4 = \tilde{s}_3,$$

$$|\widetilde{S}_1| \leq \frac{19}{720} h^5 N_4 = s_1', \quad |\widetilde{S}_2| \leq \frac{1}{90} h^5 N_4 = s_2', \quad |\widetilde{\widetilde{S}}_3| \leq \frac{1}{90} h^5 N_4 = \tilde{s}_3'.$$

With all differences expressed in terms of function values we subtract the two sets of equations (5.15) and (5.8), (5.9) for the exact and approximate values, respectively, to obtain the error equations

$$\left. \begin{aligned}
\varepsilon_j &= y_j - y(x_j) = h^2 \sum_{k=1}^{3} \sigma_{jk}(f_k - F_k) + \vartheta_j s_j \\
\varepsilon_j' &= y_j' - y'(x_j) = h \sum_{k=1}^{3} \tau_{jk}(f_k - F_k) + \vartheta_j' s_j'
\end{aligned} \right\} \quad (j = 1, 2, 3),$$

where

$$s_2 = \tilde{s}_2 + 2s_1, \quad s_3 = \tilde{s}_3 + 2\tilde{s}_2 + 5s_1, \quad s_3' = \tilde{s}_3' + s_1', \quad |\vartheta_j| \leq 1, \quad |\vartheta_j'| \leq 1.$$

Taking absolute values and using the Lipschitz condition (5.1) for the function differences $f_k - F_k$, we find that

$$\left.\begin{array}{l} |\varepsilon_j| \leq h^2 \sum_{k=1}^{3} |\sigma_{jk}| \left(K_0 |\varepsilon_k| + K_1 |\varepsilon_k'|\right) + s_j \\[2mm] |\varepsilon_j'| \leq h \sum_{k=1}^{3} |\tau_{jk}| \left(K_0 |\varepsilon_k| + K_1 |\varepsilon_k'|\right) + s_j' \end{array}\right\} \quad (j = 1, 2, 3). \quad (5.16)$$

Exactly as with the first-order case, the matrix of this system of inequalities is monotonic and the quantities Y_j, Y_j' determined from the equations

$$\left.\begin{array}{l} Y_j = h^2 \sum_{k=1}^{3} |\sigma_{jk}| \left(K_0 Y_k + K_1 Y_k'\right) + s_j \\[2mm] Y_j' = h \sum_{k=1}^{3} |\tau_{jk}| \left(K_0 Y_k + K_1 Y_k'\right) + s_j' \end{array}\right\} \quad (j = 1, 2, 3) \quad (5.17)$$

provide upper limits for the absolute values of the errors ε_j, ε_j'.

We can either solve these linear equations for the Y_j, Y_j' or derive upper bounds for the Y_j, Y_j' by the method described at the end of § 4.5 for the first-order case. The matrices

$$Y = \begin{pmatrix} Y_1 \\ Y_2 \\ Y_3 \\ Y_1' \\ Y_2' \\ Y_3' \end{pmatrix}, \quad A = \begin{pmatrix} hK_0|\sigma_{11}| & hK_0|\sigma_{12}| & hK_0|\sigma_{13}| & hK_1|\sigma_{11}| & hK_1|\sigma_{12}| & hK_1|\sigma_{13}| \\ \cdots & \cdots & \cdots & \cdots & \cdots & \cdots \\ hK_0|\sigma_{31}| & hK_0|\sigma_{32}| & \cdots & hK_1|\sigma_{31}| & \cdots & \cdots \\ K_0|\tau_{11}| & K_0|\tau_{12}| & K_0|\tau_{13}| & K_1|\tau_{11}| & K_1|\tau_{12}| & K_1|\tau_{13}| \\ \cdots & \cdots & \cdots & \cdots & \cdots & \cdots \\ K_0|\tau_{31}| & K_0|\tau_{32}| & \cdots & K_1|\tau_{31}| & \cdots & \cdots \end{pmatrix}, \quad S = \begin{pmatrix} s_1 \\ s_2 \\ s_3 \\ s_1' \\ s_2' \\ s_3' \end{pmatrix} \quad (5.18)$$

correspond here to the matrices Y, KB, S of § 4.5, and if a is the maximum value of the elements of A, the method of § 4.5 yields

$$\left.\begin{array}{l} |\varepsilon_j| \leq Y_j \leq s_j + \dfrac{h\,a}{1 - 6\,h\,a} \sum_{k=1}^{3} (s_k + s_k') \\[4mm] |\varepsilon_j'| \leq Y_j' \leq s_j' + \dfrac{h\,a}{1 - 6\,h\,a} \sum_{k=1}^{3} (s_k + s_k') \end{array}\right\} \quad (j = 1, 2, 3). \quad (5.19)$$

We now describe several methods for the main calculation.

5.4. Extrapolation methods

Suppose that approximate values y_ϱ, y_ϱ', ..., $y_\varrho^{(n-1)}$ for the exact solution $y(x_\varrho)$, $y'(x_\varrho)$, ..., $y^{(n-1)}(x_\varrho)$ at the points x_ϱ (for $\varrho = 1, ..., r$) are known and that it is required to calculate approximate values y_{r+1}, y_{r+1}', ..., $y_{r+1}^{(n-1)}$ at the next point.

If in (2.25) of Ch. I we replace n by $n - m$ and x_0 by x_r, and use the fact that $y^{(n)} = f$, we obtain a formula for the m-th derivative of the exact solution, namely

$$y^{(m)}(x_{r+1}) = \sum_{v=0}^{n-m-1} \frac{h^v}{v!}\, y^{(m+v)}(x_r) + \\ + h^{n-m} \sum_{\varrho=0}^{p} \beta_{n-m,\varrho}\, V^\varrho F(x_r) + R_{n-m,\,p+1}, \Bigg\} \quad (5.20)$$

in which we have used the notation

$$f\big(x, y(x), y'(x), \ldots, y^{(n-1)}(x)\big) = F(x). \quad (5.21)$$

An estimate for the remainder term $R_{n-m,\,p+1}$ is given in Ch. I (2.27). By neglecting this remainder term we obtain FALKNER's extrapolation formula[1] for the approximate value $y_{r+1}^{(m)}$:

$$y_{r+1}^{(m)} = \sum_{v=0}^{n-m-1} \frac{h^v}{v!}\, y_r^{(m+v)} + h^{n-m} \sum_{\varrho=0}^{p} \beta_{n-m,\varrho}\, V^\varrho f_r, \quad (5.22)$$

where $f_r = f(x_r, y_r, y_r', \ldots, y_r^{(n-1)})$. For the special case $n=1$, $m=0$, this reduces to the Adams extrapolation formula (3.6) for differential equations of the first order. For second-order equations ($n=2$, $m=0, 1$) we have

$$y_{r+1} = y_r + h\,y_r' + h^2 \Big[\tfrac{1}{2} f_r + \tfrac{1}{6} V f_r + \tfrac{1}{8} V^2 f_r + \tfrac{19}{180} V^3 f_r + \cdots\Big] \Bigg\} \quad (5.23)$$
$$y_{r+1}' = y_r' \qquad + h\Big[f_r + \tfrac{1}{2} V f_r + \tfrac{5}{12} V^2 f_r + \tfrac{3}{8} V^3 f_r + \cdots\Big].$$

A general formula for the coefficients $\beta_{n,\varrho}$ is given by (2.26) of Ch. I.

There are, however, many other ways in which we can set up extrapolation formulae. For example, by linearly combining certain formulae based on (2.23) of Ch. I we can construct a formula which involves only y and f values, i.e. all intermediate derivatives can be eliminated. Such a formula is particularly suitable when the differential equation has the form $y^{(n)} = f(x, y)$.

To derive this formula we take the set of $n-1$ equations obtained from (2.23) of Ch. I by putting $x = x_{-1}, x_{-2}, \ldots, x_{-n+1}$ in turn and multiplying by $(-1)^{q+1}\binom{n}{q+1}$ for $x = x_{-q}$, and add them. With $^{(n)}f = y$ the resulting equation reads

$$y(x_1) - \binom{n}{1} y(x_0) + \binom{n}{2} y(x_{-1}) - \cdots + (-1)^n y(x_{-n+1}) = V^n y(x_1) \\ = h^n \sum_{\varrho=0}^{p} \zeta_{n,\varrho}\, V^\varrho f_0 + \tilde{R}, \Bigg\} \quad (5.24)$$

where

$$\zeta_{n,\varrho} = P_{n,\varrho}(1) + \sum_{q=1}^{n-1} (-1)^{q+1}\binom{n}{q+1} P_{n,\varrho}(-q) \quad (5.25)$$

---

[1] FALKNER, V. M.: A method of numerical solution of differential equations. Phil. Mag. (7) **21**, 621—640 (1936).

and

$$\widetilde{R} = R_{n,p+1}(x_1) + \sum_{q=1}^{n-1} (-1)^{q+1} \binom{n}{q+1} R_{n,p+1}(x_{-q}). \qquad (5.26)$$

The first few values of the $\zeta_{n,\varrho}$ are given in Table II/27.

Table II/27. *The numbers $\zeta_{n,\varrho}$*

	$\varrho=0$	$\varrho=1$	$\varrho=2$	$\varrho=3$	$\varrho=4$	$\varrho=5$
$n=1$	1	$\dfrac{1}{2}$	$\dfrac{5}{12}$	$\dfrac{3}{8}$	$\dfrac{251}{720}$	$\dfrac{95}{288}$
$n=2$	1	0	$\dfrac{1}{12}$	$\dfrac{1}{12}$	$\dfrac{19}{240}$	$\dfrac{3}{40}$
$n=3$	1	$-\dfrac{1}{2}$	0	0	$\dfrac{1}{240}$	
$n=4$	1	-1	$\dfrac{1}{6}$	0	$-\dfrac{1}{720}$	

$$(5.27)$$

With the remainder term omitted (5.24) provides an extrapolation formula for n-th-order equations. It includes the Adams extrapolation formula (3.6) and STÖRMER's[1] extrapolation formula

$$
\begin{aligned}
y_{r+1} &= 2y_r - y_{r-1} + \\
&+ h^2\left(f_r + \frac{1}{12}\nabla^2 f_r + \frac{1}{12}\nabla^3 f_r + \frac{19}{240}\nabla^4 f_r + \frac{3}{40}\nabla^5 f_r + \cdots\right) \\
&= 2y_r - y_{r-1} + h^2\left[f_r + \frac{1}{12}(\nabla^2 f_r + \nabla^3 f_r + \nabla^4 f_r + \nabla^5 f_r) - \right. \\
&\left. - \frac{1}{240}(\nabla^4 f_r + 2\nabla^5 f_r) + \cdots\right]
\end{aligned}
\qquad (5.28)
$$

as the particular cases $n=1$ and $n=2$, respectively; the latter formula has been used a great deal on account of its convenient coefficients (particularly when fourth and higher differences are neglected).

For third-order equations the coefficients are even more convenient; in fact, since $\zeta_{32}=\zeta_{33}=0$, there are no terms in $\nabla^2 f$ and $\nabla^3 f$:

$$y_{r+1} = 3y_r - 3y_{r-1} + y_{r-2} + h^3\left(f_r - \frac{1}{2}\nabla f_r + \frac{1}{240}\nabla^4 f_r + \cdots\right). \qquad (5.29)$$

---

[1] CARL STÖRMER, the Norwegian mathematician, was born on the 3rd September, 1874 in Skien (Norway). He studied in Christiania 1892—1898, then went to Paris 1898—1900, where he became an unsalaried lecturer in 1899, and later (1902) to Göttingen. In 1903 he was appointed professor of pure mathematics in the University of Oslo.

In order to confirm his theory of the aurora borealis, STÖRMER and his colleagues spent several years calculating numerous orbits of electrons in the earth's magnetic field [STÖRMER, C.: Z. Astrophysik 1, 237—274 (1930)]. The computed orbits were reproduced very closely by E. BRÜCHE in his experimental work.

STÖRMER remarked of the calculations: "One must have ample time and patience, for the calculations are extremely long." 4500 working hours were needed for 120 orbits.

If fourth differences are neglected, we have a very simple yet accurate formula which involves only the first differences; provided we do not need higher differences for the formulae for y' and y'', which may appear explicitly in the differential equation, the difference table need not be extended beyond the first difference.

For fourth-order equations also we obtain a very simple formula when fourth differences are neglected; the formula (5.24) with $n=4$ reads

$$\left.\begin{aligned} y_{r+1} &= 4y_r - 6y_{r-1} + 4y_{r-2} - y_{r-3} + \\ &\quad + h^4\left(f_r - \nabla f_r + \frac{1}{6}\nabla^2 f_r - \frac{1}{720}\nabla^4 f_r + \cdots\right). \end{aligned}\right\} \quad (5.30)$$

5.5. Interpolation methods

Again suppose that approximate values $y_\varrho^{(m)}$ for the exact solution $y^{(m)}(x_\varrho)$ $(m=0, 1, \ldots, n-1)$ at the points x_ϱ $(\varrho=1, \ldots, r)$ are known and that it is required to calculate approximate values for $y^{(m)}(x_{r+1})$. For the exact solution, (2.36) of Ch. I with $n-m$ in place of n, x_{r+1} in place of x_0, and $y^{(n)}=f$, gives

$$\left.\begin{aligned} y^{(m)}(x_{r+1}) &= \sum_{\nu=0}^{n-m-1} \frac{h^\nu}{\nu!} y^{(m+\nu)}(x_r) + \\ &\quad + h^{n-m}\sum_{\varrho=0}^{p} \beta^*_{n-m,\varrho}\nabla^\varrho F(x_{r+1}) + R^*_{n-m,p+1}, \end{aligned}\right\} \quad (5.31)$$

where F is defined in (5.21), the $\beta^*_{n-m,\varrho}$ are given by (2.38) of Ch. I, and the remainder term is bounded by the limits (2.41) of Ch. I. Omitting the remainder term, we obtain the Adams interpolation formula for $y^{(m)}_{r+1}$:

$$y^{(m)}_{r+1} = \sum_{\sigma=0}^{n-m-1} \frac{h^\sigma}{\sigma!} y^{(m+\sigma)}_r + h^{n-m}\sum_{\varrho=0}^{p} \beta^*_{n-m,\varrho}\nabla^\varrho f_{r+1}. \quad (5.32)$$

The case $n=1$, $m=0$ has already been dealt with under the Adams interpolation method (3.11) for differential equations of the first order (§ 3.3, II). For $n=2$ we have

$$\left.\begin{aligned} y_{r+1} &= y_r + h\,y'_r + \\ &\quad + h^2\left[\frac{1}{2}f_{r+1} - \frac{1}{3}\nabla f_{r+1} - \frac{1}{24}\nabla^2 f_{r+1} - \frac{7}{360}\nabla^3 f_{r+1} - \cdots\right] \\ y'_{r+1} &= y'_r + h\left[f_{r+1} - \frac{1}{2}\nabla f_{r+1} - \frac{1}{12}\nabla^2 f_{r+1} - \frac{1}{24}\nabla^3 f_{r+1} - \cdots\right]. \end{aligned}\right\} \quad (5.33)$$

In (5.32) the unknown $y^{(m)}_{r+1}$ appears on both sides (as an argument of f_{r+1} on the right-hand side) and therefore, as in (3.14), we use the equation in an iterative form by putting the values for the ν-th iterate into the right-hand side to evaluate the $(v+1)$-th iterate in the usual way. Thus the iterative form of (5.32) is

$$y_{r+1}^{(m)\,[v+1]} = \sum_{\sigma=0}^{n-m-1} \frac{h^\sigma}{\sigma!}\, y_r^{(m+\sigma)} + h^{n-m} \sum_{\varrho=0}^{p} \beta_{n-m,\varrho}^{*}\, \nabla^\varrho f_{r+1}^{[v]}, \qquad (5.34)$$

in which the iteration superscripts are again enclosed in square brackets to distinguish them from the derivative superscripts.

Again there are occasions when we need the Lagrangian form of the equations in which the differences $\nabla^\varrho f$ are expressed in terms of the function values; corresponding to the form (3.12) of (3.11) we have the Lagrangian form

$$y_{r+1}^{(m)} = \sum_{\varrho=0}^{n-m-1} \frac{h^\varrho}{\varrho!}\, y_r^{(m+\varrho)} + h^{n-m} \sum_{\sigma=0}^{p} \beta_{n-m,\sigma,p}^{*}\, f_{r+1-\sigma} \qquad (5.35)$$

of (5.32), in which the new coefficients are given by

$$\beta_{q,\sigma,p}^{*} = (-1)^\sigma \sum_{\varrho=\sigma}^{p} \beta_{q,\varrho}^{*} \binom{\varrho}{\sigma}; \qquad (5.36)$$

the corresponding iterative form is

$$\left.\begin{aligned}
y_{r+1}^{(m)\,[v+1]} = &\sum_{\varrho=0}^{n-m-1} \frac{h^\varrho}{\varrho!}\, y_r^{(m+\varrho)} + \\
&+ h^{n-m}\left(\beta_{n-m,0,p}^{*}\, f_{r+1}^{[v]} + \sum_{\sigma=1}^{p} \beta_{n-m,\sigma,p}^{*}\, f_{r+1-\sigma}\right).
\end{aligned}\right\} \qquad (5.37)$$

Just as in § 3.3 for differential equations of the first order, other formulae of the interpolation type can be derived for equations of higher order. Thus, for instance, from (2.45) of Ch. I we have

$$\left.\begin{aligned}
y^{(m)}(x_{r+1}) = &-y^{(m)}(x_{r-1}) + 2 \sum_{\varrho=0}^{\frac{n-m}{2}-1} \frac{h^{2\varrho}}{(2\varrho)!}\, y^{(m+2\varrho)}(x_r) + \\
&+ 2h^{n-m} \sum_{\varrho=0}^{\frac{p}{2}} \beta_{n-m,\varrho}^{**}\, \nabla^{2\varrho} F(x_{r+\varrho}) + 2R_{n-m,p}^{**} \\[2ex]
y^{(m)}(x_{r+1}) = &\, y^{(m)}(x_{r-1}) + 2 \sum_{\varrho=0}^{\frac{n-m-3}{2}} \frac{h^{2\varrho+1}}{(2\varrho+1)!}\, y^{(m+1+2\varrho)}(x_r) + \\
&+ 2h^{n-m} \sum_{\varrho=0}^{\frac{p}{2}} \beta_{n-m,\varrho}^{**}\, \nabla^{2\varrho} F(x_{r+\varrho}) + 2R_{n-m,p}^{**}
\end{aligned}\right\} \quad\begin{aligned}&\text{for}\\&\text{even}\\&n-m\\[3ex]&\text{for}\\&\text{odd}\\&n-m.\end{aligned} \qquad (5.38)$$

The remainder term lies within the limits given in Ch. I (2.47); if it is omitted, we obtain the corresponding equations for the approximation

$y_{r+1}^{(m)}$, namely

$$
\left.
\begin{aligned}
y_{r+1}^{(m)} &= - y_{r-1}^{(m)} + 2 \sum_{\varrho=0}^{\frac{n-m-2}{2}} \frac{h^{2\varrho}}{(2\varrho)!} y_r^{(m+2\varrho)} + \\
&\quad + 2 h^{n-m} \sum_{\varrho=0}^{\frac{p}{2}} \beta_{n-m,\varrho}^{**} \nabla^{2\varrho} f_{r+\varrho}
\end{aligned}
\right\} \quad
\begin{aligned}
&\text{for} \\
&\text{even} \\
&n-m
\end{aligned}
$$

$$
\left.
\begin{aligned}
y_{r+1}^{(m)} &= y_{r-1}^{(m)} + 2 \sum_{\varrho=0}^{\frac{n-m-3}{2}} \frac{h^{2\varrho+1}}{(2\varrho+1)!} y_r^{(m+1+2\varrho)} + \\
&\quad + 2 h^{n-m} \sum_{\varrho=0}^{\frac{p}{2}} \beta_{n-m,\varrho}^{**} \nabla^{2\varrho} f_{r+\varrho}
\end{aligned}
\right\} \quad
\begin{aligned}
&\text{for} \\
&\text{odd} \\
&n-m,
\end{aligned}
$$

(5.39)

which describe the interpolation method known as the central-difference method. Normally these equations are used with the second sums truncated after the term with $\varrho = 1$; if the next term with $\varrho = 2$ were included, the unknown y_{r+2} would appear in addition to the unknown y_{r+1} and application of the method would be very involved[1]. Thus for $n = 2$, for example, we use the formulae

$$
\left.
\begin{aligned}
y_{r+1} &= 2 y_r - y_{r-1} + h^2 \left(f_r + \frac{1}{12} \nabla^2 f_{r+1} \right) \\
y_{r+1}' &= y_{r-1}' \qquad\quad + h \left(2 f_r + \frac{1}{3} \nabla^2 f_{r+1} \right).
\end{aligned}
\right\}
$$

(5.40)

The quantities appearing in these equations for the calculation of the values at x_{r+1} are framed in Table II/28, which also gives the required factors at the heads of the appropriate columns. The quantities in

---

[1] At the end of § 3.3, III we mentioned a method for first-order equations due to LINDELÖF which took into account the fourth difference $\nabla^4 f_{r+2}$; he applies the same idea to second-order equations, rewriting the formula in a similar way:

$$
y_{r+1} = 2 y_r - y_{r-1} + h^2 \left(f_r + \frac{1}{12} (\nabla^2 f_r + \nabla^3 f_r) \right) + \delta,
$$

where

$$
\delta = \frac{h^2}{12} \left(\nabla^4 f_{r+1} - \frac{1}{20} \nabla^4 f_{r+2} \right).
$$

If the calculation has proceeded as far as the point x_r, temporary values for the next two points are found from the formula

$$
y_{r+j+1}^{[0]} = 2 y_{r+j} - y_{r+j-1} + h^2 \left(f_{r+j} + \frac{1}{12} (\nabla^2 f_{r+j} + \nabla^3 f_{r+j}) \right) \quad \text{for } j = 0 \text{ and } 1.
$$

These are used to build up the difference table temporarily so that a value for δ can be calculated; this value of δ determines an improved value of y_{r+1} and the difference table can be filled in permanently up to the point x_{r+1}. LINDELÖF gives (loc. cit.) a further refinement of this procedure.

broken frames are those which are not needed when the differential equation does not depend explicitly on y', i.e. when it has the form $y'' = f(x, y)$.

Table II/28. *Central-difference method for differential equations of the second order.*
$$y'' = f(x, y, y')$$

		Factors for y	1		$\frac{1}{12}$
x		Factors for $h y'$	2		$\frac{1}{3}$
	y	$h y'$	$h^2 y'' = h^2 f$	$h^2 \nabla f$	$h^2 \nabla^2 f$
x_{r-2}	y_{r-2}	$h y'_{r-2}$	$h^2 f_{r-2}$		
x_{r-1}	y_{r-1}	$\boxed{h y'_{r-1}}$	$h^2 f_{r-1}$	$h^2 \nabla f_{r-1}$	$h^2 \nabla^2 f_r$
x_r	y_r	$h y'_r$	$\boxed{h^2 f_r}$	$h^2 \nabla f_r$	$\boxed{h^2 \nabla^2 f_{r+1}}$
x_{r+1}	y_{r+1}	$\boxed{h y'_{r+1}}$	$h^2 f_{r+1}$	$h^2 \nabla f_{r+1}$	

To use the equations (5.40) for the calculation of the currently next values y_{r+1} and y'_{r+1}, we can estimate the as yet unknown value $\nabla^2 f_{r+1}$ by extrapolation from the sequence of $\nabla^2 f$ values already calculated (as in § 3.4, I). If we take $\nabla^2 f_r + \nabla^3 f_r$ as the initial value $\nabla^2 f_{r+1}^{[0]}$, i.e. we assume the third differences to be constant and use $\nabla^3 f_r$ in place of $\nabla^3 f_{r+1}$, then we obtain for $y_{r+1}^{[1]}$ the value given by STÖRMER's extrapolation formula (5.28) truncated after the term in $\nabla^3 f$. Proceeding from this estimated value $\nabla^2 f_{r+1}^{[0]}$, we improve on it successively by means of the iterative scheme

$$\left. \begin{aligned} y_{r+1}^{[\nu+1]} &= 2 y_r - y_{r-1} + h^2 \left(f_r + \frac{1}{12} \nabla^2 f_{r+1}^{[\nu]} \right) \\ h y_{r+1}'^{[\nu+1]} &= h y'_{r-1} + h^2 \left(2 f_r + \frac{1}{3} \nabla^2 f_{r+1}^{[\nu]} \right) \end{aligned} \right\} \quad (\nu = 0, 1, 2, \ldots). \quad (5.41)$$

The step interval h is again to be chosen so small that

(a) the iterations converge sufficiently rapidly, say in one or two cycles, if possible,

(b) the terms following $\frac{h^2}{12} \nabla^2 f_{r+1}$ and $\frac{h}{3} \nabla^2 f_{r+1}$ in (5.40), namely $-\frac{h^2}{240} \nabla^4 f_{r+2}$ and $-\frac{h}{90} \nabla^4 f_{r+2}$, respectively, affect the approximations y_{r+1} and y'_{r+1} as little as possible.

For reference we write out the formulae (5.39) in detail for differential equations of up to the fourth order (for an equation of order m

use the first m formulae with $n = m$):

$$y^{(n-1)}_{r+1} = y^{(n-1)}_{r-1} + h\left(2f_r + \frac{1}{3}\nabla^2 f_{r+1}\right)$$

$$y^{(n-2)}_{r+1} = 2y^{(n-2)}_r - y^{(n-2)}_{r-1} + h^2\left(f_r + \frac{1}{12}\nabla^2 f_{r+1}\right)$$

$$y^{(n-3)}_{r+1} = y^{(n-3)}_{r-1} + 2h\,y^{(n-2)}_r + h^3\left(\frac{1}{3}f_r + \frac{1}{60}\nabla^2 f_{r+1}\right)$$ (5.42)

$$y^{(n-4)}_{r+1} = 2y^{(n-4)}_r - y^{(n-4)}_{r-1} + h^2 y^{(n-2)}_r + h^4\left(\frac{1}{12}f_r + \frac{1}{360}\nabla^2 f_{r+1}\right).$$

Example. For the initial-value problem

$$y''' = -y\,y'', \qquad y(0) = y'(0) = 0, \qquad y''(0) = 1,$$

which arises in boundary layer theory (Example III, § 2.6), we put

$$y'h = u, \qquad y''\frac{h^2}{2} = v, \qquad y'''\frac{h^3}{6} = w.$$

Translated into these quantities the equations (5.42) read

$$v_{r+1} = v_{r-1} + \qquad\qquad (6w_r + \nabla^2 w_{r+1}),$$

$$u_{r+1} = 2u_r - u_{r-1} + \left(6w_r + \frac{1}{2}\nabla^2 w_{r+1}\right),$$

$$y_{r+1} = y_{r-1} + 2u_r + \left(2w_r + \frac{1}{10}\nabla^2 w_{r+1}\right).$$

The values of y, y', y'' at the points $x = \pm 0{\cdot}1$, $\pm 0{\cdot}2$ are calculated from the power series[1]

$$y = \frac{1}{2!}x^2 - \frac{1}{5!}x^5 + \frac{11}{8!}x^8 - \frac{375}{11!}x^{11} + \cdots.$$

This enables us to fill in the computing scheme in Table II/29 down to the dotted line. The values of $\nabla^3 w$ are recorded so that we can more easily estimate the

Table II/29. *A non-linear third-order equation treated by the central-difference method*

x	y	$u = hy'$	$v = \dfrac{h^2}{2}y''$	$w = \dfrac{h^3}{6}y'''$ $= -\dfrac{h}{3}yv$	∇w	$\nabla^2 w$	$\nabla^3 w$	Estimate of the new value of $\nabla^2 w$
$-0{\cdot}2$	$0{\cdot}02000267$	$-0{\cdot}02000667$	$0{\cdot}00500667$	$-0{\cdot}000003338$				
$-0{\cdot}1$	$0{\cdot}00500008$	$-0{\cdot}01000042$	$0{\cdot}00500083$	$-\quad\ \ \ 833$	$+2505$			
0	0	0	$0{\cdot}005$	0	$+\ \ 833$	-1671		
$0{\cdot}1$	$0{\cdot}00499992$	$0{\cdot}00999958$	$0{\cdot}00499917$	$-0{\cdot}000000833$	$-\ \ 833$	-1667	$+\ 4$	
$0{\cdot}2$	$0{\cdot}01999733$	$0{\cdot}01999333$	$0{\cdot}00499334$	$-\quad\ \ 3328$	-2495	-1662	$+\ 5$	
$0{\cdot}3$	$0{\cdot}04497975$	$0{\cdot}02996628$	$0{\cdot}00497755$	$-\quad\ \ 7463$	-4134	-1639	$+23$	-1640
$0{\cdot}4$	$0{\cdot}07991495$	$0{\cdot}03989366$	$0{\cdot}00494698$	$-0{\cdot}000013178$	-5715	-1581	$+58$	-1580
$0{\cdot}5$	$0{\cdot}12474056$							

[1] The convergence of this series has been investigated by A. Ostrowski: Sur le rayon de convergence de la série de Blasius. C. R. Acad. Sci., Paris **227**, 580–582 (1948).

next value of $\nabla^2 w$. The procedure is completely analogous to that for a first-order equation as in Example II, § 3.5, so fewer lines of working are given.

Astronomers[1] employ a method in which columns of the sums $\Sigma f, \Sigma^2 f, \ldots$ are used in addition to the columns of the differences δf, $\delta^2 f, \ldots$ (these sum columns are formed in such a way that the $\Sigma^{n-1} f$ column contains the differences of the $\Sigma^n f$ column). In the case of a second-order differential equation $y'' = f(x, y)$, for instance, we build up the columns

x	y	$h^2 \Sigma^2 f$	$h^2 \Sigma f$	$h^2 f$	$h^2 \delta f$	$h^2 \delta^2 f$	$h^2 \delta^3 f$

This we do by using the formula

$$y_r = h^2 \left(\Sigma^2 f_r + \frac{1}{12} f_r - \frac{1}{240} \delta^2 f_r \right), \tag{5.43}$$

which follows by two successive summations from the first formula of (5.40) with the next term included, i.e.

$$\delta^2 y_r = h^2 \left(f_r + \frac{1}{12} \delta^2 f_r - \frac{1}{240} \delta^4 f_r \right)$$

in central-difference notation. Two arbitrary initial constants are introduced thereby and these may be determined as follows: if we have found y_r for $r = -1, 0, 1, 2$ by some starting procedure (cf. § 5.2), then we know also f_r for $r = -1, 0, 1, 2$ and hence $\delta^2 f_r$ for $r = 0, 1$; $\Sigma^2 f_r$ can then be calculated for $r = 0, 1$ from the equation (5.43) and this provides initial values for the two columns of Σf and $\Sigma^2 f$. If the main calculation has progressed as far as the values $y_r, f_r, \delta^2 f_{r-1}$, we first estimate values $\delta^2 f_r^{[0]}$ and $\delta^2 f_{r+1}^{[0]}$, then build $\delta^2 f_r^{[0]}$ up to $f_{r+1}^{[0]}$ and calculate $y_{r+1}^{[0]}$ from (5.43). This gives an improved value $f_{r+1}^{[1]} = f(x_{r+1}, y_{r+1}^{[0]})$ and hence also an improved value $\delta^2 f_r^{[1]}$, which can be used to make a better estimate $\delta^2 f_{r+1}^{[1]}$. These new values are then substituted in (5.43) to give an improved value $y_{r+1}^{[1]}$, and so on.

5.6. Convergence of the iteration in the main calculation

As in § 4.1 we investigate the convergence of the iterations involved in the interpolative integration formulae by first putting them in Lagrangian form with all differences expressed in terms of function values. Thus for the Adams interpolation method (5.32) we use the

---

[1] v. OPPOLZER, T. R.: Lehrbuch zur Bahnbestimmung der Kometen und Planeten, 2nd ed. Leipzig 1880. — HERRICK, S.: Step-by-step integration of $\ddot{x} = f(x, y, z, t)$ without a "corrector". Math. Tables and Other Aids to Computation 5, 61—67 (1951).

expression (5.37) for the $(\nu+1)$-th iterate. If we subtract the corresponding expression for the ν-th iterate, we obtain an expression for the iterative change

$$\delta^{(m)[\nu]} = y_{r+1}^{(m)[\nu+1]} - y_{r+1}^{(m)[\nu]},$$

namely

$$\delta^{(m)[\nu]} = h^{n-m} \beta_{n-m,0,p}^* (f_{r+1}^{[\nu]} - f_{r+1}^{[\nu-1]}).$$

From (5.36) and Ch. I (2.40) we see that the factors $\beta_{q,0,p}^*$ which occur here are given by

$$\beta_{q,0,p}^* = \sum_{\varrho=0}^{p} \beta_{q,\varrho}^* = \beta_{q,p}.$$

Using the Lipschitz condition (5.1), we have

$$|\delta^{(m)[\nu]}| \leq h^{n-m} \beta_{n-m,p} \sum_{\varrho=0}^{n-1} K_\varrho |\delta^{(\varrho)[\nu-1]}| \qquad (m = 0, 1, \ldots, n-1), \qquad (5.44)$$

and if these n inequalities are multiplied by $K_0, K_1, \ldots, K_{n-1}$, respectively, and summed, we find that the quantity

$$v^{[\nu]} = \sum_{\varrho=0}^{n-1} K_\varrho |\delta^{(\varrho)[\nu]}| \qquad (5.45)$$

satisfies the inequality

$$v^{[\nu]} \leq C v^{[\nu-1]}, \qquad (5.46)$$

where

$$C = h^n \beta_{n,p} K_0 + h^{n-1} \beta_{n-1,p} K_1 + \cdots + h \beta_{1,p} K_{n-1}. \qquad (5.47)$$

This last inequality has the same form as (4.4) in § 4.1 and by the same reasoning as used there it can be shown that the condition $C<1$ is sufficient for the convergence of $\sum\limits_{\nu=0}^{\infty} v^{[\nu]}$. Since we can assume the K_ϱ to be positive, this condition is also sufficient for the absolute convergence of each of the series $\sum\limits_{\nu=0}^{\infty} \delta^{(m)[\nu]}$ of the iteration process.

For $p=1$ and 2 this sufficient condition for convergence reads

(for $p=1$) $\frac{1}{2} h K_{n-1} + \frac{1}{6} h^2 K_{n-2} + \frac{1}{24} h^3 K_{n-3} + \cdots + h^n \beta_{n,1} K_0 < 1$,

(for $p=2$) $\frac{5}{12} h K_{n-1} + \frac{1}{8} h^2 K_{n-2} + \frac{7}{240} h^3 K_{n-3} + \cdots + h^n \beta_{n,2} K_0 < 1$.

The $\beta_{n,p}$ decrease as p increases and the restriction on h becomes correspondingly less strict.

The central-difference method (5.39) can be treated in precisely the same way and if fourth and higher differences are neglected, the corresponding sufficient condition for convergence is

$$2 \sum_{\nu=0}^{n} h^{n-\nu} |\beta^{**}_{n-\nu,1}| K_\nu < 1,$$

or, written out more fully,

$$\frac{1}{3} h K_{n-1} + \frac{1}{12} h^2 K_{n-2} + \frac{1}{60} h^3 K_{n-3} + \cdots + 2 h^n |\beta^{**}_{n,1}| K_0 < 1. \qquad (5.48)$$

5.7. Principle of an error estimate for the main calculation

Since error estimates for the various types of extrapolation and interpolation methods can all be derived in similar fashion, we restrict ourselves to the consideration of one particular method, the interpolation method characterized by formula (5.32). By subtracting (5.31) from (5.32) and expressing the differences in terms of function values as in (5.35), we find that the error

$$\varepsilon^{(m)}_r = y^{(m)}_r - y^{(m)}(x_r)$$

is given by

$$\varepsilon^{(m)}_{r+1} = \sum_{\nu=0}^{n-m-1} \frac{h^\nu}{\nu!} \varepsilon^{(m+\nu)}_r + h^{n-m} \sum_{\sigma=0}^{p} \beta^*_{n-m,\sigma,p} (f_{r+1-\sigma} - F(x_{r+1-\sigma})) - R^*_{n-m,p+1},$$

in which the notation (5.21) is used. Since the function f satisfies the Lipschitz condition (5.1), we have

$$|\varepsilon^{(m)}_{r+1}| \leq \sum_{\nu=0}^{n-m+1} \frac{h^\nu}{\nu!} |\varepsilon^{(m+\nu)}_r| + h^{n-m} \sum_{\sigma=0}^{p} \sum_{\nu=0}^{n-1} |\beta^*_{n-m,\sigma,p}| K_\nu |\varepsilon^{(\nu)}_{r+1-\sigma}| + |R^*_{n-m,p+1}|,$$

i.e.

$$|\varepsilon^{(m)}_{r+1}| - h^{n-m} \sum_{\nu=0}^{n-1} K_\nu \beta_{n-m,p} |\varepsilon^{(\nu)}_{r+1}| \leq e_m, \qquad (5.49)$$

where

$$e_m = \sum_{\nu=0}^{n-m-1} \frac{h^\nu}{\nu!} |\varepsilon^{(m+\nu)}_r| + h^{n-m} \sum_{\sigma=1}^{p} \sum_{\nu=0}^{n-1} K_\nu |\beta^*_{n-m,\sigma,p}| |\varepsilon^{(\nu)}_{r+1-\sigma}| + |R^*_{n-m,p+1}|.$$

In (5.49) we have a system of n inequalities satisfied by the n absolute errors $|\varepsilon^{(m)}_{r+1}|$ with right-hand sides e_m for which we know upper bounds provided that we know upper bounds for the earlier absolute errors $|\varepsilon^{(m)}_r|, |\varepsilon^{(m)}_{r-1}|, \ldots$. The theorem of Ch. I, § 5.5 which gives sufficient conditions for a monotonic matrix is applicable here; in (5.49), $1 - h^{n-m} K_m \beta_{n-m,p} > 0$ [since h will be chosen to satisfy the convergence condition $C < 1$ in (5.47)] and all other coefficients on the left-hand side are negative.

Upper limits $Y^{(m)}_r$ for the absolute errors $(Y^{(m)}_r \geq |\varepsilon^{(m)}_r|)$ can therefore be determined from the system of equations

$$Y^{(m)}_{r+1} - h^{n-m} \sum_{\nu=0}^{n-1} K_\nu \beta_{n-m,p} Y^{(\nu)}_{r+1} = E_m, \qquad (5.50)$$

where

$$E_m = \sum_{\nu=0}^{n-m-1} \frac{h^\nu}{\nu!} Y^{(m+\nu)}_r + h^{n-m} \sum_{\sigma=1}^{p} \sum_{\nu=0}^{n-1} K_\nu |\beta^*_{n-m,\sigma,p}| Y^{(\nu)}_{r+1-\sigma} + |R^*_{n-m,p+1}|.$$

Thus it is possible to carry out a recursive error estimate.

The step-by-step calculation of these error limits is tedious and does not provide a quick general guide to the growth of the error. However, we can also derive an independent error estimate as we did for first-order equations in § 4.4 and although its strict application is perhaps as tedious as the recursive estimate, we can use it crudely to get a rough idea of the growth of the error.

The derivation of the independent error estimate depends on the fact that systems of linear difference equations, of which (5.50) is an example, can be solved in the same way as a single linear difference equation (cf. § 4.4). We first obtain a particular solution of the inhomogeneous equations (5.50), namely $Y_r^{(m)} = $ constant $= \gamma_m$, then add to this solutions of the corresponding homogeneous equations. For these we assume the forms

$$Y_r^{(m)} = c_m z^r;$$

the constants c_m are then determined from the system of linear equations

$$c_m z^p - h^{n-m} \left(\sum_{\nu=0}^{n-1} K_\nu c_\nu \right) \left(\sum_{\sigma=0}^{p} |\beta^*_{n-m,\sigma,p}| z^{p-\sigma} \right) - z^{p-1} \sum_{\nu=0}^{n-m-1} \frac{h^\nu}{\nu!} c_{m+\nu} = 0 \left. \begin{array}{c} \\ \\ \end{array} \right\} \tag{5.51}$$
$$(m = 0, 1, \ldots, n-1),$$

which are obtained by substituting the above form in (5.50) and dividing through by z^{r+1-p}. These homogeneous equations will have a non-trivial solution only if their determinant

$$D = \begin{vmatrix} z^p - z^{p-1} - h^n K_0 \Phi_0(z) & -z^{p-1} \frac{h}{1!} - h^n K_1 \Phi_0(z) & -z^{p-1} \frac{h^2}{2!} - h^n K_2 \Phi_0(z) \cdots \\ \cdots - z^{p-1} \frac{h^{n-1}}{(n-1)!} - h^n K_{n-1} \Phi_0(z) & & \\ -h^{n-1} K_0 \Phi_1(z) & z^p - z^{p-1} - h^{n-1} K_1 \Phi_1(z) & -z^{p-1} \frac{h}{1!} - h^{n-1} K_2 \Phi_1(z) \cdots \\ \cdots - z^{p-1} \frac{h^{n-2}}{(n-2)!} - h^{n-1} K_{n-1} \Phi_1(z) & & \\ -h^{n-2} K_0 \Phi_2(z) & -h^{n-2} K_1 \Phi_2(z) & z^p - z^{p-1} - h^{n-2} K_2 \Phi_2(z) \cdots \\ \cdots - z^{p-1} \frac{h^{n-3}}{(n-3)!} - h^{n-2} K_{n-1} \Phi_2(z) & & \\ \cdots & \cdots & \cdots \\ -h K_0 \Phi_{n-1}(z) & -h K_1 \Phi_{n-1}(z) & -h K_2 \Phi_{n-1}(z) \cdots \\ \cdots z^p - z^{p-1} - h K_{n-1} \Phi_{n-1}(z) & & \end{vmatrix} \tag{5.52}$$

vanishes (printing limitations have made it necessary to spread each row of this determinant onto two lines). The $\Phi_m(z)$ introduced in the determinant to shorten the notation are polynomials in z defined by

$$\sum_{\sigma=0}^{p} |\beta^*_{n-m,\sigma,p}| z^{p-\sigma} = \Phi_m(z)$$

and depend upon n and p as well as upon m and z.

$D = 0$ is an algebraic equation for z, in general of the (np)-th degree, and has in general np roots, say

$$z_1, z_2, \ldots, z_{np}.$$

When these are all distinct, the general solution of the difference equation (5.50) is

$$Y_r^{(m)} = \gamma_m + \sum_{\nu=1}^{np} c_{m,\nu} z_\nu^r \quad (m = 0, 1, \ldots, n-1;\ r = 0, 1, 2, \ldots). \tag{5.53}$$

In the case of multiple roots the coefficients become polynomials in r; thus if $z_1 = z_2 = \cdots = z_q$, say, then we must use $(c_{m,1} + c_{m,2} r + \cdots + c_{m,q} r^{q-1}) z_1^r$ in place of

$$\sum_{\nu=1}^{q} c_{m,\nu} z_\nu^r.$$

Eventually, as r increases, the growth of the error limits is determined entirely by the root z with greatest absolute value; we can therefore use this root as a rough general guide to the rate at which the errors may be expected to grow. Even this is quite complicated to deal with in general so we now restrict ourselves to the important case of second-order differential equations. For these, equation (5.52) reads

$$z^{1-p} D = z^{p+1} - 2z^p + z^{p-1} - z\left(h^2 K_0 \Phi_0(z) + h K_1 \Phi_1(z)\right) + h^2 K_0 \Phi_0(z) + \left. + (h K_1 - h^2 K_0) \Phi_1(z) = 0. \right\} \tag{5.54}$$

For given p we can draw the curves $z = $ constant on a plane with $h^2 K_0$ and $h K$ as co-ordinates; these curves will in fact be straight lines since equation (5.54) is linear in $h^2 K_0$ and $h K_1$[1].

5.8. Instability of finite-difference methods

The rough approximations of § 4.7 can obviously be applied in similar fashion to differential equations of higher order so we will be brief; in fact we limit ourselves to a brief description of the calculations for a particular case, namely the central-difference method (5.40) for second-order equations[2].

With the same notation (perturbations η_r in y_r, η'_r in y'_r) and assumptions (in particular that f_y and $f_{y'}$ are constant) as in § 4.7 we obtain here the variation equations

$$\eta_{r+1} = 2\eta_r - \eta_{r-1} + \frac{h^2}{12} f_y(\eta_{r+1} + 10\eta_r + \eta_{r-1}) + \\ + \frac{h^2}{12} f_{y'}(\eta'_{r+1} + 10\eta'_r + \eta'_{r-1}),$$

$$\eta'_{r+1} = \eta'_{r-1} + \frac{h}{3} f_y(\eta_{r+1} + 4\eta_r + \eta_{r-1}) + \\ + \frac{h}{3} f_{y'}(\eta'_{r+1} + 4\eta'_r + \eta'_{r-1}).$$

Assuming the forms $\eta_r = p \lambda^r$, $\eta'_r = q \lambda^r$, we obtain two linear homogeneous equations for p and q; for a unique solution their determinant must

[1] Cf. Figure 6 in L. COLLATZ and R. ZURMÜHL: Z. Angew. Math. Mech. 22, 55 (1942).

[2] RUTISHAUSER, H.: Z. Angew. Math. Phys. 3, 65—74 (1952). — COLLATZ, L.: Z. Angew. Math. Phys. 4, 153—154 (1953).

vanish, and hence λ must satisfy the quartic

$$(\lambda - 1)\,[12(\lambda^2 - 1)\,(\lambda - 1) - A(\lambda + 1)\,(\lambda^2 + 10\lambda + 1)$$
$$- 4B(\lambda - 1)\,(\lambda^2 + 4\lambda + 1)] = 0,$$

where $A = h^2 f_y$, $B = h f_y'$.

Under the assumptions made in § 4.7, a perturbation $\eta(x)$ from $y(x)$ satisfies the differential equation

$$\eta'' = f_y \eta + f_{y'} \eta'.$$

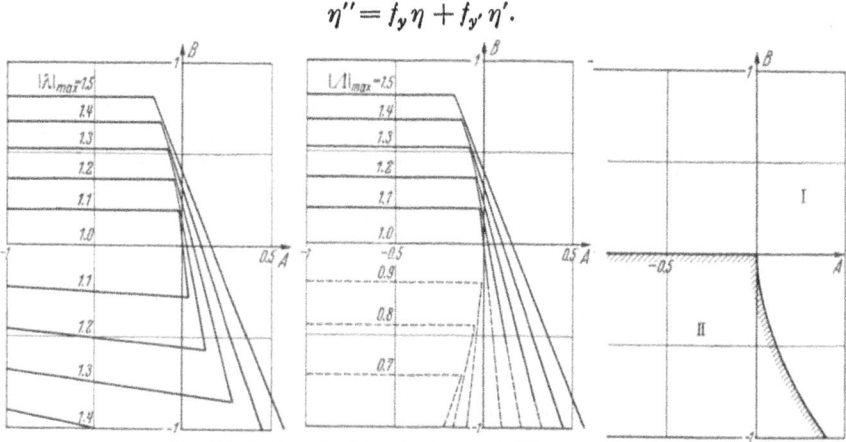

Fig. II/8. Regions of stability for the central-difference method (5.40)

This has solutions of the form $\eta(x) = e^{\mu x}$, where $\mu h = \varrho$ is a solution of the quadratic

$$\varrho^2 - B\varrho - A = 0.$$

Consequently the perturbation $\eta(x)$ increases (or decreases when $\mu < 0$) by a factor $\Lambda = e^\varrho$ over an interval of length h. If we now compare the absolute values of the roots λ with the values Λ for given A and B (Fig. II/8), we can distinguish two regions I and II in the (A, B) plane. In region I, which contains in particular the region $f_y \geqq 0$, the maximum values of $|\lambda|$ and $|\Lambda|$ coincide to a high accuracy (stable region); on the other hand, in region II there exists the danger of instability.

5.9. Reduction of initial-value problems to boundary-value problems

By raising the order of the differential equation and introducing further boundary conditions (which may be chosen with considerable arbitrariness), we can transform an initial-value problem into a boundary-value problem, which can then be treated by one of the numerous methods at our disposal[1] (cf. Ch. III). For example, by

---

[1] DE G. ALLEN, D. N., and R. T. SEVERN: The application of relaxation methods to the solution of non-elliptic partial differential equations. Quart. J. Mech. Appl. Math. **4**, 209—222 (1951).

putting $y = u'$ we can transform

$$y' = f(x, y), \qquad y(x_0) = y_0$$

into the second-order boundary-value problem

$$u'' = f(x, u'), \qquad u'(x_0) = y_0, \qquad u(x_1) = u_1,$$

where x_1 and u_1 are arbitrary.

The transformation can usually be accomplished in several ways; for instance, the initial-value problem

$$y'' = f(x), \qquad y(a) = y_a, \qquad y'(a) = y'_a$$

is equivalent to both of the boundary-value problems

$$u^{IV} = f(x); \quad u''(a) = y_a, \quad u'''(a) = y'_a, \quad u(b) = 0, \quad u'(b) = 0$$

and

$$y^{IV} = f''(x); \quad y(a) = y_a, \quad y'(a) = y'_a, \quad y''(b) = f(b), \quad y'''(b) = f'(b).$$

As yet, insufficient evidence exists for one to be able to recommend particularly any definite type of transformation.

5.10. Miscellaneous exercises on Chapter II

1. Expand the function $y(x)$ defined by

$$y' = x^2 + y^2, \qquad y(0) = -1$$

into a power series and hence calculate its values for $x = \pm 0.1, \ \pm 0.2$.

2. With the values obtained in Exercise 1 as starting values (thus with $h = 0.1$) use the Adams interpolation formula (3.11) truncated after the third difference to calculate approximations y_r at several further points.

3. Determine the "order" (in the sense of § 1.3) of DUFFING's method in § 1.5.

4. Apply the Runge-Kutta method to the following second-order problem, for which the function $f = -y^3$ does not depend on y' and the Runge-Kutta scheme therefore reduces to three rows:

Fig. II/9. Non-linear oscillations of a mass between two springs

$$y'' = -y^3; \qquad y(0) = 0.2, \qquad y'(0) = 0. \tag{5.55}$$

(This equation is satisfied by the small oscillations of a mass which is attached to two springs in the manner of Fig. II/9 and then disturbed from the position of rest[1].)

---

[1] The solution of this differential equation can be approximated in other ways, for example, by reduction to a quadrature, and can also be given in closed form in terms of elliptic functions. See, for instance, K. KLOTTER: Einführung in die Technische Schwingungslehre, 2nd ed., Vol. I, p. 138. Berlin-Göttingen-Heidelberg, Springer 1951.

The solution represents a periodic oscillation and if the period is T, it satisfies the symmetry conditions

$$y(x + T) = y(x) = y(-x) = -y\left(\frac{T}{2} - x\right).$$

Thus we need only calculate $y(x)$ for $0 \leq x \leq \dfrac{T}{4}$, i.e. as far as its first zero, for $y(x)$ is then known for all x. (Here $\frac{1}{4}T$ is approximately 10; if the quarter period is sub-divided into 5 or 10 intervals, the step interval will be $h = 2$ or $h = 1$.)

5. Calculate an approximate solution of the initial-value problem

$$y''' = -xy; \quad y(0) = 0, \quad y'(0) = 1, \quad y''(0) = 0$$

for $0 \leq x \leq 1$ using the simple extrapolation formula obtained from (5.29) by neglecting fourth and higher differences. Compare the end value with the exact value $y(1)$.

6. Use the Runge-Kutta method with $h = 0 \cdot 2$ to compute the solution for the interval $0 \leq x \leq 2$ of the problem from boundary layer theory, Example III, § 2.6:

$$y''' = -yy''; \quad y(0) = y'(0) = 0, \quad y''(0) = 1.$$

Compare the end value with the exact value given by the power series solution.

7. Apply the least-squares method of Ch. I, § 4.2 to the problem

$$y' - y = 0; \quad y(0) = 1$$

for the interval $0 \leq x \leq 1$ assuming an approximate solution of the form $w_n = \sum\limits_{\nu=0}^{n} a_\nu x^\nu$ with $a_0 = 1$. The method requires that

$$J[w_n] = \int\limits_0^1 (w_n' - w_n)^2 dx = \text{minimum}$$

and the a_ν are to be determined from the equations

$$\frac{\partial J}{\partial a_\varrho} = 0 \quad (\varrho = 1, 2, \ldots, n).$$

Compare the approximate solution for $n = 2$ and $n = 3$ with the exact solution.

5.11. Solutions of the exercises in § 5.10

1. Insertion of $y = -1 + a_1 x + a_2 x^2 + \cdots$ into the differential equation yields

$$a_1 + 2a_2 x + 3a_3 x^2 + \cdots = x^2 + 1 - 2a_1 x + x^2(a_1^2 - 2a_2) + \cdots,$$

and equating coefficients, we find that $a_1 = 1$, $a_3 = -a_1 = -1, \ldots$. To the eighth power in x the solution is

$$y = -1 + x - x^2 + \frac{4}{3} x^3 - \frac{7}{6} x^4 + \frac{6}{5} x^5 -$$

$$- \frac{37}{30} x^6 + \frac{404}{315} x^7 - \frac{3321}{2520} x^8 + \cdots,$$

which converges well for small $|x|$. The required values are given in Table II/30.

Table II/30

x	y
$-0 \cdot 2$	$-1 \cdot 2530169$
$-0 \cdot 1$	$-1 \cdot 1115133$
$0 \cdot 1$	$-0 \cdot 9088225$
$0 \cdot 2$	$-0 \cdot 8308813$

2. The results are reproduced in Table II/31, in which the values above the dotted lines are "starting values".

3. A discrepancy appears first with the term in h^4. The method is therefore of the third order[1].

---

[1] See, for instance, F. A. WILLERS: Methoden der praktischen Analysis, p. 307. Berlin and Leipzig 1928, or in the translation: Practical Analysis, p. 377. New York, Dover Publications, Inc. 1948.

Table II/31.

Adams interpolation formula (3.11) *applied to a non-linear first-order equation*

x	y	$hf(x,y)$	$h\nabla f$	$h\nabla^2 f$	$h\nabla^3 f$	$h\nabla^4 f$
−0·2	−1·2530169	0·16100514				
−0·1	−1·1115133	0·12454618	−0·03645896	0·01191278		
0	−1	0·1	−0·02454618	0·00814201	−0·00377077	
0·1	−0·9088225	0·08359583	−0·01640417	0·00584471	−0·00229730	0·00147347
0·2	−0·8308813	0·07303637	−0·01055946			0·0009192
				0·0044666	−0·0013781	0·0004730
0·3	−0·7612061	0·06694347	−0·0060929	0·0035615	−0·0009051	0·0003263
0·4	−0·6957879	0·06441208	−0·0025314	0·0029827	−0·0005788	
0·5	−0·6313749	0·06486343	+0·0004513			

4. Two calculations are carried out with $h = 1$ and $h = 2$, respectively; they are continued until y changes sign. The results are given in Tables II/32, II/33; all three lines of the Runge-Kutta calculation are given for each step up to $x = 2$ but for the remaining steps only the first line is reproduced.

Table II/32. *A non-linear oscillation computed by the Runge-Kutta method with* $h = 1$

x	y	y'	$k_j = -\frac{1}{2}y^3$	$\begin{matrix}k\\k'\end{matrix}$
0	**0·2**	**0**	−0·004	
0·5	0·199		−0·0039403	−0·0039602
1	0·1960597		— 37683	−0·0078432
1	**0·1960398**	**−0·0078432**	−0·0037671	
1·5	0·1911764		— 34936	−0·0035848
2	0·1847030		— 31506	−0·0069640
2	**0·1846118**	**−0·0148072**	−0·0031458	
3	0·1669257	−0·0202935	— 23256	
4	0·1445867	−0·0241146	— 15113	
5	0·1191990	−0·0264402	— 8468	
6	0·0920821	−0·0276418	— 3904	
7	0·0641510	−0·0281344	— 1320	
8	0·0359308	−0·0282697	— 232	
9	0·0076499	−0·0282844	— 2	
10	−0·0206345	−0·0282829	+	

Table II/33. *The same computation with* $h = 2$

x	y	$v = 2y'$	$k_j = -2y^3$	$\begin{matrix}k\\k'\end{matrix}$
0	**0·2**	**0**	−0·016	
1	0·196		−0·0150591	−0·0153727
2	0·1849409		— 126511	−0·0296292
2	**0·1846273**	**−0·0296292**	— 125868	
4	0·1446297	−0·0482341	— 60507	
6	0·0921318	−0·0552814	— 15641	
8	0·0359779	−0·0565370	— 931	
10	−0·0205907	−0·0565634		

5. The requisite starting values, separated from the following values by a dotted line in Table II/34, can be calculated by means of the power series

$$y = x - \frac{2 x^5}{5!} + \frac{2 \cdot 6 x^9}{9!} - \frac{2 \cdot 6 \cdot 10 x^{13}}{13!} + - \cdots ;$$

Table II/34. *Application of the extrapolation formula* (5.29)

x	y	$f = -xy$
$-0\cdot2$	$-0\cdot19999467$	$-0\cdot039998934$
0	0	0
$0\cdot2$	$0\cdot19999467$	$-0\cdot039998934$
$0\cdot4$	$0\cdot39982934$	$-0\cdot159931736$
$0\cdot6$	$0\cdot59870429$	$-0\cdot359225728$
$0\cdot8$	$0\cdot79454288$	$-0\cdot635634304$
1	$0\cdot98336567$	

this series is also used to calculate the exact values in Table II/35. The main calculation is carried out with the formula

$$y_{r+1} = 3 y_r - 3 y_{r-1} + y_{r-2} + \frac{h^3}{2} (f_r + f_{r-1})$$

with $h = 0\cdot2$.

Table II/35. *Exact values*

x	y
$0\cdot4$	$0\cdot399829339$
1	$0\cdot983366383$

6. Table II/36 gives the results of the calculation. The power series is used to calculate

$$y(0\cdot8) = 0\cdot3173143, \quad y(1) = 0\cdot4919304, \quad y(2) = 1\cdot78809.$$

Table II/36. *An accurate Runge-Kutta solution of the Blasius equation*

x	y	$y'h$	$y'' \dfrac{h^2}{2}$	x	y	$y'h$	$y'' \dfrac{h^2}{2}$
0	0	0	0·02	1·2	0·7003698	0·2241645	0·0150553
0·2	0·0199973	0·0399867	0·0199733	1·4	0·9388576	0·2520533	0·0127849
0·4	0·0799148	0·0797873	0·0197879	1·6	1·2028830	0·2751759	0·0103241
0·6	0·1793564	0·1189319	0·0192940	1·8	1·4875610	0·2933679	0·0078912
0·8	0·3173139	0·1566757	0·0183708	2	1·7880597	0·3068941	0·0056885
1	0·4919296	0·1920829	0·0169527				

7. For $n = 3$ we obtain the equations

$$-\frac{1}{2} + \frac{1}{3} a_1 + \frac{1}{4} a_2 + \frac{1}{5} a_3 = 0,$$

$$-\frac{2}{3} + \frac{1}{4} a_1 + \frac{8}{15} a_2 + \frac{2}{3} a_3 = 0,$$

$$-\frac{3}{4} + \frac{1}{5} a_1 + \frac{2}{3} a_2 + \frac{33}{35} a_3 = 0$$

with the solution
$$a_1 = \frac{9000}{8884}, \qquad a_2 = \frac{3780}{8884}, \qquad a_3 = \frac{2485}{8884}.$$

Table II/37 gives w_2 and w_3 for several values of x, together with their errors (by comparison with the exact solution $y = e^x$). For comparison, w_1 is also included.

Table II/37. *An application of the least-squares method*

x	$y = e^x$	$w_1(x)$	$w_2(x)$	Error $w_2 - y$	$w_3(x)$	Error $w_3 - y$
0·2	1·221 403	1·3	1·207 23	−0·014 17	1·221 869	+0·000 466
0·4	1·491 825	1·6	1·481 93	−0·009 89	1·491 202	−0·000 623
0·6	1·822 119	1·9	1·824 10	+0·001 98	1·821 427	−0·000 692
0·8	2·225 541	2·2	2·233 74	+0·008 20	2·225 970	+0·000 429
1	2·718 282	2·5	2·710 85	−0·007 43	2·718 258	−0·000 024

Chapter III

Boundary-value problems in ordinary differential equations

§ 1. The ordinary finite-difference method

The basis of the finite-difference method is the replacement of all derivatives occuring by the corresponding difference quotients; this is applicable to any problem in differential equations.

1.1. Description of the finite-difference method

We divide the interval $\langle a, b \rangle$ in which the solution $y(x)$ of the boundary-value problem is sought into n equal parts of length

$$h = \frac{b - a}{n}.$$

The points $x_i = a + ih$ are called the "pivotal points" and h is called the "pivotal interval" or merely the "interval".

We characterize the value of a function at the point $x_i = a + ih$ by a subscript i (i not necessarily integer); thus y_i denotes the value of the required function $y(x)$ at the point x_i. Further we denote an approximation to y_i by Y_i (see Fig. III/1).

The first stage of the finite-difference method is to set up a system of "finite equations" from which the Y_i can be determined. A finite equation is obtained by writing down the differential equation for a pivotal point x_i and replacing all derivatives which occur by difference quotients in some specified way; each derivative is thereby represented by a certain linear expression in the Y_i.

Thus (cf. Ch. I, § 2.1) we can replace the derivative $y'(x_i)$ by

$$\frac{Y_{i+1} - Y_{i-1}}{2h}, \tag{1.1}$$

the second derivative $y''(x_i)$ by

$$\frac{Y_{i+1} - 2Y_i + Y_{i-1}}{h^2},\tag{1.2}$$

a derivative $y^{(k)}(x_i)$ of even order k by the k-th difference quotient

$$\frac{1}{h^k}\, \Delta^k Y_{i-\frac{k}{2}},\tag{1.3}$$

and a derivative $y^{(p)}(x_i)$ of odd order p by the arithmetic mean of two p-th difference quotients:

$$\frac{1}{2}\, \frac{1}{h^p}\left(\Delta^p Y_{i-\frac{p+1}{2}} + \Delta^p Y_{i-\frac{p-1}{2}}\right).\tag{1.4}$$

The boundary conditions also are put into the form of finite equations; these must also be included in the system of equations for the Y_i.

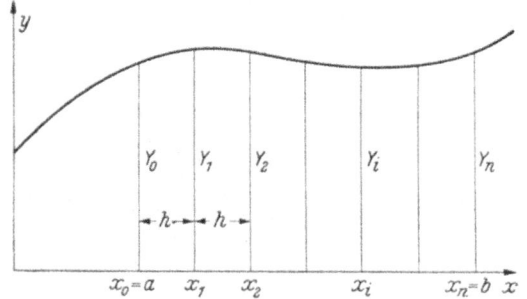

Fig. III/1. Notation used in the finite-difference method

For example, if we have a boundary-value problem of even order k with k boundary conditions at the points $x = a$ and $x = b$, we can set up a finite equation for each of the pivotal points x_i (for $i = 0, 1, \ldots, n$) and likewise for each boundary condition. In general, besides the pivotal values Y_i inside the interval $\langle a, b \rangle$, values of Y at the exterior points $x_{k/2}, x_{k/2+1}, \ldots, x_{-1}, x_{n+1}, \ldots, x_{n+k/2}$ will be involved; thus we have $n + k + 1$ equations for the same number of unknowns

$$Y_{-\frac{k}{2}}, \quad Y_{-\frac{k}{2}+1}, \ldots, \quad Y_{n+\frac{k}{2}}.$$

This system of finite equations for the Y_i is linear if and only if the boundary-value problem is linear. Nothing can be asserted about the solvability of the system without being more specific about the boundary-value problem.

Note. We can often specify finite-difference representations other than those expressed in (1.1) to (1.4) and thereby arrive at different finite equations. For example, the term $(f y')'$ can be dealt with either by differentiating the product and replacing y' and y'' by their finite-difference representations (1.1) and (1.2):

$$f'_i \frac{Y_{i+1} - Y_{i-1}}{2h} + f_i \frac{Y_{i+1} - 2Y_i + Y_{i-1}}{h^2},$$

or by replacing the two differentiations in $(fy')'$ by the finite-difference representation (1.1):

$$\frac{f_{i+1}\dfrac{Y_{i+2}-Y_i}{2h}-f_{i-1}\dfrac{Y_i-Y_{i-2}}{2h}}{2h}.$$

We can halve the interval in this last expression, which then reads

$$\frac{f_{i+\frac{1}{2}}(Y_{i+1}-Y_i)-f_{i-\frac{1}{2}}(Y_i-Y_{i-1})}{h^2}=\frac{1}{h^2}\,\Delta\,(f_{i-\frac{1}{2}}\Delta Y_{i-1}). \tag{1.5}$$

The equations which can be set up in these various ways will in general yield slightly different results.

1.2. Examples of boundary-value problems of the second order

The following worked examples are intended to familiarize the reader with the practical application of the finite-difference method and to introduce several little artifices and expedients which can often be employed with advantage. A further example will be found in Exercise 1 of § 8.11.

I. A linear boundary-value problem of the second order. Consider the bending of a strut (Fig. III/2) with flexural rigidity $EJ(\xi)$ and axial compressive

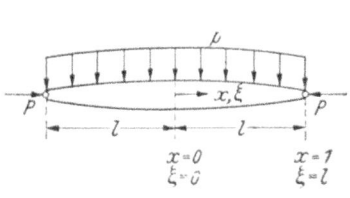

Fig. III/2. The bending of a strut

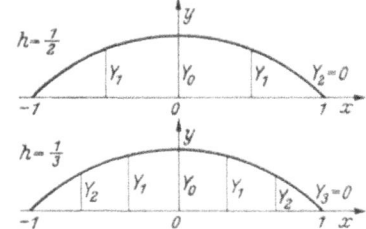

Fig. III/3. Finite-difference notation for the bending moment distribution

load P by a distributed transverse load $p(\xi)$, ξ being the co-ordinate along the axis of the strut. The bending moment distribution $M(\xi)$ satisfies the differential equation

$$\frac{d^2M}{d\xi^2}+\frac{P}{EJ(\xi)}\,M=-p(\xi).$$

We will assume here that the transverse load is a constant p but that the flexural rigidity is a variable

$$EJ(\xi)=\frac{EJ_0}{1+\left(\dfrac{\xi}{l}\right)^2}.$$

If we take $P=\dfrac{EJ_0}{l^2}$ and introduce dimensionless variables $x=\dfrac{\xi}{l}$, $y=\dfrac{M}{l^2p}$, the differential equation becomes

$$y''+(1+x^2)\,y=-1, \tag{1.6}$$

where dashes denote differentiation with respect to x. We take $M=0$ at each end, corresponding to smoothly-hinged end supports, so that the boundary conditions are

$$y(-1)=y(1)=0. \tag{1.7}$$

The interval $\langle-1,1\rangle$ is first subdivided rather coarsely into four, so that $h=\frac{1}{2}$ (Fig. III/3); then, making use of the boundary conditions

$(Y_{\pm 2}=0)$ and the symmetry $(Y_1=Y_{-1})$, we have

$$\text{(for } x=0)\qquad \frac{Y_1-2Y_0+Y_1}{h^2}+Y_0=-1,$$

$$\text{(for } x=\tfrac{1}{2})\qquad \frac{0-2Y_1+Y_0}{h^2}+\frac{5}{4}Y_1=-1.$$

With $h^2=\tfrac{1}{4}$ the solution of these two simultaneous equations is

$$Y_0=\frac{59}{61}=0\cdot967,\qquad Y_1=\frac{44}{61}=0\cdot721.$$

Thus, even though it may be rather rough, we have already obtained some idea of the bending moment distribution and with a trivial amount of computation.

To obtain more accurate values we must choose a smaller pivotal interval; thus if we take $h=\tfrac{1}{3}$ (pivotal values Y_0, Y_1, Y_2 as in Fig. III/3), which gives the system of equations

$$9(2Y_1-2Y_0)\qquad +\qquad Y_0=-1,$$

$$9(Y_2-2Y_1+Y_0)+\frac{10}{9}Y_1=-1,$$

$$9(Y_1-2Y_2)\qquad +\frac{13}{9}Y_2=-1.$$

we obtain the values

$$Y_0=\frac{53347}{56237}=0\cdot9486,\qquad Y_1=\frac{47259}{56237}=0\cdot8404,\qquad Y_2=\frac{29088}{56237}=0\cdot5172.$$

When a large number of subdivisions is used, we do not treat the difference equations as a set of simultaneous equations but take advantage of the linearity of the problem and treat them as an initial-value problem (cf. § 4.3). We start from a value $Y_0^{(1)}=1$ and calculate successively from the difference equations the values $Y_1^{(1)}$, $Y_2^{(1)}$, ..., $Y_n^{(1)}$ (for the case $h=1/n$), then repeat the procedure, starting from a different initial value $Y_0^{(2)}=0$, say, and so obtain a second set of values $Y_1^{(2)}$, $Y_2^{(2)}$, ..., $Y_n^{(2)}$. Then the linear combination

$$Y_j=\frac{Y_i^{(1)}Y_n^{(2)}-Y_i^{(2)}Y_n^{(1)}}{Y_n^{(2)}-Y_n^{(1)}}$$

of these two sets of values is the required solution; for it satisfies the difference equations on account of their linearity and also satisfies both the boundary conditions. In the example under consideration we obtain for $h=0\cdot1$ the values exhibited in Table III/1.

Improvement by h^2-extrapolation. If the finite-difference method has been applied to a particular problem with several different values of h, the results so obtained can often be improved in the following way.

We assume that at a fixed point x the error in the finite-difference method tends to zero quadratically (for the present case) with h (cf.

Table III/1. *The linear combination of two independent solutions of the difference equations*

i	$Y_i^{(1)}$	$Y_i^{(2)}$	Y_i	i	$Y_i^{(1)}$	$Y_i^{(2)}$	Y_i
0	1	0	0·933 591	6	0·641 440	−0·174 096	0·587 281
1	0·99	−0·005	0·923 923	7	0·513 688	−0·233 539	0·464 065
2	0·960 001	−0·019 950	0·894 924	8	0·368 282	−0·299 502	0·323 935
3	0·910 018	−0·044 692	0·846 617	9	0·206 836	−0·370 553	0·168 492
4	0·840 116	−0·078 946	0·779 082	10	0·031 647	−0·444 898	0
5	0·750 468	−0·122 286	0·692 510				

§ 3.3). We can then write approximately

$$y - Y \approx Ch^2, \qquad y - Y^* \approx Ch^{*2},$$

where Y and Y^* are the values calculated with the intervals h and h^*, respectively. Eliminating the constant C, we derive a new approximate value

$$\overline{Y} = \frac{Yh^{*2} - Y^*h^2}{h^{*2} - h^2} = Y + \frac{h^2}{h^{*2} - h^2}(Y - Y^*), \qquad (1.8)$$

which will in general be a better value. Table III/2 shows the result of applying this h^2-extrapolation procedure in the present example to the values at $x = 0$. One would therefore take $y(0) \approx 0.9321$ as the new approximate value.

Table III/2. h^2-extrapolation

h	Y	
$\frac{1}{2}$	0·967 2	$\overline{Y} = 0.9337$
$\frac{1}{3}$	0·948 61	$\overline{Y} = 0.932 106$
$\frac{1}{10}$	0·933 591	

II. A non-linear boundary-value problem of the second order. The boundary-value problem

$$y'' = \frac{3}{2}y^2; \qquad y(0) = 4, \qquad y(1) = 1 \qquad (1.9)$$

possesses two solutions, one of which, namely

$$y = \frac{4}{(1+x)^2}, \qquad (1.10)$$

is expressible in elementary terms while the other involves elliptic functions.

Here also we start by using a coarse subdivision of the interval to get a rough idea of the solution. With $h = \frac{1}{2}$ we have the single non-linear equation

$$\frac{4 - 2Y_1 + 1}{h^2} = \frac{3}{2} Y_1^2,$$

which gives for $y(\frac{1}{2})$ the approximation

$$Y_1 = \frac{1}{3}\left(-8 \pm \sqrt{184}\right) = \begin{cases} 1\cdot8549 \ (\text{error} +4\cdot3\%) \\ -7\cdot188. \end{cases}$$

With $h = \frac{1}{3}$ we have two non-linear equations

$$9\,(4 - 2Y_1 + Y_2) = \frac{3}{2} Y_1^2,$$

$$9\,(Y_1 - 2Y_2 + 1) = \frac{3}{2} Y_2^2$$

for the unknowns Y_1, Y_2 (see Fig. III/4). These equations represent two parabolas in the (Y_1, Y_2) plane (see Fig. III/5) and their points of intersection give the required approximate values:

$$Y_1 = 2\cdot2950 \ (\text{error} + 2\cdot0\%)$$
$$Y_2 = 1\cdot4680 \ (\text{error} + 1\cdot9\%)$$

and

$$Y_1 = -4\cdot70$$
$$Y_2 = -9\cdot72.$$

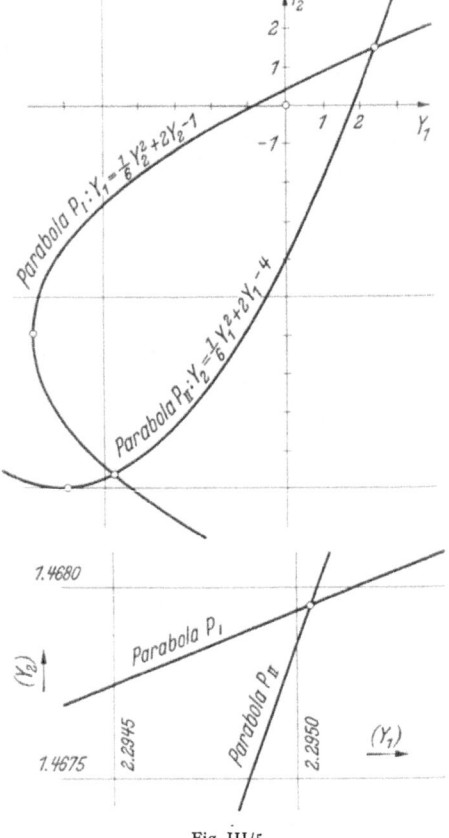

Fig. III/4   Fig. III/5

Fig. III/4. Notation for the finite-difference method applied to the non-linear problem of Example II

Fig. III/5. Solution of the algebraic equations

For finer subdivisions we treat the problem as an initial-value problem, as in Example I, by starting from $Y_0 = 4$ and a guessed value of Y_1 and calculating the remaining Y_i from the difference equation. This is repeated with several different values of Y_1 so that we can interpolate between them for a value which will give a value of Y_n satisfying the prescribed boundary condition $Y_n = 1$. The calculations with a coarse subdivision can be used to find a rough value of Y_1 for a finer subdivision.

With $h = \frac{1}{5}$ we try $Y_1 = 3$ and $Y_1 = 2 \cdot 8$ and calculate the first two Y_i columns of Table III/3 from the difference equations

$$Y_{i+1} = 0 \cdot 06\, Y_i^2 + 2 Y_i - Y_{i-1} \qquad (i = 1, 2, 3, 4). \tag{1.11}$$

Table III/3. *Solution of a non-linear problem by interpolation*

i	Y_i				
(-1)				$(6 \cdot 165\,36)$	
0	4	4	4	4	4
1	3	2·8	2·795 3	2·794 64	− 2·5138
2	2·540	2·0704	2·0594	2·057 87	− 8·6484
3	2·467	1·5980	1·5780	1·575 19	−10·2953
4	2·759	1·2788	1.2460	1·241 38	− 5·5826
5	3·509	1·0577	1·007 1	1·000 03	1

From the end values 3·509 and 1·0577 we extrapolate linearly to find the better value 2·7953 for Y_1, which we use to build up the better approximations in the third Y_i column; further interpolation yields $Y_1 = 2 \cdot 794\,64$ (from this we build up the fourth column; the extra bracketed value at x_{-1} is calculated for an error estimate in § 3.4). The last column contains the results of applying the same method to the other solution.

III. An eigenvalue problem. Consider the longitudinal vibrations of a cantilever (Fig. III/6). Let its length be l, density ϱ, modulus of elasticity E and area of cross-section $F(x)$, where the co-ordinate x is chosen along its axis with the origin at the free end. The displacement y satisfies the differential equation

$$- E(F') y' = \omega^2 \varrho F y$$

and the boundary conditions

$$y'(0) = 0, \qquad y(l) = 0.$$

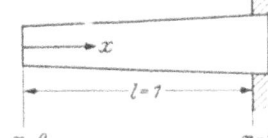

Fig. III/6. Longitudinal vibrations of a cantilever

It is required to find the natural frequencies ω. Many other physical problems, for example, the torsional vibrations of shafts, give rise to eigenvalue problems of similar form.

Let the cross-sectional area increase linearly with x from F_0 at the tip to $2 F_0$ at the base:

$$F(x) = F_0 \left(1 + \frac{x}{l}\right);$$

then with $\lambda = \omega^2 \dfrac{\varrho}{E}$ and $l = 1$ we have the fully homogeneous problem

$$- (1 + x) y'' - y' = - [(1 + x) y']' = \lambda (1 + x) y; \qquad y'(0) = 0, \qquad y(1) = 0.$$

The finite-difference method can be applied in various ways. To begin with, there are several different forms of finite equation which can be used (cf. the note in § 1.1); we choose the form

$$(1 + jh) \frac{Y_{j+1} - 2Y_j + Y_{j-1}}{h^2} + \frac{Y_{j+1} - Y_{j-1}}{2h} + \Lambda(1 + jh) Y_j = 0 \qquad (1.12)$$
$$(j = 0, 1, \ldots, n - 1),$$

where $h = 1/n$ and Λ is an approximate value for λ. Corresponding to the boundary conditions we put

$$\frac{Y_1 - Y_{-1}}{2h} = 0, \quad \text{i.e.} \quad Y_{-1} = Y_1, \quad \text{and} \quad Y_n = 0.$$

Thus we have n linear homogeneous equations for the n unknowns $Y_0, Y_1, \ldots, Y_{n-1}$. For a non-trivial solution the determinant of the coefficients must vanish. This requirement yields an algebraic equation of the n-th degree for Λ, whose n roots $\Lambda_1, \Lambda_2, \ldots, \Lambda_n$, arranged in increasing order of magnitude, are regarded as approximations to the first n eigenvalues $\lambda_1, \lambda_2, \ldots, \lambda_n$.

With $h = \frac{1}{2}$, for example, we obtain the homogeneous equations

$$(-8 + \Lambda) Y_0 + 8 Y_1 = 0,$$
$$5 Y_0 + \left(-12 + \frac{3}{2} \Lambda\right) Y_1 = 0.$$

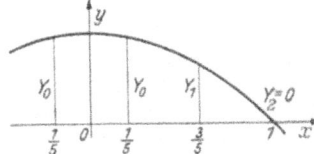

Fig. III/7. A subdivision which yields a more accurate finite-difference representation of the boundary condition $y'(0) = 0$

The condition for a non-trivial solution yields

$$0 = \begin{vmatrix} -8 + \Lambda & 8 \\ 5 & -12 + \frac{3}{2}\Lambda \end{vmatrix}$$
$$= \frac{3}{2} \Lambda^2 - 24\Lambda + 56,$$

from which we obtain

$$\Lambda = 8 \pm \frac{4}{3} \sqrt{15}, \quad \text{i.e.} \quad \begin{cases} \Lambda_1^{(2)} = 2{\cdot}836 \quad (\text{error} - 12\%) \\ \Lambda_2^{(2)} = 13{\cdot}164 \quad (\text{error} -43\%). \end{cases}$$

Here $\Lambda_k^{(m)}$ denotes the approximation to the k-th eigenvalue given by the calculation with interval $h = 1/m$.

A variation is to use a subdivision of the x axis which does not include $x = 0$ as one of its points; such a subdivision is obtained, for example, if we mark off points from $x = 1$ with $h = \frac{2}{5}$, as in Fig. III/7. The corresponding finite equations read

$$-7 Y_0 + 7 Y_1 + \Lambda \frac{24}{25} Y_0 = 0, \qquad 7 Y_0 - 16 Y_1 + \Lambda \frac{32}{25} Y_1 = 0$$

and yield the approximate values

$$\Lambda = \frac{25}{48} \left(19 \pm \sqrt{172}\right) = \begin{cases} 3{\cdot}0651 \quad (\text{error} \ -4{\cdot}8\%) \\ 16{\cdot}727 \quad (\text{error} \ -27 \ \%). \end{cases}$$

In this way we represent the boundary condition $y' = 0$ at $x = 0$ more accurately than when $x = 0$ is a pivotal point; this is shown by the fact that the above value approximates the first eigenvalue more closely than the value obtained from a

Table III/4. Solution of an eigenvalue problem (longitudinal vibrations of a cantilever) by interpolation

ϱ	Y_3	Y_2	Y_1	Y_0	Y_{-1}	$Y_{-1}-Y_1$	By interpolation	Error in $\Lambda^{(5)}$
1·88	1·990 59	2·858 46	3·490 45	3·780 41	3·630 76	0·140 31		
1·875	1·985 294	2·837 255	3·438 349	3·679 867	3·463 963	0·025 614	$\varrho_1 = 1·87387$; $\Lambda_1^{(5)} = 3·1533$	$-2·0\%$
1·874	1·984 235	2·833 020	3·427 968	3·659 899	3·430 984	0·003 016		
1·2	1·270 59	0·493 02	$-0·828 93$	$-1·667 81$	$-1·210 60$	$-0·381 67$		
1·16	1·228 235	0·386 403	$-0·934 488$	$-1·639 210$	$-0·970 608$	$-0·036 120$	$\varrho_2 = 1·15579$; $\Lambda_2^{(5)} = 21·105$	-8%
1·1558	1·223 788	0·375 418	$-0·944 777$	$-1·634 919$	$-0·944 872$	$-0·000 095$		
0	0	$-1·133 333$	0	1·339 394	0	0	$\varrho_3 = 0$; $\Lambda_3^{(5)} = 50$	-20%
$-1·1558$	$-1·223 788$	0·375 418	0·944 777	$-1·634 919$	0·944 872	0·000 095	$\varrho_4 = -1·15579$; $\Lambda_4^{(5)} = 78·9$	-35%
$-1·874$	$-1·984 235$	2·833 020	$-3·427 968$	3·659 899	$-3·430 984$	$-0·003 016$	$\varrho_5 = -1·87387$; $\Lambda_5^{(5)} = 96·8$	-52%

calculation with $h = \frac{1}{3}$ in which $x = 0$ is a pivotal point, even though this interval

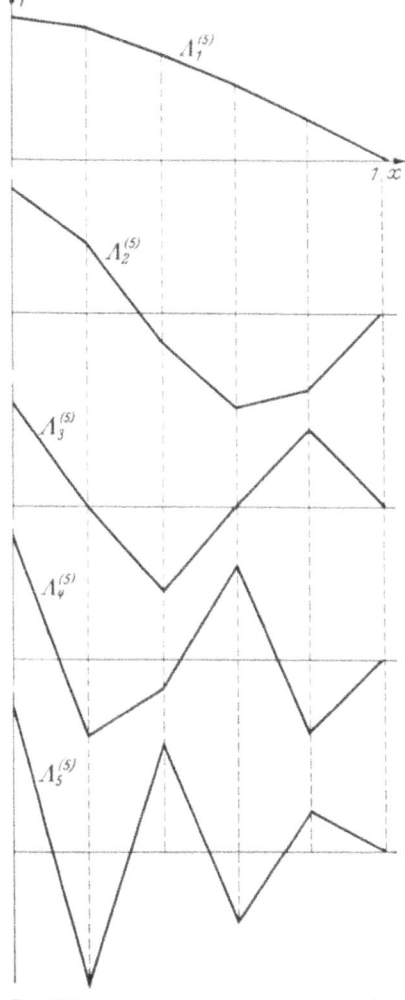

Fig. III/8. Approximations to the normal modes of vibration (longitudinal) of a cantilever

is smaller. The results for $h = \frac{1}{3}$ are

$$\Lambda_1^{(3)} = 3·0413 \ (\text{error} \ -5·5\%),$$
$$\Lambda_2^{(3)} = 18 \ (-22\%),$$
$$\Lambda_3^{(3)} = 32·96 \ (-48\%).$$

Here also it is convenient to re-duce the boundary-value problem to an initial-value problem when a large number of pivotal points are used. We estimate the value of Λ from the calculations with a coarse subdivision and then, starting from $Y_n=0$ and $Y_{n-1}=1$ (a factor is disposable here), use (1.12) to calculate $Y_{n-2}, Y_{n-3}, \ldots$. If we repeat the calculation with slightly different values of Λ, it is usually quite feasible to interpolate among them to locate the value which gives $Y_1=Y_{-1}$.

If we put $2-\Lambda h^2 = \varrho$, the equations (1.12) for $h=\frac{1}{5}$ read

$$
\begin{aligned}
1\cdot 7\,Y_3 &= Y_4\varrho \times 1\cdot 8 \\
1\cdot 5\,Y_2 &= Y_3\varrho \times 1\cdot 6 - 1\cdot 7\,Y_4 \\
1\cdot 3\,Y_1 &= Y_2\varrho \times 1\cdot 4 - 1\cdot 5\,Y_3 \\
1\cdot 1\,Y_0 &= Y_1\varrho \times 1\cdot 2 - 1\cdot 3\,Y_2 \\
0\cdot 9\,Y_{-1} &= Y_0\varrho \qquad\quad - 1\cdot 1\,Y_1.
\end{aligned}
$$

The calculations for the first two eigenfunctions using these equations in the above manner are given in Table III/4. From the form of the equations it can be seen that changing the sign of ϱ merely changes the signs of Y_3, Y_1, Y_{-1} and leaves Y_2, Y_0 unaltered. These sign changes therefore provide us with two more eigenfunctions without any further calculation; and since $\varrho = 0$ gives $Y_1 = Y_{-1} = 0$, this value yields a fifth eigenfunction (cf. the table). These approximations to the first five exact eigenfunctions are depicted in Fig. III/8.

IV. Infinite interval. As an example with a boundary condition at infinity we consider the boundary-value problem

$$
y'' = \frac{1+x}{2+x}\,y; \qquad y(0) = 1, \qquad y(\infty) = 0.
$$

Here $y(x)$ may be interpreted as the temperature difference between an infinitely long rod and its surroundings, one end $x=0$ of the rod being kept at unit temperature and the surroundings at zero temperature. The heat loss to the surroundings is assumed to be proportional to $\varphi(x)\,y$, where $\varphi = \dfrac{1+x}{2+x}$.

By means of the difference equations

$$
Y_{i+1} - 2\,Y_i + Y_{i-1} - h^2\,\frac{1+ih}{2+ih}\,Y_i = 0 \qquad (i = 1, 2, \ldots)
$$

we can express successively $Y_2, Y_3, \ldots$ as functions (linear when the differential equation is linear, as here) of Y_1; this first unknown pivotal value is then to be determined by the boundary condition $y(\infty) = 0$. Such a boundary condition can be translated into a finite condition in various ways:

1. If we replace the condition $y(\infty) = 0$ by $Y_n = 0$, we obtain an approximate value for Y_1, and hence also for $Y_2, Y_3, \ldots, Y_{n-1}$, depending on n. The values of Y_1 for a series of values of n have differences which approximate closely to a geometric sequence, as is shown in Tables III/5 and III/6 for $h=1$ and $h=\frac{1}{2}$. This facilitates the extrapolation to $n = \infty$ and indicates that the accuracy of this extrapolation, which yields the results $y(1) \approx 0\cdot 447$ and $0\cdot 444$ for $h=1$ and $\frac{1}{2}$, respectively,

Table III/5. *Solution by extrapolation from finite boundary conditions.* $h = 1$

Y_n	$Y_n = 0$ yields	Differences	Extrapolation
$Y_2 = \dfrac{8}{3} Y_1 - 1$	$Y_1 = 0\cdot375$		
		$0\cdot060$	
$Y_3 = \dfrac{19}{3} Y_1 - \dfrac{11}{4}$	$Y_1 = 0\cdot435$		
		$0\cdot010$	$Y_1 = 0\cdot4470$
$Y_4 = \dfrac{226}{15} Y_1 - \dfrac{67}{10}$	$Y_1 = 0\cdot4447$		
		$0\cdot0018$	
$Y_5 = \dfrac{1636}{45} Y_1 - \dfrac{487}{30}$	$Y_1 = 0\cdot44652$		

Table III/6. *Solution by extrapolation from finite boundary conditions.* $h = \frac{1}{2}$

Y_n	$Y_n = 0$ yields	Differences	Extrapolation
$Y_2 = 2\cdot15 Y_1 - 1$	$Y_1 = 0\cdot465$		
$Y_3 = 3\cdot6583\, Y_1 - 2\cdot1667$	$Y_1 = 0\cdot593$		
$Y_4 = 5\cdot8200\, Y_1 - 3\cdot7202$	$Y_1 = 0\cdot6392$	$0\cdot046$	$Y_1 = 0\cdot6718$
$Y_5 = 9\cdot0728\, Y_1 - 5\cdot9714$	$Y_1 = 0\cdot6582$	$0\cdot0190$	and hence
$Y_6 = 14\cdot090 Y_1 - 9\cdot3836$	$Y_1 = 0\cdot6660$	$0\cdot0078$	$Y_2 = 0\cdot4444$
$Y_7 = 21\cdot925 Y_1 - 14\cdot672$	$Y_1 = 0\cdot6692$	$0\cdot0032$	

is at least as good as the finite-difference method used. To acquire greater accuracy and confidence in the results one would repeat the calculations with smaller h. Fig. III/9 shows the approximate solution with $h = \frac{1}{2}$.

2. For large values of x the solution behaves like constant $\times e^{-x}$. (In complicated cases we put $1/x = u$ and study the behaviour of the solutions of the differential equation for small u.) Thus for large n we have

$$Y_n \approx A\, e^{-nh},$$
$$Y_{n+1} \approx A\, e^{-(n+1)h},$$

which suggests that we use

$$Y_{n+1} - e^{-h} Y_n = 0$$

Table III/7. *Method 2*

n	Y_1
2	$0\cdot665$
3	$0\cdot6681$
4	$0\cdot6702$
5	$0\cdot6710$
6	$0\cdot6713$

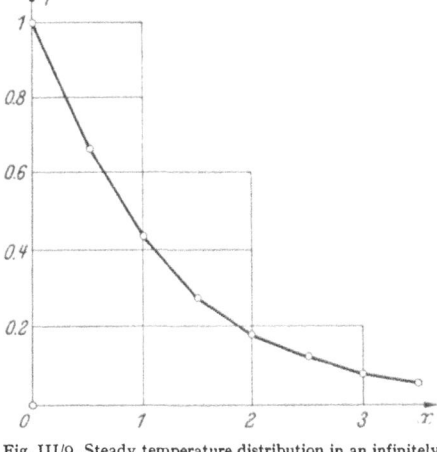

Fig. III/9. Steady temperature distribution in an infinitely long rod

as a finite boundary condition. Using the expressions for Y_n in terms of Y_1 calculated above in 1. for $h = \frac{1}{2}$, we obtain the Y_1 values of Table III/7; the extrapolated value again yields $y(1) \approx Y_2 = 0\cdot444$.

1.3. A linear boundary-value problem of the fourth order

Under the usual assumptions the transverse displacement $\eta(x)$ of an elastically embedded rail subject to a distributed transverse load (Fig. III/10) satisfies the differential equation

$$\frac{d^2}{d\xi^2}\left(E\,J(\xi)\,\frac{d^2\eta}{d\xi^2}\right) + K\eta = q(\xi).$$

Here the ξ co-ordinate is taken along the axis of the undeformed rail, $E\,J(\xi)$ denotes the flexural rigidity, K is the elastic constant of the bed material and $q(\xi)$ is the load density. For a rail with both ends freely supported, the bending moment M and the shear force Q vanish at the end points $\xi = a$ and $\xi = b$; thus the boundary conditions are

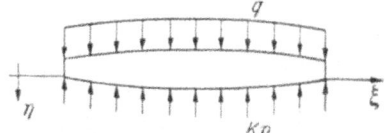

Fig. III/10. Bending of a transversely loaded, elastically embedded rail

$$\frac{d^2\eta}{d\xi^2} = \frac{d^3\eta}{d\xi^3} = 0 \quad \text{for} \quad \xi = a \quad \text{and} \quad \xi = b.$$

For the present numerical treatment of this fourth-order boundary-value problem we shall assume the following parabolic distributions of flexural rigidity and load density:

$$E\,J(\xi) = E\,J_0(2 - x^2); \quad q(\xi) = q_0(2 - x^2),$$

where $x = \xi/l$, $2l$ being the length of the rail and $x = 0$ the mid-point. Non-dimensionalizing η and K as well, we have the dimensionless set of quantities

$$x = \frac{\xi}{l}, \quad y = \frac{E\,J_0}{q_0 l^4}\eta, \quad k = \frac{l^4}{E\,J_0}K;$$

when $k = 40$ the boundary-value problem for y reads

$$\left.\begin{array}{l} [(2 - x^2)\,y'']'' + 40\,y = 2 - x^2, \\ y''(\pm 1) = y'''(\pm 1) = 0 \end{array}\right\} \qquad (1.13)$$

(dashes denote differentiation with respect to x).

We could now follow the procedure of § 1.1 and without further ado take $h = 1/n$ and write down the difference equations

$$\frac{1}{h^4}\,\Delta^2\,[(2 - (i-1)^2\,h^2)\,\Delta^2\,Y_{i-2}] + 40\,Y_i = 2 - i^2\,h^2 \qquad (i = 0, 1, 2, \ldots, n),$$

in which we put $Y_{-i} = Y_i$ because of the symmetry of the problem, and boundary conditions

$$Y_{n+1} - 2\,Y_n + Y_{n-1} = 0,$$
$$Y_{n+2} - 2\,Y_{n+1} + 2\,Y_{n-1} - Y_{n-2} = 0.$$

We would then have $n + 3$ equations for the same number of unknowns. However, it is rather more convenient to work with the equivalent system of second-order differential equations obtained by introducing the auxiliary quantity

$$v = (2 - x^2)\,y'';$$

in any case this quantity will be of interest in the calculation of the bending stresses, for it is effectively the negative of the bending moment.

The second-order system for v and y derived from (1.13) is

$$v'' + 40y - 2 + x^2 = 0, \\ (2 - x^2)\, y'' - v = 0; \left.\right\} \tag{1.14}$$

and if the symmetry of the problem is used to halve the range of x, the boundary conditions become

$$v(1) = v'(1) = y'(0) = v'(0) = 0. \tag{1.15}$$

With Y_i, V_i denoting the approximate values of y_i, v_i the finite-difference representations of these boundary conditions in the case $h = \frac{1}{2}$ read

$$V_2 = 0, \quad V_3 = V_1, \quad Y_{-1} = Y_1, \quad V_{-1} = V_1,$$

and with these relations taken into account the difference equations corresponding to the equations (1.14) reduce to

$$4\,(2V_1 - 2V_0) \quad + 40Y_0 - 2 \; = 0,$$
$$4\,(-2V_1 + V_0) \quad + 40Y_1 - \frac{7}{4} = 0,$$
$$4 \times 2V_1 \qquad\quad + 40Y_2 - 1 \; = 0,$$
$$2 \times 4\,(2Y_1 - 2Y_0) - V_0 \qquad = 0,$$
$$\frac{7}{4} \times 4\,(Y_2 - 2Y_1 + Y_0) - V_1 \; = 0$$

Solution of this system of equations yields

$$Y_0 = \frac{3726}{81\,440} = 0.045\,751\,5$$
$$Y_1 = \frac{3421}{81\,440} = 0.042\,0064$$
$$Y_2 = \frac{2666}{81\,440} = 0.032\,7358$$

$$V_0 = -\frac{488}{8144} = -0.059\,921\,4$$
$$V_1 = -\frac{315}{8144} = -0.038\,6788.$$

Again for smaller values of h it is better, as in Examples I and II, to calculate $Y_1, V_1, Y_2, V_2, \ldots$ recursively in terms of Y_0 and V_0 and derive two linear equations for Y_0 and V_0 from the conditions $V_n = 0$ and $V_{n+1} = V_{n-1}$. For $h = \frac{1}{5}$ we obtain successively

$$
\begin{array}{llll}
Y_1 = & Y_0 & + 0.01 & V_0 \\
V_1 = - & 0.8 \quad Y_0 & + & V_0 & + 0.04 \\
Y_2 = & 0.983\,674\,Y_0 & + 0.040\,408\,2\,V_0 & + 0.000\,816\,33 \\
V_2 = - & 3.2 \quad Y_0 & + 0.984 \quad V_0 & + 0.1584 \\
Y_3 = & 0.897\,782\,Y_0 & + 0.092\,2076\,V_0 & + 0.005\,076\,13 \\
V_3 = - & 7.173\,878\,Y_0 & + 0.903\,347 \quad V_0 & + 0.349094 \\
Y_4 = & 0.636\,918\,Y_0 & + 0.166\,040 \quad V_0 & + 0.017\,8504 \\
V_4 = - & 12.548\,21 \quad Y_0 & + 0.675\,162 \quad V_0 & + 0.597\,266 \\
Y_5 = & 0.005\,930\,Y_0 & + 0.259\,730 \quad V_0 & + 0.048\,191\,3 \\
V_5 = - & 19.013\,60 \quad Y_0 & + 0.181\,313 \quad V_0 & + 0.871\,278 \\
V_6 = - & 25.452\,49 \quad Y_0 & - 0.728\,104 \quad V_0 & + 1.108\,183.
\end{array}
$$

The equations $V_5 = 0$ and $V_6 = V_4$, i.e.

$$- 19 \cdot 013\,60\, Y_0 + 0 \cdot 181\,313\, V_0 + 0 \cdot 871\,278 = 0,$$
$$12 \cdot 868\,28\, Y_0 + 1 \cdot 403\,266\, V_0 - 0 \cdot 510\,917 = 0,$$

then yield

$$Y_0 = 0 \cdot 045\,331\,7 \quad \text{and} \quad V_0 = - 0 \cdot 051\,611\,5.$$

With these values of Y_0 and V_0 the remaining Y_i, V_i become

$$
\begin{aligned}
Y_1 &= 0 \cdot 044\,815\,6, & V_1 &= - 0 \cdot 047\,876\,9, \\
Y_2 &= 0 \cdot 043\,322\,4, & V_2 &= - 0 \cdot 037\,447\,3, \\
Y_3 &= 0 \cdot 041\,015\,2, & V_3 &= - 0 \cdot 022\,733\,6, \\
Y_4 &= 0 \cdot 038\,153\,4, & V_4 &= - 0 \cdot 008\,044\,1, \\
Y_5 &= 0 \cdot 035\,055\,1, & V_5 &= \quad 0.
\end{aligned}
$$

The correction (1.8) can also be applied here if we assume that the inherent errors tend to zero quadratically with the interval h. With $h = \frac{1}{5}$, $h^* = \frac{1}{2}$ it gives the improved value $Y_0 = 0 \cdot 045\,251\,7$.

1.4. Relaxation

The finite-difference method always entails the solution of a large number of simultaneous equations when the pivotal interval is small. A numerical solution of such a system of equations can be effected by the method[1] to be described now. It is applicable very generally; it can be used for linear and non-linear problems, for ordinary and partial differential equations, for the ordinary finite-difference method and also for the improved finite-difference methods (see § 2 and Ch. V, § 2).

The procedure may be outlined as follows: 1. An initial approximation is found either by estimation or by rough calculation, say by interpolation from values obtained with a coarse subdivision; 2. this initial approximation is inserted in the differential equation to discover where, i.e. at which pivotal points, the difference equation is already satisfied reasonably well and where there are still outstanding residual errors or "residuals" as they are usually called; 3. corrections are then made to the initial approximation at the points where the large residuals occur in such a way as to reduce their magnitude ("relaxation" of the initial

---

[1] This method of solving systems of linear equations had already been used by GAUSS when it was taken up by PH. L. SEIDEL: Münch. Akad. Abh. 1874, 81—108; its convergence was investigated later by R. v. MISES and H. POLLACZEK-GEIRINGER: Praktische Verfahren der Gleichungsauflösung. Z. Angew. Math. Mech. 9, 62—77 (1929), and in recent years it has been applied to a very wide range of problems and also expounded in two books by R. V. SOUTHWELL: Relaxation methods in engineering science, a treatise on approximate computation. London: Oxford University Press 1943; Relaxation methods in theoretical physics, a continuation of the treatise. London: Oxford University Press 1946, 248 pp.; further D. N. DE G. ALLEN: Relaxation methods. New York 1954, 257 pp. See also our Ch. V, § 1.6.

approximation). This relaxation usually produces new residuals at neighbouring pivotal points, but by continually decreasing the magnitude of the currently largest residual we try to approach the solution of the difference equations.

Experience is essential in making economical use of the method, for one has to appreciate the overall effect of making many varied sequences and combinations ("group relaxation") of corrections to the approximate solution to be able to "feel ones way" quickly towards the exact solution. In acquiring such experience by trying various corrections the beginner usually gets the impression that the method is difficult, for he often has to carry out a long calculation before the residuals become sufficiently small; but it should be borne in mind that the method permits the experienced "relaxer" to produce the solution very quickly to an accuracy sufficient for most technical applications.

A warning will not be out of place here, namely of the fallaciousness of the assumption that if the residuals are reduced to zero to a certain number of decimals, then the current relaxed values give the solution to the same number of decimals; particularly with a large number of equations the exact solution can differ widely from the values which ostensibly satisfy the equations to the required accuracy. In this connection it is important to note that in regions where several neighbouring residuals have the same sign an overall correction ("block relaxation") of considerable magnitude is usually needed to "liquidate" these residuals (cf. Ch. V, § 1.6).

Every computer will no doubt devise his own way of arranging the work when he has familiarized himself with the method. Consequently the layout in Tables III/8, III/9 is only offered as a suggestion.

For several classes of boundary-value problem one can determine rigorous limits for the error in a relaxation solution, cf. Ch. V, § 1.6.

Example I. A linear boundary-value problem. For the problem

$$y'' = -1 - (1 + x^2)\, y, \qquad y(\pm 1) = 0 \tag{1.16}$$

of Example I, § 1.2 we choose $h = \frac{1}{5}$ and find rough starting values for the relaxation procedure by interpolating graphically (say) between the values given by the ordinary finite-difference method with $h = \frac{1}{2}$. Thus from the results of Example I, § 1.2 we obtain the initial approximation (with $x_r = 0.2r$)

$$y_0 = 0.97, \quad y_1 = 0.94, \quad y_2 = 0.80, \quad y_3 = 0.60, \quad y_4 = 0.34.$$

The difference equations using (1.2) with $h = 1/m$ read

$$y_{k+1} - 2y_k + y_{k-1} = -h^2\big(1 + (1 + k^2 h^2)\, y_k\big),$$
$$y_{-1} = y_1, \qquad y_m = 0 \qquad (k = 0, 1, \ldots, m-1);$$

Table III/8. Examples of various relaxation procedures applied to the finite-difference method for the bent strut problem

	y_k values and corrections					z_k values					Changes z_k-y_k				
	y_0	y_1	y_2	y_3	y_4	z_0	z_1	z_2	z_3	z_4	z_0-y_0	z_1-y_1	z_2-y_2	z_3-y_3	z_4-y_4
Method 1. Starting values	0·97	0·94	0·80	0·60	0·34	0·979	0·925	0·809	0·606	0·331	9	−15	9	6	−9
Correction		−0·01				−0·01		−0·005			−10	+10	−5		
New values	0·97	0·93	0·80	0·60	0·34						−1	−5	4	6	−9
(Method 1 changes ×10⁻³)															
Method 2. Starting values	0·97	0·94	0·80	0·60	0·34	0·9794	0·9246	0·8086	0·6063	0·3311	94	−154	86	63	−89
Corrections ×10⁴	+94 / −214 / −42 / +84 / −94	−214 / −42 / +84 / −94	−42 / +84 / −94	+84 / −94	−94						−94	+47 / +107	−107 / +21	−21 / −42	+42 / +47
New values	0·9528	0·9134	0·7948	0·5990	0·3306	0·95246	0·91280	0·79464	0·59900	0·33034	−34	−60	−16	0	−16
(Method 2 starting & corrections ×10⁻⁴; new values ×10⁻⁵)															
Method 3. Starting values	0·93	0·89	0·77	0·58	0·32	0·9286	0·8885	0·7729	0·5809	0·3205	−14	−15	+29	+9	+5
Corrections ×10⁴	−14 / −44 / +14 / +32 / +42	−44 / +14 / +32 / +42	+14 / +32 / +42	+32 / +42	+42						+14	−7 / +22	+22 / −7	+7 / −16	+16 / −21
New values	0·9330	0·8944	0·7788	0·5874	0·3242	0·93306	0·89450	0·77897	0·58748	0·32434	6	10	17	8	14
Interpolation	0·93798	0·89922	0·78306	0·59056	0·32596	0·93798	0·89922	0·78306	0·59057	0·32595	0	0	0	+1	−1
(Method 3 starting & corrections ×10⁻⁴; new values & interpolation ×10⁻⁵)															

they may be written in the form

$$y_k = z_k,$$

where

$$\left.\begin{aligned} z_k = \frac{y_{k-1} + y_{k+1}}{2} + \\ + \frac{h^2}{2}\left(1 + (1 + \right. \\ \left. + k^2 h^2)\, y_k\right). \end{aligned}\right\} \quad (1.17)$$

The quantity $z_k - y_k$ represents the residual error and would normally be called the "residual" and denoted by R_k. Here it will be called the "change"[1]. This terminology arises from the close connection (cf. Ch. V, § 1.6) with the associated iteration procedure for solving such equations, i.e. the iteration defined by $y_k^{[\sigma+1]} = z_k^{[\sigma]}$; thus $z_k^{[\sigma]} - y_k^{[\sigma]}$ is the "change" $y_k^{[\sigma+1]} - y_k^{[\sigma]}$ produced by the $(\sigma+1)$-th cycle of the iteration. Calculation of residuals is equivalent to performing one cycle of the iteration procedure and noting the changes produced. If the y_k values satisfied the

Table III/9. *A relaxation solution of the system of non-linear equations arising in the finite-difference treatment of a non-linear problem*

	y_k values and corrections				z_k values				Changes $z_k - y_k$				
	y_1	y_2	y_3	y_4	z_1	z_2	z_3	z_4	z_1-y_1	z_2-y_2	z_3-y_3	z_4-y_4	
	0·50	0·85	1·05	1·1	0·478	0·842	1·052	1·105	−22	− 8	+ 2	+ 5	$\times 10^{-3}$
Correction	−0·02					−0·01							
	0·48	0·85	1·05	1·1	0·477	0·832	1·052	1·105	− 3	−18	+ 2	+ 5	
Corrections	−0·02	−0·02											
	0·46	0·83	1·05	1·1	0·467	0·821	1·042	1·105	+ 7	− 9	− 8	+ 5	
Corrections		−0·01	−0·01										
	0·46	0·82	1·04	1·1	0·462	0·815	1·037	1·100	+ 2	− 5	− 3	0	
Corrections	−0·006	−0·013	−0·012	−0·005									
	0·454	0·807	1·028	1·095	0·4550	0·8058	1·0269	1·0938	+10	−12	−11	−12	$\times 10^{-4}$
Corr. $\times 10^4$	− 14	− 50	− 60	− 44									$\times 10^{-5}$
	0·4526	0·8020	1·0220	1·0906	0·45245	0·80190	1·02187	1·09052	−15	−10	−13	− 8	
Corr. $\times 10^5$	− 60	− 90	− 100	− 66									
	0·45200	0·80110	1·02100	1·08994	0·45198	0·80106	1·02104	1·08998	− 2	− 4	+ 4	+ 4	

[1] *Translator's note.* This unusual terminology is a direct translation from the German; it is used here, in spite of its seeming awkwardness and the existence of a familiar alternative, in order that the emphasis in the German edition on the relationship with the iteration procedure may be preserved.

difference equations to the number of decimals carried, the changes would be zero.

In Table III/8 the y_k values are recorded in the columns on the left, the corresponding z_k values, calculated from them by (1.17), in the central columns and the changes $z_k - y_k$ in the columns on the right. We now describe several relaxation procedures for reducing the magnitude of these changes.

1. Point relaxation. Here we simply make such corrections in the y_k as seem appropriate, dealing with one point at a time. Thus (cf. Table III/8) we select the change $z_1 - y_1 = -0·015$ as being particularly large and try making a correction of $-0·01$ in y_1 $(=y_{-1})$. The effect of this correction is to alter z_0 by $-0·01$ (since $y_{-1} = y_1$) and z_2 by $-0·005$ but to leave its own z value z_1 unaltered since z_k is not influenced by the term in h^2 to the number of decimals carried currently; in turn the changes are altered as follows: $z_0 - y_0$ by $-0·01$, $z_1 - y_1$ by $0·01$, $z_2 - y_2$ by $-0·005$, and we can write down the new changes immediately without having to recalculate the z_k values. Continuing this procedure of making suitable corrections to the y_k, we try to decrease the magnitude of the changes as much as possible.

2. Special block relaxations. Again we first calculate the z_k values and the changes $z_k - y_k$ corresponding to the starting values for the y_k, but this time to one more decimal place. Working in fourth decimal units, we begin the relaxation by adding the correction $+94$ to y_0, which reduces the first change $z_0 - y_0$ to zero (apart from the error due to the neglect of the term in h^2, cf. method 1. above); the second change $z_1 - y_1$ is therefore altered to -107. We now reduce the remaining changes to zero successively by means of special block relaxations, i.e. relaxations with simultaneous identical corrections at neighbouring points (cf. Ch. V, § 1.6), whose effects depend on the following special property of the approximate equation $z_k - y_k = \frac{1}{2}(y_{k-1} + y_{k+1}) - y_k$: if identical corrections, say of $+2\alpha$, are made to the first r y_k values, only the r-th and $(r+1)$-th changes are altered, $z_r - y_r$ by $-\alpha$ and $z_{r+1} - y_{r+1}$ by α. Thus we remove the new change $z_1 - y_1 = -107$ by adding -214 to y_0 and y_1. At the same time $z_2 - y_2$ is altered by -107 to the new value $+21$ (see Table III/8), which can likewise be removed by adding $+42$ to y_0, y_1, y_2; and so on. By adding all the corrections we obtain new y_k values, from which we calculate new z_k values and then new changes; these are seen to have been reduced considerably, but now they all have the same sign and consequently the new y_k values still differ markedly from the required values. One would therefore repeat the process on the new y_k values with further repetitions if necessary. The method depends on the special form of the difference equation for this example but similar methods can be devised for many other types of difference equation.

3. Bracketing and interpolation. Here we aim to find two approximate solutions, one with only non-negative changes and the other with only non-positive changes. An approximation satisfying the later requirement has already been found by method 2. To find the other approximation we try new starting values $y_0 = 0 \cdot 93$, $y_1 = \cdots$ (as in Table III/8) which are likely to be too small; the changes still have varying signs but one cycle of the procedure of method 2. yields new y_k values ($y_0 = 0 \cdot 933$, $y_1 = \cdots$) for which all changes turn out positive. We obtain a new approximation (last row of Table III/8: $y_0 = 0 \cdot 93798$, $y_1 = \cdots$) with very small changes simply by interpolating between the approximations with all positive and all negative changes, respectively. The bracketing nature of this method allays the fear that the approximate values still differ considerably from the exact values, a fear which must otherwise — with the first two methods, for example — always exist.

Example II. A non-linear boundary-value problem. Suppose that we are required to find the steady temperature distribution in a homogeneous rod of length l in which, as a consequence of a chemical reaction say or some such heat-producing process, heat is generated at a rate $f(y)$ per unit time per unit length, $f(y)$ being a given function of the excess temperature y of the rod over the temperature of the surroundings. If the ends of the rod, $x = 0$ and $x = l$, are kept at given temperatures, we are to solve the first boundary-value problem

$$y'' = - c f(y); \qquad y(0) = y_0, \qquad y(l) = y_l;$$

dashes denote differentiation with respect to x, which is measured along the axis of the rod, and c is a given constant.

For this example we choose an exponential law $cf(y) = 1 + e^y$ for the heat generation and $y(0) = 0$, $y(1) = 1$ as boundary conditions.

The difference equations here are very similar in form to the corresponding equations (1.17) for Example I: with $h = 1/m$ they read

$$y_k = z_k, \quad \text{where} \quad z_k = \frac{y_{k+1} + y_{k-1}}{2} + \frac{h^2}{2} (1 + e^{y_k}), \quad (k = 1, 2, \ldots, m - 1),$$

$$y_0 = 0, \qquad y_m = 1.$$

We first get a rough idea of the solution with $h = \frac{1}{2}$, deriving from the single equation

$$e^{y_1} = 8 y_1 - 5$$

the two real solutions $y_1 = \begin{cases} 0 \cdot 9474 \\ 2 \cdot 903 \, . \end{cases}$

We shall restrict attention to the stable temperature distribution. This corresponds here to the solution with the smaller y value, from which we obtain by graphical interpolation the starting values $y_1 = 0.5$, $y_2 = \cdots$, $y_4 = 1.1$ (cf. Table III/9) for a relaxation solution with $h = \frac{1}{5}$.

The relaxation of these y_k values proceeds along essentially the same lines as for the linear boundary-value problem, the only marked difference being that on account of the non-linearity the new y_k values must be written down and from them the new z_k values calculated each time a group of corrections is made. Table III/9 needs little explanation as it is set out in the same way as Table III/8; after the first few corrections, the individual steps are omitted and sequences of corrections combined and written effectively as group corrections. By continuing the relaxation we could obtain a still more accurate solution of the difference equations but this would be rather pointless in view of the limitations of the ordinary finite-difference approximation.

§ 2. Refinements of the ordinary finite-difference method

In solving problems numerically by the "ordinary finite-difference method" described in § 1, we have seen that a very good idea of the behaviour of the solution can generally be obtained by using a large pivotal interval h with the attendant advantages of relatively short calculations and fewer unknowns. However, we have also seen that refining the subdivision ($h \to 0$) is a slowly convergent process so that to obtain accurate values one would have to use a very small pivotal interval. The amount of labour is greatly increased thereby, and consequently for accurate work the ordinary finite-difference method does not compare favourably with other methods. We therefore describe now various ways of improving the finite-difference method.

2.1. Improvement by using finite expressions which involve more pivotal values

We can speed up the convergence of the finite-difference method by replacing the derivatives by "finite expressions" which are more general than the difference quotients of § 1.1 and represent the derivatives more accurately. For example, the departure of the expression

$$\frac{1}{12h}(-y_{i+2} + 8y_{i+1} - 8y_{i-1} + y_{i-2}) \tag{2.1}$$

from $y'(x_i)$ amounts at most to $C_1 h^4$, where $C_1 = \frac{1}{30}|y^{(5)}|_{\text{max in }\langle x_{i-2},\,x_{i+2}\rangle}$, whereas the departure of the difference quotient

$$\frac{1}{2h}(y_{i+1} - y_{i-1})$$

from $y'(x_i)$ can be as large as $C_2 h^2$, where $C_2 = \frac{1}{6}|y'''|_{\text{max in }\langle x_{i-1},\,x_{i+1}\rangle}$.

Such finite expressions are easily derived by means of TAYLOR's theorem. For example, if we put

$$y'_i \approx \sum_{\varrho=-2}^{+2} c_\varrho y_{i+\varrho}, \tag{2.2}$$

where the c_ϱ are constants to be determined, and expand each term of the sum into a Taylor series:

$$y_{i+\varrho} = y_i + \varrho\, h\, y_i' + \frac{\varrho^2 h^2}{2!}\, y_i'' + \cdots,$$

we obtain for the sum the expansion

$$y_i' \approx y_i \sum_{\varrho=-2}^{2} c_\varrho + h\, y_i' \sum_{\varrho=-2}^{2} \varrho\, c_\varrho + \frac{h^2}{2!}\, y_i'' \sum_{\varrho=-2}^{2} \varrho^2 c_\varrho + \frac{h^3}{3!}\, y_i''' \sum_{\varrho=-2}^{2} \varrho^3 c_\varrho + \cdots,$$

and by solving the five linear equations

$$\sum_{\varrho=-2}^{2} c_\varrho = \sum_{\varrho=-2}^{2} \varrho^2 c_\varrho = \sum_{\varrho=-2}^{2} \varrho^3 c_\varrho = \sum_{\varrho=-2}^{2} \varrho^4 c_\varrho = 0, \quad \sum_{\varrho=-2}^{2} \varrho\, c_\varrho = \frac{1}{h}$$

for the five unknown coefficients c_ϱ we can make the expansion reduce to y_i' apart from terms in h^5 and higher powers of h. The values of c_ϱ which we obtain yield the finite expression (2.1).

2.2. Derivation of finite expressions

We now generalize the method described in § 2.1. First of all we introduce the following

Definition: *Consider the n-th order homogeneous linear differential expression*

$$L[y] = \sum_{\nu=0}^{n} f_\nu(x)\, y^{(\nu)} \tag{2.3}$$

in y with given continuous functions $f_\nu(x)$ as coefficients. Then the linear combination

$$A = \sum_{k=1}^{p} C_k\, y(x_i + \alpha_k h), \tag{2.4}$$

where C_k, α_k are constants, is called a finite expression of the r-th approximation for the differential expression $L[y]$ at the point $x = x_i$ if the factors multiplying the quantities $y^{(\nu)}(x_i)$ in the Taylor expansion of the expression A are $f_\nu(x_i)$ for $0 \leq \nu \leq n$ and zero for $n+1 \leq \nu \leq n+r$.

We then state the following

Theorem: *A finite expression of the r-th approximation (in fact, infinitely many) can be derived for any given linear differential expression $L[y]$ (2.3) at any point $x = \xi$ and for any non-negative integer r. For an arbitrary choice of $q = r+n+1$ distinct points $\xi + \alpha_k h$ $(k = 1, 2, \ldots, q)$, q quantities C_k can be determined by solving a system of linear equations so that, for every function $u(x)$ with a continuous q-th derivative,*

$$A = \sum_{k=1}^{q} C_k\, u(\xi + \alpha_k h) = L[u(\xi)] + D\, \vartheta\, \frac{h^{r+1} |u^{(q)}|_{\max}}{q!}, \tag{2.5}$$

where $|\vartheta| \leq 1$, D is a polynomial in h of at most the n-th degree and depends on the choice of the α_i but not on u, and $|u^{(q)}|_{\max}$ is the absolute maximum of the q-th derivative of u in an interval containing all the points $\xi + \alpha_k h$.

To prove this we need only apply TAYLOR's theorem to the expression on the left-hand side of (2.5). This gives

$$
\begin{aligned}
A = \quad & u(\xi)\,[\quad C_1 + \cdots + \quad C_q] + \\
& + h\,u'(\xi)\,[\alpha_1\;C_1 + \cdots + \alpha_q\;C_q] + \\
& + \frac{h^2}{2!}\,u''(\xi)\,[\alpha_1^2\;C_1 + \cdots + \alpha_q^2\;C_q] + \cdots + \\
+ \frac{h^{n+r}}{(n+r)!}\, & u^{(n+r)}(\xi)\,[\alpha_1^{n+r}C_1 + \cdots + \alpha_q^{n+r}C_q] + R,
\end{aligned}
$$

where the remainder term R may be written in the form

$$
R = \frac{h^{n+r+1}}{(n+r+1)!}\,\vartheta \cdot \big|\,u^{(n+r+1)}\big|_{\max} \sum_{\varrho=1}^{q} \big|\alpha_\varrho^{n+r+1} C_\varrho\big|
$$

with $|\vartheta| \leq 1$. Comparison of this expansion with (2.3) yields $(n+r+1)$ linear equations for the C_ϱ, namely

$$
\sum_{\varrho=1}^{q} \alpha_\varrho^k C_\varrho =
\begin{cases}
\dfrac{k!}{h^k}\,f_k(\xi) & \text{for } 0 \leq k \leq n, \\[2mm]
0 & \text{for } n+1 \leq k \leq n+r.
\end{cases}
\tag{2.6}
$$

These equations always possess a solution, for the determinant of their coefficients

$$
\begin{vmatrix}
1 & 1 & \ldots 1 \\
\alpha_1 & \alpha_2 & \ldots \alpha_q \\
\alpha_1^2 & \alpha_2^2 & \ldots \alpha_q^2 \\
\cdot\,\cdot\,\cdot\,\cdot\,\cdot\,\cdot\,\cdot\,\cdot\,\cdot\,\cdot\,\cdot\,\cdot \\
\alpha_1^{r+n} & \alpha_2^{r+n} & \ldots \alpha_q^{r+n}
\end{vmatrix}
\tag{2.7}
$$

is a Vandermonde determinant, which, being the product of the differences of the distinct numbers $\alpha_1, \ldots, \alpha_q$, is never zero[1].

Some simple finite expressions for the lower order derivatives using equidistant points $\xi + k\,h$ are listed in Table III of the appendix. From them finite expressions for any linear differential expression of up to the fourth order can be obtained by superposition[2].

---

[1] See, for instance, PERRON, O.: Algebra, Vol. I, p. 92. 3rd ed. Berlin and Leipzig 1951, or AITKEN, A. C.: Determinants and matrices, 6th ed. p. 41. London: Oliver & Boyd 1949.

[2] Closed form solutions of the system of equations (2.6) for $n=1$ and $n=2$ can be found in L. COLLATZ: Das Differenzenverfahren mit höherer Approximation für lineare Differentialgleichungen. Schr. Math. Sem. u. Inst. Angew. Math. Univ. Berlin 3, 1—34 (1935), and for general n in E. PFLANZ: Über die Bildung finiter Ausdrücke für die Lösung linearer Differentialgleichungen. Z. Angew. Math. Mech. 17, 296—300 (1937).

Expressions for non-equidistant pivotal points are given by E. PFLANZ: Allgemeine Differenzenausdrücke für die Ableitungen einer Funktion $y(x)$. Z. Angew. Math. Mech. 29, 379—381 (1949).

2.3. The finite-difference method of a higher approximation

This method follows precisely the same lines as for the ordinary finite-difference method described in § 1 (which can be regarded as a first approximation) only now the derivatives are replaced by the finite expressions of a higher approximation given in Table III of the appendix. However, on account of the extra unknowns appearing in each equation based on the higher approximation, there are always more unknowns than such equations and further equations of a lower order of approximation have to be brought in at the boundary points; these can be equations corresponding to the differential equation, for example. This device is in fact used in the following example.

Example I. For the strut problem of Example I, § 1.2, i.e.

$$y'' + (1 + x^2)\, y + 1 = 0, \qquad y(\pm 1) = 0,$$

the finite equations of the third approximation (see Table III of the appendix) for the points $x = 0$ and $x = \frac{1}{2}$ of the coarse subdivision with $h = \frac{1}{2}$ read

$$\frac{-0 + 16 Y_1 - 30 Y_0 + 16 Y_1 - 0}{12\, h^2} + 1 \cdot Y_0 + 1 = 0,$$

$$\frac{- Y_3 + 16 \cdot 0 - 30 Y_1 + 16 Y_0 - Y_1}{12 h^2} + \frac{5}{4} Y_1 + 1 = 0;$$

here we have already used the symmetry condition $Y_i = Y_{-i}$ and the boundary condition $Y_2 = 0$. To eliminate Y_3 we bring in another equation, namely the ordinary difference equation for the point $x = 1$:

$$\frac{Y_3 - 2 \cdot 0 + Y_1}{h^2} + 2 \cdot 0 + 1 = 0.$$

Solution of these equations yields

$$Y_0 = \frac{731}{787} = 0{\cdot}928\,844,$$

$$Y_1 = \frac{543}{787} = 0{\cdot}689\,962.$$

The more accurate values obtained by taking $h = \frac{1}{3}$ are

$$x = 0, \qquad Y_0 = \frac{2\,723\,933}{2\,924\,440} = 0{\cdot}931\,437,$$

$$x = \frac{1}{3}, \qquad Y_1 = \frac{2\,410\,848}{2\,924\,440} = 0{\cdot}824\,379,$$

$$x = \frac{2}{3}, \qquad Y_2 = \frac{296\,031}{584\,888} = 0{\cdot}506\,133.$$

Example II. An eigenvalue problem. For the eigenvalue problem

$$(1 + x)\, y'' + y' + \lambda(1 + x)\, y = 0, \qquad y'(0) = y(1) = 0,$$

11*

treated in Example III of § 1.2, the equation for the point $x = \frac{3}{8}$ (with $h = \frac{2}{8}$) reads

$$\frac{5}{6}(-Y_3 - 30Y_1 + 16Y_0 - Y_0) + \frac{5}{24}(-Y_3 - 7Y_0) + \frac{8}{5}\Delta Y_1 = 0$$

with the notation of Fig. III/7 ($\dot{Y}_3$ is the approximate value of $y(\frac{7}{8})$).

The unknowns Y_0 and Y_3 are eliminated by using the ordinary difference equations at the points $x = \frac{1}{8}$ and $x = 1$:

$$\frac{15}{2}(Y_1 - Y_0) + \frac{5}{4}(Y_1 - Y_0) + \frac{6}{5}\Delta Y_0 = 0,$$

$$\frac{25}{2}(Y_3 + Y_1) + \frac{5}{4}(Y_3 - Y_1) = 0, \quad \text{i.e.} \quad Y_3 = -\frac{9}{11}Y_1.$$

For $\mu = \frac{24}{25}\Lambda$ we obtain the equation

$$\begin{vmatrix} -7+\mu & 7 \\ 583 & -1275+88\mu \end{vmatrix} = 0 \quad \text{and from it} \quad \Lambda = \begin{cases} 3\cdot0968 & (\text{error} \quad -3\cdot8\%) \\ 19\cdot29 & (\text{error} \quad -16\ \%). \end{cases}$$

2.4. Basic formulae for Hermitian methods[1]

Here we consider another method of setting up finite-difference equations of greater accuracy, i.e. with truncation errors of a higher order than the ordinary finite-difference equations. The equations we obtain are often particularly convenient for differential equations of simple form, but sometimes the practical application of the method is rather complicated. *The gain in accuracy over the ordinary method·is obtained, not by including more pivotal values, as in the method of a higher approximation, but by basing the derivation of each individual difference equation on the fact that the differential equation is satisfied at several points, rather than just one as in the other methods.* This gives rise to formulae which involve the values of derivatives at more than one point in addition to the values of the function, and by analogy with HERMITE's interpolation formula we may call them Hermitian formulae and methods based upon them Hermitian methods (see footnote); as an additional justification for this terminology it may be noted that

---

[1] *Translator's note.* The German name "Mehrstellenverfahren" does not translate conveniently, for "more-point methods" would be rather misleading. The name preferred here is suggested by the use of the term "Lagrangian methods" for methods involving formulae expressed in terms of function values — by analogy with LAGRANGE's interpolation formula; here the analogy is with HERMITE'S interpolation formula, which, in addition to the values of the function, also uses the values of the derivative at several points:

$$f(x) = \sum f(x_i)\, h_i(x) + \sum f'(x_i)\, H_i(x) + R;$$

cf. HOUSEHOLDER, A. S.: Principles of numerical analysis, p. 194. New York: McGraw-Hill 1953. We do not imply that the method described here for solving differential equations is directly due to HERMITE.

HERMITE's generalization of TAYLOR's theorem (cf. Ch. I, § 2.5) provides another example of such a formula. To be more specific, the method is based on certain expressions of the form

$$P = \sum_{\nu=-p}^{p} (a_\nu y_{i+\nu} + A_\nu y_{i+\nu}^{(k)}); \tag{2.8}$$

these are derived once and for all (cf. Table III of the appendix).

We form linear combinations of the values of y and its k-th derivative at neighbouring pivotal points $x_{i+\nu}$ and determine the weighting factors a_ν, A_ν so that the coefficients of powers of h in the Taylor series for the expression P centered on the point x_i all vanish to as high a power of h as possible. Consider, for example, the case $k=1$ for the first derivative. The ordinary difference quotient (1.1) for this case yields the relation

$$y_{i+1} - y_{i-1} - 2h y_i' \approx 0,$$

and by substituting the Taylor series for y_{i+1} and y_{i-1} centered on x_i we find the error term to be proportional to $h^3 y'''$. To obtain a formula with an error term of higher order without introducing more pivotal points, it is natural to try bringing in the extra "information" provided by the values y_{i+1}' and y_{i-1}'. We take $i=0$ for simplicity and try the general form

$$P = a_{-1} y_{-1} + a_0 y_0 + a_1 y_1 + A_{-1} y_{-1}' + A_0 y_0' + A_1 y_1';$$

then with each term expanded into a Taylor series centered on x_0 we have

$$P = \begin{cases} y_0 \; (a_{-1} + a_0 + a_1) \\[4pt] + \; h y_0' \left(-a_{-1} \quad + a_1 + \frac{1}{h}[\; A_{-1} + A_0 + A_1] \right) \\[4pt] + \frac{1}{2!} h^2 y_0'' \left(\quad a_{-1} \quad + a_1 + \frac{2}{h}[-A_{-1} \quad + A_1] \right) \\[4pt] + \frac{1}{3!} h^3 y_0''' \left(-a_{-1} \quad + a_1 + \frac{3}{h}[\; A_{-1} \quad + A_1] \right) \\[4pt] + \frac{1}{4!} h^4 y_0^{IV} \left(\quad a_{-1} \quad + a_1 + \frac{4}{h}[-A_{-1} \quad + A_1] \right) + \cdots. \end{cases}$$

By putting the quantities inside the round brackets equal to zero we obtain a set of linear equations for the unknowns a_ν, A_ν. Since the equations are homogeneous, a factor remains undetermined and we therefore solve in terms of one of the unknowns, say a_1:

$$a_{-1} = -a_1, \quad a_0 = 0, \quad A_{-1} = A_1 = -\frac{h}{3} a_1 = \frac{1}{4} A_0.$$

Thus we obtain the formula

$$\frac{P}{a_1} = y_1 - y_{-1} - \frac{h}{3}(y_1' + 4y_0' + y_{-1}') = 0 + 0(h^4). \qquad (2.9)$$

We treat the general case of a derivative of arbitrary order k in a similar fashion. Thus by substituting in (2.8) the Taylor series centered on the point x_i we have

$$P = \sum_{\nu=-p}^{p} \left\{ a_\nu \left(y_i + \nu h\, y_i' + \nu^2 \frac{h^2}{2!} y_i'' + \cdots \right) + \right. \\ \left. + A_\nu \left(y_i^{(k)} + \nu h\, y_i^{(k+1)} + \nu^2 \frac{h^2}{2!} y_i^{(k+2)} + \cdots \right) \right\}, \qquad (2.10)$$

and if we define

$$b_\nu = \frac{k!}{h^k} A_\nu, \qquad (2.11)$$

the requirement that the factors multiplying y_i, y_i', y_i'', ... up to as high an order as possible, say the $(k+r)$-th, shall be zero leads to the equations

$$\sum_{\nu=-p}^{p} a_\nu \nu^\varkappa = 0 \qquad\qquad (\varkappa = 0, 1, 2, \ldots, k-1) \\ \sum_{\nu=-p}^{p} \left\{ a_\nu \nu^\varkappa + b_\nu \binom{\varkappa}{k} \nu^{\varkappa-k} \right\} = 0 \quad (\varkappa = k, k+1, \ldots, k+r) \qquad (2.12)$$

for the a_ν and b_ν.

Clearly we do not want an expression P for which all the b_ν vanish. This can be avoided by demanding that b_0, say, i.e. the coefficient corresponding to the point x_i, does not vanish. Since a factor is indeterminate in P, we can put $b_0 = 1$; then (2.12) becomes an inhomogeneous system of $(k+r+1)$ linear equations for the unknowns a_ν, b_ν. If the number of these unknowns is sufficiently large, which can be arranged merely by choosing p sufficiently large, arbitrarily many solutions will exist. [The existence of infinitely many solutions follows from the fact that to any finite expression derived as in § 2.2 there corresponds a solution of (2.12).] Hermitian expressions of the form (2.8) for the lower order derivatives (up to the fourth order) are given in Table III of the appendix.

2.5. The Hermitian method in the general case

Consider the differential equation[1,2,3]

$$y^{(n)} = f(x, y, y', y'', \ldots, y^{(n-1)}). \qquad (2.13)$$

---

[1] By using the associated GREEN's function E. J. NYSTRÖM: Zur numerischen Lösung von Randwertaufgaben bei gewöhnlichen Differentialgleichungen. Acta Math. (Stockh.) **76**, 157—184 (1944), has developed a special Hermitian method

Here we use formula (2.8) with $k = n$ and write down the equations

$$\sum_{\nu=-p}^{p} (a_\nu Y_{i+\nu} + A_\nu Y_{i+\nu}^{(n)}) = 0 \qquad (2.14)$$

for all pivotal points inside the interval, say $x_1, x_2, \ldots, x_{q-1}$.

We now substitute throughout for the $Y_\nu^{(n)}$ in terms of the lower derivatives $Y_\nu^{(n-1)}, Y_\nu^{(n-2)}, \ldots, Y_\nu', Y_\nu$ by using the differential equation. The $Y_\nu^{(s)} (1 \leq s \leq n-1)$ appear now as further unknowns; we therefore write down for each interior pivotal point the formula corresponding

for the problem $y'' = f(x, y)$, $y(x_a) = y_a$, $y(x_b) = y_b$; he gives formulae for one to four interior pivotal points (not necessarily equidistant), in particular, the following formulae for three equidistant interior points $x_i = x_a + ih$ $(i = 1, 2, 3)$, where $h = \frac{1}{4}(x_b - x_a)$:

$$y_1 = \frac{3y_0 + y_4}{4} - \frac{h^2}{480} [27 y_0'' + 332 y_1'' + 222 y_2'' + 132 y_3'' + 7 y_4''] + R_1,$$

$$y_2 = \frac{y_0 + y_4}{2} - \frac{h^2}{30} [y_0'' + 16 y_1'' + 26 y_2'' + 16 y_3'' + y_4''] \qquad + R_2,$$

$$y_3 = \frac{y_0 + 3y_4}{4} - \frac{h^2}{480} [7 y_0'' + 132 y_1'' + 222 y_2'' + 332 y_3'' + 27 y_4''] + R_3,$$

where $y_i = y(x_i)$ and R_1, R_2, R_3 are remainder terms.

[2] For second-order equations F. Stüssi: Numerische Lösung von Randwertproblemen mit Hilfe der Seilpolygongleichung. Z. Angew. Math. Phys. 1, 53—70 (1950), has derived the Hermitian formula

$$y_{i-1} - 2y_i + y_{i+1} \approx \frac{h^2}{12} (y_{i-1}'' + 10 y_i'' + y_{i+1}'')$$

purely from mechanical considerations based on the funicular polygon of graphical statics; he had already used the formula in 1935: Die Stabilität des auf Biegung beanspruchten Trägers. Abh. Internat. Vereinig. f. Brückenbau u. Hochbau 3, Zürich 1935, pp. 401—420, in particular p. 413. In numerous other works he has also derived special Hermitian formulae for ordinary differential equations of the fourth order and the biharmonic equation, always using an approach based purely on ideas from statics; see also F. Stüssi: Baustatik, Vol. I. Basel 1946. The Hermitian formula for the second derivative quoted here can also be found in Sch. E. Mikeladze: Über die Lösung von Randwertproblemen mit der Differenzenmethode. C. R. (Doklady) Acad. Sci. URSS. 28, 400—402 (1940) (Russian), where formulae of the form

$$y(a + \alpha h) \pm y(a - \alpha h) = \sum_{\lambda=0}^{n-1} \frac{1 \pm (-1)^\lambda}{\lambda!} \alpha^\lambda h^\lambda y^{(\lambda)}(a) +$$

$$+ \frac{h^n}{(n-1)!} \sum_{i=0}^{r} A_i [y^{(n)}(a + t_i h) + y^{(n)}(a - t_i h)] + \text{remainder term}$$

are used; for $n > 2$ these are of a somewhat different form from those given in the present book.

[3] A variant of the Hermitian method originally due to Numerov has been applied to the special equation $y'' = q(x) y$ (cf. G. Stracke: Bahnbestimmung der Planeten und Kometen, § 77. Berlin 1929). In this method a system of equations

to (2.14) for all derivatives which occur in the right-hand side of the differential equation. With the corresponding constants a_ν, A_ν distinguished by dashes these equations read

$$
\left.\begin{aligned}
\sum_\nu [a'_\nu Y_{i+\nu} + A'_\nu Y^{(n-1)}_{i+\nu}] &= 0, \\
\sum_\nu [a''_\nu Y_{i+\nu} + A''_\nu Y^{(n-2)}_{i+\nu}] &= 0, \\
\cdot\ \cdot\ \cdot\ \cdot\ \cdot\ \cdot\ \cdot\ \cdot\ \cdot\ \cdot\ & \\
\sum_\nu [a^{(n-1)}_\nu Y_{i+\nu} + A^{(n-1)}_\nu Y'_{i+\nu}] &= 0.
\end{aligned}\right\} \qquad (2.15)
$$

The boundary points require special attention. Here it may be necessary to use unsymmetric formulae in order to eliminate surplus unknowns. This is illustrated in Example II below in § 2.6 (see also § 2.7).

2.6. Examples of the Hermitian method

I. **Inhomogeneous problem of the second order.** Let us consider again the strut problem of Example I, § 1.2, i.e.

$$
y'' + (1 + x^2) y + 1 = 0, \qquad y(\pm 1) = 0,
$$

and use first the interval $h = \frac{1}{2}$ (Fig. III/3). From Table III of the appendix we obtain the equations

$$
Y_{i+1} - 2Y_i + Y_{i-1} - \frac{h^2}{12}(Y''_{i+1} + 10Y''_i + Y''_{i-1}) = 0 \qquad (i = 0, 1) \qquad (2.16)
$$

and from the differential equation

$$
Y''_i = -(1 + i^2 h^2) Y_i - 1.
$$

is set up for approximate pivotal values R_j of the function

$$
r(x) = y(x) - \frac{h^2}{12} y''(x) = y\left(1 - \frac{h^2}{12} q\right)
$$

instead of for approximate pivotal values Y_j of the solution $y(x)$. By TAYLOR's theorem we can show that the second central-difference of $r(x)$ has the expansion

$$
\delta^2 r(x) = r(x+h) - 2r(x) + r(x-h) = h^2 y''(x) - \frac{1}{240} h^6 y^{VI}(x) +
$$
$$
+ \text{ higher order terms.}
$$

Now $y'' = qy = qr\left(1 - \frac{h^2}{12} q\right)^{-1}$, so that if we retain only the first term on the right-hand side of the above formula, we obtain the difference equation

$$
\delta^2 R_j = R_{j+1} - 2R_j + R_{j-1} = \frac{h^2 q}{1 - \frac{h^2}{12} q} R_j
$$

to be satisfied by the approximate pivotal values of $r(x)$. STRACKE uses this formula for initial-value problems, but obviously it can also be used for boundary-value problems.

In addition we have the boundary condition $Y_2 = 0$ and symmetry condition $Y_{-i} = Y_i$, so that finally, after substitution, we have the two equations

$$2Y_1 - 2Y_0 + \frac{1}{48}\left(12 + 10Y_0 + \frac{5}{2}Y_1\right) = 0,$$

$$-2Y_1 + Y_0 + \frac{1}{48}\left(12 + Y_0 + \frac{25}{2}Y_1\right) = 0$$

for Y_0 and Y_1. Their solution is

$$Y_0 = \frac{4368}{4709} \approx 0.927585; \qquad Y_1 = \frac{3240}{4709} \approx 0.688044.$$

As in § 1.1, we treat the problem as an initial-value problem when the chosen pivotal interval is small; thus with $h = \frac{1}{5}$ we use the equations

$$300\,(2Y_1 - 2Y_0) \quad + 12 + 2.08\,Y_1 + 10 \quad Y_0 \qquad = 0,$$

$$300\,(Y_2 - 2Y_1 + Y_0) + 12 + 1.16\,Y_2 + 10.4\,Y_1 + \quad Y_0 = 0,$$

$$300\,(Y_3 - 2Y_2 + Y_1) + 12 + 1.36\,Y_3 + 11.6\,Y_2 + 1.04\,Y_1 = 0,$$

$$300\,(Y_4 - 2Y_3 + Y_2) + 12 + 1.64\,Y_4 + 13.6\,Y_3 + 1.16\,Y_2 = 0,$$

$$300\,(Y_5 - 2Y_4 + Y_3) + 12 + 2 \quad Y_5 + 16.4\,Y_4 + 1.36\,Y_3 = 0$$

to express $Y_1, Y_2, \ldots$ successively in terms of Y_0:

$$Y_1 = - \quad 0.0199309 + \quad 0.9799362\,Y_0,$$

$$Y_2 = - \quad 0.0788659 + \quad 0.9190145\,Y_0,$$

$$Y_3 = - \quad 0.1738941 + \quad 0.8154636\,Y_0,$$

$$Y_4 = - \quad 0.2990990 + \quad 0.6677412\,Y_0,$$

$$302\,Y_5 = -134.1494 \quad + 143.9457 \quad Y_0.$$

The boundary condition $Y_5 = 0$ then yields the value

$$Y_0 = 0.9319450,$$

from which we calculate the remaining Y_i:

$$Y_1 = 0.8933157,$$
$$Y_2 = 0.7776050,$$
$$Y_3 = 0.5860732,$$
$$Y_4 = 0.3231992.$$

As before (§ 1.2) we can use the values Y, Y^* calculated with two different intervals, h and h^* say, to obtain improved values $\bar{Y}$ by a formula similar to (1.8). Here, however, we assume that the error is of order h^4 (cf. § 3.3), so that the corresponding formula is

$$\overline{Y} = Y + \frac{h^4}{h^{*4} - h^4} (Y - Y^*); \tag{2.17}$$

with $h = \frac{1}{5}$, $h^* = \frac{1}{2}$ we obtain for $x = 0$

$$\overline{Y} = 0.931945 + \frac{(\frac{1}{5})^4}{(\frac{1}{2})^4 - (\frac{1}{5})^4} (0.931945 - 0.9276) = 0.932060.$$

II. An eigenvalue problem. With $h = \frac{2}{5}$ (notation as in Fig. III/7) we calculate two eigenvalues of the longitudinal vibration problem

$$-(1 + x) y'' - y' = \lambda(1 + x) y; \qquad y'(0) = y(1) = 0$$

of Example III, § 1.2. First of all we write down the Hermitian equation for the second derivative at the interior points $x = \frac{1}{5}$, $x = \frac{3}{5}$:

$$Y_1 - Y_0 - \frac{1}{75}(Y_1'' + 10 Y_0'' + Y_{-1}'') = 0,$$

$$-2 Y_1 + Y_0 - \frac{1}{75}(Y_2'' + 10 Y_1'' + Y_0'') = 0.$$

Here we have used the lower order approximation $Y_0 - Y_{-1} = 0$ for the boundary condition $y'(0) = 0$; this condition also implies that $Y_0' = -Y_{-1}'$ to the same order of approximation, but no similar relation holds for the second derivatives. These relations are used below when we write down the differential equation at the point $x = -\frac{1}{5}$. The second derivatives occuring in the Hermitian equations above are expressed in terms of the lower derivatives by writing down the differential equation for each of the four pivotal points:

$$Y_2'' + \frac{1}{2} Y_2' = 0, \qquad\qquad Y_1'' + \frac{5}{8} Y_1' + \Lambda Y_1 = 0,$$

$$Y_0'' + \frac{5}{6} Y_0' + \Lambda Y_0 = 0, \qquad Y_{-1}'' - \frac{5}{4} Y_0' + \Lambda Y_0 = 0;$$

Λ is an approximate value for λ.

We still have to eliminate the three remaining first derivatives Y_0', Y_1', Y_2' and for this we use Hermitian formulae for the first derivative — symmetric formulae for the points $x = \frac{1}{5}$, $\frac{3}{5}$ and a lower order unsymmetric formula (see Table III of the appendix) for the boundary point $x = 1$:

$$Y_1 - Y_0 - \frac{2}{15}(Y_1' + 3 Y_0') = 0,$$

$$-Y_0 - \frac{2}{15}(Y_2' + 4 Y_1' + Y_0') = 0,$$

$$-\frac{2}{3} Y_1 - \frac{2}{15}(Y_2' + Y_1') = 0.$$

The eliminations yield finally two linear homogeneous equations for Y_0 and Y_1, namely

$$(-6870 + 11\,\mu)\, Y_0 \qquad + (6995 + \mu)\, Y_1 = 0,$$
$$(4812 + \mu)\, Y_0 + (-11158 + 10\,\mu)\, Y_1 = 0,$$

where $384\,\Lambda = 5\mu$, and from the usual determinant condition for a non-trivial solution we obtain

$$\Lambda = \begin{cases} 3.1677 & (\text{error } -1.6\%), \\ 21.111 & (\text{error } -6\ \%). \end{cases}$$

2.7. A Hermitian method for linear boundary-value problems

For a "first" boundary-value problem whose differential equation does not involve y' explicitly, as in Example I of § 2.6, for instance, substitution from the differential equation into (2.14) immediately yields a set of equations for the Y_ν; on the other hand, if y' does appear in the differential equation, the set of equations derived from (2.14) and (2.15) contain the unknowns Y_ν' in addition to the Y_ν. We now describe how a set of equations for the Y_ν alone can be obtained directly for a linear boundary-value problem of the n-th order, even when the derivatives $y', \ldots, y^{(n-1)}$ all appear in the differential equation[1].

If the functions $f_k(x)$ in the linear differential equation

$$L[y] \equiv \sum_{k=0}^{n} f_k(x)\, y^{(k)} = r(x) \tag{2.18}$$

possess continuous derivatives of the k-th order, the equation may be written in the form

$$L[y] \equiv \sum_{k=0}^{n} [g_k(x)\, y]^{(k)} = r(x). \tag{2.19}$$

This may be shown very easily by construction, for if we define sets of functions $f_k^{[r]}$ successively by forming the sequence of differential expressions

$$L_1 = L - (f_n\, y)^{(n)} = \sum_{k=0}^{n-1} f_k^{[1]}\, y^{(k)},$$

$$L_2 = L_1 - (f_{n-1}^{[1]}\, y)^{(n-1)} = \sum_{k=0}^{n-2} f_k^{[2]}\, y^{(k)},$$

$$\cdots \cdots \cdots \cdots \cdots \cdots \cdots \cdots \cdots$$

$$L_n = L_{n-1} - (f_1^{[n-1]}\, y)' = f_0^{[n]}\, y = (f_0^{[n]}\, y)^{(0)},$$

which decrease in order from $(n-1)$ down to zero, then the result follows immediately by addition, and $g_k = f_k^{[n-k]}$ with $f_n^{[0]} = f_n$. For $n = 2$ we have

$$L[y] \equiv (f_2\, y)'' + [(f_1 - 2f_2')\, y]' + (f_0 - f_1' + f_2'')\, y = r(x). \tag{2.20}$$

Instead of the formula (2.15) for the lower derivatives, we use here Hermitian formulae which, though very similar, differ in that the coefficients of the derivative values (corresponding to $A_\nu', A_\nu'', \ldots, A_\nu^{(n-1)}$) are the same for each derivative and equal to those in (2.14) for the

[1] Following a somewhat differently derived method due to H. SASSENFELD: Ein Summenverfahren für Rand- und Eigenwertaufgaben linearer Differential-gleichungen. Z. Angew. Math. Mech. **31**, 240—241 (1951); presented in detail and called "Quadraturverfahren" by R. ZURMÜHL: Praktische Mathematik für Ingenieure und Physiker, 2nd ed. p. 447 et seq. Berlin-Göttingen-Heidelberg 1957.

highest derivatives; thus we derive Hermitian equations of the form

$$\sum_{\nu=-p}^{p} (c_\nu^{(k)} Y_{i+\nu} + A_\nu Y_{i+\nu}^{(k)}) = 0 \qquad (k = 1, \ldots, n-1) \qquad (2.21)$$

(we may include the equation with $k = n$ if we put $c_\nu^{(n)} = a_\nu$). The coefficients $c_\nu^{(k)}$ are determined in the usual way by making the Taylor expansions of the left-hand sides vanish to as high an order as possible; for $n = 2$, for example, we establish the formulae

$$\left. \begin{aligned} \frac{12}{h^2} (y_1 - 2y_0 + y_{-1}) - (y_1'' + 10 y_0'' + y_{-1}'') &= 0(h^4), \\ \frac{1}{2h} (-y_2 + 14 y_1 - 14 y_{-1} + y_{-2}) - (y_1' + 10 y_0' + y_{-1}') &= 0(h^4), \\ (y_1 + 10 y_0 + y_{-1}) - (y_1 + 10 y_0 + y_{-1}) &= 0. \end{aligned} \right\} \qquad (2.22)$$

[The last equation is included for convenience in writing down equation (2.23) below for specific examples.]

We now form the expression

$$\sum_{\nu=-p}^{p} A_\nu (L[y])_{x = x_{i+\nu}}$$

with $L[y]$ in the form (2.19); by virtue of (2.21) we can replace

$$\sum_{\nu=-p}^{p} A_\nu [g_k y]_{x = x_{i+\nu}}^{(k)} \quad \text{by} \quad - \sum_{\nu=-p}^{p} c_\nu^{(k)} (g_k y)_{x = x_{i+\nu}};$$

hence, using the differential equation at the points $x_{i-p}, \ldots, x_{i+p}$, we obtain

$$\sum_{\nu=-p}^{p} \sum_{k=0}^{n} c_\nu^{(k)} (g_k Y)_{x = x_{i+\nu}} = - \sum_{\nu=-p}^{p} A_\nu r(x_{i+\nu}), \qquad (2.23)$$

which provides the required linear relation between the Y_ν.

The number of equations obtained by writing this equation down for successive pivotal points x_i together with equations representing the boundary conditions must always be less than the number of unknown pivotal values and consequently further equations are needed. These can be obtained by using unsymmetric expressions for points near the boundaries; for example, in the case $n = 2$ we can use the set of formulae[1]

$$\frac{12}{h^2} (y_2 - y_1 - y_0 + y_{-1}) - (y_2'' + 11 y_1'' + 11 y_0'' + y_{-1}'') = 0(h^4),$$

$$\frac{4}{h} (y_2 + 3 y_1 - 3 y_0 - y_{-1}) - (y_2' + 11 y_1' + 11 y_0' + y_{-1}') = 0(h^4),$$

$$(y_2 + 11 y_1 + 11 y_0 + y_{-1}) - (y_2 + 11 y_1 + 11 y_0 + y_{-1}) = 0.$$

[1] These, together with formulae for differential equations of the fourth order, are to be found in R. ZURMÜHL: Praktische Mathematik für Ingenieure und Physiker, 2nd. ed. Berlin-Göttingen-Heidelberg 1957.

The boundary conditions themselves likewise require special formulae. If, for example, a boundary condition for a second-order problem specifies the value of y' or a linear combination of y and y' at a boundary, a representative finite equation can be derived from a formula such as

$$- 85 \, y_0 + 108 \, y_1 - 27 \, y_2 + 4 \, y_3 - 66 \, h \, y_0' - 18 \, h^2 \, y_0'' = 0 \, (h^5);$$

we substitute for y_0'' in terms of y_0' and y_0 by means of the differential equation and then for y_0' in terms of y_0 by means of the boundary condition (in conformity with the usual notation the y_i are to be replaced by the approximations Y_i when the remainder term is omitted).

§3. Some theoretical aspects of the finite-difference methods

3.1. Solubility of the finite-difference equations and convergence of iterative solutions

In §§ 1 and 2 we described various ways of setting up a system of linear equations for any given linear boundary-value problem. Here we consider the solubility of such a system but naturally have to restrict the generality — after all, not every linear boundary-value problem is soluble. If we focus attention on certain classes of boundary-value problems, a few additional assumptions suffice to enable us to prove the existence and uniqueness of the solution of the finite-difference equations.

We start with a simple example. For the boundary-value problem

$$\left. \begin{array}{l} - (f \, y')' + g \, y = r \, (x); \\ y \, (a) = y_a , \qquad y \, (b) = y_b \end{array} \right\} \tag{3.1}$$

the ordinary finite-difference method using (1.5) yields the equations

$$\left. \begin{array}{l} \frac{1}{h^2} \left[- f_{i+\frac{1}{2}} \, Y_{i+1} + (f_{i+\frac{1}{2}} + f_{i-\frac{1}{2}}) \, Y_i - f_{i-\frac{1}{2}} \, Y_{i-1} \right] + g_i \, Y_i - r_i = 0 \\ \hspace{6cm} (i = 1, 2, \ldots, n - 1) \\ Y_0 = y_a , \qquad Y_n = y_b . \end{array} \right\} \tag{3.2}$$

If we assume that $f(x) > 0$ and $g(x) \geq 0$, this system of equations satisfies the conditions of Theorem 2 in Ch. I, § 5.5 (in addition to the sign distribution, the weak row-sum criterion is satisfied and the matrix of coefficients does not decompose); hence the system possesses a uniquely determined solution for arbitrary values of y_a, y_b and the r_i. Further, by Theorem 3 in Ch. I, § 5.5 this solution may be computed iteratively in single or total steps, i.e. the single-step and total-step iterations converge.

We now describe another, quite different, way of showing that a unique solution of the system (3.2) exists. With this method the convergence of the single-step iteration follows at the same time. The method is also applicable to certain differential equations of higher order, for which, in general, the weak row-sum criterion is no longer satisfied. We use the fact that the system of $n-1$ equations in (3.2) (with Y_0, Y_n not included among the unknowns) is symmetric and identify it with the system of equations

$$\frac{\partial Q}{\partial Y_i} = 0 \qquad (i = 1, 2, \dots, n-1)$$

which constitute the necessary conditions for a minimum of the quadratic function

$$Q = \frac{1}{2} \sum_{\nu=0}^{n-1} f_{\nu+\frac{1}{2}} \left(\frac{Y_{\nu+1} - Y_\nu}{h} \right)^2 + \sum_{\nu=1}^{n-1} \left(\frac{1}{2} g_\nu Y_\nu^2 - r_\nu Y_\nu \right) \qquad (3.3)$$

of the $n-1$ variables $Y_1, Y_2, \dots, Y_{n-1}$ (Y_0, Y_n being fixed at the values y_a, y_b).

Now under our assumptions that $f(x) > 0$ and $g(x) \geqq 0$, the corresponding "homogeneous" quadratic form

$$Q^* = \frac{1}{2} \sum_{\nu=0}^{n-1} f_{\nu+\frac{1}{2}} \left(\frac{Y_{\nu+1} - Y_\nu}{h} \right)^2 + \frac{1}{2} \sum_{\nu=1}^{n-1} g_\nu Y_\nu^2 \quad \text{with} \quad Y_0 = Y_n = 0$$

obtained from Q by putting r_ν, Y_0, Y_n equal to zero is positive definite: obviously it cannot take negative values if the Y_ν are real and it is zero only when all the Y_ν are zero; for $Q^* = 0$ implies that

$$Y_{\nu+1} - Y_\nu = 0, \quad \text{so that} \quad Y_\nu = \text{constant},$$

and since Y_0, Y_n are zero here, the constant must be zero. The determinant of the quadratic form Q^* must therefore be positive. But this determinant is also the determinant of the system of equations (3.2) and hence that system has a unique solution for any values of r_ν, y_a, y_b. According to a well-known theorem[1], the fact that Q^* is positive definite implies also that the solution can be computed by single-step iteration.

If, instead of (3.1), we have more complicated boundary conditions, say

$$- A y(a) + y'(a) = R \qquad (3.4)$$

at $x = a$, then the corresponding finite-difference equation reads

$$- A Y_0 + \frac{Y_1 - Y_{-1}}{2h} = R.$$

---

[1] MISES, R. v., and H. POLLACZEK-GEIRINGER: Praktische Verfahren der Gleichungsauflösung. Z. Angew. Math. Mech. 9, 58—77 (1929).

If we also write down the difference equation of (3.2) with $i=0$, i.e.

$$- f_{\frac{1}{2}} Y_1 + (f_{\frac{1}{2}} + f_{-\frac{1}{2}} + g_0 h^2)\, Y_0 - f_{-\frac{1}{2}} Y_{-1} = r_0 h^2,$$

we can eliminate Y_{-1} between it and the previous equation. The resulting relation

$$(f_{\frac{1}{2}} + \alpha)\, Y_0 - f_{\frac{1}{2}} Y_1 = \beta,$$

where

$$\alpha = \frac{(g_0 h^2 + 2 h\, A\, f_{-\frac{1}{2}})\, f_{\frac{1}{2}}}{f_{\frac{1}{2}} + f_{-\frac{1}{2}}} \quad \text{and} \quad \beta = \frac{(r_0 h^2 - 2 h f_{-\frac{1}{2}} R)\, f_{\frac{1}{2}}}{f_{\frac{1}{2}} + f_{-\frac{1}{2}}},$$

between Y_0 and Y_1 can be added to the set of symmetric equations of (3.2) without disturbing the symmetry. Thus we can still identify the system with the minimum equations $\partial Q/\partial Y_i = 0$ $(i=0, 1, \ldots, n-1)$ for a quadratic function Q. The appropriate quadratic function now has the form

$$Q = \frac{1}{2} \sum_{\nu=0}^{n-1} f_{\nu+\frac{1}{2}} \left(\frac{Y_{\nu+1} - Y_\nu}{h} \right)^2 + \frac{1}{2} \alpha \left(\frac{Y_0}{h} \right)^2 + \sum_{\nu=1}^{n-1} \left(\frac{1}{2} g_\nu Y_\nu^2 - r_\nu Y_\nu \right) - \frac{\beta}{h^2} Y_0.$$

The quadratic form Q^* which results when we put $Y_n=0$, $r_\nu=0$, $R=0$ in Q will be positive definite if, in addition to the previous assumptions $f > 0$ and $g \geq 0$, we assume further that $A \geq 0$; for then $\alpha \geq 0$. This ensures also that the weak row-sum criterion is still satisfied; it will not be satisfied, however, if $g \equiv 0$ and y' is prescribed $(A=0)$ at both end-points (the "second" boundary-value problem).

In general the system of equations arising in the application of the Hermitian method of § 2.4 is not symmetric and the quadratic form approach is no longer available. In many cases, however, it can be shown that the weak row-sum criterion and the other conditions of Theorem 2 in Ch. I, § 5.5 are satisfied. As an example we choose the special boundary-value problem

$$- y'' + f(x)\, y = r(x); \qquad y(a) = y_a, \qquad y(b) = y_b.$$

With $h = \dfrac{b-a}{n}$, $x_i = a + i h$ the associated finite equations (cf. Ex. I, § 2.6) read

$$\left. \begin{array}{c} Y_{i+1} - 2 Y_i + Y_{i-1} - \dfrac{h^2}{12} (Y_{i+1}'' + 10\, Y_i'' + Y_{i-1}'') = 0 \\[2mm] Y_i'' = f_i Y_i - r_i; \qquad Y_0 = y_a, \qquad Y_n = y_b \end{array} \right\} \quad (i=1, 2, \ldots, n-1)$$

and yield the system

$$Y_{i+1}\left(1 - \frac{h^2}{12} f_{i+1}\right) - Y_i\left(2 + \frac{h^2 \cdot 10}{12} f_i\right) + Y_{i-1}\left(1 - \frac{h^2}{12} f_{i-1}\right)$$
$$= -\frac{h^2}{12} (r_{i+1} + 10\, r_i + r_{i-1}).$$

For $f(x) > 0$ and sufficiently small h, for example, $h < (12/f)^{\frac{1}{2}}$, the row-sum criterion is satisfied.

3.2. A general principle for error estimation with the finite-difference methods in the case of linear boundary-value problems

In principle we can always calculate error bounds for the finite-difference methods, for the refined methods of § 2 as well as for the ordinary method of § 1. *However, these bounds may be expected to be reasonably close to the actual error and to predict its order of magnitude correctly only when the solution is known fairly accurately; in particular, we need to have an approximate quantitive idea of the behaviour of the higher derivatives, which are critical for the error,* or if we are using maximum values of these derivatives, we must choose upper limits with as little over-estimation as possible. Although in general no error estimates exist which are both sufficiently accurate and sufficiently simple, we can at least describe a general procedure by which an error estimate is possible.

As far as we are concerned here the ordinary finite-difference method and the method of a higher approximation can be regarded as special cases of the Hermitian method of § 2.5, and we therefore consider a system of equations of Hermitian form. Thus all equations set up by the finite-difference methods for the N unknowns $Y_i, Y_i', \ldots, Y_i^{(n-1)}$ ($i = 1, \ldots, p$, say) constitute linear relations between these unknowns and the methods therefore lead to systems of equations of the form

$$A^{(\varrho)} = \sum_{\nu, k} a_{\nu, k}^{(\varrho)} Y_\nu^{(k)} = r^{(\varrho)} \qquad (\varrho = 1, 2, 3, \ldots, N). \tag{3.5}$$

We naturally assume that these equations have been set up so that the determinant of the coefficients $a_{\nu, k}^{(\varrho)}$ does not vanish, for otherwise we could not have determined an approximate solution.

We now form the same expressions $A^{(\varrho)}$ for the exact solution $y(x)$ of the boundary-value problem, whose existence and uniqueness we also assume. We then follow the usual procedure of expanding each expression into a Taylor series in the values of y and its derivatives at a point $x = x_\varrho$ and use the differential equation and boundary conditions for raising the order of the remainder terms. The resulting equations are of the form

$$\sum_{\nu, k} a_{\nu, k}^{(\varrho)} y^{(k)}(x_\nu) = r^{(\varrho)} + \vartheta_\varrho h^{n_\varrho} D_\varrho y^{(n_\varrho)}(\xi_\varrho) \qquad (\varrho = 1, 2, \ldots, N), \tag{3.6}$$

where $|\vartheta_\varrho| \leqq 1$, D_ϱ and n_ϱ are determinable quantities, and ξ_ϱ is an unknown point within the range of pivotal points involved in the ϱ-th equation.

We now derive a system of equations for the errors

$$\varphi_\nu^{(k)} = Y_\nu^{(k)} - y^{(k)}(x_\nu) \tag{3.7}$$

by subtracting equations (3.6) from (3.5):

$$\sum_{\nu, k} a_{\nu, k}^{(\varrho)} \, \varphi_{\nu}^{(k)} = - \vartheta_{\varrho} \, h^{n_{\varrho}} D_{\varrho} y^{(n_{\varrho})} (\xi_{\varrho}) . \tag{3.8}$$

Since the determinant was assumed above to be non-zero, these equations can be solved. Thus, in so far as the finite-difference method is applicable at all, the possibility of an error estimate depends on the fact that the systems (3.5) and (3.6) have the same determinant.

For the practical application of such an estimate we often put

$$\vartheta_{\varrho} \, y^{(n_{\varrho})} (\xi_{\varrho}) = \vartheta_{\varrho}^* \, |y^{(n_{\varrho})}|_{\max} , \tag{3.9}$$

then solve the equations (3.8) for the $\varphi_{\nu}^{(k)}$ and from the results derive upper bounds for the $\varphi_{\nu}^{(k)}$ using the fact that $|\vartheta^*| \leq 1$. Values for the derivatives $y^{(n_{\varrho})}$ are obtained either by expressing these derivatives in terms of lower derivatives, which are known more accurately, by differentiating the differential equation (the order of the derivatives can be reduced still further by repeated integration of the differential equation) or, less accurately, by using the differences of the Y_{ϱ} in a finite-difference derivative formula.

This is as much as can be said in the general case. That more can be proved[1] if special assumptions are made is shown by the example in the next section.

3.3. An error estimate for a class of linear boundary-value problems of the second order

In illustration of the general procedure just described in § 3.2 we derive a general error estimate for the ordinary finite-difference method applied to the special class of boundary-value problems defined by

$$L[y] = y'' + f_1 y' + f_0 y = r(x); \quad y(a) = y_0, \quad y(b) = y_n, \tag{3.10}$$

where y_0, y_n are given boundary values, f_0, f_1, r are functions with continuous second derivatives and $f_0 \leq 0$ (for positive f_0 the boundary-value problem may not possess a solution). Let the interval h be chosen so small that

$$\left| \frac{h}{2} f_1 \right| < 1 \quad \text{for} \quad a \leq x \leq b. \tag{3.11}$$

---

[1] For many boundary-value problems of monotonic type error estimates can be obtained in a quite different way which does not require knowledge of bounds for the higher derivatives; cf. L. COLLATZ: Aufgaben monotoner Art. Arch. Math. 3, 375 (1952), in which a numerical example with error limits is also given.

The corresponding boundary-value problem for the difference equations (see § 1.1) reads

$$A[Y_i] = Y_{i+1} - 2Y_i + Y_{i-1} + f_{1,i}\frac{h}{2}(Y_{i+1} - Y_{i-1}) + f_{0,i}h^2 Y_i = h^2 r_i \left.\begin{array}{c} \\ (i = 1, 2, \ldots, n-1), \\ \end{array}\right\} \quad (3.12)$$

$$Y_0 = y_0, \qquad Y_n = y_n.$$

Taylor expansion of $A[y_i]$ yields (cf. Table III of the appendix)

$$A[y_i] = h^2 L[y]_{x=x_i} + R_i, \tag{3.13}$$

where

$$|R_i| \leqq R^* = \frac{h^4}{12}\left(2|f_1 y'''|_{\max} + |y^{(4)}|_{\max}\right). \tag{3.14}$$

From (3.12) and (3.13) the error $\varphi_i = Y_i - y_i$ satisfies the difference boundary-value problem

$$A[\varphi_i] = -R_i, \qquad \varphi_0 = \varphi_n = 0. \tag{3.15}$$

This is a set of $n+1$ linear equations for the $n+1$ quantities φ_i with right-hand sides for which there exist upper limits as given by (3.14). From these equations we shall derive upper limits for the quantities $|\varphi_i|$. The matrix of coefficients satisfies the conditions of Theorem 2 in Ch. I, § 5.5 (sign distribution, weak row-sum criterion, non-decomposition), so that

$$|Y_i - y_i| = |\varphi_i| \leqq w_i, \tag{3.16}$$

where the w_i are quantities determined from equations $-A[w_i] = S_i$ with $S_i \geqq |R_i|$. Such quantities w_i can be obtained by solving the equations $-A[w_i] = 1$; for then we need only put $w_i = R^* w_i^*$, where R^* is defined in (3.14). In many cases another way is possible which avoids the necessity of solving a second system of equations. This depends on being able to find quantities w_i^* for which the numbers $-A[w_i^*] = \sigma_i$ all turn out positive (see the following numerical example); we can then write

$$|\varphi_i| \leqq w_i = \frac{w_i^* R^*}{(\sigma_i)_{\min}}. \tag{3.17}$$

Numerical example. Let us consider a simple example for which we know the exact solution so that the error limits (3.17) can be compared with the actual errors. Such an example is provided by the boundary-value problem

$$y'' - \frac{2}{x^2}y + \frac{1}{x} = 0, \qquad y(2) = y(3) = 0,$$

which has the exact solution

$$y = \frac{1}{38}\left(19x - 5x^2 - \frac{36}{x}\right).$$

The pivotal values Y_1, Y_2 corresponding to a coarse subdivision with $h = \frac{1}{3}$ are calculated from the ordinary finite-difference equations

$$\frac{-2Y_1 + Y_2}{h^2} - \frac{2 \times 9}{49} Y_1 + \frac{3}{7} = 0, \qquad \frac{Y_1 - 2Y_2}{h^2} - \frac{2 \times 9}{64} Y_2 + \frac{3}{8} = 0$$

as

$$Y_1 = \frac{217}{4932} = 0.043\,998 \quad (\text{error} -0.000\,279, \text{ i.e. } -0.6\%),$$

$$Y_2 = \frac{208}{4932} = 0.042\,174 \quad (\text{error} -0.000\,225, \text{ i.e. } -0.5\%).$$

For the error estimate we need first a value for $|y^{(4)}|_{\max}$; the maximum of the fourth difference quotient is often used for this, but in the present simple case it is better to estimate y' approximately and use the fact that

$$y^{(4)} = \frac{4}{x^4} (4y - 2xy' - x),$$

which follows from the differential equation; we find that

$$|y^{(4)}|_{\max} \approx 0.71.$$

Hence from (3.14)

$$R^* \approx \left(\frac{1}{3}\right)^4 \times \frac{0.71}{12} \approx 0.000\,73.$$

We need further a solution of the equations $-\Lambda[w_i^*] = \sigma_i$ with positive w_i^* and σ_i. Here these equations read

$$\frac{100}{49} w_1^* - w_2^* = \sigma_1, \qquad -w_1^* + \frac{65}{32} w_2^* = \sigma_2,$$

and we need look no further than the simple values $w_1^* = w_2^* = 1$; they yield $\sigma_1 = \frac{51}{49}$, $\sigma_2 = \frac{33}{32}$.

Then (3.17) yields the error estimate

$$|Y_i - y_i| \leq \frac{32}{33} \times 0.000\,73 = 0.000\,71.$$

At x_1, for example, these limits are about two and a half times greater than the actual error.

3.4. An error estimate for a non-linear boundary-value problem

In principle, the method described in § 3.2 may also be applied to non-linear boundary-value problems provided only that the solution y and its derivatives $y^{(\nu)}$ are known, or can be estimated, sufficiently accurately. Usually, of course, we will not be in a position to make the necessary estimates of y and $y^{(\nu)}$ beforehand and will have to infer their quantitive behaviour approximately from the values calculated by the finite-difference method. Consequently a rigorous error estimate is out of the question and we must be satisfied with approximate error limits. In practice, the calculation of such approximate limits is often so laborious and the results so crude that it is preferable to dispense with them altogether and rely on the indication of the order of magnitude of the error provided by the results of two or more calculations with different intervals. Thus the procedure does not warrant more explanation than is afforded by the following simple example. The problem we consider is that of Example II of § 2, i.e.

$$y'' = \tfrac{3}{2} y^2; \qquad y(0) = 4, \qquad y(1) = 1. \tag{3.18}$$

With $h = \tfrac{1}{5}$ the approximate pivotal values Y_i satisfy the equations

$$Y_{i+1} - 2 Y_i + Y_{i-1} - \tfrac{3}{2} h^2 Y_i^2 = 0 \qquad (i = 1, 2, 3, 4), \tag{3.19}$$

while the exact values y_i satisfy the inhomogeneous equations

$$y_{i+1} - 2 y_i + y_{i-1} - \tfrac{3}{2} h^2 y_i^2 = R_i \qquad (i = 1, 2, 3, 4), \tag{3.20}$$

where

$$|R_i| \leq \frac{h^4}{12} |y^{(4)}|_{\text{max in } \langle x_{i-1}, x_{i+1}\rangle}. \tag{3.21}$$

The equations for the error $\varphi_i = Y_i - y_i$ obtained by subtracting (3.19) from (3.20) can be written in the form

$$a_i \varphi_i - \varphi_{i-1} - \varphi_{i+1} = R_i \qquad (i = 1, 2, 3, 4), \tag{3.22}$$

where

$$a_i = 2 + \tfrac{3}{2} h^2 (Y_i + y_i), \qquad \varphi_0 = \varphi_5 = 0. \tag{3.23}$$

With $\alpha = a_1 a_2 - 1$, $\beta = a_3 a_4 - 1$, $D = \alpha\beta - a_1 a_4$ the explicit solution of equations (3.22) for the four unknowns φ_i reads

$$\left. \begin{aligned} D \varphi_1 &= R_1 (a_2 \beta - a_4) + R_2 \beta && + R_3 a_4 && + R_4, \\ D \varphi_2 &= R_1 \beta && + R_2 a_1 \beta && + R_3 a_1 a_4 && + R_4 a_1, \\ D \varphi_3 &= R_1 a_4 && + R_2 a_1 a_4 && + R_3 a_4 \alpha && + R_4 \alpha, \\ D \varphi_4 &= R_1 && + R_2 a_1 && + R_3 \alpha && + R_4 (a_3 \alpha - a_1). \end{aligned} \right\} \tag{3.24}$$

We complete the estimate only for the solution without zeros and further restrict attention to the point $i = 2$ (i.e. $x = 0.4$). As mentioned earlier, we have to make some approximations. Thus we replace the unknown exact solution $y(x)$ and its derivatives everywhere by the approximate values given by the finite-difference method: the $Y_i + y_i$ in the quantities a_i of (3.23) are replaced by $2 Y_i$, and pivotal values of the fourth derivative

$$y^{IV} = 3 y'^2 + \tfrac{9}{2} y^3$$

are calculated approximately by replacing the derivatives y_i' by the difference quotients $\dfrac{Y_{i+1}-Y_{i-1}}{2h}$. The approximate values of the y_i' so obtained are exhibited in Table III/10 together with the "local" maxima

Table III/10. *Quantities required in the error estimate* (3.24)

i	x_{i-1}	y_{i-1}'	$y_{\max}^{IV}$ in $\langle x_{i-1}, x_{i+1}\rangle$	Y_i	a_i
1	0	8·43	501	2·794 64	2·335
2	h	4·86	169	2·057 87	2·246
3	$2h$	3·05	68	1·575 19	2·189
4	$3h$	2·04	30	1·241 38	2·149

of y^{IV} derived from them. The latter are used in the bounds (3.21) for the R_i, and from (3.24) with $\alpha = 4\cdot25$, $\beta = 3\cdot71$, $D = 10\cdot73$, we obtain the estimate

$$|\varphi_2| \leqq 0\cdot0926.$$

This limit is about five and a half times greater than the actual error $\varphi_2 = 0\cdot01705$.

§4. Some general methods

A series of general methods, especially methods based on error distribution principles, were discussed in Ch. I, §4.2; it will therefore suffice here to refer to these methods and give examples. More general boundary-value problems for ordinary differential equations have already been formulated in Ch. I, §1.

For our approximation to the required solution $y(x)$ we choose here a function $w(x, a_1, \ldots, a_p)$ which depends on several parameters $a_1, \ldots, a_p$ and which already satisfies the boundary conditions independently of the choice of parameter values. By inserting w into the differential equation we obtain a function $\varepsilon(x, a_1, \ldots, a_p)$ which represents the residual error. We then determine the a_p in accordance with one of the principles mentioned in Ch. I, §4.2 so that this error function approximates the zero function as closely as possible.

4.1. Examples of collocation

The collocation method has a very wide range of applicability, is simple to use and requires no special previous knowledge. In many cases the method yields much more than a rough quantitative idea of the solution, in fact the accuracy of the results can sometimes be quite remarkable in view of the primitive nature of the method and the slight amount of computation involved.

Example I. For the problem of Example I, § 1.2:

$$L[y] = y'' + (1 + x^2)\, y + 1 = 0, \qquad y(\pm 1) = 0$$

we use the functions $x^{2n}(1 - x^2)$ as satisfying the boundary conditions and the symmetry about $x = 0$; thus our approximating function w is of the form

$$w = a_1(1 - x^2) + a_2(x^2 - x^4) + a_3(x^4 - x^6) + \cdots,$$

and substituted in L it yields

$$L[w] = 1 + a_1\left(-2 + (1 - x^4)\right) + a_2\left(2 - 12x^2 + x^2(1 - x^4)\right) +$$
$$+ a_3\left(12x^2 - 30x^4 + x^4(1 - x^4)\right) + \cdots.$$

If, for example, we use a two-term expression, i.e. with only two parameters a_1 and a_2, we can make $L[w] = 0$ at four symmetrically placed points, say $x = \pm\frac{1}{4}$ and $x = \pm\frac{3}{4}$; this leads to the linear equations

$$\left(\text{for } x = \frac{1}{4}\right) \quad 1 - a_1 \frac{257}{256} + a_2 \frac{5375}{4096} = 0,$$

$$\left(\text{for } x = \frac{3}{4}\right) \quad 1 - a_1 \frac{337}{256} - a_2 \frac{17881}{4096} = 0$$

with the solution $a_1 = 0.929254$, $a_2 = -0.051146$. In Table III/11 these results are compared with those obtained by using one-term and three-term expressions, respectively, with suitable collocation points.

Table III/11. *Comparison of collocation results as the number of parameters increases*

p-term expression	Collocation points x	Results: values of the a_ν		
$p = 1$	$\frac{1}{2}$	$a_1 = 0.94118$		
$p = 2$	$\frac{1}{4}$ and $\frac{3}{4}$	$a_1 = 0.92925$,	$a_2 = -0.05115$	
$p = 3$	$\frac{1}{6}$, $\frac{3}{6}$ and $\frac{5}{6}$	$a_1 = 0.932088$,	$a_2 = -0.034108$,	$a_3 = -0.030221$

Example II. Infinite interval. Another problem considered in § 1.2 was that of determining the steady temperature distribution in an infinitely long rod (Example IV); this leads to the boundary-value problem

$$L[y] = (2 + x)\, y'' - (1 + x)\, y = 0; \qquad y(0) = 1, \qquad y(\infty) = 0.$$

In order to satisfy the conditions at infinity, we choose an approximation of the form

$$y \approx w = \sum_{\nu=1}^{p} a_\nu e^{-\nu x}.$$

However, this function does not satisfy the boundary condition at $x=0$ for arbitrary values of the a_ν and we must therefore include the equation

$$\sum_{\nu=1}^{p} a_\nu = 1$$

with the collocation equations.

For $p=3$ the error function is

$$\varepsilon = L[w] = a_1 e^{-x} + a_2 (7 + 3x) e^{-2x} + a_3 (17 + 8x) e^{-3x}.$$

If, to start with, we drop the term in a_3, we can choose just one collocation point. Choosing $x=1$, we have

$$a_1 e^{-1} + a_2 10e^{-2} = 0,$$

$$a_1 + a_2 = 1,$$

and hence

$$a_1 = \frac{10}{10 - e},$$

$$a_2 = - \frac{e}{10 - e};$$

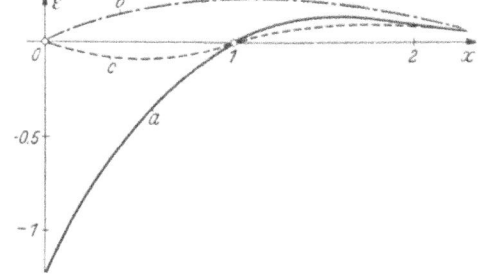

Fig. III/11. Curves of the residual error obtained by various collocations

this yields in particular

$$y(1) \approx w(1) = \frac{9}{e(10 - e)} = 0.455.$$

To get some idea of how closely the differential equation is satisfied, we can draw the corresponding ε curve (Fig. III/11, curve a). If $x=0$ is chosen as collocation point instead of $x=1$, we obtain with

$$a_1 = \tfrac{7}{6}, \qquad a_2 = - \tfrac{1}{6}$$

an ε curve (curve b in Fig. III/11) which appears to be an improvement on curve a. In spite of this, the y values are worse; for example,

$$w(1) = \frac{7e - 1}{6e^2} = 0.405.$$

When a_3 is included and $x=0$ and $x=1$, say, are used as collocation points, we have

$$a_1 = 1.3605, \qquad a_2 = -0.4768, \qquad a_3 = 0.1163;$$

the corresponding ε curve is shown in Fig. III/11 as curve c. At $x=1$ we have the good approximation

$$y(1) \approx w(1) = \frac{153e^2 - 168e + 15}{10e^2 (15 - e)(e - 1)} = 0.4418.$$

4.2. An example of the least-squares method

An example of a linear boundary-value problem is treated by the least-squares method in § 6.4, so here we confine ourselves to a non-linear example, namely the problem (1.9):

$$y'' = \tfrac{3}{2} y^2; \qquad y(0) = 4, \qquad y(1) = 1.$$

For an approximate solution we use the expression

$$\varphi = 4 - 3x + a(x - x^2),$$

which satisfies both boundary conditions for arbitrary a. As the residual error we define

$$\varepsilon = -2\varphi'' + 3\varphi^2 = 3(16 - 24x + 9x^2) + 2a(2 + 12x - 21x^2 + 9x^3) +$$
$$+ 3a^2(x^2 - 2x^3 + x^4).$$

For the least-squares method we now form

$$J[\varphi] = \int_0^1 \varepsilon^2 dx = \frac{1}{70}(42966 + 19740a + 3348a^2 + 101a^3 + a^4);$$

since a check on this calculation is desirable, we repeat it with $a = 3$:

$$\varepsilon = 3(20 - 24x^2 + 9x^4), \qquad \int_0^1 \varepsilon^2 dx = \frac{135126}{70}.$$

The least-squares equation

$$70 \frac{\partial J[\varphi]}{\partial a} = 4a^3 + 303a^2 + 6696a + 19740 = 0$$

then yields

$$a = \begin{cases} -3\cdot4671 \\ -36\cdot1 \pm 10\cdot8i, \end{cases}$$

of which only the real value is significant. With this value $a = -3\cdot4671$ the least-squares approximation to the solution at $x = \tfrac{1}{2}$ is

$$\varphi\left(\tfrac{1}{2}\right) = 1\cdot7332.$$

4.3. Reduction to initial-value problems

As a general method for the numerical solution of boundary-value problems which is very useful on occasions we may mention the technique of reducing the problem to two or more initial-value problems and treating these by one of the methods of Ch. II. (See also the examples of § 1.2.)

In the simplest case of a second-order linear differential equation

$$L[y] = r(x) \tag{4.1}$$

with boundary conditions

$$y(a) = y_a, \tag{4.2}$$

$$y(b) = y_b \tag{4.3}$$

at the points $x = a$ and $x = b$, we compute the solutions y_1 and y_2 of the two initial-value problems

$$\left. \begin{array}{llll} L[y_1] = r(x); & y_1(a) = y_a, & y_1'(a) = 0, \\ L[y_2] = 0; & y_2(a) = 0, & y_2'(a) = 1. \end{array} \right\} \quad (4.4)$$

This yields two values $y_1(b)$ and $y_2(b)$ at $x = b$. On account of the linearity of the problem, the solution of the differential equation with the initial conditions $y(a) = y_a$, $y'(a) = y_a'$ is given by

$$y(x) = y_1(x) + y_a' y_2(x).$$

Then if y_a' is calculated from

$$y_1(b) + y_a' y_2(b) = y_b,$$

the boundary condition at $x = b$ will also be satisfied and we have the required solution of the boundary-value problem.

For a fourth-order linear boundary-value problem with equation $L[y] = r(x)$ and boundary conditions

$$y(a) = y_a, \quad y'(a) = y_a', \quad y(b) = y_b, \quad y'(b) = y_b',$$

one calculates numerical solutions y_1, y_2, y_3 of the following three initial-value problems:

$$\left. \begin{array}{ll} y_1 \text{ from } & L[y_1] = r; \; y_1(a) = y_a, \; y_1'(a) = y_a', \; y_1''(a) = y_1'''(a) = 0, \\ y_2 \text{ from } & L[y_2] = 0; \; y_2(a) = y_2'(a) = 0, \; y_2''(a) = 1, \; y_2'''(a) = 0, \\ y_3 \text{ from } & L[y_3] = 0; \; y_3(a) = y_3'(a) = y_3''(a) = 0, \; y_3'''(a) = 1. \end{array} \right\} \quad (4.5)$$

The solution of the original boundary-value problem is then given by

$$y(x) = y_1(x) + y''(a) y_2(x) + y'''(a) y_3(x)$$

if $y''(a)$ and $y'''(a)$ are calculated from the two simultaneous linear equations

$$y_1(b) + y''(a) y_2(b) + y'''(a) y_3(b) = y_b,$$
$$y_1'(b) + y''(a) y_2'(b) + y'''(a) y_3'(b) = y_b'.$$

Generally, the solution $y(x)$ of an n-th-order linear boundary-value problem can always be obtained as a linear combination of the solutions $y_1, y_2, \ldots$ of at most $n+1$ initial-value problems, the constants c_ν of the combination being calculated from a system of linear equations in at most n unknowns. If the boundary conditions do not permit a convenient special choice of initial conditions such as in (4.4), (4.5), the conditions

$$L[y_k] = 0, \quad y_k^{(\nu)}(a) = \left\{ \begin{array}{ll} 0 \text{ for } \nu \neq k-1 \\ 1 \text{ for } \nu = k-1 \end{array} \right\} \quad \begin{array}{l} (\nu = 0, 1, \ldots, n-1) \\ (k = 1, 2, \ldots, n) \end{array}$$

will always suffice; the functions $y_1, y_2, \ldots, y_n$ which they define form a "fundamental system" for the homogeneous differential equation, and with the additional function y_{n+1} determined from

$$L[y_{n+1}] = r(x), \qquad y_{n+1}^{(\nu)}(a) = 0 \qquad (\nu = 0, 1, \ldots, n-1)$$

the required solution can be obtained in the form

$$y(x) = y_{n+1}(x) + \sum_{\nu=1}^{n} c_\nu y_\nu(x). \tag{4.6}$$

When we come to non-linear boundary-value problems, linear combination is no longer applicable and we resort to interpolation. Thus several different solutions $y_1, y_2, \ldots$ are computed satisfying all the boundary conditions at the point $x = a$ and we interpolate among them to find one which satisfies the boundary conditions at $x = b$.

4.4. Perturbation methods

Perturbation methods can sometimes be employed with advantage when a boundary-value problem with a known, or easily derivable solution can be found in the "neighbourhood" of the given boundary-value problem[1], i.e. such that the values of the coefficients in its differential equation differ only slightly from the corresponding values in the given equation. For simplicity we consider here only those "neighbouring" problems which have the same boundary conditions as the given problem, but perturbation methods can also be used when the boundary conditions are "disturbed" as well as the differential equation. Attention is further restricted to linear boundary conditions in the form given in Ch. I (1.7).

---

[1] Applications of perturbation methods to eigenvalue problems can be found in the following papers: MEYER ZUR CAPELLEN, W.: Methode zur angenäherten Lösung von Eigenwertproblemen mit Anwendungen auf Schwingungsprobleme. Ann. Phys. (5) **8**, 297—352 (1931). — Genäherte Berechnung von Eigenwerten. Ing.-Arch. **10**, 167—174 (1939). — RELLICH, F.: Störungstheorie der Spektralzerlegung. Math. Ann. **113**, 600—619 (1936); **114**, 677—685 (1937); **116**, 555—570 (1939); **117**, 356—382 (1940); **118**, 462—484 (1942). — NAGY, B. v. Sz.: Perturbations des transformations autoadjointes dans l'espace de Hilbert. Comm. Math. Helv. **19**, 347—366 (1946). — Perturbations des transformations linéaires fermées. Acta Sci. Math. Szeged **14**, 125—137 (1951). — SCHRÖDER, J.: Fehlerabschätzungen zur Störungsrechnung bei linearen Eigenwertproblemen mit Operatoren eines Hilbertschen Raumes. Math. Nachr. **10**, 113—128 (1953). — Fehlerabschätzungen zur Störungsrechnung für lineare Eigenwertprobleme bei gewöhnlichen Differentialgleichungen. Z. Angew. Math. Mech. **34**, 140—149 (1954) (with a summary of results in a directly applicable form). — SCHÄFKE, FR. W.: Über Eigenwertprobleme mit 2 Parametern. Math. Nachr. **6**, 109—124 (1951). — Verbesserte Konvergenz- und Fehlerabschätzungen für die Störungsrechnung. Z. Angew. Math. Mech **33**, 255—259 (1953).

We introduce a perturbation parameter ε in such a way that the differential equation

$$G(\varepsilon, x, y, y', y'', \ldots, y^{(n)}) = 0 \tag{4.7}$$

has a known solution when $\varepsilon = 0$ and reproduces the given differential equation when $\varepsilon = 1$:

$$G(1, x, y, y', \ldots, y^{(n)}) = F(x, y, y', \ldots, y^{(n)}) = 0. \tag{4.8}$$

The differential equation with $\varepsilon = 0$ we call the "undisturbed" equation and that with $\varepsilon = 1$ the "disturbed" equation.

We now assume that the solution $\varphi = y(x, \varepsilon)$ of the boundary-value problem

$$G(\varepsilon, x, y, \ldots, y^{(n)}) = 0, \quad U_\nu[y] = \gamma_\nu, \quad (\nu = 1, 2, \ldots, n) \tag{4.9}$$

may be expanded in powers of ε:

$$y(x, \varepsilon) = y_0(x) + \varepsilon y_1(x) + \varepsilon^2 y_2(x) + \cdots \tag{4.10}$$

(that this is permissible can probably not be proved with the generality considered so far). The first term $y_0(x)$ is the known solution of the undisturbed equation satisfying the boundary conditions (1.7) of Ch. I. If the remaining coefficients y_1, y_2, ... satisfy the corresponding homogeneous boundary conditions, so that

$$U_\nu[y_0] = \gamma_\nu, \quad U_\nu[y_j] = 0 \quad (\nu = 1, 2, \ldots, n; \ j = 1, 2, \ldots), \tag{4.11}$$

then the power series (4.10) will satisfy the inhomogeneous conditions.

We now replace y in the differential equation (4.9) by the series (4.10):

$$G\left(\varepsilon, x, \sum_{r=0}^{\infty} \varepsilon^r y_r, \sum_{r=0}^{\infty} \varepsilon^r y_r', \ldots, \sum_{r=0}^{\infty} \varepsilon^r y_r^{(n)}\right) = 0. \tag{4.12}$$

Expansion of the left-hand side in powers of ε using Taylor's theorem (assumed to be a valid procedure) then yields

$$G\left(\varepsilon, x, \sum_{r=0}^{\infty} \varepsilon^r y_r, \ldots, \sum_{r=0}^{\infty} \varepsilon^r y_r^{(n)}\right) = \sum_{s=0}^{\infty} \varepsilon^s G_s = 0, \tag{4.13}$$

where the coefficients G_s are differential expressions involving the functions $y_0, y_1, \ldots, y_s$ and their derivatives; for example,

$$G_0 = G(0, x, y_0, y_0', \ldots, y_0^{(n)}) = 0$$

and

$$G_1 = \left\{\frac{\partial G}{\partial \varepsilon} + y_1 \frac{\partial G}{\partial y} + y_1' \frac{\partial G}{\partial y'} + \cdots + y_1^{(n)} \frac{\partial G}{\partial y^{(n)}}\right\}_{\varepsilon=0}. \tag{4.14}$$

When we have calculated y_1, y_2, ... up to y_{s-1}, the equation $G_s = 0$ constitutes, in general, an n-th-order differential equation for y_s, from which y_s can be determined with the boundary conditions (4.11). Thus $y_1, y_2, y_3, \ldots$ can be calculated successively starting with y_1, which is calculated from y_0 by means of the linear boundary-value problem with the differential equation $G_1 = 0$ from (4.14) and the homogeneous boundary conditions from (4.11). Such a perturbation method is to be recommended only when the boundary-value problems for the y_s are of simple form.

Example. We consider again the example of the transversely loaded strut (I of § 1.2), and write the differential equation in the form

$$y'' + (1 + \varepsilon x^2) y + 1 = 0, \qquad y(\pm 1) = 0.$$

Then for $\varepsilon = 1$ we have the equation to be solved and for $\varepsilon = 0$ we have a simple differential equation with elementary solution:

$$y_0'' + y_0 + 1 = 0, \qquad y_0(\pm 1) = 0$$

with the solution

$$y_0 = \frac{\cos x}{\cos 1} - 1.$$

Substituting the power series (4.10) for y in the differential equation, we obtain

$$\sum_{n=0}^{\infty} \varepsilon^n (y_n'' + y_n) + \sum_{n=1}^{\infty} \varepsilon^n x^2 y_{n-1} + 1 = 0,$$

and hence

$$y_n'' + y_n + x^2 y_{n-1} = 0 \qquad (n = 1, 2, \ldots)$$

since the coefficients of the powers of ε must all be zero. With the boundary conditions $y_n(\pm 1) = 0$ we have a sequence of boundary-value problems from which the functions $y_1, y_2, \ldots$ can be determined successively.

The first function y_1 satisfies

$$y_1'' + y_1 = x^2 - x^2 \frac{\cos x}{\cos 1}, \qquad y_1(\pm 1) = 0,$$

and is given by

$$y_1 = x^2 - 2 + \frac{1}{12 \cos 1} \left[(3x - 2x^3) \sin x - 3 x^2 \cos x \right] + A \cos x,$$

where

$$A = \frac{1}{12 \cos^2 1} (15 \cos 1 - \sin 1).$$

The determination of further y_j is more laborious.

For the value at $x = 0$ the first approximation gives

$$y_0(0) = \frac{1 - \cos 1}{\cos 1} = 0.8508 \quad (\text{error} -9\%)$$

and the second

$$y_0(0) + y_1(0) = \frac{1}{\cos 1} \left(\frac{9}{4} - 3 \cos 1 - \frac{1}{12} \tan 1 \right) = 0.9241 \quad (\text{error} -0.8\%).$$

4.5. The iteration method or the method of successive approximations

In this method the differential equation (1.1) of Ch. I for a general boundary-value problem is put into the form

$$F_1(x, y, y', \ldots, y^{(n)}) = F_2(x, y, y', \ldots, y^{(m)}), \qquad (4.15)$$

where $m < n$ and F_1 is a convenient, simple function such that the boundary-value problem

$$F_1(x, z, z', \ldots, z^{(n)}) = r(x) \qquad (4.16)$$

with the boundary conditions (1.3) of Ch. I, which we write shortly as

$$V_\nu(z) = 0 \qquad (\nu = 1, 2, \ldots, n), \qquad (4.17)$$

can be solved readily for any right-hand side $r(x)$. If, as is usually the case, the given differential equation can be solved for the highest derivative which occurs in terms of the lower derivatives, i.e. it can be put into the form

$$y^{(n)} = \varphi(x, y, y', \ldots, y^{(n-1)}), \qquad (4.18)$$

then one obvious rearrangement of the form (4.15) has $F_1 = y^{(n)}$ and $F_2 = \varphi$.

We can now define an iterative procedure which determines a sequence of functions $y_0(x)$, $y_1(x)$, $y_2(x)$, ... in the following manner: $y_0(x)$ is chosen arbitrarily, then $y_1(x)$, $y_2(x)$, ... calculated successively as the solutions of the boundary-value problems

$$\left. \begin{array}{l} F_1\left(x, y_{k+1}(x), y'_{k+1}(x), \ldots, y^{(n)}_{k+1}(x)\right) \\ \quad = F_2\left(x, y_k(x), y'_k(x), \ldots, y^{(m)}_k(x)\right), \\ V_\nu(y_{k+1}) = 0 \end{array} \right\} \quad (k = 0, 1, 2, \ldots). \qquad (4.19)$$

With this generality nothing can be asserted about the convergence of the sequence $y_k(x)$ to the solution $y(x)$ of the boundary-value problem; it can happen that the sequence does not converge at all. When it does converge, the effectiveness of the method is often influenced considerably by the form of the rearrangement (4.15) of the given differential equation and by the choice of starting function $y_0(x)$; the method is generally more effective the closer $y_0(x)$ is to $y(x)$.

The method is often used graphically, particularly in the simple form mentioned for equations of the form (4.18). If, for example, we have a second-order equation

$$y'' = \varphi(x, y, y'), \qquad (4.20)$$

we first find a starting function $y_0(x)$, perhaps by means of the finite-difference method, then use a funicular polygon method[1] to perform

---

[1] See, for instance, E. KAMKE: Differentialgleichungen, Lösungsmethoden und Lösungen, Vol. I, 3rd ed., p. 164 et seq. Leipzig 1944.

the successive integrations which generate the sequence of functions
$y_1(x)$, $y_2(x)$, ...

If F_1, F_2, V_ν are all linear in y and its derivatives, and if certain
further assumptions are made, we can specify conditions for the convergence of the iteration.

In this case the n-th-order boundary-value problem may be written
in the form.

$$\left.\begin{array}{l} M[y] - aN[y] = r(x), \\ U_\mu[y] = \gamma_\mu \end{array}\right\} \quad (\mu = 1, 2, \ldots, n), \qquad (4.21)$$

where M and N are linear differential expressions and the $U_\mu[y] = \gamma_\mu$
are the linear boundary conditions (1.7) of Ch. I. The corresponding
iterative scheme is defined by

$$M[y_{k+1}] = aN[y_k] + r(x), \quad U_\mu[y_{k+1}] = \gamma_\mu \quad (k = 0, 1, 2, \ldots), \quad (4.22)$$

and to examine its convergence we introduce the sequence of functions
$z_k(x)$ defined by

$$z_k(x) = y_k(x) - y(x);$$

these satisfy the corresponding homogeneous iterative scheme

$$M[z_{k+1}] = aN[z_k], \quad U_\mu[z_{k+1}] = 0.$$

We now make the assumption that $z_1(x)$ possesses an eigenfunction
expansion

$$z_1(x) = \sum_{j=1}^{\infty} c_j \eta_j(x)$$

in terms of the eigenfunctions $\eta_j(x)$ of the corresponding homogeneous
eigenvalue problem

$$M[y] - \lambda N[y] = 0, \quad U_\mu[y] = 0, \qquad (4.23)$$

and also that this expansion may be differentiated term by term n
times. Then

$$z_2(x) = \sum_{j=1}^{\infty} \frac{a}{\lambda_j} c_j \eta_j(x)$$

is a solution of the boundary-value problem

$$M[z_2] = aN[z_1], \quad U_\mu[z_2] = 0,$$

and in general

$$z_{k+1}(x) = \sum_{j=1}^{\infty} \left(\frac{a}{\lambda_j}\right)^k c_j \eta_j(x). \qquad (4.24)$$

Hence the iteration converges for any c_i provided that $|a| < |\lambda_1|$,
where λ_1 is the eigenvalue of (4.23) with smallest absolute value; when
$|a| \geq |\lambda_1|$, the iteration diverges for $c_1 \neq 0$, i.e. in general.

4.6. Error estimation by means of the general iteration theorem

In many cases error estimates can be obtained by means of the general theorem on iterative processes of Ch. I, § 5; for linear boundary-value problems and also for many types of non-linear problem this theorem can be used precisely as described in Ch. I, § 5.3. This will be amplified to some extent here, although in practice the method is effective only for problems which are not far removed from linear; for otherwise the Lipschitz constant K required for the estimate can easily turn out to be greater than one. In any case, great care should be exercised in the determination of K because any increase in K will have a considerable adverse effect on the estimate on account of the factor $\frac{K}{1-K}$.

Consider the k-th-order ordinary differential equation

$$L[u] = F(x, u, u', \ldots, u^{(s)}) \tag{4.25}$$

with linear boundary conditions

$$U_\mu[u] = \gamma_\mu \qquad (\mu = 1, \ldots, k) \tag{4.26}$$

at two points a and b $(b > a)$. Here $L[u]$ is a linear homogeneous differential expression in u of order $k > s$ and F is a given function which we shall assume satisfies a Lipschitz condition of the form

$$\left. \begin{aligned} |F(x, u_1, u_1', \ldots, u_1^{(s)}) - F(x, u_2, u_2', \ldots, u_2^{(s)})| \\ \leq \sum_{\sigma=0}^{s} A_\sigma(x) |u_1^{(\sigma)} - u_2^{(\sigma)}| \end{aligned} \right\} \tag{4.27}$$

with $A_\sigma(x) \geq 0$. Further we shall assume that the boundary-value problem

$$L[u] = r(x), \qquad U_\mu[u] = 0 \qquad (\mu = 1, \ldots, k) \tag{4.28}$$

can be solved for any $r(x)$ by means of a GREEN's function $G(x, \xi)$ (cf. Ch. I, § 3.4):

$$u(x) = \int_a^b G(x, \xi)\, r(\xi)\, d\xi. \tag{4.29}$$

Finally we assume that this formula for the solution of (4.28) may be differentiated s times with respect to x, so that

$$u^{(\sigma)}(x) = \int_a^b G^{(\sigma)}(x, \xi)\, r(\xi)\, d\xi \qquad (\sigma = 0, 1, \ldots, s), \tag{4.30}$$

where

$$G^{(\sigma)}(x, \xi) = \frac{\partial^\sigma G(x, \xi)}{\partial x^\sigma}.$$

We now define a transformation T (cf. Ch. I, § 5.3) by $Tf = g$ where

$$\left.\begin{aligned}
L[g] &= F(x, f, f', \ldots, f^{(s)}) \\
U_\mu[g] &= \gamma_\mu \quad (\mu = 1, \ldots, k).
\end{aligned}\right\} \tag{4.31}$$

Then the difference $H = g_1 - g_2 = Tf_1 - Tf_2$ between two functions g_1, g_2 derived by this transformation from two functions f_1, f_2 satisfies the boundary-value problem

$$\begin{aligned}
L[H] &= F(x, f_1, f_1', \ldots, f_1^{(s)}) - F(x, f_2, f_2', \ldots, f_2^{(s)}), \\
U_\mu[H] &= 0 \quad (\mu = 1, \ldots, k),
\end{aligned}$$

and hence by (4.27)

$$\left.\begin{aligned}
|H^{(\tau)}(x)| &= \left| \int_a^b G_2^{(\tau)}(x, \xi) [F(\xi, f_1, \ldots, f_1^{(s)}) - F(\xi, f_2, \ldots, f_2^{(s)})] \, d\xi \right| \\
&\leq \int_a^b |G^{(\tau)}(x, \xi)| \sum_{\sigma=0}^s A_\sigma(\xi) |h^{(\sigma)}(\xi)| \, d\xi,
\end{aligned}\right\} \tag{4.32}$$

where $h = f_1 - f_2$. If

$$\|f\| = \underset{\text{in } a \leq x \leq b}{\text{upper limit of}} \frac{1}{W(x)} \sum_{\sigma=0}^s A_\sigma(x) |f^{(\sigma)}(x)|, \tag{4.33}$$

where $W(x)$ is a positive, or possibly non-negative, function in $a \leq x \leq b$, is used as the definition of the norm of a function possessing s continuous derivatives, then

$$\sum_{\tau=0}^s A_\tau(x) |H^{(\tau)}(x)| \leq \|h\| \sum_{\tau=0}^s A_\tau(x) \int_a^b |G^{(\tau)}(x, \xi)| \, W(\xi) \, d\xi. \tag{4.34}$$

A Lipschitz constant K for the transformation T is therefore given by

$$K = \underset{\text{in } a \leq x \leq b}{\text{upper limit of}} \frac{1}{W(x)} \sum_{\sigma=0}^s A_\sigma(x) \int_a^b |G^{(\sigma)}(x, \xi)| \, W(\xi) \, d\xi. \tag{4.35}$$

4.7. Special case of a non-linear differential equation of the second order

As an example we take a second-order differential equation in the form

$$L[u] = -u'' = F(x, u, u') \tag{4.36}$$

with the boundary conditions

$$u(a) = u_a, \quad u(b) = u_b.$$

Without loss of generality we put $a = 0$, and from Ch. I (3.26) the GREEN's function is then given by

$$G(x, \xi) = \begin{cases} \dfrac{x}{b}(b - \xi) & \text{for} \quad 0 \leq x \leq \xi, \\[2ex] \dfrac{\xi}{b}(b - x) & \text{for} \quad \xi \leq x \leq b. \end{cases} \tag{4.37}$$

For simplicity we will use constant A_0, A_1 in the Lipschitz condition (4.27). If u' does not appear explicitly in the differential equation (4.36), then $A_1 = 0$ and the eigenfunction $\sin \dfrac{\pi x}{b}$ may be chosen for $W(x)$. Since

$$\frac{\displaystyle\int_a^b G(x, \xi) \sin \dfrac{\pi \xi}{b}\, d\xi}{\sin \dfrac{\pi x}{b}} = \frac{b^2}{\pi^2}, \tag{4.38}$$

(4.35) yields the Lipschitz constant

$$K = \frac{A_0 b^2}{\pi^2}. \tag{4.39}$$

If, however, $A_1 \neq 0$, then this choice for $W(x)$ does not yield a finite upper limit in (4.35) and we must choose some other function. Let us try $W(x) = 1 - \alpha z$, where $z = \dfrac{x}{b}\left(1 - \dfrac{x}{b}\right)$; then $W > 0$ in $0 \leq x \leq b$ for $\alpha < 4$.

With

$$\left. \begin{aligned} k_0 &= \max_{0 \leq x \leq b} \frac{\displaystyle\int_0^b G(x, \xi) W(\xi)\, d\xi}{W(x)} = \max_{0 \leq z \leq \frac{1}{4}} \frac{b^2 z \left(1 - \dfrac{\alpha}{6} - \dfrac{\alpha}{6} z\right)}{2(1 - \alpha z)} \\[3ex] k_1 &= \max_{0 \leq x \leq b} \frac{\displaystyle\int_0^b G'(x, \xi) W(\xi)\, d\xi}{W(x)} = \max_{0 \leq z \leq \frac{1}{4}} b \frac{\dfrac{6 - \alpha}{12} - z + \dfrac{\alpha}{2} z^2}{1 - \alpha z} \end{aligned} \right\} \tag{4.40}$$

the Lipschitz constant (4.35) for this case is

$$K = A_0 k_0 + A_1 k_1. \tag{4.41}$$

The calculation is simplified considerably by choosing $\alpha = 0$; then $W = 1$ and we have

$$K = \frac{A_0 b^2}{8} + \frac{A_1 b}{2}. \tag{4.42}$$

As α increases, k_0 increases but k_1 decreases, so that it is often possible to obtain a sharper estimate for non-zero α; for instance, with

$$\alpha = \frac{1}{4}\left(15 - \sqrt{33}\right) \approx 2{\cdot}314$$

we have[1]

$$K = \beta \left(\frac{A_0 b^2}{2} + A_1 b \right),$$
(4.43)

where

$$\beta = \frac{9 + \sqrt{33}}{48} \approx 0.3072.$$

For the solution of the differential equation (4.36) with the boundary conditions

$$u(0) = u_0, \qquad u'(b) = u'_b$$
(4.44)

we need the GREEN's function

$$G(x, \xi) = \begin{cases} x & \text{for} \quad 0 \leq x \leq \xi, \\ \xi & \text{for} \quad \xi \leq x \leq b \end{cases}$$

instead of (4.37). In this case we choose

$$W(x) = \sin \frac{\pi x}{2b}$$

when u' does not appear in (4.36), i.e. when $A_1 = 0$; this gives the Lipschitz constant

$$K = \frac{4 A_0 b^2}{\pi^2}.$$
(4.45)

Again we must use some other function for $W(x)$ when $A_1 \neq 0$; with $W(x) = 1$ we obtain

$$K = A_0 \frac{b^2}{2} + A_1 b$$
(4.46)

and with[2] $W(x) = 3 - 6x + 4x^2$

$$K = 0.636 A_0 b^2 + \tfrac{4}{9} A_1 b.$$
(4.47)

---

[1] Cf. F. LETTENMEYER: Über die von einem Punkt ausgehenden Integralkurven einer Differentialgleichung 2. Ordnung. Dtsch. Math. 7, 56—74 (1944). LETTENMEYER considers the sequence of functions generated iteratively from [as starting funktion $u_0(x)$] the linear function which satisfies the boundary conditions; he proves convergence to the uniquely determined solution of the boundary-value for the case

$$A_0 \frac{b^2}{\pi^2} + A_1 \frac{4}{\pi^2} b < 1.$$

Comparison shows that the optimum constant b^2/π^2 appears also in our (4.39) for the case $A_1 = 0$, but that for the case $A_1 \neq 0$ our factor multiplying A_0 in (4.43) is less favourable than LETTENMEYER's while our factor multiplying A_1 is more favourable than his.

[2] For the boundary conditions (4.44) LETTENMEYER (loc. cit.) obtains the condition $\frac{4}{\pi^2} A_0 b^2 + \frac{2}{\pi} A_1 b < 1$. The optimum constant $\frac{4}{\pi^2}$ appears also in our (4.45) for the case $A_1 = 0$, as before, and for the case $A_1 \neq 0$ our factors multiplying A_0 and A_1 in (4.47) are again, as in (4.43), less and more favourable, respectively, than LETTENMEYER's.

4.8. Examples of the iteration method with error estimates

I. A linear problem. The bending of a strut (Example I of § 1.2) is again used as an example; the boundary-value problem reads

$$y'' = -1 - (1 + x^2)\, y, \qquad y(\pm 1) = 0,$$

and several iteration schemes suggest themselves.

A. The simplest is defined by

$$y''_{k+1} = -1 - (1 + x^2)\, y_k, \qquad y_{k+1}(\pm 1) = 0 \qquad (k = 0, 1, 2, \ldots).$$

We start with the function

$$y_0(x) = A(1 - x^2),$$

which satisfies the boundary conditions, and determine the constant A so that the first iterate

$$y_1(x) = \frac{15 + 14A}{30} - \frac{1 + A}{2}\, x^2 + \frac{A}{30}\, x^6$$

is of the same order of magnitude as $y_0(x)$. The condition

$$y_0(0) = y_1(0)$$

yields $A = \frac{15}{16}$ and $y_0(0) = y_1(0) = A = 0.9375$ (error 0.0054). With this value of A we have $32 y_1 = 30 - 31\, x^2 + x^6$.

A second iteration yields

$$32 y_2 = \frac{75379}{2520} - 31\, x^2 + \frac{1}{12}\, x^4 + \frac{31}{30}\, x^6 - \frac{1}{56}\, x^8 - \frac{1}{90}\, x^{10},$$

whose value at $x = 0$ is $y_2(0) = \dfrac{75379}{80640} \approx 0.93476$ (error 0.0027).

To facilitate the evaluation of the next iterate, we simplify y_2 by neglecting the two highest powers of x and approximating the remaining coefficients by simpler values:

$$y_2^* = \frac{1}{379}\, (354 - 367\, x^2 + x^4 + 12\, x^6), \quad \text{for which} \quad y_2^*(0) = \frac{354}{379} = 0.93404.$$

A further iteration then yields

$$379 y_3^* = \frac{297009}{840} - \frac{733}{2}\, x^2 + \frac{13}{12}\, x^4 + \frac{61}{5}\, x^6 - \frac{13}{56}\, x^8 - \frac{2}{15}\, x^{10},$$

whose initial value is

$$y_3^*(0) = \frac{297009}{318360} \approx 0.932934 \quad (\text{error } 0.00088).$$

B. We obtain rather better results if the term $-y$ is taken over to the left-hand side, i.e. if we use the iteration formula

$$y''_{k+1} + y_{k+1} = -1 - x^2 y_k, \qquad y_{k+1}(\pm 1) = 0 \qquad (k = 0, 1, \ldots).$$

13*

With the same starting procedure as in **A**, i.e. putting

$$y_0(x) = A(1 - x^2), \qquad y_0(0) = y_1(0), \qquad (4.48)$$

we find that

$$y_1(x) = -1 + A(26 - 13 x^2 + x^4) + \frac{1 - 14A}{\cos 1} \cos x;$$

here

$$y_1(0) = A = \frac{1 - \cos 1}{14 - 25 \cos 1} = 0.933\,505 \quad \text{(error } 0.001\,45).$$

C. The iteration can be improved still further by making a constant-coefficient approximation for the term $x^2 y$, say $\frac{1}{4} y$, and taking this over to the left-hand side as well. We obtain the slightly modified iteration

$$y_{k+1}'' + \frac{5}{4} y_{k+1} = -1 + \left(\frac{1}{4} - x^2\right) y_k, \quad y_{k+1}(\pm 1) = 0 \quad (k = 0, 1, \ldots),$$

which, with the same starting procedure as before, yields

$$y_1(x) = -\frac{4}{5} + A \frac{1761 - 1085 x^2 + 100 x^4}{125} + \frac{100 - 776A}{125 \cos \varrho} \cos \varrho x,$$

where $\varrho = \frac{1}{2}\sqrt{5}$. Compared with iterations **A** and **B** we have

$$y_1(0) = A = \frac{100(1 - \cos \varrho)}{776 - 1636 \cos \varrho} = 0.932\,456 \quad \text{(error } 0.000\,40);$$

the error in the first iterate here is less than the error in the third iterate of method **A**, so it clearly pays to give careful consideration to the choice of iteration scheme.

Error estimates. The method of error estimation is illustrated for the iteration in **A**, since the evaluation of the limits is by far the simplest in this case. We adopt the method 1. of § 5.3 in Ch. I, which is directly applicable; in particular, we introduce the norm defined in (5.27) of Ch. I, i.e.

$$\|f\| = \underset{\text{in } -1 \le x \le 1}{\text{upper limit of}} \left| \frac{f}{W} \right|,$$

where $W(x)$ is non-negative for $-1 \le x \le 1$. We apply the method to the pair of the successive iterates $y_1(x)$, $y_2(x)$, and show how great the effect of a suitable choice for the norm is.

The error estimate derived by using a non-negative eigenfunction z for W as defined in Ch. I (5.31) is very easily found with a negligible amount of calculation since the GREEN's function is not involved explicitly; the limits so obtained are, however, rather wide. For

$$z'' + \lambda z = 0, \qquad z(\pm 1) = 0$$

we have the non-negative eigenfunction $z = \cos \frac{\pi}{2} x$ and the corresponding eigenvalue is $\lambda_z = \frac{\pi^2}{4}$. With $p = 1 + x^2$, $|p|_{max} = 2$, the formula (5.33) of Ch. I yields the value $K = \frac{8}{\pi^2}$, which is rather close to 1; this gives a correspondingly large value for

$$\frac{K}{1-K} = \frac{8}{\pi^2 - 8} \approx 4\cdot28.$$

We now calculate the "distance" separating y_1 and y_2:

$$\|y_1 - y_2\| = \underset{\text{in } -1 \leq x \leq 1}{\text{upper limit of}} \left| \frac{y_1 - y_2}{\cos \frac{\pi}{2} x} \right| = 0\cdot005\,55;$$

(5.16) of Ch. I then yields

$$\|y_2 - y\| \leq 4\cdot28 \times 0\cdot005\,55 = 0\cdot0237,$$

from which we infer the error estimate

$$|y_2 - y| \leq 0\cdot0237 \cos \frac{\pi}{2} x \quad \text{for} \quad |x| \leq 1.$$

To obtain a smaller Lipschitz constant and hence a better error estimate, we use the GREEN's function method of Ch. I (5.29). The appropriate GREEN's function for this case was given in Ch. I (3.26):

$$G(x, \xi) = \begin{cases} \dfrac{(x+1)(1-\xi)}{2} & \text{for} \quad -1 \leq x \leq \xi \\ \dfrac{(\xi+1)(1-x)}{2} & \text{for} \quad \xi \leq x \leq 1. \end{cases} \tag{4.49}$$

Choosing $W(x) = 1$, i.e. using the norm $\|f\| = \underset{-1 \leq x \leq 1}{\max} |f|$, and with $N(x) = 1 + x^2$, we have

$$K = \underset{|x| \leq 1}{\max} \int_{-1}^{1} G(x, \xi)(1 + \xi^2)\, d\xi = \underset{|x| \leq 1}{\max} \frac{1}{12}(7 - 6x^2 - x^4) = \frac{7}{12}.$$

This value of K is much better than the value $8/\pi^2$ obtained above; it gives $\frac{K}{1-K} = \frac{7}{5} = 1\cdot4$ and in fact yields narrower error limits for y_1 at the point $x = 0$ than those obtained above for y_2. With

$$\|y_1 - y_2\| = \underset{|x| \leq 1}{\max} |y_1 - y_2| = \frac{221}{80\,640} \approx 0\cdot002\,74$$

(5.16) of Ch. I now gives the limits for y_2 as

$$|y_2 - y| \leq \frac{221}{57\,600} \approx 0\cdot003\,84.$$

The estimate using the GREEN's function is better still if a non-constant function is chosen for W, in this case a polynomial, say, for convenience in integration; a suitable choice here is $W = 1 - x^2$. We then have to evaluate the integral:

$$\int_{-1}^{1} G(x, \xi)(1 + \xi^2)(1 - \xi^2)\, d\xi = \frac{1}{30}(14 - 15 x^2 + x^6)$$
$$= \frac{(1 - x^2)(14 - x^2 - x^4)}{30}.$$

Substituting in (5.29) of Ch. I, we obtain

$$K = \max_{|x| \leq 1} \frac{14 - x^2 - x^4}{30} = \frac{7}{15},$$

and hence the favourable value

$$\frac{K}{1 - K} = \frac{7}{8}.$$

Finally, with $\|y_1 - y_2\| = \frac{11}{2520}$, we have the error estimate

$$|y_2 - y| \leq \frac{77}{20160}(1 - x^2) \approx 0{\cdot}003\,82\,(1 - x^2).$$

A similar calculation with $W = 1 - \frac{1}{2} x^2$ (which gives $K = \frac{21}{40}$) leads to the estimate

$$|y_2 - y| \leq 0{\cdot}003\,25\,(1 - \tfrac{1}{2} x^2).$$

At $x = 0$ this estimate exceeds the actual error $0{\cdot}002\,71$ by less than 20%.

By using a more general definition of distance (cf. the paper by J. SCHRÖDER in the references given in Ch. I, §5.2) we can obtain limits as close as $|y_2 - y| \leq 0{\cdot}003\,04\,(1 - x^2)$ without going beyond polynomials of the second degree for W.

II. Non-linear oscillations. It is required to calculate a periodic solution of

$$\left.\begin{aligned} -y'' - 6y - y^2 &= \tfrac{3}{2}\cos x, \\ y(0) - y(2\pi) = y'(0) - y'(2\pi) &= 0 \end{aligned}\right\} \tag{4.50}$$

(forced oscillations of a system with a non-linear restoring force). The problem lies sufficiently near to the linear problem obtained by neglecting the y^2 term for the iteration defined by

$$-y''_{k+1} - 6y_{k+1} = y_k^2 + \tfrac{3}{2}\cos x, \quad y_{k+1} \text{ with period } 2\pi \quad (k = 0, 1, \ldots)$$

to converge satisfactorily and allow an error estimate to be made.

The solution

$$y_0(x) = -0{\cdot}3\cos x$$

of the linear problem is an obvious choice for the starting function. A short calculation then yields

$$y_1(x) = -\frac{3}{400}(1 + 40 \cos x + 3 \cos 2x),$$

$$y_2(x) = \frac{-3}{320\,000}(805{\cdot}5 + 32\,240 \cos x +$$
$$+ 2418 \cos 2x - 240 \cos 3x - 2{\cdot}7 \cos 4x),$$

$$y_1(x) - y_2(x) = \frac{3}{320\,000}(5{\cdot}5 + 240 \cos x +$$
$$+ 18 \cos 2x - 240 \cos 3x - 2{\cdot}7 \cos 4x),$$

$$|y_1(x) - y_2(x)| \leq \frac{3}{320\,000}\,380 = 0{\cdot}003\,56.$$

For the error estimate we use the basic formula (5.29) of Ch. I with $W = 1$; the appropriate GREEN's function is given by (3.28) of Ch. I. With $n = \sqrt{6}$ we have

$$2n \sin n\pi \int_0^{2\pi} |G(x,\xi)|\, d\xi = 2 \int_0^\pi |\cos n x|\, dx = 2\,\frac{\sin(\sqrt{6}\pi) + 4}{\sqrt{6}} = 2 \times 2{\cdot}0361;$$

then from (5.25), (5.29) of Ch. I

$$K = \frac{2{\cdot}0361}{\sqrt{6}\,\sin(\sqrt{6}\pi)}\,N = 0{\cdot}8418\,N.$$

We now need a value for N, the Lipschitz constant for the function $y^2 + \frac{3}{2}\cos x$. Clearly an upper bound for $|2y|$ will suffice, and if we assume that $|y| \leq 0{\cdot}4$, i.e. we try to find a solution in the subspace F of all continuous functions $y(x)$ of period 2π whose members satisfy $|y(x)| \leq 0{\cdot}4$, then we can use the value

$$N = |2y|_{\max} = 0{\cdot}8;$$

this gives

$$K = 0{\cdot}8418 \times 0{\cdot}8 = 0{\cdot}673, \qquad \frac{K}{1-K} = 2{\cdot}06.$$

The "sphere" Σ defined by

$$|y - y_2| \leq 0{\cdot}003\,56 \times 2{\cdot}06 = 0{\cdot}007\,35$$

lies entirely in F, and hence there is a solution in this subspace, namely the function towards which the sequence $y_0, y_1, y_2, \ldots$ is converging. Incidently this shows that there are no other periodic solutions of (4.50) with $|y| \leq 0{\cdot}4$.

The fact that y lies in the sphere Σ gives us a smaller upper bound for $|y|_{\max}$:

$$|y| \leq |y_1| + |y_1 - y_2| + 0{\cdot}007\,35 \leq 0{\cdot}3409.$$

We can now repeat the above calculation with more refined values:

$$N = 2|y|_{\max} = 0.6818, \quad K = 0.8418 \times 0.6818 = 0.585,$$
$$\frac{K}{1-K} = 1.352.$$

Thus we obtain the better error estimate

$$|y - y_2| \leq 0.00356 \times 1.352 = 0.00482.$$

This in turn gives a still smaller upper bound for $|y|_{\max}$, and the process can be repeated; however, little is gained thereby, for with $K = 0.570$ we obtain $|y - y_2| \leq 0.00472$.

4.9. Monotonic boundary-value problems for second-order differential equations

Consider the boundary-value problems with the differential equation

$$L[y] = - [p(x) y'(x)]' = - F(x, y),$$

where $p(x)$ is positive and possesses a continuous derivative in the interval $a \leq x \leq b$, and with the boundary conditions

(i) $\qquad U_1[y] = y(a) = y_a, \quad U_2[y] = y(b) = y_b,$

or

(ii) $\qquad \begin{aligned} U_1[y] &= y'(a) - c\, y(a) = \gamma_a, \\ U_2[y] &= y'(b) + d\, y(b) = \gamma_b \end{aligned}$

with $c \geq 0$, $d \geq 0$. It will be shown that these problems are monotonic when $F(x, y)$ satisfies certain conditions.

Such problems were considered generally in Ch. I, § 5.6, where it was shown that they were of monotonic type if

$$L[\varepsilon] + \varepsilon A(x) \geq 0 \quad \text{with} \quad A(x) \geq 0 \quad \text{and} \quad U_\mu[\varepsilon] = 0 \quad (\mu = 1, 2) \quad (4.51)$$

implied that $\varepsilon \geq 0$. We now show by indirect proof that this is true of the above problems, excluding only the case with $A(x) \equiv 0$, $c = d = 0$. Suppose that the continuous function ε has a negative minimum value which it assumes at the point $x = \xi$.

1. If ξ is an interior point of the open interval (a, b), then $p(\xi)\, \varepsilon'(\xi) = 0$. Since ε is continuous, there is an interval (α, β) with $a \leq \alpha < \xi < \beta \leq b$ in which $\varepsilon < 0$ and hence $(p\varepsilon')' \leq 0$ from (4.51). This implies that $\varepsilon' \geq 0$ for $\alpha < x \leq \xi$ and $\varepsilon' \leq 0$ for $\xi \leq x < \beta$; hence ε is constant in (α, β), for otherwise, since $x = \xi$ is a minimum, there would exist ϱ with $\varepsilon(\varrho) > \varepsilon(\xi)$, $\xi < \varrho < \beta$ and hence also $\tilde{\varrho}$ with $\varepsilon'(\tilde{\varrho}) > 0$, $\xi < \tilde{\varrho} \leq \varrho$. It follows that ε must be constant in (a, b) also. In case (i) we therefore have $\varepsilon \equiv 0$, which contradicts the assumption of a negative minimum value. In case

(ii) we have $c\varepsilon(a) = d\varepsilon(b) = 0$ since $\varepsilon'(a) = \varepsilon'(b) = 0$; if c and d are not both zero, then again $\varepsilon \equiv 0$; if $c = d = 0$, then $A \not\equiv 0$ (since we are excluding the case $c = d = 0$, $A(x) \equiv 0$), so that there exists σ in $a < \sigma < b$ with $A(\sigma) > 0$, and since $\varepsilon' \equiv 0$ and $\varepsilon < 0$, (4.51) is contradicted at $x = \sigma$. Consequently ξ cannot be a point of the open interval (a, b).

2. The alternative, that ξ is a boundary point, say $\xi = a$, can be ruled out immediately in case (i), for it requires that $\varepsilon(a) < 0$. In case (ii) this implies that $\varepsilon'(a) \leq 0$; but, since $\xi = a$ is a minimum, we must also have $\varepsilon'(a) \geq 0$; hence $\varepsilon'(a) = 0$, and we can use the same arguments to obtain a contradiction as in 1. but with $x = a$ taking over the role played there by $x = \xi$. Therefore ε cannot be negative in $\langle a, b \rangle$.

This result may be stated as follows:

The boundary-value problem

$$T u \equiv - [p(x) u'(x)]' + F(x, u) = 0 \tag{4.52}$$

with the boundary conditions

(i)
$$u(a) = u_a, \qquad u(b) = u_b \tag{4.53}$$

or

(ii)
$$\begin{cases} u'(a) - c u(a) = \gamma_a \\ u'(b) + d u(b) = \gamma_b, \end{cases} \tag{4.54}$$

where $p(x)$ is positive and possesses a continuous derivative in (a, b), $u_a, u_b, \gamma_a, \gamma_b$ are given constants and $c \geq 0$, $d \geq 0$, is of monotonic type in a domain H of the (x, u) space [defined by $a \leq x \leq b$, $u_0(x) \leq u \leq u_1(x)$, say] when $\dfrac{\partial F}{\partial u}$ exists in H and in case (i) $\dfrac{\partial F}{\partial u} \geq 0$ and in case (ii) either $\dfrac{\partial F}{\partial u} > 0$ or $\dfrac{\partial F}{\partial u} \geq 0$ and $c^2 + d^2 > 0$.

Under these circumstances, if the boundary-value problem possesses a solution $y(x)$ in H and we have two functions y_1 and y_2 which satisfy the boundary conditions and are such that $T y_1 \leq 0 \leq T y_2$, then we know that $y_1(x) \leq y(x) \leq y_2(x)$.

Example. The problem

$$y'' = 6 x y^2, \qquad y(0) = y(1) = 1$$

is of monotonic type in the domain H defined by $0 \leq x \leq 1$, $0 \leq y \leq 1$. Suppose that the existence of a solution in H has been established, say by considering the initial-value problem

$$T y = - y'' + 6 x y^2 = 0; \qquad y(0) = 1, \qquad y'(0) = \sigma$$

with variable σ. For y we can make the approximation

$$w = 1 + \sum_{\nu=3}^{5} a_\nu (x - x^\nu),$$

which satisfies the boundary conditions for arbitrary a_ν; the term with $\nu = 2$ is omitted since $y''(0) = 0$; more terms can be added as desired, of course. The functions

$$w_1 = 1 - (x - x^3)$$

$$w_2 = 1 - 0.43\,(x - x^4)$$

give $Tw_1 \leqq 0$, $Tw_2 \geqq 0$, respectively, and hence $w_1 \leqq y \leqq w_2$; these limits may easily be improved by using more accurate expressions for w.

§5. RITZ's method for second-order boundary-value problems

For many boundary-value problems of even order it is possible to specify an integral expression $J[\varphi]$ which can be formed for all functions φ of a certain class and which has a minimum value for just that function y which solves the boundary-value problem. Consequently solution of the boundary-value problem is equivalent to minimizing $J[\varphi]$. In RITZ's method[1] the solution of this variational problem is approximated by a linear combination of suitably chosen functions. This method shares with the finite-difference method a very favoured position among the methods for the approximate solution of boundary-value problems.

5.1. EULER's differential equation in the calculus of variations

The means of formulating the required expression $J[\varphi]$ are furnished by the calculus of variations. Consider, for example, the simple variational problem of determining in the domain of all functions $\varphi(x)$ with continuous derivatives in $a \leqq x \leqq b$ and prescribed values $\varphi(a) = A$, $\varphi(b) = B$ at the end points (the domain of "admissible" functions) that function which minimizes the value of the integral

$$J[\varphi] = \int_a^b F(x, \varphi, \varphi')\,dx, \qquad (5.1)$$

in which F is a given function possessing continuous derivatives with respect to each of its arguments. It may be shown[2] that if a solution

---

[1] WALTER RITZ, born 22 February 1878 in Sion (Switzerland) in the Rhone valley, son of the artist Raphael Ritz, studied in Zürich, then in Göttingen, where in 1902 he obtained his doctor's degree under Voigt; he then worked in Leyden under H. A. Lorentz, in Paris under A. Cotton, in Tübingen under F. Paschen and in 1908 went back to Göttingen, where he died on the 7 July 1909. After a poorly healed pleurisy in 1900 his zeal for his scientific work was continually in conflict with consideration for his state of health, until eventually his health was sacrificed. (Obituary by PIERRE WEISZ in W. RITZ: Gesammelte Werke. Paris 1911.) — Habilitationsschrift von RITZ über eine neue Methode zur Lösung gewisser Variationsprobleme der mathematischen Physik. J. Reine Angew. Math. **135**, H. 1 (1908). — Ann. Phys., Lpz. **28**, 737 (1909).

[2] Cf., for example, G. GRÜSZ: Variationsrechnung, p. 11. Leipzig 1938, or R. COURANT: Differential and Integral Calculus Vol. II, p. 495 et seq. London: Blackie 1936.

$\varphi = y(x)$ which gives the integral a value not greater than that given by any other admissible function $\varphi(x)$ exists at all, then $y(x)$ must necessarily satisfy the second-order differential equation

$$-\frac{d}{dx}\left(\frac{\partial F}{\partial \varphi'}\right) + \frac{\partial F}{\partial \varphi} = 0 \tag{5.2}$$

[and also, of course, the boundary conditions $y(a) = A$, $y(b) = B$]. This is EULER's equation for the functional (5.1).

5.2. Derivation of EULER's conditions

For the application of RITZ's method to the solution of boundary-value problems we must reverse the procedure of § 5.1 and try to write the differential equation of the given boundary-value problem as the Euler equation of some variational problem; this will yield the appropriate expression $J[\varphi]$. For example, for the second-order linear boundary-value problem

$$L[y] = -\frac{d}{dx}(p\,y') + q(x)\,y = r(x) \tag{5.3}$$

with the boundary conditions

$$\left.\begin{array}{l} U_1[y] = \alpha_0\,y(a) + \alpha_1\,y'(a) = \gamma_1, \\ U_2[y] = \beta_0\,y(b) + \beta_1\,y'(b) = \gamma_2 \end{array}\right\} \tag{5.4}$$

it can be seen immediately that with

$$F = p\,\varphi'^2 + q\,\varphi^2 - 2r\,\varphi \tag{5.5}$$

the Euler equation (5.2) is identical with the given differential equation (5.3). The boundary conditions here are rather more general than in § 5.1 and $J[\varphi]$ must be modified slightly (by the addition of certain "boundary" terms), otherwise a solution of the variational problem with the new domain of admissible functions satisfying the new boundary conditions (assuming not both of α, β are zero) will not exist in general; this modification does not affect the Euler equation. We demonstrate by giving a short derivation of the Euler conditions for the more general expression

$$J[\varphi] = \int_a^b (p\,\varphi'^2 + q\,\varphi^2 - 2r\,\varphi)\,dx + A_0\,\varphi_a^2 + B_0\,\varphi_b^2 + 2A_1\,\varphi_a + 2B_1\,\varphi_b, \tag{5.6}$$

in which the more compact notation φ_a has been used for $\varphi(a)$, etc.

At first we restrict the domain of the admissible functions to those functions φ with continuous second derivatives which satisfy the boundary conditions (5.4); we shall see later that this domain can sometimes be enlarged to include functions satisfying fewer boundary conditions, or even none at all.

Since we do not here go into the question of the existence of a minimum of $J[\varphi]$ with respect to the given domain of admissible functions, we will assume that there is a smallest value of J and that there is a function Y among the admissible functions φ for which J attains this value, i.e.

$$J[Y] \leqq J[\varphi]. \tag{5.7}$$

As yet, this function Y has nothing to do with the boundary-value problem (5.3), (5.4).

If η is a function with continuous second derivatives which satisfies the homogeneous boundary conditions

$$\left. \begin{aligned} \alpha_0 \eta_a + \alpha_1 \eta_a' = 0, \\ \beta_0 \eta_b + \beta_1 \eta_b' = 0 \end{aligned} \right\} \tag{5.8}$$

corresponding to the inhomogeneous conditions (5.4), then the function $\varphi = Y + \varepsilon \eta$, where ε is an arbitrary constant, satisfies the inhomogeneous conditions (5.4) and is an admissible function; it therefore satisfies the relation (5.7). If we allow ε to run through a range of positive and negative values, $J[\varphi]$ defines a function $\Phi(\varepsilon)$ of the parameter ε which possesses a continuous derivative with respect to ε and which attains a minimum value, namely the smallest value of $J[\varphi]$, at $\varepsilon = 0$. Hence the derivative with respect to ε must vanish at this point:

$$\Phi'(0) = \left[\frac{dJ[\varphi(\varepsilon)]}{d\varepsilon} \right]_{\varepsilon=0} = \left[\frac{dJ[Y + \varepsilon\eta]}{d\varepsilon} \right]_{\varepsilon=0} = 0. \tag{5.9}$$

[δJ is often written for the expression $\varepsilon \Phi'(0)$ and is called the "first variation" of J.]

Substituting $\varphi = Y + \varepsilon \eta$ in (5.6), we obtain

$$\Phi(\varepsilon) = J[Y + \varepsilon \eta]$$
$$= \int\limits_a^b \{ p (Y' + \varepsilon\eta')^2 + q (Y + \varepsilon\eta)^2 - 2r(Y + \varepsilon\eta) \} dx + A_0 (Y_a + \varepsilon\eta_a)^2 + \cdots,$$

and hence

$$\tfrac{1}{2} \Phi'(0) = \int\limits_a^b (p Y' \eta' + q Y \eta - r\eta) dx + A_0 Y_a \eta_a + B_0 Y_b \eta_b + A_1 \eta_a + B_1 \eta_b.$$

The integral of the first term in the integrand can be transformed by integration by parts:

$$\int\limits_a^b p Y' \eta' dx = p_b Y_b' \eta_b - p_a Y_a' \eta_a - \int\limits_a^b (p Y')' \eta dx. \tag{5.10}$$

The equation $\Phi'(0) = 0$ then reads

$$\int\limits_a^b \eta [-(p Y')' + q Y - r] dx + \eta_a W_a + \eta_b W_b = 0, \tag{5.11}$$

where

$$W_a = A_0 Y_a + A_1 - p_a Y_a', \qquad W_b = B_0 Y_b + B_1 + p_b Y_b'. \tag{5.12}$$

Now equation (5.11) is to hold for any function $\eta(x)$ which possesses a continuous second derivative and satisfies the boundary conditions (5.8). This is possible only if the factor multiplying η in the integrand vanishes identically, for otherwise we can always construct a function $\eta = \eta^{**}$ satisfying the above conditions but for which (5.11) does not hold: let $\eta = \eta^*$ be a function satisfying the above conditions and for which (5.11) does hold, and let x_0 be a point in (a, b) where the factor is positive, say; by continuity it must be positive also for $|x - x_0| \leqq \delta$ with $\delta > 0$, and we put

$$\eta^{**} = \begin{cases} \eta^* + (x - x_0 - \delta)^3 (x - x_0 + \delta)^3 & \text{for} \quad |x - x_0| \leqq \delta \\ \eta^* \text{ otherwise}. \end{cases} \tag{5.13}$$

Consequently we must have

$$(-p Y')' + q Y - r = 0. \tag{5.14}$$

This differential equation, known as the Euler equation of the variational problem corresponding to (5.4), (5.6), therefore furnishes a necessary condition to be satisfied by the solution $Y(x)$ of that problem.

From (5.11) and (5.14) it follows that another necessary condition is

$$\eta_a W_a + \eta_b W_b = 0. \tag{5.15}$$

For further discussion we distinguish three cases:

1. Case $\alpha_1 \neq 0$, $\beta_1 \neq 0$. Here, given any values η_a, η_b we can calculate η_a', η_b' from (5.8). From all functions η with continuous second derivatives and any given boundary values η_a, η_b we can select one with the boundary derivative values η_a', η_b' calculated from (5.8); this function will then satisfy the homogeneous boundary conditions. Thus for any given η_a, η_b there is a function η with the boundary values η_a, η_b and satisfying all the conditions of admission. Now (5.15) can be valid for all η_a, η_b only if

$$W_a = W_b = 0.$$

These equations represent two boundary conditions [see (5.12)] which Y must necessarily satisfy.

In order that the solution of the variational problem of minimizing (5.6) shall also solve the boundary-value problem (5.3), (5.4), we have only to put

$$A_0 = -\frac{\alpha_0}{\alpha_1} p_a, \quad A_1 = \frac{\gamma_1}{\alpha_1} p_a, \quad B_0 = \frac{\beta_0}{\beta_1} p_b, \quad B_1 = -\frac{\gamma_2}{\beta_1} p_b, \tag{5.16}$$

for this ensures that the boundary conditions on Y are identical with the conditions (5.4) [we have already arranged that the differential equations (5.3) and (5.14) are identical]; p_a, p_b are here assumed to be non-zero, for otherwise the points a, b, respectively, would be singular points of the differential equation.

Since we have made $W_a = W_b = 0$ by using the values determined from (5.16), we have $\eta_a W_a + \eta_b W_b = 0$ for any function η with a continuous second derivative; thus (and this is an important new aspect) the boundary conditions can be suppressed and the solution in the so widened domain of admissible functions will satisfy the boundary conditions automatically. In this case ($\alpha_1 \neq 0, \beta_1 \neq 0$) the boundary conditions (5.4) are "suppressible" according to the definition in § 1.2 of Ch. I; the sense in which this type of boundary condition is suppressible is now explained.

2. Case $\alpha_1 = \beta_1 = 0$, $\alpha_0 \neq 0$, $\beta_0 \neq 0$. Here our domain of admissible functions is restricted to those with the fixed values

$$\varphi(a) = \frac{\gamma_1}{\alpha_0}, \qquad \varphi(b) = \frac{\gamma_2}{\beta_0} \tag{5.17}$$

at the end points, and from (5.8) we now have $\eta_a = \eta_b = 0$ for all the functions η. Consequently the expression $\eta_a W_a + \eta_b W_b$ vanishes however W_a and W_b are constituted and we may therefore choose A_0, B_0, A_1, B_1 as we please. Thus no "boundary terms" are needed in this case, for we may take all these constants to be zero and use just the integral term in $J[\varphi]$. We cannot, however, extend the domain of admissible functions as in the previous case, and hence the boundary conditions are essential. These boundary conditions are "essential" according to the definition in § 1.2 of Ch. I; and that terminology also is now explained.

3. Case in which just one of α_1, β_1 is zero. We can combine what has been said for the two previous cases; for example, if $\alpha_1 = 0$, $\beta_1 \neq 0$, we obtain the necessary conditions

$$\eta_a = 0, \qquad W_b = 0.$$

We summarize these results in the following

Theorem. *The second-order linear boundary-value problem*

$$- (p(x) y')' + q(x) y = r(x), \quad where \quad p(a) \neq 0, \quad p(b) \neq 0, \\ \left. \begin{array}{c} \alpha_0 y(a) + \alpha_1 y'(a) = \gamma_1, \\ \beta_0 y(b) + \beta_1 y'(b) = \gamma_2, \end{array} \right\} \tag{5.18}$$

including the boundary conditions, may be written as the necessary conditions to be satisfied by the solution of a variational problem $J[\varphi] = minimum$. In the case $\alpha_1 \neq 0$, $\beta_1 \neq 0$,

$$J[\varphi] = \int_a^b (p\varphi'^2 + q\varphi^2 - 2r\varphi)\, dx + R_a + R_b, \tag{5.19}$$

where

$$R_a = \frac{p(a)}{\alpha_1}\left(-\alpha_0\varphi^2(a) + 2\gamma_1\varphi(a)\right), \quad R_b = \frac{p(b)}{\beta_1}\left(\beta_0\varphi^2(b) - 2\gamma_2\varphi(b)\right), \quad (5.20)$$

and no boundary conditions need be imposed on the admissible functions φ.
In the case $\alpha_1 = 0$ *we may put* $R_a = 0$, *but we must restrict the admissible functions by demanding that they satisfy the "essential" boundary condition* $\alpha_0\varphi(a) = \gamma_1$. *Similarly, when* $\beta_1 = 0$, *we may omit the term* R_b *but must insist on the boundary condition* $\beta_0\varphi(b) = \gamma_2$. *This theorem does not, however, answer the question whether the function* $y(x)$ *is actually a solution of the variational problem so formulated or indeed whether* $J[\varphi]$ *possesses a minimum value at all* [1].

5.3. The Ritz approximation [2]

Having formulated the variational problem (5.19), (5.20) corresponding to the given boundary-value problem (5.18), we now use RITZ's method for its approximate solution. This consists in the reduction of the problem to an ordinary minimum problem by substituting a linear combination of suitable functions for $\varphi(x)$, say

$$\varphi(x) = v_0(x) + \sum_{\nu=1}^{p} a_\nu v_\nu(x), \qquad (5.21)$$

so that $J[\varphi]$ becomes an ordinary function of the unknown coefficients of the combination. If we had an "essential" boundary condition at $x = a$, say, i.e. $\alpha_0 y(a) = \gamma_1$, $\alpha_1 = 0$, then we would choose $v_0(x)$ to satisfy this boundary condition and the $v_\nu(x)$ to satisfy the corresponding homogeneous condition $v_\nu(a) = 0$; this ensures that $\alpha_0\varphi(a) = \gamma_1$ for all values of the a_ν. On the other hand, if all the boundary conditions were "suppressible", we could omit v_0 and choose the v_ν regardless of any boundary conditions, but we would have to take into account the boundary terms R_a and R_b as defined in (5.20).

On account of the linearity of the expression assumed for φ, the equations for determining the a_ν so as to minimize $J[\varphi]$ are also linear. We have

$$\left.\begin{aligned}
\frac{1}{2}\frac{\partial J[\varphi]}{\partial a_\nu} &= \int_a^b (p\,\varphi'\,v_\nu' + q\,\varphi\,v_\nu - r\,v_\nu)\,dx + \\
&+ \frac{p(a)\,v_\nu(a)}{\alpha_1}\left(-\alpha_0\varphi(a) + \gamma_1\right) + \frac{p(b)\,v_\nu(b)}{\beta_1}\left(\beta_0\varphi(b) - \gamma_2\right) = 0.
\end{aligned}\right\} \quad (5.22)$$

---

[1] See Exercise 6 of § 8.11.

[2] Error estimates for RITZ's method can be found in NICOLAS KRYLOFF: Les méthodes de solution approchée des problèmes de la physique mathématique. Mémorial. Sci. Math., Paris **49** (1931).

Of the numerous shorter papers by KRYLOFF on error estimates of this kind we mention without giving their titles just a few published in the C. R. Acad. Sci., Paris **180**, 1316—1318 (1925); **181**, 86—88 (1925); **183**, 476—479 (1926); **186**, 298—300, 422—425 (1928).

If the v_ν possess continuous second derivatives, integration by parts yields

$$\left.\begin{aligned}
\int_a^b v_\nu \{L[\varphi] - r\}\, dx - \frac{p(a)\, v_\nu(a)}{\alpha_1}\, \{U_1[\varphi] - \gamma_1\} + \\
+ \frac{p(b)\, v_\nu(b)}{\beta_1}\, \{U_2[\varphi] - \gamma_2\} = 0 \qquad (\nu = 1, 2, \ldots, p),
\end{aligned}\right\} \tag{5.23}$$

in which the notation of (5.3), (5.4) has been used. Written in full with the expression (5.21) inserted for φ the equations for the a_ν read

$$\left.\begin{aligned}
\sum_{k=1}^{p} a_k \left\{ \int_a^b v_\nu L[v_k]\, dx - \frac{p(a)\, v_\nu(a)}{\alpha_1}\, U_1[v_k] + \frac{p(b)\, v_\nu(b)}{\beta_1}\, U_2[v_k] \right\} \\
= - \int_a^b v_\nu \{L[v_0] - r\}\, dx + \frac{p(a)\, v_\nu(a)}{\alpha_1}\, \{U_1[v_0] - \gamma_1\} - \\
- \frac{p(b)\, v_\nu(b)}{\beta_1}\, \{U_2[v_0] - \gamma_2\} \qquad (\nu = 1, 2, \ldots, p).
\end{aligned}\right\} \tag{5.24}$$

When we have only "essential" boundary conditions, the boundary terms in (5.23) disappear and we are left with equations identical with GALERKIN's equations (see Ch. I, § 4.3).

Note 1. The differential equation (5.3), which is of self-adjoint form, is a special case of the general second-order linear equation

$$L[y] = f_2 y'' + f_1 y' + f_0 y = r(x). \tag{5.25}$$

However, when $f_2(x) \neq 0$ [a zero of $f_2(x)$ is a singular point of the differential equation], the general equation (5.25) may be transformed into the special form (5.3); thus if we multiply by

$$\varrho(x) = \exp \int_{x_0}^{x} \frac{f_1(\xi) - f_2'(\xi)}{f_2(\xi)}\, d\xi$$

we obtain

$$(f_2 \varrho y')' + f_0 \varrho y = r \varrho. \tag{5.26}$$

The considerations of this section are therefore also applicable to every non-singular linear differential equation of the second order.

Note 2. In principle, any second-order differential equation (including non-linear equations) may be written as the Euler equation of some variational problem[1]. In practice the derivation of the appropriate expression $J[\varphi]$ often occasions considerable difficulty; in general it requires the solution of a partial differential equation.

Note 3. The Ritz method can also be combined with other approximate methods; for example, one can demand that the approximation function (5.21) satisfies the differential equation exactly at $q(<p)$ points (combination with the collocation method); this immediately eliminates q parameters and the Ritz calculation with $p - q$ parameters is correspondingly simpler, while the accuracy

---

[1] BOLZA, O.: Vorlesungen über Variationsrechnung. Leipzig 1909. Reprinted 1949, 705 + 10 pp.

of the approximation so obtained, as shown by examples, can be just as good as that of the pure Ritz calculation with p parameters provided that the additional q conditions are suitably chosen[1].

5.4. Examples of the application of Ritz's method to boundary-value problems of the second order

Example I. A linear inhomogeneous boundary-value problem. From (5.19) the variational problem corresponding to the strut problem of § 1.2 (Example I)

$$y'' + (1 + x^2) y + 1 = 0, \qquad y(\pm 1) = 0$$

reads

$$J[\varphi] = \int_{-1}^{1} \left(\varphi'^2 - (1 + x^2) \varphi^2 - 2\varphi \right) dx = \min.,$$

where φ must satisfy the "essential" boundary conditions $\varphi(\pm 1) = 0$. The simplest form we can assume for φ is a polynomial expression satisfying the boundary conditions and exploiting the symmetry of the solution; thus we write

$$\varphi = \sum_{\nu=1}^{p} a_\nu v_\nu(x) = \sum_{\nu=1}^{p} a_\nu (1 - x^{2\nu}).$$

Such an expression with two terms $(p = 2)$:

$$\varphi = a(1 - x^2) + b(1 - x^4) \qquad \text{with} \qquad \varphi' = -2ax - 4bx^3$$

(in which we have used a, b instead of a_1, a_2) yields

$$\frac{1}{8} J[\varphi] = \frac{19}{105} a^2 + \frac{10}{45} \cdot 2ab + \frac{1244}{3465} b^2 - \frac{1}{3} a - \frac{2}{5} b. \qquad (5.27)$$

A simple yet significant check can be made on this calculation by putting $a = b = A$, so that $\varphi = A(2 - x^2 - x^4)$, and calculating J again for this function. We should get the same result, namely

$$\frac{1}{8} J = \frac{379}{385} A^2 - \frac{11}{15} A,$$

by putting $a = b = A$ in (5.27). We now calculate a and b from

$$\frac{1}{8} \frac{\partial J[\varphi]}{\partial a} = \frac{38}{105} a + \frac{20}{45} b - \frac{1}{3} = 0,$$

$$\frac{1}{8} \frac{\partial J[\varphi]}{\partial b} = \frac{20}{45} a + \frac{2488}{3465} b - \frac{2}{5} = 0,$$

obtaining

$$a = \frac{4200}{4252} = 0.98777, \qquad b = -\frac{231}{4252} = -0.05433.$$

---

[1] See, for example, TH. PÖSCHL: Über eine mögliche Verbesserung der Ritz-schen Methode. Ing.-Arch. **23**, 365—372 (1955).

The second approximation by this method is therefore

$$y_2 = \frac{1}{4252}\,(3969 - 4200\,x^2 + 231\,x^4);$$

in particular we have

$$y_2(0) = \frac{3969}{4252} = 0.933\,443.$$

If more accurate values are required, we can use a three-term expression:

$$\varphi = a\,(1 - x^2) + b\,(1 - x^4) + c\,(1 - x^6).$$

This yields

$$\frac{1}{8}\,J[\varphi] = \frac{19}{105}\,a^2 + \frac{10}{45}\,2ab + \frac{1244}{11 \times 315}\,b^2 + \frac{93}{385}\,2ac + \frac{19846}{143 \times 315}\,2bc +$$
$$+ \frac{25943}{143 \times 315}\,c^2 - \frac{1}{3}\,a - \frac{2}{5}\,b - \frac{3}{7}\,c,$$

which we can check similarly by putting $a = b = c = A$; with $\varphi = A\,(3 - x^2 - x^4 - x^6)$ we obtain

$$\frac{1}{8}\,J = \frac{3764}{1287}\,A^2 - \frac{122}{105}\,A.$$

The linear equations

$$\frac{1}{16}\,\frac{\partial J[\varphi]}{\partial a} = \frac{19}{105}\,a + \frac{10}{45}\,b + \frac{93}{385}\,c - \frac{1}{6} = 0,$$
$$\frac{1}{16}\,\frac{\partial J[\varphi]}{\partial b} = \frac{10}{45}\,a + \frac{1244}{11 \times 315}\,b + \frac{19846}{143 \times 315}\,c - \frac{1}{5} = 0,$$
$$\frac{1}{16}\,\frac{\partial J[\varphi]}{\partial c} = \frac{93}{385}\,a + \frac{19846}{143 \times 315}\,b + \frac{25943}{143 \times 315}\,c - \frac{3}{14} = 0,$$

i.e.

$$8151\,a + 10010\,b + 10881\,c - 7507.5 = 0,$$
$$10010\,a + 16172\,b + 19846\,c - 9009 = 0,$$
$$10881\,a + 19846\,b + 25943\,c - 9652.5 = 0,$$

have the solution

$$a = 0.9664778\,,$$
$$b = -0.00473781,$$
$$c = -0.02966958.$$

These values define the third approximation $\varphi = y_3$, which yields in particula

$$y_3(0) = a + b + c = 0.9320704.$$

II. An eigenvalue problem.

Corresponding to the eigenvalu problem

$$-\,[(1 + x)\,y']' = \lambda\,(1 + x)\,y, \qquad y'(0) = y(1) = 0,$$

associated with the longitudinal vibrations of a rod (Example III, § 1.2) we have from (5.19) the formal variational problem

$$J[\varphi] = \int\limits_0^1 [(1 + x)\,\varphi'^2 - \lambda\,(1 + x)\,\varphi^2]\,dx = \text{extremum}, \qquad \varphi(1) = 0.$$

Here λ is formally treated as a constant parameter (justification of this procedure is considered in §§ 8.7 and 8.8). The admissible functions φ need satisfy only the "essential" boundary condition $\varphi(1) = 0$, but to obtain better results we choose an expression for φ which also satisfies the "suppressible" boundary condition $\varphi'(0) = 0$:

$$\varphi = a(1 - x^2) + b(1 - x^3).$$

The boundary terms which should be added to the integral in $J[\varphi]$ are zero in this case. With Λ as the corresponding approximate value for λ we have

$$J[\varphi] = \frac{7}{3} a^2 + 2 \cdot \frac{27}{10} a b + \frac{33}{10} b^2 - \Lambda \left(\frac{7}{10} a^2 + 2 \cdot \frac{163}{210} a b + \frac{243}{280} b^2 \right).$$

Repeating the calculation with $-a = b = 1$, i.e. $\varphi = x^2 - x^3$, as a check, we obtain

$$J = \frac{7}{30} - \frac{13}{840} \Lambda,$$

which is also the value obtained by putting $-a = b = 1$ in $J[\varphi]$.

The equations

$$\frac{1}{2} \frac{\partial J}{\partial a} = a \left(\frac{7}{3} - \Lambda \frac{7}{10} \right) + b \left(\frac{27}{10} - \Lambda \frac{163}{210} \right) = 0,$$

$$\frac{1}{2} \frac{\partial J}{\partial b} = a \left(\frac{27}{10} - \Lambda \frac{163}{210} \right) + b \left(\frac{33}{10} - \Lambda \frac{243}{280} \right) = 0$$

have a non-trivial solution if and only if the determinant of their coefficients vanishes:

$$\begin{vmatrix} \dfrac{7}{3} - \Lambda \dfrac{7}{10} & \dfrac{27}{10} - \Lambda \dfrac{163}{210} \\[2mm] \dfrac{27}{10} - \Lambda \dfrac{163}{210} & \dfrac{33}{10} - \Lambda \dfrac{243}{280} \end{vmatrix} = 0.$$

This yields the approximations

$$\Lambda = \begin{cases} 3 \cdot 218524, \\ 25 \cdot 334 \end{cases}$$

for the first two eigenvalues.

By suppressing the second row and column of the above determinant we obtain for comparison the "first approximation" to the smallest eigenvalue; thus the single-term expression $\varphi = a(1 - x^2)$ yields

$$\frac{7}{3} - \Lambda \frac{7}{10} = 0, \quad \text{so that} \quad \Lambda = \frac{10}{3} \approx 3 \cdot 333.$$

Approximations for the corresponding eigenfunctions can be found by solving the homogeneous equations with the appropriate value of Λ

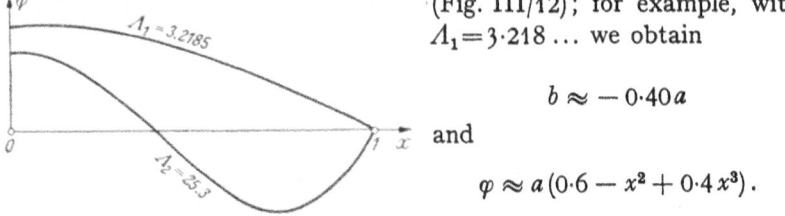

(Fig. III/12); for example, with $\Lambda_1 = 3.218\ldots$ we obtain

$$b \approx -0.40\,a$$

and

$$\varphi \approx a\,(0.6 - x^2 + 0.4\,x^3).$$

Fig. III/12. Approximate eigenfunctions

III. A non-linear boundary-value problem. For the Example II of § 1.2

$$y'' = \frac{3}{2}\,y^2; \qquad y(0) = 4, \qquad y(1) = 1$$

the corresponding variational problem according to (5.1), (5.2) is

$$J[\varphi] = \int_0^1 (\varphi'^2 + \varphi^3)\,dx = \text{extremum}, \qquad \varphi(0) = 4, \qquad \varphi(1) = 1.$$

Fig. III/13. Solution of the non-linear equations

As in (5.21), we assume a form

$$\varphi = 4 - 3x + a_1\,(x - x^2) + a_2\,(x - x^3)$$

which satisfies the boundary conditions for arbitrary a_1, a_2. Then from

$$\frac{\partial J}{\partial a_1} = \int_0^1 [2\varphi'(1 - 2x) + 3\varphi^2\,(x - x^2)]\,dx = 0,$$

$$\frac{\partial J}{\partial a_2} = \int_0^1 [2\varphi'(1 - 3x^2) + 3\varphi^2\,(x - x^3)]\,dx = 0$$

we obtain the two non-linear equations

$$1407 + 490\,a_1 + 726\,a_2 + 9\,a_1^2 + 27\,a_1 a_2 + \frac{41}{2}\,a_2^2 = 0,$$

$$1302 + 484\,a_1 + 750\,a_2 + 9\,a_1^2 + \frac{82}{3}\,a_1 a_2 + 21\,a_2^2 = 0.$$

By linear combination of these equations two new equations can be produced which do not contain terms in a_1^2 and a_2^2, respectively, and which can therefore be easily solved for a_1 in terms of a_2 and for a_2 in terms of a_1, respectively:

$$a_1 = \frac{-105 + 24 a_2 + \frac{1}{2} a_2^2}{6 - \frac{1}{3} a_2}, \qquad a_2 = \frac{2856 + 368 a_1 + \frac{9}{2} a_1^2}{129 - \frac{20}{3} a_1};$$

this facilitates the determination of a_1 and a_2, see Fig. III/13. Table III/12 gives the values obtained for a_1, a_2 and compares specimen values of the solution with those obtained by means of a single-term expression (without the term in a_2).

Table III/12. *Approximate solution of a non-linear problem by* RITZ's *method*

x	Single-term approximation		Two-term approximation	
	The everywhere positive	The other	The everywhere positive	The other
	solution		solution	
	$9 a_1 = -245 \pm \sqrt{47362}$		$a_1 = -7 \cdot 07004$	$a_1 = -32 \cdot 20$
	$a_1 = -3 \cdot 0413$	$a_1 = -51 \cdot 40$	$a_2 = 2 \cdot 72044$	$a_2 = -12 \cdot 60$
$\frac{1}{4}$	$2 \cdot 6798$ (error $+4 \cdot 7\%$)	$-6 \cdot 38$	$2 \cdot 56197$ (error $+0 \cdot 08\%$)	$-5 \cdot 75$
$\frac{1}{2}$	$1 \cdot 7397$ (error $-2 \cdot 1\%$)	$-10 \cdot 35$	$1 \cdot 7627$ (error $-0 \cdot 86\%$)	$-10 \cdot 28$
$\frac{3}{4}$	$1 \cdot 1798$ (error $-9 \cdot 7\%$)	$-7 \cdot 88$	$1 \cdot 31701$ (error $+0 \cdot 84\%$)	$-8 \cdot 43$

§6. RITZ's method for boundary-value problems of higher order

We now turn to problems in which higher order derivatives appear; we derive the Euler equations generally, including non-linear cases, but we can now be brief.

6.1. Derivation of higher order Euler equations

Consider the variational problem

$$J = J[\varphi] = \int_a^b F(x, \varphi, \varphi', \ldots, \varphi^{(m)}) \, dx + A\left(\varphi(a), \varphi'(a), \ldots, \varphi^{(m-1)}(a)\right) + \\ + B\left(\varphi(b), \varphi'(b), \ldots, \varphi^{(m-1)}(b)\right) = \text{minimum} \quad (6.1)$$

with respect to the domain of admissible functions $\varphi(x)$ which possess continuous derivatives of order $2m$ and satisfy certain prescribed linear (and linearly independent) boundary conditions

$$U_\mu[\varphi] = \gamma_\mu \qquad (\mu = 1, 2, \ldots, k) \qquad (6.2)$$

at the points a and b; these boundary conditions may not contain derivatives of order higher than $m - 1$. F is a given continuous function of $x, \varphi, \varphi', \ldots, \varphi^{(m)}$ with continuous partial derivatives of up to the $(m + 1)$-th order, and A, B are given continuous functions of the boundary values of $\varphi, \varphi', \varphi'', \ldots, \varphi^{(m-1)}$ with continuous first partial derivatives.

We assume that the minimum value of J exists and that there is a function $y(x)$ among the admissible functions $\varphi(x)$ for which J takes this minimum value:

$$J[y] \leqq J[\varphi]. \tag{6.3}$$

Then the one-parameter family of admissible functions defined by $\varphi(x) = y(x) + \varepsilon\eta(x)$, where $\eta(x)$ is any function possessing a continuous derivative of order $2m$ and satisfying the homogeneous boundary conditions

$$U_\mu[\eta] = 0 \qquad (\mu = 1, 2, \ldots, k), \tag{6.4}$$

must also satisfy the inequality (6.3). For this family of functions $J[\varphi]$ is a function $\Phi(\varepsilon)$ of the parameter ε; it can have a minimum at $\varepsilon = 0$ only if the derivative $\Phi'(0)$ vanishes.

We now evaluate $\Phi'(0)$:

$$\begin{aligned}
\Phi'(0) = \frac{d}{d\varepsilon} &\left\{ \int_a^b F(x, y + \varepsilon\eta, y' + \varepsilon\eta', \ldots, y^{(m)} + \varepsilon\eta^{(m)})\, dx + \right. \\
&+ A(x + \varepsilon\eta, \ldots, y^{(m-1)} + \varepsilon\eta^{(m-1)})_{x=a} + \\
&\left. + B(y + \varepsilon\eta, \ldots, y^{(m-1)} + \varepsilon\eta^{(m-1)})_{x=b} \right\}_{\varepsilon=0} \\
= \int_a^b &(F_\varphi\eta + F_{\varphi'}\eta' + \cdots + F_{\varphi^{(m)}}\eta^{(m)})\, dx + A_\varphi\eta(a) + \cdots + \\
&+ A_{\varphi^{(m-1)}}\eta^{(m-1)}(a) + B_\varphi\eta(b) + \cdots + B_{\varphi^{(m-1)}}\eta^{(m-1)}(b),
\end{aligned} \tag{6.5}$$

where the subscripts denote partial derivatives, for example,

$$F_{\varphi^{(i)}} = \frac{\partial F}{\partial \varphi^{(i)}}.$$

The integrals of the individual terms of the integrand can be repeatedly transformed by integration by parts until the factor η appears in place of $\eta^{(i)}$; thus we have

$$\begin{aligned}
\int_a^b F_{\varphi'}\eta'\, dx &= [\eta F_{\varphi'}]_a^b - \int_a^b \eta\, \frac{d}{dx} F_{\varphi'}\, dx, \\
\int_a^b F_{\varphi''}\eta''\, dx &= \left[\eta' F_{\varphi''} - \eta\, \frac{d}{dx} F_{\varphi''}\right]_a^b + \int_a^b \eta\, \frac{d^2}{dx^2} F_{\varphi''}\, dx
\end{aligned} \tag{6.6}$$

and so on. With all the terms transformed the condition $\Phi'(0) = 0$ becomes

$$\int_a^b \eta\left[F_\varphi - \frac{d}{dx} F_{\varphi'} + \frac{d^2}{dx^2} F_{\varphi''} - \cdots + (-1)^m \frac{d^m}{dx^m} F_{\varphi^{(m)}}\right] dx + S = 0, \tag{6.7}$$

where S is the boundary expression defined by

$$
\left.
\begin{aligned}
S = &\left[\eta\, F_{\varphi'} + \left(\eta'\, F_{\varphi''} - \eta\, \frac{d}{dx}\, F_{\varphi''} \right) + \cdots + \right. \\
&+ \left(\eta^{(m-1)} F_{\varphi^{(m)}} - \eta^{(m-2)} \frac{d}{dx}\, F_{\varphi^{(m)}} + \eta^{(m-3)} \frac{d^2}{dx^2}\, F_{\varphi^{(m)}} + \cdots + \right. \\
&+ \left. \left. (-1)^{m-1} \eta\, \frac{d^{m-1}}{dx^{m-1}}\, F_{\varphi^{(m)}} \right) \right]_a^b + A_\varphi \eta(a) + \cdots + \\
&+ A_{\varphi^{(m-1)}} \eta^{(m-1)}(a) + B_\varphi \eta(b) + \cdots + B_{\varphi^{(m-1)}} \eta^{(m-1)}(b) .
\end{aligned}
\right\} \quad (6.8)
$$

Just as in § 5.2, we infer from the arbitrariness of η that the bracketed factor of the integrand must vanish [we have only to put

$$
(x - x_0 - \delta)^{2m+1} (x - x_0 + \delta)^{2m+1}
$$

in place of the third power used in (5.13)]; this yields the Euler equation

$$
F_\varphi - \frac{d}{dx}\, F_{\varphi'} + \frac{d^2}{dx^2}\, F_{\varphi''} - \cdots + (-1)^m \frac{d^m}{dx^m}\, F_{\varphi^{(m)}} = 0, \qquad (6.9)
$$

which again represents a necessary condition to be satisfied by the solution of the variational problem. From (6.7) the boundary expression S must vanish also.

We now assume that the given functions A, B appearing in (6.1) are quadratic functions of φ and its derivatives; then the expression S is linear in the boundary values of y and its derivatives, and is linear and homogeneous in the boundary values of η and its derivatives. The k equations provided by the homogeneous boundary conditions (6.4) can be solved for k of the $2m$ boundary values $\eta(a), \eta'(a), \ldots, \eta^{(n-1)}(a)$, $\eta(b), \eta'(b), \ldots, \eta^{(m-1)}(b)$ in terms of the remaining ones. These $2m - k$ remaining boundary values of derivatives of η may be called "free boundary values"[1] and will be denoted in any order by $\eta_1, \eta_2, \ldots, \eta_{2m-k}$. If we express all boundary values of η (and its derivatives) appearing in S in terms of the free boundary values, then S can be written in the form

$$
S = \sum_{\nu=1}^{2m-k} \eta_\nu\, W_\nu[y] = 0, \qquad (6.10)
$$

where $W_\nu[y]$ is a function of the boundary values of y, y', ... Since the η_ν may be chosen arbitrarily, and since there exists a corresponding admissible function, it follows from (6.10) that

$$
W_\nu[y] = 0 \qquad (\nu = 1, 2, \ldots, 2m - k). \qquad (6.11)
$$

These equations, which must be satisfied by the solution y of the variational problem (6.1), (6.2), constitute $2m - k$ additional boundary

---

[1] KAMKE, E.: Math. Z. **48**, 70 (1942).

conditions to be associated with the $2m$-th-order differential equation (6.9), thus making up the number of boundary conditions to the required number $2m$. We assert nothing about the existence of a solution of this boundary-value problem in the present generality; we also defer discussion of the different types of boundary condition (6.2) and (6.11) till the next section.

6.2. Linear boundary-value problems of the fourth order

For illustration, and also because of their important applications, we consider in detail the fourth-order boundary-value problems with the self-adjoint differential equation

$$L[y] = (g_2 y'')'' - (g_1 y')' + g_0 y = r(x) \tag{6.12}$$

and two linearly independent non-contradictory boundary conditions at each of the points $x = a$, $x = b$. These boundary conditions are divided into "essential" and "suppressible" according to the definitions in Ch. I, §1.2. For fourth-order problems the "essential" boundary conditions are those which contain no second or third derivatives, only the first derivative and the function itself.

As the expression to be minimized in the corresponding variational problem we take

$$J[\varphi] = \int_a^b [g_2 \varphi''^2 + g_1 \varphi'^2 + g_0 \varphi^2 - 2r\varphi] \, dx + R_a + R_b, \tag{6.13}$$

where

$$R_b = A \varphi_b^2 + 2B \varphi_b \varphi_b' + C \varphi_b'^2 + 2D \varphi_b + 2E \varphi_b' \tag{6.14}$$

and R_a is defined similarly but with b replaced by a and with different constants in place of $A, B, \ldots, E$. The subscript notation here denotes argument values, e.g. $\varphi_b = \varphi(b)$, $\varphi_b' = \varphi'(b)$. The variation $\varphi = y + \varepsilon\eta$ leads to the necessary condition

$$\left. \frac{1}{2} \left(\frac{d}{d\varepsilon} J[y + \varepsilon\eta] \right) \right|_{\varepsilon=0} \\ = \int_a^b [g_2 y'' \eta'' + g_1 y' \eta' + g_0 y\eta - r\eta] \, dx + R_a^* + R_b^* = 0, \Bigg\} \tag{6.15}$$

where

$$R_b^* = A y_b \eta_b + B (y_b \eta_b' + y_b' \eta_b) + C y_b' \eta_b' + D \eta_b + E \eta_b'$$

and R_a^* is defined correspondingly. Integration by parts then yields

$$0 = \int_a^b \{L[y] - r\} \eta \, dx + S_a + S_b, \tag{6.16}$$

where

$$S_b = g_2 y'' \eta' - g_2' y'' \eta - g_2 y''' \eta + g_1 y' \eta + A y\eta + B y\eta' + \\ + B y' \eta + C y' \eta' + D \eta + E \eta',$$

in which all functions are evaluated at the point $x=b$ (S_a reads similarly). By the usual argument the integrand in (6.16) must vanish; thus y must satisfy (6.12) and we must have $S_a+S_b=0$. We now rearrange S_a and S_b in the forms

$$S_a = \eta_a W_a + \eta_a' W_a',$$
$$S_b = \eta_b W_b + \eta_b' W_b'$$

with W_a, W_a', W_b, W_b' independent of η; for instance

$$\left.\begin{aligned}
W_b &= -g_2 y''' - g_2' y'' + (g_1 + B) y' + A y + D, \\
W_b' &= g_2 y'' + C y' + B y + E,
\end{aligned}\right\} \qquad (6.17)$$

and there are analogous expressions for W_a, W_a'.

We now consider the prescribed boundary conditions, two at each of the points $x=a$, $x=b$, and distinguish the following cases:

Case I. All the boundary conditions are "essential", i.e. the values of y and y' are prescribed at $x=a$ and $x=b$. Here there are admissible functions such that W_a, W_a', W_b, W_b' do not vanish, and consequently, for S_a+S_b to vanish, we must demand that $\eta_a=\eta_a'=\eta_b=\eta_b'=0$, i.e. that the admissible functions satisfy the "essential" boundary conditions.

Case II. All boundary conditions are "suppressible". They may therefore be put into the form

$$\left.\begin{aligned}
y_b'' &= \beta_1 y_b + \beta_2 y_b' + \beta_3, \\
y_b''' &= \beta_4 y_b + \beta_5 y_b' + \beta_6
\end{aligned}\right\} \qquad (6.18)$$

with two corresponding conditions at $x=a$. [We assume that $g_2(a) \neq 0$, $g_2(b) \neq 0$, otherwise the differential equation would be singular at the end points.] For arbitrary values of $\eta_a, \eta_a', \eta_b, \eta_b'$ we can calculate $\eta_a'', \eta_a''', \eta_b'', \eta_b'''$ from the homogeneous boundary conditions corresponding to (6.18) and the analogous equations for $x=a$; with these values we can then construct a function η such that $y+\varepsilon\eta$ satisfies all the boundary conditions and is an admissible function. Consequently we must have

$$S_a + S_b = \eta_a W_a + \eta_a' W_a' + \eta_b W_b + \eta_b' W_b' = 0$$

for arbitrary values of $\eta_a, \eta_a', \eta_b, \eta_b'$, and hence $W_a = W_a' = W_b = W_b' = 0$. These four conditions must therefore be made equivalent to the suppressible boundary conditions at $x=a$ and $x=b$ [(6.18)]. We see immediately that this is not possible in general, for there are six free parameters β_i in (6.18) and only five disposable constants A, B, C, D, E in (6.17). Hence for the equivalence of the boundary conditions to be possible at all a relation must exist between the β_i, namely

$$\beta_1 + \beta_2 \frac{g_2' b}{g_2 b} + \beta_5 - \frac{g_1 b}{g_2 b} = 0; \qquad (6.19)$$

a similar relation must exist for the boundary conditions at the other boundary $x = a$.

Case III. Mixed boundary conditions. What has been said in case I about essential conditions and in case II about suppressible conditions is combined appropriately; here also only a certain class of suppressible boundary conditions can be dealt with.

Summary. *The fourth-order linear boundary-value problem*

$$\left(g_2(x)\, y''\right)'' - \left(g_1(x)\, y'\right)' + g_0(x)\, y = r(x), \tag{6.20}$$

where

$$g_2(a) \neq 0, \qquad g_2(b) \neq 0,$$

with two linearly independent non-contradictory boundary conditions at each of the points $x = a$, $x = b$, may be identified with the necessary conditions to be satisfied by the solution y of a variational problem $J[\varphi] = $ minimum, where

$$J[\varphi] = \int_a^b (g_2 \varphi''^2 + g_1 \varphi'^2 + g_0 \varphi^2 - 2r\,\varphi)\, dx + R_a + R_b, \tag{6.21}$$

under the following conditions:

(i) If all four boundary conditions are essential, then we can put $R_a = R_b = 0$ but we must specifically require that the admissible functions φ shall satisfy all the boundary conditions.

(ii) If the boundary conditions at $x = a$ are suppressible and such that when written in the form

$$\left.\begin{array}{l} y_a'' = \alpha_1 y_a + \alpha_2 y_a' + \alpha_3, \\ y_a''' = \alpha_4 y_a + \alpha_5 y_a' + \alpha_6 \end{array}\right\} \tag{6.22}$$

the relation

$$(\alpha_1 + \alpha_5)\, g_{2a} - g_{1a} + \alpha_2 g_{2a}' = 0 \tag{6.23}$$

holds, then we must put

$$\left.\begin{array}{l} R_a = - (\alpha_4 g_{2a} + \alpha_1 g_{2a}')\, \varphi_a^2 + 2\alpha_1 g_{2a} \varphi_a \varphi_a' + \\ \qquad + \alpha_2 g_{2a} \varphi_a'^2 - 2(\alpha_6 g_{2a} + \alpha_3 g_{2a}')\, \varphi_a + 2\alpha_3 g_{2a}\varphi_a'. \end{array}\right\} \tag{6.24}$$

There is no need for φ to satisfy any of the boundary condition at $x = a$.

(iii) Correspondingly, if the boundary conditions at $x = b$ are suppressible and can be written in the form

$$\left.\begin{array}{l} y_b'' = \beta_1 y_b + \beta_2 y_b' + \beta_3, \\ y_b''' = \beta_4 y_b + \beta_5 y_b' + \beta_6 \end{array}\right\} \tag{6.25}$$

with

$$(\beta_1 + \beta_5)\, g_{2b} - g_{1b} + \beta_2 g_{2b}' = 0, \tag{6.26}$$

then we must put

$$R_b = (\beta_4 g_{2b} + \beta_1 g'_{2b})\, \varphi_b^2 - 2\beta_1 g_{2b}\, \varphi_b\, \varphi'_b - \beta_2 g_{2b}\, \varphi'^2_b + \left.\begin{matrix} \\ \end{matrix}\right\}$$
$$+ 2(\beta_6 g_{2b} + \beta_3 g'_{2b})\, \varphi_b - 2\beta_3 g_{2b}\, \varphi'_b, \quad \left.\begin{matrix} \\ \end{matrix}\right\} \quad (6.27)$$

and φ need not satisfy any boundary conditions at $x = b$.

6.3. Example

For the problem of the elastically mounted rail of § 1.3, i.e.

$$[(2 - x^2)\, y'']'' + 40\, y = 2 - x^2, \quad y''(\pm 1) = y'''(\pm 1) = 0,$$

the corresponding variational problem is

$$J[\varphi] = \int_{-1}^{1} [(2 - x^2)\, \varphi''^2 + 40\varphi^2 - 2(2 - x^2)\, \varphi]\, d x = \min.$$

The function φ need not satisfy any of the boundary conditions, for they are all suppressible. The boundary terms R_a and R_b turn out to be zero in this case since all the α_i and β_i in (6.22), (6.25) are zero.

Because of the symmetry, we take

$$\varphi = \sum_{\nu=1}^{p} a_\nu\, x^{2\nu-2}, \quad \text{so that} \quad \varphi'' = \sum_{\nu=1}^{p} a_\nu (2\nu - 2)(2\nu - 3)\, x^{2\nu-4}.$$

Then for $p = 3$

$$\frac{1}{2}\, J[\varphi] = 40 a_1^2 + \frac{44}{3}\, a_2^2 + \frac{13064}{315}\, a_3^2 + \frac{40}{3}\, 2 a_1 a_2 + 8 \times 2 a_1 a_3 +$$
$$+ \frac{592}{35}\, 2 a_2 a_3 - \frac{5}{3}\, 2 a_1 - \frac{7}{15}\, 2 a_2 - \frac{9}{35}\, 2 a_3$$

(the check with $a_1 = a_2 = a_3 = a$ is used again here). The necessary conditions for a minimum

$$\frac{\partial J}{\partial a_1} = 0: \quad 40 a_1 + \frac{40}{3}\, a_2 + \qquad 8 a_3 = \frac{5}{3}$$

$$\frac{\partial J}{\partial a_2} = 0: \quad \frac{40}{3}\, a_1 + \frac{44}{3}\, a_2 + \frac{592}{35}\, a_3 = \frac{7}{15}$$

$$\frac{\partial J}{\partial a_3} = 0: \quad 8 a_1 + \frac{592}{35}\, a_2 + \frac{13064}{315}\, a_3 = \frac{9}{35}$$

yield the values

$$a_1 = \frac{143363}{40 \times 79301} = 0{\cdot}045\,195\,8$$

$$a_2 = -\frac{953}{79301} = -0{\cdot}012\,017\,5$$

$$a_3 = \frac{189}{79301} = 0{\cdot}002\,383\,3.$$

These give, for example, the approximation

$$y(1) = a_1 + a_2 + a_3 = 0{\cdot}035\,561\,7.$$

6.4. Comparison of RITZ's method with the least-squares process

Consider the boundary-value problem

$$y'' + y + x = 0, \qquad y(0) = y(1) = 0.$$

RITZ's method and the least-squares method both replace the boundary-value problem by a variational problem; here the respective variational problems read

$$J_R[\varphi] = \int\limits_0^1 (\varphi'^2 - \varphi^2 - 2x\varphi)\, dx = \min., \qquad \varphi(0) = \varphi(1) = 0$$

and

$$J_L[\varphi] = \int\limits_0^1 (\varphi'' + \varphi + x)^2\, dx = \min., \qquad \varphi(0) = \varphi(1) = 0.$$

Using the same two-term expression

$$\varphi = a_1(x - x^2) + a_2(x - x^3)$$

for both of these problems, we have

$$J_R = \frac{3}{10}\, a_1^2 + \frac{9}{10}\, a_1 a_2 + \frac{76}{105}\, a_2^2 - \frac{1}{6}\, a_1 - \frac{4}{15}\, a_2,$$

$$J_L = \frac{1}{210}\, (707 a_1^2 + 2121 a_1 a_2 + 2200 a_2^2 - 385 a_1 - 784 a_2 + 70).$$

Then the equations $\partial J/\partial a_1 = \partial J/\partial a_2 = 0$ yield for RITZ's method:

$$a_1 = \frac{8}{369}, \qquad a_2 = \frac{7}{41}.$$

and hence the approximate solution

$$y_R = \frac{1}{369}\, (71 x - 8 x^2 - 63 x^3);$$

and for the least squares method

$$a_1 = \frac{4448}{101 \times 2437}, \qquad a_2 = \frac{413}{2437},$$

$$y_L = \frac{1}{246137}\, (46161 x - 4448 x^2 - 41713 x^3).$$

Table III/13 compares some values of y_R and y_L with the exact solution

$$y = \frac{\sin x}{\sin 1} - x.$$

The numerical results and the curves in Fig. III/14 show that of the two methods RITZ's reproduces the function y more accurately while least-squares gives the more accurate values for y''.

Now consider the fourth-order boundary-value problem

$$\frac{d^4 y}{d x^4} = q(x), \qquad y(0) = y''(0) = y(l) = y''(l) = 0$$

Table III/13. *Comparison of results obtained by* Ritz's *method and the least-squares method*

	Exact solution $y(x)$	Approximate solution	
		by Ritz's method: $y_R(x)$	by the least-squares method: $y_L(x)$
$y(\tfrac{1}{2})$	0·069 75	$\frac{5}{72} = $ 0·069 44 (error −0·45%)	$\frac{134\,035}{1\,969\,096} = $ 0·068 07 (error −2·4%)
$y'(0)$	0·188 40	$\frac{71}{369} = $ 0·192 41 (error +2·1%)	$\frac{46\,161}{246\,137} = $ 0·187 54 (error −0·46%)
$y'(1)$	−0·357 91	$-\frac{134}{369} = $ − 0·363 14 (error −1·5%)	$-\frac{87\,874}{246\,137} = $ − 0·357 01 (error +0·25%)

for the deflection y of a loaded beam with constant flexural rigidity supported by smooth pin-joints a distance l apart. The corresponding Ritz variational problem reads

$$J_R[\varphi] = \int_0^l (\varphi''^2 - 2q\,\varphi)\,dx$$
$$= \min.,$$
$$\varphi(0) = \varphi(l) = 0.$$

Since $q = y^{IV}$, the integral can be transformed by two integrations by parts to give

$$J_R[\varphi] = \int_0^l (\varphi''^2 - 2\varphi\,y^{IV})\,dx$$
$$= \int_0^l (\varphi''^2 - 2\varphi''\,y'')\,dx$$
$$= \min.$$

This integral differs only by the constant

$$\int_0^l y''^2\,dy = A$$

from the integral in

$$J_R^*[\varphi] = \int_0^l (\varphi'' - y'')^2\,dx$$
$$= \min.,$$

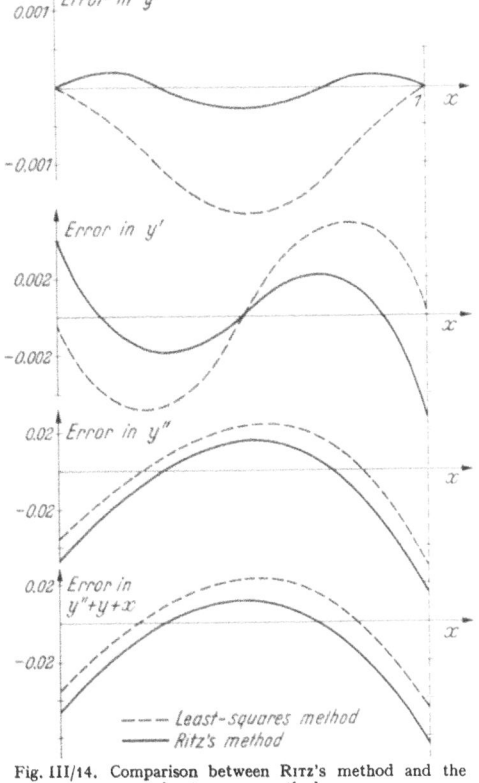

Fig. III/14. Comparison between Ritz's method and the least-squares method

which the Ritz method therefore minimizes with $J_R[\varphi]$. The corresponding least-squares variational problem reads

$$J_L[\varphi] = \int_0^l (\varphi^{IV} - q)^2\,dx = \int_0^l (\varphi^{IV} - y^{IV})^2\,dx = \min.$$

Thus RITZ's method minimizes the error integral of the second derivative while the least-squares method minimizes the error integral of the fourth derivative. Consequently least-squares gives the better approximation to the fourth derivative but its values for the function y itself are less accurate than the corresponding Ritz values; in fact it can happen that they are hopelessly wrong — C. WEBER[1] gives an example in which the approximation to the deflection even has the wrong sign throughout.

§7. Series solutions

7.1. Series solutions in general

For many boundary-value problems, especially those with a simple analytic formulation, a solution can be obtained in the form of an infinite series

$$y(x) = \sum_{\varrho=1}^{\infty} c_{\varrho} \psi_{\varrho}(x) \tag{7.1}$$

of given functions $\psi_{\varrho}(x)$ whose coefficients c_{ϱ} are initially undetermined. *The success of the method depends principally on the nature of the boundary-value problem and on a suitable choice of the system of functions ψ_{ϱ}.* In the examples of § 7.3 the assumed series converge well and in some cases the series method is actually superior to all other methods; on the other hand, there are cases in which the infinite series either converges only very slowly or does not converge at all.

A general procedure for obtaining a series solution of a linear or non-linear boundary-value problem Ch. I (1.1) with linear boundary conditions Ch. I (1.7) is now described.

Let z be a function satisfying the boundary conditions

$$U_{\nu}[z] = \gamma_{\nu} \qquad (\nu = 1, 2, \ldots, n). \tag{7.2}$$

Then for the ψ_{ϱ} we choose a system of functions which is complete in the interval considered, $\langle a, b \rangle$ say, (for example, a system of trigonometrical functions or of spherical harmonics) and whose members satisfy the homogeneous boundary conditions

$$U_{\nu}[\psi_{\varrho}] = 0 \qquad (\nu = 1, 2, \ldots, n; \ \varrho = 1, 2, \ldots). \tag{7.3}$$

The assumed completeness of the system ψ_{ϱ} permits us to express the solution y in the form

$$y(x) = z(x) + \sum_{\varrho=1}^{\infty} c_{\varrho} \psi_{\varrho}(x), \tag{7.4}$$

which satisfies the boundary conditions (7.2) for any c_{ϱ}. These coefficients are to be determined so that y satisfies the (linear or non-linear)

---

[1] WEBER, C.: Z. Angew. Math. Mech. **21**, 310—311 (1941).

differential equation (1.1) of Ch. I, i.e. $F(x, y, y', \ldots, y^{(n)}) = 0$. For this we must assume further that the expansion (7.4) may be differentiated term by term n times so that it may be substituted in the differential equation. If y is the required solution, this should yield $F \equiv 0$; therefore the coefficients in the expansion of F in the form (7.1) (this expansion also is possible because of the completeness of the ψ_ϱ) must all vanish. If the ψ_ϱ were also orthogonal, this would lead directly to the equations

$$\int_a^b F\psi_\varrho \, dx = 0 \qquad (\varrho = 1, 2, \ldots); \tag{7.5}$$

otherwise the ψ_ϱ first have to be orthogonalized. Equations (7.5) written out in detail read

$$\left. \int_a^b F\left(x, z + \sum_{\sigma=1}^\infty c_\sigma \psi_\sigma, z' + \sum_{\sigma=1}^\infty c_\sigma \psi_\sigma', \ldots, z^{(n)} + \sum_{\sigma=1}^\infty c_\sigma \psi_\sigma^{(n)}\right) \psi_\varrho(x) \, dx = 0 \right\} \tag{7.6}$$
$$(\varrho = 1, 2, \ldots).$$

Thus we have an infinite set of equations, non-linear in general, for the infinite number of unknowns c_ϱ. For numerical calculation the series (7.4) is terminated after a finite number of terms (say p); we then have a finite set of equations for $c_1, \ldots, c_p$:

$$\left. \int_a^b F\left(x, z + \sum_{\sigma=1}^p c_\sigma \psi_\sigma, \ldots, z^{(n)} + \sum_{\sigma=1}^p c_\sigma \psi_\sigma^{(n)}\right) \psi_\varrho(x) \, dx = 0 \right\} \tag{7.7}$$
$$(\varrho = 1, 2, \ldots, p).$$

For the special case of a linear second-order problem with essential boundary conditions these equations are identical with the Ritz-Galerkin equations (5.23).

A series approach[1] which is occasionally used for problems with one or more boundary conditions at infinity makes use of two series, a power series at a finite point $x = x_0$ and an asymptotic series at $x = \infty$; these expressions for the solution are joined smoothly at a suitable intermediate point $x = \alpha$.

7.2. Power series solutions

We expand all functions appearing in the differential equation in powers of x or $x - \xi$ for suitable ξ and assume a corresponding power series expansion for the solution:

$$y(x) = \sum_{k=0}^\infty a_k (x - \xi)^k. \tag{7.8}$$

---

[1] An example of the application of this technique to a non-linear problem can be found in U. T. Bödewadt: Die Drehströmung über festem Grunde. Z. Angew. Math. Mech. **20**, 241−253 (1940).

By inserting this series in the differential equation, rearranging in powers of $x - \xi$ and equating to zero the coefficients of $(x - \xi)^k$ (for $k = 0, 1, 2, \ldots$), we obtain in general an infinite set of equations for the a_ν. The n boundary conditions yield n additional equations for the a_ν. This infinite system of equations is solved approximately by retaining only a finite number of equations and the same finite number of unknowns a_ν. In many cases a formula for the successive a_ν is obtained from which convergence of the power series can be inferred and the assumption that the solution can be expressed in the form (7.8) thereby justified (cf. Example I of § 7.3).

7.3. Examples

We will show how the series method is carried out in practice by means of a few examples.

I. The form of the coefficients in the differential equation of the strut problem

$$y'' + (1 + x^2)\, y + 1 = 0, \qquad y(\pm 1) = 0$$

(Example I of § 1.2) suggests that a power series expansion would be convenient here. There are two obvious ways of proceeding. One is to put

$$y = \sum_{\nu=0}^{\infty} a_\nu x^{2\nu}, \tag{7.9}$$

in which we include only the even powers of $x^{2\nu}$ because of symmetry; we then express the a_ν successively in terms of a_0 by inserting the series in the differential equation and equating coefficients to zero; finally we determine a_0 from the boundary condition $y(1) = 0$. The other alternative is to put

$$y = \sum_{\nu=1}^{\infty} b_\nu (1 - x^{2\nu}),$$

which satisfies the boundary conditions already, then determine the b_ν from the infinite set of linear equations obtained by inserting the series in the differential equation and equating the sum of the coefficients of like powers of x to zero. Both methods lead to substantially the same results, so we will describe only the first. Thus using the series (7.9) we have

$$(1 + x^2)\, y = \sum_{\nu=0}^{\infty} a_\nu (x^{2\nu} + x^{2\nu+2}) = a_0 + \sum_{\nu=1}^{\infty} (a_\nu + a_{\nu-1})\, x^{2\nu},$$

$$y'' = \sum_{\nu=0}^{\infty} (2\nu + 2)(2\nu + 1)\, a_{\nu+1} x^{2\nu},$$

and substituting in the differential equation we obtain

$$1 + a_0 + 2a_1 + \sum_{\nu=1}^{\infty} [a_\nu + a_{\nu-1} + (2\nu + 2)(2\nu + 1)\, a_{\nu+1}]\, x^{2\nu} = 0.$$

By the uniqueness theorem for power series, this implies that the coefficients of all powers of x must vanish separately; consequently

$$1 + a_0 + 2a_1 = 0,$$
$$a_{\nu-1} + a_\nu + (2\nu + 2)(2\nu + 1) a_{\nu+1} = 0 \quad (\nu = 1, 2, 3, \ldots). \tag{7.10}$$

From this recursive formula the $a_1, a_2, a_3, \ldots$ can be expressed successively in terms of a_0 (see first column of Table III/14). We can also use this formula to show that

$$|a_\nu| \leq \frac{1 + |a_0|}{(2\nu - 1) 2\nu} \quad \text{for} \quad \nu = 1, 2, 3, \ldots.$$

Table III/14. *Calculation of the coefficients in a power series solution*

a_p	p	$\sum\limits_{\nu=0}^{p} a_\nu$	Approximate values of a_0
$a_1 = -\frac{1}{2}(1 + a_0) = -\frac{1}{2!}(1 + a_0)$	1	$\frac{1}{2!}(-1 + a_0)$	1
$a_2 = -\frac{1}{3 \times 4}(a_0 + a_1) = \frac{1}{4!}(1 - a_0)$	2	$\frac{1}{4!}(-11 + 11 a_0)$	1
$a_3 = -\frac{1}{5 \times 6}(a_1 + a_2) = \frac{1}{6!}(11 + 13 a_0)$	3	$\frac{1}{6!}(-319 + 343 a_0)$	$\frac{319}{343} = 0.93003$
$a_4 = \frac{1}{8!}(-41 + 17 a_0)$	4	$\frac{5}{8!}(-3581 + 3845 a_0)$	0.931339
$a_5 = \frac{1}{10!}(-575 - 745 a_0)$	5	$\frac{5}{10!}(-322405 + 345901 a_0)$	0.93207304
$a_6 = \frac{5}{12!}(853 - 157 a_0)$	6	$\frac{5}{12!}(-42556607 + 45658775 a_0)$	0.93205757
$a_7 = \frac{5}{14!}(14327 + 19825 a_0)$	7	$\frac{5}{12!}(-42556528.3 + 45658884 a_0)$	0.93205362

For $\nu = 1, 2$ the validity of this upper bound can be verified from Table III/14; for $\nu > 2$ it follows by complete induction: if the inequality is true for $\nu - 1$ and ν, then

$$|a_{\nu+1}| = \frac{|a_{\nu-1} + a_\nu|}{(2\nu + 1)(2\nu + 2)} \leq \frac{1 + |a_0|}{(2\nu + 1)(2\nu + 2)} \left(\frac{1}{(2\nu - 3)(2\nu - 2)} + \frac{1}{(2\nu - 1) 2\nu} \right)$$
$$\leq \frac{1 + |a_0|}{(2\nu + 1)(2\nu + 2)} \quad (\nu = 2, 3, \ldots);$$

thus it is also true for $\nu + 1$, and hence for all ν. Since $\sum\limits_{\nu=1}^{\infty} \frac{1}{(2\nu - 1) 2\nu}$ converges, the assumed power series converges absolutely and uniformly in the closed interval $-1 \leq x \leq 1$; hence the expansion is justified.

The boundary condition

$$y(1) = \sum_{\nu=0}^{\infty} a_\nu = 0$$

yields an equation for a_0. For numerical calculation the series is truncated after some term, say the term in a_p; the fourth column of Table III/14 gives the sequence of approximations to a_0 for $p = 1$, $2, \ldots, 7$.

Error estimation: The error in a_0 due to truncation of the series after the term with $v = p$ can be estimated quite easily in this example. We can write

$$a_v = \alpha_v + a_0 \beta_v, \qquad \sum_{v=0}^{p} a_v = A_p + a_0 B_p,$$

where the numbers $\alpha_v, \beta_v, A_p, B_p$ are independent of a_0 (their values can be found from columns 1 and 3 of Table III/14); then the p-th approximation to a_0 is the quotient $-A_p/B_p$, values of which are given in the fourth column of the table. To illustrate the procedure we will estimate the error in $-A_7/B_7$. It can be seen from the table that

$$|\alpha_7| < |\alpha_6|, \qquad |\beta_7| < |\beta_6|;$$

then from (7.10) we obtain successively the upper bounds

$$|\alpha_8| \leqq \frac{2|\alpha_6|}{15 \times 16} = \frac{|\alpha_6|}{120}, \qquad |\alpha_9| \leqq \frac{|\alpha_6|}{120}, \qquad |\alpha_{10}| \leqq \frac{|\alpha_6|}{60} \frac{1}{19 \times 20},$$

$$|\alpha_{11}| \leqq \frac{|\alpha_6|}{60} \frac{1}{21 \times 22}, \dots,$$

so that

$$\sum_{v=8}^{\infty} |\alpha_v| \leqq \frac{|\alpha_6|}{60} \left(1 + \frac{1}{19 \times 20} + \frac{1}{21 \times 22} + \frac{1}{23 \times 24} + \cdots \right) = \frac{|\alpha_6|}{60} \times 1{\cdot}027.$$

Precisely similar considerations yield

$$\sum_{v=8}^{\infty} |\beta_v| \leqq \frac{|\beta_6|}{60} \times 1{\cdot}027.$$

Since $|\alpha_6| = \dfrac{4265}{12!}$, the error in A_7 is at most $\dfrac{4265}{60} \times \dfrac{1{\cdot}027}{12!}$; with $-A_7 \approx 213 \times 10^6/12!$ the relative error is therefore at most $0{\cdot}35 \times 10^{-6}$. Similarly the relative error in B_7 is at most $\dfrac{785}{60} \times \dfrac{1{\cdot}027}{228 \times 10^6} < 0{\cdot}06 \times 10^{-6}$, and hence the error in a_0 is at most $0{\cdot}42 \times 10^{-6}$. Thus we have the error estimate

$$|a_0 - 0{\cdot}9320536| < 4{\cdot}2 \times 10^{-7}.$$

II. For the non-linear problem (Example II, § 1.2)

$$y'' = \tfrac{3}{2} y^2; \qquad y(0) = 4, \qquad y(1) = 1$$

we expand $y(x)$ as a power series at $x = 1$, say in powers of $s = 1 - x$:

$$y(s) = 1 + \sum_{v=1}^{\infty} a_v s^v.$$

Substitution in the differential equation yields

$$2a_2 + 6a_3 s + 12a_4 s^2 + 20a_5 s^3 + 30a_6 s^4 + 42a_7 s^5 + \cdots$$
$$= \tfrac{3}{2} + 3a_1 s + (\tfrac{3}{2} a_1^2 + 3a_2) s^2 + (3a_3 + 3a_1 a_2) s^3 + (\tfrac{3}{2} a_2^2 + 3a_1 a_3 + 3a_4) s^4 +$$
$$+ (3a_2 a_3 + 3a_1 a_4 + 3a_0) s^5 + \cdots,$$

and comparing coefficients of like powers of x we find that

$$\tfrac{3}{2} = 2a_2, \quad 6a_3 = 3a_1, \quad 12a_4 = \tfrac{3}{2}a_1^2 + 3a_2, \ldots.$$

By means of these equations we can express $a_3, a_4, \ldots$ successively in terms of a_1, so that the series for y becomes

$$y = 1 + a_1 s + \frac{3}{4} s^2 + \frac{a_1}{2} s^3 + \frac{2a_1^2 + 3}{16} s^4 +$$
$$+ \frac{3a_1}{16} s^5 + \frac{4a_1^2 + 3}{64} s^6 + \frac{a_1^3 + 6a_1}{112} s^7 + \cdots.$$

The boundary condition $y = 4$ at $s = 1$ is then an equation for a_1. If the series is terminated at the term in s^4, we obtain the quadratic equation

$$\frac{a_1^2}{8} + \frac{3}{2} a_1 + \frac{31}{16} = 4$$

with the solutions

$$a_1 = -6 \pm \sqrt{52 \cdot 5} = \begin{cases} 1 \cdot 2457 \\ -13 \cdot 246. \end{cases}$$

If two more terms are retained, we still have only a quadratic to solve, namely

$$\frac{3a_1^2}{16} + \frac{27}{16} a_1 + \frac{127}{64} = 4,$$

from which we obtain the values

$$a_1 = -\tfrac{9}{2} \pm \sqrt{31} = \begin{cases} 1 \cdot 067\,76 \\ -10 \cdot 068. \end{cases}$$

The corresponding approximations for the everywhere positive solution $y(x)$ for several values of x are given in Table III/15 with their percentage errors in brackets.

The power series for the other (partly negative) solution does not appear to converge since $y(\tfrac{1}{2}) \approx -11$ (cf. Example II, § 1.2).

§8. Some special methods for eigenvalue problems

Eigenvalue problems have already been treated several times in this chapter, but by methods, which are also applicable to ordinary boundary-value problems — actually we used several finite-difference methods and RITZ's method, but other such methods can be used. We now describe some methods designed specifically for eigenvalue problems, giving particular

Table III/15. *Approximate values obtained by applying a power series method to a non-linear problem (Example II)*

(percentages in brackets are the relative errors)

s	x	Approximate values for the everywhere positive solution, obtained by truncating the power series at the term in			other solution		
		s^3	s^4	s^4	s^4	s^4	s^4
0·25	0·75	1·4336 (+9·8%)	1·3696 (+4·9%)	1·3393 (+2·5%)	1·32365 (+1·3%)	−2·282	−2·621	−1·500
0·5	0·5	2·0313 (+14 %)	1·9120 (+11 %)	1·8502 (+4·1%)	1·8168 (+2·2%)	−4·881	−5·443	−3·632
0·75	0·25	2·8633 (+12 %)	2·7395 (+7 %)	2·6656 (+4·1%)	2·6209 (+24%)	−4·308	−4·763	−3·496

15*

prominence to the method of successive approximations or iteration method. In § 8.3 we mention an important bracketing theorem for a special class of eigenvalue problems; in many cases this theorem enables one to obtain limits for the first eigenvalue to an accuracy sufficient for most applications with very little computation.

In the following we make use of a series of results from the theory of eigenvalue problems; for proofs the reader is referred to the literature[1].

8.1. Some concepts and results from the theory of eigenvalue problems

Let the differential equation read

$$M[y] = \lambda N[y], \tag{8.1}$$

where $M[y]$ and $N[y]$ are linear homogeneous ordinary differential expressions [see Ch. I (1.6)] of the form

$$M[y] = \sum_{\nu=0}^{2m} p_\nu(x)\, y^{(\nu)}, \tag{8.2}$$

$$N[y] = \sum_{\nu=0}^{2n} q_\nu(x)\, y^{(\nu)} \tag{8.3}$$

with $m > n$. The $p_\nu(x)$ and $q_\nu(x)$ are given real functions which will here be assumed continuous. With the differential equation we associate $2m$ linear homogeneous boundary conditions [cf. Ch. I (1.7)]

$$U_\mu[y] = 0 \qquad (\mu = 1, 2, \ldots, 2m). \tag{8.4}$$

The differential expressions $M[y]$ and $N[y]$ are often assumed to be of the special "self-adjoint" form

$$\left.\begin{aligned}
M[y] &= \sum_{\nu=0}^{m} (-1)^\nu [f_\nu(x)\, y^{(\nu)}]^{(\nu)}, \\
N[y] &= \sum_{\nu=0}^{n} (-1)^\nu [g_\nu(x)\, y^{(\nu)}]^{(\nu)}.
\end{aligned}\right\} \tag{8.5}$$

The f_ν and g_ν are given functions which we assume here to be real and to possess continuous derivatives of order ν; further we assume that $f_m \neq 0$ and $g_n \neq 0$. When $n = 0$, the eigenvalue problem, then of the form

$$M[y] = \lambda g_0(x)\, y, \tag{8.6}$$

is called a "special" eigenvalue problem; when $n > 0$, it is called a "general" eigenvalue problem.

---

[1] For example, R. Courant and D. Hilbert: Methods of mathematical physics, Vol. I, 1st English ed. New York: Interscience Publishers, Inc. 1953. — Collatz, L.: Eigenwertaufgaben mit technischen Anwendungen. Leipzig 1949.

For problems in which the eigenvalue λ does not appear in the boundary conditions we make the following definitions:

A *comparison function* is any function other than the zero function which satisfies all the boundary conditions and possesses a continuous derivative of order $2m$.

An *admissible function* is any function other than the zero function which satisfies the essential boundary conditions (see Ch. I, § 1.2) and possesses a continuous derivative of order m.

The eigenvalue problem (8.1), (8.4) is called *self-adjoint* when

$$\left.\begin{aligned} \int_a^b \left(u\,M[v] - v\,M[u]\right) dx = 0, \\ \int_a^b \left(u\,N[v] - v\,N[u]\right) dx = 0 \end{aligned}\right\} \tag{8.7}$$

for any two comparison functions u, v.

The eigenvalue problem (8.1), (8.4) is called *full-definite* when

$$\int_a^b u\,M[u]\,dx > 0 \quad \text{and} \quad \int_a^b u\,N[u]\,dx > 0 \tag{8.8}$$

for any comparison function u.

Whether these conditions of self-adjointness (8.7) and full-definiteness (8.8) are satisfied in any given case can be readily determined by integration by parts[1]; for example, the eigenvalue problem

$$-y'' = \lambda e^x y; \quad y'(0) = 0, \quad y(1) + y'(1) = 0$$

is shown to be self-adjoint and full-definite, respectively, by the following two simple calculations:

$$\int_0^1 (u v'' - u'' v)\,dx = [u v' - v u']_0^1$$
$$= u(1)\,v'(1) - v(1)\,u'(1) - u(0)\,v'(0) + v(0)\,u'(0) = 0$$

for any two comparison functions u, v since

$$v'(1) = -v(1), \quad u'(1) = -u(1), \quad u'(0) = v'(0) = 0;$$

and

$$\int_0^1 (-u u'')\,dx = -[u u']_0^1 + \int_0^1 u'^2 dx = [u(1)]^2 + \int_0^1 u'^2 dx > 0$$

for any comparison function u. This last inequality sign could be replaced by an equality sign only for a function with $u' \equiv 0$ and $u(1) = 0$, i.e. for the function $u \equiv 0$, but the zero function is excluded from the domain of comparison functions.

---

[1] See L. COLLATZ: Eigenwertaufgaben, § 4. Leipzig 1949.

In eigenvalue theory an important role is played by "RAYLEIGH's quotient"

$$R[u] = \frac{\int\limits_a^b u\, M[u]\, dx}{\int\limits_a^b u\, N[u]\, dx}. \tag{8.9}$$

For full-definite problems it exists for all comparison functions u and is positive.

A self-adjoint, full-definite eigenvalue problem has the following properties:

It possesses a countably infinite sequence of positive eigenvalues $0 < \lambda_1 \leqq \lambda_2 \leqq \cdots$ with corresponding eigenfunctions $y_1, y_2, \ldots$. Its first eigenfunction y_1 is the solution of the variational problem of minimizing the Rayleigh quotient (8.9) with respect to the domain of comparison functions; the corresponding minimum value is the first eigenvalue λ_1. More generally, the $(s+1)$-th eigenvalue λ_{s+1} is the minimum value of the Rayleigh quotient (8.9) with respect to the domain of all comparison functions u which are orthogonal to the first s eigenfunctions $y_1, \ldots, y_s$ in a "generalized sense", namely in the sense that they satisfy the "generalized orthogonality relations"

$$\int\limits_a^b u\, N[y_j]\, dx = 0 \qquad (j = 1, 2, \ldots, s).$$

When multiple eigenvalues occur, they must be counted according to their multiplicity.

8.2. The iteration method in the general case

We base the description of the method on the differential equation (8.1) with the boundary conditions (8.4).

In the general case the procedure is as follows: We start with an arbitrary function $F_0(x)$ and from it determine successively a sequence of functions $F_1, F_2, \ldots$ by calculating F_k as the solution of a certain boundary-value problem which depends on F_{k-1}. This boundary-value problem is obtained from the eigenvalue problem (8.1), (8.4) by writing F_{k-1} in place of y in all terms (in the differential equation and the boundary conditions) which are multiplied by λ, writing F_k in place of y in all the remaining terms, and finally putting $\lambda = 1$; thus for the eigenvalue problem

$$y^{IV} + \lambda y'' + y = 0; \quad y'(0) = y''(0) = y'(1) = y(1) - \lambda y'''(1) = 0$$

F_k is determined from F_{k-1} as the solution of the boundary-value problem

$$F_k^{IV} + F_k = -F_{k-1}''; \quad F_k'(0) = F_k''(0) = F_k'(1) = 0, \quad F_k(1) = F_{k-1}'''(1).$$

The absence of λ from the boundary conditions simplifies the description of the procedure: starting from any function $F_0(x)$ we determine successively $F_1, F_2, \ldots$ from the boundary-value problems

$$\left.\begin{aligned} M[F_k] &= N[F_{k-1}], \\ U_\mu[F_k] &= 0 \end{aligned}\right\} \quad (k = 1, 2, \ldots). \tag{8.10}$$

If the ratios of successive terms in the sequence $F_0, F_1, F_2, \ldots$ tend to a constant (independent of x and k) as k increases, say $F_{k-1}/F_k \to K$, in which case we say the sequence converges, then one expects the functions F_k to assume more and more the form of an eigenfunction y_s with eigenvalue $\lambda_s = K$ (an eigenfunction is only determined to within a constant factor). The quotient $M[F_k]/N[F_k]$ should provide an approximation to λ_s but as it is still a function of x we use a more suitable approximate value Λ obtained by taking weighted mean values of the numerator and denominator as in RAYLEIGH's quotient (8.9):

$$\Lambda = \frac{\int\limits_a^b F_k M[F_k]\,dx}{\int\limits_a^b F_k N[F_k]\,dx} = \frac{\int\limits_a^b F_k N[F_{k-1}]\,dx}{\int\limits_a^b F_k N[F_k]\,dx}.$$

The integrals appearing here in the numerator and denominator are denoted by

$$\left.\begin{aligned} a_{2k} &= \int\limits_a^b F_k N[F_k]\,dx, \\ a_{2k+1} &= \int\limits_a^b F_{k+1} N[F_k]\,dx \end{aligned}\right\} \quad (k = 0, 1, 2, \ldots)$$

and are called SCHWARZ's constants after H. A. SCHWARZ. Λ is then the quotient of two successive Schwarz constants; if we introduce the Schwarz quotients

$$\mu_{k+1} = \frac{a_k}{a_{k+1}} \quad (k = 0, 1, 2, \ldots), \tag{8.11}$$

Λ is the Schwarz quotient μ_{2k}.

In order to say more about Λ we must make more restrictive assumptions about the eigenvalue problem[1].

8.3. The iteration method for a restricted class of problems

Here we consider only those eigenvalue problems[2] (8.1), (8.4), (8.5) which are self-adjoint and full-definite and whose boundary conditions are independent of the eigenvalue λ.

---

[1] A more general theorem has been established by H. WIELANDT: Das Iterationsverfahren bei nicht selbstadjungierten Eigenwertaufgaben. Math. Z. **50**, 93—143 (1944).

[2] Fewer restrictive assumptions are made by E. STIEFEL and H. ZIEGLER: Natürliche Eigenwertprobleme. Z. Angew. Math. Phys. **1**, 111—138 (1950).

We restrict our choice of starting function $F_0(x)$ slightly by demanding that it shall be continuous, possess continuous derivatives of order up to $2n$ and satisfy just so many of the boundary conditions (not necessarily all) that $a_0 > 0$ and

$$\int_a^b \left(F_0 N[u] - u N[F_0] \right) dx = 0 \qquad (8.12)$$

for all comparison functions u. Integration by parts readily establishes which boundary conditions should be satisfied in any given case.

Consider, for example, a special eigenvalue problem with $N[f] = g_0(x) f$: (8.12) is automatically satisfied and also $a_0 > 0$ since the full-definiteness condition implies that $g_0(x) > 0$. In this case, therefore, F_0 need not satisfy any boundary conditions.

If F_0 satisfies all the boundary conditions, then (8.12) and $a_0 > 0$ are automatically satisfied for any self-adjoint full-definite problem since F_0 is then a comparison function. Usually F_0 will be thus chosen on account of the more accurate results which may be expected; nevertheless it can sometimes be more convenient for calculation to impose a smaller (but sufficient) number of boundary conditions on F_0.

For the sequence of functions $F_0, F_1, F_2, \ldots$ derived from such an F_0 by the iteration process described in § 8.2 the Schwarz constants can be defined generally by

$$a_k = \int_a^b F_i N[F_{k-i}] \, dx \qquad (0 \le i \le k, \ k = 0, 1, 2, \ldots). \qquad (8.13)$$

Their dependence on i is only apparent, for using the iterative relation of (8.10) and the self-adjoint condition (8.7) we have

$$a_k = \int_a^b F_i M[F_{k-i+1}] \, dx = \int_a^b F_{k-i+1} M[F_i] \, dx = \int_a^b F_{k-i+1} N[F_{i-1}] \, dx$$

$$= \int_a^b F_{i-1} N[F_{k-i+1}] \, dx,$$

and consequently the integral $\int_a^b F_i N[F_{k-i}] \, dx$ depends only on the sum of the suffices $i + (k - i) = k$; for example,

$$a_2 = \int_a^b F_2 N[F_0] \, dx = \int_a^b F_1 N[F_1] \, dx = \int_a^b F_0 N[F_2] \, dx. \qquad (8.14)$$

The full-definite condition (8.8) implies that $a_1, a_2, \ldots$ are all positive, for

$$\left. \begin{aligned} a_{2k} &= \int_a^b F_k N[F_k] \, dx > 0, \\ a_{2k-1} &= \int_a^b F_k M[F_k] \, dx > 0 \end{aligned} \right\} \qquad (k = 1, 2, \ldots) \qquad (8.15)$$

and since also $a_0 > 0$, all the Schwarz quotients (8.11) are positive. The quotients μ_{2k} with even suffices may be written as Rayleigh quotients (8.9); for example,

$$\mu_2 = \frac{a_1}{a_2} = \frac{\int\limits_a^b F_1 N[F_0]\,dx}{\int\limits_a^b F_1 N[F_1]\,dx} = \frac{\int\limits_a^b F_1 M[F_1]\,dx}{\int\limits_a^b F_1 N[F_1]\,dx} = R[F_1], \qquad (8.16)$$

and generally we have
$$\mu_{2s} = R[F_s]. \qquad (8.17)$$

Under the assumptions of this section it can be shown that

$$\mu_1 \geqq \mu_2 \geqq \mu_3 \geqq \cdots \geqq \lambda_1.$$

It is also possible to prove the following bracketing theorem:

Theorem: *Let the eigenvalue problem* (8.1), (8.4), (8.5) *be such that*

(i) *the conditions of self-adjointness and full-definiteness are satisfied,*

(ii) *the eigenvalue λ does not appear in the boundary conditions,*

(iii) *the smallest eigenvalue λ_1 is simple,*

and as starting function for the iteration (8.10) *take any continuous function $F_0(x)$ with continuous derivatives of order up to $2n$ and which satisfies so many of the prescribed boundary conditions that (8.12) holds and $a_0 > 0$. If μ_k is the k-th Schwarz quotient (8.11) calculated from the iterates F_k via the Schwarz constants a_k (8.13) and if l_2 is a lower bound for the second eigenvalue such that $l_2 > \mu_{k+1}$, then the first eigenvalue λ_1 lies between the limits*

$$\mu_{k+1} - \frac{\mu_k - \mu_{k+1}}{\dfrac{l_2}{\mu_{k+1}} - 1} \leqq \lambda_1 \leqq \mu_{k+1} \qquad (k = 1, 2, \ldots). \qquad (8.18)$$

8.4. Practical application of the method

Starting from a chosen function $F_0(x)$ we have to determine each subsequent function $F_1(x), F_2(x), \ldots$ by solving a boundary-value problem (8.10). If the given differential equation is of very simple form with coefficients given formally by simple expressions, then a number of iterates F_k can be calculated. But for complicated eigenvalue problems the repeated solution of these boundary-value problems presents considerable difficulties, and it is therefore important that for the restricted class of eigenvalue problems of § 8.3 an accuracy sufficient for technical applications is generally achieved by using only $F_0(x)$ and $F_1(x)$. The calculation can be divided into three steps:

1. Find two functions F_0, F_1 with the properties: $M[F_1] = N[F_0]$, F_1 satisfies all the boundary conditions and F_0 satisfies so many that (8.12) holds and $a_0 > 0$.

In many cases two such functions F_0, F_1 can be determined by graphical integrations (when applying the method graphically, one should always integrate and never differentiate); however, we shall not go into the graphical application any further here[1].

For the special eigenvalue problems, for which $N[F_0] \equiv g_0 F_0$, we can choose F_1 arbitrarily from among the comparison functions and calculate F_0 from $g_0 F_0 = M[F_1]$. F_1 can be expressed as an arbitrary linear combination of functions ψ_i which satisfy all the boundary conditions, and the coefficients of the combination will usually be determined so that F_0 also satisfies the boundary conditions. This often leads to substantially better results, for in general the more F_0 and F_1 resemble the first eigenfunction, the more accurate the results turn out to be.

2. Calculate the Schwarz constants and quotients

$$
\left.
\begin{aligned}
a_0 &= \int_a^b F_0 N[F_0]\, dx, \quad a_1 = \int_a^b F_1 N[F_0]\, dx, \\
a_2 &= \int_a^b F_1 N[F_1]\, dx; \quad \mu_1 = \frac{a_0}{a_1}, \quad \mu_2 = \frac{a_1}{a_2}.
\end{aligned}
\right\}
\tag{8.19}
$$

We then know that

$$
\mu_1 \geqq \mu_2 \geqq \lambda_1.
$$

3. Finally we calculate a lower limit for λ_1. For this we need a lower limit l_2 for the second eigenvalue λ_2; it must be greater than μ_2, but as long as this requirement is satisfied it need not be particularly close to λ_2. In many cases such a lower limit l_2 can be obtained by comparison with an eigenvalue problem with constant coefficients (see § 8.5). Then from (8.18) with $k = 1$ we have

$$
\mu_2 - \frac{\mu_1 - \mu_2}{\dfrac{l_2}{\mu_2} - 1} \leqq \lambda_1 \leqq \mu_2.
\tag{8.20}
$$

For technical applications it is not always essential to have rigorous limits and it will often suffice to have a rough guide to the closeness of μ_2 to λ_1 so that we may know how many decimals in μ_2 can be used for λ_1. In such cases one can replace l_2 in (8.20) by an approximation to λ_2 calculated by the Ritz method or the finite-difference method, or in fact by any other suitable method. Although this procedure is not strictly valid, it can be justified to some extent on the grounds that changes in l_2 have little effect on the lower limit in (8.20) when l_2 is appreciably greater than μ_2.

---

[1] See L. COLLATZ: Eigenwertaufgaben, § 13. Leipzig 1949.

8.5. An example treated by the iteration method

For the eigenvalue problem

$$- [(1 + x) y']' = \lambda (1 + x) y, \qquad y'(0) = y(1) = 0$$

of Example III in §1.2 the conditions to be satisfied by F_0 and F_1 are

$$- [(1 + x) F_1']' = (1 + x) F_0, \qquad F_1'(0) = F_1(1) = 0.$$

Although F_0 need not satisfy any of the boundary conditions, better results may be expected if it satisfies at least the essential boundary condition $F_0(1) = 0$. We therefore take F_1 to be a polynomial of at least the third degree

$$F_1 = a_0 + a_1 x + a_2 x^2 + a_3 x^3,$$

for we have three conditions to satisfy and a constant factor is indeterminate. The boundary conditions on F_1 eliminate two of the constants; forming F_0 and putting $F_0(1) = 0$ eliminates a third; the fourth is chosen conveniently to give

$$F_1 = 3 - 5 x^2 + 2 x^3, \qquad F_0 = \frac{2(5 + 4x - 9 x^2)}{1 + x}.$$

We now calculate the Schwarz constants from (8.19):

$$a_0 = \int_0^1 (1 + x) F_0^2 \, dx = 256 \ln 2 - 121 = 56.44568,$$

$$a_1 = \int_0^1 (1 + x) F_0 F_1 \, dx = \frac{526}{30} \qquad = 17.53333,$$

$$a_2 = \int_0^1 (1 + x) F_1^2 \, dx \quad = \frac{572}{105} \qquad = 5.4476190.$$

These values yield the Schwarz quotients

$$\mu_1 = \frac{a_0}{a_1} = 3.219335, \qquad \mu_2 = \frac{a_1}{a_2} = 3.218532,$$

which constitute upper limits for λ_1.

To calculate a lower limit from (8.20) we need a lower limit l_2 for the second eigenvalue. This, as has been said before, may be fairly crude and for this example can be obtained by comparison of the given problem with a simpler problem possessing an exact solution. If the function $1 + x$, which occurs on both sides of the differential equation, is replaced by its maximum value of 2 on the right-hand side and by its minimum value of 1 on the left-hand side, we obtain the constant-coefficient eigenvalue problem

$$- y'' = 2 \lambda^* y, \qquad y'(0) = y(1) = 0,$$

whose second eigenvalue λ_2^* is given by

$$2\lambda_2^* = (\tfrac{3}{2}\pi)^2.$$

This new eigenvalue problem has the property[1] that $R^*[u] \leqq R[u]$ for all comparison functions u, where $R^*[u]$ and $R[u]$ are the Rayleigh quotients of the new and old problems, respectively. Hence $\lambda_s^* \leqq \lambda_s$ for $s = 1, 2, \ldots$ and in particular we can use λ_2^* as a lower limit for λ_2; thus we put

$$l_2 = \lambda_2^* = \tfrac{9}{8}\pi^2.$$

With this value (8.20) yields

$$3 \cdot 218\,211 \leqq \lambda_1 \leqq 3 \cdot 218\,532.$$

8.6. The enclosure theorem

The following theorem, which is proved in the theory of eigenvalues and usually known as the enclosure theorem, is often quite useful in numerical work. It is concerned with a certain class of eigenvalue problems, the "single-term" class, for which the operator $N[y]$ consists of only one term:

$$N[y] = (-1)^n [g_n(x)\, y^{(n)}]^{(n)}.$$

The differential equation is consequently of the form

$$M[y] = (-1)^n \lambda [g_n(x)\, y^{(n)}]^{(n)}. \tag{8.21}$$

We impose the usual conditions of self-adjointness and full-definiteness and, as in § 8.1, assume that the function $g_n(x)$ has the same sign throughout the fundamental interval (a, b). We assume further that the given boundary conditions are such that

$$(-1)^n \int_a^b u\,[g\,v^{(n)}]^{(n)}\, dx = \int_a^b g\,u^{(n)}\, v^{(n)}\, dx$$

for any two comparison functions u, v and any function $g(x)$ with continuous derivatives of up to the n-th order (that this condition is satisfied for any given problem can easily be verified by integration by parts).

The theorem states that if we have a function $F_0(x)$ with continuous derivatives of order up to $2n$ and a comparison function $F_1(x)$ such that $M[F_1] = N[F_0]$, cf. (8.10), and if further the function

$$\Phi(x) = \frac{F_0^{(n)}(x)}{F_1^{(n)}(x)} \tag{8.22}$$

---

[1] Cf. the comparison theorem in L. Collatz: Eigenwertaufgaben, § 8. Leipzig 1949; this deals with the case in which the differential expression M is the same in both problems but the extension to the more general case considered here is immediate.

is bounded and does not change sign in the interval (a, b), then at least one eigenvalue λ_s lies between the maximum and minimum values of Φ in the interval (a, b):

$$\Phi_{min} \leqq \lambda_s \leqq \Phi_{max}. \tag{8.23}$$

For special eigenvalue problems $(n = 0)$ the conditions to be satisfied by F_0 and F_1 can be formulated more simply[1]. It is sufficient to select any comparison function F_1 and put

$$F_0 = \frac{1}{g_0(x)} M[F_1]. \tag{8.24}$$

Then

$$\Phi(x) = \frac{M[F_1]}{g_0 F_1} = \frac{M[F_1]}{N[F_1]}, \tag{8.25}$$

and we have

$$\left(\frac{M[F_1]}{g_0 F_1}\right)_{min} \leqq \lambda_s \leqq \left(\frac{M[F_1]}{g_0 F_1}\right)_{max}. \tag{8.26}$$

The function F_0 need not satisfy any of the boundary conditions but one will usually try to choose F_1 in such a way that F_0 satisfies as many boundary conditions as possible; for by so doing one approximates an eigenfunction more closely and there is a better chance of obtaining reasonably narrow limits in (8.23).

The results can be improved, often quite substantially, by the introduction of a parameter[2] ϱ into the assumed expressions for F_0 and F_1. This parameter is then chosen so as to minimize the difference between the upper and lower limits of (8.23).

Example. Again we consider the vibration problem of § 1.2

$$M[y] = -[(1 + x) y']' = \lambda (1 + x) y, \qquad y'(0) = y(1) = 0$$

but this time we choose for F_1 the expression

$$F_1(x) = (1 + x)^q \sin[a(x - 1)].$$

The boundary condition $F_1(1) = 0$ is already satisfied for all values of the parameters q and a, which are yet to be determined.

To form $M[F_1]$ we evaluate

$$[(1 + x) F_1']' = [q^2(1 + x)^{q-1} - a^2(1 + x)^{q+1}] \sin a(x - 1) +$$
$$+ (2q + 1) a (1 + x)^q \cos a(x - 1),$$

---

[1] For the special eigenvalue problems the enclosure theorem for the first eigenvalue was proved by G. TEMPLE: The computation of characteristic numbers and characteristic functions. Proc. Lond. Math. Soc. (2) **29**, 257−280 (1929).

[2] This method is due to F. KIESZLING: Eine Methode zur approximativen Berechnung einseitig eingespannter Druckstäbe mit veränderlichem Querschnitt. Z. Angew. Math. Mech. **10**, 594−599 (1930).

and we see that for the quotient function Φ of (8.25) to remain finite the zero of the sine factor in the denominator must be cancelled by a similar zero in the numerator. This can be achieved by choosing $q = -\frac{1}{2}$, for then the cosine term disappears.

We still have to satisfy the other boundary condition $F_1(0) = 0$, so we choose the remaining parameter a to be a root of the equation

$$\tan a = -2a.$$

Let $a_1, a_2, a_3, \ldots$ be the positive roots of this transcendental equation. With these values of a and $q = -\frac{1}{2}$

$$\Phi(x) = \frac{-[(1 + x) F_1']'}{(1 + x) F_1} = a_k^2 - \frac{1}{4(1 + x)^2} \qquad (k = 1, 2, \ldots),$$

and inserting the maximum and minimum values of this function in (8.23) we obtain the limits

$$a_k^2 - \frac{1}{4} \leqq \lambda_s \leqq a_k^2 - \frac{1}{16}.$$

Thus we have obtained limits for infinitely many eigenvalues at one fell swoop. The constant difference between the upper and lower limits means that the percentage error is smaller for higher eigenvalues, and in actual fact this method is probably superior to all others for the higher eigenvalues. Several values are given in Table III/16.

Table III/16. *Bracketing of eigenvalues*

a_k	a_k	Lower	Upper
		limits for λ_s	
1·8366	3·3731	3·123	3·311
4·816	23·19	22·94	23·13
7·917	62·68	62·43	62·62
11·041	121·90	121·65	121·84
14·173	200·86	200·61	200·80
17·308	299·56	299·31	299·50

Generalizations of this bracketing theorem have been derived by H. WIELANDT[1]. One formulation, which includes all previously known bracketing theorems as special cases, is based on the Schwarz constants a_k of (8.13). In it, real numbers $b_1, b_2, \ldots, b_p$ are choosen arbitrarily and used, together with the first $p + 1$ Schwarz constants $a_0, a_1, \ldots, a_p$, to define two closed sets M_1 and M_2 of real numbers x: M_1 consists of all numbers x for which $\sum_{\nu=1}^{p} b_\nu x^\nu \geqq \frac{1}{a_0} \sum_{\nu=1}^{p} a_\nu b_\nu$, and M_2 consists of all

---

[1] WIELANDT, H.: Ein Einschließungssatz für charakteristische Wurzeln normaler Matrizen. Arch. Math. **1**, 348—352 (1949), and Fiat-Review, Naturforschung und Medizin in Deutschland 1939—1946, **2**, 98 (1948). An older formulation was given by K. FRIEDRICHS and G. HORVAY: The finite Stieltjes momentum problem. Proc. Nat. Acad. Sci., Wash. **25**, 528—534 (1939); a detailed presentation has been given by H. BÜCKNER: Die praktische Behandlung von Integralgleichungen (Ergebnisse der Angew. Mathematik, H. 1). Berlin-Göttingen-Heidelberg 1952.

numbers x for which $\sum\limits_{\nu=1}^{p} b_\nu x^\nu \leqq \dfrac{1}{a_0} \sum\limits_{\nu=1}^{p} a_\nu b_\nu$. The theorem then states that there is at least one eigenvalue whose reciprocal lies in M_1 and at least one eigenvalue whose reciprocal lies in M_2.

8.7. Three minimum principles

RITZ's method for solving variational problems has already been applied to an eigenvalue problem[1,2] (in § 5) by deriving a variational problem whose Euler conditions can be identified with the eigenvalue problem in question. Now these Euler conditions, although necessary, are not sufficient for an extremum of the variational problem, and hence conclusions about the approximate values obtained cannot be drawn solely from the fact that the eigenvalue problem can be formally identified with them. However, for certain special classes of eigenvalue problem there are other variational problems, based on certain minimum principles of eigenvalue theory, which also have solutions satisfying the eigenvalue problem, but which allow specific statements to be made about the approximate eigenvalues obtained by solving them approximately by RITZ's method. We quote the results without proof.

We now state the above-mentioned minimum principles.

A. RAYLEIGH's minimum principle. For a self-adjoint full-definite eigenvalue problem (8.1), (8.4), (8.5) with boundary conditions independent of the eigenvalue λ the smallest eigenvalue λ_1 is the minimum value of the Rayleigh quotient $R[u]$ (8.9) in the domain of all comparison functions u (cf. § 8.1).

B. KAMKE's minimum principle[3]. Here fewer restrictions are imposed on u at the expense of the generality of the eigenvalue problem; this is often of considerable advantage in practical work. This minimum principle applies to the same eigenvalue problems as RAYLEIGH's except that one additional restriction must be imposed, namely that the eigenvalue problem is "K-definite", a term which we now explain. Firstly the integrals appearing in the Rayleigh quotient (8.9) are put

---

[1] A comprehensive presentation based on the techniques of functional analysis is given by N. ARONSZAJN: Study of eigenvalue problems. The Rayleigh-Ritz and the Weinstein methods for approximation of eigenvalues. Dept. of Math. Oklahoma Agricultural and Mechanical College. Stillwater 1949. 214 pp.

[2] Error estimates for self-adjoint full-definite eigenvalue problems have been obtained by G. BERTRAM: Zur Fehlerabschätzung für das Ritzsche Verfahren bei Eigenwertaufgaben. Diss. 56 pp. Hannover 1950. Other estimates and investigations of the rate of convergence can be found in L. V. KANTOROVICH and V. I. KRYLOV: Näherungsmethoden der Höheren Analysis, pp. 226—329. Berlin 1956.

[3] KAMKE, E.: Über die definiten selbstadjungierten Eigenwertaufgaben IV. Math. Z. **48**, 67—100 (1942).

into "Dirichlet's form"

$$\left.\begin{array}{l} \int\limits_a^b u\,M[u]\,d\,x = \int\limits_a^b \sum\limits_{\nu=0}^m f_\nu[u^{(\nu)}]^2 d\,x + M_0[u], \\[2mm] \int\limits_a^b u\,N[u]\,d\,x = \int\limits_a^b \sum\limits_{\nu=0}^n g_\nu[u^{(\nu)}]^2 d\,x + N_0[u]; \end{array}\right\} \tag{8.27}$$

$M_0[u]$ and $N_0[u]$ are called the "Dirichlet boundary expressions". By integrating by parts ν times we transform a general term

$$\int\limits_a^b u(-1)^\nu [f_\nu u^{(\nu)}]^{(\nu)}\,d\,x$$

into

$$\int\limits_a^b u^{(\nu)} f_\nu u^{(\nu)}\,d\,x$$

plus terms involving only boundary values. The Dirichlet boundary expression $M_0[u]$ is then the sum of these boundary terms for all values of ν, and $N_0[u]$ is found similarly. These boundary expressions are quadratic forms in the values of $u, u', u'', \ldots$ at the boundary points $x = a$, $x = b$. If the two forms are positive definite when their variables are restricted to run through all sets of numbers satisfying the relations represented by the boundary conditions and if all the functions $f_\nu(x)$ and $g_\nu(x)$ are non-negative [previously the only restrictions had been on $f_m(x)$ and $g_n(x)$, namely $f_m \neq 0$, $g_n \neq 0$], then the eigenvalue problem is called K-definite.

It can be deduced from the self-adjointness of the eigenvalue problem that the boundary conditions are such that, by suitable linear combination, each boundary derivative of order greater than or equal to m may be expressed in terms of boundary derivatives of order $0, 1, 2, \ldots,$ $m-1$. Hence derivatives of order greater or equal to m can be eliminated from $M_0[u]$ and $N_0[u]$, which then become quadratic forms in $u(a)$, $u'(a), \ldots, u^{(m-1)}(a), u(b), u'(b), \ldots, u^{(m-1)}(b)$. Rayleigh's quotient (8.9) can therefore be put into the form

$$K[u] = \frac{\int\limits_a^b \sum\limits_{\nu=0}^m f_\nu[u^{(\nu)}]^2 d\,x + M_0[u]}{\int\limits_a^b \sum\limits_{\nu=0}^n g_\nu[u^{(\nu)}]^2 d\,x + N_0[u]}, \tag{8.28}$$

in which no boundary derivative of order greater than $m-1$ appears.

Kamke's theorem then states that λ_1 is the minimum value of the quotient $K[u]$ in the domain of all admissible functions u.

C. A minimum principle for the special eigenvalue problems[1]. For those problems in which $N[y] = g_0(x)\,y$ another minimum principle

---

[1] Collatz, L.: Z. Angew. Math. Mech. **19**, 228 (1939).

can be established under the conditions laid down in **A** for the application of RAYLEIGH's principle; it is based on the iteration method. As mentioned in § 8.4, for the special eigenvalue problems the first iterate $F_1(x)$ in the iteration process may be chosen arbitrarily from among the comparison functions; its predecessor $F_0(x)$ in the iterative sequence is then given by

$$F_0(x) = \frac{1}{g_0(x)} M[F_1].\tag{8.29}$$

Here the Rayleigh quotient (8.9) may be written as the Schwarz quotient μ_2, as in (8.16):

$$R[u] = R[F_1] = \mu_2,$$

and we know from **A** that the minimum value of this quantity in the domain of comparison functions u is the smallest eigenvalue λ_1. But since the μ_k decrease monotonically, we also have $\mu_1 \geqq \lambda_1$. Now μ_1 may easily be expressed in terms of F_1:

$$\mu_1 = \frac{\int\limits_a^b F_0 N[F_0]\,dx}{\int\limits_a^b F_1 N[F_0]\,dx} = \frac{\int\limits_a^b F_0 M[F_1]\,dx}{\int\limits_a^b F_1 M[F_1]\,dx},\tag{8.30}$$

where F_0 is given by (8.29), and if we replace F_1 by u to emphasize its arbitrary character, we have

$$\mu_1 = \frac{\int\limits_a^b \frac{1}{g_0(x)}\,(M[u])^2\,dx}{\int\limits_a^b u\,M[u]\,dx}.\tag{8.31}$$

The minimum value of this quotient in the domain of comparison functions is the smallest eigenvalue λ_1. In general $R[u]$ is a better approximation to λ_1 than $\mu_1[u]$ for the same function u but $\mu_1[u]$ can often be calculated rather more quickly. The extension of this minimum principle to the general eigenvalue problems is discussed at the end of § 8.8.

8.8. Application of RITZ's method

For any of the three minimum problems just described we can, as in Ch. I (4.1), assume a general expression for u depending on undetermined parameters and then write down the necessary minimum conditions to be satisfied by these parameters. Convenient equations are obtained if the expression for u is chosen to depend linearly on the parameters:

$$u = a_1 v_1(x) + a_2 v_2(x) + \cdots + a_p v_p(x),\tag{8.32}$$

rather than in a general (non-linear) way. The $v_1, v_2, \ldots, v_p$ will be chosen as fixed comparison functions which are linearly independent in $\langle a, b \rangle$. Then u also will be a comparison function, and will vanish identically only when $a_1 = a_2 = \cdots = a_p = 0$. This case is therefore excluded.

First let us consider RAYLEIGH's principle. Substitution of (8.32) for u in RAYLEIGH's quotient (8.9) yields

$$R[u] = \frac{\int_a^b \sum_{r=1}^p a_r v_r(x) \sum_{s=1}^p a_s M[v_s(x)] \, dx}{\int_a^b \sum_{r=1}^p a_r v_r(x) \sum_{s=1}^p a_s N[v_s(x)] \, dx}. \tag{8.33}$$

For convenience we now introduce quantities m_{rs}, n_{rs} defined by

$$\left. \begin{aligned} m_{rs} &= \int_a^b v_r(x) M[v_s(x)] \, dx, \\ n_{rs} &= \int_a^b v_r(x) N[v_s(x)] \, dx, \end{aligned} \right\} \tag{8.34}$$

which, on account of the assumed self-adjointness, possess the symmetric properties

$$m_{rs} = m_{sr} \quad \text{and} \quad n_{rs} = n_{sr}. \tag{8.35}$$

Then RAYLEIGH's quotient has the form of a quotient of two quadratic forms Q_1, Q_2 in the parameters a_r:

$$R[u] = \frac{Q_1}{Q_2},$$

where

$$Q_1 = \sum_{r,s=1}^p m_{rs} a_r a_s, \quad Q_2 = \sum_{r,s=1}^p n_{rs} a_r a_s. \tag{8.36}$$

The assumption of full-definiteness (8.8) means that the numerator and denominator in $R[u]$ can take only positive values (we have excluded the case $a_1 = a_2 = \cdots = a_p = 0$); consequently Q_1 and Q_2 are positive definite quadratic forms.

The necessary conditions for a minimum of $R[u(a_1, \ldots, a_p)]$ read

$$\frac{\partial R}{\partial a_r} = \frac{Q_2 \dfrac{\partial Q_1}{\partial a_r} - Q_1 \dfrac{\partial Q_2}{\partial a_r}}{Q_2^2} = 0 \qquad (r = 1, 2, \ldots, p).$$

The minimum value of R in this restricted class of comparison functions we denote by Λ:

$$\Lambda = \min R[u(a_1, \ldots, a_p)],$$

and use it as an approximate value for the first eigenvalue λ_1. At the same time, of course, Λ is the value of Q_1/Q_2 at the minimum and hence the equations for the a_r may be simplified to

$$\frac{\partial Q_1}{\partial a_r} - \varLambda \frac{\partial Q_2}{\partial a_r} = 0 \qquad (r = 1, 2, \ldots, p).\tag{8.37}$$

Now from (8.36)

$$\frac{\partial Q_1}{\partial a_r} = 2 \sum_{s=1}^{p} m_{rs} a_s, \qquad \frac{\partial Q_2}{\partial a_r} = 2 \sum_{s=1}^{p} n_{rs} a_s,$$

so that the equations (8.37) can be written

$$\sum_{s=1}^{p} a_s (m_{rs} - \varLambda n_{rs}) = 0 \qquad \text{for} \qquad r = 1, 2, \ldots, p.\tag{8.38}$$

These equations are called GALERKIN's equations (cf. Ch. I, § 4.3). They are p homogeneous equations for p unknowns a_s and have a non-trivial solution (i.e. one in which at least one a_s is non-zero) if and only if the determinant of their coefficients vanishes:

$$\det (m_{rs} - \varLambda n_{rs}) = \begin{vmatrix} m_{11} - \varLambda n_{11} & m_{12} - \varLambda n_{12} & \cdots & m_{1p} - \varLambda n_{1p} \\ m_{21} - \varLambda n_{21} & m_{22} - \varLambda n_{22} & \cdots & m_{2p} - \varLambda n_{2p} \\ \cdots & \cdots & \cdots & \cdots \\ m_{p1} - \varLambda n_{p1} & m_{p2} - \varLambda n_{p2} & \cdots & m_{pp} - \varLambda n_{pp} \end{vmatrix} = 0.\tag{8.39}$$

This is an algebraic equation for $\varLambda$ of the p-th degree and has p roots $\varLambda_1, \varLambda_2, \ldots, \varLambda_p$. Under the assumptions which we have made it can be shown that the p roots $\varLambda_k$ are real and positive, and that, when they are arranged in increasing order of magnitude, they furnish upper limits for the corresponding eigenvalues, i.e.

$$\varLambda_k \geq \lambda_k \qquad \text{for} \qquad k = 1, 2, \ldots, p.\tag{8.40}$$

The technique of substituting a linear expression (8.32) for u can be used in exactly the same way for the other two minimum principles of § 8.7. We derive the corresponding equations only for the third minimum principle § 8.7, **C**, but also show how they can be extended to the general eigenvalue problems. We deal first with a special eigenvalue problem, for which the differential equation reads

$$M[y] = \lambda g_0(x) y$$

and (8.31) is to be minimized. Proceeding as for RAYLEIGH's principle we substitute for u an arbitrary linear combination (8.32) of p linearly independent comparison functions $v_r(x)$, which satisfy all the boundary conditions, and obtain

$$\mu_1[u] = \frac{Q_1^{\otimes}}{Q_2^{\otimes}},$$

in which the quadratic forms $Q_1^{\otimes}$, $Q_2^{\otimes}$ are here given by

$$Q_1^{\otimes} = \int\limits_a^b \frac{1}{g_0(x)} \left(\sum_{r=1}^p a_r M[v_r(x)] \right) \left(\sum_{s=1}^p a_s M[v_s(x)] \right) dx = \sum_{r,s=1}^p m_{rs}^{\otimes} a_r a_s,$$

where

$$m_{rs}^{\otimes} = \int\limits_a^b \frac{1}{g_0(x)} M[v_r(x)] M[v_s(x)] dx, \qquad (8.41)$$

and

$$Q_2^{\otimes} = \sum_{r,s=1}^p n_{rs}^{\otimes} a_r a_s,$$

where

$$n_{rs}^{\otimes} = \int\limits_a^b v_r(x) M[v_s(x)] dx. \qquad (8.42)$$

Thus the quantities $n_{rs}^{\otimes}$ are formed in the same way as the quantities m_{rs} for RAYLEIGH's principle.

The rest of the calculation is identical with the corresponding calculation for RAYLEIGH's principle if the notation Q_1, Q_2, m_{rs}, n_{rs} is modified with a superscript $\otimes$. The equations corresponding to (8.38), i.e.

$$\sum_{s=1}^p a_s(m_{rs}^{\otimes} - \Lambda n_{rs}^{\otimes}) = 0 \qquad (r = 1, 2, \ldots, p), \qquad (8.43)$$

are GRAMMEL's equations[1], and the approximate values Λ_k are the zeros of the determinant

$$\det(m_{rs}^{\otimes} - \Lambda n_{rs}^{\otimes}) = \begin{vmatrix} m_{11}^{\otimes} - \Lambda n_{11}^{\otimes} & \ldots & m_{1p}^{\otimes} - \Lambda n_{1p}^{\otimes} \\ \cdot \cdot \cdot \cdot \cdot \cdot \cdot \cdot \cdot \cdot \cdot \cdot \cdot \cdot \\ m_{p1}^{\otimes} - \Lambda n_{p1}^{\otimes} & \ldots & m_{pp}^{\otimes} - \Lambda n_{pp}^{\otimes} \end{vmatrix}. \qquad (8.44)$$

In a practical application, particularly one in which the f_v and g_0 are not simple functions, one would normally specify the functions $M[v_r(x)]$ and calculate the $v_r(x)$ by integration, thus avoiding the numerical or graphical differentiation involved in calculating the $M[v_r(x)]$ from chosen $v_r(x)$. The determination of each $v_r(x)$ is equivalent to the first step in the iteration method.

We now turn to the general eigenvalue problem with the differential equation $M[y] = \lambda N[y]$. Clearly we must go back to the original

---

[1] GRAMMEL, R.: Ein neues Verfahren zur Lösung technischer Eigenwertprobleme. Ing.-Arch. **10**, 35—46 (1939). GRAMMEL derives the equations in a different way.

Practical examples are worked out by E. MAIER: Biegeschwingungen von spannungslos verwundenen Stäben, insbesondere von Luftschraubenblättern. Ing.-Arch. **11**, 73—98 (1940). See also R. GRAMMEL: Über die Lösung technischer Eigenwertprobleme. VDI-Forsch.-Heft, Gebiet Stahlbau **6**, 36—42 (1943).

definition of μ_1 in (8.30), which does not depend on the special form of N. Let us put

$$F_1(x) = u(x) = \sum_{r=1}^{p} a_r v_r(x);$$

then we have to determine a function F_0 which is related to F_1 by the differential equation $M[F_1] = N[F_0]$, in accordance with (8.10), and which satisfies those of the boundary conditions sufficient for (8.12) to hold. Corresponding to each of the p functions v_r we determine a function $w_r(x)$ such that $M[v_r] = N[w_r]$ and which satisfies as many of the boundary conditions as are sufficient for (8.12) to hold, i.e. so that

$$\int_a^b \left(w_r N[\varphi] - \varphi N[w_r] \right) dx = 0 \qquad (8.45)$$

for all comparison functions $\varphi(x)$. We can then put

$$F_0(x) = \sum_{r=1}^{p} a_r w_r(x).$$

With this expression substituted in (8.30) we have again a quotient of two quadratic forms $\mu_1 = Q_1^\otimes / Q_2^\otimes$ but the numerator $Q_1^\otimes$ is now defined by

$$Q_1^\otimes = \int_a^b \left(\sum_{r=1}^{p} a_r w_r(x) \right) \left(\sum_{s=1}^{p} a_s M[v_s] \right) dx = \sum_{r,s=1}^{p} m_{rs}^\otimes a_r a_s,$$

where the $m_{rs}^\otimes$ are also defined differently:

$$m_{rs}^\otimes = \int_a^b w_r M[v_s] \, dx.$$

The a_r are determined to minimize μ_1 as before, i.e. from the equations (8.43), and Λ is determined from (8.44), but the new definition of the $m_{rs}^\otimes$ is used instead of (8.41); the definition of the $n_{rs}^\otimes$ (8.42) is unaltered.

8.9. TEMPLE's quotient

The limits for the first eigenvalue λ_1 given by (8.18) in § 8.3 may be extended to higher eigenvalues, and they then provide an alternative means to the Ritz method for the approximate location of the higher eigenvalues; they can also be used in combination with the Ritz method as in the example below.

Starting with a function $F_0(x)$ which satisfies the same conditions as in § 8.3, we form the next function $F_1(x)$ in the iterative sequence generated by the iteration procedure of § 8.2 and calculate the Schwarz constants a_0, a_1, a_2 and quotients μ_1, μ_2, as in (8.19).

We now assume that $F_0(x)$ may be expanded as a uniformly convergent series of normalized eigenfunctions $y_j(x)$, i.e. normalized in the general sense:

$$\int_a^b y_i N[y_j]\, dx = \begin{cases} 0 & \text{when } i \neq j \\ 1 & \text{when } i = j. \end{cases}$$

We may then write

$$F_0(x) = \sum_{j=1}^{\infty} c_j y_j(x),$$

and assuming further that this series can be differentiated term by term sufficiently often we readily deduce that

$$F_1(x) = \sum_{j=1}^{\infty} c_j \frac{y_j(x)}{\lambda_j} \tag{8.46}$$

and

$$a_k = \sum_{j=1}^{\infty} \frac{c_j^2}{\lambda_j^k} \quad (k = 0, 1, 2). \tag{8.47}$$

The validity of this expansion can be established[1] when, for example, the eigenvalue problem (8.1), (8.4), (8.5) is self-adjoint and full-definite, $N[y]$ has the special "single-term" form

$$N[y] = (-1)^n [g_n(x) y^{(n)}]^{(n)} \quad \text{with} \quad g_n(x) > 0,$$

the boundary conditions include the equations

$$y(a) = y'(a) = \cdots = y^{(n-1)}(a) = y(b) = y'(b) = \cdots = y^{(n-1)}(b) = 0$$

and $F_0(x)$ is a comparison function.

From the Schwarz constants we now form the "Temple quotient"

$$\left. \begin{aligned} T(t) &= \frac{a_0 - t a_1}{a_1 - t a_2} = \frac{a_1}{a_2} \frac{\dfrac{a_0}{a_1} - t}{\dfrac{a_1}{a_2} - t} = \mu_2 \frac{\mu_1 - t}{\mu_2 - t} \\ &= \mu_2 \left(1 + \frac{\mu_1 - \mu_2}{\mu_2 - t}\right) = \mu_2 - \frac{\mu_1 - \mu_2}{\dfrac{t}{\mu_2} - 1}. \end{aligned} \right\} \tag{8.48}$$

Here t is an arbitrary parameter restricted only by the condition that it shall differ from μ_2. This quotient $T(t)$ can easily be expressed in

---

[1] See L. COLLATZ: Eigenwertaufgaben. Leipzig 1949. pp. 144 and 191. For formula (8.54) the assumption that $F_0(x)$ is a comparison function can be replaced by the weaker conditions specified for $F_0(x)$ in § 8.3; the proof of this is given by N. J. LEHMANN: Beiträge zur numerischen Lösung linearer Eigenwertprobleme. Z. Angew. Math. Mech. **29**, 341—356 (1949); **30**, 1—16 (1950).

terms of the expansion coefficients c_j via (8.47):

$$T(t) = \frac{a_0 - t a_1}{a_1 - t a_2} = \frac{\sum\limits_{j=1}^{\infty} c_j^2 \left(1 - \frac{t}{\lambda_j}\right)}{\sum\limits_{j=1}^{\infty} c_j^2 \frac{1}{\lambda_j}\left(1 - \frac{t}{\lambda_j}\right)},$$ (8.49)

and hence, for any eigenvalue λ_s,

$$T(t) - \lambda_s = \frac{\sum\limits_{j=1}^{\infty} c_j^2 \left(1 - \frac{\lambda_s}{\lambda_j}\right)\left(1 - \frac{t}{\lambda_j}\right)}{\sum\limits_{=1}^{\infty} c_j^2 \frac{1}{\lambda_j}\left(1 - \frac{t}{\lambda_j}\right)}.$$

Now

$$\mu_2 - t = \frac{a_1 - a_2 t}{a_2} = \frac{\sum\limits_{j=1}^{\infty} c_j^2 \frac{1}{\lambda_j}\left(1 - \frac{t}{\lambda_j}\right)}{\sum\limits_{j=1}^{\infty} \left(\frac{c_j}{\lambda_j}\right)^2},$$

and since the sum of the series which occurs both in the numerator here and in the denominator of the quotient for $T(t) - \lambda_s$ cannot be zero (we have assumed $t \neq \mu_2$), we may multiply the two quotients together and cancel:

$$(T(t) - \lambda_s)(\mu_2 - t) = \frac{\sum\limits_{j=1}^{\infty} c_j^2 \left(1 - \frac{\lambda_s}{\lambda_j}\right)\left(1 - \frac{t}{\lambda_j}\right)}{\sum\limits_{j=1}^{\infty} \left(\frac{c_j}{\lambda_j}\right)^2}.$$ (8.50)

Let the eigenvalues be arranged in ascending order of magnitude, as in § 8.1 (our assumptions imply that $\lambda_1 > 0$ but the present considerations are also valid for $\lambda_1 < 0$). Whether λ_s is simple or multiple, there is a next smaller eigenvalue λ_{s-} and a next larger eigenvalue λ_{s+} (unless $\lambda_s = \lambda_1$, when λ_{s-} does not exist). Thus we have $\lambda_{s-} < \lambda_s < \lambda_{s+}$ and no eigenvalue other than λ_s lies between λ_{s-} and λ_{s+}. When λ_s is simple, $\lambda_{s-} = \lambda_{s-1}$ and $\lambda_{s+} = \lambda_{s+1}$.

Now the numerator of the quotient on the right-hand side of (8.50) cannot be negative if t lies in the interval $\lambda_{s-} \leq t \leq \lambda_{s+}$; moreover, the denominator, whose value is a_2, is always positive since the eigenvalue problem is assumed to be full-definite. Consequently

$$(T(t) - \lambda_s)(\mu_2 - t) \geq 0 \quad \text{for} \quad \lambda_{s-} \leq t \leq \lambda_{s+}.$$ (8.51)

This implies that

and

$$\left.\begin{array}{ll} T(t) \geq \lambda_s & \text{when} \quad t < \mu_2 \\ T(t) \leq \lambda_s & \text{when} \quad t > \mu_2 \end{array}\right\}$$ (8.52)

provided always that $\lambda_{s-} \leq t \leq \lambda_{s+}$.

It can be seen immediately that the limits (8.20) for the first eigenvalue are special cases of this result: if we know that a number t, greater than μ_2, is a lower limit for λ_2, then (8.52) gives $T(t) \leq \lambda_1$, which is precisely the lower limit in (8.20); if we take the limit as $t \to -\infty$ in (8.48), we obtain $T(-\infty) = \mu_2 \geq \lambda_1$, and hence the result also includes the upper limit in (8.20).

For higher eigenvalues we can summarize thus: If we have an approximation $F_0(x)$ to an eigenfunction y_s corresponding to the eigenvalue λ_s which is sufficiently accurate that the Schwarz quotient μ_2 which it generates lies in the interval $\lambda_{s-} < \mu_2 < \lambda_{s+}$, and if also we know an upper limit L_{s-} for λ_{s-} and a lower limit l_{s+} for λ_{s+} which are both so near that

$$\lambda_{s-} \leq L_{s-} < \mu_2 < l_{s+} \leq \lambda_{s+} , \tag{8.53}$$

then λ_s is bracketed by the values $T[l_{s+}]$ and $T[L_{s-}]$:

$$T(l_{s+}) \leq \lambda_s \leq T(L_{s-}) . \tag{8.54}$$

Example. As in § 8.5, we again treat the problem

$$-[(1+x)\,y']' = \lambda\,(1+x)\,y, \quad y'(0) = y(1) = 0$$

by assuming a polynomial expression for $F_1(x)$. This time, however, we use a polynomial of higher degree with an extra arbitrary constant so that in addition to satisfying both boundary conditions on F_1 and the essential boundary condition on F_0, which yields

$$F_0(x) = \frac{2}{1+x}\{c_1(14 + 28x - 18x^2 - 24x^3) + c_2(5 + 4x - 9x^2)\}$$
$$F_1(x) = \quad c_1(11 - 14x^2 + 3x^4) \quad + c_2(3 - 5x^2 + 2x^3),$$

we can determine the ratio $c_1 : c_2$ of the remaining free constants to make $F_1(x)$ approximate the second eigenfunction. The Schwarz constants generated by this $F_0(x)$ (cf. § 8.5) are

$$a_0 = 4 \quad \{ 204 \cdot 9614\,c_1^2 + 2 \times 53 \cdot 061\,42\,c_1 c_2 + 14 \cdot 111\,42\,c_2^2\}$$
$$a_1 = \frac{2}{21}\{2619 \cdot 8 \quad c_1^2 + 2 \times 693 \cdot 2 \quad c_1 c_2 + 184 \cdot 1 \quad c_2^2\}$$
$$a_2 = \frac{1}{21}\{1623 \cdot 3 \quad c_1^2 + 2 \times \frac{1723 \cdot 3}{4} \quad c_1 c_2 + 114 \cdot 4 \quad c_2^2\} .$$

One way of determining the ratio $c_1 : c_2$ would be to put $c_1 = 1$, calculate the limits (8.54) for different values of c_2 and then interpolate for the value of c_2 which gives the narrowest limits. Here it is convenient to apply the Rayleigh-Ritz method with $F_1(x)$ as the assumed form of approximate solution; we have already calculated the expressions for a_1 and a_2, i.e. the quadratic forms Q_1 and Q_2 of § 8.8, and can write down immediately the Galerkin equations

$$(2\times 2619{\cdot}8 - \varLambda\times 1623{\cdot}3)\,c_1 + \left(2\times 693{\cdot}2 - \varLambda\times\frac{1723{\cdot}3}{4}\right)c_2 = 0,$$
$$\left(2\times 693{\cdot}2 - \varLambda\times\frac{1723{\cdot}3}{4}\right)c_1 + (2\times 184{\cdot}1 - \varLambda\times 114{\cdot}4)\ \ c_2 = 0.$$

The determinant of the coefficients is a quadratic expression in $\varLambda$, whose zeros

$$\varLambda = \begin{cases}\varLambda_1 = \ \ 3{\cdot}218\,505 \\ \varLambda_2 = 23{\cdot}189\,7\end{cases}$$

are the Rayleigh-Ritz approximations to the first and second eigenvalues. We take $\varLambda = \varLambda_2$ and solve the corresponding set of homogeneous equations to obtain

$$c_1 = 1, \quad c_2 = -3{\cdot}766\,05.$$

With these values we have

$$a_0 = 4\times 5{\cdot}441\,75, \quad a_1 = 2\times 0{\cdot}459\,85, \quad a_2 = 0{\cdot}039\,664,$$
$$\mu_1 = 23{\cdot}667, \quad\quad \mu_2 = 23{\cdot}190, \quad \mu_1 - \mu_2 = 0{\cdot}48.$$

Since μ_2 is equal to the Rayleigh-Ritz approximation $\varLambda_2$, we know immediately that $\mu_2 \geqq \lambda_2$. [This is a smaller upper limit than that given by (8.54).]

In order to calculate a lower limit for λ_2 from (8.54) we need a rough lower limit for λ_3. If, as in § 8.5, we compare with a problem having constant coefficients, we obtain a very crude lower limit

$$l_3 = \frac{25}{8}\,\pi^2 \approx 30{\cdot}8.$$

It is greater than μ_2, however, and can therefore be used in (8.54); it yields

$$T(l_3) = 21{\cdot}735 \leqq \lambda_2.$$

This limit will be improved if we can deduce a better lower limit for λ_3; for example, $l_3 = 60$ would give

$$T(60) = 22{\cdot}888 \leqq \lambda_2.$$

8.10. Some modifications to the iteration method

Theory shows that the ordinary iteration (8.10) will converge to the q-th eigenfunction if $F_0(x)$ is chosen so as to be orthogonal in the generalized sense to the first $q-1$ eigenfunctions. In practice these first $q-1$ eigenfunctions are not known exactly and therefore their components cannot be excluded completely from $F_0(x)$ and subsequent iterates. In a modification of the iteration method given by KOCH[1] the unwanted components, which would otherwise become dominant, are suppressed by subtraction at each stage of the iteration of the components of the best available approximations Y_j to the first $q-1$ eigenfunctions. Thus the direct relation (8.10) between F_k and F_{k-1} is replaced by

$$
\left.
\begin{aligned}
M[F_k^*] &= N[F_{k-1}], \\
U_\mu[F_k^*] &= 0, \\
F_k &= F_k^* - \sum_{j=1}^{q-1} \frac{Y_j \int_a^b F_k^* N[Y_j]\, dx}{\int_a^b Y_j N[Y_j]\, dx}
\end{aligned}
\right\} \quad (k = 1, 2, \ldots).
$$

Another modification[2], which occasionally improves the convergence of both this modified iteration and the ordinary iteration (8.10), consists in replacing $M[F_k]$ by $M[F_k - \vartheta F_{k-1}]$. Thus the iteration (8.10) is modified to

$$
M[F_k - \vartheta F_{k-1}] = N[F_{k-1}], \qquad U_\mu[F_k] = 0 \qquad (k = 1, 2, \ldots).
$$

Assuming that the expansion theorem is applicable we may then write

$$
F_1 = \sum_{j=1}^{\infty} c_j y_j \left(\vartheta + \frac{1}{\lambda_j}\right), \qquad F_k = \sum_{j=1}^{\infty} c_j y_j \left(\vartheta + \frac{1}{\lambda_j}\right)^k \qquad (k = 1, 2, \ldots)
$$

in place of (8.46).

One can go a step further and use a different ϑ in each cycle[3], in which case the iteration formula becomes

$$
M[F_k - \vartheta_k F_{k-1}] = N[F_{k-1}], \qquad U_\mu[F_k] = 0 \qquad (k = 1, 2, \ldots)
$$

and the expansion formula must be modified to

$$
F_k = \sum_{j=1}^{\infty} c_j y_j \left(\vartheta_1 + \frac{1}{\lambda_j}\right)\left(\vartheta_2 + \frac{1}{\lambda_j}\right) \cdots \left(\vartheta_k + \frac{1}{\lambda_j}\right) \qquad (k = 1, 2, \ldots).
$$

---

[1] KOCH, J. J.: Bestimmung höherer kritischer Drehzahlen schnell laufender Wellen. Verh. 2. Internat. Kongr. Techn. Math., Zürich 1926, pp. 213—218.
Another method is given by A. FRAENKLE: Ing.-Arch. **1**, 499—526 (1930), specifically "Methode II", p. 510 et seq. A method of minimized iterations has been devised by C. LANCZOS: An iteration method for the solution of the eigenvalue problem of linear differential and integral operators. J. Res. Nat. Bur. Stand. **45**, 255—282 (1950).

[2] Developed for integral equations by G. WIARDA: Integralgleichungen unter besonderer Berücksichtigung der Anwendungen, p. 126. Leipzig 1930.

[3] Developed for integral equations by H. BÜCKNER: Ein unbeschränkt anwendbares Iterationsverfahren für Fredholmsche Integralgleichungen. Math. Nachr. **2**, 304—313 (1949).

If approximations Λ_σ are known for, say, the first $p-1$ eigenvalues λ_σ, one can put

$$\left.\begin{aligned}\vartheta_\sigma = \vartheta_{p+\sigma} = \vartheta_{qp+\sigma} = -\frac{1}{\Lambda_\sigma} \\ \vartheta_p = \vartheta_{qp} = 0\end{aligned}\right\} \quad \text{for} \quad \sigma = 1, 2, \ldots, p-1, \quad q = 1, 2, 3, \ldots.$$

This furnishes another iteration method for the p-th eigenfunction provided that the approximations Λ_σ are sufficiently accurate.

8.11. Miscellaneous exercises on Chapter III

1. The stress distribution in a rotating disc (a steam turbine rotor, say) of thickness $\eta(r)$ and radius R may be calculated from the differential equation[1]

$$r^2 S'' + r\left(3 - \frac{r}{\eta}\eta'\right)S' - \frac{m-1}{m}\frac{r}{\eta}\eta' S + \frac{3m+1}{m}\frac{\gamma}{g}\eta r^2 \omega^2 = 0.$$

Here r is the distance from the axis of rotation (Fig. III/15), $S(r) = \eta(r)\sigma_r$, where σ_r is the radial stress, and m, γ, ω, g are constants (usual notation as in reference[1] below). If we consider a thickness profile of the form

$$\eta(r) = \frac{\eta_0}{1 + \left(\dfrac{2r}{R}\right)^2}$$

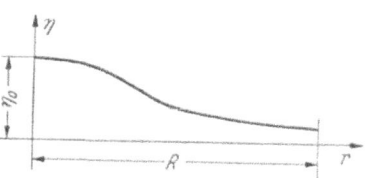

Fig. III/15. Radial section of the disc in Exercise 1

and introduce the dimensionless variables

$$x = \frac{2r}{R}, \quad y = \frac{S}{C}, \quad C = \frac{3m+1}{m}\frac{\gamma}{g}\omega^2\frac{R^2}{4}\eta_0,$$

then the equation with $m = 3$ reads

$$(1 + x^2)y'' + \left(\frac{3}{x} + 5x\right)y' + \frac{4}{3}y + 1 = 0, \tag{8.55}$$

in which the dashes now denote differentiation with respect to x.

Calculate $y(x)$ by the ordinary finite-difference method with interval $h = \frac{1}{4}$, say, for the case in which the boundary conditions are

$$y(\pm 2) = 0.6, \quad y \text{ regular at } x = 0,$$

i.e. for a solid disc with given radial stress at its edge $r = R$ $(x = 2)$.

2. Failure of the finite-difference method. Show that for the boundary-value problem

$$x y'' = c y'; \quad y(0) = 0, \quad y(1) = 1$$

the solution given by the ordinary finite-difference method is always partly of the wrong sign when $c > 2$ (when $c = 3$, for example).

---

[1] BIEZENO, C. B., and R. GRAMMEL: Technische Dynamik, 2nd ed., Vol. II, p. 25. Berlin-Göttingen-Heidelberg 1953.

3. Show that the solution of the non-linear problem

$$y'' = \frac{2 x^2}{y}, \qquad y(\pm 1) = 1$$

given by the ordinary finite-difference method is exact.

4. The boundary-value problem

$$y'' = y^2; \qquad y(0) = 0, \qquad y(2) = -2$$

possesses only complex solutions. Determine one of these solutions roughly by the ordinary finite-difference method.

5. The boundary-value problem $y'' = 0$; $y(0) = 1$, $y(1) = y'(1)$ has no solution. Can the method of § 5.2 be applied to obtain a variational problem whose Euler conditions are identical with this boundary-value problem?

6. Show that the variational problem corresponding to the boundary-value problem

$$y'' + 4 y = 2, \qquad y(\pm 1) = 0$$

according to the formulae of § 5.2 possesses neither a minimum nor a maximum. Can one still apply the Ritz method, say with a polynomial approximation of the form

$$y = \sum_{\nu=1}^{p} a_\nu (1 - x^{2\nu}).$$

7. Failure of the Ritz method. The non-linear boundary-value problem

$$y' y'' = 4 x, \qquad y(\pm 1) = 0$$

represents the Euler conditions of the variational problem

$$J[\varphi] = \int_{-1}^{1} [\tfrac{1}{6} \varphi'^3 + 4 x \varphi] \, dx = \text{extremum}$$

with the auxiliary conditions $\varphi(\pm 1) = 0$. The boundary-value problem has two solutions symmetrical about $x = 0$, namely

$$y_1 = 1 - x^2 \quad \text{and} \quad y_2 = x^2 - 1.$$

Apply Ritz's method with the approximation

$$\varphi = \sum_{\nu=1}^{p} a_\nu (1 - x^{2\nu}) \quad \text{or} \quad \varphi = \sum_{\nu=1}^{p} a_\nu \cos \nu \pi x.$$

8. Bracketing of eigenvalues. Use the enclosure theorem (§ 8.6) to calculate upper and lower limits for some of the smaller eigenvalues of the eigenvalue problem

$$- y'' = \lambda (2 - x^2) y; \qquad y(0) = 0, \qquad 2 y(1) + y'(1) = 0.$$

(This represents the vibrations of an inhomogeneous string with one end fixed and the other constrained to move transversely under an elastic force.)

9. Use Ritz's method with the two-term approximation

$$\varphi = a_1 (1 - x^2) + a_2 (1 - x^4)$$

to find the symmetric solutions of the eigenvalue problem

$$- y'' = \lambda (1 + x^2) y, \qquad y(\pm 1) = 0,$$

which represents the vibrations of an inhomogeneous string fixed at each end

10. Find upper and lower limits for the first eigenvalue of the problem of the previous exercise by using the iteration method and the theorem of § 8.3.

11. Apply RITZ's method to the eigenvalue problem

$$y^{IV} = \lambda(1 + x)\, y; \qquad y(0) = y'(0) = y''(1) = y'''(1) = 0$$

(transverse vibrations of a cantilever of variable cross-section but constant flexural rigidity) with two-term and three-term polynomial approximations.

12. What is the Euler differential equation (6.9) for the variational problem

$$J[\varphi] = \int_a^b f(x)\, \varphi^{(m-1)}\, \varphi^{(m+1)}\, dx = \min.,$$

where $m \geq 1$ and $f(x)$ is a given function of x with a continuous second derivative? No boundary conditions need be specified for this.

13. Someone asserts that in the application of RITZ's method to boundary-value problems as described in § 5 the addition of extra boundary terms to the integral, as in (5.6), and the recognition of two quite different types of boundary conditions — essential and suppressible — are quite unnecessary and that all one need do is simply to restrict the admissible functions to the domain of functions which satisfy all the boundary conditions and use merely the integral whose Euler equation coincides with the differential equation in question. Thus for the problem

$$y'' = 0; \qquad y(0) = y'(0), \qquad y(1) = 1,$$

for example, one would consider the variational problem

$$J^*[\varphi] = \int_0^1 \varphi'^2\, dx = \min.$$

subject to the auxiliary conditions $\varphi(0) = \varphi'(0)$, $\varphi(1) = 1$ and argue that the minimizing function $y(x)$ must satisfy the boundary conditions since all admissible functions satisfy them and must also satisfy the Euler equation $y'' = 0$ and therefore it must be the solution of the boundary-value problem. Comment!

8.12. Solutions

1. The singularity at $x = 0$ in (8.55) implies that for a regular solution we must have $y'(0) = 0$. Now $\lim\limits_{x \to 0} \dfrac{y'}{x} = y''(0)$, so that

$$\lim_{x \to 0}\left(\frac{3}{x} + 5x\right) y' = 3\, y''(0);$$

hence for the point $x = 0$ the differential equation (8.55) can be replaced by

$$4\,y'' + \tfrac{4}{3}\, y + 1 = 0.$$

With $Y_{-i} = Y_i$ (on account of symmetry) we obtain the difference equations

$$(x = 0) \quad 16(2Y_1 - 2Y_0) + \tfrac{4}{3}Y_0 + 1 = 0,$$

$$(x = \tfrac{1}{4}) \quad 5(Y_2 - 2Y_1 + Y_0) + \tfrac{17}{2}(Y_2 - Y_0) + \tfrac{4}{3}Y_1 + 1 = 0$$

and two more similar equations corresponding to the points $x = 1$, $x = \tfrac{3}{4}$. If we then express $Y_1, Y_2, \ldots$ successively in terms of Y_0:

$$Y_1 = 0{\cdot}958\,333\,Y_0 - 0{\cdot}031\,25,$$
$$Y_2 = 0{\cdot}874\,486\,Y_0 - 0{\cdot}094\,136,$$
$$Y_3 = 0{\cdot}801\,612\,Y_0 - 0{\cdot}148\,791,$$
$$Y_4 = 0{\cdot}742\,773\,Y_0 - 0{\cdot}192\,920,$$

the condition $Y_4 = 0{\cdot}6$ yields

$$Y_0 = 1{\cdot}067\,514,$$

and hence

$$Y_1 = 0{\cdot}991\,784,$$
$$Y_2 = 0{\cdot}839\,390,$$
$$Y_3 = 0{\cdot}706\,941.$$

2. The exact solution of the boundary-value problem is $y = x^{1+c}$ and is positive throughout the range $0 < x \leq 1$. With the pivotal interval $h = 1/n$ the difference equation corresponding to $x = h$ is

$$h\,\frac{Y_2 - 2Y_1 + Y_0}{h^2} = c\,\frac{Y_2 - Y_0}{2h},$$

which on account of the boundary condition $Y_0 = 0$ reduces to

$$Y_2 = -\frac{4}{c-2}\,Y_1.$$

Consequently Y_1 and Y_2 always have opposite signs when $c > 2$. Thus the finite-difference method yields at least one pivotal value Y_i with the wrong sign for all the values of h, however small. For the numerical example $c = 3$, $h = \tfrac{1}{2}$ the exact solution gives $y(\tfrac{1}{2}) = \tfrac{1}{16}$ and the corresponding finite-difference approximation is $Y_1 = -\tfrac{1}{4}$.

3. The exact solution $y = x^2$ is a parabola and for a parabola no error is introduced by the substitution of the second difference quotient for the second derivative. Thus, for example, with $h = \tfrac{2}{3}$ we have $y(\tfrac{1}{3}) = \tfrac{1}{9}$ and this value for Y_1 satisfies the difference equation $\dfrac{1 - 2Y_1 + Y_{-1}}{h^2} = \dfrac{1 - Y_1}{h^2} = \dfrac{2 \cdot \tfrac{1}{6}}{Y_1}$ exactly.

4. Using the pivotal interval $h = 1$ we find that $y(1) \approx -1 \pm i$. With $h = \tfrac{2}{3}$ the approximate solution with positive imaginary part is

$$y(\tfrac{2}{3}) \approx -0{\cdot}65 + 0{\cdot}87\,i,$$
$$y(\tfrac{4}{3}) \approx -1{\cdot}45 + 1{\cdot}24\,i$$

and with $h = \frac{1}{2}$ it is

$$y(\tfrac{1}{2}) \approx -0.500 + 0.728\,i,$$
$$y(1) \approx -1.070 + 1.275\,i,$$
$$y(\tfrac{3}{2}) \approx -1.760 + 1.139\,i;$$

the complex conjugates of these values give the other solution.

The more accurate approximations with $h = 2/n$, $n = 3, 4, \ldots$, which require the solution of systems of two or more equations, are obtained by interpolation. We take several values of $y(h)$ covering a range suggested by the very rough approximation $y(1) \approx -1 + i$, which indicates how the solutions for smaller h run, and for each of these values we calculate $y(2h)$, $y(3h)$, ... from the difference equations. A value of $y(h)$ for which $y(nh) + 2 \approx 0$ is then determined by two-dimensional interpolation.

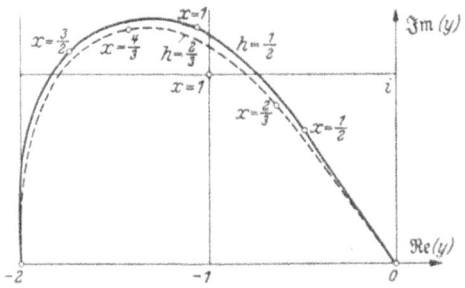

Fig. III/16. The complex solution of a real boundary-value problem

The solution can be regarded as the parametric (parameter x) representation of a curve in the complex y plane which joins the points $y = 0$ and $y = -2$. Fig. III/16 shows the approximations for the points on this curve obtained with $h = \frac{1}{2}$ and $h = \frac{2}{3}$; the calculated points are shown as small circles.

5. Yes. The required variational problem is

$$J[\varphi] = \int\limits_0^1 \varphi'^2 dx - [\varphi(1)]^2 = \text{extremum}$$

subject to the auxiliary condition $\varphi(0) = 1$. It also has no solution: with $\varphi = 1 - ax$, $J[\varphi]$ becomes $2a - 1$, which can take all values.

6. From (5.19) the corresponding variational problem is

$$J[\varphi] = \int\limits_{-1}^1 (-\varphi'^2 + 4\varphi^2 - 4\varphi)\, dx = \text{extremum}$$

subject to the auxiliary conditions $\varphi(\pm 1) = 0$. That $J[\varphi]$ has neither a maximum nor a minimum in the domain of continuously differentiable functions $\varphi(x)$ with $\varphi(\pm 1) = 0$ can be shown simply by exhibiting functions in this domain which give the integral arbitrarily large positive and negative values; such functions are

$$\varphi = a(-1 + x^2), \quad \text{which yields} \quad J[\varphi] = 8\left(\tfrac{1}{5} a^2 + \tfrac{2}{3} a\right),$$

and

$$\varphi = a(-x^2 + x^4), \quad \text{which yields} \quad J[\varphi] = -8\left(\tfrac{5}{63}a^2 - \tfrac{2}{15}a\right).$$

However, for $\varphi = y$, where

$$y = \frac{1}{2}\left(1 - \frac{\cos 2x}{\cos 2}\right)$$

is the solution of the boundary-value problem, $J[\varphi]$ has a stationary value; thus the first variation δJ still vanishes when $\varphi = y$ and consequently Ritz's method still gives significant approximations. The stationary character of $J[\varphi]$ for $\varphi = y$ can be seen clearly from the form it takes when we put $\varphi = y(x) + \varepsilon(x)$ with $\varepsilon(\pm 1) = 0$:

$$J[\varphi] = J[y] + \int\limits_{-1}^{1} (4\varepsilon^2 - \varepsilon'^2)\, dx, \quad \text{where} \quad J[y] = \tan 2 - 2.$$

For the two-term Ritz approximation $\varphi = a(1 - x^2) + b(1 - x^4)$ we have

$$\frac{1}{8} J[\varphi] = \frac{1}{5}a^2 + \frac{22}{105}2ab + \frac{44}{315}b^2 - \frac{2}{3}a - \frac{4}{5}b,$$

and from $\dfrac{\partial J}{\partial a} = 0, \dfrac{\partial J}{\partial b} = 0$ we obtain $a = \dfrac{7}{3}, b = -\dfrac{7}{11}$; these values yield the approximate solution $\varphi = \dfrac{7}{3}(1 - x^2) - \dfrac{7}{11}(1 - x^4)$ and in particular $\varphi(0) = \dfrac{56}{33} = 1\cdot696\,97$ (error $-0\cdot3\%$).

7. For both of the suggested Ritz approximations for φ it turns out that $J[\varphi] \equiv 0$; this happens, in fact, for any even function $[\varphi(x) = \varphi(-x)]$, for J is then the integral of an odd function over a symmetric interval. Consequently the a_ν cannot be calculated from the equations $\partial J/\partial a_\varrho = 0$.

8. If we try an approximation of the form

$$F_1 = f(x) \sin(g(x))$$

and calculate

$$F_1', F_1'' \quad \text{and} \quad \varPhi = -\frac{F_1''}{(2 - x^2) F_1},$$

we find that the special form

$$F_1 = (1 + bx)^{-\frac{1}{2}} \sin\left[a\left(x + \frac{bx^2}{2}\right)\right]$$

is particularly convenient. This gives

$$\varPhi = \frac{a^2 \xi^2 - \dfrac{3}{4}\dfrac{b^2}{\xi^2}}{2 - x^2}, \quad \text{where} \quad \xi = 1 + bx.$$

The boundary condition $2F_1(1) + F_1'(1) = 0$ remains to be satisfied; it implies a relation between a and b, which, if we put $\gamma = a(1 + \tfrac{1}{2}b)$, can be written

$$\frac{\tan\gamma}{\gamma} = \frac{-4(1 + b)^2}{(4 + 3b)(2 + b)}.$$

We now choose a value for b and calculate several of the smaller of the infinitely many corresponding values of γ. Each of these defines a value of a and hence a corresponding $\Phi(x)$, whose upper and lower bounds Φ_{max} and Φ_{min} give upper and lower bounds for an eigenvalue. By varying b we can easily see when the limits become narrower. Trials with $b = -0.3$ and -0.25 show that the best values are obtained with say $b = -0.285$. The limits for the first three eigenvalues calculated with this value of b are included in Table III/17.

Table III/17. *Bracketing of eigenvalues*

b	γ	a	$\Phi_{min} \leqq \lambda_s \leqq \Phi_{max}$
-0.3	2·409	2·834	$3.24 \leqq \lambda_1 \leqq 3.98$
	5·190	6·106	$15.2 \leqq \lambda_2 \leqq 18.61$
-0.25	2·385	2·726	$3.22 \leqq \lambda_1 \leqq 4.10$
	5·167	5·906	$15.24 \leqq \lambda_2 \leqq 19.5$
-0.285	2·4028	2·8021	$3.25 \leqq \lambda_1 \leqq 3.90$
	5·1832	6·0445	$15.25 \leqq \lambda_2 \leqq 18.56$
	8·1665	9·5236	$37.8 \leqq \lambda_3 \leqq 46.25$

9. Here we have

$$\frac{1}{2} J[\varphi] = \int_0^1 [\varphi'^2 - \Lambda(1 + x^2)\varphi^2]\, dx = 4\left(\frac{1}{3} a_1^2 + 2 \cdot \frac{2}{5} a_1 a_2 + \frac{4}{7} a_2^2\right) -$$
$$- 32\Lambda\left(\frac{2}{105} a_1^2 + \frac{2}{45} a_1 a_2 + \frac{92}{3465} a_2^2\right).$$

If $8\Lambda = 3\mu$, the determinant condition for the equations $\partial J/\partial a_\nu = 0$ for $\nu = 1, 2$ reads

$$\begin{vmatrix} 6\mu - 35 & 7\mu - 42 \\ 7\mu - 42 & \dfrac{92}{11}\mu - 60 \end{vmatrix} = 0,$$

which yields the quadratic

$$13\mu^2 - 712\mu + 3696 = 0.$$

The roots of this equation yield upper bounds for the first and third eigenvalues since the second corresponds to an antisymmetric eigenfunction:

$$\Lambda_{1,3} = \frac{3}{26}\left(89 \pm \sqrt{4918}\right) = \begin{cases} 2.177486 \geqq \lambda_1 \\ 18.361 \qquad \geqq \lambda_3. \end{cases}$$

10. According to (8.10) the first iterate is to be calculated here from

$$-F_1'' = (1 + x^2) F_0, \qquad F_1(\pm 1) = 0.$$

Starting from $F_0 = 1 - x^2$ we obtain

$$F_1 = \frac{1}{30}(14 - 15x^2 + x^6).$$

Then we calculate the Schwarz constants (8.13):

$$\frac{1}{2}\,a_0 = \int\limits_0^1 (1 + x^2)\,F_0^2\,dx = \frac{64}{105}\,,$$

$$\frac{1}{2}\,a_1 = \frac{3232}{35 \times 330}\,, \qquad \frac{1}{2}\,a_2 = \frac{32 \times 32\,564{\cdot}8}{900 \times 9009}$$

and from them the Schwarz quotients (8.11):

$$\mu_1 = \frac{a_0}{a_1} = \frac{220}{101} = 2{\cdot}178\,218,$$

$$\mu_2 = \frac{a_1}{a_2} = \frac{177\,255}{81\,412} = 2{\cdot}177\,259.$$

For the application of formula (8.18) we need further a lower limit l_2 for the second eigenvalue λ_2, or here a lower limit l_3 for the third eigenvalue λ_3 since the second eigenfunction is antisymmetric and we are considering here only symmetric modes — this is because we chose a symmetric starting function $F_0(x)$, which can therefore be expanded in a series of symmetric eigenfunctions (cf. § 8.9). The third eigenvalue for the comparable problem with constant coefficients $-y'' = 2\lambda y$, $y(\pm 1) = 0$ is $9\pi^2/8$ and using this value for l_3 in (8.18) we obtain the narrow limits

$$2{\cdot}177\,034 \leqq \lambda_1 \leqq 2{\cdot}177\,259.$$

11. According to (6.20), (6.21) the corresponding variational problem reads

$$J[\varphi] = \int\limits_0^1 [\varphi''^2 - \lambda(1 + x)\,\varphi^2]\,dx = \text{extremum}, \qquad \varphi(0) = \varphi'(0) = 0.$$

For the three-term approximation $\varphi = a_1 x^2 + a_2 x^3 + a_3 x^4$ the conditions $\partial J/\partial a_\nu = 0$ $(\nu = 1, 2, 3)$ yield three linear homogeneous equations whose determinant reads

$$\begin{vmatrix} 4 - \dfrac{11}{30}\,\varLambda & 6 - \dfrac{13}{42}\,\varLambda & 8 - \dfrac{15}{56}\,\varLambda \\[2ex] 6 - \dfrac{13}{42}\,\varLambda & 12 - \dfrac{15}{56}\,\varLambda & 18 - \dfrac{17}{72}\,\varLambda \\[2ex] 8 - \dfrac{15}{56}\,\varLambda & 18 - \dfrac{17}{72}\,\varLambda & \dfrac{144}{5} - \dfrac{19}{90}\,\varLambda \end{vmatrix},$$

where $\varLambda$ is an approximation for λ. With $\nu = \varLambda/1008$ the usual determinant condition reduces to

$$5175\,\nu^3 - 44\,634\,\nu^2 + 13\,554\,\nu - 90 = 0,$$

from which we obtain

$$\nu = \begin{cases} 0{\cdot}006\,792\,0 \\ 0{\cdot}307 \\ 8{\cdot}33 \end{cases} \qquad \varLambda \approx \begin{cases} 6{\cdot}8464 \\ 310 \\ 8390. \end{cases}$$

For a two-term approximation (φ without the term $a_3 x^4$) the determinant is the sub-determinant of the above determinant indicated by the dotted lines, and if $\varLambda = 840\,\mu$ the determinant condition becomes $425\,\mu^2 - 369\,\mu + 3 = 0$; this yields the values

$$\varLambda \approx \begin{cases} 6{\cdot}8944 \\ 722. \end{cases}$$

12. A possible solution y must satisfy the differential equation

$$2\,[f(x)\,y^{(m)}]^{(m)} + [f''(x)\,y^{(m-1)}]^{(m-1)} = 0,$$

as can be verified by simple manipulations with the aid of the binomial coefficients.

13. The assertion made is untrue. The Ritz method, say with the approximation

$$\varphi(x) = \tfrac{1}{2}(1+x) + \sum_{\nu=1}^{n} c_\nu (1 + x - 2\,x^{\nu+1}),$$

which satisfies all the boundary conditions, would yield solutions $\varphi_n(x)$ which converge to a limit function $\psi(x)$ as n increases, but this limit function would not be the solution of the given boundary-value problem; for it is determined by the boundary-value problem $\psi''=0$; $\psi'(0)=0$, $\psi(1)=1$, in which the suppressible boundary condition $\varphi(0)=\varphi'(0)$ has been replaced by the natural (for the suggested variational problem) boundary condition $\psi'(0)=0$, and for the solution of this boundary-value problem $\psi(0) \neq \psi'(0)$. Thus one could not recognize the approximations as false from the fact that they do not converge; for this reason it is critically important that the variational problem be formulated correctly.

The variational problem suggested in the question, namely $J^*[\varphi] = $ minimum, $\varphi(0)=\varphi'(0)$, $\varphi(1)=1$ has no solution; with φ subject to these boundary conditions $J^*[\varphi]$ has the lower bound zero, and this is assumed for the function $\varphi \equiv 1$, for which $\varphi(0) \neq \varphi'(0)$, but not for any comparison function. For the solution of the given boundary-value problem, namely $\varphi = y = \tfrac{1}{2}(1+x)$, J^* assumes the value $J^*[y] = \tfrac{1}{4}$, which is not a minimum. The comparison function

$$\varphi(x) = \begin{cases} \dfrac{1}{2\varepsilon + \varepsilon^2}\,(2\varepsilon + 2\varepsilon x - x^2) & \text{for} \quad 0 \leq x \leq \varepsilon \\ 1 & \text{for} \quad \varepsilon \leq x \leq 1 \end{cases} \quad \text{with} \quad \varepsilon > 0$$

gives J^* the value

$$\frac{\varepsilon}{3}\left(\frac{1}{1+\dfrac{\varepsilon}{2}}\right)^2,$$

and by choosing ε small enough we can make this value as close to zero as we please. As ε decreases the Ritz approximations represented by this family of comparison functions

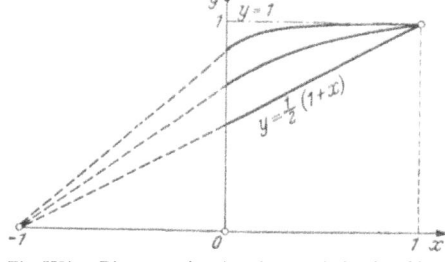

Fig. III/17. Ritz approximations for a variational problem with no solution

approximate more and more closely the function $\varphi \equiv 1$ (see Fig. III/17), yet the solution of the boundary-value problem reads $y = \tfrac{1}{2}(1+x)$.

According to § 5 the correct variational problem corresponding to the given boundary-value problem reads

$$J[\varphi] = \int_0^1 \varphi'^2 \, dx + [\varphi(0)]^2 = \text{min.},$$

where φ need satisfy only the essential boundary condition $\varphi(1) = 1$.

Chapter IV

Initial- and initial-/boundary-value problems in partial differential equations

The need for a sound theoretical foundation

In Ch. II, §§ 1.2, 1.3 some fundamental difficulties associated with the treatment of initial-value problems and error estimation for the approximate methods used were discussed with regard to ordinary differential equations. Naturally these difficulties are amplified when partial differential equations are considered; but over and above this, partial differential equations give rise to an extraordinarily large variety of phenomena and types of problem, while such essentials as the existence and uniqueness of solutions and the convergence of approximating sequences are covered by present theory only for a limited number of special classes of problems. These theoretical questions have not yet been settled in a satisfactory manner for many problems which arise in practical work. When confronted with such a problem one may be forced to rely solely on some approximate method, a finite-difference method, for example, and hope that the results obtained will be significant. Naturally such a procedure is not only unsatisfactory but even very questionable, as will be enlarged upon more precisely below; nevertheless, it is often unavoidable when a specific technical problem has to be solved and a theoretical investigation of the corresponding mathematical problem is not asked for. Consequently there is a pressing need for the accumulation of much more practical experience of approximate methods and for research into their theoretical aspects.

That an investigation of the situation is absolutely essential is revealed even by quite simple examples; they show that formal calculation applied to partial differential equations can lead to false results very easily and that approximate methods can converge in a disarmingly innocuous manner to values bearing no relation to the correct solution.

Consider, for example, the problem

$$\frac{\partial^2 u}{\partial x^2} = \frac{\partial^2 u}{\partial y^2};\quad \left\{\begin{array}{l} u(x, 0) = \cos x \\[4pt] \dfrac{\partial u(x, 0)}{\partial y} = 0, \end{array}\right\} \text{ for } |x| < \frac{\pi}{2},$$

$$u\left(\pm\frac{\pi}{2}, y\right) = \sin y \quad \text{for} \quad y \geq 0.$$

This describes the oscillations of a string of length π which is initially $(y=0)$ at rest in a displaced position defined by $u(x, 0) = \cos x$ and which is periodically excited at its ends $(x = \pm\pi/2)$.

Let the solution be expanded as a power series in the neighbourhood of the point $x = 0$, $y = 0$:

$$u(x, y) = \sum_{m, n=0}^{\infty} a_{mn} x^m y^n.$$

By inserting the series in the differential equation and equating coefficients we find immediately that the a_{mn} must satisfy the recursive formula

$$a_{m, n+2} = \frac{(m+2)(m+1)}{(n+2)(n+1)} a_{m+2, n}.$$

This relation enables us to express all the a_{mn} with $n \geq 2$ in terms of the $a_{k, 0}$ and $a_{k, 1}$; in particular,

$$a_{0, 2q} = a_{2q, 0} \quad \text{for} \quad q = 0, 1, 2, \ldots.$$

To satisfy the initial condition $u_y(x, 0) = 0$ the $a_{k, 1}$, and hence the $a_{0, 2q+1}$, must all be zero. The other initial condition $u(x, 0) = \cos x$ determines the remaining $a_{k, 0}$ $(= a_{0, k})$ with even k:

$$a_{k, 0} = \begin{cases} 0 & \text{when } k \text{ is odd} \\[6pt] \dfrac{(-1)^q}{(2q)!} & \text{when } k = 2q \text{ is even}. \end{cases}$$

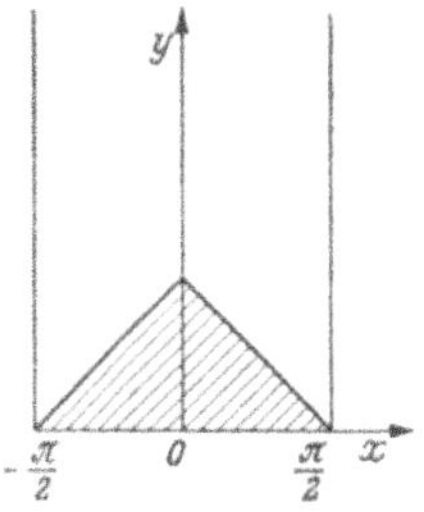

On the y axis, i.e. when $x = 0$, the power series becomes

$$\sum_{k=0}^{\infty} a_{0, k} y^k = \sum_{q=0}^{\infty} \frac{(-1)^q}{(2q)!} y^{2q} = \cos y,$$

Fig. IV/1. Region outside of which the power series solution breaks down

which converges for all y, and hence represents its sum function $\cos y$ on the whole y axis. Nevertheless, $\cos y$ is not the correct solution along the whole y axis; in fact $u(0, y)$ takes the value $\cos y$ only on that part of the y axis cut off by the characteristics $y = \frac{\pi}{2} \pm x$ emanating from the points $\left(\frac{\pi}{2}, 0\right)$, $\left(-\frac{\pi}{2}, 0\right)$, respectively (see Fig. IV/1).

It is not difficult to construct similar examples for which the power series gives the wrong solution in every interval of the y axis, even though it converges for all values of y. Such a situation arises, for instance, if in the previous example we replace the initial displacement $u(x, 0) = \cos x$ by $u(x, 0) = e^{-(1/x^2)} + c x^2$, where c is chosen so that $u\left(\pm \frac{\pi}{2}, 0\right) = 0$, and calculate the coefficients $a_{k, 0}$ from the values of the derivatives of u at the point $x = y = 0$ [in this case, of course, even $u(x, 0)$ is not represented by the power series].

It is therefore very desirable, if not essential, that an approximate treatment of a problem in partial differential equations should be coupled with theoretical substantiation of some sort. In this book we cannot develop the theory for each type of problem considered, so the reader must refer to the textbooks on partial differential equations for theoretical details[1]. We shall, however, make use of the results of the theory.

§ 1. The ordinary finite-difference method

Because of its importance for applications, this section is written very comprehensively and much is repeated from earlier sections so that the reader need refer back as little as possible.

The ordinary finite-difference method provides a simple, general method by which one can obtain a reasonable quantitative idea of the solutions of many problems in differential equations. The accuracy achieved is not usually very great but often suffices for technical problems. Refinements of the method will be discussed in § 2.

To simplify the presentation we limit ourselves for the most part to problems with two independent variables x, y, although the method may be applied in exactly the same way to problems with more than two (see the example in § 1.8).

---

[1] Of the numerous textbooks we may mention — COURANT, R., and D. HILBERT: Methoden der mathematischen Physik, 2nd ed., Vol. 1, Berlin 1931, 469 pp.; Vol. 2, Berlin 1937, 549 pp.; English edition, London 1953. — KAMKE, E.: Differentialgleichungen reeller Funktionen, 2nd ed., Leipzig 1944, 442 pp. — Differentialgleichungen, Lösungsmethoden und Lösungen, Vol. 2, Leipzig 1944, 243 pp. — FRANK, PH., and R. v. MISES: Die Differential- und Integralgleichungen der Mechanik und Physik, 2nd ed., Brunswick, Vol. 1, 1930, 916 pp.; Vol. 2, 1935, 1106 pp. — HORN, J.: Partielle Differentialgleichungen, 3rd ed. Berlin and Leipzig 1944, 228 pp. — SOMMERFELD, A.: Partielle Differentialgleichungen der Physik (Vol. 6, consisting of lectures in theoretical physics). Leipzig 1947, 332 pp. — WEBSTER, A. G., and G. SZEGÖ: Partielle Differentialgleichungen der mathematischen Physik. Leipzig and Berlin 1930, 528 pp. — COURANT, R., and K. FRIEDRICHS: Supersonic flow and shock waves, Interscience Publishers Inc. New York 1948, 464 pp. — BERNSTEIN, DOROTHY L.: Existence theorems in partial differential equations (Annals of Mathematics Study 23). Princeton 1950, 228 pp. — SAUER, R.: Anfangswertprobleme bei partiellen Differentialgleichungen. Berlin-Göttingen-Heidelberg: Springer 1952, 229 pp.

1.1. Replacement of derivatives by difference quotients

We base the finite differences on the pivotal points

$$\left.\begin{array}{l} x_i = x_0 + i\,h \\ y_k = y_0 + k\,l \end{array}\right\} \qquad (i, k = 0, \pm 1, \pm 2, \ldots), \qquad (1.1)$$

which may be defined as the nodes of a rectangular mesh made up of the "mesh lines" $x = x_i$, $y = y_k$ ($i, k = 0, \pm 1, \pm 2, \ldots$) displaced by the "mesh widths" h and l, respectively (see Fig. IV/2); (x_0, y_0) is any conveniently chosen origin for the "mesh co-ordinates" i, k. A square mesh is a rectangular mesh with $h = l$.

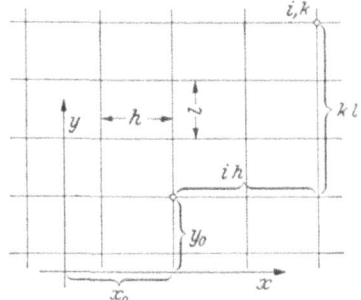

Fig. IV/2. The rectangular finite-difference mesh

Meshes other than rectangular can also be used (see Ch. V, § 2.7) but in general the resulting finite-difference expressions are more complicated and less convenient for numerical work, particularly when the chosen mesh cannot be transformed into itself by a translation[1].

Function values at mesh points, the pivotal values, will be characterized by the appropriate subscripts; thus $u_{i,k}$ will denote the value of the exact solution $u(x, y)$ at the mesh point $x = x_i$, $y = y_k$. $U_{i,k}$ will denote an approximation to the pivotal value $u_{i,k}$.

Just as for an ordinary differential equation (Ch. III, § 1), the partial differential equation is replaced by a difference equation which is derived by the substitution of an appropriate difference quotient for each derivative occurring.

This difference equation is not necessarily unique, for there are often several choices of appropriate difference quotient; for example, the derivative $(\partial u/\partial y)_{i,k}$ can be approximated by any of the following three difference quotients:

the "forward" difference quotient $\qquad \dfrac{U_{i,k+1} - U_{i,k}}{l} = \dfrac{\Delta_y U_{i,k}}{l},$ $\qquad (1.2)$

the "backward" difference quotient $\qquad \dfrac{U_{i,k} - U_{i,k-1}}{l} = \dfrac{\Delta_y U_{i,k-1}}{l},$ $\qquad (1.3)$

the "central" difference quotient $\qquad \dfrac{U_{i,k+1} - U_{i,k-1}}{2l} = \dfrac{\Delta_y}{2l}(U_{i,k-1} + U_{i,k}).$ $\qquad (1.4)$

---

[1] For the Tricomi differential equation $k(y)u_{xx} - u_{yy} = f(x, y)$ K. H. BAUERSFELD: Zum Differenzenverfahren bei Anfangswertaufgaben partieller Differentialgleichungen 2. Ordnung. Diss. Hannover, 1954, uses a rectangular mesh whose mesh width in the x direction is kept constant while its mesh width in the y direction is varied so that the mesh diagonals are chords of the characteristics.

The subscript y to the forward difference operator Δ defined in Ch. I, § 2 signifies that the operand is to be regarded as a function of y alone; thus the effects of Δ_x and Δ_y operating on any function of x and y are

$$\left.\begin{array}{l} \Delta_x g(x, y) = g(x + h, y) - g(x, y), \\ \Delta_y g(x, y) = g(x, y + k) - g(x, y) \end{array}\right\} \tag{1.5}$$

and on a discrete "mesh function" g_{ik}

$$\Delta_x g_{ik} = g_{i+1,k} - g_{ik}, \qquad \Delta_y g_{ik} = g_{i,k+1} - g_{ik}. \tag{1.6}$$

It is well known that in general the central difference quotient is a substantially better approximation to the local derivative than either the forward or the backward difference quotient. One might think therefore that it is best to use only central differences in numerical work. In fact, as is demonstrated in § 1.3, this often results in troublesome error propagation (instability).

The second partial derivative $(\partial^2 u/\partial x^2)_{i,k}$ is usually replaced by the second difference quotient

$$\frac{U_{i+1,k} - 2U_{i,k} + U_{i-1,k}}{h^2} = \frac{\Delta_x^2 U_{i-1,k}}{h^2}. \tag{1.7}$$

Generally, we can replace the partial derivative

$$\left(\frac{\partial^{m+n} u}{\partial x^m \partial y^n}\right)_{i,k} \quad \text{by} \quad \frac{\Delta_x^m \Delta_y^n U_{i-r,k-s}}{h^m l^n}, \tag{1.8}$$

in which there is a certain amount of freedom in the choice of r and s; normally r (and s correspondingly) will be $\dfrac{m}{2}$ or $\dfrac{m-1}{2}$ according as m is even or odd, although when m is odd there is also the alternative of replacing $\partial^m u/\partial x^m$ by the symmetrical expression

$$\frac{1}{2h^m}\left[\Delta_x^m\left(U_{i-\frac{m-1}{2},k} + U_{i-\frac{m+1}{2},k}\right)\right]. \tag{1.9}$$

Thus $\left(\dfrac{\partial^3 g}{\partial x^2 \partial y}\right)_{0,0}$, for example, can be replaced either by

$$\frac{1}{h^2 l}\left(g_{1,1} - 2g_{0,1} + g_{-1,1} - g_{1,0} + 2g_{0,0} - g_{-1,0}\right)$$

or by

$$\frac{1}{2h^2 l}\left(g_{1,1} - 2g_{0,1} + g_{-1,1} - g_{1,-1} + 2g_{0,-1} - g_{-1,-1}\right).$$

In this way we can find a finite-difference approximation to any partial derivative, and hence we can set up a difference equation representing the differential equation at each mesh point[1].

---

[1] A method in which only the derivatives in one fixed direction, for example, in the x direction or in the y direction, are replaced by difference quotients is given by D. R. HARTREE and J. R. WOMERSLEY: A method for the numerical or mechanical solution of certain types of partial differential equations. Proc. Roy. Soc. Lond. Ser. A **161**, 353—366 (1937). Consider, for example, the problem $u_{xx} = K u_y$ with given boundary values $u(x, 0)$, $u(0, y)$, $u(a, y)$. If we use a backward

1.2. An example of a parabolic differential equation with given boundary values

If radiation and convection are neglected, the temperature $u(x, t)$ at time t in a thin homogeneous rod at a distance x from one end satisfies the differential equation

$$\frac{\partial^2 u}{\partial x^2} = c \frac{\partial u}{\partial t} \tag{1.10}$$

(the one-dimensional heat equation), where $c = \varrho\sigma/k$ is a function of the density ϱ, specific heat σ and thermal conductivity k. Given the initial temperature distribution $u(x, 0)$ for $0 \leq x \leq a$, where a is the length of the rod, and the end temperatures $u(0, t)$, $u(a, t)$ as functions of time for $t \geq 0$, it is required to calculate the temperature distribution $u(x, t)$ along the rod at subsequent times (cf. Fig. IV/3).

With $y = t$ and $h = a/n$ we cover the region of the (x, y) plane in which we are interested with the mesh defined by

$$x_i = ih, \quad y_k = kl \quad (i = 0, 1, 2, \ldots, n; \ k = 0, 1, 2, \ldots).$$

Then if we use first of all the crude approximation with the forward difference quotient in place of the time derivative, the difference equation corresponding to the differential equation (1.10) at the point (x_i, y_k) is

$$\frac{U_{i+1, k} - 2U_{i, k} + U_{i-1, k}}{h^2} = c \frac{U_{i, k+1} - U_{i, k}}{l}. \tag{1.11}$$

This has the following advantages:

1. The mesh width l in the time direction can be chosen so that the term in $U_{i, k}$ does not appear in the difference equation (1.11):

$$l = h^2 \frac{c}{2}. \tag{1.12}$$

This simplifies the calculation considerably, for $U_{i, k+1}$ is then formed merely by taking the arithmetic mean of $U_{i+1, k}$ and $U_{i-1, k}$:

$$U_{i, k+1} = \tfrac{1}{2}(U_{i+1, k} + U_{i-1, k}). \tag{1.13}$$

difference quotient in the y direction, the equation becomes

$$\left(\frac{\partial^2 u}{\partial x^2}\right)_{i, k} = K \frac{u_{i, k} - u_{i, k-1}}{l},$$

and if the values up to the $(k-1)$-th row are already known, we have to solve a boundary-value problem for an ordinary differential equation to obtain the values $u(x, kl)$ on the k-th row. If, on the other hand, we use finite differences in the x direction, so that

$$K\left(\frac{\partial u}{\partial y}\right)_{i, k} = \frac{1}{h^2}\{u_{i+1, k} - 2u_{i, k} + u_{i-1, k}\} \quad (i = 1, \ldots, n-1),$$

we have to solve an initial-value problem for a system of ordinary differential equations of the first order. Machine methods of solution are particularly suitable here.

Table IV/1. *Finite-difference solution of the heat equation with $\partial u/\partial t$ replaced by the forward difference quotient*

k	i=0	i=1	i=2	i=3	i=4	i=5	i=6	i=7	i=8	Row-sum check S_k	Row-sum check T_k
0	0	0	0	0	0	0	0	0	0		
1	0·5	0·25	0	0	0	0	0	0	0	0·25	1·1160
2	0·8660	0·4955	0·125	0·0625	0	0	0	0	0	0·6205 / 0·6204	1·1160 / 1·6204
3	1	0·6395	0·2790	0·1551	0·0312	0·0156	0	0	0	0·9809 / 0·9810	1·6204 / 1·8470
4	0·8660	0·6316	0·3973	0·2414	0·0854	0·0466	0·0078	0·0039	0	1·2115 / 1·2114	1·8470 / 1·7114
5	0·5	0·4682	0·4365	0·2902	0·1440	0·0846	0·0252	0·0126	0	1·2304 / 1·2304	1·7112 / 1·2304
	0	0·1896	0·3792	0·2833	0·1874	0·1180	0·0486	0·0243	0	1·2304 / 1·0165	1·2304 / 1·5164
	−0·5	−0·1318	0·2364	0·2185	0·2006	0·1359	0·0712	0·0356	0	1·0164 / 0·6126	1·5164
10	−0·8660	−0·4113	0·0434	0·1103	0·1772	0·1315	0·0858	0·0429	0	0·6128 / 0·1152	−0·2532 / −0·2532
	−1	−0·5752	−0·1505	−0·0148	0·1209	0·1040	0·0872	0·0436	0	0·1152 / −0·3532	−0·8848 / −0·8848
15	−0·8660	−0·5805	−0·2950	−0·1252	0·0446	0·0592	0·0738	0·0369	0	−0·3532	−1·2192 / −1·2192
20	−0·5	−0·4264	−0·3528	−0·1929	−0·0330	0·0075	0·0480	0·0240	0	−0·6756 / −0·6756	−1·1756 / −1·1756
25	0	−0·1548	−0·3096	−0·2012	−0·0927	−0·0384	0·0158	0·0079	0	−0·7732 / −0·7730 / −0·6261	−0·7730 / −0·7730

−0·1260	−0·6260
−0·1260	−0·2794
0·5864	−0·2796
0·5866	0·1694
1·1692	0·1692
1·1692	0·5959
1·4618	0·5958
1·4620	0·8828
1·3828	0·8828
1·3826	0·9499
0·9498	0·9498
0·9496	0·7768

0	0	0	0	0	0
−0·0076	−0·0188	−0·0229	−0·0192	−0·0089	0·0050
−0·0152	−0·0376	−0·0458	−0·0384	−0·0178	0·0100
−0·0675	−0·0728	−0·0539	−0·0164	0·0290	0·0696
−0·1198	−0·1082	−0·0620	0·0056	0·0759	0·1293
−0·1489	−0·0511	0·0652	0·1682	0·2296	0·2324
−0·1780	0·0060	0·1924	0·3307	0·3833	0·3356
0·1610	0·4360	0·5962	0·5984	0·4416	0·1678
0·5	**0·8660**	**1**	**0·8660**	**0·5**	**0**
				30	35

With a little practice this averaging process can be performed in the head, so that the successive rows can be written down immediately. Build-up of rounding errors can be inhibited effectively simply by rounding to the nearest even end digit; thus, for example, we round

$$0·437\,65 \text{ to } 0·4376, \qquad 0·437\,75 \text{ to } 0·4378.$$

2. The difference equation (1.13) relates the values of U at three mesh

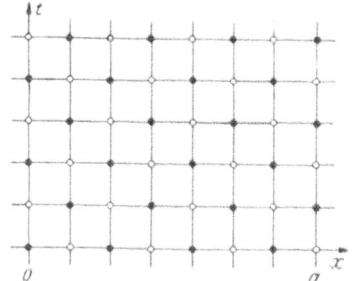

Fig. IV/3. The two interlacing sets of mesh points

points belonging to a sub-set of the whole set of mesh points. There are two such sub-sets (see Fig. IV/3), each complementary to and interlacing the other; they consist of the mesh points with $i+k$ even and odd, respectively, which are marked with black and white spots, respectively, in Fig. IV/3. The calculation need only be carried out for one of these two sub-sets.

Numerical example. Let one end $(x=2)$ of a rod of length 2 with the initial temperature distribution $u(x, 0) = 0$ be kept at this constant temperature while the temperature at the other end $(x=0)$ is varied sinusoidally with period $\frac{3}{4}c$, so that $u(2, t) = 0$ and $u(0, t) = \sin \omega t$, where $\omega = \frac{8\pi}{3c}$. Then with $h = \frac{1}{4}$ the boundary temperature $u(0, t)$ reaches its maximum value of 1 when $t = 6l$, i.e. when $k = 6$. Thus we know the boundary values printed in heavy type in Table IV/1. We can immediately proceed to calculate the values of U for one of the two sub-sets of mesh points, starting from the row $k=0$ and working downwards row by row in accordance with formula (1.13). Thus, for convenience, the time axis in the table points downwards, the opposite direction to that in Fig. IV/3.

An important row-sum check can be derived as follows. Suppose that n is even, say, (as in the numerical example, where $n=8$) and consider the sub-set of mesh points for which $i+k$ is even. Then for even k we add together the equations obtained from (1.13) by putting $i=1, 3, \ldots, n-1$ and find that

$$2 \sum_{i=1}^{\frac{n}{2}} U_{2i-1,\,k+1} = U_{0,k} + 2 \sum_{i=1}^{\frac{n}{2}-1} U_{2i,\,k} + U_{n,k}; \qquad (1.14)$$

a corresponding formula for odd k can be derived by adding the difference equations with $i=2, 4, \ldots, n-2$. If we now introduce quantities S_k and T_k defined by

$$\left. \begin{aligned} S_k = 2 \sum_{i=1}^{\frac{n}{2}-1} U_{2i,k}, \quad & T_k = U_{0,k} + S_k + U_{n,k} \\[1mm] \text{when } k=2s \text{ is even and by} & \\[3mm] \sigma_k = U_{1,k} + U_{n-1,k}, \quad S_k = \sigma_k + 2\sum_{i=1}^{\frac{n}{2}-2} U_{2i+1,k}, \quad & T_k = \sigma_k + S_k \end{aligned} \right\} \qquad (1.15)$$

when $k=2s+1$ is odd, we find that

$$T_{2s} = T_{2s+1}, \qquad S_{2s+1} = S_{2s+2}. \qquad (1.16)$$

Thus the check consists in forming S_k and T_k for each row and observing alternately whether the value of S_k or T_k is the same as for the previous row. A simpler row-sum check is possible when the calculation is carried out for both sub-meshes simultaneously.

The results obtained show that once the effect of the varying end temperature reaches any fixed point of the rod the temperature at that point oscillates with the exciting frequency but with a phase lag (more than half a wavelength near the end $x=2$) and that the amplitude decreases towards the end $x=2$. For small x and large t this behaviour is confirmed by the solution of the heat equation

$$u = e^{-ax} \sin(\omega t - ax),$$

where $a = \sqrt{c\omega/2}$ (see Exercise 1 in § 6.4).

1.3. Error propagation

Now let us try to improve upon the results obtained in § 1.2 by recalculating them from a formula in which the time derivative has been replaced by the more accurate central[1] difference quotient (1.4) instead of the forward difference quotient. Equation (1.11) is now replaced by

$$\frac{U_{i+1,k} - 2U_{i,k} + U_{i-1,k}}{h^2} = c\,\frac{U_{i,k+1} - U_{i,k-1}}{2l}.$$

Hence with the same relation between the mesh widths $l = \frac{1}{2}ch^2$ as in (1.12) we have

$$U_{i,k+1} = U_{i+1,k} - 2U_{i,k} + U_{i-1,k} + U_{i,k-1}. \tag{1.17}$$

With this formula we need starting values on the row $t = l$. We could take over the values already calculated in § 1.2, namely

$$U_{1,1} = U_{2,1} = \cdots = U_{7,1} = 0,$$

but since we are using a more accurate formula, we do not want to lose any accuracy through the propagation of errors due to poor starting values. Consequently we

Table IV/2. *Starting values for the central-difference formula* (1.17)

k'	$i'=0$	$i'=1$	$i'=2$	$i'=3$	$i'=4$	$i'=5$
0	0		0		0	
1		0		0		0
2	0·058 144		0		0	
3		0·029 072		0		0
4	0·116 096		0·014 536		0	
5		0·065 316		0·007 268		0
6	0·173 648		0·036 292		0·003 634	
7		0·104 970		0·019 963		0·001 817
8			0·062 466		0·010 890	
9				0·036 678		

calculate the values on the row $t = l$ by repeating the calculation of § 1.2 with a finer mesh having mesh widths $h' = \frac{1}{3}h$, $l' = \frac{1}{9}l$, which also satisfy the relation (1.12).

This short initial calculation need only be taken as far as is shown in Table IV/2, for we only require the last value:

$$x = 3h', \qquad y = 9l', \qquad U'_{3,9} = 0·036 678.$$

We then put $U_{1,1} = U'_{3,9}$, $U_{2,1} = \cdots = U_{7,1} = 0$ and work downwards row by row using formula (1.17), with the results shown in Table IV/3.

---

[1] The third possibility, namely of using a backward difference quotient for the time derivative, leads to the difference equation

$$\sigma(U_{i+1,k} - 2U_{i,k} + U_{i-1,k}) = U_{i,k} - U_{i,k-1}$$

and is followed up by P. LAASONEN: Über eine Methode zur Lösung der Wärmelei-tungsgleichung. Acta Math. **81**, 309—317 (1949). With this method the solutions of the finite-difference equations converge to the solution of the differential equation for arbitrary, fixed σ, but the calculation is complicated by the fact that one has to solve a system of linear equations to find the values $U_{i,k}$ on each new row.

Table IV/3. *Solution of the heat equation using the central-difference formula* (1.17)

k	i=0	i=1	i=2	i=3	i=4	i=5	i=6	i=7	i=8
0	0	0	0	0	0	0	0	0	0
1	0·25882	0·03668	0	0	0	0	0	0	0
2	0·5	0·18548	0·03668	0	0	0	0	0	0
3	0·70711	0·20244	0·11210	0·03668	0	0	0	0	0
4	0·86603	0·59979	0·05160	0·03874	0·03668	0	0	0	0
5	0·96593	−0·07951	0·64743	0·04748	−0·03462	0·03668	0	0	0
6	1	2·37217	−1·27529	0·55659	0·19008	−0·10798	0·03668	0	0
7	0·96593	−5·09914	6·12677	−2·15091	0·03383	0·47940	−0·18134	0·03668	0

Just these few rows show already that the calculation is quite useless. If we proceed any further, we obtain numbers with alternating signs and assorted magnitudes which bear no relation whatsoever to the solution of the problem. This is caused by a rapid build-up of

Table IV/4. *Propagation table for the difference equation* (1.17)

		x_{i-1}	x_i	x_{i+1}		
0	0	ε	0	0	0	
0	ε	$-2\,\varepsilon$	ε	0	0	
ε	$-4\,\varepsilon$	$7\,\varepsilon$	$-4\,\varepsilon$	ε	0	
$-6\,\varepsilon$	$17\,\varepsilon$	$-24\,\varepsilon$	$17\,\varepsilon$	$-6\,\varepsilon$	ε	
$31\,\varepsilon$	$-68\,\varepsilon$	$89\,\varepsilon$	$-68\,\varepsilon$	$31\,\varepsilon$	$-8\,\varepsilon$	
$-144\,\varepsilon$	$273\,\varepsilon$	$-338\,\varepsilon$	$273\,\varepsilon$	$-144\,\varepsilon$	$49\,\varepsilon$	
$641\,\varepsilon$	$-1096\,\varepsilon$	$1311\,\varepsilon$	$-1096\,\varepsilon$	$641\,\varepsilon$	$-260\,\varepsilon$	

errors, which is a characteristic of formula (1.17). We can analyse such a build-up of errors by assuming that, besides the error already present due to the finite-difference approximation, an additional error ε is made at some stage in the calculation, say at (x_i, y_i). The propagation of this error is easily traced by repeated application of (1.17); thus on the next row $y = y_{k+1}$ it causes additional errors ε, -2ε, ε at $x = x_{i-1}$, x_i, x_{i+1}, which in turn cause additional errors ε, -4ε, 7ε, -4ε, ε at $x = x_{i-2}, \ldots, x_{i+2}$ on the next row, and so on. These additional errors arising from the single error ε can be tabulated in a "propagation table" as in Table IV/4.

Examination of the numbers in the table reveals that the error grows approximately by a factor of four from row to row. Even if the calculation were carried out with so many guarding figures that the propagation of rounding errors could not affect the result, the intrinsic or truncation errors which are inherent in the use of the finite-difference approximations would increase so rapidly in consequence of the unstable character of the difference equation (1.17) that the method would still be quite useless. It could be argued that the presence of the boundaries inhibits the growth of the error, but if we form the propagation table for a single error ε in a boundary value, due to rounding say, we see

Table IV/5. *Propagation of an error on the boundary*

ε	0	0	0	0	0
0	ε	0	0	0	0
0	$-2\,\varepsilon$	ε	0	0	0
0	$6\,\varepsilon$	$-4\,\varepsilon$	ε	0	0
0	$-18\,\varepsilon$	$16\,\varepsilon$	$-6\,\varepsilon$	ε	0
0	$58\,\varepsilon$	$-60\,\varepsilon$	$30\,\varepsilon$	$-8\,\varepsilon$	ε
0	$-194\,\varepsilon$	$224\,\varepsilon$	$-134\,\varepsilon$	$48\,\varepsilon$	$-10\,\varepsilon$

(Table IV/5) that there is no appreciable reduction in the growth of the error, the effect of the boundaries amounting chiefly to a delay of about one row.

In contrast, the propagation table for the difference equation (1.13) (Table IV/6) shows no build-up from an error ε introduced at any point; in fact the propagated errors gradually decrease in magnitude.

Table IV/6. *Stable propagation table for* $U_{i,k+1} = \tfrac{1}{2}\,(U_{i+1,k} + U_{i-1,k})$

0		ε	0	
0	$0{\cdot}5\,\varepsilon$	$0{\cdot}5\,\varepsilon$	$0{\cdot}5\,\varepsilon$	0
	$0{\cdot}25\,\varepsilon$	$0{\cdot}5\,\varepsilon$	$0{\cdot}25\,\varepsilon$	
$0{\cdot}125\,\varepsilon$	$0{\cdot}375\,\varepsilon$	$0{\cdot}375\,\varepsilon$	$0{\cdot}375\,\varepsilon$	$0{\cdot}125\,\varepsilon$
	$0{\cdot}25\,\varepsilon$	$0{\cdot}375\,\varepsilon$	$0{\cdot}25\,\varepsilon$	
$0{\cdot}15625\,\varepsilon$	$0{\cdot}3125\,\varepsilon$	$0{\cdot}3125\,\varepsilon$	$0{\cdot}3125\,\varepsilon$	$0{\cdot}15625\,\varepsilon$
		$0{\cdot}3125\,\varepsilon$		

Summary[1]. If we replace the time derivative in the heat equation by the (crude) forward difference quotient and take $l = \tfrac{1}{2}c\,h^2$, we obtain formula (1.13), whose propagation table shows a gradual decrease in the magnitude of the propagated errors, and the calculation proceeds smoothly (that this method is usable is assured by a convergence proof in § 3.3); if, on the other hand, we replace the time derivative by the central difference quotient, which in itself is more accurate, and use the same mesh relation $l = \tfrac{1}{2}c\,h^2$, we obtain the formula (1.17), whose propagation table shows a rapid increase in the magnitude of the propagated errors and which is therefore unusable ("unstable").

---

[1] The question of error propagation is discussed further in § 3.4.

1.4. Error propagation and the treatment of boundary conditions

In many applications we are not given the boundary values of u itself; often it is the normal derivative of u or a combination of u and its normal derivative which is specified on the boundary. These types of boundary condition can be dealt with in various ways, which we will now illustrate by an example.

A physical problem. Consider an electromagnet whose core is a solid metal cylinder of specific resistance ϱ and permeability μ. The eddy-current density j and the magnetic field strength H directed along the axis satisfy the equations [1]

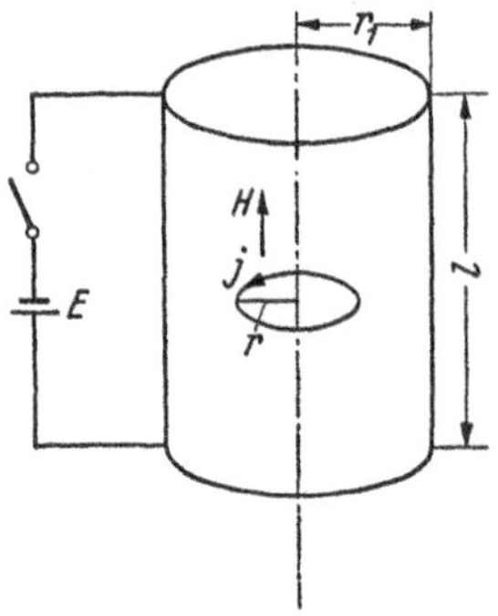

$$4\pi j = -\frac{\partial H}{\partial r}, \qquad \frac{\varrho}{r}\frac{\partial(rj)}{\partial r} = -\mu\frac{\partial H}{\partial t}, \qquad (1.18)$$

where r is the usual cylindrical co-ordinate. Elimination of j yields a second-order parabolic differential equation for H:

$$\frac{\partial^2 H}{\partial r^2} + \frac{1}{r}\frac{\partial H}{\partial r} = \frac{4\pi\mu}{\varrho}\frac{\partial H}{\partial t}. \qquad (1.19)$$

Fig. IV/4. Eddy currents in a metal cylinder

It remains to specify some initial conditions. Let a constant electromotive force E be suddenly switched into the windings circuit at time $t = 0$, there being no current in the coil and no magnetic field in the core before this time (Fig. IV/4). The subsequent eddy-current density j and magnetic field strength H are to be calculated as functions of r and t.

If the core is of radius r_1 and length l and the coil has N uniform windings of ohmic resistance R, the boundary conditions [2] can be written

$$
\left.
\begin{aligned}
t = 0: \quad & H = 0 \quad \text{for} \quad 0 \leqq r \leqq r_1, \\
r = r_1: \quad & H + \beta\frac{\partial H}{\partial r} = \frac{1}{\alpha}E \quad \text{for} \quad t > 0, \\
r = 0: \quad & H,\ \frac{\partial H}{\partial r},\ \frac{\partial^2 H}{\partial r^2} \quad \text{continuous,}
\end{aligned}
\right\} \qquad (1.20)
$$

where

$$\alpha = \frac{Rl}{4\pi N}, \qquad \beta = \frac{2\pi N^2 r_1 \varrho}{Rl}.$$

---

[1] WAGNER, K. W.: Operatorenrechnung, 2nd ed., p. 230 et seq. Leipzig 1950. The equations (1.18) follow from the field equations $\operatorname{curl} H = 4\pi j$, $\varrho\operatorname{curl} j = -\mu\dfrac{\partial H}{\partial t}$ (H is the magnetic field strength, j the current vector) by making use of the axial symmetry.

[2] If U is the induced back e.m.f. in the windings, the effective e.m.f. is $E + U$, which must be equal to the potential drop across the windings: $E + U = RI$. Now U can be calculated from the electric field strength at the surface of the core and hence expressed in terms of the eddy-current density: $U = 2\pi r_1 \varrho N j(r_1)$. Further, the line integral of the magnetic field strength yields $lH(r_1) = 4\pi NI$. Elimination of U, I and $j(r_1)$ from these three equations together with (1.18) leads to the boundary condition of (1.20) at $r = r_1$ (cf. K. W. WAGNER: see last footnote).

For our numerical example we consider the case $\beta = \frac{1}{2} r_1$. With the dimensionless variables

$$x = \frac{r}{r_1}, \qquad y = \frac{\varrho t}{4 \pi \mu r_1^2}, \qquad u = \frac{\alpha}{E} H$$

(1.19) becomes[1]

$$\frac{\partial^2 u}{\partial x^2} + \frac{1}{x} \frac{\partial u}{\partial x} = \frac{\partial u}{\partial y}, \tag{1.21}$$

and if the range of the new variable x is extended to cover the whole axial section, the boundary conditions can be put in the form

$$
\left.
\begin{aligned}
&y = 0: & &u = 0 \quad \text{for} \quad -1 \leq x \leq 1, \\[4pt]
&\begin{array}{c} x = 0: \\ \text{(point of symmetry)} \end{array} & &u, \; \frac{\partial u}{\partial x}, \; \frac{\partial^2 u}{\partial x^2} \; \text{continuous}, \\[4pt]
&x = 1: & &u + \frac{1}{2} \frac{\partial u}{\partial x} = 1, \\[4pt]
&x = -1: & &u - \frac{1}{2} \frac{\partial u}{\partial x} = 1.
\end{aligned}
\right\} \tag{1.22}
$$

The last condition may be omitted if we use the symmetry about $x = 0$ and restrict ourselves to the interval $0 \leq x \leq 1$.

Again we employ a mesh (1.1) with $y_0 = 0$ and $2l = h^2$, as in (1.12); and, after our experience in § 1.3, we replace the derivative in the time direction $\left(\frac{\partial u}{\partial y}\right)_{i,k}$ by the forward difference quotient $\frac{1}{l}(U_{i,k+1} - U_{i,k})$. The difference equation does not depend so critically on the radial derivative $\left(\frac{\partial u}{\partial x}\right)_{i,k}$ and we therefore replace this derivative by the central difference quotient $\frac{1}{2h}(U_{i+1,k} - U_{i-1,k})$. The resulting difference

---

[1] A method which is much used, particularly by electrical engineers, for the solution of initial-value and mixed initial-/boundary-value problems with linear differential equations (mostly with constant coefficients and infinite fundamental regions) is the operational method using the Laplace transformation; see, for example, G. DOETSCH: Theorie und Anwendung der Laplace-Transformation. Berlin 1937, 436 pp. — Tabellen zur Laplace-Transformation und Anleitung zum Gebrauch. Berlin and Göttingen 1947, 185 pp. — CHURCHILL, R. V.: Modern operational mathematics in engineering. New York and London 1944, 306 pp. — WAGNER, K. W.: Operatorenrechnung und Laplacesche Transformation nebst Anwendung in Physik und Technik, 2nd ed. Leipzig 1950, 489 pp. — CARSLAW, H. S., and J. C. JAEGER: Conduction of heat in solids. Oxford 1948, 386 pp., here in particular pp. 239—290, 320—338.

From a numerical point of view (for problems not soluble in closed form) the Laplace transform method is only suitable for a restricted range of problems. For some problems it leads to results which can be obtained just as well by other means and for others to series expansions whose numerical evaluation is often merely tedious. Future developments will show whether or not the frequently expressed hopes of the Laplace transform are justified.

Another method for the solution of similar problems is the "mixed" Ritz method, see Ch. V, § 5.9.

equation reads

$$U_{i,k+1} = \frac{1}{2}(U_{i+1,k} + U_{i-1,k}) + \frac{h}{4x_i}(U_{i+1,k} - U_{i-1,k}). \qquad (1.23)$$

The boundary conditions can be taken into account in several ways:

1. Choose a mesh with the boundary $x = 1$ as a mesh line, say $x = x_n$, and replace the derivative in the boundary condition by the backward difference quotient. This gives the "finite" boundary condition

$$U_{n,k} + \frac{1}{2h}(U_{n,k} - U_{n-1,k}) = 1,$$

so that

$$U_{n,k} = \frac{U_{n-1,k} + 2h}{1 + 2h}. \qquad (1.24)$$

2. Again choose a mesh with $x_n = 1$ but replace the boundary derivative by the central difference quotient. This entails keeping a column

Table IV/7. *Propagation tables for various boundary formulae at* $x = 1$

Formula (1.24)			Formula (1.25)				Formula (1.26)		
$x = \frac{1}{3}$	$x = \frac{2}{3}$	$x = 1$	$x = \frac{1}{3}$	$x = \frac{2}{3}$	$x = 1$	$x = \frac{4}{3}$	$x = \frac{2}{5}$	$x = \frac{4}{5}$	$x = \frac{6}{5}$
0	0	ε	0	0	ε	$-1{\cdot}333\,\varepsilon$	0	0	ε
0	$0{\cdot}625\,\varepsilon$	$0{\cdot}375\,\varepsilon$	0	$0{\cdot}625\,\varepsilon$	$-0{\cdot}778\,\varepsilon$	$1{\cdot}662\,\varepsilon$	0	$0{\cdot}625\,\varepsilon$	$0{\cdot}268\,\varepsilon$
$0{\cdot}469\,\varepsilon$	$0{\cdot}234\,\varepsilon$	$0{\cdot}141\,\varepsilon$	$0{\cdot}469\,\varepsilon$	$-0{\cdot}486\,\varepsilon$	$1{\cdot}229\,\varepsilon$	$-2{\cdot}125\,\varepsilon$	$0{\cdot}469\,\varepsilon$	$0{\cdot}167\,\varepsilon$	$0{\cdot}072\,\varepsilon$

of values of $U_{n+1,k}$. If the k-th row is completed, $U_{n,k+1}$ is calculated from (1.23) with $i = n$, then $U_{n+1,k+1}$ is calculated from the "finite" boundary condition

$$U_{n,k+1} + \frac{1}{2}\frac{1}{2h}(U_{n+1,k+1} - U_{n-1,k+1}) = 1,$$

which is used in the form

$$U_{n+1,k+1} = 4h(1 - U_{n,k+1}) + U_{n-1,k+1}. \qquad (1.25)$$

3. Choose a mesh for which the boundary $x = 1$ lies halfway between two mesh lines, say $x = x_n$ and $x = x_{n+1}$, and use $\frac{1}{2}(U_{n,k} + U_{n+1,k})$ as the approximation for u at $x = 1$. Replacing $\partial u / \partial x$ in the boundary condition by the central difference quotient we find that $U_{n+1,k}$ is given by

$$U_{n+1,k} = \frac{2h + U_{n,k}(1 - h)}{1 + h}. \qquad (1.26)$$

The propagation table for a boundary error depends on which of these methods we adopt. Table IV/7 shows the results of using (1.24) and (1.25) with $h = \frac{1}{3}$ and (1.26) with $h = \frac{2}{5}$ [the corresponding homogeneous equations must be used in deriving the error propagation; for example, (1.25) must be used without the term $4h \times 1$ on the right-hand side]. It can be seen that (1.25) compares unfavourably with the crude approximation represented by (1.24). In fact, if we try to calculate the solution using (1.25), the increasing randomness of the results (see Table IV/8) soon convinces us that the use of this apparently reasonable

Table IV/8. *Invalid calculation using the boundary formula* (1.25)

k	$x = 0$	$x = \frac{1}{3}$	$x = \frac{2}{3}$	$x = 1$	$x = \frac{4}{3}$
0	0	0	0	0	0
1	0	0	0	0	1·3333
2	0	0	0	0·7778	0·2963
3	0	0	0·4861	0·1728	1·5890
4		0·3646	0·0907	1·1295	−0·0820
5			0·8427	−0·0100	2·1893

method of dealing with the boundary condition has rendered the calculation quite useless. Of the other two methods (1.24) and (1.26), the latter has the advantage of the accuracy of the central difference quotient.

The other boundary $x = 0$ can also be dealt with in various ways:

1. Using a mesh with $x = 0$ as a mesh line. Since

$$\lim_{x \to 0} \frac{\partial u/\partial x}{x} = \lim_{x \to 0} \frac{\partial^2 u/\partial x^2}{1} = \left(\frac{\partial^2 u}{\partial x^2} \right)_{x=0}, \tag{1.27}$$

the differential equation (1.21) becomes

$$2 \frac{\partial^2 u}{\partial x^2} = \frac{\partial u}{\partial y}$$

as $x \to 0$. We now require a corresponding difference equation which does not lead to instability. From our experience in § 1.3 we try replacing $(\partial u/\partial y)_{0,k}$ by the forward difference quotient $\frac{1}{l}(U_{0,k+1} - U_{0,k})$, but find that the difference equation so obtained, namely

$$U_{0,k+1} = 2 U_{1,k} - U_{0,k} \tag{1.28}$$

(in which we have used the symmetry $U_{1,k} = U_{-1,k}$), propagates a boundary error with increasing magnitude (see Table IV/9). It turns out that the backward difference quotient $\frac{1}{l}(U_{0,k} - U_{0,k-1})$ is the best to use here, as can be seen from the propagation table (Table IV/9) for the resulting difference equation

$$U_{0,k} = \frac{1}{3}(2 U_{1,k} + U_{0,k-1}). \tag{1.29}$$

2. Using a mesh such that the boundary $x = 0$ lies halfway between the two mesh lines $x = -\frac{1}{2}h$ and $x = +\frac{1}{2}h$. With such a mesh we do not need any special

18*

Table IV/9. *Propagation tables for boundary formulae at* $x = 0$

Formula (1.28)			Formula (1.29)		
$x = -h$	$x = 0$	$x = h$	$x = -h$	$x = 0$	$x = h$
0	ε	0	0	ε	0
$0.25\,\varepsilon$	$-\varepsilon$	$0.15\,\varepsilon$	$0.25\,\varepsilon$	$0.333\,\varepsilon$	$0.25\,\varepsilon$
$-0.25\,\varepsilon$	$1.5\,\varepsilon$	$-0.25\,\varepsilon$	$0.083\,\varepsilon$	$0.277\,\varepsilon$	$0.083\,\varepsilon$
$0.445\,\varepsilon$	$-2\,\varepsilon$	$0.445\,\varepsilon$		$0.148\,\varepsilon$	
$2.891\,\varepsilon$					

boundary equations, we merely simplify the difference equation (1.23) by utilizing the symmetry about $x = 0$. This is in fact the method we adopt in the calculation below (Table IV/10) with $h = \frac{1}{4}$.

Having decided on the best equations to use, we can proceed to calculate $U_{i,k}$ row by row from (1.23) for $i = 1, 2, \ldots, n-1$ and from (1.26) for $i = n$.

Table IV/10. *Valid calculation of the growth of a magnetic field in a metal cylinder*

k	y	1 $\frac{1}{3}$ $\frac{2}{3}$ $i = 1$ $x = 0.125$	$\frac{2}{5}$ $\frac{3}{5}$ $i = 2$ $x = 0.375$	$\frac{3}{7}$ $\frac{4}{7}$ $i = 3$ $x = 0.625$	$\frac{3}{8} + 0.4$ $i = 4$ $x = 0.875$	$i = 5$ $x = 1.125$	Row-sum check ϱ_k	σ_k
0	0	0	0	0	0	0		
1		0	0	0	0	0.4	4.8	
2		0	0	0	0.22857	0.53714	8.50281	-0.00007
3		0	0	0.13714	0.30694	0.58416	11.82948	$+$ 05
4	0.125	0	0.09143	0.18416	0.39258	0.63555	14.74509	$+$ 01
5		0.09143	0.12277	0.27212	0.44210	0.66526	17.42304	$+$ 03
6		0.12277	0.21189	0.31437	0.49677	0.69806	19.83852	$-$ 04
7		0.21189	0.25050	0.38282	0.53362	0.72017	22.07709	$-$ 07
8	0.250	0.25050	0.32584	0.42037	0.57559	0.74535	24.11412	$-$ 17
9		0.32584	0.36375	0.47569	0.60607	0.76364	26.00493	$-$ 07
10		0.36375	0.42574	0.50914	0.64023	0.78414	27.73176	$-$ 05
11		0.42574	0.46068	0.55443	0.66628	0.79977	29.33355	$-$ 05

These equations are best written out explicitly. For $h = \frac{1}{4}$ they read

$$\bar{u}_1 = u_2, \quad \bar{u}_2 = \tfrac{1}{3}(u_1 + 2u_3), \quad \bar{u}_3 = \tfrac{1}{5}(2u_2 + 3u_4), \quad \bar{u}_4 = \tfrac{1}{7}(3u_3 + 4u_5), \quad \bar{u}_5 = \tfrac{1}{8}(2 + 3\bar{u}_4),$$

where we have used the more concise notation u_i for $U_{i,k}$ and $\bar{u}_i$ for $U_{i,k+1}$. The coefficients can be conveniently recorded at the heads of the columns as in Table IV/10.

As a row-sum check we can form

$$\varrho_k = 3(u_1 + 3u_2 + 5u_3 + 3u_4 + 4u_5)$$

and

$$\sigma_k = \varrho_k - \varrho_{k-1} - 8(1 - u_5)$$

and check that σ_k vanishes.

1.5. Hyperbolic differential equations

Using a rectangular mesh (1.1) and the approximations (1.4), (1.6) we can represent the differential equation

$$a \frac{\partial^2 u}{\partial x^2} + c \frac{\partial^2 u}{\partial y^2} + d \frac{\partial u}{\partial x} + e \frac{\partial u}{\partial y} + f u = r, \tag{1.30}$$

where a, c, d, e, f, r are given functions of x and y with $a > 0$, $c < 0$, by the difference equation

$$\left. \begin{aligned} a_{ik} \frac{U_{i+1,k} - 2U_{i,k} + U_{i-1,k}}{h^2} + d_{ik} \frac{U_{i+1,k} - U_{i-1,k}}{2h} + \\ + c_{ik} \frac{U_{i,k+1} - 2U_{i,k} + U_{i,k-1}}{l^2} + e_{ik} \frac{U_{i,k+1} - U_{i,k-1}}{2l} + f_{ik} U_{i,k} = r_{ik}. \end{aligned} \right\} \tag{1.31}$$

For the wave equation

$$\frac{\partial^2 u}{\partial x^2} = \frac{1}{\omega^2} \frac{\partial^2 u}{\partial y^2} \tag{1.32}$$

and with the mesh relation

$$h = \omega l \tag{1.33}$$

this difference equation simplifies to

$$U_{i,k+1} = U_{i+1,k} + U_{i-1,k} - U_{i,k-1}. \tag{1.34}$$

We now consider some examples of the various types of initial- and boundary-value problems which can arise.

1. Values of u and $\partial u/\partial y$ specified on the whole x axis:

$$u = G(x), \quad \frac{\partial u}{\partial y} = H(x) \quad \text{for} \quad -\infty < x < +\infty, \; y = 0. \tag{1.35}$$

In this case we know immediately the values $U_{i,0} = G(ih)$ [for a mesh (1.1) with $x_0 = y_0 = 0$]; we can calculate the values $U_{i,1}$ on the neighbouring row $k = 1$ as follows. Write (1.31) for $k = 0$ in the form

$$U_{i,1} \left(\frac{c_{i0}}{l^2} + \frac{e_{i0}}{2l} \right) + U_{i,-1} \left(\frac{c_{i0}}{l^2} - \frac{e_{i0}}{2l} \right) = r_i^*, \tag{1.36}$$

where the r_i^* are known and given by

$$r_i^* = - a_{i0} \frac{U_{i+1,0} - 2U_{i,0} + U_{i-1,0}}{h^2} - d_{i0} \frac{U_{i+1,0} - U_{i-1,0}}{2h} +$$
$$+ 2c_{i0} \frac{U_{i,0}}{l^2} - f_{i0} U_{i,0} + r_{i0};$$

from the boundary condition $\partial u/\partial y = H(x)$ we have

$$U_{i,1} - U_{i,-1} = 2l H(ih); \tag{1.37}$$

then elimination of $U_{i,-1}$ between (1.36) and (1.37) yields

$$U_{i,1} = l\,H(i\,h)\left[1 - \frac{l}{2}\frac{e_{i0}}{c_{i0}}\right] + \frac{l^2}{2}\frac{r_i^*}{c_{i0}}. \qquad (1.38)$$

We now have starting values for the difference equation (1.31), by means of which we can calculate successively the remaining values of $U_{i,k}$ for $k = 2, 3, 4, \ldots$.

2. Values of u specified on the boundary shown in Fig. IV/5 as a heavy line and values of $\partial u/\partial y$ specified on the part of that boundary which lies along the x axis:

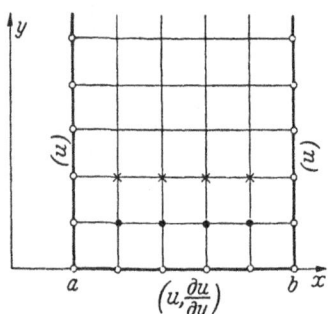

Fig. IV/5. Boundary of the solution domain for the problem (1.30), (1.39)

$$\left.\begin{aligned}
u &= G(x) \\
\frac{\partial u}{\partial y} &= H(x)
\end{aligned}\right\} \quad \begin{aligned} &\text{for } a \leqq x \leqq b, \\ &\qquad y = 0, \end{aligned}$$

$$u = \varphi(y) \quad \text{for} \quad \begin{aligned} x &= a, \\ y &> 0, \end{aligned}$$

$$u = \psi(y) \quad \text{for} \quad \begin{aligned} x &= b, \\ y &> 0. \end{aligned}$$

$$(1.39)$$

For a mesh (1.1) with

$$x_0 = a, \quad y_0 = 0, \quad h = \frac{b-a}{n},$$

where n is an integer greater than one, we know the $U_{i,k}$ at the boundary points ("white points" in the figure) and can calculate the values $U_{i,1}$ (at the "black points" in the figure) from (1.38). The remaining points are dealt with using (1.31).

3. Other types of boundary conditions often occur in applications; for example, we could have

$$\alpha\,u + \beta\,\frac{\partial u}{\partial x} = \varphi(y) \quad \text{for} \quad x = a, \quad y > 0$$

instead of $u = \varphi(y)$ as in 2. Boundary conditions of this kind can be treated by the methods of § 1.4.

1.6. A numerical example

A uniform circular membrane of radius a is fixed rigidly around its circumference and is deformed[1] into a position defined by the displacement

$$g(r) = v_0\left(1 - \frac{r^2}{a^2}\right),$$

where r is the distance from the centre in the mean plane of the membrane and v_0 is a constant (the maximum displacement of the centre of the membrane). When

---

[1] Cf. K. W. WAGNER: Operatorenrechnung, 2nd ed., p. 243. Leipzig 1950.

the membrane is released (at time $t = 0$), vibrations are set up. The displacement $v(r, t)$ at subsequent times is to be calculated. It satisfies the well-known wave equation, which we write down directly in polar co-ordinates:

$$\nabla^2 v - \frac{1}{c^2} \frac{\partial^2 v}{\partial t^2} = \frac{\partial^2 v}{\partial r^2} + \frac{1}{r} \frac{\partial v}{\partial r} - \frac{1}{c^2} \frac{\partial^2 v}{\partial t^2} = 0. \qquad (1.40)$$

The constant c denotes the wave velocity.

In the dimensionless co-ordinates

$$u = \frac{v}{v_0}, \qquad x = \frac{r}{a}, \qquad y = \frac{c\,t}{a}$$

(1.40) reads

$$\frac{\partial^2 u}{\partial x^2} + \frac{1}{x} \frac{\partial u}{\partial x} = \frac{\partial^2 u}{\partial y^2} \qquad (1.41)$$

and the boundary conditions become

$$u(x, 0) = 1 - x^2, \frac{\partial u}{\partial y} = 0 \qquad \text{on} \quad y = 0, \quad |x| \leqq 1 \quad \text{(initial conditions)},$$

$$u = 0 \text{ (no displacement)} \quad \text{on} \quad x = \pm 1, \quad y > 0 \quad \text{(the fixed boundary)}.$$

The symmetry about the y axis allows us to restrict the calculation to the half region with $x \geq 0$. Then for a square mesh with $h = l = \frac{1}{5}$ we know the boundary values printed in heavy type in Table IV/12 and can proceed using the equations

$$\left. \begin{array}{l} U_{i,1} = \dfrac{1}{2} U_{i+1, 0} \left(1 + \dfrac{1}{2i} \right) + \dfrac{1}{2} U_{i-1, 0} \left(1 - \dfrac{1}{2i} \right), \\[2mm] U_{i, k+1} = U_{i+1, k} \left(1 + \dfrac{1}{2i} \right) + U_{i-1, k} \left(1 - \dfrac{1}{2i} \right) - U_{i, k-1} \\[2mm] \qquad (i = 1, \ldots, n-1; \; k = 1, 2, \ldots). \end{array} \right\} \qquad (1.42)$$

These suffice until we come to deal with values on the y axis; these require special treatment because of the singularity at $x = 0$, or $i = 0$. At this point the differential equation (1.41) becomes

$$2 \frac{\partial^2 u}{\partial x^2} = \frac{\partial^2 u}{\partial y^2}$$

on account of (1.27), and the corresponding difference equation reads

Table IV/11. *Propagation table for* (1.43)

$i = -1$	$i = 0$	$i = 1$
0	ε	0
$\varepsilon/2$	-2ε	$\varepsilon/2$
$-\varepsilon$	7ε	$-\varepsilon$
	-20ε	

$$U_{0, k+1} = 2(U_{1, k} - U_{0, k} + U_{-1, k}) - U_{0, k-1}. \qquad (1.43)$$

However, this formula has a very unfavourable propagation table (see Table IV/11) and in fact an attempt at using it yields quite useless results after a few rows.

The simplest way of getting over this difficulty is not to use the differential equation at all, but to put a parabola through the points with $i = \pm 1, \pm 2$ and use its value at $i = 0$ for $U_{0, k}$:

$$U_{0, k} = \frac{4 U_{1, k} - U_{2, k}}{3} = U_{1, k} + \frac{1}{3} (U_{1, k} - U_{2, k}).$$

The last two columns in Table IV/12 provide a simple check which can be applied after each row has been completed. We calculate the

Table IV/12. *Finite-difference approximation for the vibrations of a circular membrane*

β_i	$\beta_0=0.5$	$\beta_1=0.75$	$\beta_2=2.333\,33$	$\beta_3=2.125$	$\beta_4=1.166\,67$		Row-sum check	
$1\pm\dfrac{1}{2i}$	$\tfrac{1}{2}\quad\tfrac{3}{2}$	$\tfrac{3}{4}\quad\tfrac{5}{4}$	$\tfrac{5}{6}\quad\tfrac{7}{6}$	$\tfrac{7}{8}$				
y \ k	$i=0$ $x=0$	$i=1$ $x=0.2$	$i=2$ $x=0.4$	$i=3$ $x=0.6$	$i=4$ $x=0.8$	$i=5$ $x=1$	σ_k	τ_k
0 \ 0	1	0·96	0·84	0·64	0·36	0		2·80
0·2 \ 1	0·92	0·88	0·76	0·56	0·28	0	4·41	2·48
0·4 \ 2	0·68	0·64	0·52	0·32	0·13	0	2·865	1·61
0·6 \ 3	0·28	0·24	0·12	0·025	0	0	0·653 12	0·385
0·8 \ 4	−0·323 75	−0·32	−0·308 75	−0·220	−0·108 13	0	−1·715 94	−0·956 88
1 \ 5	−0·941 67	−0·865 00	−0·635 00	−0·408 44	−0·192 50	0	−3·693 77	−2·100 94
\ 6	−1·187 60	−1·103 34	−0·850 55	−0·533 74	−0·249 26	0	−4·830 92	−2·736 89
\ 7	−1·052 95	−1·004 63	−0·859 68	−0·591 15	−0·274 52	0	−4·862 33	−2·729 98
\ 8	−0·736 26	−0·712 66	−0·641 86	−0·502 93	−0·268 00	0	−3·781 69	−2·125 45
\ 9	−0·333 89	−0·326 29	−0·303 48	−0·256 40	−0·165 54	0	−1·857 76	−1·051 71
2 \ 10	+0·095 11	0·090 49	0·076 64	0·056 90	0·043 65	0	0·466 09	0·267 68
\ 11	0·504 25	0·488 81	0·442 47	0·371 19	0·215 33	0	2·691 16	1·517 80
\ 12	0·849 13	0·825 34	0·753 96	0·563 04	0·281 14	0	4·327 27	2·423 48
\ 13	1·128 82	1·066 70	0·880 34	0·585 11	0·277 33	0	4·985 47	2·809 48
\ 14	1·153 62	1·059 58	0·777 45	0·494 13	0·230 83	0	4·504 87	2·561 99
3 \ 15	0·724 38	0·676 29	0·532 01	0·332 06	0·155 03	0		1·695 39

quantities

$$\sigma_k = \sum_{i=0}^{n-1} \beta_i\, U_{i,k}, \qquad \tau_k = \sum_{i=1}^{n-1} U_{i,k},$$

where

$$\beta_0 = \frac{1}{2}, \qquad \beta_1 = \frac{3}{4}, \qquad \beta_i = 2 + \frac{1}{i^2 - 1} \quad \text{for} \quad i = 2, 3, \ldots, n-2,$$

$$\beta_{n-1} = 1 + \frac{1}{2(n-2)},$$

and check that

$$\tau_{k+1} = \sigma_k - \tau_{k-1} \qquad (k = 1, 2, \ldots).$$

The numbers β_i and the coefficients $\left(1 - \dfrac{1}{2i}\right)$ and $\left(1 + \dfrac{1}{2i}\right)$ in (1.42) may be recorded conveniently at the heads of the appropriate columns.

1.7. Graphical treatment of parabolic differential equations by the finite-difference method

It is often convenient to apply the ordinary finite-difference method in graphical form[1]. For instance, the process of forming mean values

---

[1] Although it is not intended that this book should also cover graphical methods, we give here a brief description of this method for parabolic equations because of the many applications.

in the treatment of the heat equation

$$\frac{\partial^2 u}{\partial x^2} = c\,\frac{\partial u}{\partial y} \tag{1.44}$$

by the difference equation (1.13) is very easily performed graphically. The approximations $U_{i,k}$ are plotted against x, the points with equal k defining a sequence of polygons.

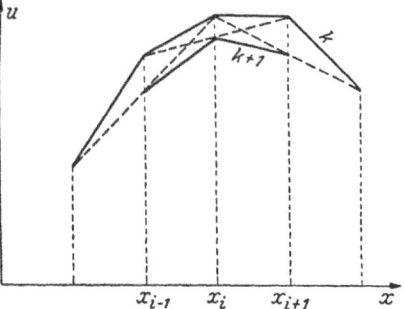

The $(k+1)$-th polygon is formed from the k-th by joining alternate points of the latter and taking the midpoints of these joins, i.e. the points of intersection with the ordinate lines, as the vertices of the next polygon (see Fig. IV/6)[1].

This construction is applied to an example in Fig. IV/7; the specific boundary values chosen are

Fig. IV/6. Graphical construction for the solution of the heat equation

$$\left.\begin{array}{l} u(x, 0) = 0 \quad \text{for} \quad -1 \leqq x \leqq 1, \\[2mm] u(1, t) = u(-1, t) = 1 - e^{-\frac{5}{c}t}, \end{array}\right\} \tag{1.45}$$

which correspond to a rod being heated from zero temperature by the application of heat to its ends so that their common temperature rises exponentially to the value 1. The graphical solution is carried out with $h = \frac{1}{5}$ and therefore $l = c/50$ in accordance with (1.12). Only the region $x \geqq 0$ is considered on account of symmetry; only the construction lines are shown in Fig. IV/7 since the sides of the polygons, which are shown in Fig. IV/6, would have confused the picture. The values of k are marked at various points; the corresponding time can be calculated from $t = k\,l$.

If we choose a mesh with $2l = c_k h^2 = c(y_k) h^2$, the more general differential equation

$$\frac{\partial^2 u}{\partial x^2} + f(x, y)\,\frac{\partial u}{\partial x} = c(y)\,\frac{\partial u}{\partial y} + a(x, y) + b(x, y)\,u \tag{1.46}$$

is represented by the difference equation

$$\left.\begin{array}{l} U_{i, k+1} = \dfrac{1}{2}\left(1 + \dfrac{h f_{i,k}}{2}\right) U_{i+1, k} + \\[4mm] \qquad\quad + \dfrac{1}{2}\left(1 - \dfrac{h f_{i,k}}{2}\right) U_{i-1, k} - \dfrac{h^2}{2}\,(a_{ik} + b_{ik}\,U_{i, k}). \end{array}\right\} \tag{1.47}$$

We can still utilize the chord construction but here we use the ordinate cut off on the line $x = \xi_{i, k} = x_i + \frac{1}{2} h^2 f_{i, k}$ instead of the line $x = x_i$ (see

---

[1] SCHMIDT, E.: Einführung in die technische Thermodynamik, 2nd ed., p. 282. Berlin 1944.

Fig. IV/8). This ordinate is $\frac{1}{2}(1+\frac{1}{2}hf_{i,k})\,U_{i+1,k}+\frac{1}{2}(1-\frac{1}{2}hf_{i,k})\,U_{i-1,k}$; we can obtain $U_{i,k+1}$ from it by laying off $\frac{1}{2}h^2(a_{ik}+b_{ik}U_{i,k})$. The construction is particularly simple when $f(x,y)$ depends only upon x, for

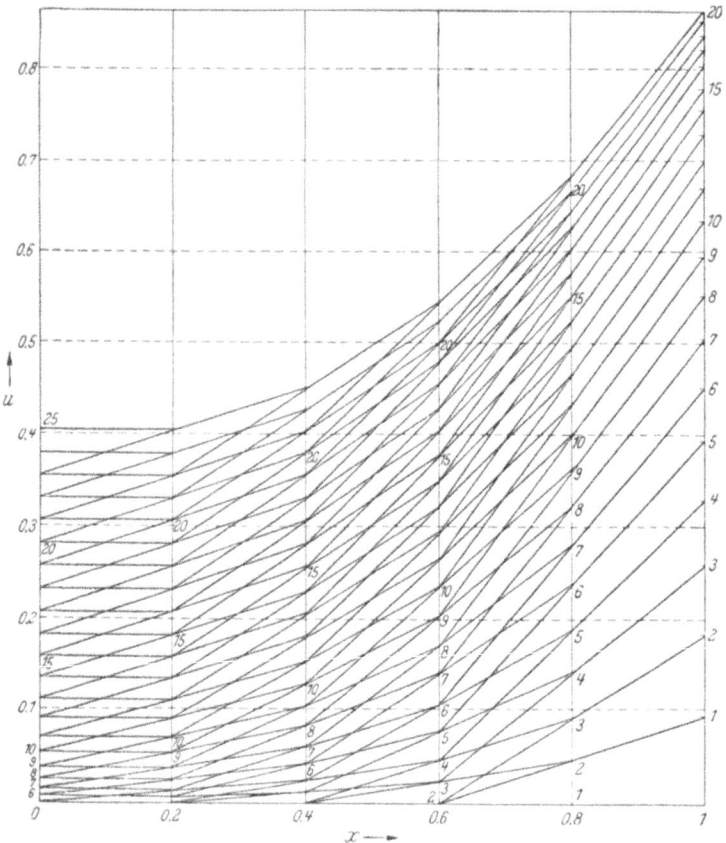

Fig. IV/7. Graphical solution for the varying temperature distribution in a heated rod

then one set of lines $x=\xi_i$ can be used for all k. The variable quantity $c(y)$ does not come into the construction but is needed when we want to find the value of $y=y_k$ corresponding to a particular value of k. This is given by

$$y_k = \frac{h^2}{2}\sum_{\varkappa=0}^{k-1}c_\varkappa.$$

Fig. IV/9 shows the construction for the example (1.21), (1.22) of § 1.4, which concerned the eddy currents in the metal core of an electromagnet. The special construction for the boundary conditions is described in the next paragraph.

If the values of u are prescribed at a boundary, say $x=\alpha$, then a mesh is chosen with $x=\alpha$ as a mesh line. On the other hand, if the boundary condition at $x=\alpha$

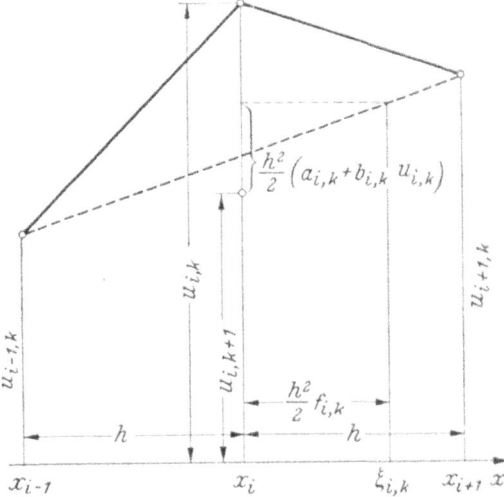

Fig. IV/8. Graphical construction for the more general differential equation (1.46)

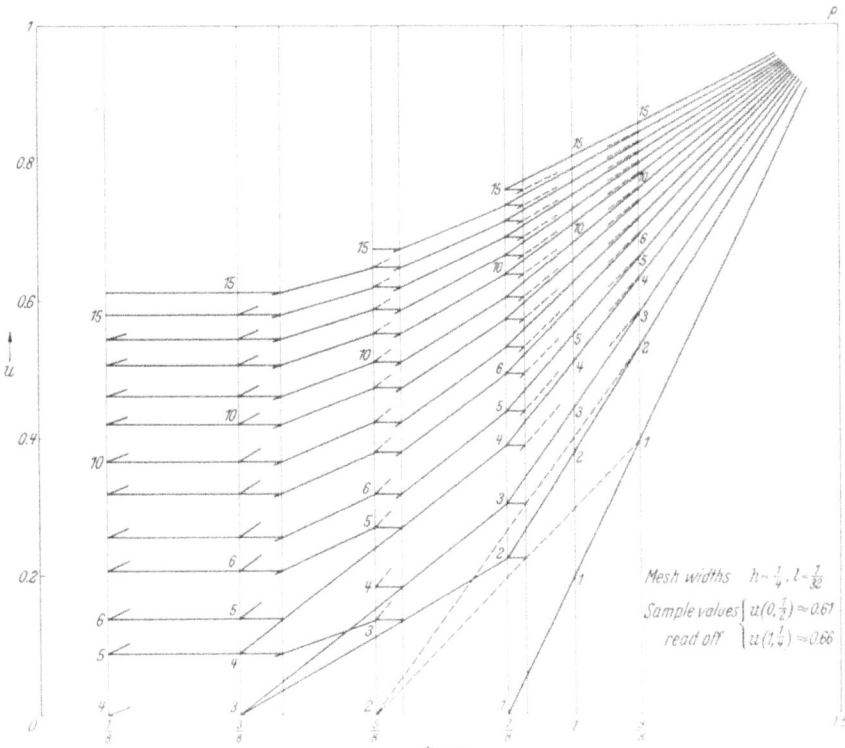

Fig. IV/9. Graphical calculation of the growth of a magnetic field in a metal cylinder

is of the form

$$u + A(y)\,\frac{\partial u}{\partial x} = B(y) \qquad (1.48)$$

with $A(y) \neq 0$, then a mesh is chosen so that $x = \alpha$ lies halfway between the two mesh lines $x = x_n$ and $x = x_{n+1}$, i.e. so that $\alpha = \frac{1}{2}(x_n + x_{n+1})$; one can then replace the x derivative by the central difference quotient, as in (1.26), which yields

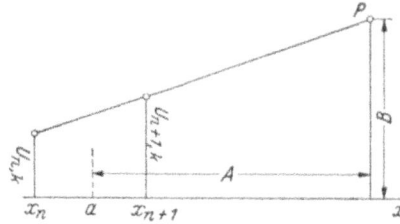

$$\left.\begin{aligned} U_{n+1,\,k}\\ = \frac{2Bh + (2A - h)\,U_{n,\,k}}{2A + h} \end{aligned}\right\} \quad (1.49)$$

This equation implies that the point P with co-ordinates $(\alpha + A,\ B)$ is collinear with the points $(x_n,\ U_{n,\,k})$ and $(x_{n+1},\ U_{n+1,\,k})$, as in Fig. IV/10;

Fig. IV/10. Graphical representation of a boundary condition involving a normal derivative

hence, knowing $U_{n,\,k}$, we can construct $U_{n+1,\,k}$. If A and B do not depend on y, P is the same point for all k (as for the example in Fig. IV/9).

If u does not occur in the boundary condition (1.48), i.e. $\partial u/\partial x$ is prescribed, then $U_{n+1,\,k}$ can be constructed from $U_{n,\,k}$ merely by drawing a line in the specified direction.

1.8. The two-dimensional heat equation

When we consider the equation of heat flow in two dimensions

$$\frac{\partial^2 u}{\partial x^2} + \frac{\partial^2 u}{\partial y^2} = c\,\frac{\partial u}{\partial z}, \qquad (1.50)$$

we find that an approximate finite-difference solution can still be obtained by an averaging procedure and hence by a simple graphical construction on a plane projection of the $(x,\ y,\ u)$ space.

We use a three-dimensional mesh

$$\left.\begin{aligned} x &= x_0 + i h_x\\ y &= y_0 + k h_y\\ z &= z_0 + l h_z \end{aligned}\right\} \quad (i,\ k,\ l = 0,\ \pm 1,\ \pm 2, \ldots) \qquad (1.51)$$

with mesh widths

$$h_x = h_y = h, \qquad h_z = c\,\varrho\,h^2. \qquad (1.52)$$

The corresponding "forward" difference equation reads

$$\left.\begin{aligned} U_{i,\,k,\,l+1} = \varrho\,(U_{i+1,\,k,\,l} + U_{i-1,\,k,\,l} + U_{i,\,k+1,\,l} + U_{i,\,k-1,\,l}) +\\ + (1 - 4\varrho)\,U_{i,\,k,\,l}. \end{aligned}\right\} \quad (1.53)$$

With $\varrho = \frac{1}{2}$, the value used in (1.13), we have the formula[1]

$$U_{i,\,k,\,l+1} = \frac{1}{2}S - U_{i,\,k,\,l}, \qquad (1.54)$$

[1] ELSER, K.: Schweizer Arch. Angew. Wiss. Techn. **10**, 341—343 (1944).

$l = 0$

0	0	0
0	ε	0
0	0	0
0	0	0

$l = 1$

0	$\frac{1}{2}\varepsilon$	0
$\frac{1}{2}\varepsilon$	$-\varepsilon$	$\frac{1}{2}\varepsilon$
0	$\frac{1}{2}\varepsilon$	0
0	0	0

$l = 2$

$\frac{1}{2}\varepsilon$	$-\varepsilon$	$\frac{1}{2}\varepsilon$	0
$-\varepsilon$	2ε	$-\varepsilon$	$\frac{1}{4}\varepsilon$
$\frac{1}{2}\varepsilon$	$-\varepsilon$	$\frac{1}{2}\varepsilon$	0
0	$\frac{1}{4}\varepsilon$	0	0

$l = 3$

$-\frac{3}{2}\varepsilon$	$\frac{21}{8}\varepsilon$	$-\frac{3}{2}\varepsilon$	$\frac{3}{8}\varepsilon$	0
$\frac{21}{8}\varepsilon$	-4ε	$\frac{21}{8}\varepsilon$	$-\frac{3}{4}\varepsilon$	$\frac{1}{8}\varepsilon$
$-\frac{3}{2}\varepsilon$	$\frac{21}{8}\varepsilon$	$-\frac{3}{2}\varepsilon$	$\frac{3}{8}\varepsilon$	0
$\frac{3}{8}\varepsilon$	$-\frac{3}{4}\varepsilon$	$\frac{3}{8}\varepsilon$	0	0

$l = 4$

$\frac{9}{2}\varepsilon$	$-\frac{13}{2}\varepsilon$	$\frac{9}{2}\varepsilon$	$-\frac{3}{2}\varepsilon$	$\frac{1}{4}\varepsilon$	0
$-\frac{13}{2}\varepsilon$	$\frac{37}{4}\varepsilon$	$-\frac{13}{2}\varepsilon$	$\frac{5}{2}\varepsilon$	$-\frac{1}{2}\varepsilon$	$\frac{1}{16}\varepsilon$
$\frac{9}{2}\varepsilon$	$-\frac{13}{2}\varepsilon$	$\frac{9}{2}\varepsilon$	$-\frac{3}{2}\varepsilon$	$\frac{1}{4}\varepsilon$	0
$-\frac{3}{2}\varepsilon$	$\frac{5}{2}\varepsilon$	$-\frac{3}{2}\varepsilon$	$\frac{3}{8}\varepsilon$	0	0

where

$$S = U_{i+1,k,l} + U_{i-1,k,l} + \left. + U_{i,k+1,l} + U_{i,k-1,l}. \right\} \quad (1.55)$$

Its error propagation, shown in the sequence of tables above, renders it practically useless; an error ε is approximately doubled at each step in the z direction.

However, putting $\varrho = \frac{1}{4}$ in (1.53) yields a usable formula[1] which is also very simple:

$$U_{i,k,l+1} = \tfrac{1}{4}S, \quad (1.56)$$

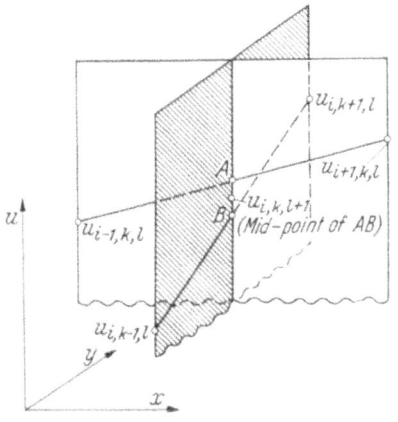

Fig. IV/11. Graphical construction for two-dimensional heat flow

where S is as defined in (1.55). The graphical construction for this formula is shown in Fig. IV/11 and a numerical example is given in Exercise 6 in § 6.4.

---

[1] Mentioned briefly by K. ELSER: see last footnote.

1.9. An indication of further problems

The finite-difference method has been applied with success to many complicated hydrodynamic and aerodynamic problems but to go into details here would be beyond the scope of this book. Suffice it to mention (in addition to the references given in § 5) the application of the finite-difference method to the systems of partial differential equations arising in fluid dynamics[1] and to boundary-value problems for differential equations of "mixed type", i.e. equations which are partly elliptic and partly hyperbolic in the considered region, there being a dividing curve along which the equation is parabolic. These latter problems are important in gas dynamics[2] (the dividing curve corresponds to the transonic region); their numerical solution needs to be treated with particular care and attention.

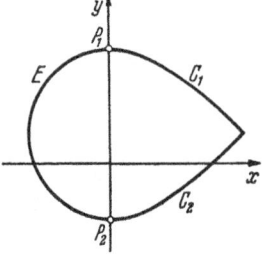

As an example which can be formulated very simply (this problem has been treated theoretically) consider the differential equation

$$- u_{xx} + f(x)\, u_{yy} = 0, \qquad (1.57)$$

Fig. IV/12.   A boundary-value problem for a differential equation of "mixed type"

where $f(x)$ is a non-decreasing function of x with the same sign as x. Then (1.57) is elliptic for $x < 0$, parabolic on the line $x = 0$, and hyperbolic for $x > 0$.

Let the boundary Γ consist of an arc E in the half-plane $x \leqq 0$ which has only its end-points P_1 and P_2 in common with the y axis, together with the two characteristics C_1 and C_2 which proceed from P_1 and P_2, respectively, into the half-plane $x \geqq 0$ (Fig. IV/12). Let u be prescribed on the boundaries E and C_1. Then u is determined uniquely inside Γ, and it should be possible to calculate u numerically.

§ 2. Refinements of the finite-difference method

The finite-difference method can be refined in exactly the same ways for partial differential equations as for ordinary differential equations (Chapter III, § 2).

---

[1] Ways of overcoming difficulties occasioned by the presence of singularities on the boundary are given by H. Görtler: Ein Differenzenverfahren zur Berechnung laminarer Grenzschichten. Ing.-Arch. **16**, 173—187 (1948). — Witting, H.: Verbesserungen des Differenzenverfahrens von H. Görtler zur Berechnung laminarer Grenzschichten. Z. Angew. Math. Phys. **4**, 376—397 (1953).

[2] The literature on the subject of differential equations of mixed type began with the pioneering work of F. Tricomi: Sulle equazioni lineari alle derivate parziali di 2° ordino di tipo misto. Mem. Real. Accad. Lincei **14**, 133—247 (1923). It has expanded so rapidly in recent years that we will content ourselves here with mentioning a few arbitrarily selected papers: Protter, M. H.: A boundary value problem for an equation of mixed type. Trans. Amer. Math. Soc. **71**, 416—429 (1951). — Bergman, St.: On solutions of linear partial differential equations of mixed type. Amer. J. Math. **74**, 444—474 (1952). — Hellwig, G.: Anfangs- und Randwertprobleme bei partiellen Differentialgleichungen von wechselndem Typus auf den Rändern. Math. Z. **58**, 337—357 (1953). — Lax, P. D.: Weak Solutions of Nonlinear Hyperbolic Equations and their Numerical Computation. Comm. Pure Appl. Math. **7**, 159—193 (1954).

2.1. The derivation of finite equations

To simplify the description we limit ourselves to linear partial differential equations in two independent variables, x, y. Such an equation of the m-th order can be written

$$L[u] = \sum_{p+q \leq m} A_{pq}(x, y) \frac{\partial^{p+q} u}{\partial x^p \partial y^q} = r(x, y), \tag{2.1}$$

where the $A_{pq}(x, y)$ and $r(x, y)$ are given functions which we shall assume to be continuous. Let sufficient starting data be given, say on the x axis and certain boundary curves, to determine uniquely a specific solution $u(x, y)$. It is required to calculate the approximations $U_{i,k+1}$ on the $(k+1)$-th row when the calculation based on the rectangular mesh (1.1) has progressed as far as the k-th row. For this purpose we derive a finite equation

$$\sum_{\iota, \varkappa} C_{\iota, \varkappa} U_{\iota, \varkappa} = r_{\iota_0, \varkappa_0} \tag{2.2}$$

corresponding to (2.1) at (x_0, y_0), in which the coefficients $C_{\iota, \varkappa}$ are determined so that the Taylor expansion of the sum

$$\Phi = \sum_{\iota, \varkappa} C_{\iota, \varkappa} u_{\iota, \varkappa} \tag{2.3}$$

in terms of u and its partial derivatives at the mesh point $(\iota_0, \varkappa_0)$ agrees with (2.1) at (x_0, y_0) to as high an order as possible, i.e. so that the order of the lowest derivative in the Taylor expansion which does not have the same coefficient as in (2.1) is as high as possible. The summation in (2.2) and (2.3) is to extend over a number of mesh points in the neighbourhood of $(\iota_0, \varkappa_0)$ the choice of which depends on the given differential equation. The number of points may be greater than is absolutely necessary for the Taylor expansion to coincide with the differential equation to some order; for then we have certain free or "surplus" $C_{\iota, \varkappa}$ at our disposal, which can be chosen so as to make the finite equations as stable as possible. A rough indication of the stability is given by the "index" J, which we now define.

Normally we will use a finite equation (2.2) which, besides the term $U_{i, k+1}$ to be calculated, involves only values $U_{\iota, \varkappa}$ with $\varkappa \leq k$. If it is written in the form

$$U_{i, k+1} = -\frac{1}{C_{i, k+1}} \sum_{\varkappa \leq k} C_{\iota, \varkappa} U_{\iota, \varkappa} + \frac{1}{C_{i, k+1}} r_{\iota_0, \varkappa_0}, \tag{2.4}$$

which expresses $U_{i, k+1}$ in terms of the remaining $U_{\iota, \varkappa}$, its index J is defined as the sum of the absolute values of the coefficients of the remaining $U_{\iota, \varkappa}$:

$$J = \frac{1}{|C_{i, k+1}|} \sum_{\varkappa \leq k} |C_{\iota, \varkappa}|. \tag{2.5}$$

Apart from the restriction $\varkappa \leq k$, the summation is over all $\iota, \varkappa$ for which $C_{\iota, \varkappa} \neq 0$.

We try to keep the index J as small as possible; this can often be done by suitably choosing surplus coefficients $C_{\iota, \varkappa}$, as suggested above, and also by moving the "centre" $(\iota_0, \varkappa_0)$ of the Taylor expansion "forwards", i.e. by choosing it near the point $(i, k+1)$.

When certain other conditions are satisfied, the condition $J \leq 1$ is sufficient for the stability and convergence of the finite-difference method, see §§ 3.3, 3.4; however, formulae with a greater index can still be usable, for instance, $J = 3$ for equation (1.34).

2.2. Application to the heat equation

To illustrate the derivation of finite equations, let us follow through the procedure for the inhomogeneous heat equation

$$L[u] = -K \frac{\partial u}{\partial y} + \frac{\partial^2 u}{\partial x^2} = r(x, y). \tag{2.6}$$

For simplicity we take the origin of the mesh co-ordinates at the mesh point $(i, k+1)$ and use this point as the centre of the Taylor

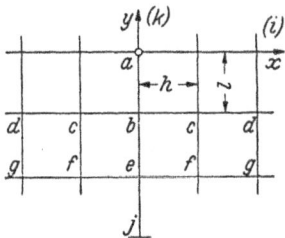

expansion; thus $\iota_0 = \varkappa_0 = 0$. We extend the summation (2.3) over the points $(\iota, \varkappa)$ with $|\iota| \leq 2, \varkappa = -1, \varkappa = -2$ and $\iota = 0, \varkappa = -3$ and denote the coefficients $C_{\iota, \varkappa}$ by a, b, c, d, e, f, g, j as indicated in Fig. IV/13, taking advantage of symmetry (thus $C_{0,0} = a$, $C_{1,-1} = C_{-1,-1} = c$, etc.). Each $u_{\iota, \varkappa}$ has now to be expanded as a Taylor series centred on the point $(0, 0)$; for example,

Fig. IV/13. Notation used in setting up a finite expression for the heat equation

$$u_{1,-1} = u_{0,0} + h u_{x,0,0} - l u_{y,0,0} + \frac{h^2}{2} u_{xx,0,0} + \cdots. \tag{2.7}$$

By similarly expanding all the terms which appear in the sum (2.3) we obtain

$$\begin{aligned}
\varPhi = u(a + b + 2c + 2d + e + 2f + 2g + j) + \\
+ l u_y(-b - 2c - 2d - 2e - 4f - 4g - 3j) + \\
+ \frac{h^2}{2} u_{xx}(2c + 8d + 2f + 8g) + \\
+ \frac{l^2}{2} u_{yy}(b + 2c + 2d + 4e + 8f + 8g + 9j) + \\
+ \frac{h^2 l}{2} u_{xxy}(-2c - 8d - 4f - 16g) + \\
+ \frac{l^3}{6} u_{yyy}(-b - 2c - 2d - 8e - 16f - 16g - 27j) +
\end{aligned} \right\} \tag{2.8}$$

$$+ \frac{h^4}{24} u_{xxxx}(2c + 32d + 2f + 32g) +$$

$$+ \frac{h^2 l^2}{4} u_{xxyy}(2c + 8d + 8f + 32g) +$$

$$+ \frac{l^4}{24} u_{yyyy}(b + 2c + 2d + 16e + 32f + 32g + 81j) +$$

$$+ \text{ terms of the 5th order,}$$

in which all the u, u_y, etc. are evaluated at the point $(0, 0)$.

This expansion is to coincide with the differential expression (2.6) to as high an order as possible, so the second bracketed factor must have the value $-K/l$, and the third must have the value $2/h^2$, while the first, fourth, and as many more as possible, must be zero. For comparison we complete the derivation for formulae correct to several different orders.

1. First-order formulae. Here we equate coefficients of u, u_y, u_{xx} only. This requires three constants, say a, b, c, and from the equations

$$a + b + 2c = 0,$$

$$-b - 2c = -\frac{K}{l},$$

$$2c = \frac{2}{h^2}$$

their values must be

$$a = -\frac{K}{l}, \quad b = \frac{K}{l} - \frac{2}{h^2}, \quad c = \frac{1}{h^2}.$$

With the notation

$$\sigma = \frac{l}{K h^2}$$

we have

$$\left. \begin{array}{l} \Phi = \frac{K}{l}\left(- u_{0,0} + (1 - 2\sigma) u_{0, -1} + \sigma(u_{1, -1} + u_{-1, -1})\right) \\ = \left(- K \frac{\partial u}{\partial y} + \frac{\partial^2 u}{\partial x^2}\right)_{0,0} + \text{remainder term of the 2nd order.} \end{array} \right\} \quad (2.9)$$

The finite equation obtained by neglecting the remainder term reads [when expressed in the form (2.4)]

$$U_{0,0} = (1 - 2\sigma) U_{0, -1} + \sigma(U_{1, -1} + U_{-1, -1}) - \sigma h^2 r_{0,0}, \quad (2.10)$$

which is the formula for the ordinary finite-difference method.

σ is positive, and for all values up to a half the index for equation (2.10) is 1; for $\sigma > \frac{1}{2}$ the index is always greater than 1. In this respect, equation (2.10) is equally favourable for all values of σ in $0 \leq \sigma \leq \frac{1}{2}$;

however, the formula obtained by putting $\sigma = \frac{1}{2}$, i.e.

$$U_{0,0} = \frac{1}{2}(U_{1,-1} + U_{-1,-1}) - \frac{1}{2}h^2 r_{0,0},$$

is more advantageous than a formula resulting from a smaller value of σ, say $\sigma = \frac{1}{3}$, which gives

$$U_{0,0} = \frac{1}{3}(U_{1,-1} + U_{0,-1} + U_{-1,-1}) - \frac{1}{3}h^2 r_{0,0},$$

because a greater value of σ means a greater value of l and hence we progress further in the time direction with the same number of computed values.

2. Second-order formulae. When we try to equate coefficients of all terms of up to the second order inclusive, we see immediately that this cannot be done with the constants a, b, c, d, since the four equations are then inconsistent. With the four constants a, b, c, e the equations are

$$a + b + 2c + e = 0, \quad -b - 2c - 2e = -\frac{K}{l},$$

$$2c = \frac{2}{h^2}, \quad b + 2c + 4e = 0,$$

and these can be solved; they yield

$$a = -\frac{3}{2}\frac{K}{l}, \quad b = 2\left(\frac{K}{l} - \frac{1}{h^2}\right), \quad c = \frac{1}{h^2}, \quad e = -\frac{K}{2l}.$$

With these values we have

$$\Phi = \frac{K}{l}\left(-\frac{3}{2}u_{0,0} + (2 - 2\sigma)u_{0,-1} + \sigma(u_{1,-1} + u_{-1,-1}) - \frac{1}{2}u_{0,-2}\right)$$

$$= (L[u])_{0,0} + \text{remainder term of the 3rd order.}$$

The corresponding finite equation reads

$$U_{0,0} = \frac{2}{3}[(2 - 2\sigma)U_{0,-1} + \sigma(U_{1,-1} + U_{-1,-1}) - \frac{1}{2}U_{0,-2} - \sigma h^2 r_{0,0}]$$

and for $0 \leq \sigma \leq 1$ has the index $J = \frac{5}{3}$. Putting $\sigma = 1$ we obtain the formula

$$U_{0,0} = \frac{1}{3}[2U_{1,-1} + 2U_{-1,-1} - U_{0,-2} - 2h^2 r_{0,0}],$$

which is stable in the sense of § 3.4. As with the ordinary finite-difference formula (1.13), this formula can be used over a sub-set of the whole set of mesh points, but here each row computed takes the solution twice as far as would a row calculated by (1.13).

3. Third-order formulae. We can equate coefficients of all terms of up to the third order inclusive by using the constants a, b, c, e, f.

The solution of the pertinent set of equations yields

$$\Phi = \frac{K}{l} \left\{ - \frac{11}{6} u_{0,0} + 3 u_{0,-1} - \frac{3}{2} u_{0,-2} + \frac{1}{3} u_{0,-3} + \right.$$

$$\left. + 2\sigma \left[- 2 u_{0,-1} + u_{0,-2} + (u_{1,-1} + u_{-1,-1}) - \frac{1}{2} (u_{1,-2} + u_{-1,-2}) \right] \right\}$$

$$= - K \frac{\partial u}{\partial y} + \frac{\partial^2 u}{\partial x^2} + \text{remainder term of the 4th order.}$$

The corresponding finite equation has the index $J = \frac{29}{11}$ for $0 \leq \sigma \leq \frac{3}{4}$; this is greater than that for either of the lower order equations and even the inclusion of the values $u_{2,-1}$ and $u_{-2,-1}$ does not improve it. [An expression with a smaller index is given in Z. Angew. Math. Mech. Bd. 16 (1936) p. 245]. A smaller index can be obtained with the Hermitian-type formulae of § 2.3.

2.3. The "Hermitian" methods

The idea of Hermitian-type formulae, which was introduced in Ch. III, § 2.4, for ordinary differential equations, can be readily extended to linear partial differential equations. Consider once more the differential equation (2.1). Instead of deriving a finite equation from the sum in (2.3), we use an expression of the form

$$\Phi = \sum_{\iota, \varkappa} C_{\iota, \varkappa} u_{\iota, \varkappa} + \sum_{\iota, \varkappa} D_{\iota, \varkappa} (L[u])_{\iota, \varkappa}, \tag{2.11}$$

where the $C_{\iota, \varkappa}$ and $D_{\iota, \varkappa}$ are constants to be determined; $(L[u])_{\iota, \varkappa}$ denotes the value taken by the differential expression $L[u]$ at the mesh point $(\iota, \varkappa)$ and the summations are extended over a number of suitably chosen mesh points in the neighbourhood of a specified point $(\iota_0, \varkappa_0)$ with which the equation is associated. As in § 2.1, the mesh points occurring in (2.11) will normally be chosen so that $C_{i,k+1} \neq 0$ and $C_{\iota,k+1} = 0$ for $\iota \neq i$; then with $(L[u])_{\iota, \varkappa} = r_{\iota, \varkappa}$ from (2.1) we use the finite equation

$$\sum_{\iota, \varkappa} C_{\iota, \varkappa} U_{\iota, \varkappa} + \sum_{\iota, \varkappa} D_{\iota, \varkappa} r_{\iota, \varkappa} = 0 \tag{2.12}$$

to calculate the approximation $U_{i, k+1}$ from the $U_{\iota, \varkappa}$ with $\varkappa \leq k$. In order that the approximations obtained by using (2.12) shall be as good as possible, the coefficients $C_{\iota, \varkappa}, D_{\iota, \varkappa}$ are determined so that the Taylor expansion of the expression (2.11) in terms of u and its partial derivatives at the point $(\iota_0, \varkappa_0)$ is zero to as high an order as possible.

Again we use the example of the inhomogeneous heat equation (2.6) to illustrate the derivation. With the same mesh points and the same notation $a, b, c, \ldots$ for the $C_{\iota, \varkappa}$ as in § 2.2 and Fig. IV/13, and with the

19*

corresponding capitals $A, B, C, \ldots$ for the $D_{i, \varkappa}$, we write

$$
\left.\begin{array}{l}
\Phi = a\,u_{0,0} + b\,u_{0,-1} + c\,(u_{1,-1} + u_{-1,-1}) + e\,u_{0,-2} + \\
\quad + f\,(u_{1,-2} + u_{-1,-2}) + A(L[u])_{0,0} + B\,(L[u])_{0,-1} + \\
\quad + C\,\{(L[u])_{1,-1} + (L[u])_{-1,-1}\}.
\end{array}\right\} \quad (2.13)
$$

Taylor expansion in terms of u and its derivatives at the point $(0, 0)$ yields

$$
\left.\begin{array}{l}
\Phi = u[a + b + 2c + e + 2f] + \\
\quad + u_y[l(-b - 2c - 2e - 4f) - K(A + B + 2C)] + \\
\quad + u_{xx}\left[\dfrac{h^2}{2}(2c + 2f) + A + B + 2C\right] + \\
\quad + u_{yy}\left[\dfrac{l^2}{2}(b + 2c + 4e + 8f) + Kl(B + 2C)\right] + \\
\quad + u_{xxy}\left[\dfrac{h^2 l}{2}(-2c - 4f) - l(B + 2C) - h^2 C K\right] + \\
\quad + u_{yyy}\left[\dfrac{l^3}{6}(-b - 2c - 8e - 16f) - \dfrac{Kl^2}{2}(B + 2C)\right] + \\
\quad + \text{terms of the 4th order.}
\end{array}\right\} \quad (2.14)
$$

If each factor in square brackets is put equal to zero, we have six homogeneous equations for eight unknowns; we therefore express six of the unknowns in terms of the remaining two, say A and f:

$$
a = \frac{5}{2}\frac{KA}{l}, \qquad b = 2(3\sigma - 1)\frac{KA}{l} + 2f, \qquad c = -3\sigma\frac{KA}{l} - f,
$$

$$
e = -\frac{1}{2}\frac{KA}{l} - 2f, \qquad \frac{KB}{l} = 2(1 - \sigma)\frac{KA}{l} + 2f, \qquad \frac{KC}{l} = \sigma\frac{KA}{l} - f
$$

$\left(\text{with the usual notation } \sigma = \dfrac{l}{K\,h^2}\right).$

Thus with $\varepsilon = \dfrac{l}{KA}\,f$ we have

$$
\left.\begin{array}{l}
\dfrac{5}{2}\,u_{0,0} + 2(3\sigma - 1 + \varepsilon)\,u_{0,-1} - (3\sigma + \varepsilon)\,(u_{1,-1} + u_{-1,-1}) - \\
\quad - \left(\dfrac{1}{2} + 2\varepsilon\right)u_{0,-2} + \varepsilon\,(u_{1,-2} + u_{-1,-2}) + \dfrac{l}{K}\,\{(L[u])_{0,0} + \\
\quad + 2(1 - \sigma + \varepsilon)\,(L[u])_{0,-1} + (\sigma - \varepsilon)\,((L[u])_{1,-1} + (L[u])_{-1,-1})\} + \\
\quad + \text{terms of the 4th order} = 0.
\end{array}\right\} \quad (2.15)
$$

It is advisable to check the results of such a Taylor expansion, say by putting $u = x^2$, y^2, y^3, etc.

If the right-hand side of the differential equation (2.6) is a constant, say r, then we obtain a very simple formula by putting $\varepsilon = 0$ and $\sigma = \frac{1}{3}$:

$$u_{0,0} = \frac{2(u_{1,-1} + u_{-1,-1}) + u_{0,-2}}{5} - \frac{6}{5}\frac{l}{K}r + \left.\begin{array}{l}\\ \\ + \text{ remainder term of the 4th order.}\end{array}\right\} \quad (2.16)$$

On account of the larger value of σ, the calculation proceeds "forwards" more rapidly with the formula

$$u_{0,0} = \frac{2(u_{1,-1} + u_{-1,-1}) + u_{1,-2} - u_{0,-2} + u_{-1,-2}}{5} - \frac{6l}{5K}r + \left.\begin{array}{l}\\ \\ + \text{ remainder term of the 4th order,}\end{array}\right\} \quad (2.17)$$

which is obtained by putting $\sigma = \frac{1}{2}$, $\varepsilon = -\frac{1}{2}$. However, (2.16) has the index $J = 1$ while for (2.17) $J = 1\cdot4$, and hence the convergence of the approximate solution to the solution of the initial-value problem is assured by § 3.3 for (2.16) but not for (2.17)[1].

2.4. An example

Let us apply the Hermitian formula (2.16) to the problem

$$\frac{\partial^2 u}{\partial x^2} - K\frac{\partial u}{\partial y} + \beta = 0 \quad \text{for} \quad |x| \leq 1, \quad y \geq 0, \qquad (2.18)$$

$$u(x, 0) = u(-1, y) = u(1, y) = 0; \qquad (2.19)$$

u can be interpreted physically as the temperature at time y of an element of a thin homogeneous rod at a distance x from the centre when heat is generated internally at a constant rate and is conducted away at the ends so as to keep the temperatures there constant at the initial uniform temperature 0 of the whole rod.

To start a calculation based on (2.16) we have to calculate the values $U_{i,1}$ on the first row by some other means. We observe that $u_{xx} = 0$ for the initial temperature distribution $u(x, 0)$; hence, from the differential equation, $u_y = \frac{\beta}{K}$ on $y = 0$, and therefore $u_{xxy} = \frac{d^2}{dx^2}\left(\frac{\beta}{K}\right) = 0$ on $y = 0$. Since the differential equation implies that $u_{xxy} = Ku_{yy}$, it follows that $u_{yy} = 0$ on $y = 0$. This means that $U_{i,1} + U_{i,-1} - 2U_{i,0} = 0$ or, since $U_{i,0} = 0$, simply that $U_{i,-1} = -U_{i,1}$. Consequently formula (2.16)

$$U_{i,k+1} = \frac{1}{5}(2U_{i+1,k} + 2U_{i-1,k} + U_{i,k-1}) + \alpha, \qquad (2.20)$$

where $\alpha = \frac{6}{5}\frac{l\beta}{K}$, becomes

$$U_{i,1} = \frac{1}{3}(U_{i+1,0} + U_{i-1,0}) + \frac{5}{6}\alpha \qquad (2.21)$$

---

[1] Further formulae can be found in J. ALBRECHT: Zum Differenzenverfahren bei parabolischen Differentialgleichungen. Z. Angew. Math. Mech. **37**, 202—212 (1957).

when $k=0$. This formula gives the required starting values and the calculation can then be continued with (2.20).

The results obtained with a mesh width $h=\frac{1}{4}$ (l is then determined from $\sigma=\dfrac{l}{Kh^2}=\dfrac{1}{3}$) are shown in Table IV/13, which gives the values

Table IV/13. *Solution of the inhomogeneous heat equation by a Hermitian method*

k	$x=-1$	$x=-\frac{3}{4}$	$x=-\frac{1}{2}$	$x=-\frac{1}{4}$	$x=0$	S_k	T_k
0	0		0		0	0	
1		0·833 33		0·833 33		3·333 33	1·666 67
2	0		1·666 67		1·666 67	5	
3		1·833 33		2·5		8·666 67	3·666 67
4	0		3·066 67		3·333 33	9·466 67	
5		2·593 33		4·06		13·306 67	5·186 67
6	0		4·274 67		4·914 67	13·464 01	
7		3·228 53		5·487 74		17·432 54	6·457 06
8	0		5·341 44		6·373 13	17·056 01	
9		3·782 28		6·783 38		21·131 32	7·564 56
10	0		6·294 55		7·701 33	20·290 43	

Steady temperature distribution:

	0		15		20		
		8·75		18·75			
	0		15		20		

of $U_{i,k}$ (apart from a factor α) for $x\leq 0$, the other half being symmetrical. The quantities formed in the row-sum check are

$S_\nu = $ sum of all $U_{i,\nu}$ in the ν-th row,

$T_\nu = $ sum of the two outermost U values in the ν-th row;

here the ν-th row means the whole ν-th row and not just the half row ($x\leq 0$) reproduced in Table IV/13. Then we should have

$$S_{2k+1}=\frac{1}{5}\left(4S_{2k}+S_{2k-1}\right)+m\alpha,$$

$$S_{2k}=\frac{1}{5}\left(4S_{2k-1}+S_{2k-2}-2T_{2k-1}\right)+(m-1)\alpha,$$

$$k=0,1,2,\ldots,\qquad m=\frac{1}{h}\ (\text{here } m=4).$$

The steady temperature distribution (the limit as $y\to\infty$) is given at the bottom of the table; it can be readily determined from the condition $U_{i,k+2}\equiv U_{i,k}$. Since the exact steady temperature distribution is parabolic, the values obtained by the approximate method are exact.

§3. Some theoretical aspects of the finite-difference methods
3.1. Choice of mesh widths

The different mesh relations (1.12) and (1.33) for the parabolic and hyperbolic differential equations, respectively, are intimately connected with the different "domains of dependence" for these equations[1]. Consider, for example, the wave equation

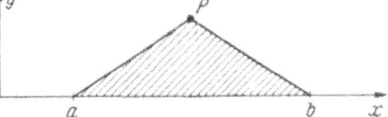

$$\frac{\partial^2 u}{\partial x^2} = \frac{1}{\omega^2} \frac{\partial^2 u}{\partial y^2} \qquad (3.1)$$

Fig. IV/14. Determinate region

with given initial values of u and $\partial u / \partial y$ on the x axis. Then the value $u(x, y)$ at a point P (Fig. IV/14) depends only on the initial data on that part of the x axis, $a \leq x \leq b$, which is cut off by the two "characteristics" through P; this is called the "domain of dependence" of the point P. Reciprocally, the segment $a \leq x \leq b$ of the x axis and the characteristics emanating from its endpoints define a region (shaded in the figures) in which u is determined completely by the initial values of u and $\partial u / \partial y$ on the segment; this may be called the "determinate region" of the initial segment.

If we apply the finite-difference method on a mesh with $h = \nu l$ and with the points $(a, 0)$, $(b, 0)$ among its mesh points, we find that the initial values in $a \leq x \leq b$ determine the $U_{i,k}$ at the mesh points in the triangle formed by the x axis and the straight lines $x \pm \nu y = \text{constant}$ which pass through the points

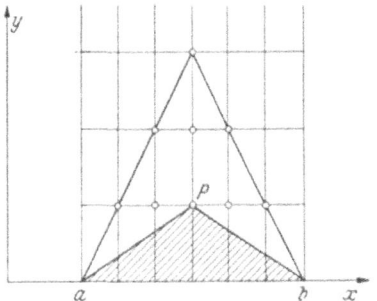

Fig. IV/15. The finite-difference determinate region

$(a, 0)$ and $(b, 0)$, respectively (these mesh points are circled in Fig. IV/15). Now if the number $1/\nu$ is chosen too large, greater than $1/\omega$ in fact, then this finite-difference "determinate region" extends beyond the actual determinate region of the initial segment $a \leq x \leq b$. This means that from initial values in $a \leq x \leq b$ the finite-difference method produces values for u at points where u is influenced by initial values outside of $a \leq x \leq b$. In this case, therefore, we cannot expect the finite-difference approximations to converge to the exact solution of the differential equation as the mesh is refined. However, convergence can be proved for $\nu = \omega$ and also an error estimate can be obtained (see § 3.2).

---

[1] See R. COURANT and D. HILBERT: Methoden der mathematischen Physik, Vol. II, p. 307. Berlin 1937, or R. COURANT and K. O. FRIEDRICHS: Supersonic flow and shock waves, p. 51. New York: Interscience Publishers Inc. 1948.

Similarly, convergence to the correct solution cannot be expected when the finite-difference method is applied to the heat equation $\partial^2 u/\partial x^2 = \partial u/\partial y$ with a constant mesh ratio $h/l = \nu$. This can be seen by considering the exact solution

$$u = \frac{1}{\sqrt{y}} e^{\frac{-x^2}{4y}}, \tag{3.2}$$

which is always positive for $y > 0$ and vanishes on the x axis except at $x = 0$. Now if $0 < a < b$, the finite-difference method would always give zero values in the large triangle of Fig. IV/15; for constant ν this is a fixed triangle and consequently the results would necessarily converge to the wrong solution.

3.2. An error estimate for the inhomogeneous wave equation

We choose here an example admitting of an explicit error estimate from which the convergence[1] of the solution of the finite-difference equations to the solution of the differential equation can be inferred directly.

Consider the initial-value problem for the inhomogeneous wave equation

$$L[u] = \frac{\partial^2 u}{\partial x^2} - \frac{1}{\omega^2} \frac{\partial^2 u}{\partial y^2} = t(x, y) \tag{3.3}$$

in which the values

$$u(x, 0) = G(x) \quad \text{and} \quad \frac{\partial u(x, 0)}{\partial y} = H(x) \tag{3.4}$$

are given on the initial line $y = 0$. The treatment which we shall give here can still be applied when the boundary conditions are more general than this.

First we write down the finite-difference equation corresponding to (3.3) for the mesh (1.1) (with $x_0 = y_0 = 0$):

$$L_{\iota, \varkappa}[U] \equiv \frac{U_{\iota+1, \varkappa} - 2U_{\iota, \varkappa} + U_{\iota-1, \varkappa}}{h^2} - \frac{U_{\iota, \varkappa+1} - 2U_{\iota, \varkappa} + U_{\iota, \varkappa-1}}{\omega^2 l^2} = t_{\iota, \varkappa}. \tag{3.5}$$

---

[1] The proof of convergence may be carried over to more general problems; see R. COURANT, K. FRIEDRICHS and H. LEWY: Über die partiellen Differenzengleichungen der mathematischen Physik. Math. Ann. **100**, 32—74 (1928). — FRIEDRICHS, K., and H. LEWY: Das Anfangswertproblem einer beliebigen nichtlinearen hyperbolischen Differentialgleichung beliebiger Ordnung in 2 unabhängigen Variablen; Existenz, Eindeutigkeit und Abhängigkeitsbereich der Lösung. Math. Ann. **99**, 200—221 (1928). — COURANT, R., and P. LAX: On nonlinear partial differential equations with two independent variables. Comm. Pure Appl. Math. **2**, 255—273 (1949). — COURANT, R., E. ISAACSON and MINA REES: On the solution of non-linear hyperbolic differential equations by finite differences. Comm. Pure Appl. Math. **5**, 243—255 (1952), in which the system

$$\sum_{i=1}^{n} \left(a_{ij} \frac{\partial u_i}{\partial x} + b_{ij} \frac{\partial u_i}{\partial y} \right) = c_j \quad (j = 1, \ldots, n)$$

is considered, where a_{ij}, b_{ij}, c_j are functions of $x, y, u_1, \ldots, u_n$.

With $h = \omega l$ this simplifies to

$$- h^2 L_{\iota, \varkappa}[U] = (U_{\iota, \varkappa+1} - U_{\iota+1, \varkappa}) - (U_{\iota-1, \varkappa} - U_{\iota, \varkappa-1}) = - h^2 t_{\iota, \varkappa}. \qquad (3.6)$$

On the initial line $y = 0$ we have

$$U_{\iota, 0} = u_{\iota, 0} = u(\iota h, 0) = G(\iota h). \qquad (3.7)$$

The values of U on the line $y = h$ are determined by the other initial condition

$$\frac{U_{\iota, 1} - U_{\iota, -1}}{2h} = H(\iota h) = \frac{\partial u(\iota h, 0)}{\partial y}: \qquad (3.8)$$

$U_{\iota, -1}$ can be eliminated between this equation and (3.6) with $\varkappa = 0$, i.e.

$$U_{\iota, 1} - U_{\iota+1, 0} - U_{\iota-1, 0} + U_{\iota, -1} = - h^2 t_{\iota, 0},$$

to obtain

$$U_{\iota, 1} = \frac{u_{\iota-1, 0} + u_{\iota+1, 0}}{2} + h H(\iota h) - \frac{h^2}{2} t_{\iota, 0}. \qquad (3.9)$$

The solution $U_{\iota, \varkappa}$ calculated from these starting values by repeated application of (3.6) can be given explicitly. From (3.6) we have

$$U_{\iota, \varkappa} - U_{\iota+1, \varkappa-1} = U_{\iota-1, \varkappa-1} - U_{\iota, \varkappa-2} - h^2 t_{\iota, \varkappa-1} = \cdots$$

$$= U_{\iota-\varkappa+1, 1} - U_{\iota-\varkappa+2, 0} - h^2 \sum_{\nu=0}^{\varkappa-2} t_{\iota-\nu, \varkappa-\nu-1},$$

and similarly

$$U_{\iota+1, \varkappa-1} - U_{\iota+2, \varkappa-2} = U_{\iota-\varkappa+3, 1} - U_{\iota-\varkappa+4, 0} - h^2 \sum_{\nu=0}^{\varkappa-3} t_{\iota-\nu+1, \varkappa-\nu-2}$$

$$\cdot \quad \cdot \quad \cdot \quad \cdot \quad \cdot \quad \cdot \quad \cdot \quad \cdot \quad \cdot \quad \cdot \quad \cdot \quad \cdot \quad \cdot \quad \cdot \quad \cdot \quad \cdot \quad \cdot \quad \cdot$$

$$U_{\iota+\varkappa-2, 2} - U_{\iota+\varkappa-1, 1} = U_{\iota+\varkappa-3, 1} - U_{\iota+\varkappa-2, 0} - h^2 t_{\iota+\varkappa-2, 1}.$$

Addition of these equations then yields

$$U_{\iota, \varkappa} = \underbrace{\sum_{\nu=1}^{\varkappa} U_{\iota-\varkappa+2\nu-1, 1}}_{\text{sum (i)}} - \underbrace{\sum_{\nu=1}^{\varkappa-1} U_{\iota-\varkappa+2\nu, 0}}_{\text{sum (ii)}} - \underbrace{h^2 \sum_{\mu=1}^{\varkappa-1} \sum_{\nu=1}^{\varkappa-\mu} t_{\iota-\nu+\mu, \varkappa-\nu-\mu+1}}_{\text{sum (iii)}}. \qquad (3.10)$$

The summations (i) and (ii) extend over the U values at the points circled in Fig. IV/16 and the summation (iii) extends over the values of t at the points marked with crosses.

We now derive an upper bound of the form constant $\times h^2$ for the absolute value of the error in the approximation U at an arbitrary but fixed point in the determinate region for the difference equations; this implies the convergence of the solution $U_{\iota, \varkappa}$ of the finite-difference equations to the solution (assumed to be differentiable sufficiently often) of the differential equation.

If M_v is the maximum value of $|\partial^v u/\partial x^v|$ and $|\partial^v u/\partial y^v|$ (for $v=3,4$) in a convex region of the (x,y) plane which includes all mesh points involved in the calculation, then it follows from (3.5) (for remainder term see Table III of the appendix) that

$$L_{\iota,\varkappa}[u] = \frac{\partial^2 u}{\partial x^2} - \frac{1}{\omega^2}\frac{\partial^2 u}{\partial y^2} + \frac{2h^2}{12}\vartheta_{\iota,\varkappa}M_4 = t_{\iota,\varkappa} + \frac{h^2}{6}\vartheta_{\iota,\varkappa}M_4,$$

where $|\vartheta_{\iota,\varkappa}|\leq 1$; hence the error $\varepsilon = U - u$ satisfies the difference equation

$$L_{\cdot,\varkappa}[\varepsilon] = -\frac{h^2}{6}\vartheta_{\iota,\varkappa}M_4 \quad \text{with} \quad |\vartheta_{\iota,\varkappa}|\leq 1. \tag{3.11}$$

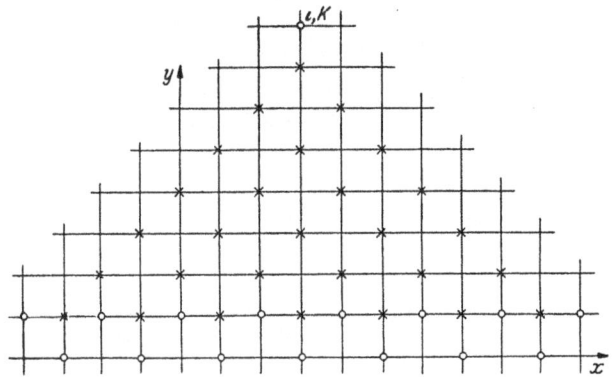

Fig. IV/16. Construction of the closed solution of the difference equations

Since on the "starting line" $y=h$ we have

$$\frac{u_{\iota,1}-u_{\iota,-1}}{2h} = \frac{\partial u(\iota h,0)}{\partial y} + \frac{h^2}{6}\vartheta_{\iota}M_3 \quad \text{with} \quad |\vartheta_{\iota}|\leq 1,$$

it follows from (3.8) that

$$\frac{\varepsilon_{\iota,1}-\varepsilon_{\iota,-1}}{2h} = -\frac{h^2}{6}\vartheta_{\iota}M_3,$$

and since $\varepsilon_{\iota,0}=0$, from (3.11) with $\varkappa=0$ we also have

$$\varepsilon_{\iota,1}+\varepsilon_{\iota,-1} = \frac{h^4}{6}\vartheta_{\iota,0}M_4;$$

we can therefore specify an upper bound for the errors on the starting line $y=h$:

$$|\varepsilon_{\iota,1}| \leq \frac{1}{2}|\varepsilon_{\iota,1}-\varepsilon_{\iota,-1}| + \frac{1}{2}|\varepsilon_{\iota,1}+\varepsilon_{\iota,-1}| \leq \frac{h^4}{12}M_4 + \frac{h^3}{6}M_3. \tag{3.12}$$

We now use formula (3.10) to solve (3.11):

$$\varepsilon_{\iota,\varkappa} = \underset{\text{(i)}}{\sum \varepsilon_{\iota+2v+1,1}} - \underset{\text{(ii)}}{\sum \varepsilon_{\iota+2v+2,0}} - \frac{h^4}{6}M_4\underset{\text{(iii)}}{\sum\sum \vartheta_{\iota\ldots,\varkappa\ldots}}.$$

The terms in sum (ii) are all zero, the terms of sum (i), $\varkappa$ in number, are each bounded as in (3.12) and the $\frac{1}{2}\varkappa(\varkappa-1)$ terms of sum (iii) are each of absolute value 1 or less; therefore

$$|U_{\iota,\varkappa}-u_{\iota,\varkappa}|=|\varepsilon_{\iota,\varkappa}|\leq\varkappa\left(\frac{h^4}{12}M_4+\frac{h^3}{6}M_3\right)+\frac{h^4}{12}M_4(\varkappa^2-\varkappa).$$

Substituting for $\varkappa=y/h$ we have

$$|U_{\iota,\varkappa}-u_{\iota,\varkappa}|\leq\frac{h^2}{12}(2y\,M_3+y^2\,M_4);\tag{3.13}$$

thus the error at a fixed point (x,y) tends to zero quadratically with h as the mesh is refined.

3.3. The principle of the error estimate for more general problems with linear differential equations

Let the Taylor expansion of the expression (2.3) coincide with the given differential expression in (2.1) up to the terms of the r-th order $(r\geq m)$ inclusive. Under the assumption that in the region considered all the partial derivatives of u of the $(r+1)$-th order exist and are bounded, say by M_{r+1} in absolute value, the remaining terms in the Taylor expansion can be collected together into the single remainder term $\vartheta D l^{r-m+1}M_{r+1}$, where $|\vartheta|\leq 1$ and D depends on the mesh and the differential equation but not on the function u; in fact D is a polynomial in l of at most the m-th degree. The estimation of the quantity M_{r+1} often causes difficulty in practical applications; to get an idea of its order of magnitude one can sometimes calculate approximate values for the partial derivatives from the differences of the numerical solution.

Using this remainder term we can write

$$\sum_{\iota,\varkappa}C_{\iota,\varkappa}u_{\iota,\varkappa}=L[u_{\iota_0\varkappa_0}]+\vartheta\,D\,l^{r-m+1}M_{r+1}.\tag{3.14}$$

The error $\varepsilon_{\iota,\varkappa}=U_{\iota,\varkappa}-u_{\iota,\varkappa}$ therefore satisfies

$$\sum_{\iota,\varkappa}C_{\iota,\varkappa}\,\varepsilon_{\iota,\varkappa}=-\vartheta D\,l^{r-m+1}M_{r+1}.\tag{3.15}$$

If we know limits for the errors in the first k rows, this formula enables us to estimate the errors in the $(k+1)$-th row; in principle, therefore, a recursive error estimate is possible. To apply it, we need limits for the errors in the "starting" rows, i.e. the rows with $\varkappa=0$, $1,\dots,s$ if the values of $\varkappa$ occurring in the finite equation (2.2) differ at most by s. If u is determined uniquely by prescribed initial data on the x axis, so that theoretically we can calculate the value of any partial derivative of u at $y=0$, then the values of $U_{\iota,\varkappa}$ in the starting rows could be calculated from the Taylor series

$$U_{\iota,\varkappa} = u_{\iota,0} + \sum_{\nu=1}^{r} \frac{l^\nu \varkappa^\nu}{\nu!} \frac{\partial^\nu u_{\iota,0}}{\partial y^\nu} \qquad (\varkappa = 0, 1, 2, \ldots, s-1)$$

(some other special starting procedure may be possible, of course); we could then use

$$|\varepsilon_{\iota,\varkappa}| \leq \frac{l^{r+1}\varkappa^{r+1}}{(r+1)!} M_{r+1} \qquad (\varkappa = 0, 1, 2, \ldots, s-1) \qquad (3.16)$$

as the required limits for the starting errors.

If the index J introduced in (2.5) is not greater than one, then in cases with $r>m$ an independent, as opposed to a recursive, error estimate can be given which implies the convergence of the finite-difference approximation to the solution of the initial-value problem. We imagine the differential equation (2.1) to have been multiplied through by such a factor that the coefficient $C_{i,k+1}$ in the finite equation (2.2) is unity; then the finite equation (2.4) reads

$$U_{i,k+1} = -\sum_{\varkappa \leq k} C_{\iota,\varkappa} U_{\iota,\varkappa} + r_{\iota_0,\varkappa_0}. \qquad (3.17)$$

Under the assumption $J \leq 1$, the corresponding equation for the error

$$\varepsilon_{i,k+1} = -\sum_{\varkappa \leq k} C_{\iota,\varkappa} \varepsilon_{\iota,\varkappa} - \vartheta D \, l^{r-m+1} M_{r+1}$$

yields the estimate

$$|\varepsilon_{i,k+1}| \leq \max \left(|\varepsilon_k|, |\varepsilon_{k-1}|, \ldots, |\varepsilon_{k-s+1}| \right) + M, \qquad (3.18)$$

where $|\varepsilon_\varkappa| = \max_\iota |\varepsilon_{\iota,\varkappa}|$ and

$$M = l^{r-m+1} D_{\max} M_{r+1}.$$

If boundary conditions are imposed which prescribe the values of u on, say, two straight lines which are used as mesh lines, then $\varepsilon=0$ there. With C denoting the largest of all the limits in (3.16), it follows from (3.18) that for $\varkappa \geq s-1$

$$\left. \begin{aligned} |\varepsilon_{\iota,\varkappa}| \leq Y_\varkappa &= C + (\varkappa - s + 1) M \\ &= \left[\frac{[(s-1) \, l]^{r+1}}{(r+1)!} + (\varkappa - s + 1) D_{\max} l^{r-m+1} \right] M_{r+1}. \end{aligned} \right\} \qquad (3.19)$$

If $r>m$, these error limits tend to zero with l at a fixed point (x, y) with $y=\varkappa l$; this is still true when $r=m$ provided that the polynomial D has the mesh width l as a factor [this is always the case, for example, when the given differential equation (2.1) has no undifferentiated term in u, i.e. $A_{0,0}=0$].

The condition $J \leq 1$ and this last condition are satisfied, for example, by formula (2.10) for the inhomogeneous heat equation provided that σ is chosen in the range $0 \leq \sigma \leq \frac{1}{2}$; thus we have shown, in particular, that the finite-difference formula (1.13) yields convergent approximations[1] and that an error estimate can be given for them.

3.4. A more general investigation of error propagation and "stability"

Various courses have been pursued in the quest for systematic means of determining the "stability" characteristics of finite-difference equations, i.e. for criteria which will indicate whether or not the calculation can be rendered useless by unfavourable error propagation. Here we describe first a very simple method[2] by applying it to the initial-/ boundary-value problem for the heat equation $u_{xx} = K u_t$ with prescribed values of $u(x, 0)$, $u(0, t)$ and $u(a, t)$.

Let the mesh widths h, l be given by $h = a/N$, $l = K \sigma h^2$, where N is an integer and σ is arbitrary for the present. We assume that the values $U_{j,0}$ on the first row have certain errors ε_j and ask whether the solution of the difference equations

$$\frac{U_{j+1,k} - 2U_{j,k} + U_{j-1,k}}{h^2} - K \frac{U_{j,k+1} - U_{j,k}}{l} = 0$$

remains bounded or not as $k \to \infty$.

Provided that α and β are related by

$$\frac{2(\cos \beta h - 1)}{h^2} - K \frac{e^{\alpha l} - 1}{l} = 0,$$

i.e.

$$e^{\alpha l} = 1 - 4\sigma \sin^2 \frac{\beta h}{2},$$

$u = e^{\alpha t} \sin \beta x$ is a solution of the difference equations.

---

[1] Convergence for the case with u prescribed on the whole x axis and with $\sigma = \frac{1}{4}$ is proved in the paper by R. COURANT, K. FRIEDRICHS and H. LEWY already referred to: Math. Ann. **100**, 32—74 (1928); for the region defined by the strip $0 \leq x \leq a$, $y \geq 0$ and under certain restrictions on the prescribed boundary values convergence is proved for $0 < \sigma \leq \frac{1}{4}$ by W. LEUTERT: J. Math. Phys. **30**, 245—251 (1952), and for $0 < \sigma < \frac{1}{2}$ by F. B. HILDEBRAND: On the convergence of numerical solutions of the heat-flow equation. J. Math. Phys. **31**, 35—41 (1952).

[2] BRIEN, G. O., M. HYMAN and S. KAPLAN: A study of the numerical solutions of partial differential equations. J. Math. Phys. **29**, 223—251 (1951). — HYMAN, M. A.: On the numerical solution of partial differential equations. Proefschrift Techn. Hogeschool, 106 pp. Delft 1953. Cf. also J. TODD: A direct approach to the problem of stability in the numerical solution of partial differential equations. Nat. Bur. Stand. Rep. No. 4260, 1955, 27 pp.

Now if we imagine the initial errors ε_j to be expressed as a finite trigonometrical series of the form

$$U_{j,0} = \varepsilon_j = \sum_{\nu=1}^{N-1} A_\nu \sin \frac{\nu j \pi}{N}$$

and replace α, β by α_ν, β_ν with $\beta_\nu = \dfrac{\nu \pi}{h N}$, then the required solution of the difference equations is

$$U_{j,k} = \sum_{\nu=1}^{N-1} A_\nu e^{\alpha_\nu k l} \sin \frac{\nu j \pi}{N}.$$

We see that the $U_{j,k}$ will remain bounded for an arbitrary initial perturbation, i.e. for arbitrary A_ν, if and only if $|e^{\alpha_\nu l}| \leqq 1$ for $\nu = 1, 2, \ldots, N-1$; this is equivalent to the condition

$$4\sigma \sin^2 \frac{\beta_\nu h}{2} \leqq 2,$$

and since $\sin^2 \frac{1}{2} \beta_\nu h \leqq 1$, we obtain as the condition for stability

$$\sigma \leqq \frac{1}{2}, \quad \text{i.e.} \quad l \leqq \frac{K}{2} h^2. \tag{3.20}$$

Equation (1.12) actually specifies the greatest possible value permitted by this condition, namely $l = \frac{1}{2} K h^2$; this value is also recommended in § 2.2 (1st-order formulae).

To avoid any possible misunderstanding, we once more state explicitly that we are concerned here only with the stability behaviour of the solution of the finite-difference equations and do not say anything about the deviation from the solution of the corresponding initial- or initial-/boundary-value problem.

Similar procedures can be carried out in several other cases. The technique required for dealing with the effects of an isolated disturbance at one mesh point is considerably more complicated than for a disturbance of the form $\sin \beta x$; nevertheless, such isolated disturbances have been investigated for fairly general differential equations[1].

However, for cases in which $J \leqq 1$, J being the index defined in (2.5), it can be seen immediately that the influence of an isolated disturbance

---

[1] This theory goes back to JOHN V. NEUMANN; it has been discussed by R. P. EDDY: Stability in the numerical solution of initial value problems in partial differential equations. Naval Ordnance Labor. Memorandum **10**, 232 (1949). — Numerous other papers have been written on the question of stability; suffice it to mention just a few: TODD, J.: A direct Approach to the Problem of Stability in the Numerical Solution of Partial Differential Equations. Comm. Pure Appl. Math. 9, 597—612 (1956). — LAX, P. D., and R. D. RICHTMYER: Survey of the Stability of Linear Finite Difference Equations. Comm. Pure Appl. Math. 9, 267—293 (1956). — MLAK, W.: Remarks on the stability problem for parabolic equations. Ann. Polonici Math. **3**, 343—348 (1957).

remains bounded; thus the finite-difference method is always stable when the formula used has $J \leq 1$. On account of the linearity of the differential equation a disturbance $\varepsilon_{i,k}$ superimposed on the solution of the differential equation is propagated according to the homogeneous equation corresponding to (2.4), i.e.

$$\varepsilon_{i,k+1} = -\frac{1}{C_{i,k+1}} \sum_{\varkappa \leq k} C_{\iota,\varkappa}\,\varepsilon_{\iota,\varkappa}. \tag{3.21}$$

As in § 2.1, the sum extends over all $\iota, \varkappa$ for which $C_{\iota,\varkappa} \neq 0$ except $\varkappa = k+1$. From the definition of J (2.5) and the fact that it does not exceed 1 we see that

$$|\varepsilon_{i,k+1}| \leq \max_{\varkappa \leq k}|\varepsilon_{\iota,\varkappa}|.$$

Therefore, if we introduce an isolated disturbance $\varepsilon_{i_0,k_0} = 1$, so that $\varepsilon_{\iota,\varkappa} = 0$ for $\varkappa < k_0$ and also for $\varkappa = k_0$ except when $\iota = i_0$, we must have $|\varepsilon_{i,k+1}| \leq 1$ for all $k \geq k_0$.

The assumption $J \leq 1$ is satisfied, for example, by equation (2.10) for the heat equation provided that σ is chosen in the range $0 \leq \sigma \leq \frac{1}{2}$; this is the same condition as (3.20).

3.5. An example: The equation for the vibrations of a beam

For illustration we select an example from the theory of the flexural vibrations of thin beams which neglects such subsidiary effects as those due to the changing inclinations of the elements of the beam. In this theory the displacement of the beam satisfies the differential equation

$$\frac{\partial^4 u}{\partial x^4} + K\frac{\partial^2 u}{\partial y^2} = 0. \tag{3.22}$$

Some typical initial and boundary conditions which arise are:
initial conditions (given initial displacement and velocity distributions):

$$u(x,0) = f_1(x), \quad \frac{\partial u(x,0)}{\partial y} = f_2(x) \quad \text{for} \quad 0 \leq x \leq a;$$

boundary conditions (smoothly pinned ends):

$$u(0,y) = \frac{\partial^2 u(0,y)}{\partial x^2} = u(a,y) = \frac{\partial^2 u(a,y)}{\partial x^2} = 0 \quad \text{for} \quad y \geq 0. \tag{3.23}$$

For a rectangular mesh (1.1) with mesh widths $h = a/N$ and l (where N is integral and greater than unity) the ordinary finite-difference method replaces the differential equation (3.22) by the difference equation

$$\left. \begin{array}{l} \dfrac{U_{i+2,k} - 4U_{i+1,k} + 6U_{i,k} - 4U_{i-1,k} + U_{i-2,k}}{h^4} + \\[2mm] \quad + \dfrac{U_{i,k+1} - 2U_{i,k} + U_{i,k-1}}{l^2}\,K = 0. \end{array} \right\} \tag{3.24}$$

As we did for the heat equation in (1.12), let us choose a convenient mesh relation; with

$$K h^4 = l^2$$

the difference equation simplifies to

$$U_{i,k+1} = - U_{i+2,k} + 4 U_{i+1,k} - 4 U_{i,k} + 4 U_{i-1,k} - U_{i-2,k} - U_{i,k-1}. \quad (3.25)$$

Once we have found values on two consecutive rows, this formula will give the values on the next row, and it would appear that we could therefore continue the solution as far as required; however, a glance at the propagation table of formula (3.25) shows immediately that it is quite unsuitable for repeated application — just the few rows given in Table IV/14 proclaim its severe instability.

Table IV/14. *The propagation table of formula* (3.25)

0	ε	0	0	0
4ε	-4ε	4ε	$-\varepsilon$	0
-40ε	49ε	-40ε	24ε	-8ε
496ε	-560ε	496ε	-337ε	172ε
-6200ε	6833ε	-6200ε		

Our choice of mesh relation implies too large a value for the mesh width l for given h. To investigate a better choice of mesh relation, let us introduce the parameter[1]

$$z = \frac{l^2}{K h^4}. \quad (3.26)$$

The formula (3.25) just considered arises as the special case $z=1$.

Suppose that at some stage of the calculation the $U_{j,k}$ values on the last two rows, say $k=p$ and $k=p+1$, have deviated from their true values by errors $\varepsilon_{j,p}$ and $\varepsilon_{j,p+1}$, respectively. Then we ask again whether the errors $\varepsilon_{j,k}$ produced in subsequent rows by using (3.24) remain bounded or not as $k \to \infty$.

It can easily be shown that equation (3.24) possesses the particular solution

$$U_{j,k} = e^{(k-p)\eta} \sin j\xi,$$

if ξ and η are related by

$$2z(\cos \xi - 1)^2 = 1 - \cosh \eta; \quad (3.27)$$

therefore with $\xi = \xi_\nu = \dfrac{\nu \pi}{N}$ ($\nu = 1, \ldots, N-1$) and corresponding values $\eta = \pm \eta_\nu$ from (3.27)

$$v_{j,k} = \sum_{\nu=1}^{N-1} \sin \frac{j\nu\pi}{N} \left(A_\nu e^{(k-p)\eta_\nu} + B_\nu e^{-(k-p)\eta_\nu} \right)$$

[1] COLLATZ, L.: Z. Angew. Math. Mech. **31**, 392–393 (1951).

is also a solution of (3.24). Moreover, it satisfies the finite-difference boundary conditions corresponding to (3.23) and by a suitable choice of A_ν and B_ν can be made to assume the values $\varepsilon_{j,p}$ and $\varepsilon_{j,p+1}$ for $k=p$ and $k=p+1$, respectively: if we obtain quantities α_ν, β_ν from

$$\varepsilon_{j,p} = \sum_{\nu=1}^{N-1} \alpha_\nu \sin\frac{j\nu\pi}{N}\,, \qquad \varepsilon_{j,p+1} = \sum_{\nu=1}^{N-1} \beta_\nu \sin\frac{j\nu\pi}{N}$$

by the formulae of harmonic analysis, then, since $\eta_\nu \neq 0$ (cos $\xi \neq 1$ and $z \neq 0$ imply that cosh $\eta \neq 1$), the A_ν and B_ν are uniquely determined by the equations $A_\nu + B_\nu = \alpha_\nu$, $A_\nu e^{\eta\nu} + B_\nu e^{-\eta\nu} = \beta_\nu$.

With these values of A_ν and B_ν we have $v_{j,k} = \varepsilon_{j,k}$, and we see that the $\varepsilon_{j,k}$ will remain bounded for an arbitrary error distribution on rows p and $p+1$, i.e. for arbitrary A_ν, B_ν, if and only if

$$|e^{\eta\nu}| = 1 \quad \text{for} \quad \nu = 1, \ldots, N-1. \tag{3.28}$$

Now let $s = 1 - \cosh\eta$ and $|\mathrm{im}\,\eta| \leq \pi$. When s is real and positive [as it is here by virtue of (3.27) with $\xi = \nu\pi/N$], the corresponding values of η lie on the imaginary axis for $0 < s \leq 2$ and on the lines im $\eta = \pm\pi$ for $s > 2$; hence $|e^\eta| = 1$ for $0 < s \leq 2$ and $|e^\eta| > 1$ for $s > 2$. From (3.27), (3.28) our stability condition therefore reads

$$2z(\cos\xi - 1)^2 \leq 2 \quad \text{or} \quad z \leq \frac{1}{\left(\cos\dfrac{\nu\pi}{N} - 1\right)^2} \quad \text{for} \quad \nu = 1, \ldots, N-1.$$

This upper bound for z depends on N, though not strongly; a simple sufficient condition for stability which is valid for all N can be obtained by taking the limit as $N \to \infty$, i.e. by replacing the cosine in the denominator by -1:

$$z \leq \tfrac{1}{4}.$$

With $z = \tfrac{1}{4}$ (3.24) becomes

$$U_{j,k+1} = -U_{j,k-1} + \tfrac{1}{4}\left(-U_{j+2,k} + 4U_{j+1,k} + \left.\begin{array}{c}\\ \\\end{array}\right. \right. \\ \left. + 2U_{j,k} + 4U_{j-1,k} - U_{j-2,k}\right). \Bigg\} \tag{3.29}$$

The index of this formula is $J = 4$ while that of (3.25) is 15.

§ 4. Partial differential equations of the first order in one dependent variable

A complete theory for the integration of partial differential equations of the first order in one dependent variable $u(x_1, x_2, \ldots, x_n)$ and n independent variables $x_1, x_2, \ldots, x_n$ has existed for many years. In this theory the integration of the partial differential equation is reduced to the integration of a system of ordinary differential equations, and hence a numerical treatment can be based on the methods described in Chapter II. Our presentation of the theory[1] here will be confined to a summary (in § 4.1) of the results needed for a numerical calculation.

In spite of the completeness of the theory, the numerical integration of the systems of ordinary differential equations which arise is very laborious and there is a need for other approximate methods which will give a quantitative idea of the solution with a moderate amount of computation. We therefore go into other possible methods of treatment, albeit briefly, in §§ 4.3 to 4.5.

Unless otherwise stated, all functions and derivatives which occur are assumed to be continuous.

4.1. Results of the theory in the general case

Consider the first-order partial differential equation

$$F(u, x_1, x_2, \ldots, x_n, p_1, p_2, \ldots, p_n) = 0 \tag{4.1}$$

for the function $u(x_1, x_2, \ldots, x_n)$ of the independent variables $x_1, x_2, \ldots, x_n$. The p_j denote[2] the partial derivatives of u

$$p_j = \frac{\partial u}{\partial x_j} \qquad (j = 1, 2, \ldots, n), \tag{4.2}$$

and F is a given function which will be assumed continuous in the arguments specified.

This differential equation is called linear (see Ch. I, § 1.3) when F is linear in u and the p_j, i.e. when it has the form

$$A u + \sum_{j=1}^{n} A_j p_j = B, \tag{4.3}$$

where A, B and the A_j are given functions of $x_1, x_2, \ldots, x_n$.

It is called quasi-linear (Ch. I, § 1.3) when F is linear only in the p_j, i.e. when it has the form

$$\sum_{j=1}^{n} A_j p_j = B, \tag{4.4}$$

where now the A_j and B are given functions of $x_1, \ldots, x_n$ and u.

---

[1] Presentations of the theory can be found, for example, in R. COURANT and D. HILBERT: Methoden der mathematischen Physik, Vol. II, in particular, pp. 51 to 122. Berlin 1937. — KAMKE, E.: Differentialgleichungen reeller Funktionen, 4. Abschnitt, 2nd ed. Leipzig 1944. — SAUER, R.: Anfangswertprobleme bei partiellen Differentialgleichungen, Ch. 2. Berlin-Göttingen-Heidelberg 1952. — DUFF, G.: Partial differential equations, Ch.s II, III. Toronto 1956.

[2] In Ch. I, § 1.3 the partial derivatives were denoted by u_j instead of p_j; by using the letter p here we conform to the notation customary in the literature.

Let initial values be prescribed for the unique determination of the function u; in fact let an $(n-1)$-dimensional manifold be defined by prescribing u and the x_j as functions of parameters $t_1, \ldots, t_{n-1}$:

$$x_j = x_j(t_1, \ldots, t_{n-1}) \quad (j = 1, \ldots, n), \qquad u = u(t_1, \ldots, t_{n-1}).$$

This manifold may be extended to a "strip manifold" C by including n functions $p_j(t_1, \ldots, t_{n-1})$ which satisfy the following conditions: the strip conditions $(n-1$ in number)

$$\frac{\partial u}{\partial t_j} = \sum_{k=1}^{n} p_k \frac{\partial x_k}{\partial t_j} \qquad (j = 1, \ldots, n-1) \tag{4.5}$$

and the "compatibility" condition

$$F(u, x_1, \ldots, x_n, p_1, \ldots, p_n) = 0. \tag{4.6}$$

To solve the initial-value problem a function $u(x_1, \ldots, x_n)$ must be determined which satisfies the differential equation (4.1) and defines a surface containing the manifold C.

An expression which is critical for the solubility of this initial-value problem is the determinant

$$\Delta = \begin{vmatrix} F_{p_1} & F_{p_2} & \cdots & F_{p_n} \\ \dfrac{\partial x_1}{\partial t_1} & \dfrac{\partial x_2}{\partial t_1} & \cdots & \dfrac{\partial x_n}{\partial t_1} \\ \cdots & \cdots & \cdots & \cdots \\ \dfrac{\partial x_1}{\partial t_{n-1}} & \dfrac{\partial x_2}{\partial t_{n-1}} & \cdots & \dfrac{\partial x_n}{\partial t_{n-1}} \end{vmatrix}. \tag{4.7}$$

We shall assume that this determinant Δ is non-zero at all points of the initial manifold C; then the existence and uniqueness of the solution u is assured (always assuming the continuity of all functions which appear).

The solution u can be built up from the "characteristic strips". These are one-dimensional strip manifolds defined by $(2n+1)$ functions

$$u(s), \quad x_j(s), \quad p_j(s), \tag{4.8}$$

of a parameter s which satisfy the system of "characteristic equations"

$$\left. \begin{aligned} \frac{du}{ds} &= \sum_{k=1}^{n} p_k \frac{\partial F}{\partial p_k}, \quad \frac{dx_j}{ds} = \frac{\partial F}{\partial p_j}, \\ \frac{dp_j}{ds} &= -\frac{\partial F}{\partial x_j} - p_j \frac{\partial F}{\partial u} \end{aligned} \right\} \quad (j = 1, \ldots, n). \tag{4.9} \tag{4.10}$$

To each set of values $t_1, \ldots, t_{n-1}$, i.e. to each point of the initial manifold C, there corresponds one such characteristic strip. The system of $(2n+1)$ ordinary differential equations (4.9), (4.10) for the $(2n+1)$ functions u, x_j, p_j, together with, say, the initial conditions that for $s = 0$ the u,

20*

x_j, p_j take the values corresponding to a point of the initial manifold C, determine the functions (4.8) and can be integrated numerically by the methods of Chapter II.

Quasi-linearity of the differential equation (and hence linearity in particular) introduces a considerable simplification: with the equation in the form (4.4) the first $(n+1)$ equations of the characteristic system (4.9), (4.10) become

$$\frac{du}{ds} = B, \quad \frac{dx_j}{ds} = A_j \qquad (j = 1, 2, \ldots, n), \tag{4.11}$$

in which the p_j no longer appear; thus we need no longer concern ourselves with the p_j and can ignore the equations (4.5), (4.6) and (4.10). Instead of the characteristic strips we have "characteristic curves" $u(s)$, $x_j(s)$, which are solutions of (4.11) and are ordinary space curves in the $(n+1)$-dimensional (u, x_j) space. The determinant (4.7) becomes

$$\Delta = \begin{vmatrix} A_1 & \cdots & A_n \\ \dfrac{\partial x_1}{\partial t_1} & \cdots & \dfrac{\partial x_n}{\partial t_1} \\ \cdots & \cdots & \cdots \\ \dfrac{\partial x_1}{\partial t_{n-1}} & \cdots & \dfrac{\partial x_n}{\partial t_{n-1}} \end{vmatrix} \tag{4.12}$$

and provided that it does not vanish, the solution $u(x_1, \ldots, x_n)$ consists simply of the totality of characteristic curves which pass through the points of the initial manifold.

4.2. An example from the theory of glacier motion

A non-linear differential equation in two independent variables which possesses the simplifying property of quasi-linearity occurs in the theory of glacier motion[1]. It furnishes a mathematical description of the conditions in a glacier, particularly

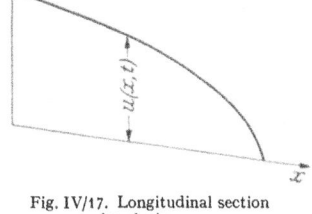

in the glacier tongue, or ablator, and appears in the form

$$[(n+1) \varkappa u^n - a] \frac{\partial u}{\partial x} + \frac{\partial u}{\partial t} = -a. \tag{4.13}$$

It pertains to a central longitudinal section of a glacier moving down a slightly inclined, straight bed.

The variables u and x refer to oblique axes in the plane of the section: $u(x, t)$ is the vertical depth of the ice at time t at a distance x along the bed (Fig. IV/17). The velocity distribution is assumed to be of the form $v = \varkappa u^n$, where $\varkappa$ depends on the slope of the bed and the exponent n is a constant lying between $\frac{1}{4}$ and $\frac{1}{3}$. The remaining symbol a

Fig. IV/17. Longitudinal section of a glacier

---

[1] See S. FINSTERWALDER: Die Theorie der Gletscherschwankungen. Z. Gletscherk. **2**, 81—103 (1907). This gives a detailed treatment of various cases of stationary glaciers, propagation of ridges, glacier shrinkage, the relative slipping of complete ice-blocks, etc.

is an ablation constant; it represents the annual melting on horizontal surfaces. The particular example which we consider here concerns a receding glacier on a very flat bed.

Let the shape of the longitudinal section of the glacier at time $t=0$ be given[1] in dimensionless variables by

$$u = 2\frac{4-x}{5-x} \quad \text{for} \quad 0 \leq x \leq 4 \tag{4.14}$$

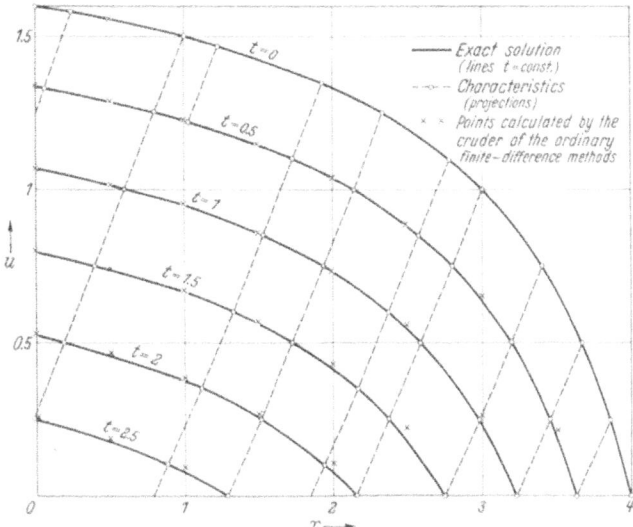

Fig. IV/18. Profile of a glacier at various times as an example of the integration of a first-order partial differential equation

and for numerical values of the parameters take $n=\frac{1}{3}$, $\varkappa=0.075$, $a=\frac{1}{2}$. Then (4.13) becomes

$$\left[0.1\sqrt[3]{u}-0.5\right]\frac{\partial u}{\partial x}+\frac{\partial u}{\partial t}=-0.5. \tag{4.15}$$

The characteristic equations (4.11) for equation (4.13) read

$$\frac{du}{ds}=-a, \quad \frac{dx}{ds}=(n+1)\varkappa u^{n}-a, \quad \frac{dt}{ds}=1. \tag{4.16}$$

Their general solution is a two-parameter family of curves in the (u, x, t) space; in this case the curves are plane and are given by

$$\left.\begin{array}{l} x+\xi = -\dfrac{\varkappa}{a}u^{n+1}+u = u - 0.15\,u^{\frac{4}{3}}, \\[2mm] u+\eta = -a\,t \qquad\qquad = -0.5\,t, \end{array}\right\} \tag{4.17}$$

---

[1] With a more sloping bed (greater $\varkappa$) and sufficiently large values of u the projections of the characteristics on the (x, t) plane slope away from the t axis instead of towards it $(dx/dt<0)$; then for the complete determination of $u(x, t)$ in the quadrant $x>0$, $t>0$, $u(0, t)$ must be prescribed in addition to $u(x, 0)$.

where ξ and η are the parameters. From this two-parameter family we select the one-parameter family of curves which pass through the points of the initial curve (4.14); the surface formed by the totality of these curves constitutes the solution of the problem. We select a set of points on the initial curve and calculate the corresponding characteristic parameters; for example, the point $t = 0$, $x = 3$, $u = 1$ yields $\xi = -2.15$, $\eta = 1$. Then the projections of these characteristics may be drawn in a (u, x) plane and graduated at (say equal) intervals in t; curves joining the points with the same values of t give the shape of the longitudinal section at various values of t (Fig. IV/18).

4.3. Power series expansions

If, as will always be assumed, the determinant (4.7) is non-zero over the given initial manifold, it is possible, in principle, to calculate the initial values of the higher partial derivatives of u. The solution can therefore be approximated by a number of terms of its Taylor series and this will give some idea of its behaviour in a neighbourhood of the initial manifold. Naturally nothing can be asserted generally about the extent of the region of convergence. Such a series expansion should always be used circumspectly, for the series can converge in certain regions to a limit function which is not the solution of the initial-value problem; examples which demonstrate this can be constructed quite easily (the example in the introduction to this chapter demonstrates the same phenomenon for a second-order equation).

The higher derivatives are calculated by repeated differentiation of the differential equation and of the initial conditions with respect to the x_j. By differentiating the differential equation and initial conditions in the form (4.5), (4.6) with respect to x_1, for instance, we obtain a system of linear equations for the n quantities $u_{11}, u_{21}, \ldots, u_{n1}$ (with the derivative notation of Ch. I, § 1.3); now the determinant of the coefficients obtained is precisely the determinant (4.7), which is assumed to be non-zero, and we can therefore solve for the derivatives u_{11}, $u_{21}, \ldots, u_{n1}$.

As an illustration consider the example of the last section § 4.2. Let us write

$$f = a - (n+1)\varkappa u^n, \quad u(x,0) = \varphi(x), \quad n(n+1)\varkappa \varphi^{n-1} = \Phi; \quad (4.18)$$

then the differential equation (4.13) reads

$$u_t = -a + f u_x$$

and the initial condition is

$$u(x,0) = \varphi(x) = 2 \cdot \frac{4-x}{5-x}.$$

Differentiation of the differential equation with respect to x and t yields the two equations

$$u_{xt} = -\Phi u_x^2 + f u_{xx},$$
$$u_{tt} = -\Phi u_x u_t + f u_{xt}$$

for the required values of the second derivatives on the initial manifold $t = 0$. Two differentiations of the initial condition yield

$$u_x = \varphi' = -\frac{2}{(5-x)^2}, \qquad u_{xx} = \varphi'' = -\frac{4}{(5-x)^3}.$$

Thus we have three linear equations for u_{xx}, u_{xt}, u_{tt}, which in this case can be solved directly by successive substitutions. For u_{tt}, for example, we obtain

$$u_{tt} = f^2 \varphi'' - 2f\Phi \varphi'^2 + a\Phi\varphi' = -\frac{1}{(5-x)^4}[8f\Phi + 4(5-x)f^2 + (5-x)^2\Phi]. \quad (4.19)$$

4.4. Application of the finite-difference method

Of the possible ways of making a rapid quantitative survey of the solution without a laborious integration of the characteristic equations, the finite-difference approach is probably the most suitable. The mesh width in the progressive direction must not be taken too large, of course, for, as in § 3.1, if the method is to make sense, the finite-difference determinate region of any part of the initial manifold must be completely contained within the corresponding determinate region outlined by the characteristics. Consequently the finite-difference method should not be used without first examining the run of the characteristics at least roughly (cf. the example).

There are usually several ways in which the differential equation can be replaced by a difference equation; we illustrate two ways — one more accurate than the other — by applying them to the example of glacier shrinking of § 4.2, i.e. the initial-value problem (4.15), (4.14). We employ the usual rectangular mesh

$$x_i = ih, \qquad y_k \text{ (here } t_k) = kl,$$

and in order to choose l suitably we first estimate the steepness of the characteristics. From (4.16) and using the fact that $u \geq 0$ (the depth of the glacier cannot be negative) and also that $0.5 - 0.1\sqrt[8]{u} > 0$, we have

$$\left|\frac{dt}{dx}\right| = \frac{1}{|(n+1)\varkappa u^n - a|} = \frac{1}{|0.5 - 0.1\sqrt[8]{u}|} \geq \frac{1}{0.5} = 2;$$

thus we can safely choose $l \leq 2h$ and for our example we take $l = h = \frac{1}{2}$.

1. The cruder method. Here the derivative with respect to x is replaced by the central difference quotient (1.4) but the derivative with respect to t is approximated crudely by the forward difference quotient

Table IV/15. *Calculation of glacier recession by the cruder finite-difference method*

	t	$i=0$, $x=0$	$i=1$, $x=0.5$	$i=2$, $x=1$	$i=3$, $x=1.5$	$i=4$, $x=2$	$i=5$, $x=2.5$	$i=6$, $x=3$	$i=7$, $x=3.5$	$i=8$, $x=4$
$U_{i,0}$	0	1·6	1·55556	1·5	1·42857	1·33333	1·2	1	0·66667	0
$\sqrt[3]{U_{i,0}}$		1·1696	1·1587	1·1447	1·1262	1·1006	1·0627	1	0·87358	0
$\tfrac{1}{2}f(U_{i,0})$		0·19152	0·19207	0·19277	0·19369	0·19497	0·19687	0·20000	0·20632	0·25
$U_{i+1,0}-U_{i-1,0}$		−0·07778	−0·10000	−0·12698	−0·16667	−0·22857	−0·33333	−0·53333	−1	−1·66667
$U_{i,1}$	0·5	1·33510	1·28635	1·22552	1·14629	1·03877	0·88438	0·64333	0·21035	−0·66667
$U_{i,2}$	1	1·06845	1·01491	0·94800	0·85938	0·73655	0·55450	0·25391	−0·32820	
$U_{i,3}$	1·5	0·79981	0·74085	0·66676	0·56657	0·42410	0·22119	−0·18882		
$U_{i,4}$	2	0·52888	0·46361	0·38080	0·26594	0·10073	−0·16351			
$U_{i,5}$	2·5	0·25520	0·18232	0·08855	−0·04507	−0·24664				
$U_{i,6}$	3	−0·02206	−0·10462	−0·21323						

(1.2). With f defined as in (4.18) we have

$$\left.\frac{U_{i,k+1}-U_{i,k}}{l} = -a + f\frac{U_{i+1,k}-U_{i-1,k}}{2h},\right\} \quad (4.20)$$

which becomes

$$U_{i,k+1} = U_{i,k} - 0.25 + \tfrac{1}{2}f(U_{i,k})\left[U_{i+1,k}-U_{i-1,k}\right]$$

when the numerical values we have chosen are inserted.

This formula enables us to calculate immediately the values $U_{i,k}$ above the heavy line in Table IV/15 [for the first row $(k=1)$ the intermediate calculation has been reproduced, but for the subsequent rows only the results are given]. By our choice of mesh ratio we have ensured that this triangle of points lies within the determinate region of the initial segment $0 \leqq x \leqq 4$, $t=0$. Actually we want to cover that part of the quadrant $x \geqq 0$, $t \geqq 0$ in which $u \geqq 0$, so we want to extend the calculation beyond the triangle as far as the boundary $u = 0$. This can be done by using unsymmetric difference quotients at the boundaries; for example, the formula

$$\left.\begin{aligned} U_{0,k+1} &= U_{0,k} - 0.25 + f(U_{0,k}) \times \\ &\times \frac{-3U_{0,k}+4U_{1,k}-U_{2,k}}{2} \end{aligned}\right\} \quad (4.21)$$

Table IV/16. Calculation of glacier recession by a more accurate finite-difference method

t		$i=0$ $x=0$	$i=1$ $x=0.5$	$i=2$ $x=1$	$i=3$ $x=1.5$	$i=4$ $x=2$	$i=5$ $x=2.5$	$i=6$ $x=3$	$i=7$ $x=3.5$	$i=8$ $x=4$
0	$U_{i,0}$	1·6	1·55555	1·5	1·42857	1·33333	1·2	1	0·66667	0
	$\frac{1}{2}u_i$	−0·26532	−0·26897	−0·27410	−0·28163	−0·29333	−0·31300	−0·35	−0·43340	
	$\frac{1}{8}u_{tt}$	−0·00072	−0·00098	−0·00140	−0·00209	−0·00333	−0·00585	−0·01188	−0·03121	
0·5	$U_{i,1}$	1·33396	1·28560	1·22451	1·14485	1·03667	0·88115	0·63812	0·20206	−0·66667
	$\sqrt[3]{U_{i,1}}$	1·1008	1·0873	1·0698	1·0461	1·0121	0·9587	0·8609	0·5868	
	$f(U_{i,1})$	0·38992	0·39127	0·39302	0·39539	0·39879	0·40413	0·41391	0·44132	
	$U_{i+1,1}-U_{i-1,1}$	−0·08399	−0·10945	−0·14075	−0·18784	−0·26370	−0·39855	−0·67909	−1·30479	
1	$U_{i,2}$	1·06725	1·01273	0·94468	0·85429	0·72817	0·53893	0·21892	−0·40916	
1·5	$U_{i,3}$	0·79597	0·73662	0·66084	0·55714	0·40736	0·16797	−0·27878		
2	$U_{i,4}$	0·52559	0·45737	0·37057	0·24841	0·06243	−0·26628			
2·5	$U_{i,5}$	0·24655	0·17105	0·07137	−0·07756	−0·32957				
3	$U_{i,6}$	−0·02987	−0·12050	−0·24342						

Table IV/17. *Error (in 5th-decimal units) in results obtained for the glacier problem by the various approximate methods*

Method	t	x = 0	x = 0·5	x = 1	x = 1·5	x = 2	x = 2·5	x = 3	x = 3·5
Ordinary finite-difference method — 1. Crude approximation using (4.20), (4.21)	0·5	+ 118	81	112	162	245	394	719	1810
	1	235	177	263	388	606	1044	2171	32934
	1·5	358	266	482	724	1187	4014	14679	
	2	512	486	804	1259	2651	10904		
	2·5	724	773	1342	2728	8466			
	3	1152	1625	3316					
Ordinary finite-difference method — 2. Better approximation using (4.22), (4.23)	0·5	4	6	11	18	35	71	198	981
	1	− 115	− 41	− 69	− 121	− 232	− 513	− 1328	24838
	1·5	− 26	− 157	− 110	− 219	− 487	− 1308	+ 5683	
	2	− 183	− 138	− 219	− 494	− 1179	+ 627		
	2·5	− 141	− 354	− 376	− 521	+ 173			
	3	− 371	+ 37	− 297					
$u^{[0]}$ obtained from a constant-coefficient approximation to the given differential equation	0·5	− 497	− 613	− 771	− 1005	− 1359	− 1933	− 2900	− 4225
	1	− 1055	− 1314	− 1680	− 2217	− 3049	− 4406	− 6553	+ 15734
	1·5	− 1682	− 2152	− 2732	− 3660	− 5112	− 7391	− 1439	
	2	− 2376	− 3018	− 3943	− 5335	− 7422	− 6078		
	2·5	− 3129	− 3997	− 5240	− 6654	− 6156			
	3	− 3785	− 4580	− 5361					
$u^{[1]}$ obtained from $u^{[0]}$ by one cycle of the iteration procedure	0·5	− 19	− 26	− 34	− 50	− 75	− 124	− 174	+ 562
	1	− 82	− 114	− 164	− 246	− 392	− 674	− 1228	+ 18863
	1·5	− 201	− 318	− 405	− 617	− 912	− 1496	+ 2410	
	2	− 391	− 556	− 820	− 1284	− 2418	− 654		
	2·5	− 624	− 903	− 1453	− 2717	− 2353			
	3	− 1178	− 1469	− 1840					

Values of the exact solution $u(x, t)$

t	x = 0	x = 0·5	x = 1	x = 1·5	x = 2	x = 2·5	x = 3	x = 3·5
0·5	1·33392	1·28554	1·22440	1·14467	1·03632	0·88044	0·63614	0·19225
1	1·06610	1·01314	0·94537	0·85550	0·73049	0·54406	0·23220	−0·65754
1·5	0·79623	0·73819	0·66194	0·55933	0·41223	0·18105	−0·33561	
2	0·52376	0·45875	0·37276	0·25335	0·07422	−0·27255		
2·5	0·24796	0·17459	0·07513	−0·07235	−0·33130			
3	−0·03358	−0·12087	−0·24639					

can be used at the boundary $i=0$ (there is a similar formula for the other boundary). This is permissible because in the region of interest ($u \geqq 0$) the characteristics have $dx/dt < 0$ and the region therefore lies within the determinate region of the initial segment $0 \leqq x \leqq 4$, $t=0$. Thus we may proceed until the $U_{i,k}$ become negative.

2. The more accurate method. Here we replace the derivatives with respect to x and t both by central difference quotients:

$$U_{i,k+1} = U_{i,k-1} - 0\cdot 5 + f(U_{i,k})\,[U_{i+1,k} - U_{i-1,k}]. \tag{4.22}$$

Starting values are now required on the first *two* rows ($k=0$, $k=1$). The values on the row $k=1$ can be found from

$$U_{i,1} - U_{i,-1} = 2l\,u_t, \qquad U_{i,1} - 2U_{i,0} + U_{i,-1} = l^2 u_{tt},$$

which yield

$$U_{i,1} = U_{i,0} + l\,u_i + \tfrac{1}{2} l^2 u_{tt} \tag{4.23}$$

(the beginning of the Taylor series). The derivative u_{tt} can be calculated in the same way as for the power series expansion of § 4.3; in fact the required formula has already been derived as (4.19). Since Φ becomes infinite when $x=4$, this formula cannot be used for the value at $x=4$, $t=1$; thus $U_{8,1}$ must be calculated by some other means, and in Table IV/16 the value found by the cruder method is taken over. Here also an unsymmetric formula [corresponding to (4.21)] must be used at the boundary:

$$U_{0,k+1} = U_{0,k-1} - 0\cdot 5 + f(U_{0,k})\,[-3U_{0,k} + 4U_{1,k} - U_{2,k}].$$

The errors in the results obtained by these two finite-difference methods are given in Table IV/17 (the exact values were calculated from the characteristics by interpolation). It can be seen that the errors are smaller for the second method than for the first, as was to be expected.

4.5. Iterative methods

An iterative process can be defined by solving the given differential equation (4.1) for one of the partial derivatives, say p_1:

$$p_1 = \frac{\partial u}{\partial x_1} = G(u, x_1, \ldots, x_n, p_2, \ldots, p_n),$$

and replacing u by $u^{[k+1]}$ on the left and by $u^{[k]}$ on the right. Thus the next approximation $u^{[k+1]}$ is determined from the current approximation $u^{[k]}$ by solving the differential equation

$$\frac{\partial u^{[k+1]}}{\partial x_1} = G\left(u^{[k]}, x_1, \ldots, x_n, \frac{\partial u^{[k]}}{\partial x_2}, \ldots, \frac{\partial u^{[k]}}{\partial x_n}\right)$$

for a function $u^{[k+1]}$ satisfying the initial conditions prescribed for u (this method may not always be suitable or even possible). In propagation-type problems the variable corresponding to time will be taken as x_1.

Accordingly, with f defined as in (4.18), the iteration formula for the differential equation (4.13) of the example in § 4.2 reads

$$\frac{\partial u^{[k+1]}}{\partial t} = - a + f(u^{[k]}) \frac{\partial u^{[k]}}{\partial x}.$$

A first approximation can be found by solving the initial-value problem

$$- a \frac{\partial u^{[0]}}{\partial x} + \frac{\partial u^{[0]}}{\partial t} = - a, \quad u^{[0]}(x, 0) = \varphi(x) = 2 \cdot \frac{4 - x}{5 - x},$$

whose differential equation represents a constant-coefficient approximation to (4.13). This simplified equation admits of a general solution expressible very simply in terms of an arbitrary function w:

$$u^{[0]}(x, t) = - a t + w(x + a t).$$

The initial condition implies that $w = \varphi$, so that finally

$$u^{[0]}(x, t) = - a t + \varphi(x + a t) = 2 - \frac{t}{2} - \frac{2}{5 - x - \frac{t}{2}}.$$

With

$$u^{[0]}_x = \frac{\partial u^{[0]}}{\partial x} = - \frac{2}{\left(5 - x - \frac{t}{2}\right)^2}$$

the equations for the next approximation $u^{[1]}$ read

$$u^{[1]}_t = - a + f(u^{[0]}) u^{[0]}_x, \quad u^{[1]}(x, 0) = \varphi(x).$$

For a given value of x, $u^{[1]}_t$ can be immediately tabulated at intervals h in t, then $u^{[1]}$ calculated by numerical integration. In Table IV/18 this is done with $h = \frac{1}{2}$ for the values of x used in the finite-difference methods in § 4.4. The values of $u^{[1]}$ for $t = 1, 2, 3, \ldots$ are calculated by SIMPSON's rule and the intermediate values for $t = \frac{1}{2}, \frac{3}{2}, \frac{5}{2}, \ldots$ by the formula

$$\int_0^h g(x) \, dx \approx \frac{h}{12} \left[5 g(0) + 8 g(h) - g(2h)\right]$$

(for the rows with $t \geq 1$ only the results are given). The errors in the values of $u^{[0]}$ and $u^{[1]}$ are also given in Table IV/17.

Table IV/18. *First cycle of the iteration procedure in the iterative solution of the glacier recession problem*

t		$x=0$	$x=0.5$	$x=1$	$x=1.5$	$x=2$	$x=2.5$	$x=3$	$x=3.5$	$x=4$
$t=0$	$u^{[0]}$	1.6	1.55556	1.5	1.42857	1.33333	1.2	1	0.66667	0
	$5-x-\tfrac{t}{2}$	5	4.5	4	3.5	3	2.5	2	1.5	1
	$f(u^{[0]})$	0.38304	0.38413	0.38553	0.38738	0.38994	0.39373	0.4	0.41264	0.5
	$u_x^{[0]}$	−0.08	−0.09877	−0.125	−0.16327	−0.22222	−0.32	−0.5	−0.88889	−2
	$u_t^{[1]}$	−0.53064	−0.53794	−0.54819	−0.56325	−0.58665	−0.62599	−0.7	−0.86679	−2.5
$t=0.5$	$u^{[0]}$	1.32895	1.27941	1.21667	1.13462	1.02273	0.86111	0.60714	0.15	
	$5-x-\tfrac{t}{2}$	4.75	4.25	3.75	3.25	2.75	2.25	1.75	1.25	
	$f(u^{[0]})$	0.39006	0.39144	0.39324	0.39570	0.39925	0.40486	0.41532	0.44687	
	$u_x^{[0]}$	−0.08864	−0.11073	−0.14222	−0.18935	−0.26446	−0.39506	−0.65306	−1.28	
	$u_t^{[1]}$	−0.53458	−0.54334	−0.55593	−0.57493	−0.60558	−0.65994	−0.77123	−1.07199	
	$u^{[1]}$	1.33373	1.28528	1.22406	1.14417	1.03557	0.87920	0.63440	0.19787	
$t=1$	$u^{[0]}$	1.05556	1	0.92857	0.83333	0.7	0.5	0.16667	−0.5	
	$u_t^{[1]}$	−0.53933	−0.55	−0.56571	−0.59020	−0.63159	−0.71032	−0.89553	−1.65874	
	$u^{[1]}$	1.06528	1.01200	0.94373	0.85304	0.72657	0.53732	0.21992	−0.46891	
$t=1.5$	$u^{[0]}$	0.77941	0.71667	0.63462	0.52273	0.36111	0.10714	−0.35		
	$u_t^{[1]}$	−0.54517	−0.55838	−0.57840	−0.61092	−0.66940	−0.79551	−1.23020		
	$u^{[1]}$	0.79422	0.73501	0.65789	0.55316	0.40311	0.16609	−0.31151		
$t=2$	$u^{[0]}$	0.5	0.42857	0.33333	0.2	0	−0.33333			
	$u_t^{[1]}$	−0.55258	−0.56932	−0.59570	−0.64129	−0.75	−1.00608			
	$u^{[1]}$	0.51985	0.45319	0.36456	0.24051	−0.05004	−0.27909			
$t=2.5$	$u^{[0]}$	0.21667	0.13462	0.02273	−0.13889	−0.39286				
	$u_t^{[1]}$	−0.56257	−0.58497	−0.62474	−0.71799	−0.87436				
	$u^{[1]}$	0.24172	0.16556	0.06060	−0.09952	−0.35483				
$t=3$	$u^{[0]}$	−0.07143	−0.16667	−0.3	−0.5					
	$u_t^{[1]}$	−0.58841	−0.62334	−0.68142	−0.78969					
	$u^{[1]}$	−0.04536	−0.13556	−0.26479	−0.47664					

4.6. Application of Hermite's formula

Another method has been suggested by PFLANZ[1]; it utilizes HERMITE'S generalization of TAYLOR'S formula (Ch. I, § 2.5). For the sake of simplicity we describe the method for a differential equation

$$F(x, y, u, p, q) = 0 \qquad (4.24)$$

in only two independent variables x, y (here $p = \partial u/\partial x$, $q = \partial u/\partial y$). Let U_k, P_k, Q_k be approximations to u, p, q at the point $x = x_k$, $y = y_k$ and U, P, Q corresponding approximations at the point x, y. Then (2.60) of Ch. I with $k = m = 1$ and x_1, y_1, x_0, y_0, h, k replaced by x, y, x_j, y_j, h_j, k_j, where $h_j = x - x_j$, $k_j = y - y_j$, yields

$$U - \frac{h_j}{2} P - \frac{k_j}{2} Q = C_j = U_j + \frac{h_j}{2} P_j + \frac{k_j}{2} Q_j. \qquad (4.25)$$

Now let x_0, y_0, u_0 and x_1, y_1, u_1 be two points on the initial curve; then U_0, P_0, Q_0, U_1, P_1, Q_1 are known. If we write down (4.25) for $j = 0$ and $j = 1$, and also demand that U, P, Q satisfy the differential equation (4.24), i.e.

$$F(x, y, U, P, Q) = 0, \qquad (4.26)$$

then we have three equations for the three unknowns U, P, Q. P and Q can be expressed linearly in terms of U by means of (4.25) with $j = 0$ and $j = 1$, so we are left with (4.26) to solve for U; this is usually done iteratively.

As long as the computational labour does not become too great the method may be extended to higher approximations by using (2.60) of Ch. I with $k = m = 2$. Approximations R, S, T to $\dfrac{\partial^2 u}{\partial x^2}, \dfrac{\partial^2 u}{\partial x \partial y}, \dfrac{\partial^2 u}{\partial y^2}$ must then be included and HERMITE's formula has to be written down for three points on the initial curve:

$$\left. \begin{array}{l} 12U - 6(h_j P + k_j Q) + h_j^2 R + 2h_j k_j S + k_j^2 T \\ = 12U_j + 6(h_j P_j + k_j Q_j) + h_j^2 R + 2h_j k_j S_j + k_j^2 T_j \end{array} \right\} \quad (j = 0, 1, 2). \qquad (4.27)$$

The system of equations for U, P, Q, R, S, T is completed by (4.26) together with the two equations

$$F_x + F_u P + F_p R + F_q S = 0, \qquad F_y + F_u Q + F_p S + F_q T = 0. \qquad (4.28)$$

§5. The method of characteristics for systems of two differential equations of the first order

In this section we describe briefly the elements of a characteristics method which has been used extensively for hydrodynamic and aerodynamic problems but mostly in graphical mode. For the details, in particular, the exploitation of various advantages afforded by graphical treatment[2], and also for more extensive problems[3], the reader is referred to the literature.

---

[1] PFLANZ, E.: Bemerkungen über die Methode von G. DUFFING zur Integration von Differentialgleichungen. Z. Angew. Math. Mech. **28**, 167—172 (1948). He also gives a numerical example.

[2] MASSAU, J.: Mémoire sur l'intégration graphique des équations aux dérivées partielles. Gent 1900; for a description of the method see in particular §§ 5, 6 Ch. III, pp. 46—58; the method is applied to numerous hydrodynamic problems.

[3] COURANT, R., and D. HILBERT: Methoden der mathematischen Physik, Vol. II, p. 303 et seq. Berlin 1937. — SAUER, R.: Charakteristikenverfahren für die ein-

When carried out numerically, the method has a certain resemblance to the finite-difference method, but unlike that method it has the important advantage (cf. § 3.1) that it builds up an accurate approximation to the true determinate region of a given initial segment.

We shall consider a system of two quasi-linear differential equations in two dependent variables $u(x, y)$ and $v(x, y)$, i.e. a system of the form

$$a_1 u_x + a_2 u_y + a_3 v_x + a_4 v_y = A,$$
$$b_1 u_x + b_2 u_y + b_3 v_x + b_4 v_y = B, \tag{5.1}$$

where a_j, b_j, A, B are given functions of x, y, u, v.

5.1. The characteristics

Let us begin by considering the problem of calculating u and v in the neighbourhood of a curve C on which their values are prescribed. Imagine curvilinear co-ordinates ξ, η introduced in such a way that one of them, say ξ, is constant on the given curve C (Fig. IV/19). We shall assume that in a neighbourhood of C the co-ordinate transformation $(x, y) \to (\xi, \eta)$ is one-one, i.e. the Jacobian of the transformation is non-zero:

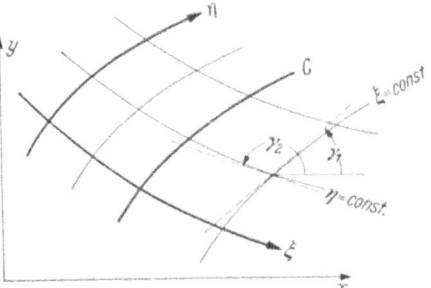

$$\Phi = \frac{\partial(\xi, \eta)}{\partial(x, y)} = \begin{vmatrix} \xi_x & \xi_y \\ \eta_x & \eta_y \end{vmatrix} \neq 0. \tag{5.2}$$

Fig. IV/19. Introduction of the characteristics

u and v are known on C and hence also are their derivatives u_η, v_η along C. We now try to determine the values of the other derivatives u_ξ, v_ξ on C. Substituting

$$u_x = u_\xi \xi_x + u_\eta \eta_x$$

dimensionale instationäre Gasströmung. Ing.-Arch. **13**, 79—89 (1942). — SCHULTZ-GRUNOW, F.: Nichtstationäre, eindimensionale Gasbewegung. Forsch.-Arb. Ing.-Wes. **13**, 125—134 (1942). — SAUER, R.: Charakteristikenverfahren für Kugel- und Zylinderwellen reibungsloser Gase. Z. Angew. Math. Mech. **23**, 29—32 (1943). — Theoretische Einführung in die Gasdynamik, p. 146 et seq. Berlin 1943. — OSWATITSCH, KL.: Über die Charakteristikenverfahren der Hydrodynamik. Z. Angew. Math. Mech. **25**, 195—208 (1947); **27**, 264—270 (1947). (A comprehensive review.) — COURANT, R., and K. FRIEDRICHS: Supersonic flow and shock waves. New York: Interscience Publishers, Inc. 1948. — SAUER, R.: Dreidimensionale Probleme der Charakteristikentheorie partieller Differentialgleichungen. Z. Angew. Math. Mech. **30**, 347—356 (1950). — Anfangswertprobleme bei partiellen Differentialgleichungen, p. 229 et seq. Berlin-Göttingen-Heidelberg 1952. — Hyperbolische Probleme der Gasdynamik mit mehr als zwei unabhängigen Veränderlichen. Z. Angew. Math. Mech. **33**, 331—336 (1953).

and corresponding expressions for u_y, v_x, v_y into (5.1) we obtain

$$
\left.\begin{aligned}
(a_1 \xi_x + a_2 \xi_y)\, u_\xi &+ (a_3 \xi_x + a_4 \xi_y)\, v_\xi \\
&= A - (a_1 \eta_x + a_2 \eta_y)\, u_\eta - (a_3 \eta_x + a_4 \eta_y)\, v_\eta, \\
(b_1 \xi_x + b_2 \xi_y)\, u_\xi &+ (b_3 \xi_x + b_4 \xi_y)\, v_\xi \\
&= B - (b_1 \eta_x + b_2 \eta_y)\, u_\eta - (b_3 \eta_x + b_4 \eta_y)\, v_\eta.
\end{aligned}\right\} \tag{5.3}
$$

Thus we have two linear equations for the unknown values of u_ξ and v_ξ on C. If the determinant of coefficients

$$
\varDelta = \begin{vmatrix} a_1 \xi_x + a_2 \xi_y & a_3 \xi_x + a_4 \xi_y \\ b_1 \xi_x + b_2 \xi_y & b_3 \xi_x + b_4 \xi_y \end{vmatrix} \tag{5.4}
$$

does not vanish, u_ξ and v_ξ may be calculated. Then in the first instance the values of u and v at neighbouring points $(\xi + \xi', \eta)$ for sufficiently small ξ' can be approximated quite crudely by the first-order terms of their Taylor series.

If, however, the determinant $\varDelta$ does vanish along C, then the values of u and v at neighbouring points cannot be so calculated. A curve $\xi(x, y) = $ constant along which $\varDelta$ vanishes identically is called a characteristic. In general this "characteristic" property depends on the particular solution u, v, so that, unless certain simplifications obtain, we can only speak of a curve as a characteristic with respect to a specific solution u, v. These characteristics form the basis of the approximate method to be described; we therefore derive now several of their properties which will be needed later. It is convenient to introduce a notation for certain determinants which will appear in most of the formulae:

$$
a_{jk} = a_j b_k - a_k b_j, \quad A_j = A b_j - B a_j \quad (j, k = 1, 2, 3, 4); \tag{5.5}
$$

thus the determinant (5.4), for instance, can be expressed in the form

$$
\varDelta = a_{13} \xi_x^2 + (a_{14} + a_{23})\, \xi_x \xi_y + a_{24} \xi_y^2. \tag{5.6}
$$

Let γ be the angle between the tangent to a characteristic $\xi = $ constant and the x axis, so that

$$
\tan \gamma = - \frac{\xi_x}{\xi_y}. \tag{5.7}
$$

The equation $\varDelta = 0$ yields a quadratic equation for $\tan \gamma$, namely

$$
a_{13} \tan^2 \gamma - (a_{14} + a_{23}) \tan \gamma + a_{24} = 0, \tag{5.8}
$$

from which we obtain the two values

$$
\tan \gamma = \frac{a_{14} + a_{23} \pm D}{2 a_{13}}, \tag{5.9}
$$

where

$$
D^2 = (a_{14} + a_{23})^2 - 4 a_{13} a_{24}. \tag{5.10}
$$

In the following we assume that $D^2 > 0$ so that there are two real values for $\tan \gamma$. The system of differential equations (5.1) is then called hyperbolic with respect to the solution u, v. For a given solution u, v we therefore have two values of $\tan \gamma$, say $\tan \gamma_1$ and $\tan \gamma_2$, at each point (x, y) and hence two corresponding ("characteristic") directions whose direction ratios satisfy the equation $\Delta = 0$. Thus we have two "direction fields", each of which generates a one-parameter family of characteristic curves. We assume for the remainder that $\xi = $ constant and $\eta = $ constant are the two families of characteristics.

The equations for the characteristics do not depend explicitly on the functions A, B and they depend on u and v only through the functions a_j, b_j. If the latter depend only on x and y, the characteristics can be determined from (5.9) without reference to a particular solution and thus constitute two families of curves which, for a given system (5.1), are fixed once and for all, and are independent of any boundary conditions which would be needed to specify a particular solution.

We now note for later use some algebraic transformations. By expressing the following determinant in terms of the second-order sub-determinants which can be formed from its first two rows, and using the fact that its value must be zero since it has at least one pair of identical rows, we obtain

$$\frac{1}{2} \begin{vmatrix} a_1 & a_2 & a_3 & a_4 \\ b_1 & b_2 & b_3 & b_4 \\ a_1 & a_2 & a_3 & a_4 \\ b_1 & b_2 & b_3 & b_4 \end{vmatrix} = a_{12} a_{34} - a_{13} a_{24} + a_{14} a_{23} = 0. \tag{5.11}$$

By virtue of this identity, (5.10) can be put in the form

$$D^2 = (a_{23} - a_{14})^2 - 4 a_{12} a_{34}. \tag{5.12}$$

The expressions (5.9) for the two values of $\tan \gamma$ differ from each other only in the sign of D; we associate them with the two families of characteristics with angles γ_1 and γ_2 (Fig. IV/19) as follows:

$$\left. \begin{aligned} \tan \gamma_1 &= -\frac{\xi_x}{\xi_y} = \frac{2 a_{24}}{a_{14} + a_{23} - D} = \frac{a_{14} + a_{23} + D}{2 a_{13}}, \\ \tan \gamma_2 &= -\frac{\eta_x}{\eta_y} = \frac{2 a_{24}}{a_{14} + a_{23} + D} = \frac{a_{14} + a_{23} - D}{2 a_{13}}. \end{aligned} \right\} \tag{5.13}$$

The equivalence of the two alternative expressions given here can be verified immediately using (5.10).

5.2. Consistency conditions

Along a characteristic, say $\xi = $ constant, there exists a certain relation between the values of u and v. This relation, which is, of course, a consequence of the characteristic condition, is established as follows.

In a region of the (x, y) plane in which u, v and their partial derivatives are continuous u_ξ and v_ξ have specific finite values satisfying the conditions (5.3), and therefore, since the determinant $\varDelta$ [(5.4)] vanishes when $\xi = $ constant is a characteristic, the right-hand sides of (5.3) must satisfy the usual consistency conditions along a characteristic; these conditions yield, in fact, a linear relation between the values of u_η and v_η; for example, the condition that the determinant corresponding to u_ξ must vanish, namely

$$\varDelta_1 = \begin{vmatrix} A - (a_1 \eta_x + a_2 \eta_y)\, u_\eta - (a_3 \eta_x + a_4 \eta_y)\, v_\eta & a_3 \xi_x + a_4 \xi_y \\ B - (b_1 \eta_x + b_2 \eta_y)\, u_\eta - (b_3 \eta_x + b_4 \eta_y)\, v_\eta & b_3 \xi_x + b_4 \xi_y \end{vmatrix} = 0, \quad (5.14)$$

yields the relation

$$\left. \begin{aligned} A_3 \xi_x + A_4 \xi_y - u_\eta (a_{13} \xi_x \eta_x + a_{14} \xi_y \eta_x + a_{23} \xi_x \eta_y + \\ + a_{24} \xi_y \eta_y) + v_\eta a_{34} \varPhi = 0, \end{aligned} \right\} \quad (5.15)$$

in which we have used the notation of (5.2) and (5.5).

The factor multiplying u_η here can be simplified using (5.7) and (5.8): with $\gamma = \gamma_1$ along $\xi = $ constant we have

$$\xi_y \left[\eta_x (- a_{13} \tan \gamma_1 + a_{14}) + \eta_y (- a_{23} \tan \gamma_1 + a_{24}) \right]$$
$$= \xi_y \left[\eta_x (- a_{13} \tan \gamma_1 + a_{14}) + \eta_y \tan \gamma_1 (- a_{13} \tan \gamma_1 + a_{14}) \right]$$
$$= \xi_y \left(\eta_x - \eta_y \frac{\xi_x}{\xi_y} \right) (- a_{13} \tan \gamma_1 + a_{14}) = \varPhi (a_{13} \tan \gamma_1 - a_{14}).$$

Then, after division by a_{34}, (5.15) becomes

$$\varPhi v_\eta - q_1 \varPhi u_\eta + \frac{A_3 \xi_x + A_4 \xi_y}{a_{34}} = 0, \quad (5.16)$$

where [from (5.12), (5.13)]

$$q_1 = \frac{a_{13} \tan \gamma_1 - a_{14}}{a_{34}} = \frac{- a_{14} + a_{23} + D}{2 a_{34}} = \frac{2 a_{12}}{- a_{14} + a_{23} - D}. \quad (5.17)$$

Now let the arc length s be introduced as parameter on the characteristic $\xi = $ constant; then

$$\frac{dv}{ds} = \frac{\partial v}{\partial \eta} \frac{d\eta}{ds} = v_\eta (\eta_x x_s + \eta_y y_s) = v_\eta x_s \left(\eta_x + \eta_y \frac{dy}{dx} \right)$$
$$= v_\eta x_s \left(\eta_x - \eta_y \frac{\xi_x}{\xi_y} \right) = - \frac{v_\eta x_s}{\xi_y} \varPhi,$$

and similarly

$$\frac{du}{ds} = - \frac{u_\eta x_s}{\xi_y} \varPhi;$$

also

$$x_s = \cos \gamma_1 \quad \text{and} \quad - \frac{\xi_x}{\xi_y} = \tan \gamma_1,$$

so that from (5.16) we have finally

$$\left(\frac{dv}{ds} - q_1\frac{du}{ds}\right)_{\xi=\text{const}} = -\frac{A_3}{a_{34}}\sin\gamma_1 + \frac{A_4}{a_{34}}\cos\gamma_1. \tag{5.18}$$

For a characteristic of the other family $\eta=$ constant, D is to be replaced by $-D$; hence we have (using the same symbol s to denote the new arc length)

$$\left(\frac{dv}{ds} - q_2\frac{du}{ds}\right)_{\eta=\text{const}} = -\frac{A_3}{a_{34}}\sin\gamma_2 + \frac{A_4}{a_{34}}\cos\gamma_2, \tag{5.19}$$

where

$$q_2 = \frac{-a_{14}+a_{23}-D}{2a_{34}} = \frac{2a_{12}}{-a_{14}+a_{23}+D}. \tag{5.20}$$

Thus along the characteristics the values of u and v must satisfy the conditions (5.18), (5.19), respectively, while along curves on which $\varDelta \neq 0$ u and v are not related in this manner.

5.3. The method of characteristics

We suppose now that the values of u and v are given on an arc of a curve K_1 which is nowhere tangent to a characteristic. We may suppose further that the values of u (say) alone are given on a contiguous arc K_2 which is likewise no-where tangent to a character-istic and which makes with K_1 at their intersection P an angle which includes just one of the characteristics passing through P (see Fig. IV/20).

Let P_1 and P_2 be two points on the curve K_1 and let the characteristic $\xi =$ constant passing through P_1 intersect the characteristic $\eta=$ constant passing through P_2 at the

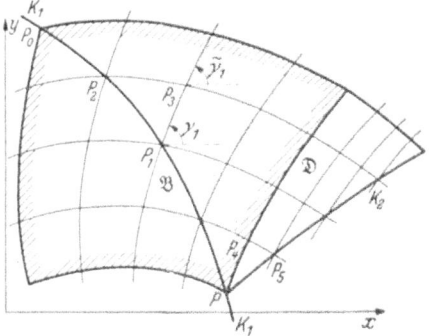

Fig. IV/20. Determinate regions

point P_3 (Fig. IV/20). We do not yet know the position of P_3, but if P_1 and P_2 are sufficiently close together we can use the point of intersection of the tangents to the characteristics at P_1 and P_2 as an approximation to it. Further, we can approximate to the equations (5.18), (5.19), which hold along the respective characteristics, by replacing the deriv-atives by difference quotients for the steps $s_1 = P_1P_3$ and $s_2 = P_2P_3$, respectively. In this way we obtain two equations for u_3 and v_3:

$$(v_3 - v_1) - q_1(u_3 - u_1) = -\frac{s_1}{a_{34}}(A_3\sin\gamma_1 - A_4\cos\gamma_1), \tag{5.21}$$

$$(v_3 - v_2) - q_2(u_3 - u_2) = -\frac{s_2}{a_{34}}(A_3\sin\gamma_2 - A_4\cos\gamma_2). \tag{5.22}$$

21*

In similar fashion we can calculate values for u and v at all points of the approximate characteristic mesh within the determinate region of the arc K_1 (the curvilinear "parallelogram" shaded in Fig. IV/20), or rather within the polygonal approximation to this region outlined by the approximate characteristics through P and P_0.

If, as mentioned at the beginning, we also have values of u prescribed on the curve K_2, we can start say at P_5 in Fig. IV/20 and use the equation corresponding to (5.22) for P_4, P_5 to calculate the missing value v_5 from the known values u_4, v_4, u_5; the next interior point can then be treated as above, and in this way, using just the one equation whenever we have to deal with a boundary point, we can proceed to fill in the region between K_2 and the enclosed characteristic through P.

We can improve somewhat on the approximation represented by (5.21), (5.22) by using one of the methods of the second chapter for the integration of (5.18), (5.19), such as one of the simple methods of § 1.5 of that chapter. We can also improve on the approximate position of P_3 by using, for example, the principle of the Euler-Cauchy method: firstly, a provisional approximation is calculated by the method described above, i.e. the intersection of the tangents at the points P_1, P_2 is used as a provisional position $\tilde{P}_3$ for the point P_3 of the characteristic mesh and corresponding values $\tilde{u}_3$, $\tilde{v}_3$ for u_3, v_3 are calculated from (5.21), (5.22); from these are calculated corresponding approximations $\tilde{\gamma}_1$, $\tilde{\gamma}_2$ to the characteristic directions at P_3 (see Fig. IV/20); then the intersection of the new straight lines through P_1 and P_2 making the angles $\frac{1}{2}(\gamma_1 + \tilde{\gamma}_1)$ and $\frac{1}{2}(\gamma_2 + \tilde{\gamma}_2)$, respectively, with the x axis may be expected to give a better approximation to the position of P_3. A simple method from which better values for u and v may be expected is to use now the mean values $\frac{1}{2}(u_1 + \tilde{u}_3)$, $\frac{1}{2}(v_1 + \tilde{v}_3)$ for u and v in the right-hand side of (5.21) and correspondingly for (5.22); effectively, we write down (5.18) for the mid-point of $\overline{P_1 P_3}$ with these mean-value approximations for the local values of u and v, and correspondingly for (5.19), and replace the derivatives by central difference quotients.

5.4. Example

We choose a particularly simple example with differential equations admitting of predetermined characteristics and for which the exact solution is known so that the error in any approximation is also known.

The current $J(x, t)$ and potential $V(x, t)$ in an electric cable satisfy the equations

$$C \frac{\partial V}{\partial t} + S V = -\frac{\partial J}{\partial x}, \quad L \frac{\partial J}{\partial t} + R J = -\frac{\partial V}{\partial x}, \qquad (5.23)$$

where t is the time, x is measured along the cable and C, R, L, S are the usual symbols for capacity, resistance, self-induction and leakage (per unit length of cable).

Let us consider the case $S = 0$. Then with $u = J, v = -CV, t = y$, $LC = \alpha^2$, $RC = \beta$ the differential equations become

$$\left.\begin{array}{l} u_x - v_y = 0 \\ -\alpha^2 u_y + v_x = \beta u. \end{array}\right\} \quad (5.24)$$

Here the matrix of the coefficients a_j, b_j, A, B in (5.1) reads

$$\begin{pmatrix} 1 & 0 & 0 & -1 & 0 \\ 0 & -\alpha^2 & +1 & 0 & \beta u \end{pmatrix}$$

and the values of the determinants (5.5) are

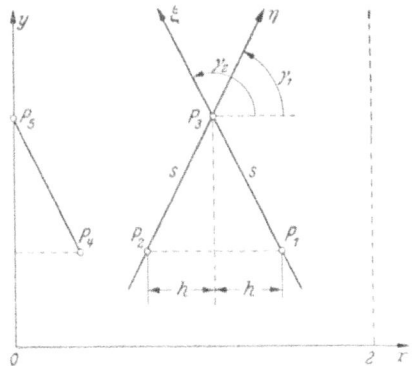

Fig. IV/21. The fixed straight-line characteristics of the example of § 5.4

$$a_{12} = -\alpha^2, \quad a_{13} = 1, \quad a_{14} = a_{23} = 0, \quad a_{24} = -\alpha^2, \quad a_{34} = 1,$$
$$A_3 = 0, \quad A_4 = \beta u.$$

From (5.10), (5.13) we have

$$D^2 = 4\alpha^2, \quad D = 2\alpha, \quad \tan \gamma_1 = \alpha, \quad \tan \gamma_2 = -\alpha,$$

so that the characteristics are the straight lines $y \pm \alpha x = $ constant; they are independent of boundary conditions and particular solutions.

If, for convenience, we measure the arc length s from the initial line $y = 0$, we have $\cos \gamma_1 = -\cos \gamma_2 = (1 + \alpha^2)^{-\frac{1}{2}}$; then, since $q_1 = \alpha, q_2 = -\alpha$ from (5.17), (5.20), the relations (5.18), (5.19) which hold along the characteristics read

$$\left.\begin{array}{l} \left(\dfrac{dv}{ds} - \alpha \dfrac{du}{ds}\right)_{\xi = \text{const}} = \dfrac{1}{\sqrt{1 + \alpha^2}} \beta u, \\[3mm] \left(\dfrac{dv}{ds} + \alpha \dfrac{du}{ds}\right)_{\eta = \text{const}} = -\dfrac{1}{\sqrt{1 + \alpha^2}} \beta u. \end{array}\right\} \quad (5.25)$$

If we write down these equations for two points P_1, P_2 lying on a parallel to the x axis a distance $2h$ apart (Fig. IV/21) and replace the derivatives by forward difference quotients, we obtain (with function values at the points P_j in Fig. IV/21 denoted by u_j, v_j)

$$v_3 - v_2 - \alpha(u_3 - u_2) = h\beta u_2, \quad v_3 - v_1 + \alpha(u_3 - u_1) = -h\beta u_1 \quad (5.26)$$

and hence the first crude approximations

$$\left.\begin{array}{l} v_3 = \tfrac{1}{2}\left[(v_1 + v_2) + (\alpha - \beta h)(u_1 - u_2)\right], \\[2mm] u_3 = \dfrac{1}{2\alpha}\left[(v_1 - v_2) + (\alpha - \beta h)(u_1 + u_2)\right]. \end{array}\right\} \quad (5.27)$$

We can approximate equations (5.25) rather more accurately by writing them down for the mid-points of $\overline{P_1 P_3}$ and $\overline{P_2 P_3}$, using the arithmetic means of the end-point values for the function values at the mid-points and replacing the derivatives by central difference quotients:

$$\left.\begin{aligned}
v_3 - v_2 - \alpha\,(u_3 - u_2) &= h\beta\,\frac{u_2 + u_3}{2}\,, \\[2mm]
v_3 - v_1 + \alpha\,(u_3 - u_1) &= -\,h\beta\,\frac{u_1 + u_3}{2}\,.
\end{aligned}\right\} \tag{5.28}$$

Solving these two equations for u_3, v_3 we obtain the approximations

$$\left.\begin{aligned}
v_3 &= \frac{1}{2}\left[(v_1 + v_2) + \left(\alpha - \frac{\beta\,h}{2}\right)(u_1 - u_2)\right], \\[2mm]
u_3 &= \frac{1}{2\alpha + h\beta}\left[(v_1 - v_2) + \left(\alpha - \frac{\beta\,h}{2}\right)(u_1 + u_2)\right].
\end{aligned}\right\} \tag{5.29}$$

We now consider a particular problem with the initial and boundary values

$$\left.\begin{aligned}
u(x, 0) &= \sin\frac{\pi}{2}\,x \\[2mm]
v(x, 0) &= 0
\end{aligned}\right\} \quad \text{for} \quad 0 \leq x \leq 2,$$

$$u(0, y) = u(2, y) = 0 \quad \text{for} \quad y \geq 0.$$

From (5.27) and an appropriate boundary formula, u and v can be calculated row by row for $y = \alpha h,\ 2\alpha h,\ 3\alpha h, \ldots$. Since u is given on the boundary, the appropriate formula to be used to complete each row at the boundaries is the second formula of (5.26); for the boundary at $x = 0$, for instance, we use

$$v_5 = v_4 - \alpha\,(u_5 - u_4) - h\beta\,u_4,$$

in which the subscripts refer to the points P_4, P_5 situated as in Fig. IV/21 and the value of u_5 is given (zero in the present example). Since symmetry (or antisymmetry in the case of v) exists about the line $x = 1$, we can restrict the calculation to the half $0 \leq x \leq 1$. The results obtained with $h = \frac{1}{4}$ for the case $\alpha = 2$, $\beta = 1$ are exhibited in Table IV/19. The value of u for each mesh point is given with the associated error (in fifth-decimal units) in brackets and the corresponding value of v immediately below. The error was obtained by comparison with the exact solution

$$u(x, y) = e^{-\frac{1}{2}y}\sin\frac{\pi}{2}\,x\left\{\cos\frac{\nu\,y}{8} - \frac{1}{\nu}\sin\frac{\nu\,y}{8}\right\}, \quad \text{where} \quad \nu = \sqrt{4\pi^2 - 1}.$$

The last column of the table is a check column containing row sums, the summations being extended over the whole row $0 \leq x \leq 2$ for u but only over the half row $0 \leq x \leq 1$ for v (the sum of all v values for $0 \leq x \leq 2$

Table IV/19. *Values of* $\left\{\begin{matrix}u\\v\end{matrix}\right.$ *obtained by using* (5.26), (5.27) *with* $h = \frac{1}{4}$

$x =$	0	0·25	0·5	0·75	1	1·25	Row sums
$y = 0$	0 0		**0·70711** 0		**1** **0**		2·41422 0
$y = 0·5$		0·309 36 (−154) 0·618 72		0·746 86 (−374) 0·25628		0·74686 −0·25628	2·11244 0·875
$y = 1$	0 1·160 10		0·371 49 (−374) 0·820 31		0·525 36 (−529) 0		1·268 34 1·980 41
$y = 1·5$		0·077 58 (−126) 0·315 26		0·187 29 (−306) 0·544 79			0·529 74 1·860 05
$y = 2$	0 1·451 02		−0·076 74 (+99) 1·026 02		−0·108 52 (+141) 0		−0·262 00 2·477 04
$y = 2·5$		−0·139 82 (+293) 1·171 37		−0·337 56 (+707) 0·485 20			−0·954 76 1·656 57
$y = 3$	0		−0·38040 (981)		−0·537 97 (1387)		−1·298 77

Table IV/20. Values of $\begin{Bmatrix} u \\ v \end{Bmatrix}$ obtained by using (5.26), (5.27) with $h = \frac{1}{8}$

$x =$	0	0·125	0·25	0·375	0·5	0·625	0·75	0·875	1	Row sums
$y = 0$	0 0		**0.38268** 0		**0.70711** 0		**0.92388** 0		**1** 0	5·02734 0
$y = 0·25$	0 0	0·17938 0·35876		0·51084 0·30415		0·76453 0·20322		0·90182 0·07136		4·71314 0·93750
$y = 0·5$	0		0·30989 (−101)		0·57260 (−188)		0·74814 (−246)		0·80978 (−266)	4·07104

Table IV/21. Values of $\begin{Bmatrix} u \\ v \end{Bmatrix}$ obtained by using (5.28), (5.29) with $h = \frac{1}{4}$

$x =$	0	0·25	0·5	0·75	1
$y = 0$	0 0		**0.70711** 0		0 1
$y = 0·5$		0·31196 (+106) 0·62692		0·75314 (+254) 0·27458	
$y = 1$	0 1·24784		0·37853 (330) 0·88235		0·53532 (467) 0
$y = 1·5$		0·08100 (216)		0·19556 (521)	

is zero and therefore does not provide a check). The check consists in verifying the relation

$$8 S_{m+1} = 7 (S_m - u_a) - 4 v_a$$

(where u_a and v_a are the first values of u and v appearing in the m-th row), which should hold between successive row sums S_m, S_{m+1} of the u values.

Table IV/20 gives the beginning of a similar calculation based on the finer characteristic mesh with $h = \frac{1}{8}$ and Table IV/21 gives the beginning of a calculation based on the coarser mesh with $h = \frac{1}{4}$ but employing the more accurate formulae (5.28), (5.29). Again the errors in the u values are given. The gain in accuracy over the first rough calculation is slight.

§ 6. Supplements

In §§ 6.1 and 6.2 we prove theorems which permit the deduction of bounds for the error in approximate solutions of various problems in parabolic differential equations. These theorems play a role for parabolic equations similar to that played by the boundary-maximum theorem for elliptic equations (Ch. V, § 3), which likewise provides a possible basis for the estimation of the error in approximate solutions. Rather less is known about hyperbolic differential equations in this respect; a theorem has been established[1] which states that under certain conditions the maximum of a function satisfying a differential inequality is assumed only on a boundary curve, but the cases covered by the theorem do not yet possess the degree of generality which has been achieved for elliptic and parabolic equations. Nevertheless, it is worthy of note that an error estimate is possible for the "mixed-type" problem with equation (1.57), which was mentioned in § 1.9 (the Tricomi problem).

6.1. Monotonic character of a wide class of initial-/boundary-value problems in non-linear parabolic differential equations

A very general estimation theorem which nevertheless admits of an elementary proof has been established by WESTPHAL[2]. Let B be the open region

$$0 < y < Y, \quad x_0(y) < x < x_1(y),$$

---

[1] AGMON, S., L. NIRENBERG and M. H. PROTTER: A maximum principle for a class of hyperbolic equations and applications to equations of mixed elliptic-hyperbolic type. Comm. Pure Appl. Math. 6, 455—470 (1953). — PROTTER, M. H.: A Boundary Value Problem for an Equation of Mixed Type. Trans. Amer. Math. Soc. 71, 416—429 (1951).

Some applications to error estimation for hyperbolic differential equations will be published shortly by the present author; cf. also L. COLLATZ: Fehlermaß-prinzipien in der praktischen Analysis. Proc. Internat. Congr. Math. Vol. I, Amsterdam 1954.

[2] WESTPHAL, H.: Zur Abschätzung der Lösungen nichtlinearer parabolischer Differentialgleichungen. Math. Z. 51, 690—695 (1949). — Similar and more general theorems can be found in the following papers: — PICONE, M.: Sul problema della

where Y is a positive constant and $x_0(y)$, $x_1(y)$ are two given continuous functions of y such that $x_0(y) < x_1(y)$ for $0 \leq y \leq Y$ (see Fig. IV/22).

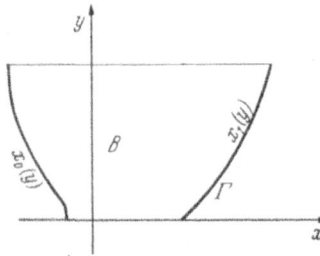

We shall need to distinguish between that part of the boundary of B which lies along the line $y = Y$ and the remaining part; we denote them by Γ_Y and Γ, respectively. Now let T be the operator defined by

$$T u = u_y - f(x, y, u, u_x, u_{xx}) \qquad (6.1)$$

Fig. IV/22. Boundaries for a class of boundary-value problems with non-linear parabolic differential equations

for functions $u(x, y)$ which are continuous in the closed region $\overline{B} = B + \Gamma + \Gamma_Y$ and for which u_y, u_x, u_{xx} exist in B. If we restrict the given function $f(x, y, u, u_x, u_{xx})$ to be monotonic non-decreasing with respect to u_{xx} for any fixed values of x, y, u, u_x, then the following theorem holds.

Theorem: *If, with the above definitions and restrictions, two functions u and v are such that $Tv < Tu$ in $B + \Gamma_Y$ and $v < u$ on Γ, then $v < u$ throughout $\overline{B}$.*

Proof: Let $w = u - v$; then $w > 0$ on Γ and we wish to show that this is true also in $B + \Gamma_Y$. Let M be the set of values of y in $0 \leq y \leq Y$ for which there is at least one value of x in $x_0(y) < x < x_1(y)$ with $w \leq 0$; we wish to show that this set is null. Suppose that it is not. If it is an infinite set, it will have a lower limit $\tilde{y}$ and there will be a sequence of values y_ν with corresponding values x_ν such that $y_\nu \to \tilde{y}$ and $w(x_\nu, y_\nu) \leq 0$; now these x_ν form a bounded infinite set, which must have a limit $\tilde{x}$, and we can choose a sub-sequence x'_ν, with corresponding values y'_ν, such that the points (x'_ν, y'_ν) converge to the limit point $\widetilde{P} = (\tilde{x}, \tilde{y})$; then by continuity $w(\widetilde{P}) \leq 0$. If M is finite, we take the smallest member of M as $\tilde{y}$ and a corresponding x value as $\tilde{x}$. Now $\widetilde{P}$ cannot lie on Γ, so we can choose a sequence of points $(\tilde{x}, y)$ with $w > 0$ which tend to $\widetilde{P}$ from below (i.e. with $y < \tilde{y}$); hence, by continuity, $w(\widetilde{P}) \geq 0$ and we must have $w(\widetilde{P}) = 0$; we see also that $w_y(\widetilde{P}) \leq 0$ (w_y is assumed to exist

propagazione del calore in un messo di frontiera conduttore, isotrope e anogenes. Math. Ann. **101**, 701−712 (1929). − MARASIMHAN: On the Asymptotic Stability of Solutions of Parabolic Differential Equations. J. Rational Mech. Anal. **3**, 303−313 (1954). − NIRENBERG, L. A.: A Strong Maximum Principle for Parabolic Equations. Comm. Pure Appl. Math. **6**, 167−177 (1953). − SZARSKI, L.: Sur la limitation et unicité des solution d'un système non-linéaire d'équations paraboliques aux derivées partielles du second ordre. Ann. Polonoci Math. **2**, 237−249 (1955). − COLLATZ, L.: Fehlerabschätzungen für Näherungslösungen parabolischer Differentialgleichungen. Anais Acad. Brasileira de Ciéncias **28**, 1−9 (1956).

in B). Similarly, $w \geq 0$ on the whole line segment $y = \tilde{y}$, $x_0(\tilde{y}) \leq x \leq x_1(\tilde{y})$ and therefore, since w_x and w_{xx} exist at $\tilde{P}$,

$$w_x(\tilde{P}) = 0, \quad w_{xx}(\tilde{P}) \geq 0.$$

In terms of u and v the situation at $\tilde{P}$ is that $u = v$, $u_x = v_{,x}$ $u_y \leq v_y$, $u_{xx} \geq v_{xx}$; consequently

$$f(x, y, u, u_x, u_{xx}) \geq f(x, y, v, v_x, v_{xx}) \quad \text{and} \quad Tu \leq Tv,$$

which contradicts our definition of u and v. Thus the set M must be null and the theorem is proved.

6.2. Estimation theorems for the solutions

In the reference cited, WESTPHAL makes among others the two following applications of the theorem in § 6.1.

Theorem 1. *With the definitions of § 6.1 let there be constants ε_1, ε_2, δ such that*

$$|Tu| \leq \varepsilon_1, \quad |Tv| \leq \varepsilon_2 \quad \text{in} \quad B + \Gamma_Y \quad \text{and} \quad |u - v| \leq \delta \quad \text{on } \Gamma; \quad (6.2)$$

further, in addition to the monotonic condition of § 6.1, let f satisfy

$$|f(x, y, u_1, u_x, u_{xx}) - f(x, y, u_2, u_x, u_{xx})| \leq C^*, \quad (6.3)$$

where C^* is a constant, for all values of x, y, u, u_x, u_{xx} which come into consideration. Then

$$|u - v| \leq \delta + (C^* + \varepsilon_1 + \varepsilon_2) y \quad (6.4)$$

in the whole of $\overline{B}$.

Proof. Again let $w = u - v$, and apply to it the operator T^* defined for fixed v by

$$T^* \varphi = \varphi_y - f(x, y, v, v_x + \varphi_x, v_{xx} + \varphi_{xx}) + f(x, y, v, v_x, v_{xx}):$$

we have

$$\begin{aligned} T^* w &= u_y - v_y - f(x, y, v, u_x, u_{xx}) + f(x, y, v, v_x, v_{xx}) \\ &= Tu - Tv + f(x, y, u, u_x, u_{xx}) - f(x, y, v, u_x, u_{xx}) \\ &\leq \varepsilon_1 + \varepsilon_2 + C^* \end{aligned}$$

in $B + \Gamma_Y$. Now the function defining $T^* \varphi$ is a monotonic non-decreasing function of φ_{xx}, so that T^* has the same monotonic property as was prescribed for T in § 6.1; moreover, the function $W = \delta + \varepsilon + (C^* + \varepsilon_1 + \varepsilon_2 + \varepsilon) y$, for which

$$T^* W = C^* + \varepsilon_1 + \varepsilon_2 + \varepsilon,$$

is such that $T^* w < T^* W$ in $B + \Gamma_Y$ and $w < W$ on Γ for any positive constant ε; consequently we can use the theorem of § 6.1 to show that $w < W$ in the whole of $\overline{B}$. Similarly $-w < W$, and hence $|w| < W$, in the whole of $\overline{B}$. (6.4) follows by letting ε tend to zero.

Theorem 2. *Let the region B of § 6.1 be such that*

$$x_0(y) \leq -A, \qquad x_1(y) \geq A > 0$$

and in addition to the monotonic condition of § 6.1 let f satisfy a Lipschitz condition of the form

$$\left.
\begin{aligned}
&|f(x, y, u', u'_x, u''_{xx}) - f(x, y, u'', u''_x, u''_{xx})| \\
&\qquad \leq M_0 |u' - u''| + M_1 |u'_x - u''_x| + M_2 |u'_{xx} - u''_{xx}|.
\end{aligned}
\right\} \tag{6.5}$$

Then for two functions u, v satisfying the differential equation $Tu = Tv = 0$ in $B + \Gamma_y$ and which are equal on that part of the boundary Γ along $y = 0$ and differ by at most δ on the remainder of Γ, we have

$$|u - v| \leq \delta e^{(M_0 + M_1 + 2M_2) y - A} \left(e^{|x|} + \tfrac{1}{2} \right) \tag{6.6}$$

in the whole of $\overline{B}$. By putting $\delta = 0$ we obtain the uniqueness theorem for the first boundary-value problem.

Proof. Again let $w = u - v$, but this time consider the operator

$$T^* \varphi = \varphi_y - M_0 |\varphi| - M_1 |\varphi_x| - M_2 \, \Phi(\varphi_{xx}),$$

where

$$\Phi(z) = \tfrac{1}{2}(z + |z|) = \begin{cases} z & \text{for } z \geq 0 \\ 0 & \text{for } z < 0; \end{cases}$$

$\Phi(z)$ is a monotonic non-decreasing function of z, so that T^* again satisfies the monotonic condition of § 6.1.

Now

$$\begin{aligned}
u_y - v_y &= f(x, y, u, u_x, u_{xx}) - f(x, y, v, v_x, v_{xx}) \\
&= f(x, y, u, u_x, v_{xx}) - f(x, y, v, v_x, v_{xx}) + \\
&\quad + f(x, y, u, u_x, u_{xx}) - f(x, y, u, u_x, v_{xx}) \\
&\leq M_0 |w| + M_1 |w_x| + M_2 \, \Phi(w_{xx}),
\end{aligned}$$

where we have used the fact that f is a monotonic non-decreasing function of its last argument. Hence $T^* w \leq 0$.

For the function

$$W = \delta e^{(M_0 + M_1 + 2M_2 + \varepsilon) y - A} \left(e^{|x|} + \tfrac{1}{2} e^{-2|x|} \right),$$

where ε is any positive constant, a simple calculation yields

$$\begin{aligned}
T^* W = \delta\, e^{(M_0 + M_1 + 2M_2 + \varepsilon) y - A} \times \\
\times \left[\varepsilon (e^{|x|} + \tfrac{1}{2} e^{-2|x|}) + \tfrac{3}{2} M_1 e^{-2|x|} + M_2 (e^{|x|} - e^{-2|x|}) \right].
\end{aligned}$$

Consequently $T^* W > 0 \geq T^* w$. Since we also have $W > w$ on Γ, it follows from the theorem of § 6.1 that $w < W$ in the whole of $\overline{B}$. By interchanging u and v we can show also that $-w < W$, and hence $|w| < W$, in the whole of $\bar{B}$. (6.6) follows by letting ε tend to zero.

GÖRTLER[1] has considered such estimation theorems generalized for the solutions of the system of partial differential equations

$$u\,u_x + v\,u_y = f(x, u, u_y, u_{yy})$$
$$u_x + v_y = 0,$$

which is of great importance in fluid dynamics.

Example. Consider the example of § 2.4 with $K = \beta = 1$:

$$Tu = u_y - u_{xx} - 1 = 0 \quad \text{in} \quad |x| \leqq 1, \quad y \geqq 0 \quad (B + \Gamma)$$

$$u(x, 0) = 0 \quad \text{for} \quad |x| \leqq 1, \quad u(\pm 1, y) = 0 \quad \text{for} \quad y \geqq 0 \quad (\text{boundary } \Gamma).$$

The function

$$v(x, y) = \tfrac{1}{2}(1 - x^2) + \sum_{n=1}^{N} a_n \cos(c_n x)\, e^{-c_n^2 y},$$

where $c_n = (n - \tfrac{1}{2})\pi$, satisfies the differential equation $Tv = 0$ and the boundary conditions $v(\pm 1, y) = 0$ (for $y \geqq 0$) for any integer N and for arbitrary constants a_n. From Theorem 1 with $\varepsilon_1 = \varepsilon_2 = C^* = 0$ it follows that $|u - v| \leqq \delta$ in the whole of $\bar{B}$, where δ is an upper bound for $|u - v|$ along the initial line segment $y = 0$, $-1 \leqq x \leqq 1$. We therefore choose the a_n so that

$$\left| \tfrac{1}{2}(1 - x^2) + \sum_{n=1}^{N} a_n \cos(c_n x) \right| \leqq \delta$$

in the interval $-1 \leqq x \leqq 1$ with δ as small as possible. For $N = 3$ we can take δ as small as 0.002 with the constants $a_1 = -0.516$, $a_2 = 0.0193$, $a_3 = -0.005$. The Hermitian finite-difference method of § 2.4 yields the approximate value $u(0, \tfrac{5}{24}) \approx \frac{7.70133}{40} = 0.19253$ at the point $i = 0$, $k = 10$ (since $l = \tfrac{1}{48}$ and $\alpha = \tfrac{1}{40}$). Here we obtain the approximation $v(0, \tfrac{5}{24}) = 0.19158$ and can assert with certainty that $|u(0, \tfrac{5}{24}) - 0.19158| \leqq 0.002$. This error estimate can be improved somewhat by the addition of a suitable constant to v.

6.3. Reduction to boundary-value problems

As for ordinary differential equations (Ch. II, § 5.9), initial- and initial-/boundary-value problems in partial differential equations may also be transformed into boundary-value problems, often in several different ways, and then dealt with by one of the several methods

[1] GÖRTLER, H.: Über die Lösungen nichtlinearer partieller Differentialgleichungen vom Reibungsschichttypus. Z. Angew. Math. Mech. **30**, 265—267 (1950).

available for the treatment of boundary-value problems (Ch. V). The transformation is effected by raising the order of the differential equation and adding further boundary conditions, which can often be chosen quite arbitrarily.

Such a treatment of the heat-flow problem

$$\frac{\partial u}{\partial t} = K \frac{\partial^2 u}{\partial x^2},$$

$$u(x, 0) = u(0, t) = 0, \quad u(L, t) = 100$$

is given in detail by ALLEN and SEVERN[1].

They put $u = \dfrac{\partial w}{\partial t} + K \dfrac{\partial^2 w}{\partial x^2}$, so that $\dfrac{\partial^2 w}{\partial t^2} = K^2 \dfrac{\partial^4 w}{\partial x^4}$, and as additional boundary conditions take

$$w(0, t) = w(L, t) = \frac{\partial w}{\partial t}(x, T) = 0,$$

where T is a suitably chosen time. This boundary-value problem for the rectangular region $0 \le x \le L$, $0 \le t \le T$ they then solve by the finite-difference method in conjunction with the relaxation technique (Ch. V, § 1.6).

6.4. Miscellaneous exercises on Chapter IV

1. The numerical example of § 1.2 concerned the heat flow in a rod at one end of which the temperature was varied sinusoidally. After a long time (for large t) the temperature distribution settles down to a regular oscillation with the imposed frequency ω. Calculate a half-cycle of this oscillating distribution by the finite-difference method [formula (1.13) with mesh width $h = \frac{1}{4}$ as in § 1.2] by introducing the values $U_{i,q}$ on some row $k = q$ (q assumed to be large and for convenience a multiple of 12) as unknowns

$$U_{2,q} = a, \quad U_{4,q} = b, \quad U_{6,q} = c,$$

expressing the $U_{i,k}$ in the strip $q < k \le q + 12$ in terms of a, b and c by means of the formula (1.13) and the boundary conditions, say $U_{0,q+2s} = \sin\left[\pi\left(1 + \frac{s}{6}\right)\right]$ and $U_{8,q+2s} = 0$ for $s = 0, 1, \ldots, 5$, and then determining a, b, c from the requirement that
$$U_{2r,q} = -U_{2r,q+12} \quad \text{for} \quad r = 1, 2, 3.$$

2. Let there be initially a linear fall in temperature from $100°$ C to $0°$ C along a thin homogeneous rod of length 10. Then let the hot end be cooled rapidly while the other end is kept at $0°$ C. Calculate approximately the temperature distribution $u(x, y)$ at several subsequent times y from the initial-/boundary-value problem

$$\frac{\partial^2 u}{\partial x^2} = 2 \frac{\partial u}{\partial y}; \quad u(x, 0) = 10x, \quad u(0, y) = 0, \quad u(10, y) = 100 e^{-0.1 y}$$

by means of the finite-difference method of § 1 with the mesh widths $h = 1$ and $\frac{1}{2}$.

3. In dealing with the heat flow in a thin homogeneous rod in § 1.2 we neglected radiation and convection effects. If we take into consideration a heat loss to the surrounding medium across the surface of the rod with the rate of loss of heat

---

[1] ALLEN, D. N. DE G., and R. T. SEVERN: The application of relaxation methods to the solution of non-elliptic partial differential equations. Quart. J. Mech. Appl. Math. 4, 209—222 (1951).

proportional to the temperature difference (or to the temperature u of the rod if we take the temperature of the surroundings as zero), then the differential equation reads

$$\frac{\partial^2 u}{\partial x^2} = \frac{\varrho\,\sigma}{K} \frac{\partial u}{\partial t} + \lambda u,$$

where ϱ, σ, K, λ are physical constants. If the rod (of length $2a$) is initially heated to a uniform temperature $u = 1$ and then allowed to cool by itself, the initial and boundary conditions are

$$u(x, 0) = 1 \quad \text{for} \quad -a \leqq x \leqq a,$$

$$u \pm K \frac{\partial u}{\partial x} = 0 \quad \text{for} \quad x = \pm a \quad \text{and} \quad t > 0.$$

Calculate the solution of this problem for $\lambda = \tfrac{1}{2}$, $a = 1$, $K = 1$. The introduction of the new variable $y = \dfrac{Kt}{\varrho\,\sigma}$ reduces the differential equation to

$$\frac{\partial^2 u}{\partial x^2} = \frac{\partial u}{\partial y} + \frac{1}{2}\,u.$$

4. Use the finite-difference method to calculate an approximate solution of the cable problem which was treated by the method of characteristics in § 5.4, reducing the system of two equations for u and v to a single equation for u and similarly eliminating v from the boundary conditions. Compare the approximate solution so obtained with the exact solution given in § 5.4 and with the other approximate solutions obtained in § 5.4.

5. Apply the ordinary finite-difference method to the problem

$$J_{tt} + \tfrac{1}{2} J_t = J_{xx}; \quad J = 0 \quad \text{for} \quad x = 0 \quad \text{and} \quad x = 1,$$

$$\left.\begin{array}{l} J = 0 \\[4pt] \dfrac{\partial J}{\partial t} = -\dfrac{dq_0}{dx} = e^{-x} \end{array}\right\} \quad \text{for} \quad t = 0, \quad \text{and} \quad 0 < x < 1$$

[transient current produced by an initially non-uniform charge distribution $q_0(x) = e^{-x}$ in an open circuit consisting of a thin uniform conductor] and incorporate a running check.

6. The free cooling of a long square prism from a uniform temperature in excess of that of the surroundings can be reduced to the problem

$$u_{xx} + u_{yy} = u_t \quad \text{in the region} \quad |x| \leqq 1, \quad |y| \leqq 1, \quad t \geqq 0,$$

$$u - \frac{\partial u}{\partial \nu} = 0 \quad \text{on the boundary} \quad |x| = 1, \quad |y| = 1, \quad t \geqq 0,$$

and $u = 1$ for $t = 0$. Apply the ordinary finite-difference method in the form of equation (1.56) of § 1.8, incorporating a current check.

7. In § 1.8 we applied the ordinary finite-difference method to the heat equation (1.50) with two space co-ordinates x, y, and with the mesh widths (1.52) we obtained the formula (1.53), which depends on the parameter ϱ. The formula was shown to be unstable for $\varrho = 1$, and the value $\varrho = \tfrac{1}{4}$ was recommended. Is the formula stable for $\varrho = \tfrac{1}{8}$?

8. Let the two-dimensional wave equation

$$u_{xx} + u_{yy} = C\,u_{tt}$$

be approximated on the mesh (1.51) with the mesh widths $h_x = h_y = h$, $h_t = (c\varrho)^{\frac{1}{2}} h_x$ by the difference equation

$$U_{i,k,l+1} = 2U_{i,k,l} - U_{i,k,l-1} + \varrho(S - 4U_{i,k,l}),$$

where S is as defined in (1.55). Is the formula stable for $\varrho = 1$ and $\varrho = \frac{1}{4}$, respectively?

9. In § 2.2 several finite expressions of a higher approximation were derived for the equation $u_{xx} - K u_y = r(x, y)$, K being a non-zero constant. With the same mesh notation as used there (mesh widths h, l with $l = \sigma K h^2$) determine the quantities c and σ in the expression

$$\Phi = U_{0,1} - U_{0,0} - \sigma(U_{1,0} - 2U_{0,0} + U_{-1,0}) + c$$

so that Φ is equal to a remainder term of the sixth order by expanding Φ by Taylor's theorem and using the relations

$$K^2 u_{yy} = \frac{\partial^4 u}{\partial x^4} - S_1, \qquad K^3 u_{yyy} = \frac{\partial^6 u}{\partial x^6} - S_2,$$

where

$$S_1 = r_{xx} + K r_y, \qquad S_2 = \frac{\partial^4 r}{\partial x^4} + K r_{xxy} + K^2 r_{yy}.$$

10. Derive in the same way as in Exercise 9 a finite expression with a sixth-order remainder term for the equation

$$\nabla^2 u = u_{xx} + u_{yy} = K u_t + r(x, y, t).$$

6.5. Solutions

1. The equations for the a, b, c read

$$1131a + 140b + 91c = 229 + 92\sqrt{3} = 388 \cdot 35$$
$$140a + 1222b + 140c = 88 + 60\sqrt{3} = 191 \cdot 92$$
$$91a + 140b + 1131c = 29 + 12\sqrt{3} = 49 \cdot 785,$$

Table IV/22. *Half-cycle of the oscillating temperature distribution*

k	$i=0$	$i=1$	$i=2$	$i=3$	$i=4$	$i=5$	$i=6$	$i=7$	$i=8$
q	0		0·3284		0·1191		0·0028		0
$q+1$		0·1642		0·2238		0·0610		0·0014	
$q+2$	−0·5		0·1940		0·1424		0·0312		0
		−0·1530		0·1682		0·0868		0·0156	
	−0·8660		0·0076		0·1275		0·0512		0
$q+5$		−0·4292		0·0676		0·0894		0·0256	
	−1		−0·1808		0·0785		0·0575		0
		−0·5904		−0·0512		0·0680		0·0288	
	−0·8660		−0·3208		0·0084		0·0484		0
		−0·5934		−0·1562		0·0284		0·0242	
$q+10$	−0·5		−0·3748		−0·0639		0·0263		0
		−0·4374		−0·2194		−0·0188		0·0132	
$q+12$	0		−0·3284		−0·1191		−0·0028		0

and since the matrix of coefficients has strongly dominant diagonal elements, they may be solved conveniently by single-step iteration; we obtain

$$a = 0 \cdot 3284, \qquad b = 0 \cdot 1191, \qquad c = 0 \cdot 0028.$$

Table IV/22 shows the corresponding temperature distribution calculated from (1.13).

2. With the mesh relation (1.12) the finite-difference formula reduces to (1.13) and the calculation consists solely in forming arithmetic means. The results for $h = 1$ and $h = 0·5$ are given in Tables IV/23, IV/24 and the temperature distribution is shown for various times in Fig. IV/23.

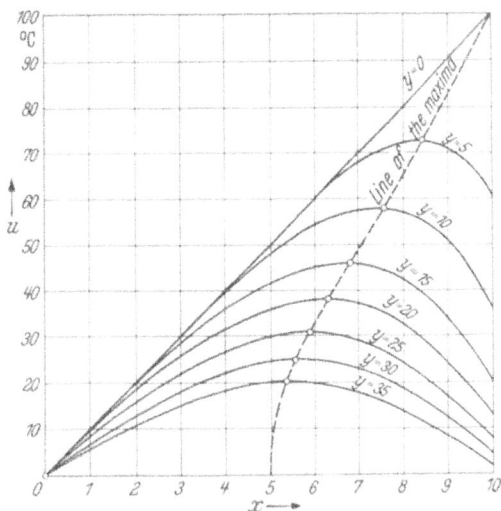

Fig. IV/23. Temperature distribution u along the rod at various times y

Table IV/23. *Finite-difference approximation for the mesh widths $h = l = 1$*

$k=y$	$x=0$	$x=1$	$x=2$	$x=3$	$x=4$	$x=5$	$x=6$	$x=7$	$x=8$	$x=9$	$x=10$
0	0		20		40		60		80		100
1		10		30		50		70		90	
2	0		20		40		60		80		81·873
3		10		30		50		70		80·937	
4	0		20		40		60		75·468		67·032
5		10		30		50		67·734		71·250	
6	0		20		40		58·867		69·492		54·881
7		10		30		49·434		64·180		62·187	
8	0		20		39·717		56·807		63·183		44·933
9		10		29·858		48·262		59·995		54·058	
10	0		19·929		39·060		54·129		57·027		36·788
11		9·965		29·495		46·594		55·578		46·907	
12	0		19·730		38·044		51·086		51·242		30·119
16	0		18·870		35·173		44·323		40·550		20·190
20	0		17·460		31·522		37·993		32·607		13·534
24	0		15·711		27·722		32·217		26·120		9·072
28	0		13·854		24·010		27·091		21·440		6·081
32	0		12·026		20·581		22·768		16·978		4·076
36	0		10·326		17·477		18·934		13·578		2·732

Table IV/24. *A section of the results for the mesh widths* $h = \frac{1}{2}$, $l = \frac{1}{4}$

k	y	x=3·5	x=4	x=4·5	x=5	x=5·5	x=6	x=6·5	x=7	x=7·5	x=8	x=8·5	x=9	x=9·5	x=10
0	0		40		50		60		70		80		90		100
1		35		45		55		65		75		85		95	
2			40		50		60		70		80		90		95·123
3		35		45		55		65		75		85		92·562	
4	1		40		50		60		70		80		88·781		90·484
5		35		45		55		65		75		84·391		89·633	
6			40		50		60		70		79·695		87·012		86·071
7		35		45		55		65		74·848		83·354		86·542	
8	2		40		50		60		69·924		79·101		84·948		81·873
12	3		40		49·996		59·926		69·432		77·249		80·397		74·082
16	4		39·995		49·945		59·658		68·481		74·802		75·654		67·032

3. We use the mesh

$$x_i = (i - \tfrac{1}{2}) h, \qquad y_k = k l,$$

with the mesh widths $h = \frac{1}{5}$ and $l = \frac{1}{50}$ [in accordance with (1.12)]. Then the finite-difference equation reads

$$U_{i,k+1} = \tfrac{1}{2} (U_{i+1,k} + U_{i-1,k}) - 0\cdot01\, U_{i,k}$$

and the boundary condition (with $n = 5$)

$$\frac{U_{n,k} + U_{n+1,k}}{2} + \frac{U_{n+1,k} - U_{n,k}}{h} = 0, \quad \text{so that} \quad U_{n+1,k} = \frac{9}{11}\, U_{n,k}.$$

We need to use this boundary condition on the initial row $y = 0$ in order to calculate the value $U_{6,0}$ which is required for the calculation of $U_{5,1}$. If we had simply put $U_{6,0} = 0$, say, this discontinuity would have made its presence felt in large fluctuations in the U values, and the unevenness of the U distribution would

Table IV/25. *Temperature in a cooling rod by the finite-difference method*

k	y	i=1 x=0·1	i=2 x=0·3	i=3 x=0·5	i=4 x=0·7	i=5 x=0·9	i=6 x=1·1	Row-sum check σ_k	$U_{4,k-1}$ minus $U_{4,k}$
0	0	1	1	1	1	1	0·81818	5	
1	0·02	0·99	0·99	0·99	0·99	0·89909	0·73562	4·85909	0·01000
2	0·04	0·9801	0·9801	0·9801	0·93465	0·85382	0·69858	4·72877	0·05535
3	0·06	0·97030	0·97030	0·94757	0·90761	0·80808	0·66115	4·60386	0·02704
		0·96060	0·94923	0·92948	0·86875	0·77630	0·63515	4·48436	0·03886
5	0·10	0·94531	0·93555	0·89970	0·84420	0·74419	0·60889	4·36895	0·02455
		0·93098	0·91315	0·88088	0·81350	0·71910	0·58835	4·25761	0·03070
		0·91276	0·89680	0·85452	0·79186	0·69373	0·56759	4·14967	0·02164
		0·89565	0·87467	0·83578	0·76621	0·67279	0·55047	4·04510	0·02565
10	0·20	0·85782	0·83557	0·79367	0·72438	0·63336	0·51820	3·84480	0·02224
		0·83812	0·81739	0·77204	0·70627	0·61496	0·50315	3·74878	0·01811
		0·81937	0·79691	0·75411	0·68644	0·59856	0·48974	3·65538	0·01983
		0·79995	0·77877	0·73413	0·66947	0·58210	0·47626	3·56442	0·01697
14	0·28	0·78136	0·75925	0·71678	0·65142	0·56704	0·46394	3·47585	0·01805

certainly have been far greater than that shown by the values obtained with no discontinuity (Table IV/25). The differences of the column $i = 4$, which are given in the last column of the table, show up this unevenness; it is quite considerable initially but gradually dies away. The row-sum check to be satisfied by the row-sums recorded in the penultimate column is

$$\sigma_{k+1} = (1 - 0.01)\,\sigma_k - \tfrac{1}{9} U_{n+1,k}, \quad \text{where} \quad \sigma_k = \sum_{i=1}^{n} U_{i,k}.$$

4. With $\alpha = 2$, $\beta = 1$ the problem reads

$$u_{xx} = 4 u_{yy} + u_y,$$

$$\left.\begin{aligned}
u(x, 0) &= \sin \frac{\pi}{2} x \\
u_y(x, 0) &= -\frac{1}{4} \sin \frac{\pi}{2} x
\end{aligned}\right\} \quad \text{for} \quad 0 \le x \le 2,$$

$$u(0, y) = u(2, y) = 0 \qquad \text{for} \quad 0 \le y.$$

With $h = \tfrac{1}{4}$, $l = \tfrac{1}{2}$ the difference equation reads

$$U_{i,k+1} = \tfrac{1}{17}[16(U_{i+1,k} + U_{i-1,k}) - 15 U_{i,k-1}] \qquad (k = 0, 1, 2, \ldots).$$

To find starting values we must proceed as in (1.35) to (1.38); we obtain the formula

$$U_{i,1} = \tfrac{1}{32}(16 U_{i+1,0} - \tfrac{15}{4} U_{i,0} + 16 U_{i-1,0}).$$

The first few rows of the calculation are given in Table IV/27, which includes a row-sum check in the last two columns. If $S_m = \sum_{j=1}^{7} U_{j,m}$ is the sum of the U values in the m-th row, then

$$\tau_m = 17 S_{m+1} - 32(S_m - U_{1,m}) + 15 S_{m-1}$$

should be zero. The τ values are very sensitive to small variations in the U values. The errors in the U values are given in brackets in fifth-decimal units. Initially their magnitude is substantially greater than that of the corresponding errors obtained by the method of characteristics in § 5.4.

Table IV/26. *Current produced by a non-uniform charge distribution*

t	$x=0$	$x=0.2$	$x=0.4$	$x=0.6$	$x=0.8$	$x=1$	σ_k	τ_k
0	0	0	0	0	0	0	0	0
0·2	0	0·155 56	0·127 36	0·104 27	0·085 37	0	0·231 63	0·472 57
0·4	0	0·121 30	0·247 46	0·202 60	0·099 31	0	0·450 06	0·670 67
0·6	0	0·094 93	0·193 25	0·235 91	0·115 71	0	0·429 16	0·639 80
0·8	0	0·074 30	0·091 20	0·110 94	0·134 83	0	0·202 14	0·411 26
1	0	0·000 96	0·001 58	0·001 82	0·000 96	0	0·003 39	0·005 32
1·2	0	−0·065 72	−0·079 86	−0·097 96	−0·120 26	0	−0·177 82	−0·363 80
1·4	0	−0·076 93	−0·157 31	−0·192 23	−0·094 16	0	−0·349 54	−0·520 64
1·6	0	−0·090 36	−0·184 09	−0·150 87	−0·074 28	0	−0·334 96	−0·499 59
2·2	0	0·092 96	0·075 32	0·061 14	0·050 59	0	0·136 47	0·280 01
2·8	0	0·046 04	0·057 05	0·068 86	0·082 89	0	0·125 90	0·254 83
3·4	0	−0·046 15	−0·094 81	−0·115 98	−0·056 73	0	−0·210 79	−0·313 67
4	0	−0·001 82	−0·003 43	−0·002 97	−0·001 82	0	−0·006 40	−0·010 04

Table IV/27. *Finite-difference approximation for the problem treated by the method of characteristics in § 5.4*

y= \ x=	0	0·25	0·5	0·75	1	Row sums S_m	τ_m
−0·5	0	0·40438	0·74719	0·97625	1·05669	5·31233	−0·00050
0	0	**0·38268**	**0·70711**	**0·92388**	1	5·02734	
0·5	0	0·30871 (−219)	0·57042 (−406)	0·74528 (−532)	0·80669 (−575)	4·05551	+0·00014
1	0	0·19921 (−386)	0·36807 (−716)	0·48092 (−934)	0·52053 (−1012)	2·61692	+0·00030
2	0	0·07403 (−481)	0·13681 (−888)	0·17873 (−1162)	0·19347 (−1256)	0·97261	−0·00030
	0	−0·04701 (−494)	−0·08688 (−915)	−0·11349 (−1193)	−0·12286 (−1293)	0·61762	−0·00009
3	0	−0·14709 (−434)	−0·27177 (−800)	−0·35511 (−1048)	−0·38434 (−1131)	−1·93228	−0·00002
	•	−0·21430 (−312)	−0·39600 (−579)	−0·51738 (−755)	−0·56004 (−820)	−2·81540	
	0	−0·24292					

Table IV/28. *Cooling of a square prism*

l	$U_{0,0,1}\left(\tfrac{1}{8}\right)$	$U_{0,1,1}\left(\tfrac{7}{8}\right)$	$U_{0,2,1}\left(\tfrac{3}{8}\right)$	$U_{0,3,1}\left(\tfrac{1}{8}\right)$	$U_{1,1,1}\left(\tfrac{3}{4}\right)$	$U_{1,2,1}\left(\tfrac{5}{8}\right)$	$U_{1,2,1}\left(\tfrac{3}{8}\right)$	$U_{2,2,1}\left(\tfrac{1}{4}\right)$	$U_{2,2,1}\left(\tfrac{1}{4}\right)$	σ_l
1	1	1	1	0·66667	1	1	0·66667	1	0·66667	5·66667
2	1	1	0·91667	0·61111	1	0·91667	0·61111	0·83333	0·55556	5·37500
3	1	0·97917	0·86111	0·57407	0·95833	0·84028	0·56019	0·73611	0·49074	5·08623
4	0·97917	0·94444	0·80845	0·53897	0·90973	0·77894	0·51929	0·65551	0·44367	4·80527
5	0·94444	0·90177	0·76032	0·50688	0·86169	0·72575	0·48383	0·61131	0·40754	4·53354
6	0·90177	0·85704	0·71504	0·47669	0·81376	0·67929	0·45286	0·56664	0·37776	4·27497
7	0·85704	0·81108	0·67308	0·44872	0·76817	0·63708	0·42472	0·52853	0·35235	4·02861
8	0·81108	0·76661	0·63349	0·42233	0·72408	0·59863	0·39909	0·49472	0·32981	3·79602
9	0·76661	0·72318	0·59655	0·39770	0·68262	0·56285	0·37523	0·46422	0·30948	3·57575
10	0·72318	0·68210	0·56164	0·37443	0·64301	0·52966	0·35310	0·43616	0·29077	3·36834
11	0·68210	0·64271	0·52896	0·35264	0·60588	0·49848	0·33232	0·41021	0·27347	3·17241
12	0·64271	0·60571	0·49808	0·33205	0·57060	0·46934	0·31289	0·38598	0·25732	2·98805
13	0·60571	0·57050	0·46911	0·31274	0·53752	0·44189	0·29459	0·36333	0·24222	2·81411
14	0·57050	0·53747	0·44176	0·29450	0·50620	0·41616	0·27743	0·34206	0·22804	2·65044
15	0·53747	0·50616	0·41606	0·27737	0·47680	0·39186	0·26124	0·32209	0·21473	2·49611
16	0·50616	0·47678	0·39181	0·26121	0·44901	0·36905	0·24603	0·30330	0·20220	2·35088
17	0·47678	0·44900	0·36902	0·24602	0·42292	0·34754	0·23169	0·28563	0·19042	

5. With $h = 0.2$ the finite-difference equations read

$$J_{i,1} = l\left(1 - \frac{l}{4}\right)e^{-ih},$$

$$J_{i,k+1} = \frac{1}{1.05}\,[J_{i+1,k} + J_{i-1,k} - 0.95\,J_{i,k-1}] \qquad (k = 1, 2, \ldots),$$

for which we deduce the check $1.05\,\tau_{k+1} = \sigma_k + \tau_k - 0.95\,\tau_{k-1}$, where $\sigma_k = J_{2,k} + J_{3,k}$ and $\tau_k = \sum_{i=1}^{4} J_{i,k}$. The results are given in Table IV/26, where only every third row is reproduced after the row $k = 8$.

6. With the mesh widths $h_x = h_y = \frac{2}{5}$, $h_z = \frac{1}{4}h_x^2 = \frac{1}{25}$ (and $x_0 = y_0 = 0$) the boundary condition is represented by $U_{i,3,l} = \frac{2}{3}U_{i,2,l}$. A convenient arrangement of the calculation is shown in Table IV/28, which includes a column for the quantities $\sigma_l = \sum_{i,k}\beta_{i,k}\,U_{i,k,l}$, where the $\beta_{i,k}$ are the coefficients $\frac{1}{4}, \frac{7}{4}, \ldots$ encircled at the heads of the columns and the summation extends over all mesh points (i, k) which occur in the table; these quantities are needed for a check, which consists in verifying that

$$\tau_{l+1} = \sigma_l,$$

where $\tau_l = \sum_{i,k}^{*} U_{i,k,l}$ over all interior mesh points occurring in the table.

7. No. One should not be deceived by the fact that the absolute maximum of the propagated errors (set out below in the manner of § 1.8) decreases at first:

$l = 0$	$l = 1$	$l = 2$	$l = 3$
$0 \quad 0$	$\frac{1}{3}\varepsilon \quad 0$	$-\frac{2}{9}\varepsilon \quad \frac{2}{9}\varepsilon \quad 0$	$\frac{4}{9}\varepsilon \quad -\frac{2}{9}\varepsilon \quad \frac{1}{9}\varepsilon \quad 0$
$\boxed{\varepsilon} \quad 0$	$\boxed{-\frac{1}{3}\varepsilon} \quad \frac{1}{3}\varepsilon$	$\boxed{\frac{5}{9}\varepsilon} \quad -\frac{3}{9}\varepsilon \quad \frac{1}{9}\varepsilon$	$\boxed{-\frac{13}{27}\varepsilon} \quad \frac{4}{9}\varepsilon \quad -\frac{1}{9}\varepsilon \quad \frac{1}{27}\varepsilon$

8. For $\varrho = 1$ the formula is unstable[1], while for $\varrho = \frac{1}{2}$ we obtain the simple formula $U_{i,k,l+1} = \frac{1}{2}S - U_{i,k,l-1}$, which is stable.

A more accurate formula using more mesh points is given by MILNE[2] on p. 144 of his book.

9. The Taylor expansion yields

$$\Phi = c - \frac{l}{K}\,r - \frac{l^2}{2K^2}\,S_1 - \frac{l^3}{6K^3}\,S_2 + u_{xx}\left(\frac{l}{K} - \sigma h^2\right) +$$

$$+ \frac{\partial^4 u}{\partial x^4}\left(\frac{l^2}{2K^2} - \sigma\,\frac{h^4}{12}\right) + \frac{l^3}{6K^3}\,\frac{\widetilde{\partial^6 u}}{\partial x^6} - \frac{\sigma h^6}{360}\,\frac{\widetilde{\partial^6 u}}{\partial x^6},$$

where all values are taken to be at $x = y = 0$ except those distinguished by wavy lines, which are to be taken at certain intermediate points. The factor multiplying u_{xx} vanishes, and so also does the factor multiplying $\partial^4 u/\partial x^4$ if we choose $\sigma = \frac{1}{6}$,

---

[1] LJUSTERNIK, L. A.: Dokl. Akad. Nauk SSSR. (N. S.) **89**, 613−616 (1953) [Russian], gives the critical value ϱ_0 (stability for $\varrho < \varrho_0$, instability for $\varrho > \varrho_0$) for several finite-difference representations of the form (1.54) for the heat equation (1.50); ϱ_0 depends on the smallest eigenvalue for the corresponding difference operator.

[2] MILNE, W. E.: Numerical solution of differential equations. New York and London 1953.

i.e. $l = \frac{1}{6} K h^2$. For this value of σ we therefore obtain the finite equation[1]

$$U_{0,1} = \frac{1}{6} (U_{-1,0} + 4 U_{0,0} + U_{1,0}) - \frac{h^2}{6} r_{0,0} - \frac{h^4}{72} (r_{xx} + K r_y)_{0,0}.$$

10. With the notation for the u values in the plane $t = t_0 = $ constant defined in Table VIII of the appendix, and using the results of that table, we find that for the mesh widths $h_x = h_y = h$, $h_t = \sigma K h^2$

$$Q = a u_a + b \sum u_b + e \sum u_e + p u(x_0, y_0, t_0 + h_t) + q$$

$$= \left\{ q - p \sigma r h^2 - \frac{p}{2} \sigma^2 h^4 (K r_t + V^2 r) \right\} + u_a [p + a + 20] + h^2 V^2 u_a [p \sigma + 6] +$$

$$+ \frac{h^4}{2} V^4 u_a [p \sigma^2 + 1] + \frac{p \sigma^3 h^6}{6} (V^6 u_a - K^2 r_{tt} - K V^2 r_t - V^4 r) +$$

$$+ \frac{h^6}{60} (V^6 u_a + 2 V^2 D u_a) + \cdots.$$

Here we have put $b = 4$, $e = 1$ in order that no term in $D u_a$ shall appear. If we put the three quantities in square brackets equal to zero, we obtain $p = -36$, $\sigma = \frac{1}{6}$, $a = 16$. Then the equation $Q = 0$ yields[2]

$$36 u(x_0, y_0, t_0 + \frac{1}{6} K h^2) = 16 u_a + 4 \sum u_b + \sum u_e - 6 h^2 r_{0,0} -$$
$$- \frac{1}{2} h^4 (K r_t + V^2 r)_{0,0} + \text{remainder term of the 6th order.}$$

This formula has index $J = 1$ and is therefore stable.

Chapter V

Boundary-value problems in partial differential equations

Many of the methods described for boundary-value problems in ordinary differential equations in Chapter III carry over without difficulty to partial differential equations. What has already been said in the introduction to Chapter IV about the need for more theoretical investigation and more practical experience applies particularly to boundary-value problems for partial differential equations and bears repeating here. For the solutions of these problems we do not possess existence and uniqueness theorems covering anything like the range desirable from the standpoint of technical applications; moreover, the diversity of problems which arise in applications is continually increasing. There is also an urgent need for existing approximate methods to be subjected to extensive practical tests and thorough theoretical investigations on a much larger scale than hitherto.

§1. The ordinary finite-difference method

The finite-difference method is applicable to boundary-value problems generally. The equations are easily set up and, if a coarse mesh is used, the solution is usually obtained to an accuracy sufficient for technical purposes

---

[1] MILNE (see last footnote) p. 122. For $r(x, y) = $ constant this formula includes the formula (2.10) with $\sigma = \frac{1}{6}$.

[2] This formula is given by MILNE p. 137 for the case $r \equiv 0$.

with a relatively short calculation. Moreover, for partial differential equations the finite-difference method may be the only practicable means of solution; for boundaries can occur of such shapes that other methods cannot cope with the boundary conditions at all, or at least, not without difficulty. If, however, the solution is required to a greater accuracy, the ordinary finite-difference method has the disadvantage of slow convergence and it is usually better to use one of the refinements of the method described in § 2 rather than repeat the calculation with a smaller mesh width.

1.1. Description of the method[1]

Although the method may be applied to problems in three or more independent variables (cf. § 1.7), we will simplify the description, as in Ch. IV, § 1, by restricting ourselves to two variables x, y.

As in Ch. IV, § 1, we introduce a rectangular mesh

$$\left. \begin{array}{l} x_i = x_0 + ih \\ y_k = y_0 + kl \end{array} \right\} \quad (i, k = 0, \pm 1, \pm 2, \ldots) \quad (1.1)$$

and characterize function values at mesh points by corresponding subscripts (for instance, $u_{i,k}$ denotes $u(x_i, y_k)$ and $U_{i,k}$ denotes an approximation to it). As in Ch. IV (1.8), we can associate with every partial derivative $\dfrac{\partial^{m+n} u}{\partial x^m \partial y^n}$ a difference quotient which involves only function values at mesh points; thus, given any differential equation for u, we can derive in this way a corresponding difference equation, which provides a relation between the approximate values $U_{i,k}$. Such an equation is written down for every mesh point in the interior of the considered region. In a similar manner we derive difference equations corresponding to the boundary conditions, until we have as many equations as unknown values $U_{i,k}$. We can say nothing about the solubility of this system of equations without being more specific about the original boundary-value problem (cf. § 1.2).

---

[1] The following is a selection of the extensive literature on the subject: RUNGE, C.: Über eine Methode, die partielle Differentialgleichungen $\nabla^2 u =$ Constans numerisch zu integrieren. Z. Math. Phys. **56**, 225—232 (1908). — RICHARDSON, L. F.: The approximate arithmetical solution by finite differences of physical problems involving differential equations with an application to the stresses in a masonry dam. Phil. Trans. Roy. Soc. Lond., Ser. A **210**, 308—357 (1911). — LIEBMANN, H.: Die angenäherte Ermittlung harmonischer Funktionen und konformer Abbildung. S.-B. Bayer. Akad. Wiss., Math.-phys. Kl. **1918**, 385—416. — HENCKY, H.: Die numerische Bearbeitung von partiellen Differentialgleichungen in der Technik. Z. Angew. Math. Mech. **2**, 58—66 (1922). — COURANT, R.: Über Randwertaufgaben bei partiellen Differentialgleichungen. Z. Angew. Math. Mech. **6**, 322—325 (1926). — COURANT, R., K. FRIEDRICHS and H. LEWY: Über die partiellen Differenzengleichungen der mathematischen Physik. Math. Ann. **100**, 32—74 (1928). — MARCUS, H.: Die Theorie elastischer Gewebe und ihre Anwendung auf die Berechnung biegsamer Platten, 2nd ed. Berlin 1932.

Since the practical derivation of the difference equations has already been treated in detail for similar cases (§ 1 of Ch. III, §§ 1, 2 of Ch. IV), there is little point in pursuing it further here; suffice it to explain the application of the method by means of the simple examples in §§ 1.5, 1.6. In §§ 1.2 to 1.4 we consider a special class of problems for which precise assertions can be made concerning convergence and error estimation.

1.2. Linear elliptic differential equations of the second order

Consider the differential equation

$$L[u] \equiv a \frac{\partial^2 u}{\partial x^2} + c \frac{\partial^2 u}{\partial y^2} + d \frac{\partial u}{\partial x} + e \frac{\partial u}{\partial y} - g u = r \qquad (1.2)$$

and let boundary values $\bar{u}$ be prescribed for $u(x, y)$ on a simple, closed, piecewise-smooth curve Γ in the (x, y) plane (Fig. V/1). This is the

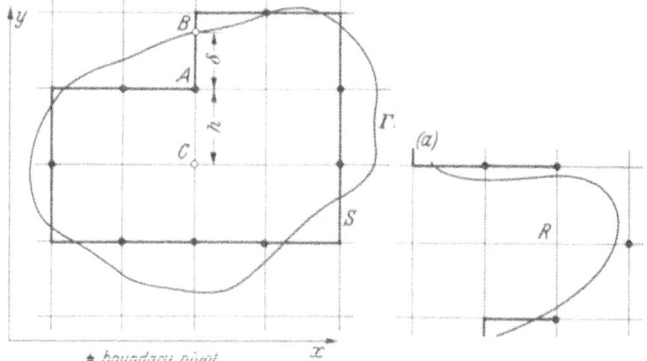

Fig. V/1. The boundary curve Γ and corresponding mesh boundary S

so-called "first" boundary-value problem, or the boundary-value problem of the first kind, for the differential equation (1.2) and the boundary Γ. We assume that the given functions a, c, d, e, g, r are continuous, and further that a, c are positive and g non-negative in a region B^* of the (x, y) plane containing the boundary Γ. The region actually bounded by Γ we denote by B and usually take this to be the closed region including the boundary Γ. B^* is as large an extension of B as is needed for the problem under consideration, and in some cases may be the improper extension $B^* = B$ (cf. § 2.1).

We first introduce some useful terminology concerning the mesh points of the basic finite-difference mesh. Two mesh points will be referred to as "neighbouring points" if they are consecutive mesh points on a mesh line. The term "pivot" will be used for a mesh point which lies within or on Γ. The pivots can be divided into "interior pivots" and "boundary pivots": an interior pivot is a pivot for which all four neighbouring points are also pivots; a boundary pivot, on the other hand, is a pivot for which at least one of its neighbouring points is not a pivot[1]. We assume that the interior pivots are "connected", i.e. given any two non-neighbouring interior pivots P and P^*, we can find further interior pivots $P_1, P_2, \ldots, P_n$ ($n \geq 1$) such that consecutive members of the sequence $P\, P_1\, P_2 \ldots P_n\, P^*$ are neighbouring points. (An assumption of "simple connectivity" is not needed for our purposes.) The boundary curve Γ may be associated with a "mesh boundary" S (Fig. V/1); this is a polygon with sides consisting only of sections of mesh lines and having the property that all mesh points which lie on them are either boundary pivots or isolated corners projecting beyond Γ[2].

We now derive the equations for the approximate pivotal values $U_{i,k}$. For every interior pivot we write down a difference equation corresponding to (1.2):

$$a_{i,k} \frac{U_{i+1,k} - 2U_{i,k} + U_{i-1,k}}{h^2} + c_{i,k} \frac{U_{i,k+1} - 2U_{i,k} + U_{i,k-1}}{l^2} +$$

$$+ d_{i,k} \frac{U_{i+1,k} - U_{i-1,k}}{2h} + e_{i,k} \frac{U_{i,k+1} - U_{i,k-1}}{2l} - g_{i,k} U_{i,k} = r_{i,k}.$$

If a square mesh ($h = l$) is used, this can be written

where
$$\left.\begin{aligned}
l_{ik}[U] &= h^2 r_{ik}, \\
l_{ik}[U] &\equiv l_{ik}^*[U] - (2a_{ik} + 2c_{ik} + g_{ik} h^2)\, U_{ik}, \\
l_{ik}^*[U] &\equiv \left(a_{ik} - \frac{h}{2} d_{ik}\right) U_{i-1,k} + \left(a_{ik} + \frac{h}{2} d_{ik}\right) U_{i+1,k} + \\
&\quad + \left(c_{ik} - \frac{h}{2} e_{ik}\right) U_{i,k-1} + \left(c_{ik} + \frac{h}{2} e_{ik}\right) U_{i,k+1}.
\end{aligned}\right\} \quad (1.3)$$

With every boundary pivot A we associate a boundary point B (i.e. a point on Γ) which lies on a mesh line through A such that the distance δ from A to B is less than the mesh width h ($0 \leq \delta < h$). We assume that h is chosen so small that C, the neighbouring point to A furthest from B (Fig. V/1), is an interior pivot. Then for the pivot A we write down the equation

$$U(A) = \frac{\delta U(C) + h\, \bar{u}(B)}{\delta + h}, \qquad (1.4)$$

---

[1] These definitions of "interior" and "boundary" pivots are convenient for the theory, at least, as far as it is developed in the following. In a practical calculation, however, it is often desirable to use function values at mesh points lying near, though outside of, the boundary curve Γ [for example, the heavily marked points in the subsidiary figure (a) of Fig. V/1] and it is then convenient to include these points as boundary pivots. As a consequence, some points which were originally boundary pivots, such as the point R in the figure, become interior pivots.

[2] Actually we do not need this "mesh boundary" for the theory presented here; for our purposes the classification of the pivots into boundary pivots and connected interior pivots is sufficient.

in which the modified notation has the obvious significance; the equation implies that graphically the values $\bar{u}(B)$, $U(A)$, $U(C)$ are collinear. If A lies on Γ, then $B = A$, $\delta = 0$ and $U(A)$ is given its required boundary value $\bar{u}(A)$.

Another approach for curved boundaries is to derive difference equations corresponding to (1.3) which make use of actual boundary points as pivotal points instead of only mesh points. For example, if $a_1 h$, $a_2 h$, $a_3 h$, $a_4 h$ are the distances from a mesh point $P_0 = (i, k)$ of four neighbouring pivotal points $P_1 = (i + a_1, k)$, $P_2 = (i, k + a_2)$, $P_3 = (i - a_3, k)$, $P_4 = (i, k - a_4)$, as in Fig. V/2, then

$$\frac{2u(P_1)}{a_1(a_1 + a_3)} + \frac{2u(P_2)}{a_2(a_2 + a_4)} + \frac{2u(P_3)}{a_3(a_3 + a_1)} + \frac{2u(P_4)}{a_4(a_4 + a_2)} -$$
$$- 2\left(\frac{1}{a_1 a_3} + \frac{1}{a_2 a_4}\right) u(P_0) = h^2(u_{xx} + u_{yy})(P_0) + O(h^3)$$

as can be verified by substituting the Taylor expansions centred on P_0 for the values of u which appear. If the differential equation is of the form $u_{xx} + u_{yy} = gu - r$, this formula without the remainder term, and accordingly with $u(P_j)$ replaced by the approximations $U(P_j)$, provides a difference equation corresponding to the point P_0. A similar application of TAYLOR's theorem can also be used to derive corresponding difference equations for the more general differential equation (1.2).

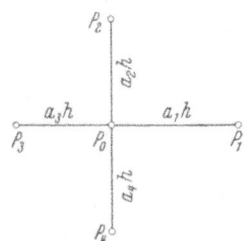

Fig. V/2. Unsymmetrical point pattern for curved boundaries

In many applications the boundary condition for the whole of the boundary Γ or some part of it Γ^* does not specify u but instead prescribes the value of a linear combination of u and its derivative $\partial u/\partial \nu$ in the direction of the inward normal ν, i.e. we can have a condition of the form

$$A_1(s)\, u + A_2(s)\, \frac{\partial u}{\partial \nu} = A_3(s) \quad \text{on} \quad \Gamma^*, \tag{1.5}$$

where s is the boundary parameter. When $\Gamma^* = \Gamma$, the problem of determining u is called the "second" boundary-value problem if $A_1(s) \equiv 0$ and the "third" boundary-value problem if $A_1(s) \not\equiv 0$ (cf. Ch. I, § 3.3); when Γ^* is properly a part of Γ, we speak of a mixed boundary-value problem. One way of deriving a difference equation to simulate a boundary condition of the type (1.5) is the following[1,2].

---

[1] See BATSCHELET, E.: Über die numerische Auflösung von Randwertproblemen bei elliptischen partiellen Differentialgleichungen. Z. Angew. Math. Phys. **3**, 165—193 (1952), where error estimates for the finite-difference method are also given; see also F. B. HILDEBRAND: Methods of applied mathematics, p. 307. New York 1952. — F. S SHAW: An introduction to relaxation methods, p. 120 et seq. New York 1953.

[2] More accurate formulae can be derived by bringing in more mesh points. A formula using 11 points is given by W. E. MILNE: Numerical solution of dif-

Consider a boundary pivot A of a sufficiently fine square mesh and let the local normal through A meet the boundary Γ in a point B which belongs to Γ^* (Fig. V/3). The line BA produced will cut a mesh line at a first point C between two neighbouring mesh points D, E. Let the associated distances be denoted as follows:

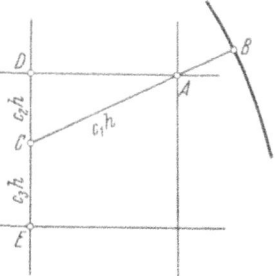

$$\overline{AC} = c_1 h, \quad \overline{DC} = c_2 h, \quad \overline{EC} = c_3 h,$$

where

$$1 \leq c_1 \leq \sqrt{2}, \quad c_2 \geq 0, \quad c_3 \geq 0,$$
$$c_2 + c_3 = 1.$$

Fig. V/3. One way of dealing with the third boundary-value problem

Then for the boundary pivot A we can write down the difference equation

$$A_1(s)\, U(A) + A_2(s)\, \frac{1}{c_1 h} \left[\frac{c_3\, U(D) + c_2\, U(E)}{c_2 + c_3} - U(A) \right] = A_3(s), \quad (1.6)$$

where s is to be given the value corresponding to the boundary point B.

Iterative solution of the system of difference equations. When a fine mesh is used, the system of difference equations becomes rather large, and in consequence an iterative solution is frequently resorted to. Thus, starting from approximations $U_{ik}^{[0]}$, we determine successive iterates $U_{ik}^{[1]}$, $U_{ik}^{[2]}$, ... from the general formulae

$$\left.\begin{aligned}(2a_{ik} + 2c_{ik} + g_{ik}\, h^2)\, U_{ik}^{[\nu+1]} &= l_{ik}^*[U^{[\nu]}] - h^2 r_{ik}, \\ U^{[\nu+1]}(A) &= \frac{\delta\, U^{[\nu]}(C) + h\, \bar{u}(B)}{\delta + h} \end{aligned}\right\} \ (\nu = 0, 1, 2, \ldots).$$

(1.7)
(1.8)

For LAPLACE's equation

$$\nabla^2 u = \frac{\partial^2 u}{\partial x^2} + \frac{\partial^2 u}{\partial y^2} = 0$$

the iteration formula (1.7) reduces to

$$\left.\begin{aligned} U_{ik}^{[\nu+1]} &= \tfrac{1}{4}(U_{i+1,k}^{[\nu]} + U_{i,k+1}^{[\nu]} + U_{i-1,k}^{[\nu]} + U_{i,k-1}^{[\nu]}) \\ &= \tfrac{1}{4} \times (\text{Sum of the } \nu\text{-th iterates at the neighbouring points}); \end{aligned}\right\} \quad (1.9)$$

ferential equations, p. 150. New York and London 1953. — Further formulae are given by J. ALBRECHT and W. UHLMANN: Differenzenverfahren für die erste Randwertaufgabe mit krummlinigen Rändern bei $\Delta u(x, y) = r(x, y, u)$. Z. Angew. Math. Mech. **37**, 212—224 (1957). — UHLMANN, W.: Differenzenverfahren für die 1. Randwertaufgabe mit krummflächigen Rändern bei $\Delta u(x, y, z) = r(x, y, z, u)$. Z. Angew. Math. Mech. **38**, 130—139 (1958). — Differenzenverfahren für die 2. und 3. Randwertaufgabe mit krummlinigen Rändern bei $\Delta u(x, y) = r(x, y, u)$. Z. Angew. Math. Mech. **38**, 236—251 (1958).

thus at each stage the value at the interior pivot (i, k) is to be replaced by the mean of the four neighbouring values (LIEBMANN's averaging procedure[1]).

Existence and uniqueness of the solution of the system of difference equations. The system of equations (1.3), (1.4) for the U_{ik} will always possess a uniquely determined solution if the mesh width h is chosen so small that $a_{ik} \pm \frac{1}{2}hd_{ik}$ and $c_{ik} \pm \frac{1}{2}he_{ik}$ are positive for all pivots (i, k). This statement follows directly from Theorem 2 of Ch. I, § 5.5, for we can easily verify that the conditions of that theorem are satisfied (the a_{ik} in § 5.5 do not have the same meaning as here). Firstly, the equations can be written down with the coefficients $2a_{ik} + 2c_{ik} + h^2 g_{ik}$ or 1 (corresponding to interior or boundary equations, respectively) on the leading diagonal; then the diagonal coefficients are positive and the non-zero off-diagonal coefficients, which can only be $-a_{ik} \pm \frac{1}{2}hd_{ik}$, $-c_{ik} \pm \frac{1}{2}he_{ik}$ or $-\dfrac{\delta}{\delta + h}$, are all negative. Secondly, all row sums of coefficients are non-negative and at least one row sum is positive: when they occur, equations of the form (1.4) give positive row sums since $1 - \dfrac{\delta}{\delta + h} > 0$; when no such equations occur, i.e. when all boundary pivots lie on the boundary curve Γ, a positive row sum will arise from one of the equations (1.3) corresponding to an interior pivot with at least one of its neighbouring pivots, say P, on Γ, for in such an equation $u(P)$ is known and is therefore part of the inhomogeneous term, so that its coefficient is not to be included in the row sum. Thirdly, the non-decomposition of the matrix of coefficients follows from the assumption that the interior pivots are connected.

The weaker conditions of Theorem 3 of Ch. I, § 5.5 are therefore also satisfied, so that the convergence of the total-step and single-step iteration procedures[2] is assured at the same time.

1.3. Principle of an error estimate for the finite-difference method

An error estimate[3] for the class of problems considered in § 1.2 has been given by GERSCHGORIN[*]; existence of the solution $u(x, y)$ is

---

[*] GERSCHGORIN, S.: Fehlerabschätzung für das Differenzenverfahren zur Lösung partieller Differentialgleichungen. Z. Angew. Math. Mech. **10**, 373—382 (1930).

[1] LIEBMANN, H.: Die angenäherte Ermittlung harmonischer Funktionen und konformer Abbildung. S.-B. Bayer. Akad. Wiss., Math.-phys. Kl. **1918**, 385—416.— WOLF, F.: Über die angenäherte numerische Berechnung harmonischer und biharmonischer Funktionen. Z. Angew. Math. Mech. **6**, 118—150 (1926).

[2] A different convergence proof based on subharmonic functions has been given by J. B. DIAZ and R. C. ROBERTS: On the numerical solution of the Dirichlet problem for Laplace's difference equation. Quart. Appl. Math. **9**, 355—360 (1952).

[3] Error estimation can be looked at in another way, which is suggested by the inexactness of applied problems. This has been discussed by R. v. MISES:

assumed and also knowledge of upper bounds M_n for the maxima in B of the absolute values of the partial derivatives $\partial^n u/\partial x^n$ and $\partial^n u/\partial y^n$ up to $n = 4$:

$$\left|\frac{\partial^n u}{\partial x^n}\right| \leqq M_n, \qquad \left|\frac{\partial^n u}{\partial y^n}\right| \leqq M_n \quad \text{in } B \text{ for } n = 1, 2, 3, 4. \qquad (1.10)$$

It will often be difficult to establish strict upper bounds M_n, and sometimes, for want of something better, one will use instead upper bounds for the corresponding difference quotients of the calculated approximations U_{ik}. Since the partial derivatives may exceed these bounds, one clearly cannot claim that the error limits so obtained are rigorous; however, they should yield an indication of the size of the error, and sometimes one may be content with this[1].

The derivation of the error estimate follows the same general pattern as for ordinary differential equations (cf. Ch. III, § 3). We write down the difference expressions corresponding to (1.3) for the exact solution $u(x, y)$ and expand them by means of TAYLOR's theorem into differential

---

On network methods in conformal mapping and in related problems. Nat. Bur. Stand., Appl. Math. Ser. **18**, 1—7 (1952). He maintains that from a physical point of view it is more logical to look for an approximate solution U such that

$$|L[U]| \leqq \delta, \qquad |\Phi_\mu[U]| \leqq \delta$$

for a given tolerance δ ($L[u] = 0$ being the given differential equation and $\Phi_\mu[u] = 0$ the associated boundary conditions) than to investigate the error $\varepsilon = U - u$ in a given approximate solution U. Thus the idea of convergence is replaced by the idea of simultaneous approximation of the differential equation and the boundary conditions.

Application of this idea to the finite-difference method is a little troublesome; having calculated approximate values at the mesh points, one still has to construct from them a function U for which $L[U]$ can be formed. If, for example, we are treating a boundary-value problem with a second-order differential equation in two independent variables x, y by the finite-difference method and have obtained approximate values U_{ik} for a function $u(x, y)$ at the mesh points (x_i, y_k), then the next step is to construct a set of polynomials, one for each cell of the mesh, which join together smoothly to form one continuous function $U(x, y)$ with continuous first and second derivatives with respect to each independent variable.

[1] This difficulty in establishing or estimating the quantities M_n which embarrasses the application of the formulae (1.26) to (1.28) can sometimes be overcome by using special properties of the problem in question; for instance, GERSCHGORIN (loc. cit.) has shown that M_4 can be calculated exactly for the torsion problem for a rectangular bar. W. WASOW [J. Res. Nat. Bur. Stand. **48** (1952) Res. Paper 2321] employs a different approach in his derivation of error estimates for the plane Laplace equation $u_{xx} + u_{yy} = 0$ in a rectangle. He considers the first boundary-value problem with $u = f(s)$ on the boundary, where $f(s)$ is a "reasonable" function, and starts by reducing the problem to one in which $f(s) = 0$ at the four corners of the rectangle; this can be done by subtracting out a suitable harmonic polynomial of the form $a_0 + a_1 x + a_2 y + a_3 x y$. Then u can be written as the sum of four functions u_j satisfying $\nabla^2 u_j = 0$ ($j = 1, 2, 3, 4$) each of which also satisfies

expressions plus remainder terms. For instance, we have

$$u_{i+1,k}+u_{i-1,k}-2u_{ik}=h^2\left(\frac{\partial^2 u}{\partial x^2}\right)_{ik}+\frac{h^4}{24}\left(\frac{\partial^4 u_{\overline{i,i-1},k}}{\partial x^4}+\frac{\partial^4 u_{\overline{i,i+1},k}}{\partial x^4}\right),$$

in which we have used the notation $\partial^n u_{\overline{i,j},k}/\partial x^n$ to denote a value of $\partial^n u/\partial x^n$ at a point on the section of mesh line between the mesh points (i, k) and (j, k). Similarly expanding the other difference expressions in (1.3), we obtain

$$\left.\begin{aligned}\frac{1}{h^2}l_{ik}[u]&=a_{ik}\left\{\left(\frac{\partial^2 u}{\partial x^2}\right)_{ik}+\frac{h^2}{24}\left(\frac{\partial^4 u_{\overline{i,i-1},k}}{\partial x^4}+\frac{\partial^4 u_{\overline{i,i+1},k}}{\partial x^4}\right)\right\}+\\&\quad+c_{ik}\{\cdots\}+d_{ik}\{\cdots\}+e_{ik}\{\cdots\}-g_{ik}u_{ik}\\&=(L[u])_{ik}+\frac{1}{h^2}R_{ik}[u]=r_{ik}+\frac{1}{h^2}R_{ik}[u],\end{aligned}\right\} \quad (1.11)$$

in which the remainder term $R_{ik}[u]$ can be expressed in the form

$$\left.\begin{aligned}R_{ik}[u]&\\=\frac{h^4}{24}&\left\{a_{ik}\left(\frac{\partial^4 u_{\overline{i,i-1},k}}{\partial x^4}+\frac{\partial^4 u_{\overline{i,i+1},k}}{\partial x^4}\right)+c_{ik}\left(\frac{\partial^4 u_{i,\overline{k,k-1}}}{\partial y^4}+\frac{\partial^4 u_{i,\overline{k,k+1}}}{\partial y^4}\right)\right\}+\\+\frac{h^4}{12}&\left\{d_{ik}\left(\frac{\partial^3 u_{\overline{i,i-1},k}}{\partial x^3}+\frac{\partial^3 u_{\overline{i,i+1},k}}{\partial x^3}\right)+e_{ik}\left(\frac{\partial^3 u_{i,\overline{k,k-1}}}{\partial y^3}+\frac{\partial^3 u_{i,\overline{k,k+1}}}{\partial y^3}\right)\right\}.\end{aligned}\right\} \quad (1.12)$$

Accordingly, since the approximations U_{ik} satisfy (1.3), the errors

$$\varepsilon_{ik}=U_{ik}-u_{ik} \quad (1.13)$$

at interior pivots (i, k) satisfy the difference equations

$$l_{ik}[\varepsilon]=-R_{ik}[u]. \quad (1.14)$$

Assuming that the derivatives of u are bounded as in (1.10) we have

$$\left|\frac{1}{h^2}R_{ik}[u]\right|\leqq\frac{h^2}{12}\{M_4(a_{ik}+c_{ik})+2M_3(|d_{ik}|+|e_{ik}|)\}=h^2 C_1,\text{ say. } (1.15)$$

the modified boundary condition on just one of the four sides of the rectangle and is zero on the other three.

With a suitable choice of co-ordinate system u_1, for example, may be expressed as a series of the form

$$u_1=\sum_{n=1}^{\infty}d_n\sin b_n x\sinh c_n y.$$

Now the finite-difference solution, which in itself is only defined at the mesh points, can also be represented by such a series, and this enables one to estimate the difference between the two solutions. If the mesh width is h, the upper bound for the absolute error finally arrived at is $(A_2 M_2^*+A_3 M_3^*)h^2$, where A_2, A_3 depend only on the size of the rectangle and M_2^*, M_3^* are upper bounds for the absolute values of the second and third derivatives, respectively, of the given boundary values along the sides of the rectangle. All these quantities can be computed from the data.

Using TAYLOR's theorem in a similar manner at the boundary pivots, we have (for a boundary pivot A and points B, C as in Fig. V/1 on a line parallel the x axis, for instance)

$$u(C) = u(B) + \frac{\partial u(B)}{\partial x}(h + \delta) + \frac{\partial^2 u(\bar{B})}{\partial x^2} \cdot \frac{(h + \delta)^2}{2}$$

$$u(A) = u(B) + \frac{\partial u(B)}{\partial x}\delta + \frac{\partial^2 u(\bar{\bar{B}})}{\partial x^2} \cdot \frac{\delta^2}{2},$$

where $\bar{B}, \bar{\bar{B}}$ are intermediate points. Elimination of $\partial u/\partial x$ yields

$$u(A) = \frac{h}{h + \delta}u(B) + \frac{\delta}{h + \delta}u(C) - R_A, \tag{1.16}$$

where

$$R_A = \frac{\delta(h + \delta)}{2} \cdot \frac{\partial^2 u(\bar{B})}{\partial x^2} - \frac{\delta^2}{2} \cdot \frac{\partial^2 u(\bar{\bar{B}})}{\partial x^2}.$$

Comparison with (1.4) reveals that the error ε is given by

$$\varepsilon(A) = \frac{\delta}{h + \delta}\varepsilon(C) + R_A \tag{1.17}$$

at the boundary pivots, and hence, using the fact that $\delta < h$, we have

$$|\varepsilon(A)| \leq \tfrac{1}{2}|\varepsilon(C)| + \tfrac{3}{2}h^2 M_2. \tag{1.18}$$

To estimate the error ε from (1.14), (1.15), (1.18), GERSCHGORIN introduces an auxiliary function W defined by

$$L[W] = -1 \quad \text{in } B, \quad W = 0 \quad \text{on } \Gamma.$$

Since we can hardly determine this function W in general, or even find limits for it, we will restrict ourselves for the remainder to a special case for which such an auxiliary function can be majorized and thereby a simple error estimate obtained. This special case is nevertheless still sufficiently general to cover many technically important problems. The additional assumption which we make is that we can find an ellipse E with axes parallel to the co-ordinate axes which has the following properties: no point of E lies inside of Γ and its semi-axes p and q are such that

$$Q = \frac{a}{p^2} + \frac{c}{q^2} - \frac{|d|}{p} - \frac{|e|}{q} > 0 \tag{1.19}$$

everywhere in B.

Let (x_c, y_c) be the centre of this ellipse and consider the auxiliary function

$$Z(x, y) = \mu + \beta\left[1 - \left(\frac{x - x_c}{p}\right)^2 - \left(\frac{y - y_c}{q}\right)^2\right] = \mu + \beta \cdot \varphi(x, y), \tag{1.20}$$

in which μ and β are positive constants to be determined. Firstly β is chosen so that

$$- \frac{1}{h^2} l_{ik}[Z] \geq h^2 C_1 \quad \text{in } B. \tag{1.21}$$

From (1.11), (1.12) we have

$$- \frac{1}{h^2} l_{ik}[Z] = - (L[Z])_{ik} - \frac{1}{h^2} R_{ik}[Z] = - (L[Z])_{ik} = \mu g + 2\beta \times$$

$$\times \left\{ \frac{a}{p^2} + \frac{c}{q^2} + d \frac{(x - x_c)}{p^2} + e \frac{(y - y_c)}{q^2} + \frac{g}{2} \left[1 - \left(\frac{x - x_c}{p} \right)^2 - \left(\frac{y - y_c}{q} \right)^2 \right] \right\}.$$

Now for the ellipse E, (1.19) holds and also

$$|x - x_c| \leq p, \quad |y - y_c| \leq q, \quad 1 - \left(\frac{x - x_c}{p} \right)^2 - \left(\frac{y - y_c}{q} \right)^2 \geq 0 \quad \text{in } B;$$

consequently

$$- \frac{1}{h^2} l_{ik}[Z] \geq 2\beta Q \geq h^2 C_1, \tag{1.22}$$

in which the last inequality determines a lower bound for β. Thus we can choose

$$\beta = \frac{h^2}{2} \left(\frac{C_1}{Q} \right)_{\text{max in } B}. \tag{1.23}$$

We now choose μ so that the result corresponding to (1.21) for boundary pivots also holds. In fact we verify that $\mu = \beta + 3 h^2 M_2$ is a suitable choice. For a typical boundary pivot A and its associated interior pivot C, we have

$$Z(A) - \frac{\delta}{h + \delta} Z(C) = \mu \frac{h}{h + \delta} + \beta \left(\varphi(A) - \frac{\delta}{h + \delta} \varphi(C) \right)$$

$$= \beta \left[\frac{h}{h + \delta} + \varphi(A) - \frac{\delta}{h + \delta} \varphi(C) \right] + 3 h^2 M_2 \frac{h}{h + \delta}.$$

Here, the last term is at least $\frac{3}{2} h^2 M_2$ and, since $0 \leq \delta \leq h$, $0 \leq \varphi(A) \leq 1$, $0 \leq \varphi(C) \leq 1$, the factor multiplying β is non-negative (in fact it lies between 0 and 2); thus the right-hand side cannot be less than the upper bound $\frac{3}{2} h^2 M_2$ of (1.17) for $|R_A|$ and we have

$$\left| \varepsilon(A) - \frac{\delta}{h + \delta} \varepsilon(C) \right| \leq \frac{3}{2} h^2 M_2 \leq Z(A) - \frac{\delta}{h + \delta} Z(C). \tag{1.24}$$

This corresponds to

$$\left| - \frac{1}{h^2} l_{ik}[\varepsilon] \right| \leq - \frac{1}{h^2} l_{ik}[Z]$$

for interior pivots, which follows from (1.14), (1.15), (1.22).

The matrix of the system of these inequalities written down for every pivot satisfies the conditions of Theorem 2 of Ch. I, § 5.5, as we have already seen in §1.2; we can therefore use the monotonic property which is assured by that theorem to obtain the error estimate

$$|\varepsilon(A)| \leqq Z(A) \quad \text{and} \quad |\varepsilon_{ik}| \leqq Z_{ik} \quad \text{in } B, \tag{1.25}$$

or, using the maximum value of Z from (1.20),

$$|U_{ik} - u_{ik}| \leqq \mu + \beta, \tag{1.26}$$

where

$$\beta = \frac{h^2}{24} \left(\frac{M_4(a+c) + 2M_3(|d| + |e|)}{\dfrac{a}{p^2} + \dfrac{c}{q^2} - \dfrac{|d|}{p} - \dfrac{|e|}{q}} \right)_{\text{max in } B} \tag{1.27}$$

and

$$\mu = \beta + 3h^2 M_2. \tag{1.28}$$

When all boundary pivots actually lie on the boundary curve Γ, no equations of the type (1.4) are needed and the estimate (1.26), (1.27) is valid with $\mu = 0$.

The special case with the differential equation

$$a(x, y) \frac{\partial^2 u}{\partial x^2} + c(x, y) \frac{\partial^2 u}{\partial y^2} = r(x, y) \tag{1.29}$$

includes many important problems. Since $d = e = 0$ for this case, the expression (1.27) for β simplifies considerably to

$$\beta = h^2 \frac{M_4 \varrho^2}{24} \tag{1.30}$$

if we choose $p = q = \varrho$ as the radius of a circle enclosing B.

1.4. An error estimate for the iterative solution of the difference equations

For the class of problems considered in § 1.2 a very simple error estimate for the iteration procedure described there can be derived by using again the monotonic property assured by Theorem 2 of Ch. I, § 5.5. In § 1.2 it has already been established that the matrix of coefficients associated with the system of equations (1.3), (1.4) satisfies the conditions of that theorem. The practical application of the consequent possibility of bracketing the solution of the difference equations is described by means of an example in § 1.6, where it is used for error estimation in connection with the relaxation method. We use it here in a different way to derive an error estimate of more general type.

Suppose that we have continued the iteration process until the numbers have settled to the accuracy of the number of decimals carried, say m. We then have approximations $U_{ik}^{[\nu]}$ for which no further iterations improve the last decimal. This, however, is no guarantee that the last decimal is correct; in fact, as is well illustrated by an example given by WOLF[1], it can still differ considerably from the corresponding figure in the exact solution U_{ik} of the system of equations (1.3), (1.4). Consequently one is interested in having an indication of how great this departure $\zeta_{ik}^{[\nu]} = U_{ik}^{[\nu]} - U_{ik}$ can be for known changes $\sigma_{ik}^{[\nu]} = U_{ik}^{[\nu+1]} - U_{ik}^{[\nu]}$.

From (1.3) and (1.7) it follows that

$$
\left.
\begin{aligned}
\sigma_{ik}^{[\nu]} &= \frac{h^2}{2a_{ik} + 2c_{ik} + g_{ik}h^2} \left[\frac{l_{ik}^*[U^{[\nu]}]}{h^2} - r_{ik} \right] - U_{ik}^{[\nu]} \\
&= \frac{h^2}{2a_{ik} + 2c_{ik} + g_{ik}h^2} \left[\frac{l_{ik}[U^{[\nu]}]}{h^2} - r_{ik} \right] \\
&= \frac{h^2}{2a_{ik} + 2c_{ik} + g_{ik}h^2} \left[\frac{l_{ik}[\zeta^{[\nu]}]}{h^2} \right];
\end{aligned}
\right\} \tag{1.31}
$$

similarly from (1.4) and (1.8) we have

$$
\left.
\begin{aligned}
\sigma^{[\nu]}(A) &= U^{[\nu+1]}(A) - U^{[\nu]}(A) = \frac{\delta}{h+\delta} U^{[\nu]}(C) + \\
&+ \frac{h}{h+\delta} \bar{u}(B) - U^{[\nu]}(A) = \frac{\delta}{h+\delta} \zeta^{[\nu]}(C) - \zeta^{[\nu]}(A).
\end{aligned}
\right\} \tag{1.32}
$$

Thus the errors $\zeta_{ik}^{[\nu]}$, $\zeta^{[\nu]}(A)$ satisfy a system of equations with the same matrix of coefficients as the system (1.14), (1.17) but with different right-hand sides. The $-\frac{1}{h^2} R_{ik}[u]$ and $-R_A$, which are bounded by $\pm h^2 C_1$ and $\pm\frac{3}{2}h^2 M_2$, respectively, are replaced here by $\frac{2a_{ik} + 2c_{ik} + g_{ik}h^2}{h^2} \sigma_{ik}^{[\nu]}$ and $\sigma^{[\nu]}(A)$, respectively. Since all the changes $\sigma_{ik}^{[\nu]}$, $\sigma^{[\nu]}(A)$ are known, we can find bounds for the new right-hand sides, and we have precisely the same situation as in § 1.3. Consequently the error estimate given there [(1.26) to (1.28)] can be carried over immediately with the substitutions just mentioned:

$$
|\zeta_{ik}^{[\nu]}| = |U_{ik}^{[\nu]} - U_{ik}| \leq \mu^* + \beta^*, \tag{1.33}
$$

where

$$
\beta^* = \frac{1}{h^2} \left(\frac{a + c + \frac{1}{2}g h^2}{\frac{a}{p^2} + \frac{c}{q^2} - \frac{|d|}{p} - \frac{|e|}{q}} \right)_{\text{max in } B} \cdot |\sigma_{ik}^{[\nu]}|_{\text{max}} \tag{1.34}
$$

and

$$
\mu^* = \beta^* + 2|\sigma^{[\nu]}(A)|_{\text{max}}. \tag{1.35}
$$

In particular, for the differential equation (1.29) we can use

$$
\beta^* = \frac{\varrho^2 |\sigma_{ik}^{[\nu]}|_{\text{max}}}{h^2}, \tag{1.36}
$$

where ϱ has the same significance as in (1.30).

---

[1] See pp. 130—131 of the paper by F. WOLF already cited in § 1.2 [Z. Angew. Math. Mech. **6**, 118—150 (1926)].

If this iterative calculation is carried out with m decimals, then for the stage at which the numbers settle we can put $|\sigma^{(\nu)}|_{\max} = \tfrac{1}{2} \times 10^{-m}$. If, for instance, $\varrho = 4$, $h = 0\cdot 1$ and $m = 4$, β^* can be as large as $0\cdot 08$ and we cannot even be sure of the first decimal. This situation worsens as h decreases, and often it is not allowed for sufficiently.

1.5. Examples of the application of the ordinary finite-difference method

I. A problem in plane potential flow. We consider the flow of an incompressible "ideal" fluid through a two-dimensional channel in which there is a right-angle bend. Let ABC and DEF be the walls of the channel as in Fig. V/4. We are to calculate the velocity distribution and streamlines within the region $ABCDEF$ assuming that the fluid flows in with unit velocity across the entrance AF and out with unit velocity across the exit CD. The velocity components v_x and v_y can be written as the partial derivatives of a stream-function Ψ or of a potential Φ:

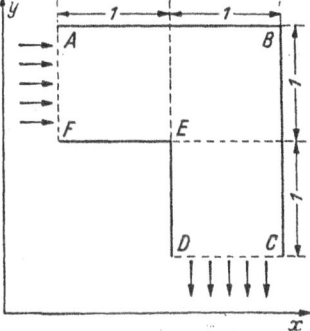

$$v_x = \frac{\partial \Phi}{\partial x} = \frac{\partial \Psi}{\partial y}, \qquad v_y = \frac{\partial \Phi}{\partial y} = -\frac{\partial \Psi}{\partial x};$$

$$\nabla^2 \Phi = \nabla^2 \Psi = 0.$$

If we work in terms of Φ, we have the "second boundary-value problem" of potential theory, for the normal derivative $\partial \Phi / \partial \nu$,

Fig. V/4. Flow through a bent channel

representing the normal component of the velocity (ν directed inwards), is prescribed on the whole of the boundary. Thus we have $\nabla^2 \Phi = 0$ within the region $ABCDEF$ and

$$\frac{\partial \Phi}{\partial \nu} = \begin{cases} 0 \text{ along } ABC \text{ and } DEF, \\ 1 \text{ along } AF, \\ -1 \text{ along } CD. \end{cases}$$

The second boundary-value problem of potential theory may be reduced to the first by the introduction of the conjugate potential function, which in this case coincides with the stream-function Ψ ($= -u$ say). (The present problem can also be formulated directly in terms of the stream-function.)

The required streamlines are therefore the lines $u = $ constant, where u is the harmonic function determined by the boundary values

$$u = \begin{cases} 0 \text{ along } ABC, \\ 1 \text{ along } DEF, \\ \text{linearly increasing along } AF \text{ and } CD. \end{cases}$$

(a) First of all a rough idea of the distribution of u values is obtained by using a coarse mesh with $h = \tfrac{1}{3}$. If we make use of the symmetry

23*

about BE, the number of unknowns reduces to nine; we denote them by $a, b, \ldots, i$ at the points indicated in Fig. V/5. We use the difference equation (1.3) to express $c, d, e, f, \ldots, i$ successively in terms of a and b:

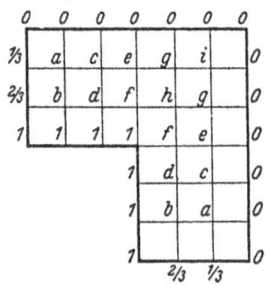

Fig. V/5. The boundary-value problem for the conjugate potential function

$$c = 4a - b - \tfrac{1}{3},$$

$$d = -a + 4b - \tfrac{5}{3},$$

$$e = 16a - 8b + \tfrac{1}{3},$$

$$\ldots\ldots\ldots\ldots\ldots;$$

then the difference equations for the two pivots on the line of symmetry yield a pair of simultaneous equations for a and b with the solution

$$a = \frac{5021}{15\,525} \quad \text{and} \quad b = \frac{10\,177}{15\,525},$$

which, substituted back, gives the approximate values

0	0	0	0	0	0	0
0·333 33	0·323 41	0·304 80	0·263 77	0·182 54	0·091 27	0
0·666 67	0·655 52	0·632 01	0·567 73	0·375 14	0·182 54	0
1	1	1	1	0·567 73	0·263 77	0

(b) We next illustrate LIEBMANN's iteration method on the equations obtained using a slightly finer mesh with $h = \tfrac{1}{4}$. The iteration is started with the following values, which were obtained by graphical interpolation from the values calculated in (a) with $h = \tfrac{1}{3}$:

0	0	0	0	0	0	0	0	0
0·25	0·25	0·24	0·22	0·19	0·15	0·10	0·05	0
0·50	0·50	0·49	0·46	0·39	0·30	0·20		0
0·75	0·75	0·74	0·71	0·65	0·48			0
1	1	1	1	1				0

We calculate new values at each interior pivot from the four neighbouring values by the averaging procedure of (1.9). The results are set out below; beneath each new value we record in brackets the change from the old value (in fourth-decimal units), which may be regarded as a correction.

0	0	0	0	0	0	0	0	0
0·25	0·2475	0·24	0·2225	0·19	0·1475	0·10	0·05	0
	(− 25)	(0)	(+ 25)	(0)	(− 25)	(0)	(0)	
0·50	0·4975	0·4850	0·4525	0·40	0·3050	0·20		0
	(− 25)	(− 50)	(− 75)	(+ 100)	(+ 50)	(0)		
0·75	0·7475	0·7375	0·7125	0·6450	0·4750			0
	(− 25)	(− 25)	(+ 25)	(− 50)	(− 50)			
1	1	1	1	1				0

By applying the same averaging procedure to these changes we obtain the changes for the next cycle, which in turn yield the changes for the following cycle (in brackets):

```
0    0         0         0         0         0         0         0      0
0 —  6(—10)  —13(+14)  —19(+ 9)  +25(— 6)  +13(+ 6) ·— 6(+10) 0(—3) 0
0 —25(—12)  —31(— 6)  +25(—27)  —19(+19)  + 6(+ 5) +25(0)           0
0 —13(— 9)  —13(—20)  —38(+ 8)  +19(—14)   0(+13)                   0
0    0         0         0         0                                0
```

One can continue averaging the changes, then calculate "best" values by extrapolation and repeat the whole process again with these as starting values. In general the convergence of the process is slow and many suggestions for improving it have been made[1]. These do not always have the desired effect; for example, an extrapolation based on a geometric series can, in fact, make matters worse. (A different type of correction calculation using relaxation is described in § 1.6.) In our example here a number of further iterations yields the following results, for which the next iteration would produce a maximum change of 0·00003:

```
0       0        0        0        0        0        0        0      0
0·25  0·24404  0·23517  0·21948  0·19217  0·15028  0·10135  0·05068 0
0·50  0·49110  0·47716  0·45064  0·39898  0·30756  0·20447          0
0·75  0·74323  0·73185  0·70701  0·64564  0·47660                   0
1     1        1        1        1                                  0
```

One normally increases the accuracy by using a smaller mesh width, but here the accuracy can also be increased by subtracting out the singularity[2]. This can be a useful expedient in many other situations in which singularities appear.

II. An equation of more general type (1.2). If a single turn of a helical spring of small angle α and radius R is deformed into a plane ring under the influence of an axial load, the stress-function Φ can be shown to satisfy the dif-

---

[1] WELLER, R., G. H. SHORTLEY and B. FRIED: J. Appl. Phys., Lancaster Pa. 9, 334—344 (1939); 11, 283—290 (1940).

[2] This device is used by S. GERSCHGORIN: Z. Angew. Math. Mech. 10, 373—382 (1930); see also W. E. MILNE: Numerical solution of differential equations, p. 221. New York and London 1953. — WOODS, L. C.: The relaxation treatment of singular points in Poissons equation. Quart. J. Mech. Appl. Math. 6, 163—185 (1953), deals with three types of singularities: 1. A logarithmic singularity in u at a point P, which can be subtracted out. 2. A discontinuity in u at P; for instance, there may be a jump in the given boundary values at a point P on the boundary [for plane potential problems one can subtract out the function $a\varphi$, where (r, φ) are polar co-ordinates centred at P and a is suitably chosen]. 3. The derivatives of u are unbounded at P although u is continuous there (this is the case in the above example). With suitable α, β, m one can subtract the function $(\alpha \sin m\varphi + \beta \cos m\varphi) r^m$.

ferential equation[1]

$$\frac{\partial^2 \Phi}{\partial x^2} + \frac{\partial^2 \Phi}{\partial y^2} + \frac{3}{R-y}\frac{\partial \Phi}{\partial y} - 2G\lambda = 0$$

and the boundary condition

$$\Phi = 0 \quad \text{on } \Gamma,$$

where Γ is the boundary of the cross-section in the (x, y) plane (containing the axis of the spring) (Fig. V/6). G is the modulus of rigidity and $\lambda = \dfrac{\sin\alpha\cos\alpha}{R}$.

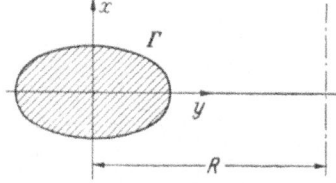

Fig. V/6. Cross-section of a coil of a helical spring

The shear-stress components are given by

$$\tau_y = \left(\frac{R}{R-y}\right)^2 \frac{\partial \Phi}{\partial x},$$

$$\tau_x = -\left(\frac{R}{R-y}\right)^2 \frac{\partial \Phi}{\partial y}.$$

We consider the special case with the rectangular cross-section

$$|x| \leq \tfrac{1}{2}, \qquad |y| \leq 1$$

and $R = 5$. Then, in terms of a new dependent variable u defined by

$$\Phi = -2G\lambda u,$$

the problem reads

$$u_{xx} + u_{yy} + \frac{3}{5-y}u_y + 1 = 0, \qquad u = 0 \text{ on } \Gamma.$$

(a) For a coarse mesh with $h = \tfrac{1}{2}$ there are only the three unknown u values a, b, c at the points indicated in Fig. V/7. They satisfy the difference equations

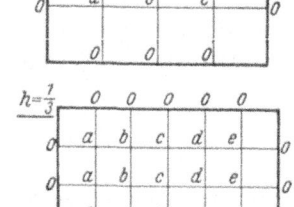

Fig. V/7. Notation for pivotal values

$$4(b - 4a) + \frac{3}{5 + \tfrac{1}{2}}\frac{b - 0}{1} + 1 = 0,$$

$$4(c + a - 4b) + \frac{3}{5}\frac{c - a}{1} + 1 = 0,$$

$$4(b - 4c) + \frac{3}{5 - \tfrac{1}{2}}\frac{0 - b}{1} + 1 = 0$$

with the solution

$$a = \frac{1379}{16 \times 929} = 0.09277,$$

$$b = \frac{99}{929} = 0.10657,$$

$$c = \frac{1259}{16 \times 929} = 0.08470.$$

(b) For a mesh with $h = \tfrac{1}{4}$ we have, on account of symmetry, just five unknown values a, b, c, d, e as in Fig. V/7. If we express four of these unknowns in terms of the other, say e, by means of the difference equations, then the last equation determines e:

$$d = 3.391304e - \tfrac{1}{8} \times 1.130435,$$
$$c = 10.151783e - \tfrac{1}{8} \times 4.918261,$$
$$b = 29.70435\ e - \tfrac{1}{8} \times 16.12367,$$
$$a = 86.07586\ e - \tfrac{1}{8} \times 48.54253,$$
$$7680.677e - \tfrac{1}{8} \times 4388.762 = 0.$$

[1] BIEZENO, C. B., and R. GRAMMEL: Technische Dynamik, 2nd ed., Vol. I. p. 351. 1953.

Therefore

$$a = 0·071\,28,$$
$$b = 0·094\,39,$$
$$c = 0·098\,25,$$
$$d = 0·089\,71,$$
$$e = 0·063\,489\,2,$$
$$\tfrac{9}{8}c = 0·110\,53.$$

The last value $\tfrac{9}{8}c$ corresponds to the centre of the rectangle.

(c) For the finer mesh with $h = \tfrac{1}{4}$ the difference equations are best solved by iteration. With mesh points $(x_i, y_k) = (ih, kh)$ the equations can be put in the form

$$U_{i,k} = \frac{1}{4}\,(U_{i+1,k} + U_{i-1,k} + U_{i,k+1} + U_{i,k-1}) + \frac{3}{32}\,\frac{1}{5-y}\,(U_{i+1,k} - U_{i-1,k}) + \frac{1}{64}\,.$$

As in Example I, we estimate starting values for the U_{ik} from graphs of the results of the calculation with $h = \tfrac{1}{2}$. For convenience we iterate only on the changes. The results are exhibited in Table V/1, whose last row gives the factors $\dfrac{3}{32}\cdot\dfrac{1}{5-y}$, which can be calculated once and for all.

Table V/1. *Results of the iterative solution of the difference equations with* $h = \tfrac{1}{4}$

i	x	$k=-4$ $y=-1$	$k=-3$	$k=-2$	$k=-1$	$k=0$ $y=0$	$k=1$	$k=2$	$k=3$	$k=4$ $y=1$
−2	−0·6	0	0	0	0	0	0	0	0	0
−1	−0·25	0	0·05244	0·07447	0·08289	0·08420	0·08010	0·06912	0·04622	0
0	0	0	0·06794	0·09798	0·10968	0·11151	0·10576	0·09050	0·05936	0
$\dfrac{3}{32}\,\dfrac{1}{5-y}$			0·016304	0·017046	0·017857	0·018750	0·019736	0·020833	0·022059	

III. A differential equation of the fourth order. If a non-uniformly loaded square plate of side $2A$ has two opposite sides firmly clamped and the other two smoothly hinged (Fig. V/8), the deflection u will satisfy

$$\nabla^4 u = \frac{p}{N} = \left(\frac{x}{A}\right)^2 \frac{p_0}{N}\,,$$

$$u = \frac{\partial u}{\partial x} = 0 \quad \text{for} \quad |x| = A,$$

$$u = \nabla^2 u = 0 \quad \text{for} \quad |y| = A.$$

Let p_0 and N be constants.

Using the symmetry of the problem, we can reduce the number of unknown values for the mesh width $h = \tfrac{1}{2}A$ to four, namely the a, b, c, d as shown in the figure. In setting up the difference equations, we find that values outside of the square are needed. As indicated in the figure, these can be expressed in terms of the values inside by means of the boundary conditions; for example, the values $-c$, $-d$ follow from the difference equations representing the boundary condition $\nabla^2 u = 0$. Let us put

$$k = \frac{h^4}{2} \frac{p_0}{N} = \frac{q}{128};$$

then, using Table VI of the appendix, we can immediately write down the set of difference equations:

$$20a - 16b - 16c + 8d = 0,$$
$$20b - 8a - 16d + 4c + 2b = 2k,$$
$$20c - 8a - 16d + 4b = 0,$$
$$20d - 8b - 8c + 2a + 2d = 2k.$$

The solution is

$$a = \frac{310}{497} k = 0.00487\,q,$$

$$b = \frac{257}{497} k = 0.00405\,q,$$

$$c = \frac{227}{497} k = 0.00357\,q,$$

$$d = \frac{193}{497} k = 0.00304\,q.$$

If we choose $h = \frac{2}{5}A$, we obtain somewhat more accurate results with still only four unknowns (see Fig. V/8). Thus from the difference equations

$$6\alpha - 5\beta - 5\gamma + 2\delta = s,$$
$$-5\alpha + 13\beta + 2\gamma - 7\delta = 9s,$$
$$-5\alpha + 2\beta + 11\gamma - 7\delta = s,$$
$$2\alpha - 7\beta - 7\gamma + 20\delta = 9s,$$

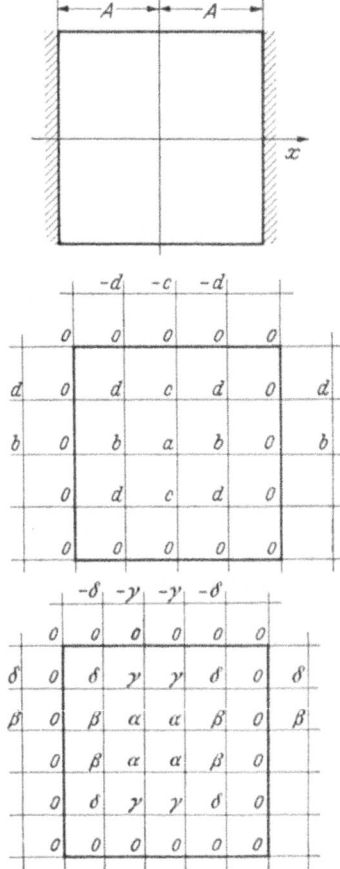

Fig. V/8. Notation for pivotal values

where $s = \dfrac{h^4}{25} \cdot \dfrac{p_0}{N} = \dfrac{16}{15625}\,q$, we obtain the values

$$\alpha = \frac{6755}{1522} s = 0.00454\,q, \qquad \beta = \frac{4693}{1522} s = 0.00315\,q,$$

$$\gamma = \frac{4383}{1522} s = 0.00294\,q, \qquad \delta = \frac{3186}{1522} s = 0.00214\,q.$$

1.6. Relaxation with error estimation

The practical application of the ordinary finite-difference method in conjunction with the relaxation procedure[1] will be explained by means

---

[1] See, for example, G. SHORTLEY, R. WELLER, P. DARBY and E. H. GAMBLE: Numerical solution of axisymmetrical problems, with applications to electrostatics

of an example. We consider the problem of determining the steady temperature distribution $u(x, y)$ in a homogeneous square plate of side A whose edges are kept at the constant temperatures $u = 0$ and $u = 1$ as specified in Fig. V/9; u satisfies the potential equation $\nabla^2 u = 0$.

As usual we begin with a coarse mesh so as to get a rough idea of the solution from which we can estimate starting values for the next step. With $h = \frac{1}{3}A$ we have two unknown values a, b (see Fig. V/9); they satisfy the difference equations

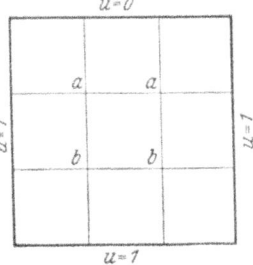

$$4a = a + b + 1, \qquad 4b = a + b + 1 + 1,$$

from which we obtain $a = \frac{5}{8}$, $b = \frac{7}{8}$. These values, rounded to 0·63 and 0·88, are taken over for the first approximation on the finer mesh with $h = \frac{1}{6}A$. Thus we have first approximations to the values g, j in Fig. V/9; we estimate the remaining values by graphical interpolation, dealing with f, i, k first. These starting values are recorded in the upper-left quadrants at the intersections in Table V/2, Stage (i); each intersection corresponds to a pivot. The numbers in the upper-right quadrants are the arithmetic means

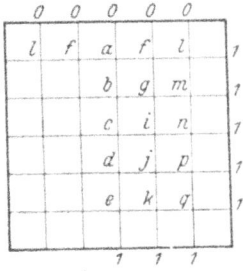

Fig. V/9. Notation for the temperature distribution in a square plate

of the values in the four neighbouring upper-left quadrants, i.e. the next approximation in the iteration of §1.2; for the point a, for instance, this is

$$0·25 \times (0·39 + 0 + 0·39 + 0·59) = 0·3425.$$

In the lower-right quadrants we enter the changes thus produced, expressing them in fourth-decimal units (for example, at a we have $0·3425 - 0·32 = 225 \times 10^{-4}$).

Now if the difference equation for a particular pivot is satisfied by the starting values, the corresponding change must be zero[1]; con-

---

and torsion. J. Appl. Phys. **18**, 116—129 (1947). A detailed description, which also describes many little artifices, is given by L. Fox: Some improvements in the use of relaxation methods for the solution of ordinary and partial differential equations. Proc. Roy. Soc. Lond., Ser. A **190**, 31—59 (1947). Comprehensive treatments are to be found in the books by SOUTHWELL which were cited earlier and also in F. S. SHAW: An introduction to relaxation methods. New York: Dover Publ., Inc. 1953, 396 pp. A theory is developed by G. TEMPLE: The general theory of relaxation methods applied to linear systems. Proc. Roy. Soc. Lond., Ser. A **169**, 476—500 (1939).

[1] Translator's note: These changes produced in the starting values by one step of the iteration procedure are, in fact, identical with (or at least proportional to) what would in the customary relaxational parlance be called the "residuals" corresponding to the starting values; see Ch. III, § 1.4.

Table V/2. *Various stages in the solution of the difference equations by relaxation*

Stage (i)

	0	0	0
0·32	0·3425	0·39 0·3775	0·56 0·5375
	+225	−125	−225
0·59	0·5825	0·63 0·63	0·76 0·7625
	−75	0	+25
0·75	0·755	0·78 0·78	0·86 0·865
	+50	0	+50
0·87	0·8625	0·88 0·88	0·92 0·925
	−75	0	+50
0·94	0·9425	0·95 0·945	0·96 0·9675
	+25	−50	+75
1	1	1	1

Stage (ii)

	0	0	0
0·3316	0·3313	0·3716	0·3714 0·5316	0·5317
0·5820	0·5819	0·6224	0·6222 0·7552	0·7550
0·7512	0·7512	0·78	0·7799 0·8660	0·8660
0·8628	0·8627	0·88	0·8797 0·9288	0·9288
0·9396	0·9393	0·9472	0·9472 0·9692	0·9690

(corrections: −3, −2, +1; −1, −2, −2; −1, 0, 0; −1, −3, 0; −3, 0, −2)

Stage (iii)

	0	0	0
0·33064	0·33064	0·37084	0·37082 0·53132	0·53134
0·58088	0·58085	0·62132	0·62130 0·75452	0·75452
0·75012	0·75011	0·77896	0·77894 0·86544	0·86543
0·86164	0·86164	0·87888	0·87885 0·92824	0·92825
0·93868	0·93869	0·94656	0·94656 0·96868	0·96870

(corrections: 0, −2, +1, +2; 0, −3, −2, +1; −3, −2, −1, −2; −1, −1, 0, +1; +1, 0, 0, +2)

Table V/3. *Numerical details of the relaxation procedure (all numbers expressed in fourth-decimal units)*

(1) Relaxation of the starting values

Corrections	Changes
+200 −120 −240	[+225] −200 −60 / [−125] +50 −60 +120 / [−225] +240 −30
— —	[−75] +50 +15 / [0] −30 / [+25] −60 +15
+60 —	[+50] −60 −15 / [0] +15 +15 / [+50] −60 +15
−60 —	[−75] +15 +60 / [0] −15 +15 −5 / [+50] +15 −60 +20
— −20 +80	[+25] −15 −10 / [−50] +20 +20 / [+75] +15 −80 −5

(2) Relaxation of the improved values

Corrections	Changes
−60 −60 −60 / −24 −4 +16	[−15] +30 −16 +3 +1 / [−15] +15 −6 +4 +4 / [−15] +30 −6 +3 +1
−60 −60 −60 / −20 −16 +12	[−20] +30 +4 −12 −4 / [−30] +15 +3 −1 +16 −5 / [−20] +30 +4 −12 −4
−48	[+5] −15 +7 +3 / [+30] −15 −4 −12 / [+5] −15 +7 +3
−12 +28	[+25] −28 +3 / [−5] +7 +2 −3 / [+25] −28 +3
−4 −8 +12	[+5] +7 −12 −2 / [−10] +3 +8 −1 / [+5] +7 −12 −2
	[−35] +15 +24 −2 −5 / [−10] +15 −6 −8 +20 −12 / [0] −1 −12 +12 / [0] +4 +4 −5

sequently we relax the starting values, i.e. make corrections to them, so as to make the changes as small as possible. If at the pivot a, for example, we had taken the value 0·34 instead of 0·32, a correction of $+200 \times 10^{-4}$, the change at a would have been only 25 instead of 225 (in 10^{-4} units), but the changes at the four neighbouring pivots would have been greater by $+50$. The procedure is most conveniently carried out in two relaxation tables, one for the corrections and the other for the changes, as in Table V/3. The numbers which are framed in the arrays of changes are the changes current at the start of each table, the other numbers being the alterations produced by the corrections.

For the relaxation of the starting values we note that the correction of $+200$ at a alters the changes by the amounts underlined in the table, i.e. -200 at a and $+50$ at b and f. Similarly we make corrections at all pivots where particularly large changes appear, recording these corrections and the corresponding amounts by which they alter the changes in the appropriate columns. In doing this we must remember that on account of symmetry we are correcting two values at once, except on the line of symmetry, so that any change which is affected by both must be altered accordingly; thus the correction of -120 at f alters the changes at g and l by -30, but the change at a by -60.

It is convenient to review the situation now by calculating the net effect on the changes so far and start again in a fresh table with these new changes as starting changes; these are framed in the table and it can be seen that their maximum absolute value is now 35 instead of 225. We continue the relaxation from these improved values as in the table, making corrections which eventually bring the maximum absolute value of the changes down to 3×10^{-4}; at this stage we must increase the number of decimals in order to be able to make further improvements. The current values of u are easily calculated by adding the corrections to the starting values; for example, for a we have

$$0·32 + 0·0200 - 0·0084 = 0·3316.$$

As a check we carry out the averaging process of the Liebmann iteration on these improved u values and thus make an independent calculation of the new changes [Table V/2, Stage (ii)]. The corresponding results after repeating the procedure with an extra decimal are given in Table V/2, Stage (iii)[1].

---

[1] An "over-relaxation" in which, for the solution of the system of linear equations

$$\sum_{k=1}^{n} a_{jk} x_k = r_j \qquad (j = 1, \ldots, n),$$

one determines a sequence of approximations $x_j^{[\nu]}$ generally by

$$x_j^{[\nu+1]} = x_j^{[\nu]} - \omega \left\{ \sum_{k=1}^{j-1} a_{jk} x_k^{[\nu+1]} + \sum_{k=j}^{n} a_{jk} x_k^{[\nu]} - r_j \right\}$$

is investigated by D. M. YOUNG: Iterative methods for solving partial difference

A point of computational technique may be mentioned here. It is a fact that the magnitudes of the changes can be reduced easily in a region where the changes alternate in sign, but not in a region where they have the same sign. A useful technique for dealing with this latter situation is provided by the so-called "block relaxation", in which one makes identical corrections at a block of points[1]. The general effect achieved by this is typified by the following example, in which corrections of 4α are made at points in a rectangular block:

Corrections						Effect on the changes				
						α	α	α	α	α
4α	4α	4α	4α	4α	α	-2α	$-\alpha$	$-\alpha$	$-\alpha$	-2α	α
4α	4α	4α	4α	4α	α	$-\alpha$	0	0	0	$-\alpha$	α
4α	4α	4α	4α	4α	α	-2α	$-\alpha$	$-\alpha$	$-\alpha$	-2α	α
						α	α	α	α	α

With this characteristic pattern in mind it is easy to determine for any particular case which points to include in the block and what value to take for the correction 4α. An example occurs in Table V/3, (2), in which the changes at the pivots a, b, f, g, l, m are all negative; the block correction of -60 produces the underlined alterations in the table of changes.

A comparison of Stages (ii) and (iii) of Table V/2 reveals that in the values of u the third decimal has altered even though the changes in Stage (ii) are confined to the fourth decimal. The possibility of this "ill-conditioning" has already been mentioned in § 1.4; it can become dangerously bad, particularly when a large number of pivots are used.

Error estimation with relaxation. From § 1.4 and Theorem 2 of Ch. I, § 5.5 we have the following bracketing rule for the solutions of the difference equations of § 1.2:

equations of elliptic type. Trans. Amer. Math. Soc. **76**, 92—111 (1954). — On the Solution of Linear Systems by Iteration. Proc. Symp. Appl. Math. **6**, 283—298 (1956). He gives a rule for the suitable choice of the "relaxation factor" ω. Previously L. F. RICHARDSON: Phil. Trans. Roy. Soc. Lond., Ser. A **210**, 307—357 (1911), had used a relaxation factor which even varied from step to step. — See also L. F. RICHARDSON: Phil. Trans. Roy. Soc. Lond., Ser. A **242**, 439—491 (1950). Over-relaxation is also considered by D. N. DE G. ALLEN: La méthode de libération des liaisons ... . Colloques Internat. Centre Nat. Rech. Sci. **14** (Méthodes de calcul dans les problèmes de méchaniques) 11—34. Marseilles and Paris 1949.

[1] For a description of block relaxation see R. V. SOUTHWELL: Relaxation methods in theoretical physics, p. 55. Oxford 1946. Numerous applications of block relaxation are to be found in the literature; see, for example, D. C. GILLES: The use of interlacing nets for the application of relaxation methods to problems involving two dependent variables, with a foreword by W. G. BICKLEY. Proc. Roy. Soc. Lond., Ser. A **193**, 407—433 (1948). — DUSINBERRE, G. M.: Numerical analysis of heat flow, 227 pp. New York-Toronto-London 1949; on p. 65 of this book a large number of point patterns for block relaxation are reproduced.

If we have a set of approximate values for the solution of the difference equations of § 1.2 for which the changes are everywhere non-positive (or everywhere non-negative), then each of these values is greater than or equal to (or, respectively, less than or equal to) the corresponding exact value of the solution of the difference equations. (Whether they are greater or less than the corresponding values of the exact solution of the boundary-value problem is another matter).

A set of changes with no variations in sign can usually be achieved quite quickly with block relaxation. Table V/4 shows the relaxations required to bring about all non-positive and all non-negative changes, respectively, for the approximations obtained in our present example at Stage (iii) (Table V/2, upper-left quadrants); all numbers are expressed in fifth-decimal units. In the first half A. the effect of the corrections is given in detail: the framed numbers are the starting changes taken from Table V/2, Stage (iii); the underlined numbers are the alterations produced by the block of five corrections of $+4$ in the third column; combination of all the alterations yields the non-positive new changes. The corresponding approximate solution, which is everywhere greater than or equal to the exact solution of the difference equations, is exhibited in Table V/5 (same arrangement as in Table V/2). The second half B. of Table V/4 gives the corrections necessary to produce all non-negative changes; the corresponding approximate values, which are consequently lower limits, are exhibited in Table V/6. Thus for the value b, for example, we have the limits

$$0 \cdot 58064 \leqq b \leqq 0 \cdot 58088.$$

In actual fact, although we have not proved it here, the numbers in the upper-right quadrants of Tables V/5 and V/6, i.e. the next Liebmann iterates, also provide upper and lower limits, respectively, so that the more accurate result

$$0 \cdot 58066 \leqq b \leqq 0 \cdot 58085$$

also holds[1].

A generalized form of block relaxation[2] which is usually more effective than the ordinary form for large blocks is obtained by choosing the values of the corrections so that they represent a discrete sub-harmonic function in the sense that the value at each point of the block is greater than or equal to the arithmetic mean of the values at the four neighbouring points, the values outside of the block being zero.

---

[1] For proof see L. COLLATZ: Einschließungssätze bei Iteration und Relaxation. Z. Angew. Math. Mech. **32**, 76—84 (1952).

[2] The "Scheibenrelaxation" of E. STIEFEL: Über einige Methoden der Relaxationsrechnung. Z. Angew. Math. Phys. **3**, 1—33 (1952).

Table V/4. *Relaxation of an approximate solution to obtain changes with no variations in sign*

A. Non-positive changes (for an upper bound)			B. Non-negative changes	
Corrections	Alterations produced	New changes	Corrections	New changes
0 0 +4	[0] [−2]+1 [+2]−3	0 −1 −1	−12 −12 − 4	0 +1 0
0 0 +4	[−3] [−2]+1 [0]−2	−3 −1 −2	−24 −20 −12	+2 0 +3
0 0 +4	[−1]+1 [−2]+1 [−1]−2	0 −1 −3	−24 −24 −12	0 +3 0
+4 0 +4	[0]−3 [−3]+1 [+1]−2 +1 +1	−3 0 −1	−20 −20 − 8	+2 +2 0
+4 +4 +4	[+1]−3 [0]+1 [+2]−3 +1 +1 +1 +1 −4	0 −2 0	− 8 − 8 − 4	0 0 +2

Table V/5. *Upper bounds for the solution of the difference equations*

	0		0		0	
0·33064	0·33064	0·37084	0·37083	0·53136	0·53135	1
0·58088	0 0·58085	0·62132	−1 0·62131	0·75456	−1 0·75454	1
0·75012	−3 0·75012	0·77896	−1 0·77895	0·86548	−2 0·86545	1
0·86168	0 0·86165	0·87888	−1 0·87888	0·92828	−3 0·92827	1
0·93872	−3 0·93872	0·94560	0 0·94558	0·96872	−1 0·96872	1
1	0	1	−2	1	0	

Table V/6. *Lower bounds for the solution of the difference equations*

	0		0		0	
0·33052	0·33052	0·37072	0·37073	0·53128	0·53128	1
0·58064	0 0·58066	0·62112	+1 0·62112	0·75440	0 0·75443	1
0·74988	+2 0·74988	0·77872	0 0·77875	0·86532	+3 0·86532	1
0·86144	0 0·86146	0·87868	+3 0·87870	0·92816	0 0·92816	1
0·93860	+2 0·93860	0·94648	+2 0·94648	0·96864	0 0·96866	1
1	0	1	0	1	+2	

This leaves considerable freedom of choice, and a variety of such group relaxations can easily be constructed; the following is a typical example:

Corrections					Effect on the changes			
					α	2α	2α	α	
4α	8α	8α	4α	α	0	-2α	-2α	0	α
8α	12α	12α	8α	2α	-2α	-2α	-2α	-2α	2α
8α	12α	12α	8α	2α	-2α	-2α	-2α	-2α	2α
4α	8α	8α	4α	α	0	-2α	-2α	0	α
					α	2α	2α	α	

We conclude this section on relaxation by mentioning a phenomenon which manifests itself in systematic relaxation procedures such as the modification of the Liebmann iteration (1.9) in which one deals with the pivots in a fixed cycle and always uses the latest improved values at the neighbouring pivots. To simplify the description we confine ourselves to two equations in two unknowns:

$$\left.\begin{array}{l} a_{11}x_1 + a_{12}x_2 = r_1 \\ a_{21}x_1 + a_{22}x_2 = r_2. \end{array}\right\} \quad (1.37)$$

Consider the iteration procedure defined by

$$x_1^{[\nu+1]} = \frac{1}{a_{11}}(r_1 - a_{12}x_2^{[\nu]}),$$

$$x_2^{[\nu+1]} = \frac{1}{a_{22}}(r_2 - a_{21}x_1^{[\nu+1]})$$

$$(\nu = 0, 1, 2, \ldots)$$

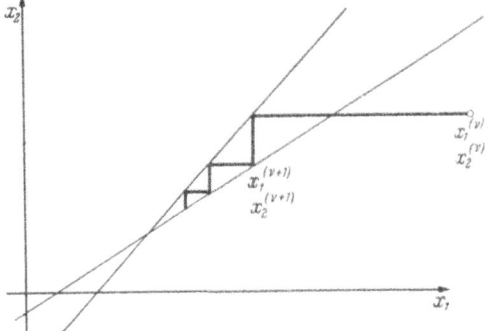

Fig. V/10. A drawback of systematic relaxation procedures due to the occurence of "cages"

(the single-step iteration method of Ch. I, § 5.5) and assume that it converges. The procedure admits of a simple geometrical representation if the two lines defined by (1.37) are drawn in the (x_1, x_2) plane, as in Fig. V/10. The zigzag line connects the sequence of points $(x_1^{[\nu]}, x_2^{[\nu]})$, $(x_1^{[\nu+1]}, x_2^{[\nu]})$, $(x_1^{[\nu+1]}, x_2^{[\nu+1]})$, ... and we see immediately that for k greater than a certain value k_0 (here $k_0 = 0$) the points $(x_1^{[k]}, x_2^{[k]})$ all remain in a fixed "wedge", which STIEFEL (see last footnote) calls a "Käfig", i.e. a "cage". When the angle of the wedge is small, the convergence of the sequence of points to the solution, i.e. to the point of intersection, is slow. Similar "cages" occur also with larger systems of equations; they cause the situation in which substantial improvements are obtained from the first stage of a systematic relaxation procedure (getting into the cage) but only moderate improvements from subsequent stages.

1.7. Three independent variables (spatial problems)

For most problems with more than two independent variables solution by finite differences is very tedious, for as soon as one starts to refine the mesh the number of unknown pivotal values increases very rapidly. Nevertheless one can easily think of examples for which the finite-difference method is the only really feasible numerical method. It may be remarked that the extension to more than two independent variables introduces no new fundamental difficulties.

We choose for illustration an example possessing a high degree of symmetry; this permits a considerable reduction in the number of unknowns. Suppose that a homogeneous material occupies the cube $|x| \leq 1$, $|y| \leq 1$, $|z| \leq 1$ and that two opposite vertices are maintained at the temperatures $+1$ and -1, respectively, while the rest of the surface is insulated. Then the steady temperature distribution satisfies the boundary-value problem

$\nabla^2 u = 0$ inside the cube,

$\dfrac{\partial u}{\partial v} = 0$ for $|x| = 1$, $|y| = 1$, $|z| = 1$,

$u = 1$ for $x = y = z = 1$ and $u = -1$ for $x = y = z = -1$.

For a three-dimensional square mesh with mesh width $h = \frac{2}{3}$ we have the nine unknown pivotal values $a, b, \ldots, k$ as shown in Fig. V/11. The boundary condition $\partial u/\partial v = 0$ is satisfied by introducing symmetrical function values at neighbouring points outside of the cube, as shown in Fig. V/11 for a few typical points. It is expedient in setting up the difference equations corresponding to $\nabla^2 u = 0$ (formula in Table VI of the appendix) to make a sketch as in Fig. V/11. We obtain the equations

$$-6a + b + 4d + 1 = 0,$$
$$a - 6b + c + 4e = 0,$$
$$b - 3c + 2f = 0,$$
$$a - 3d + e + i = 0,$$
$$b + d - 6e + f + g + 2k = 0,$$
$$c + 2e - 7f - 2g = 0,$$
$$e - f - 3g - k = 0,$$
$$d - 2i + k = 0,$$
$$2e - g + i - 8k = 0,$$

which yield

$$a = \frac{1367}{5147} = 0.2656, \quad b = \frac{415}{5147} = 0.0807, \quad c = \frac{199}{5147} = 0.0387,$$

$$d = \frac{660}{5147} = 0.1282, \quad e = \frac{231}{5147} = 0.0450, \quad f = \frac{91}{5147} = 0.0177,$$

$$g = \frac{12}{5147} = 0.0023, \quad i = \frac{382}{5147} = 0.0743, \quad k = \frac{104}{5147} = 0.0202.$$

In some cases it is possible to transform the spatial problem into a plane problem. For example, TRANTER[1] reduces the first boundary-value problem for the equation

$$\nabla^2 u + f(x, y, z) = 0 \qquad (1.38)$$

in a cylinder, i.e. with the boundary conditions

$u = h_1(x, y)$ for $z = 0$
$u = h_2(x, y)$ for $z = \pi$
$u = g(x, y, z)$ on the curved surface of a cylinder parallel to the z axis,

to a set of first boundary-value problems in the region of the (x, y) plane representing the cross-section of the cylinder by means of finite Fourier transforms

$$v(n) = v(n, x, y) = \int_0^\pi u \sin nz \, dz$$
$$(n = 1, 2, 3, \ldots).$$

If (1.38) is multiplied by $\sin nz$ and integrated with respect to z from 0 to π, integration by parts yields a two-dimensional partial differential equation for $v(n)$:

$$\left. \begin{aligned} &\left(\frac{\partial^2}{\partial x^2} + \frac{\partial^2}{\partial y^2} - n^2\right) v(n) + \\ &+ n\left[h_1 - (-1)^n h_2\right] + \\ &+ \int_0^\pi f \sin nz \, dz = 0. \end{aligned} \right\} \quad (1.39)$$

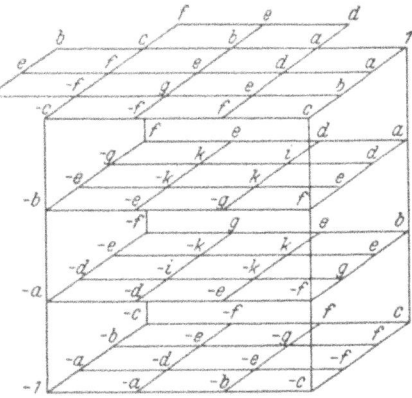

Fig. V/11. The finite-difference method for a three-dimensional temperature distribution

On the boundary of the cross-section $v(n)$ must take the value

$$\int_0^\pi g \sin nz \, dz.$$

This boundary-value problem is solved approximately for several values of n by the finite-difference method and relaxation, then u is determined by Fourier synthesis.

TRANTER also considers the mixed boundary-value problem with the normal derivative prescribed on the plane ends of the cylinder; for this he uses the cosine transform

$$v(n) = \int_0^\pi u \cos nz \, dz.$$

The appropriate transforms for the corresponding problems with a semi-infinite cylinder are

$$v(\xi) = \int_0^\infty u \begin{Bmatrix} \sin \\ \cos \end{Bmatrix} \xi z \, dz.$$

This technique of reducing the number of independent variables by means of integral transforms (Fourier, Laplace and allied transforms all have their particular applications) can be used in a variety of problems.

---

[1] TRANTER, C. J.: The combined use of relaxation methods and Fourier transforms in the solution of some three-dimensional boundary value problems. Quart. J. Mech. Appl. Math. 1, 281—286 (1948); a numerical example is given.

1.8. Arbitrary mesh systems

A formula which approximates a differential equation to a certain order of accuracy by a relation between pivotal values (i.e. function values at mesh points) generally requires more of these pivotal values when the mesh is chosen to be other than rectangular[1] (here we are again restricting ourselves to the plane). For example, in the case of the second-order differential equation

$$L[u] = A\,u_{xx} + 2B\,u_{xy} + C\,u_{yy} + 2D\,u_x + 2E\,u_y + 2F u = r(x, y) \qquad (1.40)$$

(with constant $A, B, \ldots, F$) the corresponding relation, centred on a mesh point P_0 say, must in general involve also the values at five neighbouring points $P_1, \ldots, P_5$; only in special cases does the number reduce to four. If we denote the expression in the pivotal values by

$$Q = \sum_{\nu=0}^{5} a_\nu u_\nu,$$

where u_ν is the value of u at the point P_ν with co-ordinates x_ν, y_ν, and expand these function values by TAYLOR's theorem at the point P_0 (which, for simplicity, we take to be the origin $x_0 = 0$, $y_0 = 0$):

$$Q = u_0 \sum_{\nu=0}^{5} a_\nu + \left(\frac{\partial u}{\partial x}\right)_0 \sum_{\nu=0}^{5} a_\nu x_\nu + \left(\frac{\partial u}{\partial y}\right)_0 \sum_{\nu=0}^{5} a_\nu y_\nu + \frac{1}{2}\left(\frac{\partial^2 u}{\partial x^2}\right)_0 \sum_{\nu=0}^{5} a_\nu x_\nu^2 + \cdots,$$

then by comparing with the given differential expression $L[u]$ we obtain (with an arbitrary constant ϱ) the six equations

$$\sum_{\nu=0}^{5} a_\nu x_\nu^2 = \varrho A, \qquad \sum_{\nu=0}^{5} a_\nu x_\nu y_\nu = \varrho B, \qquad \sum_{\nu=0}^{5} a_\nu y_\nu^2 = \varrho C,$$

$$\sum_{\nu=0}^{5} a_\nu x_\nu = \varrho D, \qquad \sum_{\nu=0}^{5} a_\nu y_\nu = \varrho E, \qquad \sum_{\nu=0}^{5} a_\nu = \varrho F$$

for the a_ν. Since a_0 appears in the last equation only, the other equations provide five linear equations for $a_1, \ldots, a_5$. We now ask in what circumstances just four neighbouring values, say $u_1, \ldots, u_4$ (so that $a_5 = 0$), will be sufficient. With $a_5 = 0$ the first five equations constitute five homogeneous equations for $a_1, a_2, a_3, a_4, \varrho$ and will possess a non-trivial solution if and only if the determinant vanishes:

$$\begin{vmatrix} x_1^2 & x_2^2 & x_3^2 & x_4^2 & A \\ x_1 y_1 & x_2 y_2 & x_3 y_3 & x_4 y_4 & B \\ y_1^2 & y_2^2 & y_3^2 & y_4^2 & C \\ x_1 & x_2 & x_3 & x_4 & D \\ y_1 & y_2 & y_3 & y_4 & E \end{vmatrix} = 0.$$

There are therefore five constants $\alpha, \beta, \gamma, \delta, \varepsilon$, not all zero, such that

and
$$\left.\begin{array}{l} \alpha x_j^2 + \beta x_j y_j + \gamma y_j^2 + \delta x_j + \varepsilon y_j = 0 \quad \text{for} \quad j = 1, 2, 3, 4 \\ \alpha A + \beta B + \gamma C + \delta D + \varepsilon E = 0 \end{array}\right\} \qquad (1.41)$$

This means that the four points P_1, P_2, P_3, P_4 lie with P_0 on the conic

$$\alpha x^2 + \beta x y + \gamma y^2 + \delta x + \varepsilon y = 0,$$

where $\alpha, \beta, \gamma, \delta, \varepsilon$ must satisfy the last equation of (1.41).

---

[1] See R. v. MISES: On network methods in conformal mapping and in related problems. Nat. Bur. Stand., Appl. Math. Ser. **18**, 1—7 (1952).

For the equation $V^2 u = r(x, y)$, in which $A = C$, $B = D = E = 0$, we have $\gamma = -\alpha$ and the conic is a rectangular hyperbola; the degenerate case of a pair of orthogonal straight lines is the conic used with a rectangular mesh.

1.9. Solution of the difference equations by finite sums

We present the method[1] by describing its application to the first boundary, value problem of potential theory for a rectangular region: $V^2 u = 0$ in $0 \leq x \leq a$ $0 \leq y \leq b$ and u prescribed on the boundary. For a mesh with the points

$$x_j = j h, \qquad y_k = k l, \qquad h = \frac{a}{M+1}, \qquad l = \frac{b}{N+1}$$

$$(j = 0, 1, \ldots, M+1; \ k = 0, 1, \ldots, N+1)$$

the difference equations [(1.3) with $a_{i,k} = c_{i,k} = 1$, $d_{i,k} = e_{i,k} = g_{i,k} = r_{i,k} = 0$] have particular solutions of the form

$$\lambda^k \sin \frac{n \pi j h}{a},$$

where λ, $\varrho = \dfrac{n \pi h}{a}$ and $r = \dfrac{l}{h}$ are related by the equation

$$2 r^2 (\cos \varrho - 1) + \left(\lambda + \frac{1}{\lambda} - 2 \right) = 0.$$

For given r and ϱ this yields two values for λ:

$$\left. \begin{matrix} \lambda_1 \\ \lambda_2 \end{matrix} \right\} = \left\{ \begin{matrix} \lambda_{1,n} \\ \lambda_{2,n} \end{matrix} \right\} = \mu \pm \sqrt{\mu^2 - 1},$$

where $\mu = 1 - r^2 (\cos \varrho - 1)$.

Consequently the finite sum

$$u' = \sum_{n=1}^{M} \sin \frac{n \pi j h}{a} \, (P_n' \lambda_{1,n}^k + Q_n' \lambda_{2,n}^k), \tag{1.42}$$

in which P_n', Q_n' are constants to be determined, is a solution of the difference equations. It vanishes on the boundaries $x = 0$ and $x = a$ and by suitable choice of P_n', Q_n' can be made to take the prescribed boundary values of u on the boundaries $y = 0$ and $y = b$. If u'' is the corresponding finite sum which vanishes on $y = 0$ and $y = b$ and takes the prescribed boundary values on $x = 0$ and $x = a$, then $u' + u''$ will be the required solution of the difference equations.

To determine P_n', Q_n' we first put $y = 0$ and obtain from (1.42)

$$u(j h, 0) = \sum_{n=1}^{M} (P_n' + Q_n') \sin \frac{n \pi j h}{a}.$$

Thus the quantities $P_n' + Q_n'$ are the coefficients in a harmonic analysis of the given function $u(j h, 0)$ and can be determined by any of the well-known methods (RUNGE's scheme, etc.). Similarly, putting $y = b$, we have

$$u(j h, b) = \sum_{n=1}^{M} (P_n' \lambda_{1,n}^{N+1} + Q_n' \lambda_{2,n}^{N+1}) \sin \frac{n \pi j h}{a},$$

and the quantities $P_n' \lambda_{1,n}^{N+1} + Q_n' \lambda_{2,n}^{N+1}$ can be determined in the same way. We then have two equations for P_n' and Q_n'. The u values at the individual mesh points can be calculated from the difference equations once the values on two consecutive rows have been obtained by evaluating the finite sums.

---

[1] HYMAN, M. A.: Non-iterative numerical solution of boundary value problems. Appl. Sci. Res. B 2, 325—351 (1952).

Similar finite-sum solutions can also be used for the second and third boundary-value problems and for differential equations of higher order with constant coefficients (the biharmonic equation, for example). The method can be carried over to problems in three or more dimensions by using multiple sums.

1.10. Simplification of the calculation by decomposition of the finite-difference equations[1]

Solution of the difference equations corresponding to a linear boundary-value problem with a differential equation of the form (1.2) can often be simplified as follows. Firstly, in addition to assuming the existence of a solution, we assume that the linear difference equation associated with a boundary pivot P as a representation of the boundary condition involves, in addition to the value $U(P)$ (which must occur), only the approximate values U at neighbouring pivots. Those pivots P_i $(i = 1, 2, \ldots, n)$ for which the value of U is not given immediately by the boundary condition are now divided (chequer-wise) into two sets A $(i = 1, 2, \ldots, p$; we can take $p \leq \frac{1}{2}n)$ and B $(i = p + 1, \ldots, n)$ such that no two neighbouring points belong to the same set [the values $U(P_i)$ are classified accordingly as A values or B values]. Then, multiplying each of the difference equations by a suitable constant if need be, we can put the system of equations into the form

$$-\alpha U_i + \sum_{k=p+1}^{n} b_{ik} U_k + s_i = 0 \quad (i = 1, 2, \ldots, p), \tag{1.43}$$

$$\sum_{k=1}^{p} b_{ik} U_k - \alpha U_i + s_i = 0 \quad (i = p + 1, \ldots, n). \tag{1.44}$$

In matrix notation this can be written

$$\left(\begin{array}{c|c} -\alpha I_p & B_1 \\ \hline B_2 & -\alpha I_{n-p} \end{array} \right) \left(\begin{array}{c} x_A \\ x_B \end{array} \right) + \left(\begin{array}{c} s_A \\ s_B \end{array} \right) = 0,$$

in which the matrix and vectors are partitioned into sub-matrices and sub-vectors, respectively, by lines between the p-th and $(p + 1)$-th rows and columns, and obvious notations are used for these partitions; for instance, I_p is the p-rowed unit matrix and x_A is the vector of unknowns in set A, etc. We now eliminate the unknowns of set B by pre-multiplying the system by the matrix $\left(\begin{array}{c|c} \alpha I_p & B_1 \\ \hline 0 & I_{n-p} \end{array} \right)$; this transforms the original matrix into the "decomposed" matrix $\left(\begin{array}{c|c} -\alpha^2 I_p + B_1 B_2 & 0 \\ \hline B_2 & -\alpha I_{n-p} \end{array} \right)$. In this way we obtain a "reduced system" for the unknowns in set A:

$$-\alpha^2 U_i + \sum_{k=1}^{p} \beta_{ik} U_k + \sigma_i = 0, \quad \beta_{ik} = \sum_{j=p+1}^{n} b_{ij} b_{jk}, \quad \sigma_i = \alpha s_i + \sum_{j=p+1}^{n} b_{ij} s_j$$

$$(i = 1, 2, \ldots, p),$$

i.e.

$$(-\alpha^2 I_p + B_1 B_2) x_A + \sigma_A = 0, \quad \sigma_A = \alpha s_A + B_1 s_B. \tag{1.45}$$

Thus we only have to solve p equations; the remaining unknowns are expressed explicitly in terms of the solution of these equations by (1.44). Another advantage

---

[1] The material for this section was kindly made available to me by Herr J. SCHRÖDER [for a detailed presentation see J. SCHRÖDER: Z. Angew. Math. Mech. **34**, 241—253 (1954)].

of this method is that far fewer iteration cycles are needed to solve the reduced system by single-step (or total-step) iteration than to solve similarly the original system to the same accuracy in the A values[1].

For differential equations of the form $\nabla^2 u - gu = r$, where g is a non-negative constant, use of a square mesh (mesh width h) usually permits a simple, direct derivation of the reduced system which does not involve the matrix multiplications $B_1 B_2$, $B_1 s_B$. One way of doing this is to use special stencils[2] which can be constructed by superimposing ordinary stencils. However, if U is known at all the boundary pivots, the reduced system can be written down immediately, for the coefficients are given by the following simple formulae:

$$\left.
\begin{aligned}
&\alpha = 4 + g h^2 \\
&\beta_{ik} = \text{number of pivots } P_j \text{ which are neighbours to both} \\
&\qquad P_i \text{ and } P_k \\
&\text{(in particular, } \beta_{ii} = \text{number of pivots } P_j \text{ which are neighbours to } P_i) \\
&\sigma_i = \alpha s_i + \sum_j s_j, \text{ where the sum extends over all } j \text{ for} \\
&\qquad \text{which } P_i \text{ and } P_j \text{ are neighbouring points.}
\end{aligned}
\right\} \quad (1.46)$$

Clearly the matrix of the system is symmetric in this case.

The technique can also be applied with advantage to eigenvalue problems. If, for instance, the differential equation reads $\nabla^2 u - gu + \lambda u = 0$, where g is again a non-negative constant, and if with a square mesh (mesh width h) $U = 0$ at all boundary pivots, then by the same procedure as above we can put the difference equations into the form (1.43), (1.44) [with $s_i = 0$ and $\alpha = 4 + (g - \Lambda) h^2$, where Λ is an approximate eigenvalue] and set up the corresponding reduced system (1.45) for which the β_{ik} may be found from the formula of (1.46). By virtue of the equation

$$\det \left(\begin{array}{c|c} -\alpha I_p & B_1 \\ \hline B_2 & -\alpha I_{n-p} \end{array} \right) = (-1)^{n-p} \alpha^{n-2p} \det(-\alpha^2 I_p + B_1 B_2), \quad (1.47)$$

the "latent roots" α of the matrix $B = \left(\begin{array}{c|c} 0 & B_1 \\ \hline B_2 & 0 \end{array} \right)$ may be calculated using the determinant of the reduced system. They are obviously symmetrically placed about $\alpha = 0$, and if (x_A / x_B) is a latent vector corresponding to α, then $(x_A / -x_B)$ is a latent vector corresponding to $-\alpha$. In the iterative calculation of the greatest latent root and its corresponding latent vector the same accuracy is achieved in half the number of cycles if the matrix of the reduced system is used rather than that of the original system.

Example: Let us find a function $u(x, y)$ which satisfies the equation $\nabla^2 u + \lambda u = 0$ inside the region illustrated in Fig. V/12 and vanishes on the boundary. Using the heavily lined square mesh with mesh width $h = 1$, we have seven pivots P_i (marked 1, 2, ..., 7 in the figure); set A comprises the pivots P_1, P_2, P_3 and set B the remainder ($n = 7$, $p = 3$). The numbers β_{ik} ($i, k = 1, 2, 3$) and b_{ik} ($i = 4, 5, 6, 7$;

---

[1] See SCHRÖDER's paper.

[2] Following MILNE we apply the term "stencil" to an array of coefficients set out in a pattern corresponding to the points whose associated approximate values they are to multiply. Thus the ordinary stencil here is
$$\begin{array}{|c|c|c|} \hline & 1 & \\ \hline 1 & -\alpha & 1 \\ \hline & 1 & \\ \hline \end{array}.$$

$k = 1, 2, 3$) are found from the formula of (1.46) (for example: P_1 and P_2 have two neighbouring pivots in common so that $\beta_{12} = \beta_{21} = 2$) and from the ordinary difference equations, respectively:

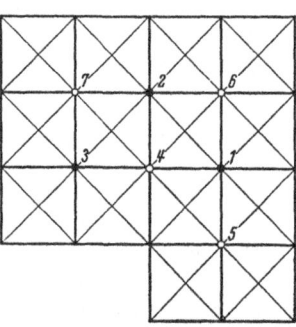

$\beta_{i,k}$	β_i	3 2 1	6
		2 3 2	7
		1 2 2	5

$b_{i,k}$	b_i	1 1 1	3
	$=$	1 0 0	1
		1 1 0	2
		0 1 1	2

Fig. V/12. Reduction of the system of difference equations

As a check one may calculate the numbers $\beta_i = \sum_{k=1}^{3} \beta_{ik}$ and $b_i = \sum_{k=1}^{3} b_{ik}$ and verify that $\beta_i = \sum_{j=4}^{3} b_{ji} b_j$.

From (1.47) the latent roots α are given by $\alpha (\alpha^6 - 8\alpha^4 + 12\alpha^2 - 3) = 0$. One pair of roots is $\alpha_2 = 1 \cdot 2523$ and $\alpha_6 = -\alpha_2$; these give $\varLambda_2 = 2 \cdot 7477$ and $\varLambda_6 = 5 \cdot 2523$. A solution of the reduced system corresponding to α_2 and α_6 is

$$U_1 = -1 \cdot 3168, \qquad U_2 = 0 \cdot 4426, \qquad U_3 = 1.$$

The ordinary difference equations (1.44) (with $s_i = 0$, $\alpha = \alpha_2, \alpha_6$) then yield the following values for the remaining U_i:

$$U_4 = \pm 0 \cdot 1004, \qquad U_5 = \mp 1 \cdot 0515, \qquad U_6 = \mp 0 \cdot 6981, \qquad U_7 = \pm 1 \cdot 1519.$$

Mesh refinement: For the type of eigenvalue problem just considered ($\nabla^2 u - gu + \lambda u = 0$, $g = \text{constant} \geqq 0$, $U = 0$ at the boundary pivots for a square mesh) one may seek to improve a calculated eigenvector x and corresponding approximate eigenvalue $\varLambda = g + \dfrac{1}{h^2}(4 - \alpha)$ by a perturbation calculation starting from the known quantities x and α (this method of improving calculated values by means of a perturbation calculation may also be applied to the corresponding inhomogeneous boundary-value problem). We make two further assumptions: the latent root α is simple and non-zero, and the boundary $\varGamma$ consists of complete sides or diagonals of the squares of the "basic mesh", i.e. the mesh with which the known approximations x and α were calculated. The diagonals of the squares determine a new mesh, the "diagonal mesh", which has at least $2n$ pivots and which takes over directly the old pivots $P_1, \ldots, P_n$ as the new set A^* of non-neighbouring pivots. Since the Laplace operator is unaltered by a rotation of the co-ordinate axes, the coefficients in the difference equations are similar except for the smaller mesh width $h/\sqrt{2}$ and the corresponding reduced system can be written

$$(-\alpha^{*2} I_n + B_1^* B_2^*)\, x_{A^*}^* = 0, \qquad \alpha^* = 4 + \frac{h^2}{2}(g - \varLambda^*),$$

where asterisked quantities refer to the diagonal mesh. It can be shown that

$$A_0 = \tfrac{1}{2} B^2 + 2B$$

does not differ greatly from $B_1^* B_2^*$, so we put

$$B_1^* B_2^* = A_0 + \varepsilon A_1,$$

in which we shall take $\varepsilon = 1$, and solve the equation $(-\alpha^{*2} I_n + A_0 + \varepsilon A_1) x_{A*}^* = 0$ by a perturbation method: we put $x_{A*}^* = \varphi_0 + \varepsilon \varphi_1 + \varepsilon^2 \varphi_2 + \cdots$ with $\varphi_0 = x$, $\alpha^{*2} = \mu_0 + \varepsilon \mu_1 + \varepsilon^2 \mu_2 + \cdots$ with $\mu_0 = \frac{1}{2}\alpha^2 + 2\alpha$ and equate coefficients of powers of ε. Using the fact that A_0 is symmetrical we find that

$$\mu_0 + \mu_1 = 2\alpha + r, \quad \text{where} \quad r = \frac{(R\,\varphi_0,\,\varphi_0)}{(\varphi_0,\,\varphi_0)}, \quad R = B_1^* B_2^* - 2B = \left(\begin{array}{c|c} R_1 & 0 \\ \hline 0 & R_2 \end{array}\right);$$

here (y, z) denotes the inner product of two n-dimensional vectors y and z.

For the above example the diagonal mesh provides twenty-one pivots as against the seven pivots of the basic mesh. We have

$$R_1 = \begin{pmatrix} 4 & 1 & 0 \\ 1 & 4 & 1 \\ 0 & 1 & 4 \end{pmatrix}, \quad R_2 = \begin{pmatrix} 4 & 1 & 1 & 1 \\ 1 & 4 & 0 & 0 \\ 1 & 0 & 4 & 0 \\ 1 & 0 & 0 & 4 \end{pmatrix},$$

and for $\alpha = \alpha_2$

$$\mu_0 + \mu_1 = 6.4363, \quad \mu_0 + \mu_1 + \mu_2 = 6.4746 \quad \text{and} \quad \mu_0 + \mu_1 + \mu_2 + \mu_3 = 6.4670.$$

These three numbers lead to the approximate values $\Lambda_2^* \approx 2.9260$, 2.9110, 2.9139, respectively (the exact solution for the calculation with twenty-one pivots is $\Lambda_2^* = 2.9135$), as compared with the value $\Lambda_2 = 2.7477$ from the basic mesh.

§2. Refinements of the finite-difference method

The derivation and application of the formulae for the methods to be described here are for the most part so similar to the corresponding considerations in Ch. III, § 2 and Ch. IV, § 2 that we can be brief and place more emphasis on the generality of the methods.

2.1. The finite-difference method to a higher approximation in the general case

Let a function $u(x_1, x_2, \ldots, x_n)$ be defined in a simply connected, closed region B^* of the n-dimensional co-ordinate space by an m-th-order linear differential equation

$$L[u] = \sum_{\alpha_1 + \cdots + \alpha_n \leq m} A_{\alpha_1, \alpha_2, \ldots, \alpha_n} \frac{\partial^{\alpha_1 + \cdots + \alpha_n} u}{\partial x_1^{\alpha_1} \ldots \partial x_n^{\alpha_n}} = t(x_1, \ldots, x_n) \qquad (2.1)$$

together with linear boundary conditions. For example, we might prescribe on certain hypersurfaces Γ_μ values of u or of $\partial u/\partial x_\varrho$ or, more generally, of a linear combination of the partial derivatives of u of up to the $(m-1)$-th order. These surfaces need not lie on the boundary of the region of definition B^*; it is often convenient to take into consideration values of the function u outside of the "boundary" surfaces on which the boundary conditions are given.

For the numerical calculation of u from the given data we introduce a system of pivotal points P_j $(j = 1, 2, \ldots, N)$ which all lie in B^* (they need not be arranged in a regular pattern) and seek approximations U_j to the values $u(P_j)$ of the exact solution u at the points P_j. For the

solution of this problem we set up a system of linear equations for the
U_j. Such an equation is obtained by forming the sum

$$\sum_{\varrho=1}^{N} C_\varrho u(P_\varrho)$$

and determining the constants C_ϱ so that it approximates the differential
expression of (2.1) at a point P_j. We say the equation is written for, or
corresponds to, the point P_j.

Each term $C_\varrho u(P_\varrho)$ of the sum is expanded by TAYLOR's theorem
about the point P_j:

$$\left.\begin{aligned}
u(P_\varrho) &= u(P_j) + \sum_{\nu=1}^{n} \left(x_\nu(P_\varrho) - x_\nu(P_j)\right) \frac{\partial u(P_j)}{\partial x_\nu} + \\
&\quad + \frac{1}{2!} \sum_{\nu,\mu=1}^{n} \left(x_\nu(P_\varrho) - x_\nu(P_j)\right)\left(x_\mu(P_\varrho) - x_\mu(P_j)\right) \frac{\partial^2 u(P_j)}{\partial x_\nu \partial x_\mu} + \\
&\quad + \cdots + \text{terms of the } r\text{-th order} + \text{remainder term}
\end{aligned}\right\} \quad (2.2)$$

and the whole sum $\sum C_\varrho u(P_\varrho)$ then rearranged in terms of $u(P_j)$ and the
partial derivatives of u at the point P_j:

$$\sum_{\varrho=1}^{N} C_\varrho u(P_\varrho) = \sum_{\alpha_1+\cdots+\alpha_n \leq r} B_{\alpha_1,\ldots,\alpha_n} \frac{\partial^{\alpha_1+\cdots+\alpha_n} u(P_j)}{\partial x_1^{\alpha_1} \ldots \partial x_n^{\alpha_n}} + \text{remainder term}. \quad (2.3)$$

The new coefficients $B_{\alpha_1,\ldots,\alpha_n}$ depend linearly on the C_ϱ. If h is the
smallest non-zero number among the values of the quantities
$|x_j(P_k) - x_j(P_l)|$, and M_{r+1} is the maximum in B^* of the absolute values
of all the $(r+1)$-th partial derivatives of u, then the remainder term
can be expressed in the form $\vartheta D_j h^{r+1} M_{r+1}$, where $|\vartheta| \leq 1$ and D_j depends
on the positions of the points P_ϱ but not on u; when the disposition of
the points P_j is known, numerical limits can be given for the D_j.

Our object is to make the expression on the left-hand side of (2.3)
as accurate a representation of the differential expression $L[u]$ in (2.1)
as possible; we can then write approximately

$$\sum_{\varrho=1}^{N} C_\varrho u(P_\varrho) \approx L[u(P_j)] = t(P_j)$$

and take

$$\sum_{\varrho=1}^{N} C_\varrho U_\varrho = t(P_j) \quad (2.4)$$

as one of the equations for the U_j.

To achieve our object we try to make the coefficients of $\dfrac{\partial^{\alpha_1+\cdots+\alpha_n} u}{\partial x_1^{\alpha_1} \ldots \partial x_n^{\alpha_n}}$
in (2.1) and (2.3) agree for all derivatives of up to as high an order as
possible. First of all we must have

$$A_{\alpha_1,\ldots,\alpha_n} = B_{\alpha_1,\ldots,\alpha_n}(C_\varrho), \quad (2.5)$$

i.e. we must have agreement for $\alpha = \alpha_1 + \alpha_2 + \cdots + \alpha_n = 0, 1, 2, \ldots, m$ at least. This minimum requirement yields the special case of the ordinary finite-difference method, for which (2.3) becomes

$$\sum_{\varrho=1}^{N} C_\varrho u(P_\varrho) = L[u(P_j)] + \vartheta D_j\, h^{m+1} M_{m+1}.$$

The linear equations given by (2.5) admit of simple solutions[1] in which many of the C_ϱ are zero and the few non-zero C_ϱ correspond to points P_ϱ lying near P_j.

Here we are more interested in higher approximations, i.e. those for which the order of the remainder term is greater than $m+1$:

$$\sum_{\varrho=1}^{N} C_\varrho u(P_\varrho) = L[u(P_j)] + \vartheta D_j\, h^{r+1} M_{r+1}, \tag{2.6}$$

where $r > m$. To this end, we add to the equations (2.5) for the C_ϱ the further equations

$$B_{\alpha_1, \ldots, \alpha_n} = 0 \quad \text{for} \quad m < \alpha \leqq r.$$

Insertion in (2.4) of a set of numbers C_ϱ which satisfy these equations yields one of the desired equations for the U_j.

Further equations of a higher approximation can be derived by finding a different set of numbers C_ϱ which satisfy the necessary equations[2] or by using a different point P_i as "centre" of the Taylor expansions. In addition, linear equations for the U_j may be derived in the same way from the boundary conditions.

One endeavours to set up as many equations for the U_j as will yield a system of linear equations with a non-zero determinant.

2.2. A general principle for error estimation

If we have such a system of equations for the U_j, then in principle we can obtain estimates for the errors

$$\varepsilon_j = U_j - u(P_j).$$

If we assume that the solution u possesses partial derivatives of up to and including the $(r+1)$-th order, then insertion of u into (2.6) yields

$$\sum_{\varrho=1}^{N} C_\varrho u(P_\varrho) = L[u(P_j)] + \vartheta D_j\, h^{r+1} M_{r+1} = t(P_j) + \vartheta D_j\, h^{r+1} M_{r+1};$$

---

[1] Provided that the distribution of pivotal points in B^* is sufficiently dense the number of C_ϱ at our disposal will be far greater than the number of equations to be satisfied.

[2] Strictly we ought to write $C_\varrho^{(j,\varkappa)}$, $\vartheta^{(j,\varkappa)}$, $D^{(j,\varkappa)}$, say, for even with the same point P_j several sets of values for the C_ϱ are possible. Since misunderstanding is unlikely, we use the simpler notation with fewer indices.

then subtracting from (2.4) we find that the errors ε_j satisfy the equation

$$\sum_{\varrho=1}^{N} C_\varrho \varepsilon_\varrho = - \vartheta D_j \, h^{r+1} M_{r+1}. \tag{2.7}$$

Thus, apart from the different right-hand sides, the errors ε_j satisfy the same system of equations as the approximate values U_j. Since the determinant of the coefficients was assumed to be non-zero, we can solve the system of equations (2.7) for the ε_ϱ.

For the finite-difference methods described here for the numerical calculation of the solution u of a linear boundary-value problem to a higher approximation, the theoretical possibility of being able to obtain error estimates follows from the possibility of being able to apply the method in the first place, provided that u is assumed to be differentiable sufficiently often in the region B^*.

Uniqueness of the solution u need not be assumed: what we estimate is the departure of our approximate pivotal values from the values of a solution whose partial derivatives of the $(r+1)$-th order are bounded absolutely in B^* by the constant M_{r+1} in (2.7). As regards the difficulties which attend the estimation of M_{r+1}, cf. the remarks made in Ch. IV, § 3.3.

2.3. Derivation of finite expressions

The Taylor expansion method of § 2.1 provides a technique for setting up a finite expression to represent any given differential expression. Here we describe an operator method which shows that there exist finite expressions of an arbitrarily high order of approximation, i.e. expressions with remainder terms of order $r+1$ for any prescribed r.

We select a typical term

$$\frac{\partial^{\alpha_1 + \alpha_2 + \cdots + \alpha_n} u}{\partial x_1^{\alpha_1} \partial x_2^{\alpha_2} \dots \partial x_n^{\alpha_n}} \tag{2.8}$$

and take our pivotal points at the nodes of a square mesh of mesh width h. We can then use the displacement operator E_i which transforms a function $u(x_1, x_2, \dots, x_n)$ into $u(x_1, x_2, \dots, x_{i-1}, x_i+h, x_{i+1}, \dots, x_n)$. This operator[1] obeys the laws of ordinary algebra and also commutes with the differential operators $\partial^{\alpha}/\partial x_i^{\alpha}$. Raised to an integral power p (positive, negative or zero) it has the effect

$$E_i^p u(x_1, x_2, \dots, x_i, \dots, x_n) \equiv u(x_1, x_2, \dots, x_{i-1}, x_i+ph, x_{i+1}, \dots, x_n). \tag{2.9}$$

In the equations between operators which we write in the following we imagine the operations to be performed on rational integral functions

---

[1] STEFFENSEN, J. F.: Interpolation, p. 4 et seq. and p. 178 et seq. Baltimore 1927. — BRUWIER, L.: Sur une équation aux dérivées et aux différences mêlées. Mathesis **47**, 103—104 (1933).

$v(x_1, x_2, \ldots, x_n)$ which are of degree not greater than r in each of the independent variables. This avoids having to take the remainder terms into account each time.

According to § 2.2 of Chapter III there exists for each individual operator $\partial^{\alpha_i}/\partial x_i^{\alpha_i}$ a finite operator [1]

$$\frac{\partial^{\alpha_i}}{\partial x_i^{\alpha_i}} = \sum_{\varrho = \left[-\frac{r}{2}\right]}^{\left[\frac{r}{2}\right]} A_\varrho^{(i)} E_i^\varrho \tag{2.10}$$

for any given integer r.

If we multiply this equation by $\partial^{\alpha_k}/\partial x_k^{\alpha_k}$ $(k \neq i)$, then, since $E_i^\varrho v$ is also a rational integral function of $x_1, x_2, \ldots, x_n$ of degree not greater than r in each of the independent variables, we can replace

$$\frac{\partial^{\alpha_k}}{\partial x_k^{\alpha_k}} E_i^\varrho v \quad \text{by} \quad \sum_\sigma A_\sigma^{(k)} E_k^\sigma E_i^\varrho v;$$

consequently

$$\frac{\partial^{\alpha_i}}{\partial x_i^{\alpha_i}} \frac{\partial^{\alpha_k}}{\partial x_k^{\alpha_k}} = \left\{ \sum_\varrho A_\varrho^{(i)} E_i^\varrho \right\} \left\{ \sum_\sigma A_\sigma^{(k)} E_k^\sigma \right\}$$

and

$$\prod_{i=1}^n \left[\sum_{\varrho = \left[-\frac{r}{2}\right]}^{\left[\frac{r}{2}\right]} A_\varrho^{(i)} E_i^\varrho \right] - \frac{\partial^{\alpha_1 + \cdots + \alpha_n}}{\partial x_1^{\alpha_1} \ldots \partial x_n^{\alpha_n}} \equiv 0 \tag{2.11}$$

for any one of our functions $v(x_1, \ldots, x_n)$.

If the product in (2.11) is now multiplied out and applied to a function u, and then each term expanded by TAYLOR's theorem with a remainder term of the $(r+1)$-th order, exactly as in § 2.1, we have

$$\prod_{i=1}^n \left[\sum_\varrho A_\varrho^{(i)} E_i^\varrho \right] u - \frac{\partial^{\alpha_1 + \cdots + \alpha_n} u}{\partial x_1^{\alpha_1} \ldots \partial x_n^{\alpha_n}}$$
$$= \sum_{\beta_1 + \cdots + \beta_n \leq r} B_{\beta_1, \ldots, \beta_n} \frac{\partial^{\beta_1 + \cdots + \beta_n} u}{\partial x_1^{\beta_1} \ldots \partial x_n^{\beta_n}} + \vartheta D_j h^{r+1} M_{r+1}$$

(notation D_j, M_{r+1} as in § 2.1). The constants $B_{\beta_1, \beta_2, \ldots, \beta_n}$ are the same not only for all our rational integral functions, but also for all functions u which possess partial derivatives of up to and including the $(r+1)$-th order. Now take u to be any rational integral function of (total) degree not greater than r; then the right-hand side must be zero since these functions are included among those for which (2.11) is valid. Further, $M_{r+1} = 0$ for these functions, so if we put $u = $ constant, we deduce that $B_{0,0,\ldots,0} = 0$; similarly we can put $u = x_i$, $u = x_i x_k$,

---

[1] The customary notation $[x]$ is used for the greatest integer which is not greater than x; for example, $[2.5] = 2$, $[2] = 2$, $[-2.5] = -3$.

$u = x_i x_k x_l, \ldots$, from which it follows that all $B_{\beta_1, \ldots, \beta_n} = 0$ for $\beta_1 + \beta_2 + \cdots + \beta_n \leqq r$. Thus for all functions u with continuous partial derivatives of the $(r + 1)$-th order we have

$$\prod_{i=1}^{n} \left[\sum_{\varrho = \left[-\frac{r}{2}\right]}^{\left[\frac{r}{2}\right]} A_{\varrho}^{(i)} E_i^{\varrho} \right] u = \frac{\partial^{\alpha_1 + \cdots + \alpha_n} u}{\partial x_1^{\alpha_1} \ldots \partial x_n^{\alpha_n}} + \vartheta D_j h^{r+1} M_{r+1}. \quad (2.12)$$

Although this operator method[1] shows the existence of finite expressions of an arbitrarily high order of approximation, the Taylor expansion method of § 2.1 is often more profitable in practice, for it can often be used to derive simpler expressions involving fewer points[2]. Several finite expressions for the frequently occurring differential expressions $\nabla^2 u$ and $\nabla^4 u$ are given for two independent variables in Table VI of the appendix (also for three independent variables for $\nabla^2 u$).

2.4. Utilization of function values at exterior mesh points

With the finite-difference methods of a higher approximation it often happens that a finite equation written down for a mesh point near the boundary involves approximate values at mesh points which lie outside

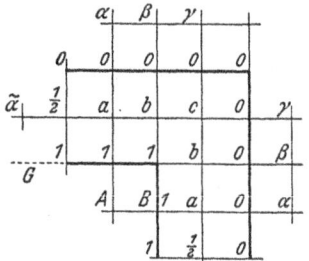

Fig. V/13. Utilization of exterior mesh points

of the fundamental region, i.e. the region enclosed by the boundaries on which the boundary values are prescribed. To render possible the elimination of these values one usually employs finite equations of a lower approximation on the boundary. This is illustrated in the following examples.

I. Consider the problem of Example I, § 1.5, which was to determine a potential function with given values on the boundaries shown in Fig. V/4. If we take $h = \frac{1}{2}$ and make use of the symmetry of the problem, we have just three unknown values a, b, c at the points indicated in Fig. V/13, but finite equations of a higher approximation written down for these points will involve function values outside of the fundamental region — in the simplest case, the values $\alpha, \beta, \gamma, \tilde{\alpha}$, A, B as in the figure.

The unknown value B is included because of the singular behaviour at the re-entrant corner. The section of the u surface in the vertical plane containing the line $\beta b B$ will have a discontinuous slope at the

---

[1] Operators are used by W. G. BICKLEY: Finite-difference formulae for the square lattice. Quart. J. Mech. Appl. Math. **1**, 35—42 (1948).

[2] Examples for comparison of the methods can be found in L. COLLATZ: Schr. Math. Sem. u. Inst. Angew. Math. Univ. Berlin **3**, 18 (1935).

corner as shown by the heavy line in Fig. V/14. Consequently the true value of 1 at B is not consistent with a polynomial representation, whereas a fictitious extrapolated value can be made so.

Using the finite expression for $\nabla^2 u$ given in Table VI of the appendix we obtain the equations

$$-60a + 16(b + 1 + \tfrac{1}{2} + 0) -$$
$$- (c + A + \tilde{a} + \alpha) = 0,$$
$$-60b + 16(c + 1 + a + 0) -$$
$$- (0 + B + \tfrac{1}{2} + \beta) = 0,$$
$$-60c + 16(0 + b + b + 0) -$$
$$- (\gamma + a + a + \gamma) = 0.$$

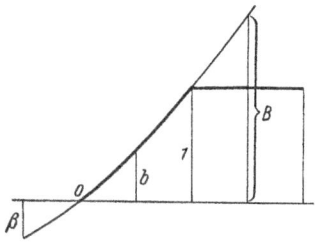

Fig. V/14. Introduction of a fictitious value near a singularity

To eliminate the extra unknowns we express them in terms of a, b and c by means of the ordinary difference equations corresponding to the boundary points:

$$-4 \times 0 + 0 + a + 0 + \alpha = 0, \quad \text{yields} \quad \alpha = -a,$$
and similarly we obtain $\beta = -b, \quad \gamma = -c,$
$$-4 \times \tfrac{1}{2} + a + 1 + \tilde{a} + 0 = 0, \quad \text{yields} \quad \tilde{a} = 1 - a,$$
$$-4 \times 1 + 1 + A + 1 + a = 0, \quad \text{yields} \quad A = 2 - a.$$

Similarly we put $B = 2 - b$ as though the value 1 on the line G in Fig.V/13 extended beyond the re-entrant corner.

With these substitutions we find that

$$-57a + 16b - c = -21,$$
$$32a - 116b + 32c = -27,$$
$$-a + 16b - 29c = 0,$$

and hence

$$a = \frac{2571}{5308} = 0\cdot48436, \qquad b = \frac{2265}{5308} = 0\cdot42671, \qquad c = \frac{1161}{5308} = 0\cdot21873.$$

If more accurate values are needed, the calculation can be repeated with a smaller mesh width h; the larger system of equations which arises can be solved iteratively.

II. A boundary-value problem whose physical background is sketched in § 4.2 is the following:

$$\nabla^2 u = -1, \quad \frac{\partial u}{\partial \nu} = u \quad \text{on the boundary of the square } |x| \leq 1, \ |y| \leq 1,$$

where ν is the inward normal. To illustrate the method, we use the rather large mesh width $h = \frac{2}{3}$; using the symmetry we have the five unknowns a, b, c, d, e (Fig. V/15). We describe three methods.

A. The ordinary finite-difference method.

The ordinary difference equations for a, b and c are

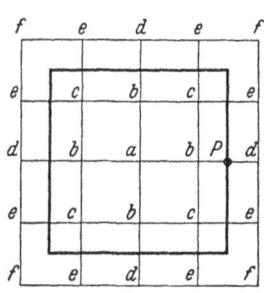

$$-4a + 4b + h^2 = 0,$$
$$-4b + a + 2c + d + h^2 = 0,$$
$$-4c + 2b + 2e + h^2 = 0;$$

to these we add the equations corresponding to the boundary conditions:

$$\frac{b-d}{h} = \frac{b+d}{2},$$

i.e.

Fig. V/15. Notation for the example treated by several finite-difference methods

$$b(2-h) = d(2+h),$$

and similarly

$$c(2-h) = e(2+h).$$

Solving these equations we obtain

$$a = \frac{53}{28}\,h^2 = \frac{53}{63} = 0.8413, \qquad b = \frac{46}{63} = 0.7302, \qquad c = \frac{40}{63} = 0.6349,$$

$$d = \frac{1}{2}\,b = 0.3651, \qquad e = \frac{1}{2}\,c = 0.3175.$$

B. A finite equation of higher approximation could have been written down for the point a. If we therefore replace the first of the equations in A by the equation

$$-60a + 64b - 4d + 12h^2 = 0$$

and use the remaining equations as they are, the solution of the new system of equations is

$$a = \frac{388}{459} = 0.8453, \qquad b = \frac{336}{459} = 0.7102, \qquad c = \frac{292}{459} = 0.6362,$$

$$d = \frac{1}{2}\,b, \qquad e = \frac{1}{2}\,c.$$

These values are worse than those obtained in A, which is rather surprising at first sight. This demonstrates the unadvisability of using a very accurate equation at one point when very crude approximations are retained at others (here at the boundary). We therefore approximate the boundary conditions more accurately in the next method.

C. Method of a higher approximation.

Here we approximate the boundary condition $\partial u/\partial v = u$ at the point P, say, in Fig. V/15 by putting a parabola through the points a, b, d.

At P this parabola has the ordinate and derivative values

$$u_P = \frac{3d + 6b - a}{8} \quad \text{and} \quad -(u_\nu)_P = \frac{d - b}{h},$$

which substituted in the boundary condition yield

$$\frac{3d + 6b - a}{8} + \frac{d - b}{h} = 0.$$

With $h = \frac{2}{3}$ we have the more accurate finite boundary condition

$$15d = 6b + a, \quad \text{and similarly} \quad 15e = 6c + b. \tag{2.13}$$

From these two equations, together with the main equations used in B:

$$-60a + 64b - 4d + 12h^2 = 0,$$

$$-4b + a + 2c + d + h^2 = 0,$$

$$-4c + 2b + 2e + h^2 = 0,$$

we obtain

$$a = \frac{5787}{7092} = 0 \cdot 8160, \quad b = \frac{4983}{7092} = 0 \cdot 7026, \quad c = \frac{4307}{7092} = 0 \cdot 6073,$$

$$d = 0 \cdot 3354, \quad e = 0 \cdot 2898.$$

These values are much better than those found in A and B.

III. In Example II of § 1.5 the boundary-value problem

$$u_{xx} + u_{yy} + \frac{3}{5 - y} u_y + 1 = 0,$$

$u = 0$ for $|x| = \frac{1}{2}$ and for $|y| = 1$

was treated by the ordinary finite-difference method for the mesh widths $h = \frac{1}{2}, \frac{1}{3}, \frac{1}{4}$. Here we apply a method of higher approximation for $h = \frac{1}{2}$. With the notation of Fig. V/16 the finite equations for the three interior points read

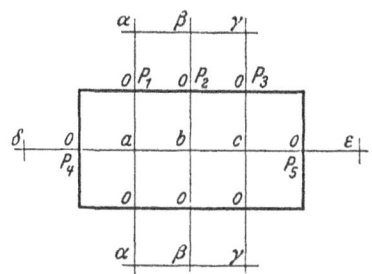

Fig. V/16. Notation for the higher approximation solution of Example III

$$\left.\begin{array}{c} \dfrac{-c + 16b - 30a - \delta}{3} + \dfrac{-2\alpha - 30a}{3} + \dfrac{6}{11} \dfrac{-c + 8b + \delta}{6} + 1 = 0, \\[2mm] \dfrac{16c - 30b + 16a}{3} + \dfrac{-2\beta - 30b}{3} + \dfrac{3}{5} \dfrac{8c - 8a}{6} + 1 = 0, \\[2mm] \dfrac{-\varepsilon - 30c + 16b - a}{3} + \dfrac{-2\gamma - 30c}{3} + \dfrac{2}{3} \dfrac{-\varepsilon - 8b + a}{6} + 1 = 0, \end{array}\right\} \tag{2.14}$$

in which appear the unknown values $\alpha, \beta, \gamma, \delta, \varepsilon$ outside of the rectangle.

These extra values are eliminated by using again the ordinary difference equations corresponding to the differential equation for the points $P_1, \ldots, P_5$:

$$\frac{\alpha + a}{h^2} + 1 = \frac{\beta + b}{h^2} + 1 = \frac{\gamma + c}{h^2} + 1 = 0,$$

$$\frac{a + \delta}{h^2} + \frac{1}{2}(a - \delta) + 1 = \frac{c + \varepsilon}{h^2} + \frac{3}{4}(\varepsilon - c) + 1 = 0,$$

by means of which the $\alpha, \ldots, \varepsilon$ can be expressed in terms of a, b, c.

Substituting these values in (2.14) we obtain three linear equations for a, b, c which have strongly dominant diagonal coefficients and are best solved by iteration:

$$-4394a + 1400b - 98c + 285{\cdot}5 = 0,$$
$$136a - 580b + 184c + 35 = 0,$$
$$-76a + 1520b - 6508c + 431 = 0.$$

Their solution is

$$a = 0{\cdot}098\,78,$$
$$b = 0{\cdot}11249,$$
$$c = 0{\cdot}091\,34.$$

2.5. Hermitian finite-difference methods (Mehrstellenverfahren)

These methods employ formulae of the Hermitian type discussed in Ch. III, § 2 and Ch. IV, § 2.3; their derivation here is very similar, though rather more general.

We consider a boundary-value problem as in § 2.1, and with the same notation as used there seek approximate values U_j for the values $u(P_j)$ of the exact solution at the pivotal points P_j distributed (not necessarily in a regular pattern) throughout the region B^*.

We write the differential equation (not necessarily linear) in the form

$$f(x_1, \ldots, x_n, u, L_1[u], L_2[u], \ldots, L_p[u]) = 0, \tag{2.15}$$

where $L_1[u], L_2[u], \ldots, L_p[u]$ are given linear differential expressions in u as in (2.1). For each of these differential expressions, and also for any further linear differential expressions $L_{p+1}[u], \ldots, L_q[u]$ which may occur in the boundary conditions, we write down expressions of the form

$$\Phi_s[u] = \sum_{\varrho=1}^{N} C_{\varrho, s}\, u(P_\varrho) + \sum_{\varrho=1}^{N} D_{\varrho, s}\, (L_s[u])_{P_\varrho} \quad (s = 1, \ldots, q) \tag{2.16}$$

and expand them by TAYLOR's theorem in terms of the values of u and its partial derivatives at the point P_j. The constants $C_{\varrho, s}$ and $D_{\varrho, s}$ are to be chosen so that the expansion vanishes identically to as high an order as possible. For most practical calculations one will naturally choose the pivotal points in a regular pattern and use expressions Φ_s for which only the $C_{\varrho, s}$ and $D_{\varrho, s}$ corresponding to "mesh points" near

P_j are non-zero. The Taylor expansion and determination of the coefficients are illustrated in detail in Ch. IV, §§ 2.2 and 2.3 for the differential expression

$$L[u] = \frac{\partial^2 u}{\partial x^2} - k\,\frac{\partial u}{\partial y}.$$

Expressions of the form (2.16) are given for the individual derivatives

$$\frac{\partial^\nu u}{\partial x_\sigma^\nu}$$

(for small ν) in Table III of the appendix and for the operators ∇^2 and ∇^4 in Table VI[1].

An approximate equation

$$\varPhi_s[U_j] = \sum_{\varrho=1}^{N} C_{\varrho,s}\,U_\varrho + \sum_{\varrho=1}^{N} D_{\varrho,s}\,U_{\varrho,s} = 0 \qquad \left(\begin{matrix} j = 1, \ldots, N \\ s = 1, \ldots, q \end{matrix}\right), \qquad (2.17)$$

where $U_{\varrho,s}$ denotes an approximate value of $(L_s[u])_{P_\varrho}$, is now written down for each point P_j and for each expression $L_s[u]$, making $N \times q$ equations in all. We have $N \times (q+1)$ unknowns U_ϱ, $U_{\varrho,s}$, so N further equations are required. These are obtained by writing down the boundary conditions for those points P_j lying on or near the boundary and the differential equation (2.15) with $(L_s[u])_{P_j}$ replaced by $U_{j,s}$ for those points P_j lying "further inside". Whether the differential equation or a boundary condition is written down for any specific point will be decided on the basis of the particular problem under consideration.

For a linear differential equation $L[u] = r(x_1, \ldots, x_n)$ a more general form[2] of the expression (2.16) can be used, namely

$$\varPhi[u] = \sum_{\varrho=1}^{N} C_\varrho\,u(P_\varrho) + \sum_{\varrho=1}^{N} D_\varrho\,(L[u])_{P_\varrho} + \sum_{\varrho=1}^{N}\sum_{\mu=1}^{M} E_{\varrho,\mu}\,(M_\mu[L[u]])_{P_\varrho},$$

where the M_μ are chosen operators and the $E_{\varrho,\mu}$ are constants to be determined in the usual way (formally the second sum is included in the double sum and could therefore be omitted). In the corresponding approximate equation we replace $(M_\mu[L[u]])_{P_\varrho}$ by $(M_\mu[r])_{P_\varrho}$, which is known. The inclusion of these extra terms offers the possibility in many cases of achieving a higher order approximation without involving any extra pivotal points. Naturally a high accuracy formula of this type will only be used when the solution possesses continuous derivatives of

---

[1] See also L. COLLATZ: Das Mehrstellenverfahren bei Plattenaufgaben. Z. Angew. Math. Mech. 30, 385—388 (1950) and R. ZURMÜHL: Behandlung der Plattenaufgabe nach dem verbesserten Differenzenverfahren. Z. Angew. Math. Mech. 37, 1—16 (1957).

[2] COLLATZ, L.: Z. Angew. Math. Mech. 31, 232 (1951).

a sufficiently high order. As an example we mention the use of the expression

$$\Phi[u] = 20u_a - 4\sum u_b - \sum u_c + \frac{h^2}{5}(34\,V^2 u_a - \sum V^2 u_b) + $$
$$+ \frac{h^4}{30}(17\,V^4 u_a + \sum V^4 u_b) = O(h^8)$$

for the two-dimensional Poisson equation, in which $L[u] = u_{xx} + u_{yy}$; it is based on a square mesh with mesh width h and makes use of just one operator $M_1 = V^2$. The notation used is as follows: a is an arbitrary mesh point P, $u_a = u(P)$, $\sum u_b$ is the sum of the u values at the four neighbouring points and $\sum u_c$ is the sum of the u values at the four mesh points at a distance $\sqrt{2}h$ from P; and similarly for $\sum V^2 u_b$, etc.[1].

Other ways of improving on the accuracy of the ordinary finite-difference method are considered by Woods[2]. For the differential equation $V^2 u = r(x, y)$, which he treats in detail, he suggests a method which is effectively an iterative solution of the system of equations obtained from the first Hermitian formula in Table VI of the appendix. This formula is easily derived by Taylor expansion, and with the same notation as above can be written in the form

$$4u_a - \sum u_b + h^2 r_a = D[u] + O(h^6),$$

where

$$D[u] = \frac{1}{6}(\sum u_c - 2\sum u_b + 4u_a) - \frac{h^2}{12}(\sum r_b - 4r_a).$$

First of all, approximations U' are obtained from the ordinary difference equations

$$4U'_a - \sum U'_b + h^2 r_a = 0$$

(together with the boundary conditions). With these values U' the quantities $D[U']$ are calculated and then corrections v' obtained from the inhomogeneous difference equations

$$4v'_a - \sum v'_b = D[U']$$

---

[1] A special formula of this type, namely

$$20u_a - 4\sum u_b - \sum u_c + 6h^2 V^2 u_a - \frac{h^4}{2} V^4 u_a = O(h^6),$$

was given as early as 1934 by SH. MIKELADZE: Sur l'intégration numérique d'équations différentielles aux dérivées partielles. Bull. Acad. Sci. URSS. **6**, 819—841 (1934) (Summaries in French and Russian); cf. also equation (2.21) of § 2.7. Further formulae of this type can be found in SH. E. MIKELADZE: Über die numerische Lösung der Differentialgleichung $u_{xx} + u_{yy} + u_{zz} = \varphi(x, y, z)$. C. R. (Doklady) Acad. Sci. URSS. **14**, 177—179 (1937), also 181—182.

[2] Woods, L. C.: Improvements to the accuracy of arithmetical solutions to certain two-dimensional field problems. Quart. J. Mech. Appl. Math. **3**, 349—363 (1950).

(with homogeneous boundary conditions), so that the new approximations $U'' = U' + v'$ satisfy the equations

$$4U''_a - \sum U''_b + h^2 r_a = D[U'].$$

If need be, further corrections v'', ... can be calculated in the same way.

2.6. Examples of the use of Hermitian formulae

In order that the methods may be compared, we use the same examples as in § 2.4 (and in the same order).

I. From Table VI of the appendix the Hermitian equations for
$\nabla^2 u = 0$ centred on the points a, b, c (notation as in Fig. V/13) can be
written down immediately:

$$40a - 8(b + 1 + \tfrac{1}{2} + 0) - 2(1 + 1 + 0 + 0) = 0,$$
$$40b - 8(c + 1 + a + 0) - 2(b + 1 + 0 + 0) = 0,$$
$$40c - 8(0 + b + b + 0) - 2(0 + 1 + 0 + 0) = 0.$$

Here no exterior pivotal values appear, in contrast to § 2.4. We obtain

$$a = \frac{40}{83} = 0.48193, \qquad b = \frac{34}{83} = 0.40964, \qquad c = \frac{71}{332} = 0.21386.$$

II. With the notation of Fig. V/15 the Hermitian equations read

$$40a - 32b - 8c = \frac{16}{3},$$
$$40b - 8(a + 2c + d) - 4(b + e) = \frac{16}{3},$$
$$40c - 16(b + e) - 2(a + 2d + f) = \frac{16}{3};$$

to these we must add the equations (2.13), corresponding to the boundary
conditions, and also the analogous equation

$$15f = 6e + d.$$

Solution of this system of equations yields

$$a = \frac{147177}{176484} = 0.8339,$$

$$b = 0.7199, \qquad c = 0.6234, \qquad d = 0.3436, \qquad e = 0.2974, \qquad f = 0.1418.$$

III. For the boundary-value problem

$$u_{xx} + u_{yy} + \frac{3}{5 - y} u_y + 1 = 0 \qquad (2.18)$$

with $u = 0$ on the sides of the rectangle $|x| \leq \frac{1}{2}$, $|y| \leq 1$ we use a different notation from § 2.4. With the mesh width $h = \frac{1}{2}$ we have the unknown function values U_2, U_3, U_4 (see Fig. V/17) and the unknown values U_1', U_2', U_3', U_4', U_5' of the partial

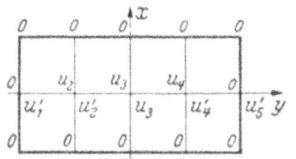

Fig. V/17. Notation for an example treated by the Hermitian method

derivative in the y direction; the corresponding values of $V^2 u$, which we denote by $V_1^2, \ldots, V_5^2$, can be expressed in terms of the U_j' by means of (2.18):

$$V_1^2 = -1 - \tfrac{1}{2} U_1', \qquad V_2^2 = -1 - \tfrac{6}{11} U_2',$$
$$V_3^2 = -1 - \tfrac{3}{5} U_3', \qquad V_4^2 = -1 - \tfrac{3}{5} U_4',$$
$$V_5^2 = -1 - \tfrac{3}{4} U_5',$$

while for $|x| = \frac{1}{2}$ we have $V^2 u = -1$ since $u_y = 0$. Thus the Hermitian equations read

$$\left. \begin{aligned} 40\,U_2 - 8\,U_3 - 2 - \tfrac{12}{11} U_2' + \tfrac{1}{4}\left(-4 - \tfrac{3}{5} U_3' - \tfrac{1}{2} U_1'\right) &= 0, \\ 40\,U_3 - 8\,U_2 - 8\,U_4 - 2 - \tfrac{6}{5} U_3' + \tfrac{1}{4}\left(-4 - \tfrac{3}{5} U_4' - \tfrac{6}{11} U_2'\right) &= 0, \\ 40\,U_4 - 8\,U_3 - 2 - \tfrac{6}{5} U_4' + \tfrac{1}{4}\left(-4 - \tfrac{3}{4} U_5' - \tfrac{3}{5} U_3'\right) &= 0. \end{aligned} \right\} \quad (2.19)$$

According to the general procedure described in § 2.5 one would now write down a set of Hermitian equations for the U_i'; from Table III of the appendix such an equation centred on $y = \frac{1}{4}$, for example, would be

$$- U_3 - \tfrac{1}{6}\left(U_5' + 4\,U_4' + U_3'\right) = 0.$$

However, in the present simple case it is more convenient, and at the same time more accurate, to put a fourth-order interpolation polynomial through the five function values U_i and use its derivatives at the points $y = -1$, $-\frac{1}{2}$, 0, $\frac{1}{2}$, 1 to express the U_i' in terms of the U_i. Thus, evaluating the derivative of the polynomial

$$U(y) = \left[-\tfrac{4}{3} U_2\, y(2y - 1) + U_3(4y^2 - 1) - \tfrac{4}{3} U_4\, y(2y + 1)\right](y^2 - 1)$$

for these values of y, we obtain

$$U_1' = 8\,U_2 - 6\,U_3 + \tfrac{8}{3} U_4, \qquad U_2' = -\tfrac{4}{3} U_2 + 3\,U_3 - U_4, \qquad U_3' = -\tfrac{4}{3} U_2 + \tfrac{4}{3} U_4,$$
$$U_4' = U_2 - 3\,U_3 + \tfrac{4}{3} U_4, \qquad U_5' = -\tfrac{8}{3} U_2 + 6\,U_3 - 8\,U_4.$$

Substitution of these expressions in (2.19) yields a system of equations for the U_i:

$$27072\,U_2 - 6945\,U_3 + 368\,U_4 - 1980 = 0,$$
$$-3138\,U_2 + 19845\,U_3 - 4822\,U_4 - 1485 = 0,$$
$$-228\,U_2 - 1845\,U_3 + 14068\,U_4 - 1080 = 0$$

which may be solved conveniently by iteration since it has strongly dominant diagonal coefficients. The solution is

$$U_2 = 0{\cdot}10098, \qquad U_3 = 0{\cdot}11346, \qquad U_4 = 0{\cdot}09329.$$

IV. We mention here an interesting theorem due to PÓLYA[1]. Let a bounded, simply connected region B in the (x, y) plane be covered by a square mesh of mesh width h and let B' be a region contained within B and bounded only by sections

---

[1] PÓLYA, G.: Sur une interprétation de la méthode des différences finies, qui peut fournir des bornes supérieures ou inférieures. C. R. Acad. Sci., Paris 235, 995−997 (1952).

of mesh lines. If λ_k is an eigenvalue of $\nabla^2 u + \lambda u = 0$ in B with $u = 0$ on the boundary Γ of B, λ'_k the corresponding eigenvalue for the region B' and Λ_k the corresponding eigenvalue of the particular difference equations

$$U_{i+1,j} + U_{i,j+1} + U_{i-1,j} + U_{i,j-1} - 4 U_{i,j} +$$

$$+ \frac{\lambda h^2}{12} (6 U_{i,j} + U_{i+1,j} + U_{i+1,j+1} + U_{i,j+1} + U_{i-1,j} + U_{i-1,j-1} + U_{i,j-1}) = 0$$

for the region B', then $\lambda_k \leq \lambda'_k \leq \Lambda_k$ for all k for which Λ_k exists. The corresponding result for the ordinary difference equations does not necessarily hold.

2.7. Triangular and hexagonal mesh systems

There are many types of boundary for which it is expedient to use a non-rectangular mesh. For illustration we select a mesh of equilateral triangles (Fig. V/18) and describe the derivation of finite expressions for the operators ∇^2 and ∇^4.

Let O be a general mesh point. We start by expanding the function values at the neighbouring mesh points $P, Q, \ldots$ into Taylor series centred on O; for example, with the notation $\alpha = \frac{1}{2}h$, $\beta = \frac{1}{2}\sqrt{3}\,h$, where h is the mesh width (here the length of the sides of the equilateral triangles), the value at P (Fig. V/18) has the expansion

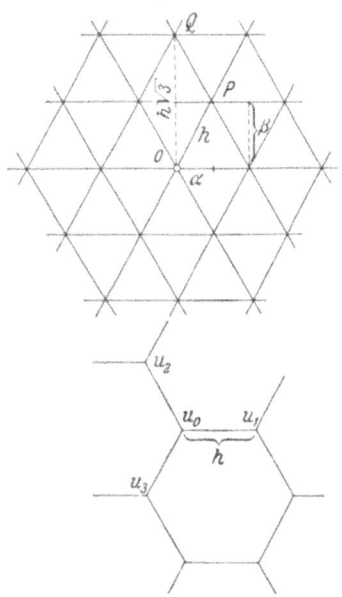

Fig. V/18. Triangular and hexagonal meshes

$$u_P = u_0 + \alpha u_x + \beta u_y +$$

$$\left. \begin{array}{l} + \frac{1}{2!} (\alpha^2 u_{xx} + 2\alpha\beta u_{xy} + \beta^2 u_{yy}) + \\[2mm] + \frac{1}{3!} (\alpha^3 u_{xxx} + 3\alpha^2\beta u_{xxy} + 3\alpha\beta^2 u_{xyy} + \beta^3 u_{yyy}) + \\[2mm] + \frac{1}{4!} (\alpha^4 u_{xxxx} + 4\alpha^3\beta u_{xxxy} + 6\alpha^2\beta^2 u_{xxyy} + 4\alpha\beta^3 u_{xyyy} + \\[2mm] \qquad\qquad + \beta^4 u_{yyyy}) + \cdots. \end{array} \right\} \quad (2.20)$$

If we write down the corresponding expansions for the other five mesh points at a distance h from O and add all six together, most terms cancel out because of the symmetry and we are left with the result

$$\sum u_P = 6u_0 + \frac{3h^2}{2} \nabla^2 u_0 + \frac{3h^4}{32} \nabla^4 u_0 + \cdots. \qquad (2.21)$$

Similarly, addition of the corresponding Taylor expansions for the six function values at the mesh points at a distance $\sqrt{3}\,h$ from O yields the formula

$$\sum u_Q = 6u_0 + \frac{9h^2}{2}\,\nabla^2 u_0 + \frac{27h^4}{32}\,\nabla^4 u_0 + \cdots, \qquad (2.22)$$

which can also be derived immediately from (2.21) by replacing h by $\sqrt{3}\,h$.

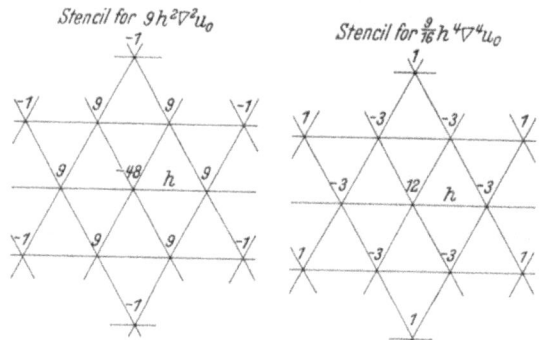

Fig. V/19. Stencils for $\nabla^2 u$ and $\nabla^4 u$

Thus as a first approximation for $\nabla^2 u_0$ we have the formula

$$\sum u_P - 6u_0 = \frac{3h^2}{2}\,\nabla^2 u_0 + O(h^4) \qquad (2.23)$$

and as a second approximation the formula

$$-48u_0 + 9\sum u_P - \sum u_Q = 9h^2\,\nabla^2 u_0 + O(h^6), \qquad (2.24)$$

which is obtained by eliminating $\nabla^4 u_0$ between (2.21) and (2.22).

Elimination of $\nabla^2 u_0$ between (2.21) and (2.22) yields an approximation for $\nabla^4 u_0$:

$$12u_0 - 3\sum u_P + \sum u_Q = \tfrac{9}{16}h^4\,\nabla^4 u_0 + O(h^6). \qquad (2.25)$$

These results are best visualized as stencils of coefficients laid over the corresponding pattern of mesh points as in Fig. V/19; such stencils are also helpful in setting up the system of difference equations.

For a mesh of regular hexagons ("honeycomb pattern") we obtain in the same way

$$u_1 + u_2 + u_3 = 3u_0 + \frac{3h^2}{4}\,\nabla^2 u_0 + O(h^3), \qquad (2.26)$$

where the function values u_0, u_1, u_2, u_3 correspond to the points indicated in Fig. V/18.

2.8. Applications to membrane and plate problems

Example I. Consider a uniform plate in the shape of a trapezium (sides of lengths $A, A, A, 2A$ and angles of $60°$ and $120°$ as in Fig. V/20) with its long edge clamped and its other edges freely supported. If it is subjected to a uniformly distributed trans-
verse load of intensity p
per unit area and its
flexural rigidity is N,
the transverse displace-
ment u satisfies the differential equation

$$\nabla^4 u = \frac{p}{N}$$

and the boundary con-
ditions

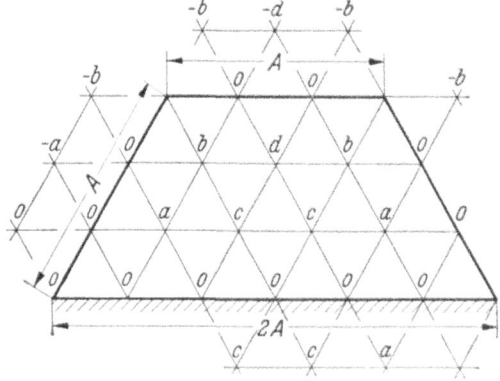

Fig. V/20. Notation for the trapezoidal plate

$$u = \frac{\partial u}{\partial v} = 0 \text{ along the clamped edge } (v \text{ in the direction}$$
$$\text{of the inward normal}),$$

$u = \nabla^2 u = 0$ along the freely supported edges.

From (2.25) the finite equations for a triangular mesh with mesh width $h = \frac{1}{3}A$ read

$$12a - 3b - 3c + d = k = \frac{A^4}{144} \cdot \frac{p}{N},$$
$$-3a + 10b - 2c - 3d = k,$$
$$-3a - 2b + 10c - 3d = k,$$
$$2a - 6b - 6c + 11d = k,$$

where a, b, c, d are the approximate function values as indicated in Fig. V/20.

In these equations we have used the values at exterior mesh points inferred from the boundary conditions (symmetrical values about the long clamped edge and antisymmetrical values about the freely supported edges — see Fig. V/20).

Solution of the equations yields

$$a = \frac{58}{241} k, \qquad b = c = \frac{95}{241} k, \qquad d = \frac{115}{241} k,$$

i.e.

$$a = 0.001\,671 \frac{p A^4}{N}, \qquad b = c = 0.002\,737 \frac{p A^4}{N}, \qquad d = 0.003\,314 \frac{p A^4}{N}.$$

A finer mesh with $h = \frac{1}{4}A$ is used in Exercise 9 of § 6.7.

Example II. The normal modes of vibration of a homogeneous membrane stretched over a frame in the shape of a regular hexagon H with sides of length L (Fig. V/21) are given by the solutions of the eigenvalue problem

$$\nabla^2 u = - \lambda u \text{ in } H,$$

$$u = 0 \text{ on the boundary of } H.$$

A. Triangular mesh with $h = \frac{1}{3}L$. Here we have nineteen mesh points inside of H. For the corresponding function values we adopt the nota-

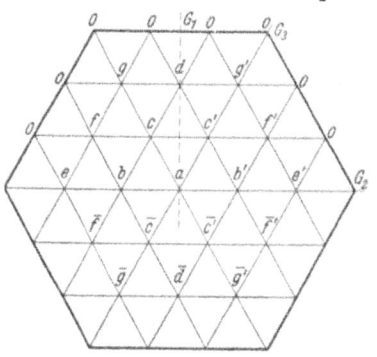

Fig. V/21. Hexagonal membrane

tion $a, b, c, \ldots, g, b', \ldots, \bar{g}'$ as in Fig. V/21, in which the same letter is associated with points lying symmetrically about the lines G_1 and G_2 (Fig. V/21) and dashes and bars distinguish on which sides of the lines they lie; for example, the image of c in G_1 is c' and in G_2 is $\bar{c}$, and the image of $\bar{c}$ in G_1 is $\bar{c}'$. A general calculation would lead to an algebraic equation of the nineteenth degree for the approximation Λ to the eigenvalue λ, but by postulating the various symmetries which can occur we can achieve our object with equations of a very much lower degree.

A1. Symmetry about G_1 and G_2. All bars and dashes can be omitted since $c = c' = \bar{c} = \bar{c}'$ etc. and we have just seven equations for $a, b, \ldots, g$ and Λ. With $\nu = 6 - \frac{2}{3}h^2\Lambda$ these read [from (2.23)]

$$\nu a = 2b + 4c$$
$$\nu b = a + 2c + e + 2f,$$
$$\nu c = a + b + c + d + f + g,$$
$$\nu d = 2c + 2g,$$
$$\nu e = b + 2f,$$
$$\nu f = b + c + e + g,$$
$$\nu g = c + d + f.$$

We can take out the factors $(\nu + 1)$ and $(\nu^2 - 3)$ from the determinant of coefficients $D(\nu)$ by combining several rows and columns, and the condition on ν for a non-trivial solution becomes

$$D(\nu) = (\nu + 1)(\nu^2 - 3)(\nu^4 - 2\nu^3 - 15\nu^2 + 24) = 0.$$

Table V/7. *The normal modes of vibration of a hexagonal membrane*

Meshes		Triangular meshes				Various hexagonal mesh systems					
Number of interior mesh points		7		19		6		12		24		4	
Mesh width		$h=\frac{1}{2}L$		$h=\frac{1}{3}L$		$h=\frac{1}{2}L$		$h=\frac{1}{3}L$		$h=\frac{1}{4}L$		$h=\frac{1}{\sqrt{3}}L$	
Symmetry about	Nodal lines	v	ΛL^2	v	ΛL^2	v	ΛL^2	v	ΛL^2	v	ΛL^2	v	ΛL^2
G_1, G_2 full symm.		$1+\sqrt{7}$	6.28	4.8715	6.77	2	5.33	$1+\sqrt{2}$	7.030	2.6750	6.933	$\sqrt{3}$	5.072
G_1 / G_2		1	13.3	3.355	15.87	1	10.67	$\frac{1+\sqrt{5}}{2}$	16.58	2.214	16.76	0	12
G_1, G_2 / —		-1	18.7	$\sqrt{3}$	25.6	-1	21.3	$\frac{\sqrt{5}-1}{2}$	28.58	1.68	28.2		
G_1, G_2 full symm.		$1-\sqrt{7}$	20.4	1.23	28.6			$1-\sqrt{2}$	41.0	1.55	31.0	$-\sqrt{3}$	18.9
G_2		-2	21.3	$\sqrt{2}-1$	33.5	-2	26.7	$-1+\sqrt{2}$	31.0	1.21	38.2		
G_1				0	36								
G_1 / G_2				-0.476	38.8			$\frac{1-\sqrt{5}}{2}$	43.4	1	42.7		
G_1, G_2 / —				-1	42			$\frac{-1-\sqrt{5}}{2}$	55.4	0.54	52.5		
G_1, G_2 full symm.				-1.63	45.8					-1.21	90		

The largest root of this equation gives the "gravest" mode (smallest value of $\varLambda$); this corresponds to the special case of full symmetry (six axes of symmetry), in which the displacements are all of the same sign and nodal lines are absent. The corresponding values are given in the first row of Table V/7. The subsequent rows show the corresponding results for the higher eigenvalues in increasing order of magnitude; sketches of the corresponding nodal lines are also given. These lines of zero displacement can be obtained approximately from the

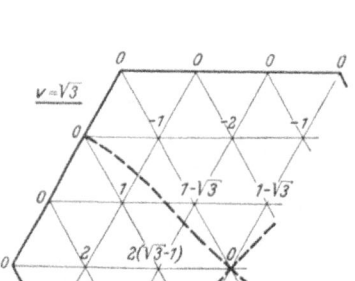

Fig. V/22. A higher mode with nodal line

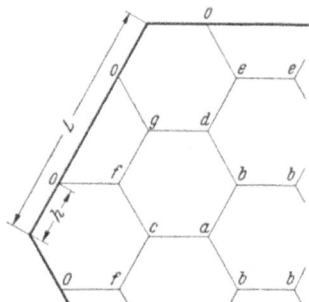

Fig. V/23. Hexagonal mesh for the calculation of normal modes

corresponding approximate mode form given by the ratios $a:b:c:d:e:f:g$; these are easily calculated for each ν value from the homogeneous equations, say by putting one of the non-zero values equal to 1. This is shown in Fig. V/22 for $\nu = \sqrt{3}$.

A2. Symmetry about G_1, antisymmetry about G_2. Here the dashes can be omitted, as before, but omission of the bars is to be accompanied by a reversal of sign; thus, for example, $c = c' = -\bar{c} = -\bar{c}'$. Further $a = b = e = 0$, and we are left with just four equations for c, d, f, g:

$$\nu c = c + d + f + g,$$
$$\nu d = 2c + 2g,$$
$$\nu f = c + g,$$
$$\nu g = c + d + f.$$

The determinantal condition now leads to the quartic

$$\nu(\nu^3 - \nu^2 - 7\nu - 3) = 0.$$

A3, A4. Antisymmetry about G_1, symmetry and antisymmetry, respectively, about G_2. The procedure is the same as in A1 and A2; the number of unknowns is reduced to five and three, respectively, with the results shown in Table V/7. For the higher modes some uncertainty exists in the ordering of the eigenvalues; this can only be removed by increasing the accuracy of the calculation.

B. Hexagonal mesh. The results for several different mesh widths are exhibited in Table V/7. It will suffice to describe the beginning of the calculation with $h = \frac{1}{4}L$, for which the mesh system and the notation $a, b, \ldots, g$ for the function values are indicated in Fig. V/23. Again we postulate the various symmetries, and taking the case of symmetry about G_1 and G_2 as an example we have from (2.26) the equations

$$\nu a = 2b + c,$$
$$\nu b = a + b + d,$$
$$\dots\dots\dots\dots\dots,$$

where $\nu = 3 - \frac{3}{4}h^2 \Lambda$. The values $c, d, \dots, g$ can easily be expressed successively in terms of a and b:

$$c = 2\nu \frac{a}{2} - 2b,$$

$$f = (\nu^2 - 1) \frac{a}{2} - \nu b,$$

$$g = (\nu^3 - 3\nu) \frac{a}{2} + (-\nu^2 + 2) b,$$

$$\dots\dots\dots\dots\dots\dots\dots\dots\dots\dots\dots$$

and the last two homogeneous equations then lead directly to the equation for ν:

$$(\nu + 1)(\nu^3 - 3\nu^2 - \nu + 5)(\nu^3 - 4\nu + 2) = 0.$$

A hexagonal mesh can be fitted into the hexagon H in various other ways, as indicated in Table V/7.

Example III. If the membrane of Example II is replaced by a homogeneous plate, the differential equation governing the vibrations becomes

$$\nabla^4 u = \lambda u;$$

the simplest associated boundary conditions are $u = 0$, $\nabla^2 u = 0$ at a freely supported edge and $u = 0$, $\partial u/\partial \nu = 0$ at a clamped edge. If all the edges are freely supported, the eigenfunctions, or mode forms, are the same as for the membrane, for a solution v of $\nabla^2 v = -\lambda v$ with $v = 0$ on the boundary satisfies $\nabla^4 v = \lambda^2 v$ and also the boundary conditions $v = \nabla^2 v = 0$. For a clamped plate, on the other hand, a new calculation is required. We have to use mesh points outside of the hexagon, and because of the clamping, the associated function values are related to the interior values in symmetrical fashion about the edges, as in Example I. With $h = \frac{1}{2}L$, $\nu = 12 - \frac{9}{16}h^4 \Lambda$ and the function values as in Fig. V/24, we have from (2.25) the equations

$$\nu a = 3(b + c + c' + b' + \bar{c}' + \bar{c}),$$
$$\nu b = 3(c + a + \bar{c}) - 2b - c' - \bar{c}',$$
$$\dots\dots\dots\dots\dots\dots\dots\dots\dots\dots\dots$$

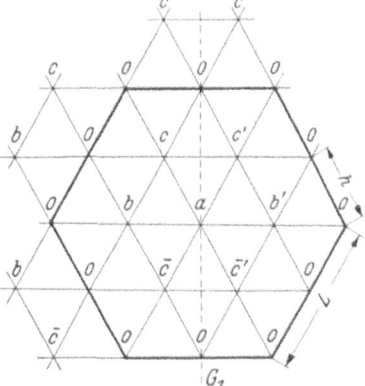

Fig. V/24. Notation for a clamped hexagonal plate

As before, these can be simplified very considerably by postulating the various symmetries that can occur. Symmetry about the line G_1 leads to the values $\nu = 1$, $\pm \sqrt{55}$, 2, -4 and antisymmetry to the values $\nu = 2$, -4, -10; the corresponding values of ΛL^4 are 104, 284, 284, 455, 455, 524, 626.

§ 3. The boundary-maximum theorem and the bracketing of solutions

The boundary-maximum theorem for second-order elliptic differential equations owes its importance to the fact that for many cases of first and third boundary-value problems it provides a basis from which one may be able to obtain in a simple manner rigorous error limits for approximate solutions, and hence also upper and lower bounds for the exact solution. The simple nature of the calculations which are involved is demonstrated by the examples in § 3.4.

3.1. The general boundary-maximum theorem

The error estimates of the following sections are based on the boundary-maximum theorem. This theorem was first established for the potential equation but is valid in much more general cases and will therefore be presented generally here.

Consider the second-order linear differential equation

with
$$\left. \begin{aligned} L[u] &= r \\ L[u] &= - \sum_{j,k=1}^{n} a_{jk} u_{jk} - \sum_{j=1}^{n} b_j u_j + c u, \end{aligned} \right\} \tag{3.1}$$

whose dependent variable $u(x_1, \ldots, x_n)$ is a function of the n independent variables $x_1, \ldots, x_n$ in a closed region $B + \Gamma$ of the n-dimensional space, B being an open region and Γ its set of boundary points. The differential expression may be written more simply as

$$L[u] = - a_{jk} u_{jk} - b_j u_j + c u \tag{3.2}$$

if we make use of the summation convention which is customary in tensor calculus; in this convention the summation signs are omitted and summation from 1 to n understood over each Latin subscript which appears twice in a term. The subscripts attached to a and b serve merely to distinguish different functions but those attached to u (and v, w, z in the following) denote as in (1.11) the partial derivatives

$$u_j = \frac{\partial u}{\partial x_j}, \qquad u_{jk} = \frac{\partial^2 u}{\partial x_j \partial x_k}.$$

We assume here that the given functions a_{jk}, b_j, c, r are continuous in $B + \Gamma$.

The differential equation is said to be elliptic when the matrix a_{jk} (which can be taken as symmetric: $a_{jk} = a_{kj}$) is positive definite in $B + \Gamma$. The theorem applies to elliptic differential equations with $c \geqq 0$. Let the open region B be bounded (and connected) and let its boundary Γ form a closed, connected, piecewise smooth[1], $(n-1)$-dimensional hypersurface. Then we can state

---

[1] COURANT, R., and D. HILBERT: Methoden der mathematischen Physik, Vol. II, p. 228. Berlin 1937.

The boundary-maximum theorem: *If a non-constant function* $w(x_1, \ldots, x_n)$ *with continuous partial derivatives of up to and including the second order is such that* $L[w] \leqq 0$ *in* B, *where* L *is the differential operator of* (3.2) *with* (a_{jk}) *positive definite and* $c \geqq 0$, *and if further (this condition may be omitted for the case* $c = 0$) w *is not everywhere negative in* $B + \Gamma$, *then the maximum value* M *attained by* w *in* $B + \Gamma$ *is assumed only on the boundary* Γ.

Proof (indirect)[1]: We derive a contradiction from the assumption that there is a point P_1 inside B with $w(P_1) = M$. Since w is not constant, there must be a second point P_2 in B with $w(P_2) < M$. There is a curve C joining P_1 and P_2 and a positive number $\varrho > 0$ such that every hypersphere of radius ϱ with a point P on C as centre lies entirely within B. The continuity of w implies the existence of a point P_3 on C such that $w(P_3) = M$ and $w(P) < M$ for all points P on C lying between P_2 and P_3. We now choose a point P_4 on C lying between P_2 and P_3 and sufficiently near P_3 that the hypersphere S_1 with P_4 as centre and with P_3 on its surface lies entirely within B. We therefore have a hypersphere S_1 with $w(P_4) < M$ at its centre P_4 and $w(P_3) = M$ at a point P_3 on its surface. We can now shrink this hypersphere until we arrive at a hypersphere S_2, centre P_4, with $w < M$ at all points inside S_2 and with at least one point P_5 on its surface where $w(P_5) = M$. Then the hypersphere S_3 with $\overline{P_4 P_5}$ as a diameter will have $w = M$ at only one point P_5 on its surface; elsewhere on its surface and in its interior $w < M$. Let P_6 be the centre of S_3 and R its radius. We now define a fourth hypersphere S_4 with centre P_5 and radius $\varrho' < R$ and also so small that S_4 lies completely within B. Then, with the origin of the co-ordinates $x_1, \ldots, x_n$ transferred temporarily to the point P_6 and with the distance from it denoted by r ($r^2 = x_j x_j$), positive constants δ and α can be chosen so that the auxiliary function $z = w + \delta(e^{-\alpha r^2} - e^{-\alpha R^2})$ is less than M everywhere on the surface $\varPhi$ of S_4; for on the part of $\varPhi$ which lies outside of S_3 the bracketed factor is negative and on the remainder, which is a closed set, $w \leqq M' < M$, so that, for given α, δ can be chosen so small that

---

[1] Cf. E. Hopf: Elementare Bemerkungen über die Lösungen partieller Differentialgleichungen 2. Ordnung vom elliptischen Typus. S.-B. Preuss. Akad. Wiss., Phys.-Math. Kl., pp. 147—152, Berlin 1927. — Courant, R., and D. Hilbert: Methoden der mathematischen Physik, Vol. II, p. 275. Berlin 1937. — Bateman, H.: Partial differential equations of mathematical physics, p. 135. New York 1944. — The theorem can be proved very simply for the special case of the potential equation — one way makes use of Poisson's integral, cf. for example Ph. Frank and R. v. Mises: Die Differentialgleichungen und Integralgleichungen der Mechanik und Physik, 2nd ed., Vol. I, pp. 691—692. Brunswick 1930. "A priori" limits for the solutions of boundary-value problems are derived by G. Fichera: Methods of functional linear analysis in mathematical physics. Proc. Internat. Math. Congr. Amsterdam 1954.

$z < M$ there also. Further, $z(P_5) = M$ from the definition of z. Consequently we have a function z which takes the value M at the centre of S_4 and is less than M everywhere on the surface of S_4; it must therefore have a relative maximum with $z \geq M$ for at least one point P_7 inside of S_4.

So far α has been an arbitrary positive number; we now choose it so that $L[z] < 0$ at P_7. We have

$$L[z] = L[w] + \delta\{L[e^{-\alpha r^2}] - c e^{-\alpha R^2}\}$$

and

$$e^{\alpha r^2}\{L[e^{-\alpha r^2}] - c e^{-\alpha R^2}\} = -4\alpha^2 a_{jk} x_j x_k + 2\alpha(a_{jj} + b_j x_j) + c - c e^{\alpha(r^2 - R^2)}.$$

Here α^2 is multiplied by a negative factor and the last term is non-positive since $c \geq 0$; therefore by choosing α large enough we can ensure that the expression in curly brackets takes only negative values (excluding zero) in the whole of S_4. Since $L[w] \leq 0$, we also have $L[z] < 0$ everywhere in S_4, in particular at P_7.

Now at P_7, z has a relative maximum; at P_7, therefore, $z_j = 0$ ($j = 1, \ldots, n$) and the negatives of the second partial derivatives $-z_{jk}$ form a positive definite or semi-definite matrix:

$$-z_{jk} \xi_j \xi_k \geq 0$$

for arbitrary real numbers $\xi_1, \ldots, \xi_n$. Consequently

$$L[z]_{P_7} = -a_{jk} z_{jk} + c z,$$

where both the a_{jk} and the $-z_{jk}$ are the coefficients of positive (semi-) definite quadratic forms, so that

$$-a_{jk} z_{jk} \geq 0$$

from a result due to FEJER[1]. This implies immediately that $L[z]_{P_7} \geq 0$ for the case $c = 0$; and since the conditions of the theorem require that $M \geq 0$ when $c \geq 0$, we have $z(P_7) \geq M \geq 0$, and hence $L[z]_{P_7} \geq 0$, also for the general case $c \geq 0$. Since z was chosen above so that $L[z]_{P_7} < 0$, we have arrived at the desired contradiction.

Corollary[2]. *The corresponding boundary-minimum theorem, in which the inequalities satisfied by $L[w]$ and M are reversed (i.e. have the signs*

---

[1] FEJER, L.: Über die Eindeutigkeit der Lösung der linearen partiellen Differentialgleichungen zweiter Ordnung. Math. Z. **1**, 70—79 (1918).

[2] The analogous theorem to the boundary-maximum theorem for the corresponding difference equations has been proved by T. S. MOTZKIN and W. WASOW: On the approximation of linear elliptic differential equations by difference equations with positive coefficients. J. Math. Phys. **31**, 253—259 (1953); this result can be proved very easily, cf. R. COURANT and D. HILBERT: Methoden der mathematischen Physik, Vol. II, p. 275. 1937.

$\leq$, $\geq$ *interchanged) and M is the minimum value attained by w, is also valid (the boundary-maximum theorem has only to be applied to the function* $-w$*).*

A further consequence of the boundary-maximum theorem is the following:

If two functions w_1 *and* w_2 *with continuous partial derivatives of up to and including the second order both vanish on the boundary* Γ *and are such that* $|L[w_1]| \leq L[w_2]$ *in B, then* $|w_1| \leq w_2$ *in B.*

This can be seen by writing $w = w_1 - w_2$, $w^* = -w_1 - w_2$, so that $L[w] \leq 0$ and $L[w^*] \leq 0$. The maximum of w cannot be negative since there are points where $w = 0$ (on the boundary Γ); thus w satisfies the conditions of the boundary-maximum theorem, and hence $M = 0$ since $w \equiv 0$ on Γ; consequently $w \leq 0$. Similarly $w^* \leq 0$, and therefore $|w_1| \leq w_2$.

3.2. General error estimation for the first boundary-value problem

Consider now the first boundary-value problem associated with the equation (3.1), in which u takes prescribed values $\tilde{u}$ on the boundary Γ:

$$u = \tilde{u} \quad \text{on } \Gamma; \tag{3.3}$$

for simplicity we will assume that the given function $\tilde{u}$ is continuous on Γ. The tilde will be used to denote boundary values for other functions also; for example, $\tilde{v}$ will denote the value of v on the boundary.

Let v be an approximate solution of the boundary-value problem which satisfies at least one of the equations (3.1), (3.3). We distinguish two cases (cf. Ch. I, § 4.1).

Case 1. v satisfies the differential equation $L[u] = r$ exactly but need not take the prescribed boundary values. The error function $w = v - u$ then satisfies the boundary-value problem $L[w] = 0$, $\tilde{w} = \tilde{v} - \tilde{u}$, in which the boundary values $\tilde{w}$ are known for a given approximate solution v. Since $L[w] = 0$, w satisfies both of the conditions $L[w] \leq 0$ and $L[w] \geq 0$, which appear in the boundary-maximum theorem and its corollary, respectively. If $w \geq 0$ somewhere in $B + \Gamma$, then by the boundary-maximum theorem w assumes its maximum value only on the boundary and $w_{max} = \tilde{w}_{max} \geq 0$. Similarly, if $w \leq 0$ somewhere in $B + \Gamma$, then by the corollary w assumes its minimum value only on the boundary and $w_{min} = \tilde{w}_{min} \leq 0$. Consequently, if $w = 0$ somewhere in $B + \Gamma$, then we have $\tilde{w}_{min} \leq w \leq \tilde{w}_{max}$, where $\tilde{w}_{min} \leq 0$, $\tilde{w}_{max} \geq 0$. If $w \neq 0$ in $B + \Gamma$, i.e. if w does not change sign in $B + \Gamma$ (w being a continuous function), then for $w > 0$, say, we have $0 < w \leq \tilde{w}_{max}$.

We can distinguish these various cases according to the behaviour of $\tilde{w}$ instead of w; this is more convenient for application and leads to the

Error estimate. *For the boundary-value problem* (3.1), (3.3) *with* $c \geq 0$ *and* (a_{ik}) *positive definite let* v *be an approximate solution which satisfies the differential equation* $L[u] = r$ *exactly. The boundary values* $\tilde{w} = \tilde{v} - \tilde{u}$ *of the error function* $w = v - u$ *are determinable. If* $\tilde{w}$ *takes the value zero somewhere on* Γ, *then*

$$\tilde{w}_{\min} \leq w \leq \tilde{w}_{\max} \tag{3.4}$$

in B. *If* $\tilde{w}$ *does not take the value zero somewhere on* Γ, *i.e.* $\tilde{w}$ *is everywhere positive or everywhere negative on* Γ, *then* w *is everywhere positive or everywhere negative, respectively, in* B *and*

$$|w| \leq |\tilde{w}|_{\max}. \tag{3.5}$$

In the case $c = 0$, (3.4) *holds independently of the behaviour of* $\tilde{w}$.

From the point of view of practical analysis the situation described here deserves mention as being particularly favourable for error estimation. Error estimates for other problems involving differential equations often turn out to be far more intricate and much less precise than those obtained here. If in fact we were specifically aiming to find constant error bounds valid over the whole of the region B, we could not do better than the estimates (3.4) and (3.5), for the error actually reaches the specified limits on the boundary Γ and the constants are therefore best-possible.

Case 2. The approximate solution v satisfies the boundary condition (3.3) but not necessarily the differential equation (3.1). This situation arises, for instance, when RITZ's method has been used (cf. § 5) and is therefore of particular interest. The residual function $L[v] - r = L[v] - L[u]$ is now not necessarily zero. We suppose that we can construct two auxiliary functions q_1 and q_2 such that

$$L[v] + L[q_1] \leq r \quad \text{and} \quad L[v] + L[q_2] \geq r \tag{3.6}$$

everywhere in $B + \Gamma$. One way of doing this is to evaluate $L[v]$, then choose suitable functions q_1^* and q_2^*, form $L[q_1^*]$ and $L[q_2^*]$ and try to find constants c_1, c_2 such that the inequalities are satisfied in $B + \Gamma$ by $q_1 = c_1 q_1^*$ and $q_2 = c_2 q_2^*$. Such functions q_1, q_2 can always be specified when a function z is known with $L[z] \leq -A < 0$ in B; for the Laplace expression

$$L[u] = -\sum_{j=1}^{n} \frac{\partial^2 u}{\partial x_j^2}$$

we can use, for example,

$$z = \frac{A}{2n} \sum_{j=1}^{n} x_j^2 = q_1^* = q_2^*.$$

The boundary values of the functions $v + q_1 - u$ and $v + q_2 - u$ are known; let M_1 be the maximum of $v + q_1 - u$ and M_2 the minimum of $v + q_2 - u$ on the boundary. Now put

$$\mu_1 = \begin{cases} M_1 & \text{for the case} \quad c = 0 \\ \max(M_1, 0) & \text{for the case} \quad c \geqq 0, \end{cases}$$

$$\mu_2 = \begin{cases} M_2 & \text{for the case} \quad c = 0 \\ \min(M_2, 0) & \text{for the case} \quad c \geqq 0. \end{cases}$$

Then, since $L[v + q_1 - u] \leqq 0$ and $L[v + q_2 - u] \geqq 0$ in $B + \Gamma$, the boundary-maximum theorem and its corollary yield

$$v + q_1 - u \leqq \mu_1, \quad v + q_2 - u \geqq \mu_2,$$

from which we obtain upper and lower bounds for u:

$$v + q_1 - \mu_1 \leqq u \leqq v + q_2 - \mu_2. \tag{3.7}$$

A somewhat simpler estimate can be derived if a function Z can be found with

$$L[Z] \geqq A > 0 \quad \text{in } B, \quad Z = 0 \quad \text{on } \Gamma. \tag{3.8}$$

Then with

$$\varrho = \max_{\text{in } B} |L[v] - r|$$

we have

$$|L[v - u]| = |L[v] - r| \leqq \varrho \leqq \frac{\varrho}{A} L[Z] = L\left[\frac{\varrho}{A} Z\right],$$

and by the consequence of the boundary-maximum theorem mentioned at the end of § 3.1

$$|v - u| \leqq \frac{Z}{A} \max_{\text{in } B} |L[v] - r|. \tag{3.9}$$

3.3. Error estimation for the third boundary-value problem

We consider a mixed boundary-value problem in which Γ is composed of two parts Γ_1 and Γ_2, but do not exclude the degenerate cases in which one part coincides with Γ and the other is null. Let the boundary value $u = A_4$ be prescribed on Γ_1 and let the boundary condition on Γ_2 be of the form[1]

$$A_1 u + A_2 L^*[u] = A_3, \tag{3.10}$$

---

[1] GRÜNSCH, H. J.: Eine Fehlerabschätzung bei der 3. Randwertaufgabe der Potentialtheorie. Z. Angew. Math. Mech. **32**, 279−281 (1952). − See also C. PUCCI: Maggioriazione della suluzione di un problema al contorno, di tipo misto, relativo a una equazione a derivate parzioli, lineare, del secondo ordine. Atti Acad. Naz. Lincei, Ser. VIII **13**, 360−366 (1953). − Bounds for Solutions of Laplace's Equation Satisfying Mixed Conditions. J. Rational Mech. Anal. **2**, 299−302 (1953).

where A_4 and A_1, A_2, A_3 are given functions on Γ_1 and Γ_2, respectively, and

$$L^*[u] = A\, \frac{\partial u}{\partial \sigma},$$

as defined in Ch. I, § 3.3, with $A > 0$ and σ denoting the conormal; further let

$$\frac{A_2}{A_1} < 0 \quad \text{on } \Gamma_2.$$

The other assumptions made in Ch. I, § 3.3 for the formulation of the third boundary-value problem are also made here.

Now let v be an approximate solution which satisfies the differential equation (3.1) exactly but the boundary conditions only approximately:

$$\left.\begin{aligned} v &= A_4^* \quad \text{on } \Gamma_1, \\ A_1 v + A_2 A\, \frac{\partial v}{\partial \sigma} &= A_3^* \quad \text{on } \Gamma_2. \end{aligned}\right\} \tag{3.11}$$

We now define a boundary function γ along the whole of the boundary Γ as the defect in the satisfaction of the boundary conditions, with the mixed condition normalized so that the coefficient of u is unity. Thus the error function $w = v - u$, which satisfies the homogeneous differential equation corresponding to (3.1):

$$L[w] = 0,$$

is subject to the boundary conditions

$$\left.\begin{aligned} w &= A_4^* - A_4 = \gamma \quad \text{on } \Gamma_1, \\ w + \frac{A_2 A}{A_1}\, \frac{\partial w}{\partial \sigma} &= \frac{A_3^* - A_3}{A_1} = \gamma \quad \text{on } \Gamma_2. \end{aligned}\right\} \tag{3.12}$$

Again w satisfies both of the conditions $L[w] \leq 0$ and $L[w] \geq 0$ of § 3.1 since $L[w] = 0$. If $w \geq 0$ somewhere in $B + \Gamma_1 + \Gamma_2$, then by the boundary-maximum theorem w assumes its non-negative maximum value M only on the boundary $\Gamma = \Gamma_1 + \Gamma_2$, say at a point Q.

If Q is on Γ_1, then $w(Q) = \gamma(Q) = M$, and consequently $M \leq \gamma_{\max}$, where $\gamma_{\max}$ denotes the maximum value of γ on Γ; if Q is on Γ_2, then $w(Q) \leq \gamma(Q)$, for the maximum property of $w(Q) = M$ implies that

$$\frac{\partial w}{\partial \sigma} \leq 0$$

at Q (it was shown in Ch. I, § 3.3 that for an elliptic differential equation the conormal points into the interior): in both cases $w \leq \gamma_{\max}$.

Similarly, if $w \leq 0$ somewhere in $B + \Gamma$, then $w \geq \gamma_{\min}$. In the case where the coefficient c in the differential equation (3.1) is zero these inequalities hold irrespective of the behaviour of w. Summarizing we have the

Error estimate. *For the (mixed) boundary-value problem* (3.1), (3.10) *with* $c \geqq 0$, (a_{jk}) *positive definite and* $\dfrac{A_2}{A_1} < 0$, *let v be an approximate solution which satisfies the differential equation. The boundary function γ associated with the error function $w = v - u$ can then be determined from* (3.12). *If γ takes both positive and negative values, or if γ takes the value zero, then*

$$\gamma_{\min} \leqq w \leqq \gamma_{\max} \tag{3.13}$$

in B. If γ has a fixed sign, then w has the same fixed sign and

$$|w| \leqq |\gamma|_{\max} \tag{3.14}$$

in B. For the case $c = 0$, (3.13) *holds independently of whether the sign of γ varies or not.*

3.4. Examples

It is often essential to know particular solutions of the differential equation if an error estimate is to be readily worked out. The derivation of particular solutions usually presents no difficulty when the differential equation is separable, i.e. when solutions in Bernoulli product form can be obtained by solving ordinary differential equations [error estimation and the derivation of particular solutions are treated in more detail in an example given by the author in Z. Angew. Math. Mech. **32**, 207 (1952)]. There are also methods of obtaining particular solutions in general cases[1].

I. The torsion problem for a beam of square cross-section (with sides of length 2) leads to the boundary-value problem

$$\nabla^2 u = 0 \text{ in } B, \qquad \text{i.e. for} \quad |x| \leqq 1, \ |y| \leqq 1,$$

$$u = x^2 + y^2 \text{ on } \Gamma, \qquad \text{i.e. for} \quad |x| = 1 \text{ and for } |y| = 1.$$

Let us assume for v an expression of the form

$$v = a_0 + a_1 v_1 + a_2 v_2,$$

where

$$v_1 = x^4 - 6x^2 y^2 + y^4 = \text{re}(x + iy)^4,$$

$$v_2 = x^8 - 28 x^6 y^2 + 70 x^4 y^4 - 28 x^2 y^6 + y^8 = \text{re}(x + iy)^8.$$

It is expedient to make fairly accurate sketches of the behaviour on the boundary of the functions v_1, v_2, u, and also of several approximate solutions v — here, on account of symmetry, we need only consider the

---

[1] BERGMAN, ST.: Operatorenmethoden in der Gasdynamik. Z. Angew. Math. Mech. **32**, 33—45 (1952).

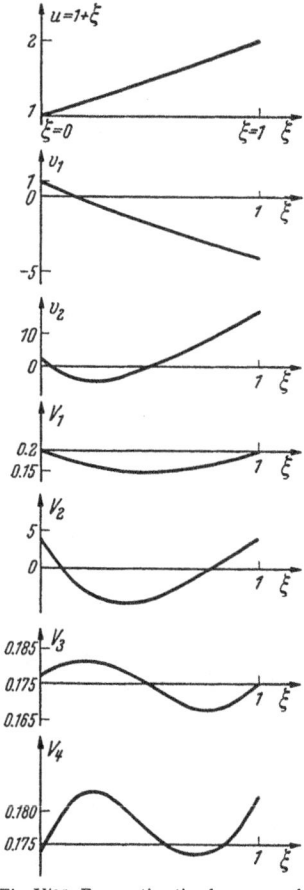

Fig. V/25. Error estimation by means of
the boundary-maximum theorem

line $y = 1$ (cf. Fig. V/25, in which $\xi = x^2$) then it is easy[1] to see how v can be ad justed by the addition of $c_1 v_1$ or $c_2 v_2$ respectively, so as to approximate u mor closely (aiming at a minimum maximum error) and to see what would be suitabl values for c_1, c_2. Each adjustment i followed up numerically with the aid, sa) of a table of function values at equidistan points, in which can be included a simpl row-sum check as in Table V/8. If onl a_0 and $a_1 v_1$ are used, we obtain

$$\varphi_1 = 1 \cdot 175 - 0 \cdot 2 v_1 (x, y),$$

and for the error along $y = 1$ we have

$$|1 \cdot 175 - 0 \cdot 2 v_1 - (x^2 + 1)| \leqq 0 \cdot 025;$$

then from the error estimate of § 3.2 Case I

$$|u - \varphi_1| \leqq 0 \cdot 025$$

holds in B also. If v_2 is used as well, w obtain

$$\varphi_2 = 1 \cdot 1786 - 0 \cdot 2 v_1 + 0 \cdot 006 V_2 + 0 \cdot 0019 v_1$$

where $V_2 = v_2 + 3 v_1$, with $|\varphi_2 - u| \leqq 0 \cdot 00$ on the boundary, and hence $|\varphi_2 - u| \leqq 0 \cdot 00$ also in B; in particular, we obtain th limits

$$1 \cdot 1736 \leqq u(0, 0) \leqq 1 \cdot 1836$$

Table V/8. *Tabular values of various approximating profiles*

	$\xi = 0$	$\xi = 0 \cdot 2$	$\xi = 0 \cdot 3$	$\xi = 0 \cdot 5$	$\xi = 0 \cdot 7$	$\xi = 0 \cdot 8$	$\xi = 1$	
$u = 1 + \xi$ (with $\xi = x^2$)	1	$1 \cdot 2$	$1 \cdot 3$	$1 \cdot 5$	$1 \cdot 7$	$1 \cdot 8$	2	
$v_1 = 1 - 6 \xi + \xi^2$	1	$-0 \cdot 16$	$-0 \cdot 71$	$-1 \cdot 75$	$-2 \cdot 71$	$-3 \cdot 16$	$- 4$	$-$
$v_2 = 1 - 28 \xi + 70 \xi^2 - 28 \xi^3 + \xi^4$	1	$-2 \cdot 0224$	$-1 \cdot 8479$	$1 \cdot 0625$	$6 \cdot 3361$	$9 \cdot 4736$	16	
$V_1 = u + 0 \cdot 2 v_1 - 1$	$0 \cdot 2$	$0 \cdot 168$	$0 \cdot 158$	$0 \cdot 150$	$0 \cdot 168$	$0 \cdot 168$	$0 \cdot 2$	
$V_2 = v_2 + 3 v_1$	4	$-2 \cdot 5024$	$-3 \cdot 9779$	$-4 \cdot 1875$	$-1 \cdot 7939$	$-0 \cdot 0064$	4	$-$
$V_3 = V_1 - 0 \cdot 006 V_2$	$0 \cdot 176$	$0 \cdot 18301$	$0 \cdot 18187$	$0 \cdot 175125$	$0 \cdot 16876$	$0 \cdot 16804$	$0 \cdot 176$	
$V_4 = V_3 - 0 \cdot 0019 v_1$	$0 \cdot 1741$	$0 \cdot 18331$	$0 \cdot 18322$	$0 \cdot 17845$	$0 \cdot 17391$	$0 \cdot 17404$	$0 \cdot 1836$	

[1] A fairly experienced computer will usually achieve the desired result ver quickly in this way; an automatic procedure could be used, of course, and the *a* determined by TREFFTZ's method (cf. § 6 — the present example is in fact treate(by this method in § 6.5) or one of the error distribution principles, but muc more computation would be involved.

for the value at the centre of the square (the error in the mean value $1 \cdot 1786$ is less than $\frac{1}{2}\%$).

II. For the heat conduction problem (cf. § 4.2)

$$\nabla^2 u = -1 \text{ in } B, \quad \text{i.e. for} \quad |x| \leq 1, \ |y| \leq 1$$

$$u = \frac{\partial u}{\partial v} \text{ on } \Gamma, \quad \text{i.e. for} \quad |x| = 1 \text{ and for } |y| = 1,$$

we use an approximate solution of the form

$$v = -\frac{x^2 + y^2}{4} + a_0 + a_1 v_1 + a_2 v_2$$

with the same v_1 and v_2 as in the last example.

Then on $y = 1$ we have (with $\xi = x^2$)

$$\gamma = \left(v + \frac{\partial v}{\partial y}\right)_{y=1} = -\frac{3 + \xi}{4} + a_0 + a_1 V_1 + a_2 V_2,$$

where

$$V_1 = 5 - 18\xi + \xi^2, \quad V_2 = 9 - 196\xi + 350\xi^2 - 84\xi^3 + \xi^4.$$

As in the previous example, we try to choose a_0, a_1, a_2 so as to make the absolute value of the error $|\gamma|$ as small as possible. As in Example I, the following results were obtained with the aid of graphs and tabulated values of the various functions involved.

With $a_2 = 0$, i.e. using v_1 only, we obtain

$$\left| -\frac{3+\xi}{4} + 0 \cdot 8217 - 0 \cdot 0147 V_1 \right| \leq 0 \cdot 0019,$$

while if we include v_2, we effectively gain another figure:

$$\left| -\frac{3+\xi}{4} + 0 \cdot 821\,66 - 0 \cdot 014\,436\,V_1 + 0 \cdot 000\,063\,1\,V_2 \right| \leq 0 \cdot 000\,19.$$

According to (3.13) we obtain from this last result the error estimate

$$\left| -\frac{x^2 + y^2}{4} + 0 \cdot 821\,66 - 0 \cdot 014\,436\,v_1 + 0 \cdot 000\,063\,1\,v_2 - u \right| \leq 0 \cdot 000\,19,$$

which is valid in the whole of $B + \Gamma$; in particular, we have the useful limits

$$0 \cdot 821\,47 \leq u(0, 0) \leq 0 \cdot 821\,85,$$

$$0 \cdot 557\,10 \leq u(0, 1) \leq 0 \cdot 557\,48,$$

$$0 \cdot 380\,22 \leq u(1, 1) \leq 0 \cdot 380\,60.$$

3.5. Upper and lower bounds for solutions of the biharmonic equation

Here we describe briefly, making use of the definitions and results already given in Ch. I, § 3.3, a fundamental way[1] of deriving limits for the solution u of the "clamped plate problem"

$$V^4 u = p \text{ in } B,$$
$$u = f, \quad u_\nu = g \text{ on } \Gamma$$

with region B, boundary Γ and inward normal ν as in Ch. I, §§ 3.2, 3.5. The existence of a solution to such a boundary-value problem will be assumed.

Then the problem

$$V^4 u^* = 0 \text{ in } B,$$
$$u^* = -\varrho, \quad u_\nu^* = -\varrho_\nu \text{ on } \Gamma$$

will also possess a solution u^*; here we select an arbitrary interior point (x_0, y_0) of B and take ϱ to be the "fundamental solution" which is singular at (x_0, y_0) as defined in Ch. I (3.40). According to Ch. I (3.41) we have

$$8\pi u(x_0, y_0) = J_1 + J_2,$$

where

$$J_1 = \iint_B \varrho \, V^4 u \, dx \, dy + \int_\Gamma [u_\nu V^2 \varrho - u(V^2 \varrho)_\nu] \, ds$$

is known but

$$J_2 = \int_\Gamma [\varrho(V^2 u)_\nu - \varrho_\nu V^2 u] \, ds \tag{3.15}$$

is not (at least, not until the problem has been solved).

Now let v, w, v^*, w^* be functions satisfying

$$\left. \begin{array}{ll} V^4 v = p, \quad V^4 v^* = 0 \text{ in } B, \\ w = f, \quad w_\nu = g, \quad w^* = -\varrho, \quad w_\nu^* = -\varrho_\nu \text{ on } \Gamma. \end{array} \right\} \tag{3.16}$$

Then from Ch. I (3.36), (3.39) we have

$$(D[u - w, \, u^* - w^*])^2 \leqq D[u - w] \, D[u^* - w^*] \leqq D[v - w] \, D[v^* - w^*],$$

---

[1] DIAZ, J. B., and H. J. GREENBERG: Upper and lower bounds for the solution of the first biharmonic boundary-value problem. J. Math. Phys. **27**, 193—201 (1948). Another way, which is based almost entirely on mechanical concepts, is described by P. FUNK and E. BERGER: Eingrenzung für die größte Durchbiegung einer gleichmäßig belasteten quadratischen Platte. Federhofer-Girkmann Festschrift 1952, pp. 199—204. — Limits for further problems from the theory of elasticity, including limits for the derivatives of the solution, are given by J. L. SYNGE: Upper and lower bounds for the solutions of problems of elasticity. Proc. Roy. Irish Acad., Sect. A **53**, 41—46 (1950). — Pointwise bounds for the solution of certain boundary value problems. Proc. Roy. Soc. Lond., Ser. A **208**, 170—175 (1951). — MAPLE, C. G.: The Dirichlet Problem: Bounds at a point for the Solution and its Derivatives. Quart. Appl. Math. **8**, 213—228 (1950). — COOPERMAN, PH.: An extension of the method of Trefftz for finding local bounds on the solutions of boundary value problems and on their derivatives. Quart. Appl. Math. **10**, 359—373 (1953). — NICOLOVIUS, R.: Abschätzung der Lösung der ersten Platten-Randwertaufgabe nach der Methode von Maple-Synge. Z. Angew. Math. Mech. **37**, 344—349 (1957).

where D is the integral operator defined in Ch. I (3.34). Now the function appearing to the second power on the left-hand side can be expressed in terms of $u(x_0, y_0)$ and known functions. We have

$$D[u - w, u^* - w^*] = D[w, w^*] - D[u, w^*] + D[u - w, u^*],$$

in which the last two terms can be evaluated by the formula (3.37) of Ch. I; with $\varphi = u^*$, $\psi = u - w$ we see that

$$D[u - w, u^*] = 0$$

and with $\varphi = u$, $\psi = w^*$ that

$$D[u, w^*] = \iint_B w^* \nabla^4 u \, dx \, dy - J_2.$$

Substituting for J_2 from the expression for $8\pi u(x_0, y_0)$ [Ch. I (3.41)] we obtain

$$D[u - w, u^* - w^*] = D[w, w^*] - \iint_B w^* \nabla^4 u \, dx \, dy + 8\pi u(x_0, y_0) - J_1.$$

With

$$W = J_1 + \iint_B w^* \nabla^4 u \, dx \, dy - D[w, w^*]$$

we therefore have the estimate

$$(8\pi u(x_0, y_0) - W)^2 \leqq D[v - w] D[v^* - w^*]. \tag{3.17}$$

To use this estimate, we have to find four functions v, w, v^*, w^* satisfying (3.15). A suitable function for w^* is probably the most difficult to find. Diaz and Green-Berg recommend that outside of a small circle C, centre (x_0, y_0), w^* be chosen as $-\varrho$, and inside of C as a polynomial in r such that w^* possesses continuous second derivatives everywhere, including on C.

Another fundamental approach has been demonstrated by Miranda[1]. If the boundary (possibly consisting of several separate curves) possesses a continuously turning tangent, and if the boundary functions f, g and also $f' = \dfrac{df}{ds}$ (s denoting the arc length along the boundary) are continuous, then the following estimates hold for the homogeneous equation $(p = 0)$:

$$\sqrt{u_x^2 + u_y^2} \leqq Q = K_1 [\max_\Gamma |g| + \max_\Gamma |f'|] + K_2 \max_\Gamma |f|,$$
$$|u(P)| \leqq \delta Q + \max_\Gamma |f|,$$

in which P is an interior point of B, δ is the distance of P from the boundary Γ and K_1, K_2 are constants depending only on the geometry of the region B; the calculation of numerical values for these constants is rather laborious. This yields a fundamental error estimate for an approximate solution of the clamped plate problem which satisfies the differential equation exactly.

§4. Some general methods

Various general methods have already been discussed in Ch. I, §§ 4 and 5; it suffices here to refer to these methods and give examples. More general boundary-value problems for partial differential equations were formulated in Ch. I, § 1.3.

---

[1] Miranda, C.: Formule di maggiorazione e teorema di esistenza per le funzioni biarmoniche di due variabili. Giorn. Mat. Battaglini **78**, 97–118 (1948/49).

4.1. Boundary-value problems of monotonic type for partial differential equations of the second and fourth orders

As in the case of ordinary differential equations, problems of monotonic type can be treated effectively by making use of their monotonic property. Let the differential equation read

$$T[u] = L[u] + F(x_j, u) = 0, \tag{4.1}$$

as in Ch. I (5.52). We consider first equations of the second order and focus attention on those equations for which $L[u]$ has the form (3.2) with coefficients satisfying the conditions laid down in § 3.1. Let the boundary condition be

$$\text{either (case I)} \qquad u = \gamma \quad \text{on } \Gamma \tag{4.2}$$

$$\text{or (case II)} \quad A_1 u + A_2 L^*[u] = A_3 \text{ on } \Gamma, \tag{4.3}$$

where γ, A_1, A_2, A_3 are continuous functions of position on Γ with $\dfrac{A_2}{A_1} < 0$, and L^* has the same significance as in Ch. I, § 3.3. We can then state the following

Theorem. *If the function $F(x_j, u)$ in the differential equation (4.1) is such that $\partial F/\partial u$ exists and is non-negative in a domain H of the $(x_1, \ldots, x_n, u)$ space which is convex with respect to u, then the boundary-value problems (4.1), (4.2) and (4.1), (4.3) with the differential expression $L[u]$ of (3.2) [satisfying the conditions of § 3.1] and the boundary conditions (4.2) and (4.3), respectively, [satisfying the above assumptions] are of monotonic type. This implies that if a solution u exists and lies in H, then it is bracketed by any two functions u_1 and u_2 which satisfy the boundary conditions and are such that $T[u_1] \leqq 0 \leqq T[u_2]$:*

$$u_1 \leqq u \leqq u_2 \text{ in } B. \tag{4.4}$$

The theorem still holds when, as in § 3.3, (4.2) is prescribed on a part Γ_1 of the boundary Γ and (4.3) is prescribed on the remaining part Γ_2.

To prove the theorem we have only to show (see Ch. I, § 5.6) that

$$L[\varepsilon] + \varepsilon A(x_j) \geqq 0 \text{ in } B,$$

where $A(x_j) \geqq 0$, together with the appropriate homogeneous boundary condition (case I: $\varepsilon = 0$ on Γ, case II: $A_1 \varepsilon + A_2 L^*[\varepsilon] = 0$), implies that $\varepsilon \geqq 0$. For this we use the boundary-maximum theorem of § 3.1.

Suppose that the minimum value of ε is negative and let $x_j = \xi_j$ be a point at which this minimum value is attained.

1. If ξ_j is an interior point of B, then since ε is continuous, there must be a neighbourhood of ξ_j in which $\varepsilon < 0$; in this neighbourhood we

therefore have $L[\varepsilon] \geq 0$. But according to the boundary-maximum theorem this means that there cannot be a negative minimum value in the interior. Consequently ξ_j cannot be an interior point of B.

2. Neither can it be a boundary point; for in case I $\varepsilon = 0$ on the boundary and in case II a negative minimum on the boundary leads to a contradiction as follows. At ξ_j on $\varGamma$ we must have $\dfrac{\partial \varepsilon}{\partial \sigma} \geq 0$, for σ, being in the direction of the conormal, points inwards (for elliptic equations). Therefore $\dfrac{A_2}{A_1} L^*[\varepsilon] \leq 0$ at ξ_j, which, since we have assumed that $\varepsilon(\xi_j) < 0$, is incompatible with the boundary conditions.

Thus we must have $\varepsilon \geq 0$ and the theorem is proved. It is applied to a numerical example in § 4.3.

The monotonic character of some fourth-order boundary-value problems can be demonstrated by means of this theorem. Consider, for example, the following simple problem which arises in the standard treatment of the flexure of a freely supported flat plate:

$$\nabla^4 u \equiv \frac{\partial^4 u}{\partial x^4} + 2 \frac{\partial^4 u}{\partial x^2 \partial y^2} + \frac{\partial^4 u}{\partial y^4} = r(x, y) \text{ in } B, \tag{4.5}$$

$$u = \gamma_1(s), \qquad \nabla^2 u = \gamma_2(s) \text{ on } \varGamma, \tag{4.6}$$

where s is the arc length along the boundary curve $\varGamma$ of a simply-connected plane region B and r, γ_1, γ_2 are given functions. Here we define the operator T by

$$T v = \nabla^4 v - r(x, y)$$

and restrict its domain of operation to those functions v in B which possess fourth-order partial derivatives and satisfy the boundary conditions. This operator also is monotonic, for it can be readily shown that

$$\nabla^4 \varepsilon \geq 0 \text{ in } B \quad \text{and} \quad \nabla^2 \varepsilon = \varepsilon = 0 \text{ on } \varGamma$$

implies $\varepsilon \geq 0$: we put $-\nabla^2 \varepsilon = \zeta$, so that $-\nabla^2 \zeta \geq 0$ in B, $\zeta = 0$ on $\varGamma$; then, according to the above theorem, $\zeta \geq 0$ and consequently $-\nabla^2 \varepsilon \geq 0$ in B, $\varepsilon = 0$ on $\varGamma$; using the above theorem again we have $\varepsilon \geq 0$.

4.2. Error distribution principles. Boundary and interior collocation

For these methods (see Ch. I, § 4) we use as an approximation to the required solution $u(x_1, \ldots, x_n)$ some function $w(x_1, \ldots, x_n, a_1, \ldots, a_p)$ which depends on p parameters a_ϱ and which, for arbitrary values of these parameters, satisfies either the differential equation or the boundary conditions, whichever is the more convenient. Insertion of w into the boundary conditions or the differential equation, respectively, yields residual error functions $\varepsilon(x_j, a_\varrho)$, and by choosing the a_ϱ suitably we make these error functions approximate the zero function as closely as possible in the sense of one of the principles mentioned in Ch. I, § 4. We recall that methods of this sort may be conveniently classified into boundary, interior and mixed methods, as in Ch. I, § 4.1. For several important classes of boundary-value problems we have means for deriving bounds for the error of an approximation function in a quite simple manner, cf. § 3 and § 4.1; it is therefore very useful to possess in the procedures based on the error distribution principles methods of a parallel simplicity for the calculation of approximation functions.

Example. Consider the problem

$$\nabla^2 u = \frac{\partial^2 u}{\partial x^2} + \frac{\partial^2 u}{\partial y^2} = -1 \quad \text{in the interior of } B,$$

$$u = \frac{\partial u}{\partial v} \quad \text{on the boundary of } B,$$

where B is the square $|x| \leq 1$, $|y| \leq 1$ (Fig. V/26) and v is in the direction of the inward normal; u could be interpreted physically as the tem-

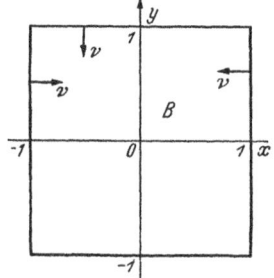

perature distribution (measured as the excess over the temperature of the surroundings) in the cross-section of a square wire carrying a current and for which the amount of heat lost to the surroundings is proportional to the excess temperature at the surface.

Fig. V/26. Boundary and inward normals for a "third" boundary-value problem

For our approximation function we can assume an expression [Ch. I (4.10)] which satisfies either the differential equation or the boundary conditions. One way of obtaining an expression which satisfies the boundary conditions is to write down a polynomial expression satisfying any symmetry conditions that exist but whose coefficients $a_{\mu\nu}$ are otherwise arbitrary:

$$w = a_{00} + a_{20}(x^2 + y^2) + a_{22}x^2y^2 + a_{40}(x^4 + y^4) + a_{42}(x^4y^2 + x^2y^4) + \cdots,$$

and then determine these coefficients from the boundary conditions (cf. § 4.3). In this way we obtain, for example,

$$w = a_1 v_1 + a_2 v_2 = a_1[9 - 3(x^2 + y^2) + x^2 y^2] + \\ + a_2[30 - 5(x^2 + y^2) - 3(x^4 + y^4) + x^2 y^2(x^2 + y^2)]. \quad \} \quad (4.7)$$

For an expression which satisfies the differential equation we require a particular solution, together with a series of functions satisfying the homogeneous equation (and also any symmetries which exist). In our present case the latter can be obtained, for example, as the real parts of the functions $(x+iy)^{4\varrho}$; then with $-\frac{1}{4}(x^2+y^2)$ satisfying the in-homogeneous equation we have

$$w = -\tfrac{1}{4}(x^2+y^2) + a_0 + a_1(x^4 - 6x^2y^2 + y^4) + \\ + a_2(x^8 - 28x^6y^2 + 70x^4y^4 - 28x^2y^6 + y^8). \quad \Big\} \quad (4.8)$$

We now illustrate several methods for determining the a_ϱ.

I. Collocation[1]. A. We use first the expression (4.7) to illustrate collocation as an "interior" method. Substitution in the differential expression yields

$$\nabla^2 w = a_1[-12 + 2(x^2+y^2)] + \\ + a_2[-20 - 36(x^2+y^2) + 2(x^4+y^4) + 24x^2y^2], \quad \Big\} \quad (4.9)$$

which is to be put equal to -1 at the collocation points.

The choice of collocation points is a matter of some uncertainty; although normally some attempt at a uniform distribution would be made, we have a virtually unrestricted choice, but no reliable guide to the effect this choice has on the results. In the case of a single-term expression (only $a_1 \neq 0$) the coefficient a_1 is given by

$$a_1 = \frac{1}{2(6 - x^2 - y^2)}$$

for an arbitrary position (x, y) of the single collocation point, and will vary between $\frac{1}{12}$ (for $x=y=0$) and $\frac{1}{8}$ (for $x=y=1$) as the collocation point moves about the square; thus we can obtain values for $w(0, 0)$ lying anywhere in the range from $0\cdot75$ to $1\cdot125$, which is a rather large variation.

For a two-term expression with a_1 and a_2 we have to dispose two collocation points. On account of the symmetries which exist, collocation actually occurs in general at sixteen points; we try to choose the two arbitrary points so that the full set of sixteen collocation points are distributed as uniformly as possible. If, for example, we choose the points $x=\frac{1}{2}$, $y=\frac{1}{4}$ and $x=1$, $y=\frac{1}{2}$, we obtain the two equations

$$a_1\frac{91}{8} + a_2\frac{3935}{128} = 1, \qquad a_1\frac{19}{2} + a_2\frac{455}{8} = 1,$$

which yield the values

$$a_1 = \frac{446}{6057} = 0\cdot07363, \qquad a_2 = \frac{32}{6057} = 0\cdot005283.$$

For $u(0, 0)$ these give the approximate value $w(0, 0) = \dfrac{4974}{6057} = 0\cdot82120$.

Table V/9 exhibits the results for various positions of the collocation points.

---

[1] See also R. A. FRAZER, W. P. JONES and SYLVIA W. SKAN: Approximations to functions and to the solutions of differential equations. Rep. and Mem. No. 1799 (2913), Aero. Res. Comm. 1937, 33 pp.

Table V/9. *Collocation for the temperature distribution in a square wire carrying a current*

Expression		Co-ordinates of the collocation points	Position	Results
Interior collocation — Expression (4.7)	only a_1	$x = y = \frac{1}{2}$		$a_1 = \frac{1}{11}$ $w(0,0) = \frac{9}{11} = 0{\cdot}8182$
	only a_1	$x = y = \frac{2}{3}$		$a_1 = \frac{9}{92}$ $w(0,0) = \frac{81}{92} = 0{\cdot}8804$
	only a_1	$x = \frac{3}{4},\ y = \frac{1}{4}$		$a_1 = \frac{4}{43}$ $w(0,0) = \frac{36}{43} = 0{\cdot}8372$
	with a_1, a_2	$x = \frac{1}{2},\ y = \frac{1}{4}$ $x = \frac{3}{4},\ y = \frac{1}{2}$		$a_1 = 0{\cdot}07400$ $a_2 = 0{\cdot}005148$ $w(0,0) = \frac{4080}{4973} = 0{\cdot}82043$
	with a_1, a_2	$x = \frac{1}{2},\ y = \frac{1}{4}$ $x = 1,\ y = \frac{1}{2}$		$a_1 = 0{\cdot}073634$ $a_2 = 0{\cdot}005283$ $w(0,0) = \frac{4974}{6057} = 0{\cdot}82120$
Boundary collocation — Expression (4.8)	only a_0	$x = 1,\ y = \frac{1}{2}$		$a_0 = \frac{13}{16}$ $w(0,0) = a_0 = 0{\cdot}8125$
	with a_0, a_1	$x = 1,\ y = \frac{1}{4}$ $x = 1,\ y = \frac{3}{4}$		$a_0 = w(0,0) = \frac{14615}{17792}$ $= 0{\cdot}821431$ $a_1 = -\frac{2}{139} = -0{\cdot}01439$
	with a_0, a_1, a_2	$x = 1,\ y = 0$ $x = 1,\ y = \sqrt{\frac{1}{2}}$ $x = 1,\ y = 1$		$a_0 = w(0,0) = \frac{12846}{15635}$ $= 0{\cdot}821618$ $a_1 = -0{\cdot}014400$ $a_2 = 0{\cdot}000064$

B. Greater convenience and less uncertainty are afforded by the use of the other expression (4.8), which satisfies the differential equation. This is because the collocation points now lie only on the boundary and it is much easier to ensure some degree of uniformity in their distribution.

On account of symmetry we need only consider the part of the boundary along $x = 1$, where $\dfrac{\partial w}{\partial v} = -\dfrac{\partial w}{\partial x}$; we have

$$\left(w + \frac{\partial w}{\partial x}\right)_{x=1} = -\frac{3 + y^2}{4} +$$
$$+ a_0 + a_1(5 - 18y^2 + y^4) + a_2(9 - 196y^2 + 350y^4 - 84y^6 + y^8),$$

which in accordance with the boundary conditions is to approximate the zero function as closely as possible.

First of all, let us put $a_1 = a_2 = 0$; then for a general boundary collocation point $(1, y)$

$$a_0 = \frac{3 + y^2}{4}.$$

Now $|y|$ lies between 0 and 1, so a_0 can vary from 0·75 to 1; for $y = \frac{1}{2}$ we have $a_0 = w(0, 0) = 0\cdot8125$.

If we include the term in a_1, we can demand that the boundary condition be satisfied at $y = \frac{1}{4}$ and $y = \frac{3}{4}$, for example; then from

$$a_0 + a_1 \frac{993}{256} = \frac{49}{64}, \qquad a_0 - a_1 \frac{1231}{256} = \frac{57}{64}$$

we obtain

$$a_0 = w(0, 0) = 0\cdot821\,43,$$
$$a_1 = -0\cdot01439.$$

The results for a three-parameter expression are also given in Table V/9.

II. GALERKIN's method. We use the same example to illustrate GALERKIN's method as an interior method; thus we take the expression $w = a_1 v_1 + a_2 v_2$ of (4.7) as an approximation function. Here we determine a_1 and a_2 from the equations [see Ch. I (4.13)]

$$\int_0^1 \int_0^1 v_\varrho (V^2 w + 1)\, dx\, dy = 0 \qquad (\varrho = 1, 2), \tag{4.10}$$

where $V^2 w$ is given by (4.9). The evaluation of the integrals leads to the equations

$$\frac{6}{5} a_1 + \frac{6782}{1575} a_2 = \frac{1}{9}, \qquad \frac{6782}{1575} a_1 + \frac{74\,144}{4725} a_2 = \frac{2}{5}.$$

The first approximation $(a_2=0)$ can be found from the first equation by putting $a_2=0$:

$$a_1 = \frac{5}{54}, \quad w(0,0) = \frac{5}{6} = 0\cdot833\ldots.$$

If a_2 is taken into account, we obtain

$$a_1 = \frac{39305}{536397} = 0\cdot0732760, \quad a_2 = 0\cdot0053831, \quad w(0,0) = 0\cdot820978.$$

4.3. The least-squares method as an interior and a boundary method

After the general description in Ch. I, § 4.2 we restrict ourselves here to just two examples.

I. Interior method. Let us consider again the example of § 4.2:

$$\left.\begin{array}{ll} V^2 u = -1 & \text{in the interior} \\[2mm] \dfrac{\partial u}{\partial v} = u & \text{on the boundary} \end{array}\right\} \text{ of the square } |x|\leq1,\ |y|\leq1. \quad (4.11)$$

We start with the simplest case of a one-parameter approximation. We need a function $\varphi(x,y)$ which satisfies the boundary conditions. Taking into account the symmetries of the problem, we put

$$\varphi = c_1 + c_2(x^2+y^2) + c_3 x^2 y^2;$$

then on the part of the boundary $x=1$, $|y|\leq1$, for example, we have

$$-\frac{\partial \varphi}{\partial v} = \frac{\partial \varphi}{\partial x} = 2c_2 x + 2c_3 x y^2,$$

and the boundary condition yields

$$\left(\varphi - \frac{\partial \varphi}{\partial v}\right)_{x=1} = c_1 + 3c_2 + (c_2 + 3c_3)\, y^2 = 0;$$

consequently

$$c_2 = -3c_3, \quad c_1 = -3c_2 = 9c_3,$$

and we take

$$\varphi = a\left(9 - 3(x^2+y^2) + x^2 y^2\right).$$

Then

$$V^2\varphi = a\left(-12 + 2(x^2+y^2)\right),$$

and inserting this in the expression to be minimized according to the minimum principle of Ch. I (4.5), namely

$$J[\varphi] = \iint_B (V^2\varphi + 1)^2 dx\, dy, \quad (4.12)$$

and evaluating the integral, we obtain

$$J = 4\left(1 - \frac{64}{3} a + 112 \times \frac{46}{45} a^2\right).$$

The condition

$$\frac{1}{4}\frac{\partial J}{\partial a} = -\frac{64}{3} + 224 \times \frac{46}{45}\, a = 0$$

yields $a = \frac{15}{161}$ and hence the approximate solution

$$\varphi = \frac{15}{161}\left(9 - 3\left(x^2 + y^2\right) + x^2 y^2\right).$$

This gives the following values at the key points of the cross-section:

$$\varphi(0,0) = \frac{135}{161} = 0\cdot8385 \quad (\text{error } + 2\cdot1\%),$$

$$\varphi(0,1) = \frac{90}{161} = 0\cdot5590 \quad (\text{error } + 0\cdot3\%),$$

$$\varphi(1,1) = \frac{60}{161} = 0\cdot3727 \quad (\text{error } - 1\cdot4\%).$$

Proceeding to the second approximation, we put

$$\varphi = c_1 + c_2\left(x^2 + y^2\right) + c_3 x^2 y^2 + c_4\left(x^4 + y^4\right) + c_5\left(x^2 y^4 + x^4 y^2\right).$$

With the relations between the c_i required by the boundary conditions the expression takes the form

$$\varphi = 9\alpha + (-3\alpha + 15\beta)\left(x^2 + y^2\right) + (\alpha - 10\beta) x^2 y^2 - 9\beta\left(x^4 + y^4\right) + 3\beta\left(x^2 y^4 + x^4 y^2\right);$$

this function satisfies the boundary conditions for arbitrary values of α and β and therefore provides a two-parameter approximation function. With $2\alpha = a_1 = a$, $2\beta = a_2 = b$ it yields

$$\nabla^2\varphi + 1 = 1 + aA_1 + bA_2,$$

where

$$A_1 = -6 + x^2 + y^2, \qquad A_2 = 30 - 64\left(x^2 + y^2\right) + 3\left(x^4 + y^4\right) + 36 x^2 y^2.$$

From the conditions $\partial J/\partial a_i = 0$ (for $i = 1, 2$) we obtain the equations

$$\iint_B A_1\, dx\, dy + a \iint_B A_1^2\, dx\, dy + b \iint_B A_1 A_2\, dx\, dy = 0,$$

$$\iint_B A_2\, dx\, dy + a \iint_B A_1 A_2\, dx\, dy + b \iint_B A_2^2\, dx\, dy = 0,$$

where

$$\alpha_{10} = \iint_B A_1\, dx\, dy = -4 \times \frac{16}{3}, \qquad \alpha_{20} = \iint_B A_2\, dx\, dy = -4 \times \frac{112}{15},$$

$$\alpha_{11} = \iint_B A_1^2\, dx\, dy = 4 \times \frac{1288}{45}, \qquad \alpha_{12} = \alpha_{21} = \iint_B A_1 A_2\, dx\, dy = 4 \times \frac{9776}{315},$$

$$\alpha_{22} = \iint_B A_2^2\, dx\, dy = 4 \times \frac{788192}{1575}.$$

We can check the calculation of these integrals by evaluating $J[\varphi]$ for $a = b = 1$: we have

$$\nabla^2\varphi + 1 = 25 - 63\left(x^2 + y^2\right) + 36 x^2 y^2 + 3\left(x^4 + y^4\right)$$

and hence

$$\iint\limits_B (\nabla^2 \varphi + 1)^2 \, dx \, dy = 4 \times \frac{99\,143}{175};$$

this value should be equal to

$$\iint\limits_B (1 + A_1 + A_2)^2 \, dx \, dy = 4 + 2\alpha_{10} + 2\alpha_{20} + \alpha_{11} + 2\alpha_{12} + \alpha_{22}.$$

The equations

$$\alpha_{10} + \alpha_{11} a + \alpha_{12} b = 0,$$
$$\alpha_{20} + \alpha_{21} a + \alpha_{22} b = 0$$

read

$$-\frac{16}{3} + \frac{1288}{45} a + \frac{9776}{315} b = 0,$$

$$-\frac{112}{15} + \frac{9776}{315} a + \frac{788\,192}{1575} b = 0,$$

i.e.

$$-210 + 1127a + 1222b = 0,$$
$$-735 + 3055a + 49262b = 0,$$

and yield

$$a = \frac{175}{4} \frac{2999}{719237} = 0.182424,$$

$$b = \frac{175}{4} \frac{593}{719237} = 0.0036071.$$

The approximate values at the key points of the cross-section are now

$$\varphi(0, 0) = \tfrac{9}{2} a \qquad = 0.820909 \ (\text{error} -0.08\%),$$
$$\varphi(0, 1) = 3(a + b) = 0.558094 \ (\text{error} +0.16\%),$$
$$\varphi(1, 1) = 2a + 4b = 0.379277 \ (\text{error} -0.19\%).$$

Upper and lower bounds for the solution. It may be mentioned here that for this problem upper and lower bounds for the solution, though admittedly fairly coarse, can be obtained very simply by direct use of its monotonic property — which is assured by the theorem of § 4.1. If a function φ satisfying the boundary conditions is such that $\nabla^2 \varphi + 1 \leq 0$ or ≥ 0 throughout the whole region, then φ is respectively an upper or lower bound for the solution u; thus, for instance, in the approximation function which we have just been using we want to choose the constants a, b so that $\nabla^2 \varphi + 1$ does not change sign (naturally we will also try to keep its magnitude as small as possible). First of all we construct a linear combination of the functions A_1 and A_2 which is as near constant as possible. It can be seen by evaluating these functions at the "key" points that the best combination (up to a constant factor) is $43 A_1 + A_2$; we have

$$-246 \leq 43 A_1 + A_2 \leq -228$$

and hence

$$\frac{1}{228} (43 A_1 + A_2) + 1 \leq 0 \leq \frac{1}{246} (43 A_1 + A_2) + 1.$$

Consequently φ is an upper bound for u when $\alpha = \dfrac{43}{456}$, $\beta = \dfrac{1}{456}$ and a lower bound when $\alpha = \dfrac{43}{492}$, $\beta = \dfrac{1}{492}$. The mean of these two bounds gives an approximation with known error limits; at the centre of the square, for example, we have $u(0, 0) = 0\cdot8177 \pm 0\cdot0311$.

II. Boundary method. We choose a different problem here, namely

$$\nabla^2 u = x^2 - 1, \qquad u = 0 \quad \text{for} \quad |x| = 1 \quad \text{and for} \quad |y| = \tfrac{1}{2};$$

u can be interpreted as the transverse displacement of a membrane stretched over a rectangular frame and distorted by a non-uniformly distributed transverse load.

Table V/10. *Values of the individual contributions a_{mn}*

$a_{00} = \dfrac{3}{2}$	$a_{10} = \dfrac{5}{6}$	$a_{20} = \dfrac{7}{10}$	$a_{30} = \dfrac{9}{14}$	$a_{40} = \dfrac{11}{18}$
$a_{01} = \dfrac{7}{24}$	$a_{11} = \dfrac{1}{8}$	$a_{21} = \dfrac{11}{120}$	$a_{31} = \dfrac{13}{168}$	$a_{41} = \dfrac{5}{72}$
$a_{02} = \dfrac{11}{160}$				

The function $\varphi = w_0 + \sum\limits_{\nu=1}^{p} a_\nu w_\nu$ with

$$w_0 = \frac{x^4}{12} - \frac{x^2}{2}, \qquad w_\nu = \operatorname{re}(x + i y)^{2\nu}$$

satisfies the differential equation exactly for arbitrary a_ν and exhibits the symmetries possessed by the problem. We restrict attention to the first quadrant and denote

$$\int\limits_0^{\frac{1}{2}} F(1, y)\, dy + \int\limits_0^1 F(x, \tfrac{1}{2})\, dx \qquad \text{by} \qquad \int\limits_\Gamma F\, ds.$$

Then the least-squares requirement reads

$$J[\varphi] = \int\limits_\Gamma \varphi^2\, ds = \text{minimum}, \tag{4.13}$$

or

$$\frac{1}{2}\frac{\partial J}{\partial a_\nu} = 0 = \int\limits_\Gamma \varphi\, w_\nu\, ds \qquad (\nu = 1, 2, \ldots, p). \tag{4.14}$$

To expedite the derivation of the equations for the a_ν from (4.14), we construct a table of values of the quantities

$$a_{mn} = \int\limits_\Gamma x^{2m} y^{2n}\, ds;$$

Table V/10 gives sufficient values for the case $p = 2$, for which we obtain the equations

$$\frac{3}{2} a_1 + \frac{13}{24} a_2 = \frac{43}{120}, \qquad \frac{13}{24} a_1 + \frac{83}{160} a_2 = \frac{2435}{120 \times 84}$$

with the solution

$$a_1 = \frac{16\,643}{60 \times 2443} = 0 \cdot 1135\,, \qquad a_2 = \frac{848}{2443} = 0 \cdot 3471\,.$$

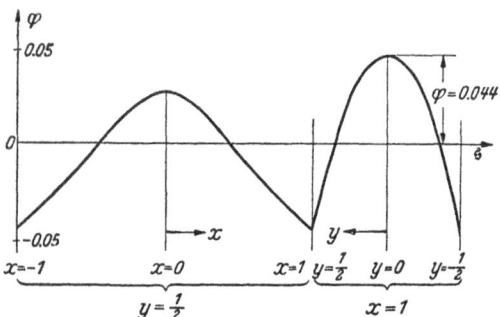

Fig. V/27. The error distribution on the boundary

The boundary values of the corresponding approximation φ are shown graphically in Fig. V/27; because of the zero boundary condition this graph represents the error distribution on the boundary. Since $u - \frac{1}{12}x^4 + \frac{1}{2}x^2$ is a harmonic function, we infer from the boundary-maximum theorem of § 3.1 that the maximum absolute error in φ on the boundary, about $0 \cdot 044$, is not exceeded in the interior: $|\varphi - u| \leqq 0 \cdot 044$ in the whole rectangle.

4.4. Series solutions

A method which is often used for the solution of boundary-value problems is to find a series expansion for the solution u of the form

$$u = \sum_{m=1}^{\infty} a_m \mathit{\Psi}_m \quad \text{say, or} \quad u = \sum_{m,n=1}^{\infty} a_{mn} \mathit{\Psi}_{mn}\,, \qquad (4.15)$$

where $\mathit{\Psi}_m$ or $\mathit{\Psi}_{mn}$ are chosen as known functions of the independent variables and the a_m or a_{mn}, respectively, are constants to be determined. Naturally triple sums and higher multiple sums can also be used as occasion demands.

What was said in Ch. III, § 7.1 concerning series solutions for ordinary differential equations is obviously also valid here; for partial differential equations the effectiveness of a series solution depends to a particularly large extent on a propitious choice of the functions $\mathit{\Psi}$ and on what is involved in carrying out the method.

Ordinarily we choose the functions $\mathit{\Psi}$ so that they satisfy either

      (i) the differential equation,

or (ii) the boundary conditions,

or (iii) at least some of the boundary conditions,

then seek to determine the expansion coefficients a_m (or a_{mn}) so that the series satisfies the differential equation or the boundary conditions when inserted. This usually leads to an infinite system of equations for infinitely many unknowns a_m; such a system is normally solved in an approximate manner by putting all the unknowns after the first N equal to zero and determining the remaining unknowns from the first N equations. A thorough treatment of the subject of series solutions would have to cover such a wide range of phenomena that we content ourselves here with a few examples which demonstrate the procedure involved in the application of the method; for the rest, reference can be made to the literature, where numerous examples[1] of series expansions are to be found[2].

4.5. Examples of the use of power series and related series

I. For the heat conduction problem (cf. § 4.2)

$$\left. \begin{array}{l} \nabla^2 u = -1 \text{ in the interior} \\ \partial u/\partial v = u \text{ on the boundary} \end{array} \right\} \quad \begin{array}{l} \text{of the square } |x| \leqq 1, \quad |y| \leqq 1, \\ (v = \text{inward normal}), \end{array} \left. \right\} \quad (4.16)$$

which has already been treated by many different methods, the power series method is superior to all others in shortness and accuracy. We make use of the symmetries of the problem and instead of the general double series

$$u = \sum_{m, n=0}^{\infty} a_{mn} x^m y^n$$

---

[1] For instance, in A. Sommerfeld: Partielle Differentialgleichungen der Physik. Leipzig 1947. — Frank, Ph., and R. v. Mises: Differential- und Integralgleichungen der Mechanik und Physik. Brunswick 1930 and 1935. See also our Ch. IV, introduction and § 4.3.

[2] A method for the solution of boundary-value problems based on the Bergman "kernel function" associated with the given region and the given differential equation is given by St. Bergman: Kernel functions and conformal mapping. Surveys of the Amer. Math. Soc. 5 (1950); see also St. Bergman and M. Schiffer: Kernel functions and elliptic differential equations in mathematical physics, Part B. New York: Academic Press Inc. 1953. The kernel function can be obtained by expansion in terms of the functions of a complete orthonormal system. For the potential equation in a simply-connected region of the complex z plane, for example the system of polynomials obtained from the powers of z by orthogonalization with respect to the given region is suitable. The kernel function is defined as the difference between the corresponding Green and Neumann functions. There are simple rules for modifying the kernel function when the differential equation or the region is altered. This method has been used at the National Bureau of Standards, Washington, to solve boundary-value problems for the potential equation in doubly-connected regions bounded by concentric rectangles. Experience must show whether it is best to work in terms of orthonormal systems or to tabulate the kernel function for several different regions and a given differential equation, say the potential equation.

use the simple series

$$u = -\tfrac{1}{4}(x^2 + y^2) + \sum_{\varrho=0}^{\infty} a_{2\varrho} \, \mathrm{re}\, (x + i\, y)^{4\varrho}$$

$$= -\tfrac{1}{4}(x^2 + y^2) + a_0 + a_2(x^4 - 6x^2 y^2 + y^4) +$$

$$+ a_4(x^8 - 28 x^6 y^2 + 70 x^4 y^4 - 28 x^2 y^6 + y^8) +$$

$$+ a_6(x^{12} - 66 x^{10} y^2 + 495 x^8 y^4 - 924 x^6 y^6 + \cdots) + \cdots.$$

The first term $-\tfrac{1}{4}(x^2 + y^2)$ satisfies the inhomogeneous potential equation and the subsequent terms, being the real parts of analytic functions, satisfy the homogeneous equation; therefore $\nabla^2 u = -1$ for arbitrary a_ϱ. Only powers $(x + i y)^{4\varrho}$ which satisfy all the symmetries of the problem have been used. The a_ϱ will now be determined so that the boundary condition $\partial u / \partial v = u$ is satisfied. We form

$$(u)_{x=1} = a_0 - \tfrac{1}{4}(1 + y^2) + a_2(1 - 6y^2 + y^4) +$$

$$+ a_4(1 - 28 y^2 + 70 y^4 - 28 y^6 + y^8) +$$

$$+ a_6(1 - 66 y^2 + 495 y^4 - 924 y^6 + \cdots) + \cdots,$$

$$\left(\frac{\partial u}{\partial x}\right)_{x=1} = -\tfrac{1}{2} + a_2(4 - 12 y^2) + a_4(8 - 168 y^2 + 280 y^4 - 56 y^6) +$$

$$+ a_6(12 - 660 y^2 + 3960 y^4 - 5544 y^6 + \cdots) + \cdots.$$

The sum of these two expressions is to vanish identically, so the coefficient multiplying each power y^{2k} in the sum must be put equal to zero; this yields the following infinite system of equations for the unknowns a_ϱ:

$$a_0 + 5a_2 + \quad 9a_4 + \quad 13 a_6 + \cdots = \quad \tfrac{3}{4} \quad \text{(coefficient of } y^0),$$

$$18 a_2 + 196 a_4 + \quad 726 a_6 + \cdots = -\tfrac{1}{4} \quad (\quad ,, \quad ,, \; y^2),$$

$$a_2 + 350 a_4 + 4455 a_6 + \cdots = \quad 0 \quad (\quad ,, \quad ,, \; y^4),$$

$$84 a_4 + 6468 a_6 + \cdots = \quad 0 \quad (\quad ,, \quad ,, \; y^6).$$

We solve these equations approximately in the customary fashion: for the ϱ-th approximation we solve the first ϱ equations for the first ϱ unknowns with the remaining unknowns put equal to zero. In the second approximation, for example, we solve the equations

$$a_0 + 5 a_2 = \tfrac{3}{4}, \qquad 18 a_2 = -\tfrac{1}{4},$$

and obtain

$$a_0 = \frac{59}{72}, \qquad a_2 = -\frac{1}{72}.$$

Table V/11 gives the first four approximations for the coefficients of the power series, together with the values calculated from them for the key points of the square. The convergence of the successive approximations is good.

Table V/11. *First four approximations by the power series method*

	1st approximation	2nd approximation	3rd approximation	4th approximation
a_0	$\dfrac{3}{4}=0{\cdot}75$	$\dfrac{59}{72}=\;0{\cdot}81944$	$\dfrac{20053}{24416}=\;0{\cdot}82131$	$\dfrac{1283427}{1562176}=0{\cdot}821\,564$
a_2	0	$-\dfrac{1}{72}=-0{\cdot}01389$	$-\dfrac{25}{1744}=-0{\cdot}01432$	$-\dfrac{22495}{1562176}=-0{\cdot}01440$
a_4	0	0	$\dfrac{1}{24416}=\;0{\cdot}000041$	$\dfrac{77}{1562176}=0{\cdot}000\,0493$
a_6	0	0	0	$\dfrac{-1}{1562176}=-0{\cdot}00000064$
$u\,(0,0)$	0·75	0·819444	$\dfrac{20053}{24416}=0{\cdot}821\,306$	$\dfrac{1283427}{1562176}=0{\cdot}821\,5636$
$u\,(0,1)$	0·50	0·555555	$\dfrac{13600}{24416}=0{\cdot}557012$	$\dfrac{870464}{1562176}=0{\cdot}557\,2125$
$u\,(1,1)$	0·25	0·375	$\dfrac{9261}{24416}=0{\cdot}379\,300$	$\dfrac{593615}{1562176}=0{\cdot}3799924$

II. We now illustrate the opposite procedure on the plate problem of § 1.5

$$\left. \begin{array}{l} \nabla^4 u = x^2, \\ u = u_x = 0 \quad \text{for} \quad |x| = 1, \quad u = \nabla^2 u = 0 \quad \text{for} \quad |y| = 1, \end{array} \right\} \quad (4.17)$$

i.e. we use a series which already satisfies all the boundary conditions:

$$u = (1 - x^2)^2 (1 - y^2) \sum_{m, n=0}^{\infty} a_{mn}\, x^{2m}\, \eta_n(y)$$

with

$$\eta_n(y) = (4n + 5)\, y^{2n} - (4n + 1)\, y^{2n+2}.$$

Again we choose a series which also satisfies the symmetries of the problem.

The expressions $\nabla^4 \Psi_{mn}$ are now worked out for the first few quantities $\Psi_{mn} = (1 - x^2)^2 (1 - y^2)\, x^{2m}\, \eta_n(y)$, but only as far as the highest powers which we intend to consider. We will illustrate the procedure with a quite crude three-term approximation involving only a_{00}, a_{01}, a_{10}. For this we need only those terms of $\nabla^4 \Psi_{mn}$ which are $O(x^2 + y^2)$, and by substituting in the differential equation and equating coefficients we obtain:

$$\text{absolute term:} \qquad 10 a_{00} - 20 a_{01} - 12 a_{10} = 0,$$

$$\text{factor multiplying } x^2: \quad -14 a_{00} + 46 a_{01} + 100 a_{10} = \frac{1}{24},$$

$$\text{factor multiplying } y^2: \quad -10 a_{00} + 140 a_{01} + 14 a_{10} = 0.$$

Solution of this system of equations yields

$$a_{00} = \frac{35}{24 \times 2487}, \qquad a_{01} = \frac{-1}{48 \times 2487}, \qquad a_{10} = \frac{15}{12 \times 2487}.$$

The approximate value

$$u\,(0, 0) \approx 5 a_{00} = 0{\cdot}00293$$

corresponding to this three-term approximation is still very crude and more terms must be taken into consideration if any real accuracy is to be attained.

4.6. Eigenfunction expansions

The set of expansion functions Ψ is often chosen as the set of eigen-functions of the associated eigenvalue problem. Consider a linear in-homogeneous boundary-value problem of the form

$$L[u^*] = r^* \quad \text{in } B,$$
$$U_\mu[u^*] = v_\mu \quad \text{on } \Gamma \quad (\mu = 1, \ldots, k), \Big\} \tag{4.18}$$

where, as usual, Γ is the boundary of a two(or more)-dimensional region B, L and the U_μ are given linear differential expressions, r^* and the v_μ are given functions of position, and u^* is to be determined. We suppose that the boundary conditions $U_\mu[u^*] = v_\mu$ render the determination unique. First of all we reduce the problem to one with homogeneous boundary conditions: if w^* is any function which satisfies the inhomogeneous boundary conditions, and also possesses derivatives of a sufficiently high order (so that $L[w^*]$ exists), then the function $u = u^* - w^*$ satisfies a linear boundary-value problem with homogeneous boundary conditions, namely

$$L[u] = r = r^* - L[w^*] \quad \text{in } B,$$
$$U_\mu[u] = 0 \quad \text{on } \Gamma \quad (\mu = 1, \ldots, k). \Big\} \tag{4.19}$$

We now suppose that the eigenvalue problem

$$L[\psi] = \lambda \psi \quad \text{in } B,$$
$$U_\mu[\psi] = 0 \quad \text{on } \Gamma \quad (\mu = 1, \ldots, k) \Big\} \tag{4.20}$$

possesses a complete system of eigenfunctions $\psi_1, \psi_2, \ldots$ corresponding to the eigenvalues $\lambda_1, \lambda_2, \ldots$, and expand the function r in (4.19) into a uniformly convergent series of these eigenfunctions (we assume that this is possible):

$$r = \sum_{\varrho=1}^{\infty} c_\varrho \psi_\varrho. \tag{4.21}$$

Then

$$u = \sum_{\varrho=1}^{\infty} \frac{c_\varrho}{\lambda_\varrho} \psi_\varrho \tag{4.22}$$

is the solution of the inhomogeneous boundary-value problem (4.19), assuming that $L[u]$ may be formed term by term for this series (4.22).

Example. The boundary-value problem which arises in the problem of the torsion of prisms (cf. Example I, § 3.4) is customarily reduced to one with homogeneous boundary conditions by putting $w^* = x^2 + y^2$; apart from a constant factor we obtain

$$\nabla^2 u = -1 \quad \text{in } B,$$
$$u = 0 \quad \text{on } \Gamma.$$

Let us consider a prism of rectangular section bounded by the lines $x = 0$, $x = a$, $y = 0$, $y = b$.

The eigenfunctions of the associated eigenvalue problem

$$\nabla^2 u = \lambda u \quad \text{in } B, \qquad u = 0 \quad \text{on } \Gamma$$

are given by

$$\Psi_{m,n} = \sin \frac{m \pi x}{a} \sin \frac{n \pi y}{b} \qquad (m, n = 1, 2, \ldots)$$

and the corresponding eigenvalues by

$$\lambda_{m,n} = -\left(\frac{m^2}{a^2} + \frac{n^2}{b^2} \right) \pi^2.$$

The eigenfunction expansion of the function $r = -1$ is here a double Fourier expansion

$$-1 = \sum_{m,n=1}^{\infty} c_{m,n} \sin \frac{m \pi x}{a} \sin \frac{n \pi y}{b} \quad \text{for} \quad 0 < x < a,\ 0 < y < b, \quad (4.23)$$

so that the coefficients $c_{m,n}$ are Fourier coefficients. They are obtained by the well-known method of multiplying by $\Psi_{p,q}$ and integrating over the fundamental region:

$$-\int_0^a \sin \frac{p \pi x}{a} d x \int_0^b \sin \frac{q \pi y}{b} d y = c_{p,q} \int_0^a \sin^2 \frac{p \pi x}{a} d x \int_0^b \sin^2 \frac{q \pi y}{b} d y = c_{p,q} \frac{a b}{4}.$$

This yields

$$c_{p,q} = \begin{cases} 0 & \text{when at least one of } p, q \text{ is even,} \\ -\dfrac{16}{\pi^2 p q} & \text{when both } p \text{ and } q \text{ are odd.} \end{cases}$$

The solution (4.22) therefore reads

$$u = \frac{16}{\pi^4} \sum_{p,q \text{ odd}} \frac{\sin \dfrac{p \pi x}{a} \sin \dfrac{q \pi y}{b}}{p q \left(\dfrac{p^2}{a^2} + \dfrac{q^2}{b^2} \right)}, \qquad (4.24)$$

where p, q run through all pairs of odd positive integers.

Let us evaluate the value at the centre point for the special case of a square section. From (4.24)

$$u\left(\frac{a}{2}, \frac{a}{2} \right) = \frac{16 a^2}{\pi^4} \sum_{p,q \text{ odd}} \frac{(-1)^{\frac{p+q}{2} - 1}}{p q (p^2 + q^2)}$$

$$= \frac{8 a^2}{\pi^4} \sum_{p,q \text{ odd}} \frac{(-1)^{\frac{p+q}{2} - 1}}{\left(\dfrac{p+q}{2} \right)^4 - \left(\dfrac{p-q}{2} \right)^4} = \frac{8 a^2}{\pi^4} S,$$

$$(4.25)$$

where

$$S = \sum_{k=1}^{\infty} (-1)^{k+1} A_k \quad \text{with} \quad A_k = \sum_{\substack{j=1,3,5,\ldots \\ j \text{ odd}}}^{2k-1} \frac{1}{k^4 - (k-j)^4};$$

for example,

$$A_4 = \frac{2}{4^4 - 3^4} + \frac{2}{4^4 - 1^4} \quad \text{and} \quad A_5 = \frac{2}{5^4 - 4^4} + \frac{2}{5^4 - 2^4} + \frac{1}{5^4}.$$

This rearrangement of the double series (permissible because of the absolute convergence), in which terms with the same $p+q$, and therefore with the same sign, are grouped together into the finite sums A_k converts it into an alternating

Table V/12. *Evaluation of the alternating series for the value at the centre*

k	Terms $(-1)^{k+1} A_k$	Partial sums s_k	First smoothing s_k'	Second smoothing s_k''
1	1	1		
2	-0.13333333	0.86666667	0.91077873	
3	0.04311491	0.90978158	0.89418493	0.89926310
4	-0.01927171	0.89050987	0.89790383	0.89667191
5	0.01030412	0.90081399	0.89669503	0.89711961
6	-0.00617112	0.89464227	0.89718455	0.89700470
7	0.00399741	0.89863968	0.89695466	0.89704204
8	-0.00274267	0.89589701	0.89707429	0.89702754
9	0.00196644	0.89786345	0.89700691	0.89703385
10	-0.00145973	0.89640372	0.89704729	0.89703084
11	0.00111455	0.89751827	0.89702187	0.89703239
12	-0.00087105	0.89664722	0.89703853	
13	0.00069417	0.89734139		

simple series S in the A_k; moreover, these A_k are easy to calculate, since the denominators are the differences of the fourth powers 1, 16, 81, ... Table V/12 gives the calculated values for k up to 13 of the quantities A_k, the partial sums $s_k = \sum_{i=1}^{k} (-1)^{i+1} A_i$ and the quantities s_k', s_k'' which are obtained from the s_k by smoothing:

$$s_k' = \tfrac{1}{4}(s_{k-1} + 2s_k + s_{k+1}), \quad s_k'' = \tfrac{1}{4}(s_{k-1}' + 2s_k' + s_{k+1}')$$

(this smoothing process gains three more decimals). We deduce from these figures that to six decimals

$$S = 0.897032,$$

and hence

$$u\left(\frac{a}{2}, \frac{a}{2}\right) = 0.073671 3a^2.$$

For the special case of a square of side $a = 2$ (cf. Example I, § 3.4) we have

$$u(1, 1) = 0.294685.$$

For this example the series method would appear to be far superior to all other methods.

§5. The Ritz method

5.1. The Ritz method for linear boundary-value problems of the second order

As in the case of ordinary differential equations (Ch. III, § 5), it is also possible for many boundary-value problems of even order in partial differential equations to specify an integral expression $J[\varphi]$ which can be formed for a certain class of functions φ and which assumes its minimum value just for that function u which satisfies the boundary-value problem, and therefore to find an approximate solution of the boundary-value problem by inserting an approximation function for φ in $J[\varphi]$ and determining the parameters to make J a minimum for this restricted class of functions.

For a boundary-value problem of the second order in two independent variables we consider the variational problem [corresponding to (5.6) in Ch. III]

$$J[\varphi] = \iint_B [A\,\varphi_x^2 + 2B\,\varphi_x\varphi_y + C\,\varphi_y^2 + F\,\varphi^2 - 2r\,\varphi]\,dx\,dy + \left. \right\}$$
$$+ \int_\Gamma (S\,\varphi^2 + 2T\,\varphi)\,ds = \text{extremum,} \quad \left. \right\} \quad (5.1)$$

and investigate the range of boundary-value problems which can be covered by it. B is assumed to be a bounded, simply-connected, closed region with a piecewise-smooth boundary curve Γ, along which the arc length s is measured anti-clockwise from some fixed point. The extremum, which we may suppose for definiteness to be a minimum, shall be relative to the set of values of J obtained by letting φ run through the domain of continuous functions which possess continuous partial derivatives of up to and including the second order and which also satisfy on Γ a certain linear boundary condition of the form

$$a_1(s)\,\varphi + a_2^*(s)\,\varphi_x + a_3^*(s)\,\varphi_y = a_4(s), \quad (5.2)$$

which will be specified more precisely later. We assume that there is a smallest value of J and a corresponding function u among the admissible functions φ for which J assumes this smallest value. The question of the existence of a minimum of J will not be pursued further here. Thus we suppose that

$$J[u] \leqq J[\varphi]. \quad (5.3)$$

If $\eta(x, y)$ is any function with continuous second partial derivatives which satisfies the homogeneous boundary condition corresponding to (5.2), i.e.

$$a_1\eta + a_2^*\,\eta_x + a_3^*\,\eta_y = 0 \quad \text{on } \Gamma, \quad (5.4)$$

then $\varphi = u + \varepsilon\eta$ is an admissible function for any value of ε, and $J[u + \varepsilon\eta] = \Phi(\varepsilon)$ is a function of ε which assumes its minimum value for $\varepsilon = 0$ and whose derivative with respect to ε must therefore vanish at $\varepsilon = 0$:

$$\Phi'(0) = \left(\frac{\partial J[u + \varepsilon \eta]}{\partial \varepsilon}\right)_{\varepsilon=0} = 0. \tag{5.5}$$

By inserting $\varphi = u + \varepsilon \eta$ in (5.1) and calculating the derivative in (5.5) we obtain

$$\tfrac{1}{2}\Phi'(0) = J[u, \eta] + \iint\limits_B r u \, dx \, dy - \int\limits_\Gamma T u \, ds = 0,$$

where $J[u, \eta]$ is the integral of the corresponding bilinear expression:

$$\left.\begin{array}{l} J[\varphi, \psi] = \iint\limits_B [A\,\varphi_x\psi_x + B\,(\varphi_x\psi_y + \varphi_y\psi_x) + C\,\varphi_y\psi_y + F\,\varphi\psi - \\ \qquad - r\,(\varphi + \psi)]\,dx\,dy + \int\limits_\Gamma [S\,\varphi\psi + T(\varphi + \psi)]\,ds; \end{array}\right\} \tag{5.6}$$

transformation of the double integral by means of GREEN'S formula [Ch. I (3.4)] then yields

$$\tfrac{1}{2}\Phi'(0) = \iint\limits_B \eta\,\{L[u] - r\}\,dx\,dy - \int\limits_\Gamma \eta\,L^*[u]\,ds + \int\limits_\Gamma \eta\,(Su + T)\,ds = 0, \tag{5.7}$$

in which $L[u], L^*[u]$ are the linear differential expressions defined in Ch. I, (3.3), (3.5):

$$L[u] = -\frac{\partial}{\partial x}(A\,u_x + B\,u_y) - \frac{\partial}{\partial y}(B\,u_x + C\,u_y) + F\,u,$$

$$L^*[u] = (A\,u_x + B\,u_y)\cos(\nu, x) + (B\,u_x + C\,u_y)\cos(\nu, y).$$

Now (5.7) can hold for all functions $\eta(x, y)$ satisfying the present conditions only if the factor $L[u] - r$ vanishes identically in B. This follows by the same argument as was used in Ch. III, § 5.2: if $L[u] - r$ were not equal to zero, but were positive say, at some interior point (x_0, y_0) of B, then by continuity it would also be positive in a small neighbourhood $(x - x_0)^2 + (y - y_0)^2 \leq \delta^2$ of (x_0, y_0), and for any η function which satisfies (5.7), say $\eta = \eta^*(x, y)$, we could specify another, $\eta = \eta^{**}$, which does not:

$$\eta^{**} = \begin{cases} \eta^* + [(x - x_0)^2 + (y - y_0)^2 - \delta^2]^3 & \text{for } (x - x_0)^2 + (y - y_0)^2 \leq \delta^2 \\ \eta^* & \text{otherwise.} \end{cases}$$

Consequently u must satisfy the differential equation

$$L[u] = r(x, y), \tag{5.8}$$

called the Euler equation of the variational problem, and (5.7) reduces to

$$\int\limits_\Gamma \eta\,L^{**}[u]\,ds = 0, \tag{5.9}$$

where

$$L^{**}[u] = Su + T - L^* = Su + T - (A\,u_x + B\,u_y)\cos(v, x) - \left.\vphantom{\begin{matrix}1\\1\end{matrix}}\right\} \quad (5.10)$$
$$- (B\,u_x + C\,u_y)\cos(v, y).$$

If our assumption of the existence of a minimizing function u is not to be contradicted, this boundary condition (5.9) must be compatible with the boundary condition (5.2) (with φ replaced by u); we see immediately that this is not always possible, and proceed to discuss the conditions under which it is.

5.2. Discussion of various boundary conditions

We distinguish two cases according as u is or is not prescribed on the boundary $\varGamma$.

Case I. First boundary-value problem. Here the boundary values of u are prescribed:

$$u(s) = g(s) \quad \text{on } \varGamma. \quad (5.11)$$

The corresponding homogeneous boundary condition for η is therefore $\eta = 0$ on $\varGamma$ and the necessary condition (5.9) is automatically satisfied; thus no condition is imposed on $L^{**}[u]$ and we may choose $S = T = 0$.

Case II. Third boundary-value problem. Here the value of a linear combination of u and its first partial derivatives as in (5.2) is prescribed on the boundary $\varGamma$. Consequently there are functions η which are not zero on $\varGamma$, and therefore in (5.9) we must have $L^{**}[u] = 0$; this condition is therefore to be identifiable with the boundary condition (5.2). Let us denote the direction cosines of the inward normal v by

$$\alpha = \cos(v, x) = -\cos(s, y), \quad \beta = \cos(v, y) = \cos(s, x) \quad (5.12)$$

and express φ_x and φ_y in terms of the normal and tangential derivatives φ_v and φ_s:

$$\varphi_x = \alpha\,\varphi_v + \beta\,\varphi_s, \left.\vphantom{\begin{matrix}1\\1\end{matrix}}\right\} \quad (5.13)$$
$$\varphi_y = \beta\,\varphi_v - \alpha\,\varphi_s.$$

Then the two boundary conditions (5.2) and $L^{**}[u] = 0$, which are to represent the same condition, read

$$a_1\varphi + (-a_3^*\alpha + a_2^*\beta)\,\varphi_s + (a_2^*\alpha + a_3^*\beta)\,\varphi_v - a_4 = 0, \quad (5.14)$$

$$Su + [(C-A)\,\alpha\beta + B\,(\alpha^2 - \beta^2)]\,u_s - [A\,\alpha^2 + 2B\,\alpha\beta + C\,\beta^2]\,u_v + T = 0. \quad (5.15)$$

If we regard the differential equation as prescribed, so that A, B, C are fixed, we have only two quantities S and T at our disposal in (5.15), and it is clear that they cannot be chosen to make (5.15) equivalent to (5.14) unless the coefficients in the given boundary condition are such that the ratio of the coefficients of the partial derivatives in each of these two equations are the same:

$$\frac{a_2}{a_3} = \frac{-a_3^*\alpha + a_2^*\beta}{a_2^*\alpha + a_3^*\beta} = \frac{(A-C)\alpha\beta - B(\alpha^2-\beta^2)}{A\alpha^2 + 2B\alpha\beta + C\beta^2}. \tag{5.16}$$

We exclude the case when the denominator vanishes identically (on the boundary), for then the boundary condition provides a relation between u and u_s, i.e. a differential equation for the boundary value $u(s)$, and by integration of this differential equation the boundary-value problem can be reduced to a first boundary-value problem. The denominator can still vanish on parts of the boundary Γ without vanishing everywhere on it. As this case will not be pursued any further, we will assume here that $A\alpha^2 + 2B\alpha\beta + C\beta^2 \neq 0$ on Γ. The results of the last two sections can then be summarized in the following

Theorem: *The boundary-value problem*

$$L[u] = -\frac{\partial}{\partial x}(A u_x + B u_y) - \frac{\partial}{\partial y}(B u_x + C u_y) + F u = r(x, y) \text{ in } B, \tag{5.17}$$

$$a_1(s) u + a_2(s) u_s + a_3(s) u_v = a_4(s) \quad \text{on } \Gamma \tag{5.18}$$

with $a_3 \neq 0$ may be written as the necessary Euler conditions for a variational problem of the form

$$\left. \begin{aligned} J[\varphi] = \iint_B [A\,\varphi_x^2 + 2B\,\varphi_x\varphi_y + C\,\varphi_y^2 + F\,\varphi^2 - 2r\,\varphi]\,dx\,dy + \\ + \int_\Gamma (S\,\varphi^2 + 2T\,\varphi)\,ds = \text{extremum} \end{aligned} \right\} \tag{5.19}$$

if and only if the coefficients satisfy the condition

$$\frac{a_2}{a_3} = \frac{(A-C)\alpha\beta - B(\alpha^2-\beta^2)}{A\alpha^2 + 2B\alpha\beta + C\beta^2}, \tag{5.20}$$

where

$$\alpha = \cos(v, x), \qquad \beta = \cos(v, y).$$

S and T are necessarily related to the other coefficients by

$$\left. \begin{aligned} S &= -(A\alpha^2 + 2B\alpha\beta + C\beta^2)\frac{a_1}{a_3}, \\ T &= +(A\alpha^2 + 2B\alpha\beta + C\beta^2)\frac{a_4}{a_3}, \end{aligned} \right\} \tag{5.21}$$

but the functions φ admitted for comparison need not be restricted to satisfy any boundary conditions.

For the corresponding first boundary-value problem, i.e. for the case $a_2 = a_3 = 0$, $a_1 \neq 0$, the values of S and T are irrelevant, and may be taken to be zero, but it is now essential that the admissible functions φ all satisfy the boundary condition: $\varphi(s) = a_4/a_1$ on Γ.

5.3. A special class of boundary-value problems

We specialize the results of the last section by taking $A = C$ and $B = 0$; this leaves the class of differential equations of the form

$$- \frac{\partial}{\partial x} [A(x, y) u_x] - \frac{\partial}{\partial y} [A(x, y) u_y] + F(x, y) u = r(x, y) \quad \text{in } B, \quad (5.22)$$

which occur frequently in applications. To satisfy the condition (5.19) we must also take $a_2 = 0$; thus we restrict ourselves to boundary conditions of the form

$$a_1(s) u + a_3(s) u_v = a_4(s) \quad \text{on } \Gamma, \quad (5.23)$$

which also occur frequently in practice.

Then for the case $a_3 \neq 0$ the corresponding variational problem reads (since $\alpha^2 + \beta^2 = 1$)

$$\left. \begin{aligned} J[\varphi] = \iint_B [A(\varphi_x^2 + \varphi_y^2) + F \varphi^2 - 2r\varphi] \, dx \, dy + \\ + \int_\Gamma \frac{A}{a_3} (-a_1 \varphi^2 + 2a_4 \varphi) \, ds = \text{extremum}, \end{aligned} \right\} \quad (5.24)$$

for which φ need not satisfy any boundary conditions.

For the case $a_3 = 0$, $a_1 \neq 0$ the boundary integral in (5.24) may be omitted, but the admissible functions φ must all take the boundary value a_4/a_1 on Γ.

5.4. Example

Once more we consider the heat conduction problem

$$\left. \begin{aligned} \nabla^2 u = -1 \quad \text{in the interior} \\ \partial u / \partial v = u \quad \text{on the boundary} \end{aligned} \right\} \text{ of the square } |x| \leq 1, \; |y| \leq 1. \quad (5.25)$$

From § 5.3 the corresponding variational problem reads

$$J[\varphi] = \iint_B [\varphi_x^2 + \varphi_y^2 - 2\varphi] \, dx \, dy + \int_\Gamma \varphi^2 \, ds = \text{extremum}, \quad (5.26)$$

for which φ need not satisfy any boundary conditions. Using polynomial approximation functions which already satisfy the symmetry conditions, we take for our first approximation the two-parameter expression

$$\varphi = a_1 + a_2 (x^2 + y^2).$$

With

$$\varphi_x = 2a_2 x, \qquad \varphi_y = 2a_2 y.$$

we obtain

$$\iint\limits_{B} [\varphi_x^2 + \varphi_y^2 - 2\varphi]\, dx\, dy = -8a_1 - \frac{16}{3} a_2 + \frac{32}{3} a_2^2,$$

$$\int\limits_{\Gamma} \varphi^2\, ds = 8a_1^2 + \frac{64}{3} a_1 a_2 + \frac{224}{15} a_2^2;$$

hence

$$J[\varphi] = 8\left[a_1^2 + \frac{8}{3} a_1 a_2 + \frac{16}{5} a_2^2 - a_1 - \frac{2}{3} a_2 \right]. \qquad (5.27)$$

Table V/13. *Results for the heat conduction problem obtained by Ritz's method*

Point	Two-parameter expression		Three-parameter expression	
	Approximate value	Error	Approximate value	Error
$x=0,\ y=0$	$\dfrac{13}{16} \approx 0{\cdot}8125$	$-1{\cdot}1\%$	$\dfrac{139}{168} \approx 0{\cdot}827\,38$	$+0{\cdot}7\%$
$x=0,\ y=1$	$\dfrac{37}{64} \approx 0{\cdot}5781$	$+3{\cdot}8\%$	$\dfrac{47}{84} \approx 0{\cdot}559\,52$	$+0{\cdot}4\%$
$x=1,\ y=1$	$\dfrac{11}{32} \approx 0{\cdot}3438$	$-9{\cdot}5\%$	$\dfrac{8}{21} \approx 0{\cdot}380\,95$	$+0{\cdot}25\%$

From the two extremum conditions

$$\frac{1}{8} \frac{\partial J}{\partial a_1} = 0 = 2a_1 + \frac{8}{3} a_2 - 1,$$

$$\frac{1}{8} \frac{\partial J}{\partial a_2} = 0 = \frac{8}{3} a_1 + \frac{32}{5} a_2 - \frac{2}{3}$$

we obtain $a_1 = \dfrac{13}{16}$, $a_2 = -\dfrac{15}{64}$, and hence the approximate solution

$$\varphi = \frac{1}{64} [52 - 15(x^2 + y^2)].$$

Several numerical values, together with their errors, are given in Table V/13.

For a better approximation we use the three-parameter expression

$$\varphi = a_1 + a_2(x^2 + y^2) + a_3 x^2 y^2.$$

J is now given by

$$\frac{1}{8} J[\varphi] = a_1^2 + \frac{8}{3} a_1 a_2 + \frac{16}{5} a_2^2 + \frac{2}{3} a_1 a_3 + \frac{88}{45} a_2 a_3 + \frac{7}{15} a_3^2 - a_1 - \frac{2}{3} a_2 - \frac{1}{9} a_3$$

and the extremum conditions

$$\frac{1}{8} \frac{\partial J}{\partial a_1} = 2\, a_1 + \frac{8}{3} a_2 + \frac{2}{3} a_3 - 1 = 0,$$

$$\frac{1}{8} \frac{\partial J}{\partial a_2} = \frac{8}{3} a_1 + \frac{32}{5} a_2 + \frac{88}{45} a_3 - \frac{2}{3} = 0,$$

$$\frac{1}{8} \frac{\partial J}{\partial a_3} = \frac{2}{3} a_1 + \frac{88}{45} a_2 + \frac{14}{15} a_3 - \frac{1}{9} = 0$$

yield

$$a_1 = \frac{139}{168}, \quad a_2 = -\frac{15}{56}, \quad a_3 = \frac{5}{56}.$$

The new approximate solution is

$$\varphi = \frac{1}{168} [139 - 45(x^2 + y^2) + 15 x^2 y^2];$$

numerical values at key points are compared with the first approximation in Table V/13.

In this example J has been worked out as a quadratic function of the a_ϱ for illustration; this is not essential, for the differentiations with respect to the parameters can be performed before the integral is evaluated, i.e. under the integral sign, and in fact the calculation can often be shortened thereby.

5.5. A differential equation of the fourth order

For the plate problem[1] of § 1.5

$$\nabla^4 u = \frac{p}{N} = x^2 \quad \text{in } B: |x| \leqq 1, \quad |y| \leqq 1, \tag{5.28}$$

$$u = u_x = 0 \quad \text{for } |x| = 1, \quad u = \nabla^2 u = 0 \quad \text{for } |y| = 1 \tag{5.29}$$

we consider an integral expression of the form[2]

$$J[\varphi] = \iint_B \left[(\nabla^2 \varphi)^2 - 2 \frac{p}{N} \varphi \right] dx\, dy. \tag{5.30}$$

To investigate the associated variational problem we proceed exactly as in §5.1, and can therefore be brief. We put $\varphi = u + \varepsilon \eta$ and form $\Phi(\varepsilon) = J[u + \varepsilon \eta]$; then (5.5), which here reads

$$\Phi'(0) = \left(\frac{\partial J}{\partial \varepsilon} \right)_{\varepsilon=0} = 2 \iint_B \left(\nabla^2 u\, \nabla^2 \eta - \frac{p}{N} \eta \right) dx\, dy = 0, \tag{5.31}$$

provides a necessary condition to be satisfied by u if it is to minimize $J[\varphi]$.

If we apply GREEN's formula [Ch. I (3.8)] with $\varphi = \nabla^2 u$, $\psi = \eta$, i.e.

$$\iint_B (\nabla^2 u\, \nabla^2 \eta - \eta\, \nabla^4 u)\, dx\, dy = \int_\Gamma \left(\eta \frac{\partial \nabla^2 u}{\partial \nu} - \nabla^2 u \frac{\partial \eta}{\partial \nu} \right) ds, \tag{5.32}$$

to the first term in the double integral of (5.31), the necessary extremum

---

[1] More general problems of stressed elastic bodies have been considered by WEBER; he obtains bounds for the displacements by making use of two minimum principles. C. WEBER: Eingrenzung von Verschiebungen mit Hilfe der Minimalsätze. Z. Angew. Math. Mech. **22**, 126—130 (1942). — Eingrenzung von Verschiebungen und Zerrungen mit Hilfe der Minimalsätze. Z. Angew. Math. Mech. **22**, 130—136 (1942).

[2] By way of further literature on variational problems for double integrals involving second derivatives we may mention A. R. FORSYTH: Calculus of variations, 656 pp. Cambridge 1927, in particular Ch. XI, pp. 567—600.

condition (5.31) becomes

$$\iint_B \eta \left(V^4 u - \frac{p}{N} \right) dx\,dy + \int_\Gamma \left(\eta \frac{\partial V^2 u}{\partial \nu} - V^2 u \frac{\partial \eta}{\partial \nu} \right) ds = 0. \quad (5.33)$$

The definitions of boundary curve Γ, arc length s and inward normal ν follow those in Ch. I, § 3.1.

As before, the factor multiplying η in the double integral must vanish, and hence the minimizing function u must satisfy the given differential equation. Thus the double integral vanishes. In order that the boundary integral shall also vanish we must demand that $\eta = 0$ everywhere on the boundary and that $\partial \eta / \partial \nu = 0$ on the parts of the boundary where the plate is clamped; the second boundary condition on the rest of the boundary, $V^2 \eta = 0$, may be suppressed, since the fact that we do not demand the vanishing of η_ν on this part of the boundary automatically ensures that the minimizing function u satisfies $V^2 u = 0$ here; we say that $V^2 u = 0$ is a "natural" boundary condition for the variational problem $J[\varphi] = \min$. Thus, as in the case of ordinary differential equations, the admissible functions φ must satisfy the essential boundary conditions, but need not satisfy the suppressible boundary conditions. Unless the latter are also natural boundary conditions, as in the present example, $J[\varphi]$ must be modified by the addition of suitable boundary integral expressions so as to obtain a variational problem for which the suppressible boundary conditions are natural, otherwise they will not be satisfied by the minimizing function u.

As a family of admissible functions for the present example we could use, for instance,

$$\varphi = (1 - x^2)^2 (1 - y^2) (a_1 + a_2 x^2 + a_3 y^2 + a_4 x^4 + a_5 x^2 y^2 + a_6 y^4);$$

these satisfy all the essential boundary conditions and also the appropriate symmetry conditions. To illustrate the method we will use the one-parameter family obtained by putting all parameters but a_1 equal to zero; the integral in (5.30) becomes

$$J[\varphi] = \frac{128}{35} \left(\frac{64}{5} a_1^2 - \frac{a_1}{9} \right), \quad (5.34)$$

and the condition $\partial J / \partial a_1 = 0$ leads to

$$a_1 = \varphi(0, 0) = \frac{5}{1152} \approx 0 \cdot 00435.$$

5.6. Direct proof of two minimum principles for a biharmonic boundary-value problem

Consider the boundary-value problem

$$V^4 u = p(x, y) \quad \text{in } B, \quad (5.35)$$

$$u = f(s), \quad u_\nu = g(s) \quad \text{on } \Gamma \quad (5.36)$$

with region B, boundary Γ, inward normal ν as in § 3.5, and assume the existence of a solution u.

Now let v and w be functions satisfying the differential equation and boundary conditions, respectively:

$$\nabla^4 v = p \text{ in } B; \quad w = f, \quad w_\nu = g \text{ on } \Gamma. \tag{5.37}$$

Then from (3.38) of Ch. I we have

$$D[v - w] = D[v - u] + D[u - w], \tag{5.38}$$

and hence

$$D[v - u] \leqq D[v - w] \tag{5.39}$$

and

$$D[u - w] \leqq D[v - w]. \tag{5.40}$$

Each of these two inequalities contains a minimum principle.

I. Consider first (5.39). On account of (5.38) we have equality if and only if $D[u - w] = 0$. Now from the definition of $D[\varphi]$, Ch. I (3.34), $D[\varphi] = 0$ implies that $\nabla^2 \varphi = 0$; further $\nabla^2 (u - w) = 0$ in B with $u - w = 0$ on the boundary Γ implies that $u - w \equiv 0$ in B. Consequently

$$D[v - u] = D[v - w] \quad \text{if and only if} \quad u = w.$$

With each side expanded as in (3.35) of Ch. I, (5.39) here reads

$$D[v] - 2D[u, v] + D[u] \leqq D[v] - 2D[v, w] + D[w]. \tag{5.41}$$

If we put $\varphi = v$, $\psi = u - w$ in (3.37) of Ch. I, the boundary integral vanishes on account of (5.37) and we have

$$D[u - w, v] = D[u, v] - D[w, v] = \iint_B (u - w) \, \nabla^4 v \, dx \, dy$$

$$= \iint_B (u - w) \, \nabla^4 u \, dx \, dy.$$

Hence (5.41) may be written

$$\iint_B [(\nabla^2 u)^2 - 2u \, \nabla^4 u] \, dx \, dy \leqq \iint_B [(\nabla^2 w)^2 - 2w \, \nabla^4 u] \, dx \, dy. \tag{5.42}$$

This result can be stated as the following

Theorem: *If w runs through the class of functions which possess continuous partial derivatives of the second order and satisfy the boundary conditions in (5.37), i.e. $w = f$, $w_\nu = g$ on Γ, then the integral*

$$J[w] = \iint_B [(\nabla^2 w)^2 - 2w \, p] \, dx \, dy \tag{5.43}$$

assumes its minimum value when and only when $w = u$, where u is the solution of the boundary-value problem (5.35), (5.36).

II. The treatment of the second inequality (5.40) is quite analogous. On account of (5.38) we have equality if and only if $D[u-v]=0$, i.e. $V^2[u-v]=0$. If we put $\varphi=u-v$, $\psi=w$ in (3.37) of Ch. I, the double integral on the right-hand side disappears and we have

$$D[u-v,w]=D[u,w]-D[v,w]=\int_\Gamma \left[w\{V^2(u-v)\}_\nu - w_\nu V^2(u-v)\right]ds.$$

The inequality (5.40), which can be written in the form

$$D[u]-2D[u,w]+D[w]\leqq D[v]-2D[v,w]+D[w],$$

therefore yields

$$\iint_B (V^2u)^2\,dx\,dy - 2\int_\Gamma \left[w(V^2u)_\nu - w_\nu V^2u\right]ds$$
$$\leqq \iint_B (V^2v)^2\,dx\,dy - 2\int_\Gamma \left[w(V^2v)_\nu - w_\nu V^2v\right]ds.$$

Thus we may state the

Theorem[1]. *If v runs through the class of functions which possess continuous partial derivatives of the fourth order and satisfy the differential equation $V^4v=p$ of (5.37), then the integral*

$$J[v]=\iint_B (V^2v)^2\,dx\,dy + 2\int_\Gamma \left[g\,V^2v - f(V^2v)_\nu\right]ds \qquad (5.44)$$

assumes a minimum value for all functions of the form $u+\varphi$, where u is the solution of the boundary-value problem (5.35), (5.36) and φ is any admissible function with $V^2\varphi=0$.

Example. We choose an example similar to one considered by WEGNER, (loc. cit.). For a uniformly loaded rectangular plate whose edges are clamped and have lengths in the ratio $1:2$ we have

$$V^4u=1 \text{ in } B,$$
$$u=\frac{\partial u}{\partial \nu}=0 \text{ on } \Gamma,$$

where Γ is the boundary of the rectangle B with $|x|\leqq 2$, $|y|\leqq 1$.

Let us assume for V^2v an expression of the form

$$V^2v=\frac{x^2+y^2}{4}+\sum_{\nu=0}^{p} c_\nu \varphi_\nu \quad \text{with} \quad \varphi_\nu=\mathrm{re}\,(x+iy)^{2\nu};$$

---

[1] Established for homogeneous boundary conditions $f=g=0$ by U. WEGNER: Ein neues Verfahren zur Berechnung der Spannungen in Scheiben. Forsch.-Arb. Ing.-Wes. **13**, 114—149 (1942), and for inhomogeneous boundary conditions by J. B. DIAZ and H. J. GREENBERG: Upper and lower bounds for the solution of the first biharmonic boundary-value problem. J. Math. Phys. **27**, 193—201 (1948). WEGNER also gives numerical examples.

then $\nabla^4 v = 1$ for arbitrary c_ν. Since the boundary conditions are homogeneous, the boundary integral in (5.44) does not appear, and the conditions $\partial J/\partial c_\nu = 0$ (for $\nu = 0, 1, \ldots, p$) read for $p = 1$

$$2c_0 + 2\ c_1 + \frac{5}{6} = 0,$$

$$2c_0 + \frac{226}{45} c_1 + \frac{3}{2} = 0,$$

with the solution

$$c_0 = -\frac{40}{204} \approx -0.196, \qquad c_1 = -\frac{45}{204} \approx -0.221;$$

for $p = 2$ the calculation yields

$$c_0 = -\frac{1850}{3 \times 3671} \approx -0.168, \qquad c_1 = -\frac{3945}{4 \times 3671} \approx -0.269,$$

$$c_2 = \frac{100}{3671} \approx 0.0272.$$

In this way we obtain directly an approximation for $\nabla^2 u$ instead of for u. Consequently the method is very suitable when the stress distribution is of most interest, but less so when the displacement has to be calculated.

5.7. More than two independent variables

We procede along the same lines[1] as in §§ 5.1—5.3, but because of the greater length and complexity of the formulae, we restrict ourselves to a simple case, namely that for which the differential equation is of the self-adjoint form (5.49); and further we present this simple case more concisely.

Corresponding to (5.1) we consider the variational problem

$$\left.\begin{aligned} J[\varphi] = \int_B \left(\sum_{i,k=1}^m A_{ik} \frac{\partial \varphi}{\partial x_i} \frac{\partial \varphi}{\partial x_k} + q\,\varphi^2 - 2r\,\varphi \right) d\tau + \\ + \int_\Gamma (K\,\varphi^2 + 2M\,\varphi)\,dS = \text{extremum}, \end{aligned}\right\} \tag{5.45}$$

where B is a given, finite, simply-connected, (say) closed region of the $(x_1, x_2, \ldots, x_m)$ space bounded by a closed surface Γ which is made up of a finite number of "faces", each with a continuously turning tangent plane; $d\tau = dx_1 dx_2 \ldots dx_m$ denotes, as in Ch. I, § 3.2, the volume element in B, and dS the surface element on Γ; A_{ik}, q, r and K, M are given continuous functions of position in B and on Γ, respectively; further the A_{ik} are symmetric ($A_{ki} = A_{ik}$) and possess continuous first partial derivatives. The domain of admissible functions will be restricted to

---

[1] See also R. WEINSTOCK: Calculus of variations, 326 pp. New York-Toronto-London 1952.

those functions $\varphi(x_1, x_2, \ldots, x_m)$ which are continuous and have continuous partial derivatives of up to and including the second order in B and which satisfy such boundary conditions on the boundary Γ as are found to be necessary. Again we assume that there is an admissible function $u(x_1, x_2, \ldots, x_m)$ for which J attains its minimum value, and then derive conditions which must be satisfied by this function. Such a condition is that for any family of admissible functions of the form $\varphi = u + \varepsilon\eta(x_1, x_2, \ldots, x_m)$, where ε is a parameter, the function $\Phi(\varepsilon) = J[u + \varepsilon\eta]$ must have a minimum value when $\varepsilon = 0$; hence the condition $\Phi'(0) = 0$ of (5.5) must be satisfied.

Here we have

$$\left.\begin{aligned}
\frac{1}{2}\,\Phi'(0) &= \frac{1}{2}\left\{\frac{\partial}{\partial\varepsilon}\,J[u + \varepsilon\eta]\right\}_{\varepsilon=0} \\
&= \int_B \left\{\sum_{i,k=1}^{m} A_{ik}\frac{\partial u}{\partial x_i}\frac{\partial\eta}{\partial x_k} + qu\eta - r\eta\right\}d\tau + \int_\Gamma \{Ku\eta + M\eta\}\,dS = 0.
\end{aligned}\right\} \quad (5.46)$$

Now according to the formulae (3.13) to (3.15) of Ch. I

$$\int_B \left\{\sum_{i,k=1}^{m} A_{ik}\frac{\partial u}{\partial x_i}\frac{\partial\eta}{\partial x_k} + qu\eta\right\}d\tau = \int_B \eta L[u]\,d\tau - \int_\Gamma \eta L^*[u]\,dS, \quad (5.47)$$

where $L[u]$ and $L^*[u]$ are defined, as in Ch. I (3.12), (3.15), by

$$L[u] = -\sum_{i,k=1}^{m} \frac{\partial}{\partial x_i}\left(A_{ik}\frac{\partial u}{\partial x_k}\right) + qu,$$

$$L^*[u] = \sum_{i,k=1}^{m} A_{ik}\frac{\partial u}{\partial x_k}\cos(\nu, x_i) = A\frac{\partial u}{\partial\sigma};$$

we can therefore rewrite (5.46) in the form

$$\int_B \eta\{L[u] - r\}\,d\tau + \int_\Gamma \eta\{Ku + M - L^*[u]\}\,dS = 0. \quad (5.48)$$

By the usual argument it can be deduced from the arbitrariness of η that the factor multiplying η in the space integral must vanish inside B; thus as the Euler differential equation we obtain

$$L[u] = -\sum_{i,k=1}^{m} \frac{\partial}{\partial x_i}\left(A_{ik}\frac{\partial u}{\partial x_k}\right) + qu = r. \quad (5.49)$$

The necessary condition (5.48) then reduces to

$$\int_\Gamma \eta\{Ku + M - L^*[u]\}\,dS = 0. \quad (5.50)$$

For further discussion we select two special cases (not all boundary conditions can be dealt with by this method anyway).

5.8. Special cases

Case I. First boundary-value problem. Here, since the value of u is prescribed on the boundary Γ, say $u = g$, we must demand that $\eta = 0$ on Γ, i.e. that all admissible functions φ satisfy the boundary condition $\varphi = g$ on Γ. Then the condition (5.50) is satisfied for arbitrary K and M; in particular, we can put $K = M = 0$, i.e. we can omit the boundary integral in (5.45).

Case II. Third boundary-value problem. Suppose that we require

$$L^*[u] + qu + w = 0 \quad \text{on } \Gamma,$$

where q, w are given functions of position. Here we minimize J for functions φ which are not prescribed on the boundary, so that the minimizing function u must satisfy $Ku + M - L^*[u] = 0$ on Γ. Thus if we choose $K = -q$, $M = -w$, the required boundary condition will be satisfied, and the admissible functions φ need not be restricted by any boundary conditions.

For the particular differential equations with

$$A_{ik} = \delta_{ik} p = \begin{cases} p & \text{for} \quad i = k \\ 0 & \text{for} \quad i \neq k, \end{cases}$$

where p is a given function of position in B, the above boundary condition reads

$$p u_\nu + q u + w = 0,$$

where u_ν is the derivative in the direction of the inward normal.

Example. For the three-dimensional problem

$$u_{xx} + u_{yy} + u_{zz} = 0,$$

$$\frac{\partial u}{\partial \nu} = 0 \quad \text{for} \quad |x| = 1, \quad |y| = 1, \quad |z| = 1.$$

$$u = \sigma \quad \text{for} \quad x = y = z = \sigma \quad \text{with} \quad \sigma = \pm 1$$

(temperature distribution in a cube C as in § 1.7) the volume integral in (5.45) reduces to the Dirichlet integral

$$J[\varphi] = \iiint_C (\text{grad } \varphi)^2 \, dx \, dy \, dz \tag{5.51}$$

and φ need only satisfy the boundary condition $\varphi = \sigma$ at the corners $x = y = z = \sigma$ with $\sigma = \pm 1$. The simplest family of admissible functions which also satisfy the symmetries of the problem may be defined by

$$\varphi = a(x + y + z) + (1 - 3a) x y z.$$

Using this expression we have

$$\varphi_x = a + (1 - 3a) y z,$$
$$\tfrac{1}{8} J[\varphi] = 6a^2 - 2a + \tfrac{1}{3},$$

and from $\partial J/\partial a = 0$ we obtain $a = \tfrac{1}{6}$; the approximate solution therefore reads

$$\varphi = \tfrac{1}{6}(x + y + z) + \tfrac{1}{2} x y z.$$

5.9. The mixed Ritz expression

As in § 5.7, let $J[\varphi]$ be an integral expression which is minimized with respect to a certain domain of admissible functions φ by the function u which is the solution of a given (say) linear boundary-value problem with the differential equation

$$L[x_1, x_2, \ldots, x_n, u, u_{x_1}, u_{x_2}, \ldots] = 0$$

and the boundary conditions $U_\varkappa[u] = \gamma_\varkappa$ $(\varkappa = 1, \ldots, k)$. The admissible functions φ may have to satisfy certain boundary conditions (cf. § 5.7, for example), which we will denote by

$$V_\mu[\varphi] = v_\mu \qquad (\mu = 1, \ldots, m). \tag{5.52}$$

The basis of the ordinary Ritz method is the minimization of J with respect to a p-parameter family of admissible functions defined by an expression of the form

$$\varphi(x_\varrho) = w_0(x_\varrho) + \sum_{\nu=1}^{p} a_\nu w_\nu(x_\varrho), \tag{5.53}$$

where w_0 satisfies the inhomogeneous boundary conditions $V_\mu[w_0] = v_\mu$ and the w_ν for $1 \leq \nu \leq p$ satisfy the corresponding homogeneous conditions $V_\mu[w_\nu] = 0$. Now if there is an independent variable, x_n say, derivatives with respect to which do not occur in the boundary conditions, then φ, as defined in (5.53), would still satisfy the boundary conditions even if the a_ν were allowed to depend on x_n, and were thus regarded as functions of x_n to be determined[1]. This defines a wider class of admissible functions, which lies somewhere between the p-parameter family (5.53) (an ordinary Ritz expression, leading to a system of linear equations for the a_ν) and the class of all admissible functions [with the a_ν depending on all the x_ϱ, so that the sum in (5.53) can be replaced by a single unknown function $a(x_\varrho)$; this leads to the Euler differential equation of the variational problem].

If we insert this "mixed" expression, in which the a_ν depend on a single variable x_n, the integral expression J becomes a functional of the p functions $a_\nu(x_n)$; we write

$$J = \tilde{J}[a_\nu(x_n)].$$

---

[1] KANTOROVICH, L.: Sur une méthode directe de la solution approximative du problème du minimum d'un intégral double [Russian]. Leningrad Bull. Ac. Sci. 7, 647—652 (1933) (Jb. Fortschr. Math. 59, 1149). The method has been used by HILLEL PORITZKY: The reduction of the solution of certain partial differential equations to ordinary differential equations. Trans. 5th Intern. Congr. Appl. Mech., Cambridge (Mass.) 1938, pp. 700—707, by N. S. SEMENOW: Biegung von Rechtecksplatten. Zbl. Mech. 11, 12 (1939) and also by E. METTLER: Allgemeine Theorie der Stabilität erzwungener Schwingungen elastischer Körper. Ing.-Arch. 17, 418—449 (1949), in particular p. 420 et seq. and p. 445 et seq.

The requirement that $\tilde{J}$ be minimized leads to a system of Euler equations consisting of p ordinary differential equations for the p unknown functions $a_\nu(x_n)$.

The method is also applicable to initial-/boundary-value problems (Ch. IV); if the time t is one of the independent variables, it can be used as x_n, and sometimes one can then draw conclusions about the behaviour of the solution as t increases indefinitely[1].

Examples. I. Consider the first boundary-value problem

$$\nabla^2 u = r(x,y,z) \quad \text{in } B,$$

$$u = 0 \quad \text{on the boundary } \Gamma \text{ of } B.$$

According to § 5.3 the corresponding variational problem reads

$$J[\varphi] = \int_B \{(\operatorname{grad}\varphi)^2 + 2r\varphi\}\, d\tau = \min.$$

Let $\varphi = a(x)\, w(x,y,z)$, where w (or the product aw if w does not depend on x) vanishes on the boundary Γ. Then $J[\varphi] = J[a(x)w] = \tilde{J}[a]$. We derive an Euler differential equation for the minimizing function $A(x)$ by the usual procedure of putting $a(x) = A(x) + \varepsilon\eta(x)$. We must have

$$\frac{1}{2}\left(\frac{\partial \tilde{J}[A + \varepsilon\eta]}{\partial \varepsilon}\right)_{\varepsilon=0} = \int_B \{(\operatorname{grad}(A\,w),\ \operatorname{grad}(\eta\,w)) + r\eta\,w\}\, d\tau = 0.$$

By using formula (3.17) of Ch. I we can separate out η as a factor:

$$\int_B \eta\,w\{-\nabla^2(A\,w) + r\}\, d\tau - \int_\Gamma \eta\,w\,\frac{\partial(A\,w)}{\partial\nu}\, dS = 0.$$

Now w (or the product aw), and hence also ηw, vanishes on the boundary, so that the boundary integral is zero. The remaining volume integral can be written as a repeated integral of the form

$$\int \eta(x)\left[\iint_{B^*(x)} w\{-\nabla^2(A(x)w) + r\}\, dy\, dz\right] dx,$$

where $B^*(x)$ is the cross-section of B in the plane $x = \text{constant}$. We see that, on account of the arbitrariness of $\eta(x)$, the double integral over the cross-section $B^*(x)$ must vanish for each x in B. This yields an ordinary differential equation of the second order for $A(x)$:

$$A'' \iint_{B^*} w^2\, dy\, dz + 2A' \iint_{B^*} w\,\frac{\partial w}{\partial x}\, dy\, dz + A \iint_{B^*} w\nabla^2 w\, dy\, dz - \iint_{B^*} r\,w\, dy\, dz = 0,$$

where dashes denote derivatives with respect to x.

In the general formulation with $w = 0$ everywhere on Γ the boundary conditions for $A(x)$ are that it remains finite at the end-points, and the uniqueness depends on the fact that the differential equation is singular at the end-points; it may be more convenient, however, to choose w so that A must vanish at the end-points. With $A(x)$ so determined $A(x)w(x,y,z)$ is the desired approximate solution.

---

[1] An example is given by F. WEIDENHAMMER: Der eingespannte, axial pulsierend belastete Stab als Stabilitätsproblem. Z. Angew. Math. Mech. **30**, 235–237 (1950).

II. Example in which the fundamental region extends to infinity:

$$\nabla^2 u = 0 \text{ in } B; \quad u(x, \pm 1) = 0, \quad u(0, y) = 1 - y^2,$$

where B is the region $x \geq 0$, $|y| \leq 1$. Here we have to minimize the Dirichlet integral:

$$J = \iint_B (\varphi_x^2 + \varphi_y^2) \, dx \, dy = \min.$$

We first put $\varphi(x, y) = (1 - y^2) f(x)$; then with dashes denoting differentiation with respect to x the new functional $J = \tilde{J}[f]$ is given by

$$\frac{1}{2} \tilde{J}[f] = \int_0^\infty \left(\frac{8}{15} f'^2 + \frac{4}{3} f^2 \right) dx.$$

If $f = F(x)$ minimizes this integral, we have by the usual linear variation method

$$0 = \frac{1}{4} \left(\frac{\partial \tilde{J}[F + \varepsilon \eta]}{\partial \varepsilon} \right)_{\varepsilon = 0} = \int_0^\infty \left(\frac{8}{15} F' \eta' + \frac{4}{3} F \eta \right) dx$$

$$= \int_0^\infty \eta \left(-\frac{8}{15} F'' + \frac{4}{3} F \right) dx + \left[\frac{8}{15} \eta F' \right]_0^\infty.$$

By the usual arguments depending on the arbitrariness of η the factor multiplying η in the integral must vanish, and also we must have $\eta(0) = F'(\infty) = 0$. Thus $F(x)$ must satisfy the Euler equation

Table V/14. *Approximations for* $u(x, 0)$

x	One-term expression	Two-term expression
0·25	0·6735	0·687 44
0·5	0·4536	0·467 63
0·75	0·3055	0·316 73
1	0·2057	0·214 14

$$-\frac{8}{15} F'' + \frac{4}{3} F = 0$$

with the boundary conditions $F(0) = 1$, $F'(\infty) = 0$. Hence $F(x) = e^{-kx}$, where $k = \sqrt{2 \cdot 5}$, and

$$\varphi(x, y) = (1 - y^2) e^{-\sqrt{2 \cdot 5} \, x} .$$

Table V/14 gives a few sample values for comparison with the next approximation.

A two-term expression

$$\varphi(x, y) = (1 - y^2) f_1(x) + (1 - y^2)^2 f_2(x)$$

leads in exactly the same way to the pair of simultaneous ordinary differential equations

$$-\frac{8}{15} F_1'' - \frac{16}{35} F_2'' + \frac{4}{3} F_1 + \frac{16}{15} F_2 = 0,$$

$$-\frac{16}{35} F_1'' - \frac{128}{315} F_2'' + \frac{16}{15} F_1 + \frac{128}{105} F_2 = 0$$

with the boundary conditions $F_1(0) = 1$, $F_2(0) = 0$, $F_1'(\infty) = F_2'(\infty) = 0$. It follows that

$$F_1(x) = a\,e^{k_1 x} + (1 - a)\,e^{k_2 x}$$
$$F_2(x) = b\,e^{k_1 x} - b\,e^{k_2 x},$$

where k_1 and k_2 are the negative roots of $k^4 - 28k^2 + 63 = 0$; if we put $k^2 = \varrho$, we have

$$\varrho_{1,2} = 14 \pm \sqrt{133} = \begin{cases} 25 \cdot 533 \\ 2 \cdot 4674 \end{cases}$$

and $a = \dfrac{7 - \varrho_2}{2\sqrt{133}}$, $b = \dfrac{7 - \varrho_1}{16}\,a$. Some values of the approximation function when $y = 0$ are given in Table V/14.

§6. The Trefftz[1] method

While the Ritz method is, in the sense of Ch. I, §4.1, an interior method, the Trefftz method[2] belongs to the category of boundary methods, and therefore possesses over the Ritz method the advantages already mentioned in Ch. I, §4.1. Moreover, error estimation by means of the boundary-maximum theorem (§3) is in many ways more simple for the Trefftz method than for the Ritz method; see §6.3 and the example in §6.5.

6.1. Derivation of the Trefftz equations

Consider the first boundary-value problem with the differential equation

$$L[u] = -\frac{\partial}{\partial x}(A\,u_x + B\,u_y) - \frac{\partial}{\partial y}(B\,u_x + C\,u_y) + F\,u = r(x, y) \text{ in } B \quad (6.1)$$

---

[1] ERICH TREFFTZ, born 21 February 1888, son of a Leipzig merchant, studied in Aachen, Göttingen and Strasbourg. He recieved much stimulation from his uncle, CARL RUNGE, whom he assisted at Göttingen and accompanied to New York when RUNGE went there as exchange professor in the Columbia University. He obtained his doctor's degree at Strasbourg in 1913 with work instigated by v. MISES. After being wounded in the first World War he went to Aachen, where he became ordinary professor of applied mathematics in 1919. In 1922 he accepted a professorship in applied mechanics at Dresden, a post which he held until his death from a malignant disease on the 21 January 1937. In his obituary [Z. Angew. Math. Mech. 17, 1 (1937)] L. PRANDTL wrote of him: "His domestic happiness, the pleasure he took in his work, the devotion of his students and the affection of his friends made him a happy man." His work in the field of mechanics was concerned chiefly with hydrodynamics, the theory of vibrations and elasticity [see R. GRAMMEL: Das wissenschaftliche Werk von Erich Trefftz. Z. Angew. Math. Mech. 18, 1—11 (1938)].

[2] TREFFTZ, E.: Ein Gegenstück zum Ritzschen Verfahren. Verh. Kongr. Techn. Mech., Zürich 1926, pp. 131—137. — Konvergenz und Fehlerschätzung beim Ritzschen Verfahren. Math. Ann. 100, 503—521 (1928). In these papers the method is presented for the potential equation and the biharmonic equation; we shall restrict ourselves to second-order equations but of the more general type (6.1) and (6.21).

and the boundary condition

$$u = g(s) \text{ on } \Gamma, \tag{6.2}$$

where s is the arc length along the boundary Γ of a region B as in § 5.1, F, r, g are given continuous functions, A, B, C are given functions with continuous first partial derivatives and $A > 0, C > 0, AC - B^2 > 0, F \geqq 0$.

Now suppose that we know a particular solution $\overline{w}$ of the inhomogeneous equation (6.1) and a number of linearly independent solutions $w_1, w_2, \ldots, w_m$ of the corresponding homogeneous equation:

$$L[\overline{w}] = r, \quad L[w_1] = L[w_2] = \cdots = L[w_m] = 0. \tag{6.3}$$

Then

$$W = \overline{w} + \sum_{\sigma=1}^{m} a_\sigma w_\sigma \tag{6.4}$$

is also a solution of (6.1), and the a_σ are at our disposal for making W approximate the solution u of the boundary-value problem as closely as possible.

The obvious way of choosing the a_σ would be to employ one of the error distribution principles of Ch. I, § 4.2; by the least-squares method, for example, we would demand that

$$J^* = \int_\Gamma [W(s) - g(s)]^2 ds = \min., \tag{6.5}$$

which on the insertion of the expression (6.4) for W yields the following symmetric system of linear equations for the a_ϱ:

$$\sum_{\varrho=1}^{m} a_\varrho \int_\Gamma w_\varrho w_\sigma ds = \int_\Gamma (g - \overline{w}) w_\sigma ds \quad (\sigma = 1, 2, \ldots, m). \tag{6.6}$$

TREFFTZ's method of determining the a_ϱ, however, is based on quite a different principle and makes further use of the differential equation. As with RITZ's method, we define the integral expression

$$J[\varphi, \psi] = \iint_B [A\, \varphi_x \psi_x + B(\varphi_x \psi_y + \varphi_y \psi_x) + C\, \varphi_y \psi_y + F\, \varphi \psi] dx\, dy \tag{6.7}$$

and put

$$J[\varphi, \varphi] = J[\varphi]. \tag{6.8}$$

As addition law we have

$$J[\varphi + \psi] = J[\varphi] + 2J[\varphi, \psi] + J[\psi]. \tag{6.9}$$

Now $J[\varphi]$ is a measure of the total deviation of the function φ from a constant (from zero in cases where $F > 0$ somewhere in B); for $J[\varphi] \geqq 0$, and $J[\varphi] = 0$ if and only if ψ_x, φ_y and $F\varphi^2$ vanish identically in B. TREFFTZ's method is to demand that the error ε in the linear combination (6.4)

$$\varepsilon = W - u = \overline{w} + \sum_{\sigma=1}^{m} a_\sigma w_\sigma - u \tag{6.10}$$

shall be as small as possible over the region B when assessed in terms of the measure J, i.e.

$$J[\varepsilon] = J\left[\overline{w} + \sum_{\sigma=1}^{m} a_\sigma w_\sigma - u\right] = \min.$$ (6.11)

In the cases when J measures the deviation from a constant which need not necessarily be zero, for example, when L is the Laplace operator (with $A = C = 1$, $B = F = 0$), this minimization leaves an additive constant free in W, which has to be determined by some other means; see the example in § 6.5.

The necessary minimum conditions

$$\frac{\partial J\left[\overline{w} + \sum_{\varrho=1}^{m} a_\varrho w_\varrho - u\right]}{\partial a_\sigma} = 2J\left[\overline{w} + \sum_{\varrho=1}^{m} a_\varrho w_\varrho - u, w_\sigma\right] = 0$$ (6.12)

$$(\sigma = 1, 2, \ldots, m)$$

yield a system of linear equations for the a_ϱ whose coefficients are double integrals over the region B. These integrals may be transformed into boundary integrals by means of GREEN's formula (3.4) of Ch. I. With $\varphi = \overline{w} + \sum_{\varrho=1}^{m} a_\varrho w_\varrho - u$, $\psi = w_\sigma$ this formula reads

$$J\left[\overline{w} + \sum_\varrho a_\varrho w_\varrho - u, w_\sigma\right] - \iint_B \left(\overline{w} + \sum_\varrho a_\varrho w_\varrho - u\right) L[w_\sigma]\, dx\, dy +$$

$$+ \int_\Gamma \left(\overline{w} + \sum_\varrho a_\varrho w_\varrho - u\right) L^*[w_\sigma]\, ds = 0 \qquad (\sigma = 1, 2, \ldots, m).$$

Here the first term vanishes by virtue of (6.12) and the second by virtue of (6.3); only the boundary integral remains, and since $u(s) = g(s)$ is known on the boundary, we have

$$\sum_{\varrho=1}^{m} a_\varrho \int_\Gamma w_\varrho L^*[w_\sigma]\, ds = \int_\Gamma (g - \overline{w}) L^*[w_\sigma]\, ds \qquad (\sigma = 1, 2, \ldots, m).$$ (6.13)

These are TREFFTZ's equations.

6.2. A maximum property

$\overline{w}$, chosen as a solution of the inhomogeneous differential equation (6.1), is a fixed function independent of the a_ϱ; consequently so also is $\overline{w} - u$. If the a_ϱ could be chosen ideally we would have $J\left[\sum_{\varrho=1}^{m} a_\varrho w_\varrho\right] = J[\overline{w} - u]$; we show that by TREFFTZ's method $J\left[\sum_{\varrho=1}^{m} a_\varrho w_\varrho\right]$ approximates

$J[\overline{w} - u]$ from below. From (6.10) and (6.9) we have

$$J\left[\sum_\varrho a_\varrho w_\varrho\right] = J[\varepsilon + u - \overline{w}] = J[\overline{w} - u - \varepsilon]$$
$$= J[\overline{w} - u] - 2J[\overline{w} - u, \varepsilon] + J[\varepsilon].$$

Now $\overline{w} - u = \varepsilon - \sum_\varrho a_\varrho w_\varrho$, so that, since $J[\varphi, \psi]$ is a bilinear functional,

$$J[\overline{w} - u, \varepsilon] = J[\varepsilon] - \sum_\varrho a_\varrho J[w_\varrho, \varepsilon];$$

but from (6.12) $J[w_\varrho, \varepsilon] = J[\varepsilon, w_\varrho] = 0$, so that

$$J[\overline{w} - u, \varepsilon] = J[\varepsilon].$$

Consequently

$$J\left[\sum_\varrho a_\varrho w_\varrho\right] = J[\overline{w} - u] - J[\varepsilon], \tag{6.14}$$

and since $J[\varepsilon]$ is non-negative,

$$J\left[\sum_{\varrho=1}^m a_\varrho w_\varrho\right] \leq J[\overline{w} - u]. \tag{6.15}$$

If RITZ's method is used to find an approximation for $\overline{w} - u$, the integral expression $J[\overline{w} - u]$ is approximated from above; for $\zeta = \overline{w} - u$ is defined by the homogeneous differential equation $L[\zeta] = 0$ and the known boundary values

$$\zeta(s) = \overline{w}(s) - g(s),$$

and is the solution of the variational problem

$$J[\varphi] = \text{min.}, \qquad \varphi = \overline{w}(s) - g(s) \quad \text{on } \varGamma$$

(assuming, as in § 5.1, that a unique solution of this problem exists). Thus by using a Ritz expression

$$\zeta \approx \vartheta = \vartheta_0 + \sum_{\varrho=1}^p b_\varrho \vartheta_\varrho,$$

where $\vartheta_0(s) = \overline{w}(s) - g(s)$, $\vartheta_\varrho(s) = 0$ $(1 \leq \varrho \leq p)$ on the boundary, but otherwise the ϑ_ϱ are chosen arbitrarily from the class of continuous functions with continuous partial derivatives, we can determine from the necessary minimum conditions $\dfrac{\partial J[\vartheta]}{\partial b_\varrho} = 0$ $(\varrho = 1, \ldots, p)$ an approximation ϑ such that $J[\vartheta] \geq J[\zeta]$. By using both TREFFTZ's and RITZ's method we can therefore bracket[1] the value of the integral $J[\overline{w} - u]$:

$$J\left[\sum_{\varrho=1}^m a_\varrho w_\varrho\right] \leq J[\overline{w} - u] \leq J[\vartheta]. \tag{6.16}$$

---

[1] Another estimate is given by J. B. DIAZ: On the estimation of torsional rigidity and other physical quantities. Proc. 1st Nat. Congr. Appl. Mech. 1953, pp. 259–263, and a further estimate in connection with the finite-difference method by G. PÓLYA: C. R. Acad. Sci., Paris 235, 995–997 (1952).

For many problems — the torsion problem[1] for a shaft of constant cross-section for example, — the value of this integral can be of greater interest than values of the dependent variable.

6.3. Special case of the potential equation

For the first boundary-value problem for the inhomogeneous potential equation

$$- V^2 u = r(x, y) \text{ in } B, \qquad u = g(s) \text{ on } \Gamma \qquad (6.17)$$

the differential expression $L[u]$ of (6.1) has

$$A \equiv C \equiv 1, \qquad B \equiv F \equiv 0$$

and the boundary expression $L^*[u]$ reduces to

$$L^*[\psi] = \frac{\partial \psi}{\partial \nu}.$$

If $\overline{w}$ and $w_1, \ldots, w_m$ are solutions of the inhomogeneous and homogeneous differential equation, respectively, i.e.

$$- V^2 \overline{w} = r; \qquad V^2 w_1 = V^2 w_2 = \cdots = V^2 w_m = 0,$$

the unknowns a_ϱ in the approximate solution

$$u \approx \overline{w} + \sum_{\sigma=1}^{m} a_\sigma w_\sigma$$

are determined in TREFFTZ's method from the equations

$$\sum_{\varrho=1}^{m} a_\varrho \int_\Gamma w_\varrho \frac{\partial w_\sigma}{\partial \nu} ds = \int_\Gamma (g - \overline{w}) \frac{\partial w_\sigma}{\partial \nu} ds \qquad (\sigma = 1, 2, \ldots, m). \quad (6.18)$$

In this special case the integral expressions (6.7), (6.8) read

$$J[\varphi, \psi] = \iint_B (\varphi_x \psi_x + \varphi_y \psi_y) \, dx \, dy = \iint_B \operatorname{grad} \varphi \operatorname{grad} \psi \, dx \, dy,$$

$$J[\varphi] = \iint_B (\varphi_x^2 + \varphi_y^2) \, dx \, dy = \iint_B \operatorname{grad}^2 \varphi \, dx \, dy,$$

and we see that TREFFTZ's method requires that the mean square of the gradient of the error function, grad ε, where

$$\varepsilon = \overline{w} + \sum_{\sigma=1}^{m} a_\sigma w_\sigma - u, \qquad (6.19)$$

shall be as small as possible.

---

[1] The bracketing of the dependent variable is considered by C. WEBER, who makes use of auxiliary problems [Z. Angew. Math. Mech. 22, 126—136 (1942)].

Error estimate. With the a_ϱ calculated from (6.18) inserted in (6.19) we form the error function $\varepsilon(s)$ along the boundary Γ. If $\varepsilon_{\min}$ and $\varepsilon_{\max}$ are the smallest and largest values of ε on the boundary Γ, then from the boundary-maximum theorem of § 3.1 we know that ε lies within these limits also in B:

$$\varepsilon_{\min} \leq \overline{w} + \sum_{\sigma=1}^{m} a_\sigma w_\sigma - u \leq \varepsilon_{\max}. \tag{6.20}$$

6.4. More than two independent variables

As with RITZ's method, we can also extend TREFFTZ's method to self-adjoint equations of the form (5.49) in higher dimensions:

$$L[u] = - \sum_{i,k=1}^{m} \frac{\partial}{\partial x_i} \left(A_{ik} \frac{\partial u}{\partial x_k} \right) + q u = r \tag{6.21}$$

with $A_{ik} = A_{ki}$, but here we must also demand that the quadratic form

$$Q = \sum_{i,k=1}^{m} A_{ik} z_i z_k \tag{6.22}$$

shall be positive definite for all x_r in B and that $q \geq 0$. Apart from this, the assumptions we make concerning the region B, boundary Γ and functions A_{ik}, q, r are the same as in § 5.7. For the present we restrict ourselves to the first boundary-value problem with $u = g$ prescribed on the boundary Γ.

The derivation of the Trefftz equations follows exactly the same lines as in § 6.1. From a particular solution $\overline{w}$ of the inhomogeneous differential equation (6.21) and p linearly independent solutions $w_1, \ldots, w_p$ of the corresponding homogeneous equation we form the function

$$W = \overline{w} + \sum_{\varrho=1}^{p} a_\varrho w_\varrho, \tag{6.23}$$

which satisfies the inhomogeneous equation (6.21) for arbitrary a_ϱ.

Now we introduce the integral expression [(3.14) of Ch. I]

$$J[\varphi, \psi] = \int_B \left\{ \sum_{i,k=1}^{m} A_{ik} \frac{\partial \varphi}{\partial x_i} \frac{\partial \psi}{\partial x_k} + q \varphi \psi \right\} d\tau \tag{6.24}$$

and put

$$J[\varphi] - J[\varphi, \varphi]. \tag{6.25}$$

From our assumptions concerning the coefficients A_{ik} and q we have $J[\varphi] \geq 0$, and $J[\varphi] = 0$ only if $\partial\varphi/\partial x_i \equiv 0$, i.e. $\varphi = \text{constant}$ (and only if $\varphi \equiv 0$ when $q > 0$ somewhere in B). Thus, as before, we can use $J[\varepsilon]$

as a measure of the deviation of the error function

$$\varepsilon = W - u = \bar{w} + \sum_{\varrho=1}^{p} a_{\varrho} w_{\varrho} - u \qquad (6.26)$$

from a constant (which may be zero) and demand that this measure shall be as small as possible:

$$J[\varepsilon] = J\left[\bar{w} + \sum_{\varrho=1}^{p} a_{\varrho} w_{\varrho} - u\right] = \min. \qquad (6.27)$$

This yields the necessary conditions

$$\frac{\partial J}{\partial a_{\mu}} = 2J\left[\bar{w} + \sum_{\varrho=1}^{p} a_{\varrho} w_{\varrho} - u, w_{\mu}\right] = 2J[\varepsilon, w_{\mu}] = 0 \quad (\mu = 1, \ldots, p), \quad (6.28)$$

which may be transformed into

$$\int_{B} \varepsilon L[w_{\mu}] d\tau - \int_{\Gamma} \varepsilon L^{*}[w_{\mu}] dS = 0 \qquad (\mu = 1, \ldots, p) \qquad (6.29)$$

by means of the formulae (3.13) to (3.15) of Ch. I. Since $L[w_{\mu}] = 0$, we obtain the system of linear equations

$$\sum_{\varrho=1}^{p} a_{\varrho} \int_{\Gamma} w_{\varrho} L^{*}[w_{\mu}] dS = \int_{\Gamma} (g - \bar{w}) L^{*}[w_{\mu}] dS \qquad (\mu = 1, \ldots, p), \quad (6.30)$$

from which the a_{ϱ} are to be determined.

If $q = 0$ and the matrix A_{ik} in (6.21) is the unit matrix (δ_{ik}), we have $L[u] = -\nabla^{2} u$ and $L^{*}[\psi] = \partial\psi/\partial\nu$.

The argument in § 6.2 leading to the result (6.15) applies equally well for the higher-dimensional measure J of (6.24), (6.25), so that here also the Trefftz approximation function is such that

$$J\left[\sum_{\varrho=1}^{p} a_{\varrho} w_{\varrho}\right] \leqq J[\bar{w} - u]. \qquad (6.31)$$

6.5. Example

The torsion problem for a shaft of square cross-section leads to the boundary-value problem

$$\nabla^{2} u = -1 \quad \text{for} \quad |x| \leqq 1, |y| \leqq 1,$$
$$u = 0 \quad \text{for} \quad |x| = 1 \quad \text{and for} \quad |y| = 1.$$

Here solutions $\bar{w}, w_{\mu}$ satisfying the inhomogeneous and homogeneous differential equation, respectively, and also satisfying the symmetries of the problem, are

$$\bar{w} = -\tfrac{1}{4}(x^2 + y^2)$$
$$w_1 = x^4 - 6x^2 y^2 + y^4 = \mathrm{re}\,(x + i\,y)^4,$$
$$w_2 = x^8 - 28x^6 y^2 + 70x^4 y^4 - 28x^2 y^6 + y^8 = \mathrm{re}\,(x + i\,y)^8.$$

The function $w = $ constant, which also satisfies the homogeneous equation, may be disregarded for the present (cf. the remarks in § 6.1 about $J[\varphi]$ measuring the deviation from a constant). The equations (6.18) for determining the constants a_μ can be written down immediately; because of the symmetries we need only take the boundary integral over a half-side, say $x = 1$, $0 \leq y \leq 1$, and the equations for $p = 2$ read

$$\sum_{\varrho=1}^{2} a_\varrho \int_0^1 \left(w_\varrho \frac{\partial w_\mu}{\partial x}\right)_{x=1} dy = -\int_0^1 \left(\bar{w}\,\frac{\partial w_\mu}{\partial x}\right)_{x=1} dy \qquad (\mu = 1, 2).$$

With the integrals evaluated, for example,

$$\int_0^1 \left(\bar{w}\,\frac{\partial w_1}{\partial x}\right)_{x=1} dy = -\int_0^1 (1 + y^2)(1 - 3y^2)\,dy = \frac{4}{15},$$

they read

$$\frac{48}{7}a_1 - \frac{1664}{99}a_2 + \frac{1}{3} = 0,$$

$$-208a_1 + \frac{15232}{13}a_2 - 11 = 0,$$

from which we obtain

$$a_1 = -\frac{3619}{79856},$$

$$a_2 = \frac{429}{319424}.$$

As mentioned previously, we have a disposable additive constant, so the Trefftz approximation can be written

$$w = C - \frac{1}{4}(x^2 + y^2) - \frac{14476}{319424}(x^4 - 6x^2 y^2 + y^4) +$$

$$+ \frac{429}{319424}(x^8 - 28x^6 y^2 + 70x^4 y^4 - 28x^2 y^6 + y^8);$$

C will be determined so that the boundary values of w approximate the zero boundary value as closely as possible. On the boundary we have

$$(w)_{x=1} - C = -\gamma + f(y), \qquad \text{where} \qquad \gamma = \frac{93903}{319424}$$

and

$$319424 f(y) = -5012 y^2 + 15554 y^4 - 12012 y^6 + 429 y^8,$$

so that to determine C we need to find the variation σ in the function $f(y)$ over the interval $0 \leq y \leq 1$; for then, with $C = \gamma + \frac{1}{2}\sigma - f_{max}$, the boundary values of w will lie between $\frac{1}{2}\sigma$ and $-\frac{1}{2}\sigma$. With the aid of the graph in Fig. V/28 we find that $\sigma = \dfrac{1150}{319424}$, and hence $C = 0.295\,434$.

According to the error estimate of §6.3, the absolute error in w cannot exceed $\frac{1}{2}\sigma$ any-where in the square. In particular, for $u(0, 0)$ we have the strict limits

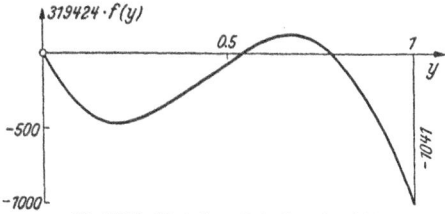

$$C - \frac{\sigma}{2} = 0.2936 \leq u(0, 0)$$

$$\leq 0.2972 = C + \frac{\sigma}{2}.$$

Fig. V/28. Variation of the function $f(y)$.

6.6. Generalization to the second and third boundary-value problems

Here we describe a different way[1] of deriving the Trefftz equations (6.30) which is applicable to the second and third boundary-value problems also. Consider again the differential equation (6.21) but let the boundary condition be of the more general type

$$A_1 u + A_2 L^*[u] = A_3 \tag{6.32}$$

on the boundary Γ, where A_1, A_2, A_3 are given functions of position on Γ and $L^*[u] = A \dfrac{\partial u}{\partial \sigma}$ is as defined by (3.15) of Ch. I.

As in §6.1, we determine an approximation of the form

$$u \approx W = \overline{w} + \sum_{\varrho=1}^{p} a_\varrho w_\varrho,$$

where $\overline{w}$ and $w_1, \ldots, w_p$ are solutions of the inhomogeneous and homogeneous differential equation, respectively, as in (6.3), but here we apply GREEN's theorem in the form (3.16) of Ch. I directly to the functions $\varphi = u - \overline{w}$ and $\psi = w_\varrho$. Since $L[u - \overline{w}] = L[w_\varrho] = 0$, the volume integrals disappear and we have

$$\int_{\Gamma} \{(u - \overline{w}) L^*[w_\varrho] - w_\varrho L^*[u - \overline{w}]\} \, dS = 0 \qquad (\varrho = 1, \ldots, p). \tag{6.33}$$

---

[1] A similar approach is considered by M. PICONE: Sur le calcul de la déformation d'un solide élastique encastré. 7th Internat. Congr. Appl. Mech., London 1948, and G. FICHERA: Risultati concernenti la risoluzioni delle equazioni funzionali lineari dovuti all' Insituto Nazionale per le applicazioni del calcolo. Mem. Accad. Naz. Lincei, Sci. Fis. Math., Ser. VIII 3, Sez. I, Fasc. 1, Rome 1950. — See also J. ALBRECHT: Eine einheitliche Herleitung der Gleichungen von TREFFTZ und GALERKIN. Z. Angew. Math. Mech. 35, 193—195 (1955).

First of all let us consider the first boundary-value problem again. Here $u - \overline{w}$ is known on the boundary, but not $L^*[u - \overline{w}]$. We replace $u - \overline{w}$ in $L^*[u - \overline{w}]$ by the approximation $\sum\limits_{\varrho=1}^{p} a_\varrho w_\varrho$ and obtain thereby a system of p linear equations for the a_ϱ:

$$\int\limits_{\Gamma} (u - \overline{w}) \, L^*[w_\varrho] \, dS - \sum\limits_{\tau=1}^{p} a_\tau \int\limits_{\Gamma} w_\varrho L^*[w_\tau] \, dS = 0 \qquad (\varrho = 1, \ldots, p).$$

On account of the relation

$$\int\limits_{\Gamma} \{w_\varrho L^*[w_\tau] - w_\tau L^*[w_\varrho]\} \, dS = 0,$$

which follows from (3.16) of Ch. I with $\varphi = w_\varrho$, $\psi = w_\tau$, this system of linear equations is precisely the same as (6.30).

. Now consider the general boundary condition (6.32), firstly with $A_1 \neq 0$ on Γ. Then on Γ we have

$$u - \overline{w} = - \frac{A_2}{A_1} L^*[u - \overline{w}] + A_3^*,$$

where

$$A_3^* = \frac{A_3}{A_1} - \overline{w} - \frac{A_2}{A_1} L^*[\overline{w}],$$

and (6.33) can be written

$$\int\limits_{\Gamma} \left(L^*[u - \overline{w}] \left\{ \frac{A_2}{A_1} L^*[w_\varrho] + w_\varrho \right\} - A_3^* L^*[w_\varrho] \right) dS = 0 \qquad (\varrho = 1, \ldots, p). \quad (6.34)$$

Here we can again replace $L^*[u - \overline{w}]$ by $\sum\limits_{\tau=1}^{p} a_\tau L^*[w_\tau]$ to obtain a system of linear equations for the p unknowns a_τ. For the cases in which $A_2 \neq 0$ we could also have expressed $L^*[u - \overline{w}]$ in terms of $u - \overline{w}$; this would have led to a different system of equations for the a_τ.

If $A_1 \equiv 0$ on Γ, then $L^*[u]$ is known on Γ; by replacing $u - \overline{w}$ by $\sum\limits_{\tau=1}^{p} a_\tau w_\tau$ in the first term of the integrand in (6.33) we obtain the equations

$$\sum\limits_{\tau=1}^{p} a_\tau \int\limits_{\Gamma} w_\tau L^*[w_\varrho] \, dS = \int\limits_{\Gamma} w_\varrho L^*[u - \overline{w}] \, dS \qquad (\varrho = 1, \ldots, p). \quad (6.35)$$

If Γ consists of two parts Γ_1 and Γ_2 on which $A_1 \neq 0$ and $A_1 \equiv 0$ respectively, we obtain a system of equations for the a_ϱ by splitting up the integral of (6.33) into one along Γ_1 and one along Γ_2, and using the transformation which leads to (6.34) in the former and that which leads to (6.35) in the latter. Again no volume integrals have to be evaluated; all the coefficients are expressed in terms of boundary integrals.

6.7. Miscellaneous exercises on Chapter V

1. **Transformer field.** Calculate a solution of LAPLACE's equation $\nabla^2 u = 0$ in the annular region between two concentric squares (Fig. V/29) on the boundaries of which u is constant at a different value on each, say 0 on the outer boundary and 1 on the inner. This is a problem for which the finite-difference method is particularly suitable; with other methods, series expansions, for instance, difficulty is experienced in dealing with the boundary conditions.

2. **Torsion of an I-section girder.** Let the cross-section of a girder be composed of seven squares of side a arranged as in Fig. V/30 to form an I-shaped region.

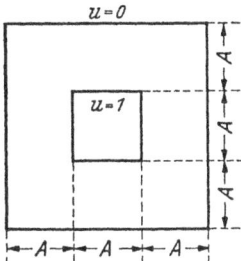

Fig. V/29. Transformer field

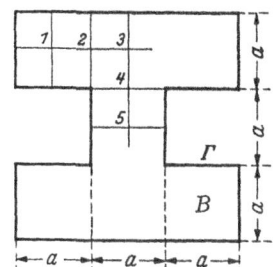

Fig. V/30. Torsion of a girder of I section

The Prandtl stress function satisfies $\nabla^2 u = -2$ inside of this region and on the boundary $u = 0$. Solve this problem approximately by the finite-difference method with

(a) the mesh width $h = a/2$ (cf. Fig. V/30),
(b) the mesh width $h = a/4$ (cf. Fig. V/35),
(c) the mesh width $h = a/2$ again, but by the Hermitian method of § 2.5.

3. Solve the torsion problem of § 6.5, i.e.

$$\nabla^2 u = -1 \quad \text{for} \quad |x| \leq 1, \ |y| \leq 1,$$
$$u = 0 \quad \text{for} \quad |x| = 1 \quad \text{and for} \quad |y| = 1,$$

by assuming a power series solution of the form

$$u = \sum_{j,k=0}^{\infty} a_{jk} x^j y^k.$$

Calculate in particular $u(0, 0)$.

4. For comparison, solve the problem of the last exercise again, this time using a series expansion of the form

$$u = \sum_{\substack{m,n=0 \\ m \geq n}}^{\infty} (1 - x^2)(1 - y^2) b_{mn}(x^{2m} y^{2n} + x^{2n} y^{2m}),$$

which already satisfies the boundary conditions and symmetries of the problem. Substitute this series in the differential equation and equate coefficients.

5. For the same problem as in 3. and 4. compare the ordinary finite-difference method with the Hermitian method for $h = \frac{1}{2}$.

29*

6. **Normal modes of vibration of the air in a room.** The amplitude of the vibration satisfies the equation[1]

$$\frac{\partial^2 v}{\partial x^2} + \frac{\partial^2 v}{\partial y^2} + \frac{\partial^2 v}{\partial z^2} = -\varkappa v,$$

in which the constant $\varkappa$ is related to the natural frequency and represents an eigenvalue parameter to be determined. The boundary conditions are that $v = 0$ on fixed walls and $\partial v/\partial \nu = 0$ (ν being the inward normal) on the boundaries of the room across which the air can move freely.

Consider here a rectangular room which has fixed walls with dimensions $a = 4\,\text{m}$, $b = 6\,\text{m}$, $c = 4\,\text{m}$ as in Fig. V/31 and with two open

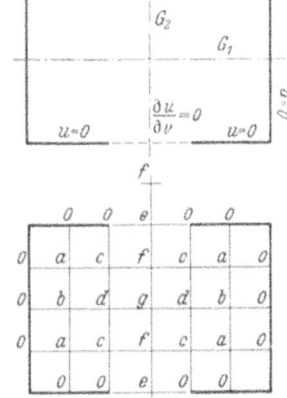

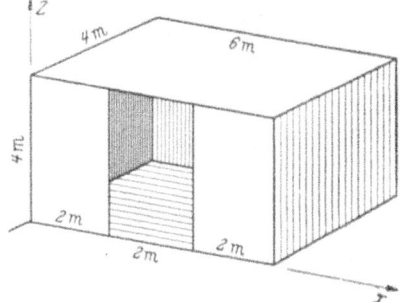

Fig. V/31. Natural acoustic frequencies of a room

doorways, one in the centre of each of the long walls and measuring $4\,\text{m} \times 2\,\text{m}$. If the system of co-ordinates x, y, z is chosen as in Fig. V/31, the problem can be reduced to

$$\nabla^2 u = \frac{\partial^2 u}{\partial x^2} + \frac{\partial^2 u}{\partial y^2} = -\lambda u$$

by assuming solutions of the form $v = u(x, y)\sin(n z/c)$; λ is a new eigenvalue parameter.

Determine the first few eigenvalues λ approximately by means of the ordinary finite-difference method with $h = 1\,\text{m}$.

7. One end-face of a solid cylinder of radius 1 (one unit of length) and length 2 is kept at the temperature $u = 1$ and the other at the temperature $u = -1$ (Fig. V/32), while from the curved surface heat is lost to the surroundings at a rate given by the boundary condition

$$\frac{\partial u}{\partial \nu} = \alpha u.$$

Use the finite-difference method to calculate approximately the steady temperature distribution ($\nabla^2 u = 0$) in the cylinder for $\alpha = 1$ and $\alpha = 5$.

---

[1] See, for instance, A. G. WEBSTER and G. SZEGÖ: Partielle Differentialgleichungen der mathematischen Physik, p. 36. Leipzig and Berlin 1930.

8. Find Ritz approximations, using one term and two terms, respectively, for the solution of the problem

$$u_{xx} + u_{yy} + \frac{3}{5-y} u_y + 1 = 0, \quad u = 0 \quad \text{for} \quad |x| = \frac{1}{2} \quad \text{and for} \quad |y| = 1$$

of Example II in § 1.5 (shear stress in a helical spring).

9. Work out Example I of § 2.8 (concerning a loaded trapezoidal plate) with the smaller mesh width $h = A/4$.

10. A homogeneous membrane stretched over a square frame Γ (length of side 1) has an elastic thread running through it along a diagonal D, which divides the square into two triangles B_1, B_2 (Fig. V/33). The amplitude distribution $u(x, y)$ for a normal mode of vibration satisfies the equations[1]

$$\nabla^2 u + \lambda u = 0 \quad \text{in } B_1 \text{ and } B_2,$$
$$u = 0 \quad \text{on } \Gamma,$$
$$\frac{\partial u}{\partial v_1} + \frac{\partial u}{\partial v_2} + \alpha \frac{\partial^2 u}{\partial s^2} + \lambda \varrho u = 0 \quad \text{on } D,$$

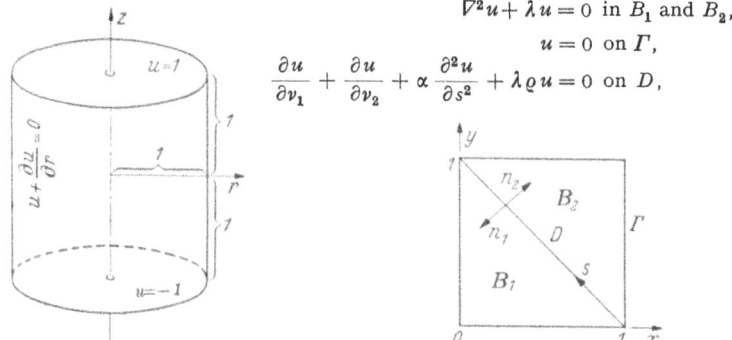

Fig. V/32. A potential problem with axial symmetry   Fig. V/33. Membrane containing an elastic thread

where s denotes the arc length along D and v_1, v_2 are the inward normals from D into B_1, B_2, respectively; α is proportional to the tension in the thread and ϱ to its density. With $\alpha = 6$, $\varrho = 1$ determine the fundamental mode approximately by means of

(a) the ordinary finite-difference method,
(b) the Ritz method or the collocation method.

11. In Example II of Ch. III, § 1.4 the steady temperature distribution in a rod in which heat was generated according to an exponential law was obtained by relaxation. A corresponding two-dimensional problem can be formulated as follows:

$$\nabla^2 u + \frac{1 + e^u}{2} = 0 \quad \text{for} \quad |x| \leqq 1, |y| \leqq 1,$$
$$u = 0 \quad \text{for} \quad |x| = 1 \quad \text{and for} \quad |y| = 1.$$

(a) Find an approximate temperature distribution by the ordinary finite-difference method.

(b) How do other methods such as Ritz's, Galerkin's, least-squares and collocation compare when applied to this non-linear problem?

---

[1] Courant, R.: Über die Anwendung der Variationsrechnung in der Theorie der Eigenschwingungen und über neue Klassen von Funktionalgleichungen. Acta Math. **49**, 1—68 (1926), here p. 60.

6.8. Solutions

1. Because of the symmetry we need only consider that part of the annular region which lies in one half-quadrant. The ordinary finite-difference method yields the following values:

for $h = A/2$

	0	0	0
	0·4167	0·2083	0
	1	0·4167	0
	1	0·4583	0

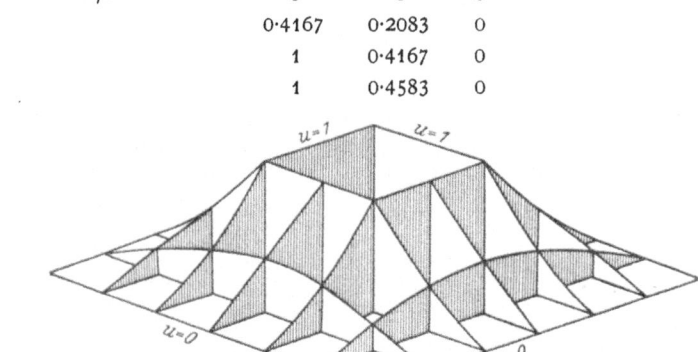

Fig. V/34. Potential distribution in a transformer coil

for $h = A/4$

		0	0	0
		0·099 86	0·049 93	0
	0·303 53	0·201 70	0·099 86	0
	0·472 05	0·303 53	0·147 82	0
1	0·640 57	0·392 56	0·187 89	0
1	0·697 65	0·438 27	0·211 17	0
1	0·711 75	0·451 70	0·218 51	0

Assuming that the error in these results is $O(h^2)$ we can extrapolate to obtain the new approximate values

	0	0	0
	0·3845	0·1995	0
	1	0·3845	0
	1	0·4495	0

Values obtained by the higher approximation method of §§ 2.3, 2.4 with $h = A/2$ are worse at as many points as they are better than those obtained by the ordinary finite-difference method with the same mesh width (Fig. V/34):

	0	0	0
	0·405 16	0·212 03	0
	1	0·405 16	0
	1	0·462 03	0

The more accurate formulae are based on the inclusion of higher order terms in the Taylor expansion, whose validity is restricted by the presence of the singularities at the corners of the inner square.

2. (a) Mesh width $h = a/2$. The five unknown function values $U_1, U_2, \ldots, U_5$ (numbered as in Fig. V/30) satisfy the difference equations

$$2h^2 = 4U_1 - U_2 = 4U_2 - U_1 - U_3 = 4U_3 - 2U_2 - U_4 = 4U_4 - U_3 - U_5 = 4U_5 - 2U_4,$$

from which we obtain

$$U_1 = \frac{251}{334} h^2 = 0.18787 a^2,$$

$$U_2 = \frac{336}{334} h^2 = 0.25150 a^2,$$

$$U_3 = \frac{425}{334} h^2 = 0.31811 a^2,$$

$$U_4 = \frac{360}{334} h^2 = 0.26946 a^2,$$

$$U_5 = \frac{347}{334} h^2 = 0.25973 a^2.$$

Fig. V/35. Numbering of the points of the finer mesh

(b) Mesh width $h = a/4$. Let the unknown function values $U_1, U_2, \ldots$ correspond to the points marked 1, 2, ... in Fig. V/35. The calculation can easily be carried out exactly (without iteration) by putting $U_1 = \alpha$, $U_2 = \beta$, $U_3 = \gamma$ and expressing

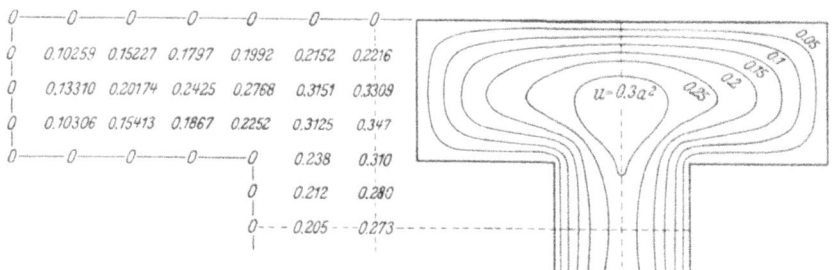

Fig. V/36. Results of the finite-difference approximation for the shear stress in a twisted girder

the remaining U_ν in terms of these first three unknowns. This is straightforward as far as U_{17}. Then the difference equation for the point 16

$$4U_{16} - 2U_{13} - U_{17} = 2h^2$$

yields a relation between α, β and γ. This relation can be used to reduce the size of the coefficients of α, β, γ at each step in the continuation of the calculation down the web, i.e. in the successive expressions for U_{18} to U_{26}. With the further relations implied by the conditions $U_{21} = U_{25}$, $U_{22} = U_{26}$ we have a system of equations for α, β, γ, which yields

$$\alpha = 0.8207286 \times 2h^2,$$

$$\beta = 1.0647835 \times 2h^2,$$

$$\gamma = 0.8244528 \times 2h^2.$$

Fig. V/36 shows the approximate values of U_ν/a^2, together with lines of shearing stress calculated from them by interpolation.

(c) The equations for the Hermitian method read

$$24h^2 = 40U_1 - 8U_2 = 40U_2 - 8U_1 - 8U_3 - 2U_4$$
$$= 40U_3 - 16U_2 - 8U_4 = 40U_4 - 8U_3 - 8U_5 - 4U_2 = 40U_5 - 16U_4,$$

and have the solution

$$U_1 = \frac{1544{\cdot}7}{7582}\, a^2 = 0{\cdot}203\,73\,a^2, \qquad U_4 = \frac{2226}{7582}\, a^2 = 0{\cdot}293\,59\,a^2,$$

$$U_2 = \frac{2037}{7582}\, a^2 = 0{\cdot}268\,66\,a^2, \qquad U_5 = \frac{2027{\cdot}7}{7582}\, a^2 = 0{\cdot}267\,44\,a^2.$$

$$U_3 = \frac{2397{\cdot}3}{7582}\, a^2 = 0{\cdot}316\,18\,a^2,$$

Most of the (c) values lie nearer to the (b) values than to the (a), i.e. nearer to the values which are to be regarded as the more accurate; since the amount of computation involved in (c) is only a little greater than that in (a), the Hermitian method appears to be quite favourable in this case.

3. We retain only those terms in the assumed power series solution which satisfy the symmetries of the problem, and arrange them in the form

$$u = \sum_{n=0}^{\infty} \sum_{\varrho=0}^{n} b_{n,\varrho}\, x^{2n-2\varrho}\, y^{2\varrho} \qquad \text{with} \qquad b_{n,\varrho} = b_{n,n-\varrho}.$$

Substituting in the differential equation and equating coefficients, we obtain

for $n = 1$: $2b_{1,0} + 2b_{1,1} = -1$, which (since $b_{1,0} = b_{1,1}$) yields $b_{1,0} = -\tfrac{1}{4}$,

and for $n > 1$:

$$b_{n,\sigma+1} = -b_{n,\sigma}\, \frac{(2n - 2\sigma)(2n - 2\sigma - 1)}{(2\sigma + 2)(2\sigma + 1)}$$

i.e.

$$b_{n,\sigma} = (-1)^{\sigma} \binom{2n}{2\sigma} b_{n,0} \qquad (\sigma = 0, 1, \ldots, n).$$

Thus $b_{n,\sigma} = 0$ for n odd, and with $b_{2m,0} = c_m$ we have

$$u(x, y) = -\frac{1}{4}(x^2 + y^2) + \sum_{m=0}^{\infty} c_m \sum_{\sigma=0}^{2m} (-1)^{\sigma} \binom{4m}{2\sigma} x^{4m-2\sigma} y^{2\sigma},$$

which satisfies the differential equation for arbitrary c_m [the factors multiplying the c_m are the real parts of the functions $(x + iy)^{4m}$].

For $x = 1$ we must have $u(1, y) \equiv 0$, i.e. all the coefficients of y^0, y^2, y^4, ... in $u(1, y)$ must vanish. This yields an infinite system of linear equations for the c_m:

coefficient of y^0 :	$\tfrac{1}{4} =$	$c_0 +$	$c_1 +$	$c_2 +$	$c_3 +$	$c_4 +$	$c_5 + \cdots$
,, ,, y^2 :	$-\tfrac{1}{4} =$		$6c_1 +$	$28c_2 +$	$66c_3 +$	$120c_4 +$	$190c_5 + \cdots$
,, ,, y^4 :	$0 =$		$c_1 +$	$70c_2 +$	$495c_3 +$	$1820c_4 +$	$4845c_5 + \cdots$
,, ,, y^6 :	$0 =$			$28c_2 +$	$924c_3 +$	$8008c_4 +$	$38760c_5 + \cdots$
,, ,, y^8 :	$0 =$			$c_2 +$	$495c_3 +$	$12870c_4 +$	$125970c_5 + \cdots$
,, ,, y^{10} :	$0 =$				$66c_3 +$	$8008c_4 +$	$184756c_5 + \cdots$

. . . .

which is solved approximately as in § 4.5. Table V/15 shows the results of the first few approximations up to the sixth; the approximate value of $u(0, 0)$ is given by c_0. The convergence is seen to be very good.

Table V/15. *Results of the approximations of order $\nu = 1, \ldots, 6$*

ν	c_0	c_1	c_2	c_3	c_4	c_5
1	0·25					
2	0·291 667	−0·041 667				
3	0·294 005 1	−0·044 642 9	0·000 638			
4	0·294 432 8	−0·045 230 3	0·000 822 4	−0·000 024 92		
5	0·294 565 6	−0·045 441 8 5	0·000 891 1	−0·000 039 6	0·000 001 5	
6	0·294 619 5	−0·045 496 2	0·000 921 6	−0·000 047 6	0·000 002 77	−0·000 000 102 9

4. It is a help in forming $\nabla^2 u$ to make a short table of the Laplacians $\nabla^2 s_{m,n}$ of the symmetric functions $s_{m,n} = x^{2m} y^{2n} + x^{2n} y^{2m}$ which occur (Table V/16); the product multiplying $b_{m,n}$ can be expanded as $s_{m,n} - s_{m+1,n} - s_{m,n+1} + s_{m+1,n+1}$. Such a table can be very useful for various other purposes also.

Table V/16. *Laplacians of various symmetric functions*

φ	$\nabla^2 \varphi$
$x^2 + y^2$	4
$x^4 + y^4$ $x^2 y^2$	$12(x^2 + y^2)$ $2(x^2 + y^2)$
$x^6 + y^6$ $x^4 y^2 + x^2 y^4$	$30(x^4 + y^4)$ $2(x^4 + y^4) + 24 x^2 y^2$
$x^8 + y^8$ $x^6 y^2 + x^2 y^6$ $x^4 y^4$	$56(x^6 + y^6)$ $2(x^6 + y^6) + 30(x^4 y^2 + x^2 y^4)$ $12(x^4 y^2 + x^2 y^4)$
$x^{10} + y^{10}$ $x^8 y^2 + x^2 y^8$ $x^6 y^4 + x^4 y^6$	$90(x^8 + y^8)$ $2(x^8 + y^8) + 56(x^6 y^2 + x^2 y^6)$ $12(x^6 y^2 + x^2 y^6) + 60 x^4 y^4$

Substituting the series in the differential equation and equating coefficients, we obtain with $2b_{mm} = \beta_m$ the following infinite system of equations:

coefficient of $x^0 y^0$: $-1 = -4\beta_0 + 4b_{10}$

,,   ,, $(x^2 + y^2)$: $0 = \quad 2\beta_0 - 16b_{10} + 12b_{20} + 2\beta_1$

,,   ,, $(x^4 + y^4)$: $0 = \qquad\qquad 2b_{10} - 32b_{20} - 2\beta_1 + 30b_{30} + 2b_{21}$

,,   ,, $x^2 y^2$: $0 = \qquad\qquad 24b_{10} - 24b_{20} - 24\beta_1 \qquad\qquad + 24b_{21}$

,,   ,, $(x^6 + y^6)$: $0 = \qquad\qquad\qquad 2b_{20} \qquad\qquad - 58b_{30} - 2b_{21}$

,,   ,, $(x^4 y^2 + x^2 y^4)$: $0 = \qquad\qquad\qquad 30b_{20} + 12\beta_1 - 30b_{30} - 54b_{21}$

The dotted lines mark off the successive finite systems from which the successive approximations are calculated (as in § 4.5). The corresponding approximate values

Table V/17. *Results of the first four approximations by a series method*

Approximation	β_0	Error in β_0	b_{10}	b_{20}	β_1	b_{30}	b_{21}
1st	$\dfrac{1}{4} = 0\cdot25$	-15 %	—	—	—	—	—
2nd	$\dfrac{2}{7} = 0\cdot28571$	$-3\cdot0$ %	$0\cdot03571$	—	—	—	—
3rd	$\dfrac{7}{24} = 0\cdot291667$	$-1\cdot0$ %	$0\cdot041667$	0	$0\cdot041667$	—	—
4th	$\dfrac{164}{559} = 0\cdot293381$	$-0\cdot44\%$	$0\cdot043381$	$-0\cdot000447$	$0\cdot056351$	$-0\cdot000447$	$0\cdot01252$

are given in Table V/17, which also gives the error in β_0, the approximation for $u(0, 0)$.

5. The approximations for the function values a, b, c as defined in Fig. V/37 are exhibited in Table V/18. Here the trivial increase in computational work entailed by the Hermitian method is handsomely repaid by the considerable gain in accuracy.

Fig. V/37. Notation for the torsion problem for a square shaft

6. For the amplitude distribution which is symmetric about both axes of symmetry G_1, G_2 we have seven unknown values $a, b, \ldots, g$ as in Fig. V/33. By expressing all unknowns in terms of a, b and $\nu = 4 - \Lambda h^2$ we obtain finally

$$\nu^7 - 14\nu^5 + 37\nu^3 - 16\nu = 0,$$

which has the roots

$$\nu = 0, \quad \pm0\cdot7332, \quad \pm1\cdot6699, \quad \pm3\cdot2671.$$

The corresponding approximate eigenvalues Λ_i are given in Table V/19, together with the corresponding results for the antisymmetric modes.

Table V/18. *Comparison of the ordinary and Hermitian finite-difference methods*

Ordinary finite-difference method		Hermitian method	
Equations	Results	Equations	Results
$4a - 4b = h^2 = \dfrac{1}{4}$	$a = \dfrac{9}{32} = 0\cdot28125$	$40a - 32b - 8c = 12h^2 = 3$	$a = \dfrac{1272}{4312} = 0\cdot29499$
$-a + 4b - 2c = \dfrac{1}{4}$	$b = \dfrac{7}{32} = 0\cdot21875$	$-8a + 36b - 16c = 3$	$b = \dfrac{990}{4312} = 0\cdot22959$
$-2b + 4c = \dfrac{1}{4}$	$c = \dfrac{11}{64} = 0\cdot1719$	$-2a - 16b + 40c = 3$	$c = \dfrac{783}{4312} = 0\cdot18159$
	Error in a: $-4\cdot5\%$		Error in a: $+0\cdot1\%$

7. Take cylindrical co ordinates r, z as in Fig. V/32 and use a mesh in the (r, z) plane as in Fig. V/38 with $h = \frac{2}{5}$. With the notation of Fig. V/38 the boundary condition $\partial u/\partial r = -u$ corresponds to

$$g = e\,\frac{2-h}{2+h} = \frac{2}{3}\,e, \qquad k = \frac{2}{3}\,f.$$

Table V/19. *Results for the natural acoustic frequencies of a room*

Mode symmetric about	G_1, G_2	G_1	G_2	—
and antisymmetric about	—	G_2	G_1	G_1, G_2
Approximate eigenvalues $\varLambda$	$\begin{aligned}\varLambda_1 &= 0\cdot7329\\ \varLambda_4 &= 2\cdot330\\ \varLambda_6 &= 3\cdot267\\ \varLambda_9 &= 4\\ \varLambda_{12} &= 4\cdot733\\ \varLambda_{14} &= 5\cdot670\\ \varLambda_{17} &= 7\cdot27\end{aligned}$	$\begin{aligned}\varLambda_2 &= 1\cdot586\\ \varLambda_8 &= 3\cdot586\\ \varLambda_{10} &= 4\cdot414\\ \varLambda_{16} &= 6\cdot414\end{aligned}$	$\begin{aligned}\varLambda_3 &= 1\cdot864\\ \varLambda_7 &= 3\cdot338\\ \varLambda_{11} &= 4\cdot662\\ \varLambda_{15} &= 6\cdot136\end{aligned}$	$\begin{aligned}\varLambda_5 &= 3\\ \\ \varLambda_{13} &= 5\end{aligned}$

As $r \to 0$ the differential equation

$$\nabla^2 u = u_{rr} + \frac{1}{r}\,u_r + u_{zz} = 0$$

tends to $2u_{rr} + u_{zz} = 0$. See Table V/20 for equations and results.

In the case with greater heat loss at the boundary, i.e. with $\partial u/\partial r = -5u$, we obtain temperatures which depart much more from the linear distribution $u = z$.

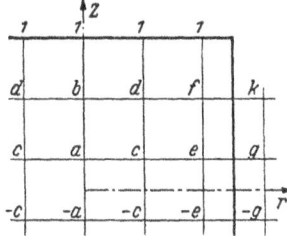

Fig. V/38. Finite-difference mesh in a cylinder

Table V/20. *Approximate temperature distribution in a cylinder*

Equations for the finite-difference method for the case $\dfrac{\partial u}{\partial r} = -u$	Results		
	for the case $\dfrac{\partial u}{\partial r} = -u$		for the case $\dfrac{\partial u}{\partial r} = -5u$
$\begin{aligned}7a- \ \ b- \ \ 4c &= 0\\ -a+6b -4d &= 1\\ -a +10c-2d- \ 3e &= 0\\ - \ \ b- \ 2c+8d - \ 3f &= 2\\ - \ 9c +50e-12f &= 0\\ -9d-12e+38f &= 12\end{aligned}$	$\begin{aligned}&\uparrow z\\ &{-}1\!\!-\!\!-\!\!1\!\!-\!\!-\!\!1\!\!-\\ &\ \ \ \big\vert\\ 0\cdot5621\ \ &0\cdot5480\ \ 0\cdot4927\\ \ \ \ \big\vert\\ 0\cdot1760\ \ &0\cdot1720\ \ 0\cdot1492\\ \ \ \ \big\vert \quad\quad\ \ r\\ -\big\vert-\!\cdot\!-\!\cdot\!-\!\cdot\!-\!\cdot\!\to\end{aligned}$		$\begin{aligned}&\uparrow z\\ &{-}1\!\!-\!\!-\!\!-\!\!1\!\!-\!\!-\!\!-\!\!1\!\!-\\ &\ \ \ \big\vert\\ 0\cdot514\ \ &0\cdot487\ \ 0\cdot365\\ \ \ \ \big\vert\\ 0\cdot154\ \ &0\cdot141\ \ 0\cdot094\\ \ \ \ \big\vert \quad\quad\ \ r\\ -\big\vert-\!\cdot\!-\!\cdot\!-\!\cdot\!-\!\cdot\!\dashv\end{aligned}$

8. First of all the differential equation must be put into the self-adjoint form (5.17):

$$(5-y)^{-3}\,\frac{\partial^2 u}{\partial x^2} + \frac{\partial}{\partial y}\left[(5-y)^{-3}\,\frac{\partial u}{\partial y}\right] + (5-y)^{-3} = 0.$$

Then from (5.20) the corresponding variational problem reads

$$J[\varphi] = \iint_B (5-y)^{-3}\,[\varphi_x^2 + \varphi_y^2 - 2\varphi]\,dx\,dy = \min.,$$

where B is the rectangle given in the problem and the admissible functions φ are to vanish on the boundary. We assume the two-parameter expression

$$\varphi = (1-y^2)\,(1-4x^2)\,(5-y)^3\,(a+by),$$

Table V/21. *Shear stress in a helical spring by Ritz's method*

One-parameter expression	Two-parameter expression		y	Approximate values for $u(0, y)$	
Solution	Equations	Solution		1-param. exp.	2-param. exp.
$a = \dfrac{7}{7264}$	$17025a - 1931b = \dfrac{525}{32}$	$a = 0.0010185$	-0.5	0.12025	0.09691
			0	0.12046	0.12732
$= 0.0009637$	$1931a + 4065b = 0$	$b = 0.0004838$	0.5	0.06586	0.08614

in which b is to be put equal to zero for a one-parameter approximation; the factor $(5 - y)^3$ serves to simplify the calculation of the integrals which occur. See Table V/21 for the results.

In Fig. V/39 these results are compared with those obtained by the ordinary finite-difference method with mesh width $h = \frac{1}{4}$. The finite-difference method

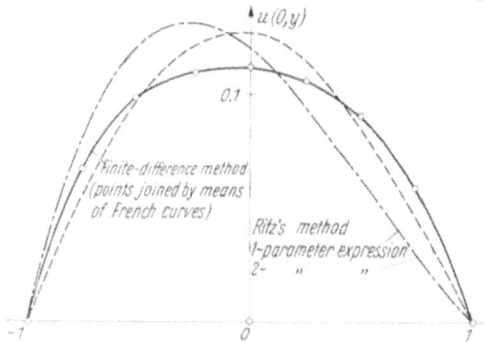

yields a curve for $u(0, y)$ which has a noticeably flat peak; further terms are needed in the Ritz approximation in order to be able to reproduce a curve of this shape.

Fig. V/39. Various approximations by Ritz's method and the finite-difference method

9. With function values denoted by $a, b, \ldots$ as in Fig. V/40 and

$$\frac{pA^4}{N} = \varrho, \qquad k = \frac{9}{16}h^4\frac{p}{N}.$$

the equations in matrix form read $A\,x = r$ with

$$A = \begin{Bmatrix} 12 & -3 & 0 & -3 & 1 & 0 & 0 & 0 \\ -3 & 11 & -3 & -3 & -3 & 1 & 1 & 0 \\ 0 & -3 & 10 & 1 & -3 & -3 & 0 & 1 \\ -3 & -3 & 1 & 13 & -3 & 0 & -3 & 1 \\ 1 & -3 & -3 & -3 & 12 & -2 & -2 & -3 \\ 0 & 1 & -3 & 0 & -2 & 8 & 1 & -3 \\ 0 & 1 & 0 & -3 & -2 & 1 & 10 & -3 \\ 0 & 0 & 2 & 2 & -6 & -6 & -6 & 12 \end{Bmatrix}, \quad x = \begin{Bmatrix} a \\ b \\ c \\ d \\ e \\ f \\ g \\ i \end{Bmatrix}, \quad r = \begin{Bmatrix} k \\ k \\ k \\ k \\ k \\ k \\ k \\ k \end{Bmatrix}.$$

Solution of these equations yields the following pivotal values, which are set out in their mesh positions:

$$c = 0.001840\varrho, \qquad f = 0.002419\varrho,$$
$$b = 0.001665\varrho, \qquad e = 0.002820\varrho, \qquad i = 0.003131\varrho,$$
$$a = 0.000711\varrho, \qquad d = 0.001385\varrho, \qquad g = 0.001730\varrho.$$

10. (a) First take $h = \frac{1}{3}$. Let the function values be denoted by a, b as in Fig. V/41, let Λ be an approximate value for λ and put $\mu = \Lambda h^2$; then corresponding

to the differential equation we have

$$2b - 4a + \mu a = 0.$$

To approximate the condition along the diagonal D we could replace $\partial u/\partial v_1$ at the point b by $\dfrac{0-b}{h\sqrt{2}}$, but this would give a very crude approximation; instead

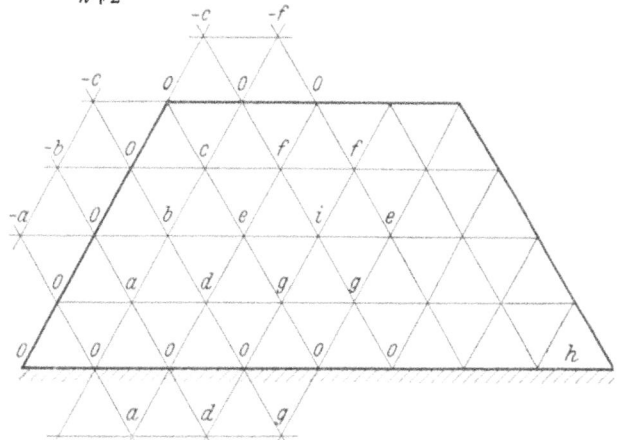

Fig. V/40. Notation for the finer mesh on the trapezoidal plate

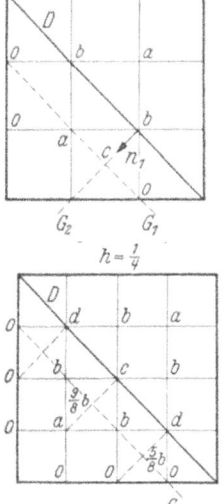

we introduce an intermediate point c (Fig. V/41) and use for $\partial u/\partial v_1$ the derivative at b of the parabola through the ordinates 0, c, b, along the line G_2. The ordinate value c can be obtained by parabolic interpolation along the line G_1; thus $c = \frac{3}{4}a$ and we have

$$\left(\frac{\partial u}{\partial v_1}\right)_b = \frac{-3b + 4c - 0}{h\sqrt{2}} = \frac{-3b + 3a}{\frac{1}{3}\sqrt{2}}.$$

This yields

$$2 \cdot \frac{-3b + 3a}{\frac{1}{3}\sqrt{2}} + 6 \cdot \frac{b - 2b + 0}{\left(\frac{1}{3}\sqrt{2}\right)^2} + \Lambda b = 0$$

as an approximation to the condition along D.

As characteristic equation we have

$$\begin{vmatrix} -4 + \mu & 2 \\ \sqrt{2} & -3 - \sqrt{2} + \mu \end{vmatrix} = 0,$$

and from it we obtain

$$\mu = \begin{cases} 2\cdot513 \\ 5\cdot902 \end{cases} \quad \text{and} \quad \Lambda = \begin{cases} 22\cdot61 \\ 53\cdot11. \end{cases}$$

Fig. V/41. Notation and extra pivotal points

The corresponding solutions of the homogeneous equations are respectively

$$a = 1, \quad b = 0\cdot744,$$

and

$$a = 1, \quad b = -0\cdot951.$$

Now take $h = \frac{1}{4}$, let the function values be denoted by a, b, c, d as in Fig. V/41 and introduce intermediate values along the line G_3 (calculated by parabolic interpolation as for $h = \frac{1}{3}$). In matrix form the equations read

$$A x = \varkappa x$$

with $\Lambda h^2 = \mu = \varkappa + 4$ and

$$
x = \begin{Bmatrix} a \\ b \\ c \\ d \end{Bmatrix}, \quad
A = \begin{Bmatrix}
0 & 2 & 0 & 0 \\
1 & 0 & 1 & 1 \\
-\dfrac{\sqrt{2}}{4} & \dfrac{9}{8}\sqrt{2} & -2 - \dfrac{3}{4}\sqrt{2} & 6 \\
0 & \dfrac{5}{8}\sqrt{2} & 3 & -2 - \dfrac{3}{4}\sqrt{2}
\end{Bmatrix}.
$$

Here it is convenient to determine the first few eigenvalues by means of bracketing theorems[1]; this avoids having to set up and solve an algebraic equation of the fourth degree. We obtain

$$\varkappa_1 = 2\cdot483; \quad \Lambda_1 = 24\cdot27; \quad a = 4\cdot025, \quad b = 5, \quad c = 4\cdot93, \quad d = 3\cdot465,$$
$$\varkappa_2 = 0\cdot62; \quad \Lambda_2 = 54\cdot1; \quad a = 10, \quad b = 3\cdot1, \quad c = -4\cdot9, \quad d = -3\cdot23.$$

The fundamental mode is depicted in Fig. V/42, in which the restraining influence of the thread can be discerned. The thread has this effect because $\alpha = 6$ is large compared with $\varrho = 1$; if α had been small and ϱ large it would have had the opposite effect.

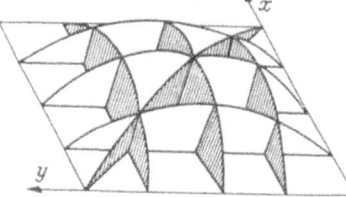

Fig. V/42. Fundamental mode of vibration of a membrane containing a thread

(b) The Ritz variational problem reads

$$J = \iint\limits_{B_1 + B_2} (\varphi_x^2 + \varphi_y^2 - \lambda \varphi^2)\, dx\, dy +$$
$$+ 6 \int\limits_D \left\{ \left(\frac{d\varphi}{ds}\right)^2 - \lambda \varphi^2 \right\} ds = \text{extremum}$$

with φ restricted to be zero on the boundary and continuous along D. To illustrate the method we find a very rough approximation using the one-parameter expressions $\varphi = axy$ in B_1, $\varphi = a(1-x)(1-y)$ in B_2. From $\partial J/\partial a = 0$ we obtain the approximation $\Lambda_1 = \dfrac{15\,(35 - 3\sqrt{2})}{17} = 27\cdot14$.

11. (a) With the function values denoted by a, α as in Fig. V/43 we obtain the approximate values of Table V/22.

We use these approximations to find rough starting values for an iterative solution of the equations for a finer mesh with $h = \frac{1}{2}$ and function value notation a, b, c as in Fig. V/43.

The equations for the ordinary finite-difference method read

$$
\left.
\begin{aligned}
8(\ 4a - 4b \quad\ ) &= 1 + e^a, \\
8(-a + 4b - 2c) &= 1 + e^b, \\
8(\quad -2b + 4c) &= 1 + e^c.
\end{aligned}
\right\} \tag{6.36}
$$

[1] See, for instance, L. Collatz: Eigenwertaufgaben mit technischen Anwendungen, p. 291 et seq. Leipzig 1949.

Table V/22. *Rough approximations for a non-linear heat flow problem*

Ordinary finite-difference method		Hermitian method	
$h=1$	$h=\tfrac{2}{3}$	$h=1$	$h=\tfrac{1}{3}$
$8a = 1 + e^a$ $a = \begin{cases} 0\cdot2925 \\ 3\cdot204 \end{cases}$	$9\alpha = 1 + e^\alpha$ $\alpha = \begin{cases} 0\cdot2544 \\ 3\cdot382 \end{cases}$	$10a = 2 + e^a$ $a = \begin{cases} 0\cdot3406 \\ 3\cdot495 \end{cases}$	$99\alpha = 14 + 10e^\alpha$ $\alpha = \begin{cases} 0\cdot2743 \\ 3\cdot506 \end{cases}$

For the "cool" solution (with the small u values) a suitable iteration process is to solve these equations with the current approximations substituted in the right-hand side. The calculation is best performed by inverting the matrix of coefficients

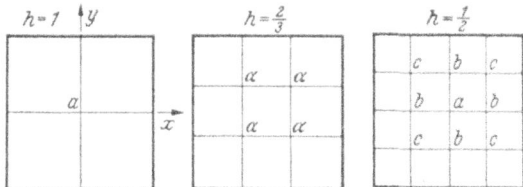

Fig. V/43. Finite-difference meshes for a non-linear problem

on the left-hand side, so that the $(n + 1)$-th approximation is given explicitly in terms of the n-th approximation by

$$
\left.\begin{aligned}
a_{n+1} &= 6A + 2B + 2C, \\
b_{n+1} &= 2A + 2B + 2C, \\
c_{n+1} &= A + B + 3C,
\end{aligned}\right\} \quad \text{where} \quad
\left\{\begin{aligned}
128A &= 1 + e^{a_n}, \\
32B &= 1 + e^{b_n}, \\
64C &= 1 + e^{c_n}.
\end{aligned}\right.
$$

Table V/23. *Iterative solution of the non-linear finite-difference equations*

n	a_n	b_n	c_n	e^{a_n}	e^{b_n}	e^{c_n}	$2A$	$2B$	$2C$
0	0·34	0·26	0·19	1·405	1·297	1·209	0·037 58	0·1436	0·069 03
1	0·325	0·250	0·194	1·384	1·284	1·214	0·037 25	0·142 75	0·069 19
2	0·3237	0·2492	0·1938	1·3822	1·2830	1·2138	0·037 222	0·142 688	0·069 181
3	0·323 53	0·249 09	0·193 73						

Successive stages of the iteration are shown in Table V/23. We infer that to four decimals the solution of the finite-difference equations is $a = 0\cdot3235$, $b = 0\cdot2491$, $c = 0\cdot1937$.

The h^2-extrapolation of Ch. III (1.8) yields

$$u(0, 0) \approx 0\cdot334$$

as a better approximation for the temperature at the centre.

For the "hot" solution (with the much larger u values) we start from rough values approximately ten times as great: $a_0 = 3\cdot4$, $b_0 = 2\cdot6$, $c_0 = 1\cdot9$. In this case, neither the above iteration process nor the corresponding inverse process, in which the current approximation is inserted in the left-hand sides of (6.36) instead of the right-hand sides and the next approximation obtained by solving for the exponentials, is found to be suitable. We therefore employ the relaxation method of § 1.6, using the change produced by the inverse iteration just mentioned as the measure of the residual error, i.e. we define the residuals for approximations a, b, c as the differences between these values and those obtained as the next approximations a', b', c' by inserting a, b, c in the left-hand sides of (6.36), solving for the exponentials and taking natural logarithms. We make corrections to the starting values so as to reduce the magnitude of the changes $a - a', \ldots$. The decrease in the sum S of the absolute values of these changes is a rough measure of the effectiveness of each step. The calculation is shown in Table V/24, which is set out in the same way as Table III/9 for the corresponding one-dimensional problem, with the addition of a column for S. So many steps were needed because the original starting values a_0, b_0, c_0 were rather poor approximations. To accelerate the process a group relaxation was used in the last line of the table; an appropriate group correction α, β, γ can be calculated fairly accurately by virtue of the fact that the exponentials can be approximated closely by linear functions for small α, β, γ. We want to choose these corrections so that the values $a + \alpha, b + \beta, c + \gamma$ satisfy (6.36). Now $32(a + \alpha) - 32(b + \beta) = 1 + e^{a+\alpha}$, for instance, can be written $e^{a'} + 32\alpha - 32\beta = e^{a'} \cdot e^{(a-a')+\alpha}$; since $(a - a')$ and α may be expected to be of the same order of magnitude, we expand the last exponential to obtain $32\alpha - 32\beta = e^{a'}[(a - a') + \alpha]$, and similarly for the other equations. With the values from the penultimate row of the table we have

$$32\alpha - 32\beta \qquad = 80.28(\quad 0.015 + \alpha),$$
$$-8\alpha + 32\beta - 16\gamma = \ 6.52(-0.015 + \beta),$$
$$-16\beta + 32\gamma = \ 2.84(\quad 0.006 + \gamma)$$

and hence

$$\alpha = -0.0163,$$
$$\beta = -0.0131,$$
$$\gamma = -0.0066.$$

(b) Application of the Ritz method does not introduce any fundamental difficulties. A corresponding variational problem can be specified:

$$J[\varphi] = \iint_Q [\varphi_x^2 + \varphi_y^2 - \varphi - e^\varphi]\, dx\, dy = \text{extremum}$$

where Q is the square $|x| \leq 1$, $|y| \leq 1$ and φ is to vanish on the boundary of Q; and also an appropriate approximation function:

$$\varphi = (1 - x^2)(1 - y^2)[a_0 + a_1(x^2 + y^2) + a_2 x^2 y^2 + \cdots].$$

Our difficulties start, however, when we come to evaluate the integrals which arise — even with a one-parameter approximation (only $a_0 \neq 0$) they are quite formidable. By and large, one may say that any method which entails integration over the square is undesirable here. A method which does not suffer from this drawback is the collocation method, but in this method one is faced with the uncertainty regarding the choice of collocation points.

Applying this method with the approximation

$$u \approx w = (1 - x^2)(1 - y^2)[a_0 + a_1(x^2 + y^2)],$$

we determine the parameters so that

$$\nabla^2 w = a_0[-4 + 2(x^2 + y^2)] + a_1[4 - 16(x^2 + y^2) + 24 x^2 y^2 + 2(x^4 + y^4)]$$

is equal to $-\frac{1}{2}(1 + e^w)$ at the collocation points.

Table V/24. Relaxational treatment of the non-linear finite-difference equations for the "hot" temperature distribution

	Starting values			32(a−b)−1 =			New values			Changes			Absolute sum S
	a	b	c	e^a	e^b	e^c	a'	b'	c'	a−a'	b−b'	c−c'	
	3·4	2·6	1·9	24·6	24·6	18·2	3·20	3·20	2·90	0·20	−0·60	−1·00	1·80
Corrections	0·1 / 3·5	2·6	1·9	3·2 / 27·8	−0·8 / 23·8	18·2	3·33	3·17	2·90	0·17	−0·57	−1·00	1·74
Corrections	3·5	2·6	−0·1 / 1·8	27·8	1·6 / 25·4	−3·2 / 15·0	3·33	3·24	2·71	0·17	−0·64	−0·91	1·72
Corrections	3·5	−0·2 / 2·4	−0·3 / 1·5	6·4 / 34·2	−1·6 / 23·8	−6·4 / 8·6	3·53	3·17	2·15	−0·03	−0·77	−0·65	1·45
Corrections	0·6 / 4·1	2·4	1·5	19·2 / 53·4	−4·8 / 19·0	8·6	3·98	2·94	2·15	0·12	−0·54	−0·65	1·31
Corrections	4·1	−0·2 / 2·2	−0·2 / 1·3	6·4 / 59·8	−3·2 / 15·8	−3·2 / 5·4	4·09	2·76	1·69	0·01	−0·56	−0·39	0·96
Corrections	0·2 / 4·3	2·2	1·3	6·4 / 66·2	−1·6 / 14·2	5·4	4·19	2·65	1·69	0·11	−0·45	−0·39	0·95
Corrections	4·3	2·2	−0·2 / 1·1	6·4 / 72·6	−3·2 / 11·0	−3·2 / 2·2	4·29	2·40	0·79	0·01	−0·40	+0·31	0·72
Corrections	4·3	−0·05 / 1·95	1·1	1·6 / 74·2	−1·6 / 9·4	0·8 / 3·0	4·31	2·24	1·10	−0·01	−0·29	0	0·30
Corrections	0·1 / 4·4	1·95	1·1	3·2 / 77·4	−0·8 / 8·6	3·0	4·35	2·15	1·10	0·05	−0·20	0	0·25
Corrections	4·4	−0·1 / 1·85	−0·05 / 1·05	3·2 / 80·6	−2·4 / 6·2	3·0	4·39	1·82	1·10	0·01	+0·03	−0·05	0·09
Corrections	4·4	0·01 / 1·86	1·05	−0·32 / 80·28	+0·32 / 6·52	−0·16 / 2·84	4·385	1·875	1·044	0·015	−0·015	0·006	0·036
Corrections	α 4·3837	β 1·8469	γ 1·0434	80·1776	6·3368	2·8384	4·3842	1·8464	1·0433	−0·0005	0·0005	0·0001	0·0011

For the first approximation (only $a_0 \neq 0$) we have one collocation point to choose. One would normally choose it roughly in the middle of the half-quadrant, say at $x = \frac{1}{2}$, $y = \frac{1}{4}$ (see Fig. V/44); this yields

$$e^\beta = 9.6\beta - 1 \quad \text{with} \quad \beta = \frac{45}{64} a_0,$$

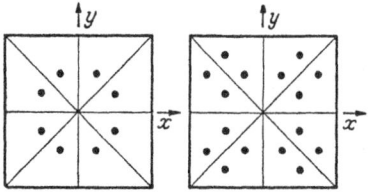

Fig. V/44. Collocation points for a non-linear heat flow problem

from which we obtain the values

$$\beta = \begin{cases} 0.2360 \\ 3.475 \end{cases}$$

and

$$w(0,0) = a_0 = \begin{cases} 0.3356 \\ 4.94. \end{cases}$$

How strongly dependent the results are on the choice of collocation points and how uncertain they are in consequence is shown in Table V/25; for comparison, results are given for some positions which obviously would not normally be chosen.

Table V/25. *Uncertainty in the results*

Collocation point	Value for a_0
$x = 0$, $y = 0$,	0.29 and 3.2
$x = 1$, $y = 0$,	0.5 —
$x = \frac{2}{3}$, $y = \frac{1}{3}$,	0.382 and 7.61
$x = 1$, $y = 1$,	∞ —

For the two-parameter approximation (including a_1) with $x = \frac{1}{2}$, $y = \frac{1}{4}$ and $x = \frac{3}{4}$, $y = \frac{1}{2}$ as collocation points (Fig. V/44) it is convenient to write $a_0 = 13\mu - 5\nu$, $a_1 = 16(\mu - \nu)$, so that the collocation equations become

$$36\mu - 9\nu = \frac{1}{2}\left(1 + e^{\frac{45}{8}\mu}\right),$$

$$-47\mu + 66\nu = \frac{1}{2}\left(1 + e^{\frac{21}{8}\nu}\right).$$

The first equation can be solved immediately for ν in terms of μ, and the second for μ in terms of ν, so that it is a simple matter to represent them graphically; their points of intersection are

$$\begin{cases} \mu = 0.0438 \\ \nu = 0.0485, \end{cases} \text{so that} \begin{cases} a_0 = 0.327 \\ a_1 = 0.075, \end{cases} \text{and} \begin{cases} \mu = 0.64 \\ \nu = 0.50, \end{cases} \text{so that} \begin{cases} a_0 = 5.8 \\ a_1 = -2.2. \end{cases}$$

Chapter VI

Integral and functional equations

§1. General methods for integral equations

1.1. Definitions

An equation for a function $u(x_1, x_2, \ldots, x_n)$ of n independent variables $x_1, x_2, \ldots, x_n$, in the simplest case for a function $y(x)$, is called an integral equation when it involves an integral with the function u appearing in its integrand and with at least one of the arguments of u among its variables of integration. When the equation also involves somewhere a derivative of u, it is called an integro-differential equation.

A special role is played by the linear integral equations

$$\lambda \int_a^b K(x, \xi)\, y(\xi)\, d\xi = f(x) \tag{1.1}$$

and

$$y(x) - \lambda \int_a^b K(x, \xi)\, y(\xi)\, d\xi = f(x) \tag{1.2}$$

of the first and second kinds, respectively, (written down here for the special case of one independent variable x). In these equations the "kernel" $K(x, \xi)$ is a given, say continuous function of x and ξ, $f(x)$ a given continuous function of x, and $y(x)$ the function to be determined. The interval of integration (a, b) may extend to infinity in one or both directions.

The theorems and methods described in the following for integral equations with one independent variable x are all equally valid for integral equations with several (finitely many) independent variables, in which the integration, instead of being taken over the fundamental interval $a \leq x \leq b$, is taken over the corresponding fundamental region.

If $f(x) \equiv 0$, the integral equation (1.2) is called homogeneous; otherwise inhomogeneous. For the inhomogeneous integral equation λ is to be regarded as a given parameter, while for the homogeneous equation it is essentially an eigenvalue parameter since the integral equation then presents an eigenvalue problem, in which ordinarily the object is to determine those values of λ, the *eigenvalues*, for which the integral equation possesses "non-trivial" solutions (i.e. solutions which do not vanish identically), correspondingly called the *eigenfunctions*.

The integral equation (1.2) is said to be "regular" when the interval (a, b) is finite and the kernel $K(x, \xi)$ is bounded and continuous[1];

---

[1] The theorems for regular integral equations may easily be carried over to cases in which fewer assumptions are made about the kernel. The definitions of regular and singular integral equations used here follow those in PH. FRANK and R. v. MISES: Differential- und Integralgleichungen der Mechanik und Physik, 2nd ed., Vol. 1, p. 535. Brunswick 1930.

"singular" integral equations, for which the fundamental interval is infinite or the kernel violates the conditions of boundedness and continuity, also occur frequently in applications (cf. § 3).

If the upper limit of integration in (1.1) or (1.2) is replaced by the variable x, we obtain a "Volterra integral equation"; corresponding to (1.2), for example, we have the Volterra integral equation of the second kind

$$y(x) - \lambda \int_a^x K(x, \xi) y(\xi) d\xi = f(x).$$

For regular integral equations there exists a complete theory[1].

A variety of applied problems give rise also to integro-differential equations; we shall frequently consider the following type of linear integro-differential equation:

$$M[y(x)] - \lambda \int_a^b K(x, \xi) N[y(\xi)] d\xi = f(x), \tag{1.3}$$

with boundary conditions

$$U_\mu[y] = \gamma_\mu \qquad (\mu = 1, 2, \ldots, m). \tag{1.4}$$

Here $M[y]$ and $N[y]$ are linear differential expressions in y as in Ch. III, (8.2), (8.3), $M[y]$ being of order $m \geq 0$, and $U_\mu[y] = \gamma_\mu$ are n given linear boundary conditions for y of the form (1.7), (1.8) of Ch. I If M and N are both of order zero, so that no derivatives appear in equation (1.3), then we have an ordinary integral equation of the form (1.2) and the boundary conditions drop out.

Occasionally initial- and boundary-value problems in differential equations are reduced to integral equations[2], cf. Examples I, II of § 1.3 but the importance of the connection between the two kinds of formulation would appear to lie primarily in the theoretical field; for the numerical solution of a differential equation problem one generally prefers to apply the methods of Ch.s II to V directly rather than to first transform the problem into an integral equation.

---

[1] See, for instance, PH. FRANK and R. v. MISES: (see last footnote). - COURANT, R., and D. HILBERT: Methods of mathematical physics, 1st English ed Vol. I. New York: Interscience Publishers, Inc. 1953. — HAMEL, G.: Integral gleichungen, 2nd ed. Berlin 1949. — SCHMEIDLER, W.: Integralgleichungen mit Anwendungen in Physik und Technik, Vol. I, Lineare Integralgleichungen, 611 pp Leipzig 1950.

[2] See, for instance, W. SCHMEIDLER: (see last footnote) pp. 328—360. — COLLATZ, L.: Eigenwertaufgaben, pp. 90—109. Leipzig 1949.

1.2. Replacement of the integrals by finite sums

As for ordinary and partial differential equations, the finite-difference method is also a very generally applicable method for integral and integro-differential equations. In this method the derivatives are replaced by finite expressions as in Ch. III, § 1 and Ch. IV, § 1 and the integrals by sums derived from a quadrature formula of a suitable kind.

We explain the method with reference to a not necessarily linear integral equation of the form

$$\int_a^b \Phi\big(x, \xi, y(x), y(\xi)\big)\, d\xi = 0 \tag{1.5}$$

with finite interval (a, b).

Let us use a quadrature formula

$$\int_a^b F(x)\, dx = \sum_{\nu=1}^n A_\nu F(x_\nu) + R, \tag{1.6}$$

where the x_ν are chosen pivotal points in $\langle a, b \rangle$, the A_ν are appropriate weighting factors and R denotes the remainder term. Let Y_j be an approximation for $y(x_j)$. If we write down (1.5) for $x = x_j$ and replace the integral by a finite sum in accordance with the quadrature formula (1.6) without the remainder term, we obtain an equation for the approximate values Y_j:

$$\sum_{\nu=1}^n A_\nu \Phi(x_j, x_\nu, Y_j, Y_\nu) = 0 \qquad (j = 1, 2, \ldots, n).$$

If such an equation is written down for $j = 1, 2, \ldots, n$, we obtain a system of n (in general non-linear) equations for the same number of unknowns Y_j.

For the linear integral equation (1.2), for instance, the finite equations read

$$Y_j - \lambda h \{\tfrac{1}{2} K_{j,0} Y_0 + \sum_{k=1}^{n-1} K_{j,k} Y_k + \tfrac{1}{2} K_{j,n} Y_n\} = f_j \qquad (j = 0, 1, \ldots, n) \tag{1.7}$$

when the quadrature formula used is the trapezium rule with the pivotal points $x_j = a + j h$ $\left(\text{pivotal interval } h = \dfrac{b-a}{n}\right)$. Here we have written $K_{j,k}$ and f_j for $K(x_j, x_k)$ and $f(x_j)$, respectively. Thus we have $n+1$ linear equations for the Y_j.

If the integral equation is homogeneous, i.e. $f(x) \equiv 0$, then the linear equations (1.7) are also homogeneous and have a non-trivial solution if and only if the determinant of the coefficients vanishes. This yields an algebraic equation for λ of (in general) the $(n+1)$-th degree. The roots $\Lambda_1, \Lambda_2, \ldots, \Lambda_{n+1}$ of this algebraic equation are used as approximate values[1] for the first $n+1$ eigenvalues λ_ν.

If SIMPSON's rule or some other more accurate quadrature formula[2] is used, one should take care that it is applied only to intervals in which the integrand is differentiable sufficiently often; for an integrand with, say, a discontinuous derivative — as is the case, for example, with certain GREEN's functions for differential equation problems — SIMPSON's rule can naturally yield worse results than the trapezium rule if applied indiscriminately.

1.3. Examples

I. Inhomogeneous linear integral equation of the second kind. In the first boundary-value problem of potential theory a solution $u(x, y)$ of the potential equation is to be determined which takes given boundary values $g(t)$ on the boundary Γ of a region B. Let the boundary curve Γ be defined in terms of the boundary parameter t by

$$x = \xi(t), \qquad y = \eta(t).$$

The solution may[3] be written in the form

$$u(x, y) = \int_\Gamma \mu(t) \frac{d\vartheta}{dt} dt, \tag{1.8}$$

---

[1] WIELANDT, H.: Proc. Internat. Congr. Math. Amsterdam 1954, Vol. II, p. 391, and: Error bounds for eigenvalues of symmetric integral equations. Proc. Symp. Appl. Math. VI, New York-Toronto-London 1956, pp. 261—282, gives an error estimate for the approximate values $\Lambda_\nu^{(n)}$ (for the eigenvalues λ_ν) obtained by using the quadrature formula (1.6) in the case of a real, symmetric, square-integrable (cf. § 2.3) kernel $K(x, \xi)$. If we imagine the $\Lambda_\nu^{(n)}$ as the first $n+1$ members of an infinite sequence whose remaining members are all zero, and if the λ_ν are numbered suitably, then there is a number C depending only on the kernel and on the quadrature formula such that (for the interval $\langle a, b \rangle = \langle 0, 1 \rangle$) the estimate

$$\left| \frac{1}{\lambda_\nu} - \frac{1}{\Lambda_\nu^{(n)}} \right| \leqq C$$

holds for all ν; for the trapezium rule we can put

$$C = \frac{0 \cdot 54}{n-1} \sup_{x, \xi} \left| \frac{\partial K(x, \xi)}{\partial x} \right|$$

and for SIMPSON's rule

$$C = \frac{0 \cdot 75}{(n-1)^2} \sup_{x, \xi} \left| \frac{\partial^2 K(x, \xi)}{\partial x^2} \right|.$$

[2] GAUSS's and CHEBYSHEV's quadrature formulae are recommended by E. J. NYSTRÖM: Über die praktische Auflösung von linearen Integralgleichungen und Anwendungen auf Randwertaufgaben der Potentialtheorie. Commentationes physico-mathematicae. Acta Soc. Sci. Fenn. 4, Nr. 15, 1—52. Helsingfors 1928. Error estimates are given by L. V. KANTOROVICH and V. I. KRYLOV: Näherungsmethoden der höheren Analysis, pp. 94—155. Berlin 1956.

[3] NYSTRÖM, E. J.: (see last footnote).

where ϑ is the angle (see Fig. VI/1) defined by

$$\vartheta = \tan^{-1} \frac{\eta(t) - y}{\xi(t) - x}$$

and μ satisfies the integral equation of the second kind

$$\pi \mu(s) + \int_\Gamma K(s,t)\,\mu(t)\,dt = g(s) \quad \text{with} \quad K(s,t) = \frac{\partial}{\partial t} \tan^{-1} \frac{\eta(t) - \eta(s)}{\xi(t) - \xi(s)}. \quad (1.9)$$

If, for example, Γ is the ellipse

$$x = a \cos t, \quad y = b \sin t, \quad a \geq b, \quad (1.10)$$

the formula for the kernel yields

$$K(s,t) = \frac{ab}{a^2 + b^2 - (a^2 - b^2)\cos(s+t)}. \quad (1.11)$$

Let us consider the particular example of the steady temperature distribution in a long cylinder of elliptical cross-section (Fig. VI/2) at the surface of which

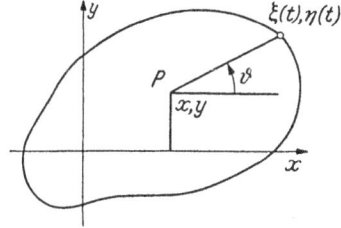

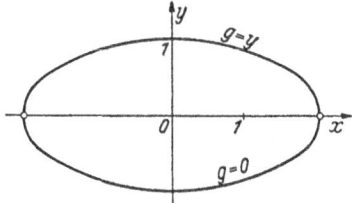

Fig. VI/1. The first boundary-value problem of potential theory

Fig. VI/2. Cross-section of an elliptical cylinder with the prescribed values for the temperature on the boundary

[the boundary of the ellipse (1.10) with $a = 2$, $b = 1$] the temperature $g(s)$ is prescribed as follows: $g(s) = 0$ for $y \leq 0$, $g(s) = y$ for $y \geq 0$ (linearly increasing temperature). Suppose that we are required to calculate approximately the temperature $u(0, 0)$ at the centre of the cross-section. According to (1.9) we have to solve the integral equation

$$\pi \mu(s) + \int_{-\pi}^{\pi} \frac{2}{5 - 3\cos(s+t)}\,\mu(t)\,dt = \begin{cases} \sin s & \text{for} \quad 0 \leq s \leq \pi, \\ 0 & \text{for} \quad -\pi \leq s \leq 0 \end{cases}$$

(there are, of course, numerous other methods for the solution of this problem, cf. Ch. V).

With the pivotal interval $h = \dfrac{\pi}{4}$ there are nine unknown pivotal values $\mu\left(\dfrac{\pi}{4}\,j\right)$ in the interval $-\pi \leq s \leq \pi$. We can reduce this number by using the symmetry of the problem; if μ_j denotes the corresponding approximate pivotal value, we have

$$\mu_j = \mu_{4-j}.$$

If the integral is evaluated approximately by the trapezium rule, we obtain for the approximations $\mu_2, \mu_1, \mu_0, \mu_{-1}, \mu_{-2}$ the linear equations

$$4\mu_2 + 0.25 \quad \mu_2 + 0.5617\mu_1 + 0.8 \quad \mu_0 + 1.3895\mu_{-1} + \quad \mu_{-2} = \frac{4}{\pi} \approx 1.2732,$$

$$4\mu_1 + 0.2808\mu_2 + 0.65 \quad \mu_1 + 0.9756\mu_0 + 1.4 \quad \mu_{-1} + 0.6948\mu_{-2} = \frac{4}{\pi\sqrt{2}} \approx 0.9003,$$

$$4\mu_0 + 0.4 \quad \mu_2 + 0.9756\mu_1 + 1.25 \quad \mu_0 + 0.9756\mu_{-1} + 0.4 \quad \mu_{-2} = 0,$$

$$4\mu_{-1} + 0.6948\mu_2 + 1.4 \quad \mu_1 + 0.9756\mu_0 + 0.65 \quad \mu_{-1} + 0.2808\mu_{-2} = 0,$$

$$4\mu_{-2} + \quad \mu_2 + 1.3895\mu_1 + 0.8 \quad \mu_0 + 0.5617\mu_{-1} + 0.25 \quad \mu_{-2} = 0,$$

with the solution (obtainable conveniently by iteration)

$$\mu_2 = 0.3422, \quad \mu_1 = 0.2329, \quad \mu_0 = -0.0396, \quad \mu_{-1} = -0.1048, \quad \mu_{-2} = -0.1354.$$

These values yield the approximation

$$u(0,0) = \int_{-\pi}^{\pi} \frac{\mu(t)}{2\cos^2 t + \frac{1}{2}\sin^2 t}\, dt \approx \frac{\pi}{4}\left[2(\mu_2 + \mu_{-2}) + 1.6(\mu_1 + \mu_{-1}) + \mu_0\right] = 0.4547$$

for the temperature at the centre of the cross-section.

II. An eigenvalue problem. In Exercise 11 of Ch. III, § 8.11 the problem representing the flexural vibrations of a cantilever, namely

$$y^{IV} = \lambda(1+x)\,y, \quad y(0) = y'(0) = y''(1) = y'''(1) = 0, \qquad (1.12)$$

was treated by RITZ's method. Here it will first be transformed into an integral equation with the aid of its GREEN's function $G(x, \xi)$. Since the problem

$$y^{IV}(x) = r(x), \quad y(0) = y'(0) = y''(1) = y'''(1) = 0$$

is equivalent to

$$y(x) = \int_0^1 G(x, \xi)\, r(\xi)\, d\xi, \quad \text{where} \quad G(x, \xi) = \begin{cases} \dfrac{x^2}{6}(3\xi - x) & \text{for} \quad x \leqq \xi, \\[2mm] \dfrac{\xi^2}{6}(3x - \xi) & \text{for} \quad x \geqq \xi, \end{cases}$$

(1.12) may be replaced by the homogeneous integral equation of the second kind

$$y(x) = \lambda \int_0^1 K(x, \xi)\, y(\xi)\, d\xi \quad \text{with} \quad K(x, \xi) = G(x, \xi)(1 + \xi).$$

This kernel is unsymmetric, but it could easily be made symmetric by introducing $s(x) = y(x)\sqrt{1 + x}$; however, for the following calculation it is more convenient to work with the unsymmetric kernel and so avoid the square roots.

We illustrate several ways of replacing the integral by a finite sum by using different quadrature formulae.

1. Using the trapezium rule. This is the simplest method, but also the crudest, and accordingly the results are not at all accurate.

First of all let us use just three pivotal points $x = 0, \frac{1}{2}, 1$. It is expedient to record the values of the kernel neatly in an array as in Table VI/1.

If Λ, y_1, y_2 are approximate values for λ, $y(\frac{1}{2})$, $y(1)$, and if $\Lambda = 6/\varrho$, the approximate equations corresponding to the integral equation read

(for $x = \frac{1}{2}$) $\varrho y_1 = \frac{1}{4} \left[0 + 2 \times \frac{3}{8} y_1 + \frac{5}{4} y_2 \right]$,

(for $x = 1$) $\varrho y_2 = \frac{1}{4} \left[0 + 2 \times \frac{15}{16} y_1 + 4 y_2 \right]$.

Table VI/1. *Values of* $6K(x, \xi)$

	$x=0$	$x=\frac{1}{2}$	$x=1$
$\xi = 1$	0	$\frac{5}{4}$	4
$\xi = \frac{1}{2}$	0	$\frac{3}{8}$	$\frac{15}{16}$
$\xi = 0$	0	0	0

Table VI/2. *Values of* $6K(x, \xi)$

	$x=0$	$x=\frac{1}{3}$	$x=\frac{2}{3}$	$x=1$
$\xi = 1$	0	$\frac{48}{81}$	$\frac{168}{81}$	$\frac{324}{81} = 4$
$\xi = \frac{2}{3}$	0	$\frac{25}{81}$	$\frac{80}{81}$	$\frac{140}{81}$
$\xi = \frac{1}{3}$	0	$\frac{8}{81}$	$\frac{20}{81}$	$\frac{32}{81}$
$\xi = 0$	0	0	0	0

With $4\varrho = \tau$ the determinant of this homogeneous system of equations for y_1, y_2 reads

$$\begin{vmatrix} \frac{3}{4} - \tau & \frac{5}{4} \\ \frac{15}{8} & 4 - \tau \end{vmatrix} = \tau^2 - \frac{19}{4} \tau + \frac{21}{32} ;$$

ts zeros are $\tau = \frac{1}{8} \left(19 \pm \sqrt{319} \right)$, so that $\lambda \approx \Lambda = \frac{24}{\tau} = \begin{cases} 5\cdot 21 \\ 168\cdot 5. \end{cases}$

Although the method using the trapezium rule is not to be recommended on account of its low accuracy, we apply it again with the smaller pivotal interval $ = \frac{1}{3}$ and $x = 0, \frac{1}{3}, \frac{2}{3}, 1$; in any case we shall need the corresponding values of the kernel (Table VI/2) when we apply the three-eights rule in 2. below and also we introduce a matrix notation which will be needed later.

If y_j is an approximation for $y(\frac{1}{3} j)$, the corresponding homogeneous equations may be written concisely in the matrix form

$$A z = \sigma z, \tag{1.13}$$

where

$$z = \begin{pmatrix} y_1 \\ y_2 \\ y_3 \end{pmatrix}, \quad A = \begin{pmatrix} 8 & 25 & 24 \\ 20 & 80 & 84 \\ 32 & 140 & 162 \end{pmatrix}, \quad \sigma = 243 \varrho = 243 \times \frac{6}{\Lambda} .$$

The required values of σ are the latent roots of the matrix A; we obtain

$$\sigma = \begin{cases} 242\cdot 3 \\ 6\cdot 78 \\ 0\cdot 95, \end{cases} \quad \text{and hence} \quad \Lambda = \begin{cases} 6\cdot 01 \\ 215 \\ 1530. \end{cases}$$

2. **Using more accurate formulae.** By using better quadrature formulae we can obtain more accurate values with the same amount of computation as in 1. not counting the calculation of the kernel values, which is the same in each case.

With the pivotal points $x = 0, \frac{1}{2}, 1$ we use SIMPSON's rule. Then the approximate equations corresponding to the integral equation become (with $\varrho = 6/\varLambda$)

$$(\text{for } x = \tfrac{1}{2}) \qquad \varrho\, y_1 = \frac{1}{6}\left[1 \times 0 + 4 \times \frac{3}{8}\, y_1 + \frac{5}{4}\, y_2\right],$$

$$(\text{for } x = 1) \qquad \varrho\, y_2 = \frac{1}{6}\left[1 \times 0 + 4 \times \frac{15}{16}\, y_1 + 4 y_2\right].$$

From the usual determinant condition, which here reads

$$\begin{vmatrix} \dfrac{3}{2} - \tau & \dfrac{5}{4} \\[2mm] \dfrac{15}{4} & 4 - \tau \end{vmatrix} = \tau^2 - \frac{11}{2}\,\tau + \frac{21}{16} = 0,$$

where $\tau = 6\varrho$, we obtain

$$\tau = \begin{cases} \dfrac{21}{4} \\[2mm] \dfrac{1}{4} \end{cases} \quad \text{and} \quad \varLambda = \frac{36}{\tau} = \begin{cases} \dfrac{48}{7} = 6\text{·}8571 \\[2mm] 144; \end{cases}$$

the value $\varLambda_1 = \dfrac{48}{7}$ in particular is a very good approximation.

With the pivotal points $x = 0, \frac{1}{3}, \frac{2}{3}, 1$ one can use the "three-eighths rule"

$$\int_0^{3h} f(x)\, dx \approx \frac{3h}{8}\,[f(0) + 3f(h) + 3f(2h) + f(3h)];$$

this yields a matrix equation of the form (1.13) with

$$A = \begin{pmatrix} 8 & 25 & 16 \\ 20 & 80 & 56 \\ 32 & 140 & 108 \end{pmatrix}, \qquad \sigma = 216\varrho = 216\,\frac{6}{\varLambda},$$

and we obtain

$$\sigma = \begin{cases} 189\text{·}16 \\ 5\text{·}91 \\ 0\text{·}93, \end{cases} \qquad \varLambda = \begin{cases} 6\text{·}851 \\ 219 \\ 1390. \end{cases}$$

Naturally with so few pivotal points we cannot expect much accuracy from the approximations for the higher eigenvalues.

III. An eigenvalue problem for a function of two independent variables. Consider the integral equation

$$u(x, y) = \lambda \int_{-1}^{1} \int_{-1}^{1} |x\xi + y\eta|\, u(\xi, \eta)\, d\xi\, d\eta. \tag{1.14}$$

Here, instead of an ordinary integral, we have a double integral to approximate by a finite sum. For this there are again crude formulae and more accurate formulae at our disposal. For illustration of the method we confine ourselves to the simple formula

$$\int_{-1}^{1} \int_{-1}^{1} \varphi(\xi, \eta) \, d\xi \, d\eta \approx \frac{4}{n^2} \sum_{\nu, \mu = 1}^{n} \varphi(a_\nu, a_\mu), \qquad (1.15)$$

in which the a_ν are a set of distinct points in the interval $(-1, 1)$.

If we first take $n = 2$ and $a_1 = -k$, $a_2 = k$, and denote the pivotal values of u by a, b, c, d as in Fig. VI/3, we obtain the four approximate equations

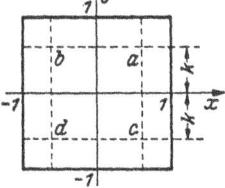

Fig. VI/3. Notation for the eigenvalue problem of Example III

$$\nu a = 2a + 2d = \nu d,$$

$$\nu b = 2b + 2c = \nu c,$$

where $\nu = 1/(\Lambda k^2)$, Λ being an approximation for λ.

The determinant condition reduces to $\nu^2 (4 - \nu)^2 = 0$, so if we exclude $\nu = 0$ (which would give $\Lambda = \infty$) we have the double root $\nu = 4$; the corresponding eigenfunction is characterized by $a = d$, $b = c$, with a, b otherwise arbitrary, and the approximate eigenvalue Λ is $1/(4k^2)$, where k is yet to be chosen. The choice $k = \frac{1}{2}$ gives $\Lambda = 1$; but a better choice is $k = \frac{1}{3}\sqrt{3}$, for (1.15) then becomes CHEBYSHEV's quadrature formula; the corresponding value of Λ is 0.75.

For the larger number of pivotal points defined by $n = 3$, $a_1 = -k$, $a_2 = 0$, $a_3 = k$ we exclude the case $1/\Lambda = 0$ from the start and make use of the consequent central symmetry about the point $x = y = 0$. We are left with

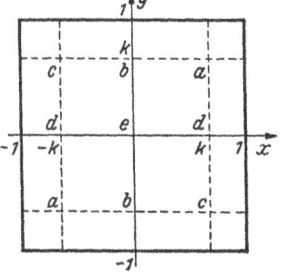

Fig. VI/4. Notation for nine pivotal points (assuming central symmetry)

five unknown function values a, b, c, d, e as indicated in Fig. VI/4; these must satisfy the equations

$$\varrho a = 2a + b \qquad + d$$

$$\varrho b = a + b + c$$

$$\varrho c = b + 2c + d$$

$$\varrho d = a + c + d$$

$$\varrho e = 0,$$

where $\varrho = \dfrac{9}{8\Lambda k^2}$. Fig. VI/5 shows the corresponding eigenvalues and eigenfunctions for the choice $k = \frac{2}{3}$; the simple calculation involved can

n = 3

$\varrho = \dfrac{3+\sqrt{17}}{2}$
$\Lambda = 0.7107$

$$\begin{pmatrix} 1 & 0.78 & 1 \\ 0.78 & 0 & 0.78 \\ 1 & 0.78 & 1 \end{pmatrix}$$

$\varrho = 2$
$\Lambda = \dfrac{81}{64} = 1.266$

$$\begin{pmatrix} -1 & 0 & 1 \\ 0 & 0 & 0 \\ 1 & 0 & -1 \end{pmatrix}$$

$\varrho = 1$
$\Lambda = \dfrac{81}{32} = 2.531$

$$\begin{pmatrix} 0 & 1 & 0 \\ -1 & 0 & -1 \\ 0 & 1 & 0 \end{pmatrix}$$

$\varrho = \dfrac{3-\sqrt{17}}{2}$
$\Lambda = -4.507$

$$\begin{pmatrix} 1 & -1.28 & 1 \\ -1.28 & 0 & -1.28 \\ 1 & -1.28 & 1 \end{pmatrix}$$

n = 4

$\sigma = 11+\sqrt{181}$
$\Lambda = 0.6543$

$$\begin{pmatrix} 1.245 & 1 & 1 & 1.245 \\ 1 & 0.415 & 0.415 & 1 \\ 1 & 0.415 & 0.415 & 1 \\ 1.245 & 1 & 1 & 1.245 \end{pmatrix}$$

$\sigma = 7+\sqrt{29}$
$\Lambda = 1.2919$

$$\begin{pmatrix} -2.516 & -1 & 1 & 2.516 \\ -1 & -0.839 & 0.839 & 1 \\ 1 & 0.839 & -0.839 & -1 \\ 2.516 & 1 & -1 & -2.516 \end{pmatrix}$$

$\sigma = 6$
$\Lambda = \dfrac{8}{3} = 2.6667$

$$\begin{pmatrix} 0 & 1 & 1 & 0 \\ -1 & 0 & 0 & -1 \\ -1 & 0 & 0 & -1 \\ 0 & 1 & 1 & 0 \end{pmatrix}$$

$\sigma = -2$
$\Lambda = -8$

$$\begin{pmatrix} 0 & -1 & 1 & 0 \\ 1 & 0 & 0 & -1 \\ -1 & 0 & 0 & 1 \\ 0 & 1 & -1 & 0 \end{pmatrix}$$

$\sigma = 11-\sqrt{181}$
$\Lambda = -6.5210$

$$\begin{pmatrix} -1.445 & 1 & 1 & -1.445 \\ 1 & -0.482 & -0.482 & 1 \\ 1 & -0.482 & -0.482 & 1 \\ -1.445 & 1 & 1 & -1.445 \end{pmatrix}$$

$\sigma = 7-\sqrt{29}$
$\Lambda = 9.908$

$$\begin{pmatrix} 0.72 & -1 & 1 & -0.72 \\ -1 & 0.24 & -0.24 & 1 \\ 1 & -0.24 & 0.24 & -1 \\ -0.72 & 1 & -1 & 0.72 \end{pmatrix}$$

Fig. VI/5. Approximate eigenvalues and eigenfunctions for the eigenvalue problem with two independent variables (Example III)

be performed very quickly. Also shown are the approximations for
$n = 4$, $a_1 = -3k$, $a_2 = -k$, $a_3 = k$, $a_4 = 3k$ with $k = \frac{1}{4}$ [here $\sigma = 1/(\Lambda k^2)$],

which are likewise obtainable with only a short calculation. (The values are still fairly crude; cf. the values calculated in § 1.5.)

If k is chosen so that the error in the quadrature formula (1.15) is of as high an order as possible, it turns out that the approximations for the lower eigenvalues are improved, but that those for the higher eigenvalues are worsened. For $n = 3$ the choice $k = \frac{1}{2}\sqrt{2}$ (CHEBYSHEV'S quadrature formula) yields for Λ the approximations $\frac{9}{16}(\sqrt{17} - 3) = 0.632$, $\frac{9}{8} = 1.125$, $-\frac{9}{16}(\sqrt{17} + 3) = -4.007$; with $n = 4$ and $k = \frac{1}{15}\sqrt{15}$ we obtain for Λ the values $\frac{1}{4}(\sqrt{181} - 11) = 0.613$, 1.211, 2.5, -6.113, -7.5, 9.289.

As in Ch. V, § 2.8, Example II, the computational work is greatly reduced if we postulate the various symmetries and anti-symmetries which can exist.

IV. A non-linear integral equation. Consider the equation

$$\int_0^1 \frac{[y(x) + y(\xi)]^2}{1 + x + \xi}\, d\xi = 1. \tag{1.16}$$

If we first make a very rough approximation to the integral by replacing the integrand by a constant, say its value at the midpoint $\xi = \frac{1}{2}$, we find that for $x = \frac{1}{2}$

$$\tfrac{1}{2}[2y(\tfrac{1}{2})]^2 \approx 1; \quad\text{so that}\quad y(\tfrac{1}{2}) \approx \pm\sqrt{\tfrac{1}{2}} \approx \pm 0.707.$$

If $y(x)$ is a solution of the integral equation, then obviously so is $-y(x)$.

By using the trapezium rule for the integral we obtain the two equations

$$\begin{aligned}
&\text{(for } x = 0) &&4y_0^2 + \tfrac{1}{2}(y_0 + y_1)^2 = 2,\\
&\text{(for } x = 1) &&\tfrac{1}{2}(y_0 + y_1)^2 + \tfrac{4}{3}y_1^2 = 2
\end{aligned}$$

for the approximate values y_0, y_1 of $y(0)$, $y(1)$. From these equations it follows immediately that $3y_0^2 = y_1^2$; for the remainder we restrict attention to the solution corresponding to $y_1 = \sqrt{3}\, y_0$ with the positive sign. For this solution we obtain

$$y_0 = \sqrt{\frac{2}{33}(6 - \sqrt{3})} = 0.5086 \quad\text{and}\quad y_1 = y_0\sqrt{3} \approx 0.8809.$$

By using SIMPSON'S rule (with three approximate pivotal values y_0, y_1, y_2 at the points $x = 0, \frac{1}{2}, 1$) we obtain the three non-linear equations

$$\begin{aligned}
&\text{(for } x = 0) &&4y_0^2 + \tfrac{8}{3}(y_0 + y_1)^2 + \tfrac{1}{2}(y_0 + y_2)^2 = 6,\\
&\text{(for } x = \tfrac{1}{2}) &&\tfrac{8}{3}(y_0 + y_1)^2 + 8y_1^2 + \tfrac{8}{5}(y_1 + y_2)^2 = 6,\\
&\text{(for } x = 1) &&\tfrac{1}{2}(y_0 + y_2)^2 + \tfrac{8}{5}(y_1 + y_2)^2 + \tfrac{4}{3}y_2^2 = 6.
\end{aligned}$$

These are best solved iteratively by expressing the principal term (underlined) in each equation in terms of the remaining ones, i.e. using the iterative scheme defined by

$$y_0^{[\nu+1]} + y_1^{[\nu+1]} = [\tfrac{3}{8}\{6 - 4y_0^{[\nu]\,2} - \tfrac{1}{2}(y_0^{[\nu]} + y_2^{[\nu]})^2\}]^{\frac{1}{2}}$$

for the first equation and similarly for the others, the bracketed superscripts $[\nu]$ and $[\nu+1]$ characterizing as usual the values obtained by the ν-th and $(\nu+1)$-th cycles of the iteration procedure.

The way in which the calculation is carried out is evident from Table VI/3, in which several cycles of the iteration are reproduced. The starting values

Table VI/3. *Iterative solution of the equations obtained by using* SIMPSON'S *rule*

ν	$y_0^{[\nu]}$	$y_1^{[\nu]}$	$y_2^{[\nu]}$	$y_0^{[\nu]}+y_1^{[\nu]}$	$y_0^{[\nu]}+y_2^{[\nu]}$	$y_1^{[\nu]}+y_2^{[\nu]}$	$y_0^{[\nu+1]}+y_1^{[\nu+1]}$	$y_1^{[\nu+1]}$	$y_1^{[\nu+1]}+y_2^{[\nu+1]}$
0	0·5	0·7	0·9	1·2	1·4	1·6	1·228	0·7085	1·569
1	0·519	0·709	0·861	1·228	1·380	1·569	1·221	0·7080	1·594
2	0·513	0·708	0·886	1·221	1·399	1·594	1·220	0·7062	1·576
3	0·514	0·706	0·870						
4	0·514	0·706	0·877	1·220	1·391	1·583	1·2210	0·7076	1·5825
5	0·5134	0·7076	0·8749						

y_0, y_1, y_2 are taken from the results obtained above with a smaller number of pivotal points. As shown in the table, the calculation can be shortened by estimating new starting values after a few cycles of the iteration.

1.4. The iteration method

With integral equations of the form

$$y(x) = \int_a^b G(x, \xi, y(x), y(\xi))\, d\xi \qquad (1.17)$$

one can choose arbitrarily a function $F_0(x)$ and from it calculate a sequence of functions $F_1, F_2, \ldots$ according to the iterative formula

$$F_{n+1}(x) = \int_a^b G(x, \xi, F_n(x), F_n(\xi))\, d\xi \qquad (n = 0, 1, 2, \ldots). \qquad (1.18)$$

For the linear integral equation (1.2) this formula[1] reads

$$F_{n+1}(x) = f(x) + \lambda \int_a^b K(x, \xi) F_n(\xi)\, d\xi \qquad (n = 0, 1, 2, \ldots), \qquad (1.19)$$

and the sequence $F_n(x)$ converges to the solution of the integral equation if the kernel $K(x, \xi)$ satisfies the conditions laid down in § 2.3 and

---

[1] For cases in which the variation of $K(x, \xi)$ with ξ is small C. WAGNER: On the numerical evaluation of Fredholm integral equations with the aid of the Liouville-Neumann series. J. Math. Phys. **30**, 232—234 (1952), gives a correction which can be used to improve the values at the current stage of the iteration.

$|\lambda| \leq |\lambda_1|$, where λ_1 is the smallest in absolute value of the eigenvalues of the homogeneous integral equation[1] corresponding to (1.2).

For eigenvalue problems [equation (1.2) with $f(x) \equiv 0$] the method parallels the iteration method in Ch. III, § 8 for eigenvalue problems in ordinary differential equations. Here the iterative procedure is defined by

$$F_{n+1}(x) = \int_a^b K(x, \xi) F_n(\xi) \, d\xi \qquad (n = 0, 1, 2, \ldots), \qquad (1.20)$$

and from the successive iterates we calculate the Schwarz constants and Schwarz quotients

$$a_k = \int_a^b F_0(x) F_k(x) \, dx, \qquad \mu_{k+1} = \frac{a_k}{a_{k+1}} \qquad (k = 0, 1, 2, \ldots). \qquad (1.21)$$

For symmetric kernels $K(x, \xi)$ we can also write

$$a_k = \int_a^b F_j(x) F_{k-j}(x) \, dx \qquad (0 \leq j \leq k; \ k = 0, 1, 2, \ldots). \qquad (1.22)$$

The μ_k then provide successive approximations to one of the eigenvalues (care should be exercised in dealing with homogeneous integral equations with unsymmetric kernels, for they need not possess real eigenvalues). Error estimates have been established[2] for problems with symmetric kernels; if, in addition, the kernels are positive definite, the estimates for the smallest eigenvalue of Ch. III (8.18) are also valid here, cf. § 2.3.

1.5. Examples of the iteration method

I. An eigenvalue problem. Consider again the integral equation (1.14), for which approximate solutions were obtained in § 1.3; these solutions suggest that

$$F_0(x, y) = |x| + |y|$$

should be a reasonable starting function.

According to (1.20) we have now to calculate

$$F_1(x, y) = \int_{-1}^1 \int_{-1}^1 |x\xi + y\eta| \, (|\xi| + |\eta|) \, d\xi \, d\eta. \qquad (1.23)$$

---

[1] A detailed and comprehensive presentation can be found in H. Bückner: Die praktische Behandlung von Integralgleichungen. (Ergebnisse der Angew. Math., H. 1.) Berlin-Göttingen-Heidelberg 1952.

[2] Collatz, L.: Schrittweise Näherungen bei Integralgleichungen und Eigenwertproblemen. Math. Z. **46**, 692—708 (1940). — Iglisch, R.: Bemerkungen zu einigen von Herrn Collatz angegebenen Eigenwertabschätzungen bei linearen Integralgleichungen. Math. Ann. **118**, 263—275 (1941).

In evaluating this integral we first suppose that $y \geqq x > 0$; in this half-quadrant we have

$$\frac{1}{2} F_1 = \int_0^1 \int_0^1 (x\xi + y\eta)(\xi + \eta)\, d\xi\, d\eta + \int_{-1}^0 \int_{-\frac{x\xi}{y}}^1 (x\xi + y\eta)(-\xi + \eta)\, d\xi\, d\eta +$$

$$+ \int_0^1 \int_{-\frac{x\xi}{y}}^0 (x\xi + y\eta)(\xi - \eta)\, d\xi\, d\eta = \frac{7}{6} y + \frac{x^2}{4y} + \frac{x^3}{12 y^2} \,.$$

F_1 is determined in the other seven half-quadrants of the fundamental region by symmetry considerations. Then from (1.21) we have

$$a_0 = \int_{-1}^1 \int_{-1}^1 F_0^2\, dx\, dy = 4 \int_0^1 \int_0^1 (x+y)^2 dx\, dy = 8 \times \frac{7}{12} \,,$$

$$a_1 = \int_{-1}^1 \int_{-1}^1 F_0 F_1\, dx\, dy = 8 \int_0^1 \int_0^y 2(x+y)\left(\frac{7}{6} y + \frac{x^2}{4y} + \frac{x^3}{12 y^2}\right) dx\, dy = 8 \times \frac{29}{30} \,,$$

and similarly from (1.22)

$$a_2 = \int_{-1}^1 \int_{-1}^1 F_1^2\, dx\, dy = 8 \times \frac{2713}{1680} \,.$$

Forming the corresponding Schwarz quotients, we obtain for the first eigenvalue the approximations

$$\mu_1 = \frac{a_0}{a_1} = \frac{35}{58} \approx 0{\cdot}603\,45 \quad \text{and} \quad \mu_2 = \frac{a_1}{a_2} = \frac{1624}{2713} \approx 0{\cdot}598\,599 \,.$$

II. A non-linear integral equation. In order to apply the iteration method to the integral equation

$$\int_0^1 \frac{[y(x) + y(\xi)]^2}{1 + x + \xi}\, d\xi = 1 \,, \tag{1.24}$$

we must first put it in the form (1.17). Naturally a general rule which will always be effective in producing the required form cannot be laid down for non-linear cases, but here we can derive a quadratic equation for $y(x)$ by taking it outside of the integral sign:

$$[y(x)]^2 \varphi_0(x) + 2 y(x)\, \varphi_1(x) + \varphi_2(x) = 1 \,, \tag{1.25}$$

where

$$\varphi_\nu(x) = \int_0^1 \frac{[y(\xi)]^\nu}{1 + x + \xi}\, d\xi \qquad (\nu = 0, 1, 2) \,.$$

If we define

$$\varphi_{v,n} = \int\limits_0^1 \frac{[F_n(\xi)]^v}{1+x+\xi}\, d\xi,$$

where $F_n(x)$ is the n-th approximation, and substitute it for $\varphi_v(x)$ in the solution of (1.25), we obtain as the next approximation

$$F_{n+1}(x) = \frac{1}{\varphi_{0,n}}\left[-\varphi_{1,n} + \sqrt{\varphi_{1,n}^2 + \varphi_{0,n}(1-\varphi_{2,n})}\right]. \qquad (1.26)$$

As starting function we choose $F_0(x) = \text{constant} = \delta$; then with $L(x) = \ln\dfrac{2+x}{1+x}$ we obtain $F_1(x) = \dfrac{1}{\sqrt{L(x)}} - \delta$. We now determine δ so that $F_0(x)$ and $F_1(x)$ are of the same order of magnitude, say by demanding that they take the same value at the mid-point $x = \frac{1}{2}$; this yields $\delta = \dfrac{1}{2\sqrt{L(\frac{1}{2})}} \approx 0.6996$. Values of $F_1(x)$ for this value of δ are given in Table VI/4.

Table VI/4. *Specimen values of the iterates*

x	$F_0(x)$	$F_1(x)$	$F_1^*(x)$	$F_2^*(x)$
0	0·6996	0·5015	0·5	0·51852
0·5	0·6996	0·6996	0·7	0·71135
1	0·6996	0·8708	0·9	0·87969

It is now convenient to round off $F_1(x)$ to $F_1^*(x) = 0.5 + 0.4x$ and apply the next iteration step to $F_1^*(x)$, for which the integrations are much simpler:

$$\varphi_{0,1}^*(x) = L(x),$$

$$\varphi_{1,1}^*(x) = 0.4 + (0.1 - 0.4x)L(x),$$

$$\varphi_{2,1}^*(x) = 0.16(2-x) + (0.1 - 0.4x)^2 L(x).$$

We then calculate $F_2^*(x)$ from (1.26); specimen values are reproduced in Table VI/4.

III. An error estimate for a non-linear equation. As an example of an equation for which the existence of and bounds for the solution can be deduced directly from the general iteration theorem of Ch. I, § 5.2, consider the non-linear equation

$$\int\limits_0^1 \sqrt{x + y(\xi)}\, d\xi = y(x)$$

(the application of the theorem does not always proceed so smoothly as for this equation). The operator T is here the integral operator defined by the left-hand side of the equation, and the equation is already of the form $Ty(x) = y(x)$ to which the theorem applies.

We must first define a norm, then determine a Lipschitz constant K as in Ch. I (5.8) for the operator T. By TAYLOR's theorem we have

$$T f_1(x) - T f_2(x) = \int_0^1 \frac{[f_1(\xi) - f_2(\xi)] \, d\xi}{2 \sqrt{x + f_1(\xi) + \vartheta(\xi) [f_2(\xi) - f_1(\xi)]}},$$

where the values of ϑ lie in the range $0 < \vartheta < 1$. Now let us introduce as norm

$$\| f(x) \| = \max_{\langle 0, 1 \rangle} | f(x) |$$

for a continuous function $f(x)$ in $\langle 0, 1 \rangle$, and as the sub-space F of Ch. I, § 5.2 let us choose the sub-space of functions $f(x)$ with $0 < m \le f(x)$ for fixed m; then

$$\| T f_1 - T f_2 \| \le K \| f_1 - f_2 \| \quad \text{if} \quad K = \max_{\langle 0, 1 \rangle} \int_0^1 \frac{d\xi}{2 \sqrt{x + m}} = \frac{1}{2 \sqrt{m}}.$$

Next we use one of the finite-sum methods of §§ 1.2, 1.3 to get a rough quantitative idea of a possible solution. The approximate values y_0, y_1 for $y(0), y(1)$ given by the approximate equations $\frac{1}{2} (\sqrt{y_0} + \sqrt{y_1}) = y_0$, $\frac{1}{2} (\sqrt{1 + y_0} + \sqrt{1 + y_1}) = y_1$ are $y_0 = 1.17$, $y_1 = 1.53$. We therefore try $m = 1$ and choose $u_0(x) = a + bx$, where $a = 1.17$ and $b = 0.36$, as starting function. One step of the iteration procedure yields

$$u_1(x) = T u_0(x) = \frac{2}{3b} [(x + a + b)^{\frac{3}{2}} - (x + a)^{\frac{3}{2}}];$$

in particular, $u_1(0) = 1.1610$. By determining the maximum difference between u_0 and u_1 we find that $\| u_1 - u_0 \| \le 0.01$, and hence, since the choice $m = 1$ gives $K = \frac{1}{2}$ and $\frac{K}{1 - K} = 1$, the "sphere" Σ of Ch. I (5.13) consists of those functions $h(x)$ with $| h - u_1 | \le 0.01$; since this implies that $h(x) \ge 1$, the sphere certainly lies in F, and consequently, by the general theorem of Ch. I, § 5.2, there exists precisely one function $y(x)$ which satisfies the integral equation and the condition $y(x) \ge 1$, and this function lies in the strip $| y(x) - u_1(x) | \le 0.01$.

1.6. Error distribution principles

The error distribution principles of Ch. I, § 4, employed for the approximate solution of differential equations in Ch. III, § 4 and Ch. V, § 4, can be employed similarly for the approximate solution of integral equations.

We shall consider general integral equations in the form (1.5) and linear integral equations in the form (1.2). Let us assume for $y(x)$ the approximation

$$y(x) \approx w(x) = \sum_{j=1}^{p} a_j v_j(x).\tag{1.27}$$

Insertion of the approximation function w into the integral equation yields a residual error function $\varepsilon(x)$; for the general equation (1.5)

$$\varepsilon(x) = \varepsilon(x, a_1, \ldots, a_p) = \int_a^b \Phi(x, \xi, w(x), w(\xi)) \, d\xi\tag{1.28}$$

and for the linear equation (1.2)

$$\varepsilon(x) = \varepsilon(x, a_1, \ldots, a_p) = w(x) - \lambda \int_a^b K(x, \xi) \, w(\xi) \, d\xi - f(x).$$

Again there are several principles on which the determination of the a_j can be based; these were listed and described in detail in Ch. I, § 4.2. Here we mention only the least-squares method and collocation.

1. Least-squares method. Here we demand that

$$J = J(a_1, a_2, \ldots, a_p) = \int_a^b \varepsilon^2 \, dx = \min.,\tag{1.29}$$

and obtain in the necessary minimum conditions

$$\frac{\partial J}{\partial a_\varrho} = 0 \qquad (\varrho = 1, 2, \ldots, p)$$

p (in general non-linear) equations for the a_ϱ.

For the linear integral equation (1.2) we obtain the system of linear equations

$$\sum_{k=1}^{p} b_{jk} a_k = r_j \qquad (j = 1, \ldots, p)$$

with

$$b_{jk} = \int_a^b \left[v_j(x) - \lambda \int_a^b K(x, \xi) \, v_j(\xi) \, d\xi \right] \left[v_k(x) - \lambda \int_a^b K(x, \xi) \, v_k(\xi) \, d\xi \right] dx$$

and

$$r_j = \int_a^b f(x) \left[v_j(x) - \lambda \int_a^b K(x, \xi) \, v_j(\xi) \, d\xi \right] dx.$$

2. Collocation. Here we demand that ε shall vanish at p points x_ϱ in the interval $\langle a, b \rangle$:

$$\varepsilon(x_\varrho) = 0 \qquad (\varrho = 1, 2, \ldots, p; \; a \leqq x_1 < x_2 < \cdots < x_p \leqq b),$$

which similarly yields p (in general non-linear) equations for the a_ϱ.

Example. The non-linear integral equation

$$\int_0^1 \sqrt{x + \xi} \, [y(\xi)]^2 \, d\xi = y(x)$$

31*

has the trivial solution $y(x) \equiv 0$, but also possesses a non-trivial solution. For this non-trivial solution we assume an approximation of the form

$$y \approx w = \sqrt{a_1 + a_2 x},$$

thus departing somewhat from the usual procedure represented by (1.27); it is, however, a more obvious choice in this case. The residual error function which arises is given by

$$\varepsilon(x) = \int_0^1 \sqrt{x+\xi}\,[a_2(x+\xi) + a_1 - a_2 x]\,d\xi - \sqrt{a_1 + a_2 x}$$

$$= \tfrac{2}{5} a_2 \left(\sqrt{x+1}^5 - \sqrt{x}^5\right) + \tfrac{2}{3}(a_1 - a_2 x)\left(\sqrt{x+1}^3 - \sqrt{x}^3\right) - \sqrt{a_1 + a_2 x}.$$

If we use first of all a single-parameter approximation with $a_2 = 0$, and determine a_1 in accordance with the collocation principle from $\varepsilon = 0$ at $x = \frac{1}{2}$, we find that $\sqrt{a_1} = \dfrac{3\sqrt{2}}{3\sqrt{3}-1} \approx 1\cdot011$.

If we incorporate both parameters a_1 and a_2, and use $x = \frac{1}{4}$ and $x = \frac{3}{4}$ as collocation points, a_1 and a_2 are to be determined from the system of non-linear equations

$$\varepsilon\left(\tfrac{1}{4}\right) = a_2 \frac{25\sqrt{5}-1}{80} + \left(a_1 - \frac{a_2}{4}\right)\frac{5\sqrt{5}-1}{12} - \sqrt{a_1 + \frac{1}{4}a_2} = 0,$$

$$\varepsilon\left(\tfrac{3}{4}\right) = a_2 \frac{49\sqrt{7}-9\sqrt{3}}{80} + \left(a_1 - \frac{3a_2}{4}\right)\frac{7\sqrt{7}-3\sqrt{3}}{12} - \sqrt{a_1 + \frac{3}{4}a_2} = 0.$$

To solve these equations we put $a_2 = a_1\gamma$. Then after dividing them through by $\sqrt{a_1}$ we can immediately eliminate $\sqrt{a_1}$ between them and obtain an equation for γ:

$$\sqrt{\frac{4+\gamma}{4+3\gamma}} = \frac{0\cdot848362 + 0\cdot474181\gamma}{1\cdot110342 + 0\cdot592910\gamma}.$$

This has the solution $\gamma = 1\cdot8643$, which yields $a_1 = 0\cdot4886$, $a_2 = 0\cdot9108$; specimen values of the corresponding approximation for $y(x)$ are

x	0	0·5	1
$w(x)$	0·6990	0·9716	1·1830

1.7. Connection with variational problems

In order to extend the technique of Ch. III, §§ 5, 6 and Ch. V, § 5 to integral equations, we must be able to find a variational problem whose Euler equation can be identified with the given integral equation. First of all we derive the Euler equation for a general class of variational problems, namely

$$J = J[u] = \int_a^b\!\!\int_a^b F\big(x, \xi, u(x), u(\xi)\big)\,dx\,d\xi + \int_a^b G\big(\xi, u(\xi)\big)\,d\xi = \text{extr.} \quad (1.30)$$

with respect to the domain of continuous functions $u(x)$ in the interval $\langle a, b \rangle$; we shall assume here that the given functions $F(x, \xi, u_1, u_2)$ and $G(\xi, u)$ are continuous in all their arguments, x, ξ, u_1, u_2 and ξ, u, respectively. We derive here only the necessary Euler equation which must be satisfied by a solution $y(x)$ if it exists, and assume that the variational problem (1.30) does, in fact, possess such a solution with the property that

$$J[y] \leqq J[u].$$

In the usual way we consider any one-parameter family of admissible functions of the form

$$u = y + \varepsilon \eta,$$

where ε is the parameter and η an admissible function. For this family of functions $J[y + \varepsilon \eta] = \Phi(\varepsilon)$ is a function of ε which must have a minimum value for $\varepsilon = 0$. A necessary condition that y shall minimize J is therefore that

$$\left(\frac{d\Phi}{d\varepsilon} \right)_{\varepsilon=0} = \left(\frac{d}{d\varepsilon} J[y + \varepsilon \eta] \right)_{\varepsilon=0} = 0$$

for any admissible function η.

Let us denote the partial derivatives of F with respect to u_1 and u_2 and of G with respect to u by subscripts thus

$$F_j = \frac{\partial F}{\partial u_j} \quad (j = 1, 2), \qquad G_u = \frac{\partial G}{\partial u}.$$

Taylor expansion of F yields

$$F\big(x, \xi, y(x) + \varepsilon \eta(x), y(\xi) + \varepsilon \eta(\xi)\big) = F\big(x, \xi, y(x), y(\xi)\big) +$$
$$+ \varepsilon \eta(x) F_1\big(x, \xi, y(x), y(\xi)\big) + \varepsilon \eta(\xi) F_2\big(x, \xi, y(x), y(\xi)\big) + \varepsilon^2 \ldots,$$

and hence

$$\left(\frac{d}{d\varepsilon} J[y + \varepsilon \eta] \right)_{\varepsilon=0} = \int_a^b \int_a^b [\eta(x) F_1 + \eta(\xi) F_2] \, dx \, d\xi + \int_a^b \eta(\xi) G_u d\xi.$$

Now in the second term in the integrand of the first integral we can interchange x and ξ, so that $\eta(x)$ will appear as a factor in both terms; further we can write x in place of ξ in the second integral. Our necessary condition can therefore be written

$$\int_a^b \eta(x) \, S \, dx = 0, \tag{1.31}$$

where

$$S = \int_a^b \big[F_1(x, \xi, y(x), y(\xi)) + F_2(\xi, x, y(\xi), y(x)) \big] \, d\xi + G_u\big(x, y(x)\big). \tag{1.32}$$

Since (1.31) is to hold for any continuous function $\eta(x)$, S must vanish identically; thus $S = 0$ is the necessary condition (Euler equation) for the variational problem (1.30).

We now consider the inverse problem of finding a suitable variational problem for a given integral equation. We try to bring the integral equation

$$g\left(x, y(x)\right) + \int_a^b h\left(x, \xi, y(x), y(\xi)\right) d\xi = 0 \tag{1.33}$$

into the form $S = 0$. First of all we can put

$$g(x, y) = G_u(x, y),$$

so that

$$G(x, y) = \int_{y_0}^y g(x, t)\, dt, \tag{1.34}$$

in which y_0 can be chosen arbitrarily. Sometimes it may be expedient to multiply the integral equation (1.33) through by a suitable function $s(x)$ beforehand.

In order to bring (1.33) into the form $S = 0$ we have yet to satisfy

$$h\left(x, \xi, y(x), y(\xi)\right) = F_1\left(x, \xi, y(x), y(\xi)\right) + F_2\left(\xi, x, y(\xi), y(x)\right). \tag{1.35}$$

The question of the existence of a function F related to a given function h by (1.35) will not be pursued further here; we content ourselves with listing in Table VI/5 some typical possibilities obtained by calculating h from some simple forms for F.

Table VI/5. *Some corresponding integrand functions*

$F(x, \xi, y(x), y(\xi))$	$h(x, \xi, y(x), y(\xi))$
$\frac{1}{2}K(x, \xi)\, y(x)\, y(\xi)$	$\frac{1}{2}[K(x, \xi) + K(\xi, x)]\, y(\xi)$
$K(x, \xi)\, y(x)$	$K(x, \xi)$
$K(x, \xi) \int_0^{y(x)} \varphi(u)\, du$	$K(x, \xi)\, \varphi(y(x))$
$K(x, \xi)\, \varphi(y(x)\, y(\xi))$	$[K(x, \xi) + K(\xi, x)]\, \varphi'(y(x)\, y(\xi))\, y(\xi)$
$K(x, \xi)\, \varphi(y(x) + y(\xi))$	$[K(x, \xi) + K(\xi, x)]\, \varphi'(y(x) + y(\xi))$

By referring to this table, which can easily be extended, and using (1.34) one can find a corresponding variational problem (1.30) for many cases of integral equations of the form (1.33); for example, for the linear integral equation

$$y(x) - \lambda \int_a^b K(x, \xi)\, y(\xi)\, d\xi = f(x) \tag{1.36}$$

with a symmetric kernel $K(x, \xi) = K(\xi, x)$ we obtain the variational problem

$$J[u] = \tfrac{1}{2}\lambda \int_a^b\int_a^b K(x, \xi)\, u(x)\, u(\xi)\, dx\, d\xi +$$
$$+ \int_a^b \left(f(\xi)\, u(\xi) - \tfrac{1}{2}[u(\xi)]^2\right) d\xi = \text{extremum.}\right\} \tag{1.37}$$

Examples. I. A linear inhomogeneous integral equation arising in a measurement problem.

Suppose that when an attempt is made to measure an intensity distribution $y(x)$, say along the section $-1 \le x \le 1$ of the x axis, the reading $f(x)$ differs from the true value $y(x)$ because of the influence of neighbouring elements. Let the influence at the point x of an element $d\xi$ at the point ξ be $K(x, \xi)\, y(\xi)\, d\xi$; normally $K(x, \xi)$ will depend only on the distance $|x - \xi|$, and here we will assume that

$$K(x, \xi) = \begin{cases} 1 - (x - \xi)^2 & \text{for } |x - \xi| \le 1 \\ 0 & \text{for } |x - \xi| \ge 1. \end{cases} \tag{1.38}$$

Then the reading taken at the point x will be

$$y(x) + \int_{-1}^1 K(x, \xi)\, y(\xi)\, d\xi = f(x).$$

Thus we have an integral equation for the required true distribution $y(x)$ in terms of the observed (measured) distribution $f(x)$. For this example we will take

$$f(x) = \frac{1}{1 + x^2}.$$

(1.36), (1.37) yield immediately the corresponding variational problem

$$J[u] = \frac{1}{2}\int_{-1}^1\int_{-1}^1 K(x, \xi)\, u(x)\, u(\xi)\, dx\, d\xi + \frac{1}{2}\int_{-1}^1 u^2(x)\, dx - \int_{-1}^1 \frac{u(x)}{1 + x^2}\, dx = \text{extr.}$$

Taking account of the symmetry, we assume the two-term approximation

$$u = a + b\, x^2, \tag{1.39}$$

and instead of first forming J, we form the expressions $\partial J/\partial a$, $\partial J/\partial b$ directly; differentiation of the double integral with respect to a, for instance, yields

$$\int_{-1}^1\int_{-1}^1 K(x, \xi)\, (2a + b(x^2 + \xi^2))\, dx\, d\xi$$
$$= 2\int_0^1\int_{\xi-1}^1 (2a + b(x^2 + \xi^2))\,(1 - (x - \xi)^2)\, dx\, d\xi = \frac{13}{3}a + \frac{6}{5}b$$

We obtain for a and b the equations

$$\frac{\partial J}{\partial a} = \frac{25}{6}a + \frac{19}{15}b - \frac{\pi}{2} = 0,$$

$$\frac{\partial J}{\partial b} = \frac{19}{15}a + \frac{73}{60}b - \frac{4 - \pi}{2} = 0,$$

with the solution

$$a = 0{\cdot}4428,$$
$$b = -0{\cdot}2164.$$

This parabolic approximation is quite crude, however; in Fig. VI/6 it is compared with the results obtained by the summation methods of § 1.2 with $h = \frac{1}{4}$ and $h = \frac{1}{3}$; these results show that the curve of the solution $y(x)$ is bell-shaped, and it cannot therefore be followed closely by a parabola, which has no points of inflexion.

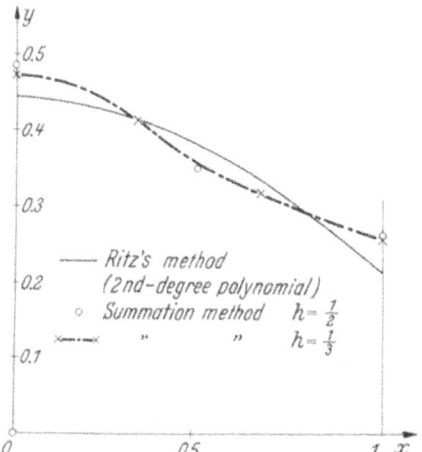

Fig. VI/6. Some approximate solutions of an inhomogeneous integral equation obtained by Ritz's method and summation methods

II. A non-linear integral equation (already treated by collocation in § 1.6).

There are various ways of setting up a corresponding variational problem for the equation

$$\int_0^1 \sqrt{x + \xi}\, y^2(\xi)\, d\xi = y(x);$$

one could, for example, multiply the equation through by $y(x)$, thereby bringing the integrand into the form given in Table VI/5, right-hand column, last row but one; we see from the left-hand column that this is generated by (1.35) from the function

$$F = \tfrac{1}{4}\sqrt{x + \xi}\, y^2(x)\, y^2(\xi).$$

Here, however, it is somewhat simpler to write the integral equation as

$$\int_0^1 \sqrt{x + \xi}\, z(\xi)\, d\xi = \sqrt{z(x)}$$

by introducing $z(x) = y^2(x)$; then according to (1.33), (1.34), (1.36), (1.37) a corresponding variational expression reads

$$J[u] = \int_0^1\!\!\int_0^1 \tfrac{1}{2}\sqrt{x + \xi}\, u(x)\, u(\xi)\, dx\, d\xi - \tfrac{2}{3}\int_0^1 u^{\frac{3}{2}}(x)\, dx.$$

For the simple linear approximation $u(x) = a + bx$ the necessary minimum conditions read

$$\frac{\partial J}{\partial a} = \int_0^1\!\!\int_0^1 \tfrac{1}{2}\sqrt{x + \xi}\,\{2a + b(x + \xi)\}\, dx\, d\xi - \int_0^1 \sqrt{a + bx}\, dx = 0,$$

$$\frac{\partial J}{\partial b} = \int_0^1\!\!\int_0^1 \tfrac{1}{2}\sqrt{x + \xi}\,\{a(x + \xi) + 2b\xi x\}\, dx\, d\xi - \int_0^1 x\sqrt{a + bx}\, dx = 0.$$

On evaluating the integrals we obtain the two non-linear equations

$$\alpha a + \beta b = \frac{2}{3b}\left(\sqrt{a+b}^3 - \sqrt{a}^3\right)$$

$$\beta a + \gamma b = \frac{2}{5b^2}\left(\sqrt{a+b}^5 - \sqrt{a}^5\right) - \frac{2a}{3b^2}\left(\sqrt{a+b}^3 - \sqrt{a}^3\right),$$

where

$$\alpha = \frac{8}{15}\left(2\sqrt{2} - 1\right) \approx 0.97515,$$

$$\beta = \frac{4}{35}\left(4\sqrt{2} - 1\right) \approx 0.53221,$$

$$\gamma = \frac{16}{135}\left(\sqrt{2} + 1\right) \approx 0.28613,$$

and these can easily be reduced to a single equation for the ratio $c = b/a$:

$$\frac{\alpha + \beta c}{\beta + \gamma c} = \frac{\frac{2}{3}c\left(\sqrt{1+c}^3 - 1\right)}{\frac{2}{5}\left(\sqrt{1+c}^5 - 1\right) - \frac{2}{3}\left(\sqrt{1+c}^3 - 1\right)};$$

this has the solution $c = 1.8861$, yielding $a = 0.4861$, $b = 0.9168$. Specimen values of the corresponding approximate solution u are

x	0	0.5	1
u	0.6972	0.9719	1.1844

which agree well with the results of § 1.6.

1.8. Integro-differential equations and variational problems

Here we can be brief because of the similarity with the last section and with Ch. III, § 6.1.

Suppose that the expression

$$\left.\begin{aligned}J[u] = \int_a^b\int_a^b & F\left(x, \xi, u(x), u(\xi), u'(x), u'(\xi), \ldots, u^{(n)}(x), u^{(n)}(\xi)\right) dx\,d\xi + \\ & + \int_a^b G\left(\xi, u(\xi), u'(\xi), \ldots, u^{(n)}(\xi)\right) d\xi + \sum_{\nu=0}^{n-1}\left(a_\nu u^{(\nu)}(a) + \right. \\ & \left. + b_\nu u^{(\nu)}(b)\right) + \sum_{\mu,\nu=0}^{n-1}\left(a_{\mu\nu}u^{(\mu)}(a)\,u^{(\nu)}(a) + b_{\mu\nu}u^{(\mu)}(b)\,u^{(\nu)}(b)\right)\end{aligned}\right\} \quad (1.40)$$

is to be minimized. F and G are given functions continuous in each of their arguments and with as many continuous partial derivatives as are needed in the following; and a_ν, b_ν, $a_{\mu\nu}$, $b_{\mu\nu}$ are given constants.

Here we can prescribe certain boundary conditions

$$U_\mu[u(a), u(b), u'(a), u'(b), \ldots, u^{(n-1)}(a), u^{(n-1)}(b)] = \gamma_\mu \\ (\mu = 1, 2, \ldots, k; \ k \leqq 2n), \tag{1.41}$$

which we will assume to be linear in the boundary values specified. Let there exist a solution $y(x)$ of this variational problem. Then the usual procedure of considering a one-parameter family of admissible functions $u = y + \varepsilon\eta$ leads to the necessary condition

$$0 = \left(\frac{d}{d\varepsilon} J[y + \varepsilon\eta]\right)_{\varepsilon=0} \\ = \int_a^b \int_a^b \sum_{\nu=0}^n \left(F_{1\nu}\eta^{(\nu)}(x) + F_{2\nu}\eta^{(\nu)}(\xi)\right) dx\, d\xi + \int_a^b \sum_{\nu=0}^n G_\nu \eta^{(\nu)}(\xi)\, d\xi + \Psi \tag{1.42}$$

for any function η satisfying the homogeneous boundary conditions

$$U_\mu[\eta] = 0 \qquad (\mu = 1, 2, \ldots, k), \tag{1.43}$$

where in (1.42) we have introduced the notation

$$F_{1\nu} = \frac{\partial F}{\partial u^{(\nu)}(x)}, \qquad F_{2\nu} = \frac{\partial F}{\partial u^{(\nu)}(\xi)}, \qquad G_\nu = \frac{\partial G}{\partial u^{(\nu)}}, \tag{1.44}$$

$$\Psi = \sum_{\nu=0}^{n-1} [a_\nu \eta^{(\nu)}(a) + b_\nu \eta^{(\nu)}(b)] + \sum_{\mu,\nu=0}^{n-1} [a_{\mu\nu}(y^{(\mu)}(a)\eta^{(\nu)}(a) + \\ + y^{(\nu)}(a)\eta^{(\mu)}(a)) + b_{\mu\nu}(y^{(\mu)}(b)\eta^{(\nu)}(b) + y^{(\nu)}(b)\eta^{(\mu)}(b))]. \tag{1.45}$$

Further, let a tilde embellishing a function symbol signify that x and ξ have been interchanged at every place where they occur in that function, so that, for example,

$$\tilde{F}_{1\nu} = F_{1\nu}(\xi, x, u(\xi), u(x), u'(\xi), u'(x), \ldots, u^{(n)}(\xi), u^{(n)}(x)),$$

and put

$$F_{1\nu} + \tilde{F}_{2\nu} = \Phi_\nu.$$

Then by interchanging x and ξ in the single integral, and also in the F_2 terms of the double integral, we can write (1.42) in the form

$$0 = \int_a^b \int_a^b \sum_{\nu=0}^n \eta^{(\nu)}(x) \Phi_\nu\, dx\, d\xi + \int_a^b \sum_{\nu=0}^n \eta^{(\nu)}(x) G_\nu\, dx + \Psi. \tag{1.46}$$

We can now transform it further by integration by parts. We have, for example,

$$\int_a^b \int_a^b \eta'(x) \Phi_1\, dx\, d\xi = \left[\eta(x) \int_a^b \Phi_1\, d\xi\right]_a^b - \int_a^b \eta(x) \left(\int_a^b \frac{d}{dx} \Phi_1\, d\xi\right) dx;$$

and similarly transforming all other terms (integrating the term in $\eta^{(\nu)}$ by parts ν times), we obtain

$$
\begin{aligned}
= \int_a^b \eta(x) &\left\{ \int_a^b \left(\Phi_0 - \frac{d}{dx}\Phi_1 + \frac{d^2}{dx^2}\Phi_2 - \cdots \right) d\xi + \left(G_0 - \frac{d}{dx}G_1 + \frac{d^2}{dx^2}G_2 - \cdots \right) \right\} dx + \\
&+ \left[\eta(x) \left\{ \int_a^b \left(\Phi_1 - \frac{d}{dx}\Phi_2 + \frac{d^2}{dx^2}\Phi_3 - \cdots \right) d\xi + \left(G_1 - \frac{d}{dx}G_2 + \frac{d^2}{dx^2}G_3 - \cdots \right) \right\} \right]_a^b + \\
&+ \left[\eta'(x) \left\{ \int_a^b \left(\Phi_2 - \frac{d}{dx}\Phi_3 + \frac{d^2}{dx^2}\Phi_4 - \cdots \right) d\xi + \left(G_2 - \frac{d}{dx}G_3 + \frac{d^2}{dx^2}G_4 - \cdots \right) \right\} \right]_a^b + \\
&+ \left[\eta''(x) \{ \cdots \} \right]_a^b + \cdots + \left[\eta^{(n-1)}(x) \left\{ \int_a^b \Phi_n\, d\xi + G_n \right\} \right]_a^b + \Psi.
\end{aligned} \tag{1.}
$$

On account of the arbitrariness of η, the factor multiplying $\eta(x)$ in the integral must vanish; this yields the integro-differential equation

$$
\int_a^b \left(\sum_{\nu=0}^n (-1)^\nu \frac{d^\nu}{dx^\nu} \Phi_\nu \right) d\xi + \sum_{\nu=0}^n (-1)^\nu \frac{d^\nu}{dx^\nu} \left(\frac{\partial G}{\partial u^{(\nu)}(x)} \right) = 0; \tag{1.48}
$$

The integral with respect to x therefore drops out of (1.47), leaving a sum which is linear and homogeneous in the boundary values $\eta(a)$, $\eta'(a), \ldots, \eta^{(n-1)}(a), \eta(b), \eta'(b), \ldots, \eta^{(n-1)}(b)$. In general, k of these $2n$ boundary values can be expressed in terms of the remaining ones by means of the boundary conditions (1.43). These $2n - k$ remaining boundary values may be called "free boundary values", and will be denoted in any order by $\eta_1, \eta_2, \ldots, \eta_{2n-k}$. If all boundary values of η appearing in (1.47) are now expressed in terms of the free boundary values, the sum remaining in (1.47) assumes the form

$$
\sum_{\nu=1}^{2n-k} \eta_\nu W_\nu[y] = 0. \tag{1.49}
$$

Since the η_ν may be chosen arbitrarily, we must have

$$
W_\nu[y] = 0 \qquad (\nu = 1, 2, \ldots, 2n - k). \tag{1.50}
$$

These equations constitute $2n - k$ further boundary conditions to be satisfied necessarily by the solution y of the variational problem.

Thus y must satisfy the integro-differential equation (1.48), which in general is of the $2n$-th order, together with the $2n$ boundary conditions (1.41), (1.50). In general this determines y uniquely.

The following two possibilities should be noted: that integrals over the fundamental interval can occur in the boundary conditions (1.50) derived from (1.47), and that boundary values of y can appear in the integro-differential equation (1.48) if integration by parts is applied to the integral.

Example. The vertical oscillations of a suspension bridge satisfy the integro-differential equation[1]

$$(E I y'')'' - H y'' + k \int_{-l}^{l} y(\xi) \, d\xi = m \, \omega^2 y$$

together with certain boundary conditions, which, if the bridge is assumed to be pinned at each end, read

$$y(\pm l) = y''(\pm l) = 0.$$

Here, EI is the flexural rigidity of the stiffening girder, H the horizontal component of the tension in the cable, m the mass per unit length of the oscillating parts, ω the frequency of the oscillations, y the deflection due to the oscillations, $2l$ the span and k a certain constant.

For this problem a corresponding variational problem reads[2]

$$J[u] = \int_{-l}^{l} (E I u''^2 + H u'^2 - m \, \omega^2 u^2) \, dx + k \left(\int_{-l}^{l} u(\xi) \, d\xi \right)^2 = \text{extr.}$$

with respect to the domain of admissible functions u satisfying the essential boundary conditions $u(\pm 1) = 0$.

1.9. Series solutions

Integral equations frequently arise whose form suits them for treatment by a series expansion method of one kind or another — a power series or trigonometric series method, for example —, the appropriate series to use depending on the particular form of the integral equation. For instance, trigonometric series are very effective for linear integral equations of the form (1.2) with "convolution-type kernels", i.e. integral equations of the form

$$y(x) = f(x) + \lambda \int_{a}^{b} K(x - \xi) \, y(\xi) \, d\xi, \tag{1.51}$$

in which the kernel is a function only of the difference $x - \xi$, provided also that K is periodic with period $b - a$:

$$K(x - \xi) = K(x - \xi + b - a).$$

We imagine the functions $K(x)$, $f(x)$ and $y(x)$ to be expanded as trigonometric series written in complex form:

$$\left. \begin{aligned} K(x) &= \sum_{\nu=-\infty}^{\infty} k_\nu \, e^{2\pi i \frac{\nu x}{b-a}}, \qquad f(x) = \sum_{\nu=-\infty}^{\infty} f_\nu \, e^{2\pi i \frac{\nu x}{b-a}}, \\ y(x) &= \sum_{\nu=-\infty}^{\infty} y_\nu \, e^{2\pi i \frac{\nu x}{b-a}} \end{aligned} \right\} \tag{1.52}$$

---

[1] Klöppel, K., and H. Lie: Lotrechte Schwingungen von Hängebrücken. Ing.-Arch. **13**, 211—266 (1942).

[2] A numerical example can be found in L. Collatz: Eigenwertaufgaben, pp. 244, 377. Leipzig 1949.

and insert these series into the integral equation (1.51); assuming their convergence to be sufficiently strong that they may be multiplied together and integrated term by term we obtain

$$\sum_{\nu=-\infty}^{\infty} (y_\nu - f_\nu) \, e^{2\pi i \frac{\nu x}{a-b}} = \lambda \sum_{\nu=-\infty}^{\infty} k_\nu e^{2\pi i \frac{\nu x}{b-a}} \sum_{\mu=-\infty}^{\infty} y_\mu \int_a^b e^{2\pi i \frac{\mu-\nu}{b-a} \xi} \, d\xi.$$

Now

$$\int_a^b e^{2\pi i \frac{k\xi}{b-a}} d\xi = \begin{cases} b-a & \text{for} \quad k = 0 \\ 0 & \text{for} \quad \text{integral } k \neq 0, \end{cases}$$

so that the only non-zero terms in the double sum are those with $\mu = \nu$, and by equating coefficients of $e^{2\pi i \frac{\nu x}{b-a}}$ we find that

$$y_\nu - f_\nu = \lambda (b - a) \, k_\nu \, y_\nu,$$

or

$$y_\nu = \frac{f_\nu}{1 - \lambda (b - a) \, k_\nu}. \tag{1.53}$$

Thus, provided that the denominator in (1.53) does not vanish, i.e. provided that λ is not one of the eigenvalues $\dfrac{1}{(b-a) \, k_\nu}$, we know the Fourier coefficients y_ν and hence also, from (1.52), the solution $y(x)$.

When the fundamental interval is infinite, we use FOURIER's integral theorem instead of the Fourier expansions (1.52) (see textbooks on integral equations)[1].

For linear integral equations of the form (1.2) the solution $y(x)$ is sometimes[2] expanded as a series of functions which are orthogonal over the fundamental interval, i.e. functions $z_1(x), z_2(x), \ldots$ with

$$\int_a^b z_j(x) \, z_k(x) \, dx = \begin{cases} 0 & \text{for} \quad j \neq k, \\ s_j & \text{for} \quad j = k. \end{cases} \tag{1.54}$$

The series

$$y(x) = \sum_{k=1}^{\infty} a_k z_k(x) \tag{1.55}$$

---

[1] An application to the integro-differential equation

$$y(x) = \int_{-\infty}^{+\infty} K(|x - \xi|) \, (p(\xi) - c \, y^{IV}(\xi)) \, d\xi$$

[for which boundary conditions are replaced by the condition that y shall be integrable in $(-\infty, \infty)$], which arises in the investigation of the effects of elasticity in railway track mountings, can be found in M. E. REISSNER: On the theory of beams resting on a yielding foundation. Proc. Nat. Acad. Sci., Wash. **23**, 328–333 (1937).

[2] Various series expansions are used by D. ENSKOG: Eine allgemeine Methode zur Auflösung von linearen Integralgleichungen. Math. Z. **24**, 670–683 (1926).

is inserted into the integral equation (1.2) and the resulting equation

$$\sum_{k=1}^{\infty} a_k z_k(x) - \lambda \int_a^b \sum_{k=1}^{\infty} a_k K(x, \xi) z_k(\xi) d\xi = f(x)$$

multiplied by $z_j(x)$, then integrated over the fundamental interval $\langle a, b \rangle$. Assuming that term-by-term integration is permissible (which has to be verified separately in individual cases), we obtain the following infinite system of equations for a countably infinite number of unknowns a_j:

$$a_j s_j - \lambda \sum_{k=1}^{\infty} \varkappa_{jk} a_k = r_j \qquad (j = 1, 2, \ldots), \tag{1.56}$$

where

$$\varkappa_{jk} = \int_a^b \int_a^b K(x, \xi) z_j(x) z_k(\xi) d x d\xi, \qquad r_j = \int_a^b f(x) z_j(x) d x. \tag{1.57}$$

This infinite system is usually solved approximately by retaining only the first p equations and solving them for the first p unknowns $a_1, \ldots, a_p$ with the remaining $a_r (r > p)$ put equal to zero. The values so obtained may be called the p-th approximation and will be denoted by $a_1^{(p)}, \ldots, a_p^{(p)}$; they are calculated from the finite system of equations

$$a_j^{(p)} s_j - \lambda \sum_{k=1}^{p} \varkappa_{jk} a_k^{(p)} = r_j \qquad (j = 1, 2, \ldots, p). \tag{1.58}$$

For the corresponding eigenvalue problem, i.e. (1.2) with $f(x) \equiv 0$, we have $r_j \equiv 0$ and we calculate the p-th approximations $\Lambda_1^{(p)}, \Lambda_2^{(p)}, \ldots, \Lambda_p^{(p)}$ to the eigenvalues λ_j as the roots of the algebraic equation obtained by putting the determinant of (1.58) equal to zero[1]:

$$\begin{vmatrix} s_1 - \lambda \varkappa_{11} & -\lambda \varkappa_{12} \ldots & -\lambda \varkappa_{1p} \\ -\lambda \varkappa_{21} & s_2 - \lambda \varkappa_{22} \ldots & -\lambda \varkappa_{2p} \\ \cdot \quad \cdot \quad \cdot \quad \cdot & \cdot \quad \cdot \quad \cdot \quad \cdot & \cdot \quad \cdot \quad \cdot \\ -\lambda \varkappa_{p1} & -\lambda \varkappa_{p2} \ldots & s_p - \lambda \varkappa_{pp} \end{vmatrix} = 0. \tag{1.59}$$

---

[1] LÖSCH, F.: Zur praktischen Berechnung der Eigenwerte linearer Integralgleichungen. Z. Angew. Math. Mech. **24**, 35—41 (1944). Under the assumption that $K(x, \xi)$ is continuous and can be expanded uniformly as a series of the form

$$K(x, \xi) = \sum_{\nu=1}^{\infty} \varphi_\nu(x) z_\nu(\xi) \quad \text{with} \quad \varphi_\nu(x) = \int_a^b K(x, \xi) z_\nu(\xi) d\xi,$$

LÖSCH proves that the $\Lambda_j^{(p)}$ tend to the eigenvalues λ_j of the integral equation in "position and order" and that when the eigenvalues of the integral equation are all simple, the suitably normalized approximating functions

$$\sum_{j=1}^{p} a_j^{(p)} z_j(x)$$

converge uniformly in $a \leq x \leq b$ to the corresponding eigenfunction of the integral equation. Symmetry of the kernel is not assumed.

The method is also applicable to linear integro-differential equations of the form (1.3) with the boundary conditions (1.4). Here we use a set of functions $z_0(x), z_1(x), z_2(x), \ldots$ with $z_0(x)$ satisfying the inhomogeneous boundary conditions (1.4) and the other $z_j(x)$ satisfying the corresponding homogeneous conditions:

$$U_\mu[z_0] = \gamma_\mu \qquad \{\mu = 1, 2, \ldots, m$$
$$U_\mu[z_j] = 0 \qquad \{j = 1, 2, \ldots.$$

Inserting

$$y = z_0 + \sum_{k=1}^{\infty} a_k z_k \tag{1.60}$$

into the integro-differential equation (1.3), we obtain (again assuming that the various operations may be performed term by term)

$$\left.\begin{aligned}
\sum_{k=1}^{\infty} a_k M[z_k] - \lambda \int_a^b \sum_{k=1}^{\infty} a_k K(x, \xi)\, N[z_k(\xi)]\, d\xi &= \varphi(x) \\
&= f(x) - M[z_0(x)] + \lambda \int_a^b K(x, \xi)\, N[z_0(\xi)]\, d\xi.
\end{aligned}\right\} \tag{1.61}$$

The constants a_k could be determined approximately by one of the general methods of § 1.6 — collocation, for example — but we can also use a method allied to the method described above for the special case $M[y] = N[y] = y$. Let $w_1(x), w_2(x), \ldots$ be a system of functions which is complete over the interval $\langle a, b \rangle$. Then multiplication of (1.61) by $w_j(x)$ and integration over the interval $\langle a, b \rangle$ yields

$$\sum_{k=1}^{\infty} (m_{jk} - \lambda n_{jk})\, a_k = r_j \qquad (j = 1, 2, \ldots), \tag{1.62}$$

where

$$\left.\begin{aligned}
m_{jk} &= \int_a^b w_j(x)\, M[z_k(x)]\, dx, \\
n_{jk} &= \int_a^b \int_a^b K(x, \xi)\, w_j(x)\, N[z_k(\xi)]\, dx\, d\xi, \\
r_j &= \int_a^b \varphi(x)\, w_j(x)\, dx.
\end{aligned}\right\} \tag{1.63}$$

Again we have [in (1.62)] an infinite system of equations for the unknowns a_j and can obtain approximate solutions as above by retaining only the first p equations and the first p unknowns.

For the corresponding eigenvalue problem, in which $f(x), \gamma_\mu, z_0(x)$, $\varphi(x), r_j$ are all identically zero, it is to be recommended that the $z_k(x)$ and $w_j(x)$ be so chosen that $m_{jk} = 0$ for $j \neq k$; then with $\lambda = 1/\varkappa$ the equation for the approximate eigenvalues reduces to the secular equation

$$
\begin{vmatrix}
n_{11} - \varkappa\, m_{11} & n_{12} & \cdots\, n_{1p} \\
n_{21} & n_{22} - \varkappa\, m_{22} \cdots n_{2p} \\
\cdot \;\; \cdot \;\;\; \cdot \;\;\; \cdot & \cdot \;\;\; \cdot \;\;\; \cdot \;\;\; \cdot \;\;\; \cdot \;\;\; \cdot \;\;\; \cdot \\
n_{p1} & n_{p2} & \cdots\, n_{pp} - \varkappa\, m_{pp}
\end{vmatrix} = 0.
$$

This can be achieved, for example, by choosing the $z_k(x)$ as the eigenfunctions of the eigenvalue problem

$$
M[z_k] = \lambda z_k, \qquad U_\mu[z_k] = 0
$$

and putting

$$
w_j(x) = z_j(x).
$$

1.10. Examples

I. An inhomogeneous integro-differential equation. Consider the equation

$$
\frac{1+x}{2}\, y'' + 10 \int\limits_0^1 \frac{y(\xi)\,d\xi}{1+x+\xi} + 1 = 0
$$

with the boundary conditions

$$
y(0) = y(1) = 0.
$$

For the series satisfying the boundary conditions we take

$$
y(x) = \sum_{k=1}^{\infty} a_k(x - x^{k+1}).
$$

For a two-parameter approximation we retain only the first two terms, putting $a_k = 0$ for $k \geq 3$; then insertion into the equation yields

$$
a_1(-1-x) + a_2(-3x - 3x^2) + 10 \int\limits_0^1 \frac{a_1(\xi - \xi^2) + a_2(\xi - \xi^3)}{1+x+\xi}\, d\xi + 1 = 0. \tag{1.64}
$$

Here, suitable functions for the $w_j(x)$ are the powers $x^{j-1}\,(j = 1, 2, \ldots)$; thus one equation for a_1, a_2 is obtained by integrating (1.64) over the interval $\langle 0, 1 \rangle$ and the other by first multiplying (1.64) by x, then integrating over $\langle 0, 1 \rangle$. Before writing these equations down, we evaluate separately the more complicated integrals which occur:

$$
J_{11} = \int\limits_0^1\!\!\int\limits_0^1 \frac{(\xi - \xi^2)\,dx\,d\xi}{1+x+\xi} = \frac{1}{6}\,(8 + 32 \ln 2 - 27 \ln 3) \qquad = 0.0863630,
$$

$$
J_{12} = \int\limits_0^1\!\!\int\limits_0^1 \frac{(\xi - \xi^3)\,dx\,d\xi}{1+x+\xi} = \frac{1}{4}\left(-\frac{23}{6} - 8 \ln 2 + 9 \ln 3\right) \qquad = 0.1272500,
$$

$$J_{21} = \int_0^1 \int_0^1 \frac{x(\xi - \xi^2)\,dx\,d\xi}{1 + x + \xi} = \frac{1}{4}\left(\frac{9}{2} + 8\ln 2 - 9\ln 3\right) \qquad = 0 \cdot 0394167,$$

$$J_{22} = \int_0^1 \int_0^1 \frac{x(\xi - \xi^3)\,dx\,d\xi}{1 + x + \xi} = \frac{1}{60}\left(-\frac{49}{2} - 88\ln 2 + 81\ln 3\right) = 0 \cdot 0581774$$

(a useful check on the order of magnitude of the value obtained in each case can be made by considering the range of values assumed by the integrand).

The two equations for a_1 and a_2 read

$$a_1\left(-\tfrac{3}{2} + 10 J_{11}\right) + a_2\left(-\tfrac{5}{2} + 10 J_{12}\right) + 1 = 0,$$

$$a_1\left(-\tfrac{5}{6} + 10 J_{21}\right) + a_2\left(-\tfrac{7}{4} + 10 J_{22}\right) + \tfrac{1}{2} = 0,$$

and have the solution

$$a_1 = 2 \cdot 7135, \qquad a_2 = -0 \cdot 59206.$$

As an approximation for the mid-point value, for example, we obtain

$$y\left(\tfrac{1}{2}\right) = 0 \cdot 4563.$$

II. A non-linear integral equation. For the integral equation

$$\int_0^1 \frac{[y(x) + y(\xi)]^2}{1 + x + \xi}\,d\xi = 1,$$

already considered in §§ 1.3, 1.5, it is expedient to move the origin to the centre of the fundamental interval before using power series expansions; we introduce the new variables $u = x - \tfrac{1}{2}$, $v = \xi - \tfrac{1}{2}$, $z(u) = y(u + \tfrac{1}{2})$, thus transforming the integral equation into

$$\int_{-\frac{1}{2}}^{\frac{1}{2}} \frac{[z(u) + z(v)]^2}{2 + u + v}\,dv = 1,$$

and then insert the power series

$$z(u) = a_0 + a_1 u + a_2 u^2 + \cdots, \qquad \frac{1}{2 + u + v} = \frac{1}{2} - \frac{u + v}{4} + \frac{(u + v)^2}{8} - + \cdots.$$

If only terms of at most the second degree in u and v after multiplying out the series in the integrand are retained, the integration yields

$$1 = \int_{-\frac{1}{2}}^{\frac{1}{2}} \ldots dv \approx 2a_0^2 + (2a_0 a_1 - a_0^2) u + \left(\frac{a_0^2}{2} + \frac{a_1^2}{2} + 2a_0 a_2 - a_0 a_1\right)\left(u^2 + \frac{1}{12}\right).$$

By equating coefficients we obtain the three equations

$$1 = 2a_0^2,$$

$$0 = 2a_0 a_1 - a_0^2,$$

$$0 = \frac{a_1^2}{2} + 2a_0 a_2 - a_0 a_1 + \frac{a_0^2}{2},$$

with the solution

$$a_1 = \frac{1}{2} a_0, \qquad a_2 = -\frac{1}{16} a_0, \qquad a_0 = \pm \sqrt{\frac{1}{2}};$$

and hence

$$z \approx \pm \sqrt{\frac{1}{2}} \left(1 + \frac{1}{2} u - \frac{1}{16} u^2\right), \qquad y \approx \pm \sqrt{\frac{1}{2}} \left(\frac{47}{64} + \frac{9}{16} x - \frac{1}{16} x^2\right).$$

For the positive solution, for example, we obtain the approximations 0·5192, 0·7071, 0·8729 for $y(0)$, $y(\frac{1}{2})$, $y(1)$, respectively.

If we had retained only the linear terms in the integrand, we would still have obtained a fairly useful approximation, namely

$$z \approx \pm \sqrt{\frac{1}{2}} \left(1 + \frac{1}{2} u\right), \qquad y \approx \pm \sqrt{\frac{1}{2}} \left(\frac{3}{4} + \frac{1}{2} x\right).$$

§2. Some special methods for linear integral equations

2.1. Approximation of kernels by degenerate kernels

The method to be described now relates to linear integral equations of the form (1.2) or to the more general case of linear integro-differential equations of the form (1.3) with the boundary conditions (1.4).

In this method the kernel $K(x, \xi)$ is approximated by kernels $K_n(x, \xi)$ of the form

$$K_n(x, \xi) = \sum_{j=1}^{n} A_j(x) B_j(\xi). \tag{2.1}$$

A kernel which can be written in this way as a finite sum of functions in which the variables x and ξ can be separated, i.e. confined to separate factors $A_j(x)$ and $B_j(\xi)$, is called a "degenerate kernel". Any continuous kernel may be approximated to any degree of accuracy by a degenerate kernel. In practice one often uses polynomials or trigonometric functions for the functions $A_j(x)$, $B_j(\xi)$. A degenerate approximation $K_n(x, \xi)$ to a kernel $K(x, \xi)$ can also be determined[1] from the equation

$$\begin{vmatrix} K_n(x, \xi) & K(x, \xi_1) & K(x, \xi_2) & \dots K(x, \xi_n) \\ K(x_1, \xi) & K(x_1, \xi_1) & K(x_1, \xi_2) & \dots K(x_1, \xi_n) \\ \cdot & \cdot & \cdot & \cdot & \cdot & \cdot & \cdot & \cdot & \cdot \\ K(x_n, \xi) & K(x_n, \xi_1) & K(x_n, \xi_2) & \dots K(x_n, \xi_n) \end{vmatrix} = 0, \tag{2.2}$$

where the x_j and ξ_j are selected points of the fundamental interval; the degenerate kernel K_n then coincides with the given kernel K along the $2n$ lines $x = x_j$ and $\xi = \xi_j$:

$$K_n(x, \xi_j) = K(x, \xi_j) \quad \text{and} \quad K_n(x_j, \xi) = K(x_j, \xi) \quad \text{for} \quad j = 1, 2, \dots, n.$$

[1] BATEMAN, H.: On the numerical solution of linear integral equations. Proc. Roy. Soc. Lond., Ser. A **100**, 441—449 (1922).

For a degenerate kernel the solution of the integro-differential equation (1.3) with the boundary conditions (1.4) may be reduced to the solution of a set of ordinary boundary-value problems and of a system of linear equations.

We therefore replace the kernel $K(x, \xi)$ in the integro-differential equation (1.3) by the degenerate approximation $K_n(x, \xi)$. Then by interchanging the order of the summation and integration, and taking the functions $A_j(x)$ outside of the integral sign, we obtain

$$M[y(x)] = f(x) + \lambda \sum_{j=1}^{n} c_j A_j(x), \tag{2.3}$$

where

$$c_j = \int_a^b B_j(\xi) N[y(\xi)] \, d\xi. \tag{2.4}$$

Now let $z(x)$ be the solution of the boundary-value problem

$$M[y(x)] = f(x), \quad U_\mu[y] = \gamma_\mu \quad (\mu = 1, 2, \dots, m)$$

and $z_k(x)$ (for $k = 1, \dots, n$) the solution of the boundary-value problem

$$M[y(x)] = A_k(x), \quad U_\mu[y] = 0 \quad (\mu = 1, 2, \dots, m)$$

(we assume that these $n + 1$ boundary-value problems possess unique solutions); then the boundary-value problem (2.3), (1.4) has the solution

$$y(x) = z(x) + \lambda \sum_{k=1}^{n} c_k z_k(x). \tag{2.5}$$

We now insert this expression for y into (2.4), and obtain a system of equations for the c_j:

$$c_j = \int_a^b B_j(\xi) N[z(\xi)] \, d\xi + \lambda \sum_{k=1}^{n} c_k \int_a^b B_j(\xi) N[z_k(\xi)] \, d\xi.$$

If we introduce the Kronecker symbol

$$\delta_{jk} = \begin{cases} 0 & \text{for} \quad j \neq k, \\ 1 & \text{for} \quad j = k \end{cases}$$

and define

$$a_{jk} = \int_a^b B_j(\xi) N[z_k(\xi)] \, d\xi, \quad r_j = \int_a^b B_j(\xi) N[z(\xi)] \, dx, \tag{2.6}$$

these equations can be written more compactly as

$$\sum_{k=1}^{n} (\delta_{jk} - \lambda a_{jk}) c_k = r_j \quad (j = 1, 2, \dots, n). \tag{2.7}$$

32*

Thus the c_k can be calculated provided that the determinant

$$\varDelta = \det\left(\delta_{jk} - \lambda\, a_{jk}\right) \tag{2.8}$$

does not vanish, and with these c_k values (2.5) gives the solution of the integro-differential equation with the degenerate kernel (2.1).

If we have a completely homogeneous problem with $f(x) \equiv 0$, $\gamma_\mu \equiv 0$ (an eigenvalue problem), then $z(x) \equiv 0$ and $r_j \equiv 0$, and the zeros of the determinant provide approximations $\varLambda_j$ for the eigenvalues λ_j.

If the degenerate kernel $K_n(x, \xi)$ defined by (2.2) is used, the solution of the integral equation (1.2) of the second kind can be expressed in the form[1]

$$y(x) = f(x) + \lambda \int_a^b G(x, \xi, \lambda)\, f(\xi)\, d\xi,$$

where the solving kernel $G(x, \xi, \lambda)$ is to be determined from the equation

$$\begin{vmatrix} G(x,\xi,\lambda) & K(x,\xi_1) & K(x,\xi_2)\ldots K(x,\xi_n) \\ K(x_1,\xi) & A_{11} & A_{12} & \ldots A_{1n} \\ \cdot & \cdot \quad \cdot \quad \cdot \quad \cdot \quad \cdot \quad \cdot \quad \cdot \quad \cdot \quad \cdot \\ K(x_n,\xi) & A_{n1} & A_{n2} & \ldots A_{nn} \end{vmatrix} = 0$$

with

$$A_{jk} = K(x_j, \xi_k) - \lambda \int_a^b K(x_j, t)\, K(t, \xi_k)\, dt.$$

2.2. Example

For the problem

$$\frac{1+x}{2}\, y'' + 10 \int_0^1 \frac{y(\xi)\, d\xi}{1+x+\xi} + 1 = 0, \qquad y(0) = y(1) = 0, \tag{2.9}$$

considered in § 1.10, let the kernel $K(x, \xi) = \dfrac{1}{1+x+\xi}$ be approximated first of all quite crudely by a constant A; the solution $\eta(x)$ of the approximate integro-differential equation can then be given in closed form. The equation reads

$$\eta'' = \frac{C}{1+x} \qquad \text{with} \qquad C = -2 - 20A \int_0^1 \eta(\xi)\, d\xi$$

and can be integrated immediately; using the boundary conditions $\eta(0) = \eta(1) = 0$, we obtain

$$\eta = C\left[(1+x)\ln(1+x) - 2x\ln 2\right].$$

---

[1] BATEMAN, H.: (see last footnote) p. 443.

To determine C we calculate

$$\int_0^1 \eta(\xi)\,d\xi = C(\ln 2 - \tfrac{3}{4}) = -C \times 0 \cdot 056853 = -C\,\delta, \text{ say},$$

insert this value for the integral into the equation defining C and find that

$$C = \frac{2}{20 A\,\delta - 1} \approx \frac{-2}{1 - 1\cdot1371 A}.$$

If we take $A = \tfrac{1}{2}$ as being a reasonable average value for the kernel, we have $C = -4\cdot6353$, and hence $\eta(x)$ is known; in particular, $\eta(\tfrac{1}{2})$ $= 0\cdot3938$.

Now let us approximate the kernel $K(x, \xi) = \dfrac{1}{1 + x + \xi}$ more accurately by using a linear function $A + B(x + \xi)$. The corresponding approximate solution $\eta(x)$ is now to be determined from

$$(1 + x)\,\eta'' = -2 - 20 \int_0^1 [A + B(x + \xi)]\,\eta(\xi)\,d\xi = C + D(1 + x),$$

$$\eta(0) = \eta(1) = 0,$$

and again we have a problem which can be solved in closed form:

$$\eta = C[(1 + x)\ln(1 + x) - 2x \ln 2] + D\,\frac{x^2 - x}{2}.$$

To determine C and D we calculate

$$\int_0^1 \eta(\xi)\,d\xi = -C\,\delta - \frac{1}{12}D, \qquad \int_0^1 \xi\,\eta(\xi)\,d\xi = -\frac{1}{36}C - \frac{1}{24}D$$

(with δ as defined above) and insert these values in the equation defining C and D; then by equating coefficients (of powers of x) we obtain two linear equations for C and D:

$$(1 - 20 A\,\delta - \tfrac{5}{9}B)\,C + (1 - \tfrac{5}{3}A - \tfrac{5}{6}B)\,D = -2,$$

$$- 20 B\,\delta\,C + (1 - \tfrac{5}{3}B)\,D = 0.$$

The values of C and D depend on the choice of values for A and B. Suitable values for A and B could be determined by the method of least squares, but here we use values $\left(A = 0\cdot8,\ B = -\dfrac{4}{15}\right)$ which were obtained simply by means of a graph; they give $C = -7\cdot6386$, $D = 1\cdot6035$ and $\eta(\tfrac{1}{2}) = 0\cdot4485$.

Naturally one could use still more accurate approximations for the kernel $K(x, \xi)$.

2.3. The iteration method for eigenvalue problems

The theory of the method of successive approximations, or iteration method, of Ch. III, § 8.3 applies also to the linear homogeneous integral equation

$$y(x) = \lambda \int_a^b K(x, \xi) \, y(\xi) \, d\xi \qquad (2.10)$$

with symmetric kernel $K(x, \xi)$ when this kernel is square-integrable and continuous in the mean[1]. Starting from an arbitrarily chosen, continuous function $F_0(x)$, one determines further functions $F_n(x)$ from the iteration formula (1.20) and calculates the corresponding Schwarz constants a_n and quotients μ_n defined by (1.21), (1.22). One can then make use of the results stated below.

We arrange the eigenvalues of (2.10) in order of increasing absolute value, treating each multiple eigenvalue, say of multiplicity m, as a set of m coincident eigenvalues:

$$0 \leq |\lambda_1| \leq |\lambda_2| \leq |\lambda_3| \leq \cdots,$$

and for the following assume that $0 < |\lambda_1| < |\lambda_2|$.

Then we may state:

1. If, besides being symmetric, square-integrable and continuous in the mean, the kernel is "positive definite", i.e. the eigenvalues are all positive, then the μ_n decrease monotonically:

$$\mu_1 \geq \mu_2 \geq \cdots \geq \lambda_1 > 0, \qquad (2.11)$$

and

$$0 \leq \mu_{n+1} - \lambda_1 \leq \frac{\mu_n - \mu_{n+1}}{\dfrac{l_2}{\mu_{n+1}} - 1}, \qquad (2.12)$$

where l_2 is a lower limit for the second eigenvalue, but is greater than μ_{n+1}:

$$\lambda_2 \geq l_2 > \mu_{n+1}.$$

2. In the more general cases for which the condition of positive definiteness of the kernel is relaxed, the μ_n need not form a monotonic sequence;

---

[1] That is when the integrals

$$\int K(x, \xi) \, dx, \quad \int K^2(x, \xi) \, dx, \quad \iint K(x, \xi) \, dx \, d\xi, \quad \iint K^2(x, \xi) \, dx \, d\xi$$

over the fundamental interval all exist and are bounded (square-integrability) and

$$\lim_{x \to x_1} \int [K(x, \xi) - K(x_1, \xi)]^2 \, d\xi = 0$$

(continuity in the mean); both of these conditions are satisfied when $K(x, \xi)$ is continuous. See G. HAMEL: Integralgleichungen, 2nd ed., p. 68. Berlin 1949.

now, however, the products $\mu_{2r-1}\mu_{2r}$ decrease monotonically:

$$\mu_1\mu_2 \geq \mu_3\mu_4 \geq \mu_5\mu_6 \geq \cdots \geq \lambda_1^2 \geq 0, \tag{2.13}$$

and the corresponding limits for the first eigenvalue[1] take the form

$$0 \leq \mu_{2n+1}\mu_{2n+2} - \lambda_1^2 \leq \frac{\mu_{2n-1}\mu_{2n} - \mu_{2n+1}\mu_{2n+2}}{\dfrac{l_2^2}{\mu_{2n+1}\mu_{2n+2}} - 1}, \tag{2.14}$$

where l_2 is a number such that

$$\lambda_2^2 \geq l_2^2 > \mu_{2n+1}\mu_{2n+2}.$$

3. *The enclosure theorem also holds[2]: If the function*

$$G(x) = \frac{F_0(x)}{F_1(x)} \tag{2.15}$$

lies between finite limits $G_{\min}$ and $G_{\max}$ and does not change sign in the fundamental interval, then $G_{\min}$ and $G_{\max}$ enclose at least one eigenvalue λ_k of (2.10):

$$G_{\min} \leq \lambda_k \leq G_{\max}. \tag{2.16}$$

Sometimes a suitable value for l_2 or a direct estimate for λ_1 can be obtained from the relation[3]

$$\sum_{\nu=1}^{\infty} \frac{1}{\lambda_\nu^2} = \int_a^b\int_a^b [K(x,\xi)]^2\,dx\,d\xi = k;$$

for example, we can deduce that

$$\frac{1}{\lambda_2^2} \leq k - \frac{1}{\lambda_1^2}, \quad \text{i.e.} \quad \lambda_2 \geq \left(k - \frac{1}{\lambda_1^2}\right)^{-\frac{1}{2}}.$$

Occasionally one also uses the "iterated kernels", in terms of which the above relation and allied results can be expressed; these "iterated kernels" are defined in § 3.1, where another use for them is described.

The various modifications to the iteration method which were described in Ch. III, §§ 8.9, 8.10 can be carried over immediately to integral equations; in the literature, in fact, they are often established for integral equations first.

---

[1] Similar results for the higher eigenvalues and other generalizations are given by L. COLLATZ: Math. Z. **46**, 692—708 (1940), and R. IGLISCH: Ann. Ann. **118**, 263—275 (1941).

[2] COLLATZ, L.: Math. Z. **47**, 395—398 (1941).

[3] See, for instance, G. HAMEL: Integralgleichungen, 2nd ed., p. 68. Berlin 1949.

§3. Singular integral equations

Applied problems frequently lead to integral and integro-differential equations for which the fundamental region is infinite or for which the kernel $K(x, \xi)$ for equations of the form (1.2), (1.3) is unbounded. A numerical treatment of such singular equations must, of course, take into account the nature of the singularity, but so many different kinds of situation can arise that a unified description seems hardly possible; it must suffice here to select a few typical cases and use them to illustrate the application of some of the methods which have been used hitherto.

Two of the most frequently occurring types of singular kernel $K(x, \xi)$ can be written in the forms

$$\frac{H(x,\xi)}{(x-\xi)^\alpha}, \qquad H(x,\xi)\ln|x-\xi|,$$

where $H(x, \xi)$ is continuous in the fundamental region $a \leqq (x, \xi) \leqq b$.

3.1. Smoothing of the kernel

For an integral equation of the form (1.2) a "smoother" kernel can be obtained by utilizing the smoothing property of integration in the following way[1]. We multiply the equation by $K(v, x)$ and integrate with respect to x over the fundamental interval, then make use of the original equation to obtain

$$\frac{y(v) - f(v)}{\lambda} = \int\limits_a^b K(v, x)\, y(x)\, dx$$

$$= \lambda \int\limits_a^b \int\limits_a^b K(v, x)\, K(x, \xi)\, y(\xi)\, dx\, d\xi + \int\limits_a^b K(v, x)\, f(x)\, dx;$$

this also is an integral equation of the form (1.2), and can be written

$$y(v) = \lambda^2 \int\limits_a^b K_2(v, \xi)\, y(\xi)\, d\xi + f_2(v), \qquad (3.1)$$

where

$$K_2(v, \xi) = \int\limits_a^b K(v, x)\, K(x, \xi)\, dx, \quad f_2(v) = f(v) + \lambda \int\limits_a^b K(v, x)\, f(x)\, dx. \quad (3.2)$$

In general the new kernel K_2 will be "smoother" than K. In particular, for a singular symmetric kernel of the form

$$K(x, \xi) = \frac{H(x,\xi)}{(x-\xi)^\alpha} \qquad (3.3)$$

[1] See G. HAMEL: Integralgleichungen, 2nd ed., p. 140. Berlin 1949; or P. MORSE and H. FESHBACH: Methods of theoretical physics, Part I, p. 922. New York-Toronto-London: McGraw-Hill 1953.

with $0<\alpha\leq\frac{1}{2}$ the new kernel K_2 will no longer be singular. For the case $\frac{1}{2}<\alpha<1$, K_2 will still be singular, but we can repeat the process on equation (3.1) and so on; after a finite number of such "smoothing steps" we will have a kernel K_n which is no longer singular. These kernels $K_2, K_3, \ldots$ are known as "iterated kernels".

It should be noted that for the case with $\alpha=1$ in (3.3) this smoothing process does not help, for the iterated kernels are again of the form (3.3). This important case will be dealt with in greater detail in § 3.2.

3.2. Singular equations with Cauchy-type integrals

A number of applied problems give rise to singular integral and integro-differential equations of the first and second kinds of the forms (the integral notation will be explained later)

$$\lambda\oint_a^b \frac{H(x,\xi)}{x-\xi}\, N[y(\xi)]\, d\xi = f(x) \tag{3.4}$$

and

$$y(x) - \lambda\oint_a^b \frac{H(x,\xi)}{x-\xi}\, N[y(\xi)]\, d\xi = f(x), \tag{3.5}$$

respectively[1], where $H(x,\xi)$ and $f(x)$ are given continuous functions and $N[y]$ is a linear differential expression in y of the n-th order. In the simplest case $n=0$ we have an equation of the form

$$\lambda\oint_a^b \frac{H(x,\xi)\,y(\xi)}{x-\xi}\, d\xi = f(x).$$

An integral of the type considered here, usually known as a Cauchy integral, is improper and in general does not converge. In order to attach a meaning to it, we define its Cauchy principal value, distinguished by a C superimposed on the integral sign, as the limit

$$\oint_a^b \frac{H(x,\xi)}{x-\xi}\, d\xi = \lim_{\varepsilon\to 0}\left\{\int_a^{x-\varepsilon} \frac{H(x,\xi)}{x-\xi}\, d\xi + \int_{x+\varepsilon}^b \frac{H(x,\xi)}{x-\xi}\, d\xi\right\},$$

provided that this limit exists. This is the value to be understood in the above equations.

The difficulties which attend the solution of integral equations with these "Cauchy kernels" can often be surmounted by making use of certain singular integrals whose Cauchy principal values are known; the formulae (3.8), (3.15) below, for example, are used to this end.

---

[1] For the one-dimensional singular Fredholm integral equation with Cauchy-type kernel a complete theory has been given by N. J. MUSKHELISHVILI: Singular integral equations, 447 pp. Groningen 1953.

Consider the integral equation

$$\frac{1}{2\pi} \oint_{-a}^{a} \frac{y(\xi)\,d\xi}{\xi - x} = f(x). \tag{3.6}$$

First we transform it by the introduction of the new variables

$$x = - a \cos \varphi, \qquad \xi = - a \cos \psi$$

into

$$\frac{1}{2\pi} \oint_{0}^{\pi} \frac{y(- a \cos \psi) \sin \psi}{\cos \varphi - \cos \psi}\, d\psi = f(- a \cos \varphi). \tag{3.7}$$

If we now assume for the function $y(- a \cos \psi) \sin \psi = g(\psi)$ appearing in the numerator of the integrand an expansion of the form

$$g(\psi) = y(- a \cos \psi) \sin \psi = \tfrac{1}{2} b_0 + \sum_{n=1}^{\infty} b_n \cos n \psi,$$

we can evaluate the integral in (3.7) by using the formula[1]

$$\oint_{0}^{\pi} \frac{\cos n \psi}{\cos \psi - \cos \varphi}\, d\psi = \pi\, \frac{\sin n \varphi}{\sin \varphi} \qquad (n = 0, 1, 2, \ldots;\ 0 < \varphi < \pi); \tag{3.8}$$

(3.7) then becomes

$$-\frac{1}{2} \sum_{n=1}^{\infty} b_n \frac{\sin n \varphi}{\sin \varphi} = f(- a \cos \varphi). \tag{3.9}$$

We now assume that the function $f(- a \cos \varphi) \sin \varphi$ may be expanded as a uniformly convergent Fourier series of the form

$$f(- a \cos \varphi) \sin \varphi = - \tfrac{1}{2} \sum_{n=1}^{\infty} b_n \sin n \varphi \qquad (0 < \varphi < \pi); \tag{3.10}$$

the b_n (except b_0) are thereby determined, and in

$$y(\xi) = \frac{1}{\sqrt{1 - \left(\frac{\xi}{a}\right)^2}} \left(\frac{1}{2} b_0 + \sum_{n=1}^{\infty} b_n \cos n \psi \right), \tag{3.11}$$

where the $\cos n \psi$ can be expressed in known fashion as polynomials in $\cos \psi$ (and hence in ξ), we have a family of solutions of the integral equation (3.6). We note that the constant b_0 can be chosen arbitrarily and that the solution $y(\xi)$ can become infinite like $(a^2 - \xi^2)^{-\frac{1}{2}}$ at the end-points $\xi = \pm a$.

---

[1] See, for example, G. HAMEL: Integralgleichungen, 2nd ed., p. 145. Berlin 1949.

Other singular integral equations, such as

$$\oint_{-a}^{+a} \frac{y(\xi)\, d\xi}{x^2 - \xi^2} = f(x),$$ (3.12)

for example, may be solved in a similar way[1].

On account of its importance in the theory of conformal transformations, we will consider one more particular integral equation, namely

$$f(\varphi) = -\frac{1}{2\pi} \oint_0^{2\pi} y(\vartheta) \cot \frac{\vartheta - \varphi}{2}\, d\vartheta.$$ (3.13)

If we substitute the Fourier series

$$y(\vartheta) = \frac{1}{2} a_0 + \sum_{n=1}^{\infty} (a_n \cos n\,\vartheta + b_n \sin n\,\vartheta)$$ (3.14)

for $y(\vartheta)$ in the integral and use the formulae

$$\left.\begin{aligned}\oint_0^{2\pi} \cos n\,\vartheta \cot \frac{\vartheta - \varphi}{2}\, d\vartheta &= -2\pi \sin n\,\varphi \\[2mm]\oint_0^{2\pi} \sin n\,\vartheta \cot \frac{\vartheta - \varphi}{2}\, d\vartheta &= 2\pi \cos n\,\varphi\end{aligned}\right\} \quad (n = 0, 1, 2, \ldots), \quad (3.15)$$

the integral equation (3.13) becomes

$$f(\varphi) = \sum_{n=1}^{\infty} (a_n \sin n\,\varphi - b_n \cos n\,\varphi).$$ (3.16)

If we assume that the given function $f(\varphi)$ may be expanded as a uniformly convergent Fourier series of the form (3.16) — this implies in particular that $f(\varphi)$ must satisfy the condition

$$\int_0^{2\pi} f(\varphi)\, d\varphi = 0$$

—, then the unknowns a_n and b_n (except a_0) are determined immediately by Fourier analysis of $f(\varphi)$. With these values in (3.14) we again obtain solutions $y(\vartheta)$ in which a constant a_0 may be chosen arbitrarily. With the aid of numerical tables and formulae for trigonometric interpolation

[1] HAMEL, G.: Integralgleichungen, 2nd ed., p. 148. Berlin 1949.

WITTICH reduces this solution[1] to a form suitable for numerical evaluation.

For singular integral equations which cannot be solved in closed form or by series expansions one tries to remove the singularity by some means or other in order that a quadrature formula may be used to evaluate the integral. A device which is often used for this purpose with kernels of Cauchy type is the following[2]. Consider the Cauchy integral

$$J(x) = \oint_a^b \frac{f(x,\xi)}{x-\xi}\,d\xi$$

and assume that $f(x, \xi)$ possesses a continuous derivative with respect to ξ. When x lies nearer a than b, say, so that $a < x < b$, $h = x - a < b - x$, we can write

$$J = J_1 + J_2,$$

where

$$J_1 = \int_a^x \frac{f(x,\xi)}{x-\xi}\,d\xi + \int_x^{x+h} \frac{f(x, 2x-\xi)}{x-\xi}\,d\xi$$

and

$$J_2 = \int_x^{x+h} \frac{f(x,\xi)-f(x,2x-\xi)}{x-\xi}\,d\xi + \int_{x+h}^b \frac{f(x,\xi)}{x-\xi}\,d\xi.$$

---

[1] THEODORSEN, T., and J. E. GARRICK: General potential theory. Nat. Adv. Comm. Aeronaut. Rep. Nr. 452 (1933) 1—35. Practical methods for conformal transformation are not dealt with in the present book because the subject lies at the very edge of the field with which the book is concerned; however, we can at least give a selection of references: TREFFTZ, E.: Eine neue Methode zur Lösung der Randwertaufgabe partieller Differentialgleichungen. Math. Ann. 79 (1919). — FRANK, PH., and R. v. MISES: Differential- und Integralgleichungen der Mechanik und Physik, Vol. I, pp. 729—734. Brunswick 1930. — HEINHOLD, J.: Ein Schmiegungsverfahren der konformen Abbildung. S.-B. Bayer. Akad. Wiss., Math.-nat. Kl. 1948, 203—222 (where further literature is mentioned), also Z. Angew. Math. Mech. 30, 286—287 (1950). — WITTICH, H.: Konforme Abbildung einfach zusammenhängender Gebiete. Z. Angew. Math. Mech. 25/27, 131—132 (1947). — LÖSCH, F.: Auftrieb und Moment eines unsymmetrischen Doppelflügels. Luftf.-Forschg. 17, 22—31 (1940) (doubly-connected region). — WITTICH, H.: Bemerkungen zur Druckverteilungsrechnung nach THEODORSEN-GARRICK. Jb. dtsch. Luftf.-Forschung 1941. — WARSCHAWSKI, S. E.: Recent results in numerical methods of conformal mapping. Proc. Symp. Appl. Math. VI, pp. 219—250. New York-Toronto-London 1956. — Experiments in the computation of conformal maps (edited by JOHN TODD). Nat. Bur. Stand., Appl. Math. Ser. 42, 61 pp. (1955). — KANTOROVICH, L. V., and V. I. KRYLOV: Näherungsmethoden der Höheren Analysis, pp. 330—508. Berlin 1956.

[2] A clear graphical explanation of the device and an application to a technical problem can be found in C. WEBER: Randverformung der Halbebene durch eine Normalbelastung, „Ein Ei des Kolumbus ?". Z. Angew. Math. Mech. 30, 240—242 (1950).

Now the integrals in J_1 combine to form one integral over the interval $a \leq \xi \leq 2x - a$ with an integrand whose numerator is symmetric and denominator antisymmetric about the mid-point $\xi = x$; hence, by the definition of the Cauchy principal value, $J_1 = 0$. The evaluation of J_2 presents no difficulties since the two integrands now remain bounded, the limit

$$\lim_{\xi \to x} \frac{f(x, \xi) - f(x, 2x - \xi)}{x - \xi} = -2 \left(\frac{\partial f(x, \xi)}{\partial \xi} \right)_{\xi = x}$$

existing by virtue of the assumption that $f(x, \xi)$ is differentiable with respect to ξ.

Example. For the integral equation

$$\frac{1}{2\pi} \oint_{-1}^{1} \frac{y(\xi) \, d\xi}{\xi - x} = |x|$$

(3.10) requires that the function $|\cos \varphi| \sin \varphi$ be expanded as a Fourier sine series:

$$-2 |\cos \varphi| \sin \varphi = \sum_{n=1}^{\infty} b_n \sin n \varphi;$$

by the usual rules of Fourier analysis we obtain

$$b_{2k} = 0, \quad b_{2k-1} = (-1)^{k-1} \frac{8}{\pi (2k-3)(2k+1)} \qquad (k = 1, 2, \ldots).$$

From (3.11) it follows that

$$y(\xi) = \frac{1}{\sqrt{1 - \xi^2}} \left(\frac{1}{2} b_0 + \sum_{n=1}^{\infty} b_n \cos n \, \psi \right), \quad \text{where} \quad \xi = -\cos \psi,$$

is a solution of the integral equation. The constant b_0 is arbitrary and $(1 - \xi^2)^{-\frac{1}{2}}$ must be a solution of the corresponding homogeneous equation. If we omit the arbitrary multiples of this function which may be added to any solution, we are left with the particular solution

$$y(\xi) = \frac{1}{\sqrt{1 - \xi^2}} \frac{8}{\pi} \left(\frac{\cos \psi}{(-1) \cdot 3} - \frac{\cos 3\psi}{1 \cdot 5} + \frac{\cos 5\psi}{3 \cdot 7} - \frac{\cos 7\psi}{5 \cdot 9} + - \cdots \right)$$

$$= \frac{8}{\pi} \frac{1}{\sqrt{1 - \xi^2}} g, \text{ say.}$$

The sum function of this Fourier series, which we have denoted by g, may be given in closed form[1]. If, however, graphical accuracy will suffice, one can quickly plot a few terms of the Fourier series to obtain

---

[1] By putting $e^{i\psi} = z$, $\cos n \psi = \operatorname{re} z^n$ and summing the power series which results, one can show that

$$g(\psi) = -\frac{\cos \psi}{2} \left(1 + \sin \psi \ln \tan \left(\frac{\pi}{4} - \frac{\psi}{2} \right) \right) \quad \text{for} \quad 0 \leq \psi \leq \frac{\pi}{2}.$$

a graph of g against ψ, transfer this graph onto a different abscissa scale to represent g as a function of ξ, then scale up the ordinate values in the appropriate manner to obtain finally a graph of $y(\xi)$ against ξ (see Fig. VI/7).

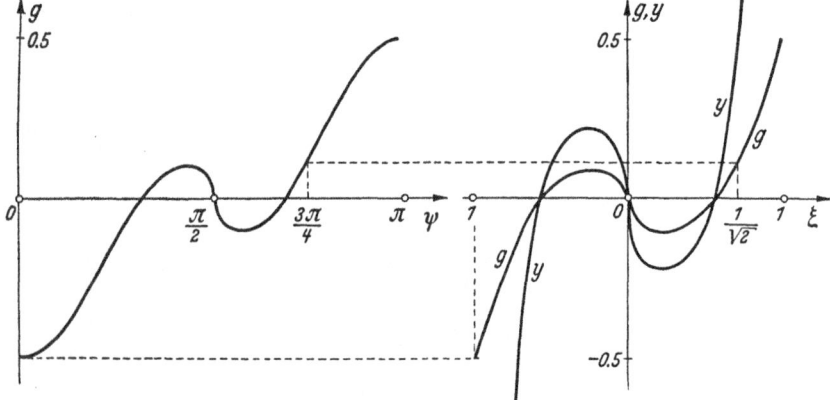

Fig. VI/7. Graphical evaluation of a Fourier series solution of a singular integral equation

3.3. Closed-form solutions

Some singular integral equations may be solved in closed form with the aid of inversion formulae[1]. Several examples of such formulae are listed in Table VI/6, in which for simplicity $f(x)$, say, is regarded as a

Table VI/6. *Inversion formulae for some well-known singular integral equations*

Integral equation	Solution
$f(x) = \dfrac{1}{2\pi} \displaystyle\oint_{-\pi}^{\pi} \left(1 + \cot\dfrac{1}{2}\,(x-\xi)\right) y(\xi)\,d\xi$	$y(\xi) = \dfrac{1}{2\pi} \displaystyle\oint_{-\pi}^{\pi} \left(1 + \cot\dfrac{1}{2}\,(x-\xi)\right) f(x)\,dx$
$f(x) = \sqrt{\dfrac{2}{\pi}} \displaystyle\int_{0}^{\infty} \cos(x\xi)\, y(\xi)\,d\xi$	$y(\xi) = \sqrt{\dfrac{2}{\pi}} \displaystyle\int_{0}^{\infty} \cos(x\xi)\, f(x)\,dx$
$f(x) = \displaystyle\int_{0}^{x} \dfrac{y(\xi)\,d\xi}{\sqrt{x-\xi}}$	$y(\xi) = \dfrac{1}{\pi}\dfrac{d}{d\xi} \displaystyle\int_{0}^{\xi} \dfrac{f(x)\,dx}{\sqrt{\xi-x}}$
$f(x) = \dfrac{1}{2\pi} \displaystyle\oint_{-1}^{1} \dfrac{y(\xi)\,d\xi}{x-\xi}$	$\sqrt{1-\xi^2}\,\pi y(\xi) = 2\displaystyle\oint_{-1}^{1} \dfrac{f(x)\sqrt{1-x^2}}{x-\xi}\,dx - \displaystyle\int_{-1}^{1} y(x)\,dx$

[1] See, for example, W. Magnus and F. Oberhettinger: Formeln und Sätze für die speziellen Funktionen der mathematischen Physik, 2nd ed., p. 186. Berlin 1948.

given continuous function; actually, continuity may be replaced by a weaker condition, but for this and other details we must refer the reader to the literature.

3.4. Approximation of the kernel by degenerate kernels

The method of § 2.1 can also be applied to equations with singular kernels if these kernels have known expansions in terms of their eigenfunctions, for we can then truncate these expansions to obtain degenerate approximations to the given singular kernels.

Thus, for example, the kernel

$$K(\varphi, \psi) = \ln |\cos \varphi - \cos \psi|,$$

with a logarithmic singularity, has the expansion

$$K(\varphi, \psi) = \sum_{n=0}^{\infty} \frac{y_n(\varphi) \, y_n(\psi)}{\lambda_n}$$

with eigenfunctions $y_n(\varphi)$ given by

$$y_0 = \frac{1}{\sqrt{\pi}}, \qquad y_n(\varphi) = \sqrt{\frac{2}{\pi}} \cos n\varphi \qquad (n = 1, 2, \ldots)$$

(these form an orthonormal system) and eigenvalues λ_n by

$$\lambda_0 = -\frac{1}{\pi \ln 2}, \qquad \lambda_n = -\frac{n}{\pi} \qquad (n = 1, 2, \ldots).$$

For PRANDTL's circulation equation

$$\alpha_i(y) = \frac{1}{4\pi\nu} \int_{-\frac{b}{2}}^{\frac{b}{2}} \frac{d\Gamma}{dy'} \frac{dy'}{y - y'}$$

(an explanation of the notation can be found in the literature[1]), which

---

[1] The following is only a short selection from the extensive literature on the subject: K. SCHRÖDER: Über die Prandtlsche Integro-Differentialgleichung der Tragflügeltheorie. S.-B. Preuss. Akad. Wiss., Math.-nat. Kl. 16, 1—35 (1939). —

SÖHNGEN, H.: Die Lösungen der Integralgleichung $g(x) = \frac{1}{2\pi} \int_{-a}^{a} \frac{f(\xi)\,d\xi}{x - \xi}$ und deren

Anwendung in der Tragflügeltheorie. Math. Z. 45, 245—264 (1939). — MULTHOPP, H.: Die Berechnung der Auftriebsverteilung von Tragflügeln. Luftf.-Forschg. 15, 153—169 (1938). — SCHWABE, M.: Luftf.-Forschg. 15, 170—180 (1938), gives a computing scheme for MULTHOPP's method for the calculation of the lift distribution over a wing.

occurs in wing theory, this method of replacing the kernel by a degenerate kernel leads to a system of equations first derived by MULTHOPP, but in a different way[1].

§ 4. Volterra integral equations

4.1. Preliminary remarks

If the upper limit in the integral equation is variable, it is called a Volterra integral equation; in particular,

$$f(x) = \int_a^x K(x, \xi)\, y(\xi)\, d\xi \tag{4.1}$$

is called a Volterra integral equation of the first kind and

$$y(x) = f(x) + \int_a^x K(x, \xi)\, y(\xi)\, d\xi \tag{4.2}$$

a Volterra integral equation of the second kind. $y(x)$ is to be determined for given functions $f(x)$ and $K(x, \xi)$ which we assume to be continuous for $a \leq x \leq b$ and $a \leq x \leq \xi \leq b$, respectively.

In general, differentiation with respect to x transforms a Volterra integral equation of the first kind into one of the second kind; thus we obtain

$$f'(x) = K(x, x)\, y(x) + \int_a^x \frac{\partial K(x, \xi)}{\partial x}\, y(\xi)\, d\xi,$$

which can be written in the form (4.2) by dividing through by $K(x, x)$:

$$y(x) = \frac{f'(x)}{K(x, x)} - \int_a^x \frac{\partial K(x, \xi)}{\partial x}\, \frac{y(\xi)}{K(x, x)}\, d\xi. \tag{4.3}$$

For this transformation to be legitimate we must assume that $f(x)$ and $K(x, \xi)$ are differentiable with respect to x and that $K(x, x) \neq 0$.

Under the assumptions of continuity made above, it may be shown that the Volterra integral equation of the second kind (4.2) always possesses one and only one continuous solution $y(x)$[2].

---

[1] That MULTHOPP's equations could be derived by using the principle mentioned here was recognized by K. JAECKEL: Praktische Auswertung singulärer Integralgleichungen. Lecture. Hannover 1949.

[2] See, for example, G. HAMEL: Integralgleichungen, 2nd ed., p. 32 et seq. Berlin 1949.

Volterra integral equations which occur in applications often have kernels of convolution type, i.e. kernels which are functions of the difference $\delta = x - \xi$ alone: $K(x, \xi) = K(x - \xi) = K(\delta)$. For such cases WHITTAKER[1] recommends that $K(\delta)$ be approximated by a sum of kernel functions of similar form, i.e. of convolution type, but for which the integral equation can be solved in closed form. He exhibits several classes of such kernel functions. Here we mention only the class of kernels of the form $K(\delta) = \sum_{\nu=1}^{n} q_\nu e^{p_\nu \delta}$; for this kernel the solution of (4.2) (with $a = 0$) reads

$$y(x) = f(x) + \int_0^x \widetilde{K}(x - s) f(s) \, ds,$$

where

$$\widetilde{K}(x) = \sum_{\nu=1}^{n} \frac{(\alpha_\nu - p_1)(\alpha_\nu - p_2) \ldots (\alpha_\nu - p_n)}{(\alpha_\nu - \alpha_1)(\alpha_\nu - \alpha_2) \ldots (\alpha_\nu - \alpha_{\nu-1})(\alpha_\nu - \alpha_{\nu+1}) \ldots (\alpha_\nu - \alpha_n)} e^{p_\nu x},$$

the $\alpha_1, \alpha_2, \ldots, \alpha_n$ being the roots (assumed to be distinct) of the equation

$$\sum_{\nu=1}^{n} \frac{q_\nu}{x - p_\nu} = 1.$$

4.2. Step-by-step numerical solution

Perhaps the most obvious method for the numerical solution of (4.2) is to replace the integral by a finite sum[2]. Let

$$x_i = \xi_i = a + ih, \quad y_i = y(x_i), \quad f_i = f(x_i), \quad K_{il} = K(x_i, \xi_l), \quad (4.4)$$

where h is a suitably chosen pivotal interval, and let Y_i be an approximation to be calculated for y_i. Then from (4.2) it follows first of all that $f_0 = y_0 = Y_0$. For $x = x_1$ we evaluate the integral by the trapezium rule (the use of a more accurate quadrature formula is described in § 4.3) and obtain the equation

$$Y_1 = f_1 + \frac{h}{2}(K_{10} Y_0 + K_{11} Y_1);$$

thus we can calculate

$$Y_1 = \frac{f_1 + \frac{h}{2} K_{10} Y_0}{1 - \frac{h}{2} K_{11}}. \tag{4.5}$$

---

[1] WHITTAKER, E. T.: On the numerical solution of integral equations. Proc. Roy. Soc. Lond., Ser. A **94**, 367−383 (1918).

[2] Cf. J. R. CARSON: Electrical circuit theory and the operational calculus, p. 145. New York: McGraw-Hill 1926. A similar method is used by A. HUBER: Eine Näherungsmethode zur Auflösung Volterrascher Integralgleichungen. Mh. Math. Phys. **47**, 240−246 (1939); he replaces the function to be determined by piecewise-linear functions. Cf. also V. I. KRYLOV: Application of the Euler-Laplace formula to the approximate solution of integral equations of Volterra type. Trudy Mat. Inst. Steklov **28**, 33−72 (1949) [Russian; reviewed in Zbl. Math. **41**, 79 (1952), also in Math. Rev. **12**, 540 (1951)].

For $x = x_n$ we obtain in the same way the equation

$$Y_n = f_n + \frac{h}{2}(K_{n0}Y_0 + 2K_{n1}Y_1 + 2K_{n2}Y_2 + \cdots + 2K_{n,n-1}Y_{n-1} + K_{nn}Y_n),$$

and hence we can calculate

$$Y_n = \frac{1}{1 - \frac{h}{2}K_{nn}}\left(f_n + \frac{h}{2}K_{n0}Y_0 + h\sum_{i=1}^{n-1}K_{ni}Y_i\right) \tag{4.6}$$

in terms of the values $Y_0, Y_1, \ldots, Y_{n-1}$ calculated in the previous steps.

Example. Linear transmission systems. Suppose that we have an arbitrary linear transmission system (whose physical nature we may leave unspecified), i.e. we imagine a system with an input and output for which the associated input and output quantities (functions of time) are related by a linear transformation. If we denote the input function by $y(t)$ and the output function by $S(t)$, this transformation may be expressed in the form[1]

$$S(t) = y(t)A(0) - \int_0^t y(\xi)\frac{d}{d\xi}A(t-\xi)\,d\xi \tag{4.7}$$

or, equivalently,

$$S(t) = \frac{d}{dt}\int_0^t y(t-\xi)A(\xi)\,d\xi; \tag{4.8}$$

here $A(t)$ is the response to a unit step function, i.e. the output $S(t)$ produced when $y(t) = 0$ for $t < 0$ and $y(t) = 1$ for $t > 0$. The physical nature of the input and output quantities may or may not be the same; in a mechanical system, for example, the input might be a force and the output a displacement, and in an electrical system both quantities might be currents.

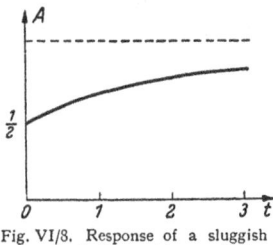

Fig. VI/8. Response of a sluggish measuring instrument to a unit step function input

For our example we will consider a sluggish measuring instrument whose immediate response to a unit step function input is a deflection of only a half a unit, which then gradually increases to the correct unit deflection according to the law (see Fig. VI/8)

$$A(t) = 1 + \frac{e^{-t} - 1}{2t}. \tag{4.9}$$

The question arises what is the time distribution of the quantity to be measured $y(t)$ if a specific time distribution of the reading $S(t)$ has

---

[1] See, for example, K. W. WAGNER: Operatorenrechnung, 2nd ed., p. 14. Leipzig 1950.

)een recorded. Equation (4.7) is then a Volterra integral equation of
he second kind for $y(t)$. We will consider the case $S(t) \equiv 1$; thus we
ısk how the quantity to be measured $y(t)$ must vary in order that the
leflection may remain constant.

Integration of (4.8) from 0 to t yields

$$t = \int_0^t y(t - \xi) A(\xi) \, d\xi; \tag{4.10}$$

his is a Volterra integral equation of the first kind in a form which
:an be obtained by a trivial change in the variable of integration for

Table VI/7. *Step-by-step solution of a Volterra integral equation*

t	$A(t)$	Calculation with $h=0.2$		Calculation with $h=0.1$	
		Approximation Y	Error	Approximation Y	Error
0	0·5	2		2	
0·1	0·524 187 0			1·903 252	
0·2	0·546 827 0	1·812 692	−0·009	1·822 052	+0·0005
0·3	0·568 030 3			1·744 497	
0·4	0·587 900 0	1·683 48	+0·003	1·680 822	+0·0009
0·5	0·606 530 7			1·618 255	
0·6	0·624 009 7	1·558 94	−0·008	1·568 44	+0·0012
0·7	0·640 418 1			1·517 57	
0·8	0·655 830 6	1·483 37	+0·006	1·478 77	+0·0015
0·9	0·670 316 5			1·437 00	
1	0·683 939 7	1·396 3	−0·009	1·406 98	+0·0018

ıny kernel of convolution type. If $A_n = A(nh)$ and Y_n is the approxima-
ion for $y(nh)$, the equation derived by using the trapezium rule to
:valuate the integral reads

$$t_n = nh = \frac{h}{2} (Y_0 A_n + 2 Y_1 A_{n-1} + 2 Y_2 A_{n-2} + \cdots + 2 Y_{n-1} A_1 + Y_n A_0);$$

ıence

$$Y_n = \frac{1}{A_0} \left(2n - Y_0 A_n - 2 \sum_{\nu=1}^{n-1} Y_\nu A_{n-\nu} \right). \tag{4.11}$$

Γhe values obtained by means of this formula with the two different
)ivotal intervals $h=0.2$ and $h=0.1$ are given in Table VI/7. Better
⁄alues can be obtained by using a more accurate quadrature formula —
ƃIMPSON's rule, for example, cf. § 4.3.

33*

4.3. Method of successive approximations (iteration method)

Starting from a continuous function $y_0(x)$ we construct a sequence of functions[1] $y_n(x)$ according to the iteration formula

$$y_{n+1}(x) = f(x) + \int_a^x K(x, \xi)\, y_n(\xi)\, d\xi \qquad (n = 0, 1, 2, \ldots). \quad (4.12)$$

Under the continuity assumptions made in §4.1, the sequence converges uniformly to the solution $y(x)$ of (4.2) in every finite interval $\langle a, b^*\rangle$ with $b^* < b$. If we denote pivotal values of $y_n(x)$ by

$$y_{(n)k} = y_n(x_k)$$

and evaluate the integral by SIMPSON's rule, the iteration formula (4.12) reads

$$\left.\begin{aligned}
y_{(n+1),2k} = f_{2k} + \frac{h}{3}\,(&K_{2k,0}\,y_{(n)0} + 4K_{2k,1}\,y_{(n)1} + \\
&+ 2K_{2k,2}\,y_{(n)2} + 4K_{2k,3}\,y_{(n)3} + 2K_{2k,4}\,y_{(n)4} + \cdots + \\
&+ K_{2k,2k}\,y_{(n)2k}) \qquad (k = 1, 2, 3, \ldots;\ n = 0, 1, 2, \ldots).
\end{aligned}\right\} \quad (4.13)$$

With this formula we can calculate the values $y_{(n+1)\nu}$ for even ν, but for the next cycle of the iteration we also need the values for odd ν; these can be obtained by interpolating between the "even" values, say by putting a parabola through three consecutive points:

$$y_{(n)\nu} = \tfrac{1}{8}\,(3\,y_{(n),\nu-1} + 6\,y_{(n),\nu+1} - y_{(n),\nu+3}) \quad (4.14)$$

or, rather more accurately, by putting a cubic through four consecutive points:

$$y_{(n)1} = \frac{1}{16}\,(5\,y_{(n)0} + 15\,y_{(n)2} - 5\,y_{(n)4} + y_{(n)6}), \quad (4.15)$$

$$\left.\begin{aligned}
y_{(n)\nu} = \frac{1}{16}\,(-\,y_{(n),\nu-3} + 9\,y_{(n),\nu-1} + 9\,y_{(n),\nu+1} - y_{(n),\nu+3}) \\
(\nu = 3, 5, 7, \ldots).
\end{aligned}\right\} \quad (4.16)$$

If the solution is taken as far as the point $x = x_{2p}$, the value $y_{(n),2p-1}$, must, of course, be calculated from the formula corresponding to (4.15) and not from (4.16).

Example. We consider the same problem as in § 4.2, i.e. the problem (4.7), (4.9) concerning a sluggish measuring instrument. With $S(t) \equiv 1$, $A(0) = \tfrac{1}{2}$ and $A'(t) = \tfrac{1}{2}t^{-2}\,[1 - (1 + t)\,e^{-t}]$ the iteration formula (4.12) reads

$$y_{n+1}(t) = 2 - 2\int_0^1 y_n(\xi)\, A'(t - \xi)\, d\xi. \quad (4.17)$$

For the calculation implied by (4.13) it is convenient to make a separate table of the various multiples of the pivotal values of $A'(t)$ which will actually be needed. These are reproduced in Table VI/8 for the chosen pivotal interval $h = 0{\cdot}1$.

---

[1] An application can be found in K. ZOLLER: Die Entzerrung bei linearen physikalischen Systemen. Ing.-Arch. **15**, 1—18 (1944).

Table VI/8. *Multiples of $A'(t)$ needed for* SIMPSON's *rule*

t	$A'(t)$	$2A'(t)$	$4A'(t)$
0	0·25		
0·1	0·2339420		0·9357680
0·2	0·2190388	0·4380775	
0·3	0·2052018		0·8208070
0·4	0·1923498	0·3846996	
0·5	0·1804080		0·7216321
0·6	0·1693075	0·3386149	
0·7	0·1589847		0·6359387
0·8	0·1493812	0·2987623	
0·9	0·1404430		0·5617720
1	0·1321206		

Table VI/9. *Iterative solution with integrals evaluated by* SIMPSON's *rule with* $h = 0·1$

t	Simpson sum	$y_1(t)$	$y_2(t)$	$y_3(t)$	$y_4(t)$	$y_5(t)$
0		2	2	2	2	2
0·1		1·90327	1·90570	1·905678	1·9056807	1·9056798
0·2	2·8096	1·81269	1·821848	1·821544	1·8215509	1·8215506
0·3		1·7279	1·74758	1·746590	1·7466253	1·7466246
0·4	5·2740	1·6484	1·68200	1·679808	1·6799182	1·6799139
0·5		1·5739	1·6242	1·62008	1·620334	1·620323
0·6	7·4406	1·5040	1·5734	1·56672	1·567216	1·567186
0·7		1·4383	1·5290	1·51884	1·519711	1·519650
0·8	9·3499	1·3767	1·4904	1·47590	1·477304	1·477192
0·9		1·3188	1·4569	1·4373	1·43940	1·439207
1	11·0364	1·2642	1·4280	1·4023	1·40539	1·405090

Table VI/10. *Iterative solution with integrals evaluated by* SIMPSON's *rule with* $h = 0·2$

t	$y_1(t)$	$y_2(t)$	$y_3(t)$	$y_4(t)$	$y_5(t)$	$y_6(t)$
0	2	2	2	2	2	2
0·2	1·8130	1·82252	1·822411	1·822416	1·822404	
0·4	1·6484	1·68197	1·679733	1·679821	1·6798168	1·6798184
0·6	1·5038	1·5730	1·56616	1·566639	1·566618	
0·8	1·3767	1·4903	1·47588	1·477292	1·477183	1·477191
1	1·264	1·4277	1·40198	1·40510	1·40480	
1·2	1·165	1·3824	1·34198	1·34780	1·34712	1·347187
1.4	1·076	1·351	1·2923	1·30202	1·30069	
1·6	0·998	1·329	1·2493	1·26439	1·26208	1·262379

We have $y_n(0) = 2$. Starting from the quite crude approximation $y_0(x) \equiv 2$ we obtain the values shown in Table VI/9 for the first five cycles of the iteration procedure. For the first iterate $y_1(t)$ we also give the value of the integral calculated by SIMPSON's rule (the Simpson sum), and indicate the interpolated values by indenting them.

For comparison we give in Table VI/10 the values obtained by a calculation performed with the double step $h = 0.2$.

4.4. Power series solutions

If the kernel $K(x, \xi)$ and the inhomogeneous term $f(x)$ are analytic functions of simple form, it is often convenient to calculate the solution $y(x)$ for small values of $x - a$ by means of its Taylor series

$$y(x) = \sum_{\nu=0}^{\infty} \frac{y^{(\nu)}(a)}{\nu!} (x - a)^{\nu}. \tag{4.18}$$

We will illustrate the procedure for the integral equation of the second kind (4.2). First we express each derivative $y^{(\nu)}(x)$ in terms of lower derivatives by repeated differentiation of (4.2); for the first two derivatives, for example, we obtain

$$\left.\begin{aligned}
y'(x) &= f'(x) + K(x, x)\, y(x) + \int_a^x \frac{\partial K(x, \xi)}{\partial x}\, y(\xi)\, d\xi, \\
y''(x) &= f''(x) + \frac{dK(x, x)}{dx}\, y(x) + K(x, x)\, y'(x) + \\
&\quad + \frac{\partial K(x, x)}{\partial x}\, y(x) + \int_a^x \frac{\partial^2 K(x, \xi)}{\partial x^2}\, y(\xi)\, d(\xi),
\end{aligned}\right\} \tag{4.19}$$

where we have used the convention that

$$\frac{dK(x, x)}{dx} = \left(\frac{\partial K(x, \xi)}{\partial x} + \frac{\partial K(x, \xi)}{\partial \xi}\right)_{\xi=x}$$

and

$$\frac{\partial K(x, x)}{\partial x} = \left(\frac{\partial K(x, \xi)}{\partial x}\right)_{\xi=x}.$$

We now put $x = a$ in (4.2) and (4.19) and obtain a set of equations from which we can calculate successively the derivatives of $y(x)$ at $x = a$ which are needed in (4.18):

$$\left.\begin{aligned}
y(a) &= f(a), \\
y'(a) &= f'(a) + K(a, a)\, y(a), \\
y''(a) &= f''(a) + K(a, a)\, y'(a) + \\
&\quad + \left[2\frac{\partial K(x, \xi)}{\partial x} + \frac{\partial K(x, \xi)}{\partial \xi}\right]_{\substack{x=a \\ \xi=a}} \times y(a),
\end{aligned}\right\} \tag{4.20}$$

. . . .

§5. Functional equations

Functional equations can appear in so many different forms that there would be little sense in setting out to describe generally applicable methods; the more fruitful approach here is to look at particular equations to see what methods suggest themselves. In any case most of the methods mentioned in §§ 1 and 2 for the solution of integral equations can be readily adapted for the solution of more general equations, and we therefore limit ourselves to a few examples.

5.1. Examples of functional equations

Any equation which expresses a property possessed by one or more functions or by a class of functions may be called a functional equation[1]. In such an equation might appear, for example, a function $u(x, y)$, its partial derivatives, its values at points other than (x, y), say $u(x+h, y+k)$, integrals with integrands involving u, etc. Thus differential equations, integral equations, integro-differential equations, difference equations, in fact all the equations dealt with in this book, are functional equations. Sometimes, however, the term functional equation is used in a more restricted sense applying only to those cases in which the arguments of the function to be determined are not the same throughout the equation. If we restrict the meaning of the term in this way, we may, for example, refer to an equation in which the arguments of the unknown function are not the same throughout and in which derivatives of the unknown function appear as a functional-differential equation.

We now select just a few examples from the very many different types of functional equation which can occur; at the same time we indicate possible ways of dealing with them to show that there is a corresponding diversity of methods for their solution.

A very simple functional-differential equation is

$$y'_{(x)} = y_{(x-1)}$$

(whenever there is a possibility of the argument being misread as a factor, as here, we write it as a subscript). Here, $y(x)$ in the interval $\langle 0, 1 \rangle$, say, may be chosen arbitrarily from among the differentiable functions with $y'(1) = y(0)$, then $y(x)$ in the intervals $\langle 1, 2 \rangle$, $\langle 2, 3 \rangle$, ... determined successively by integration in accordance with the differential equation. Thus further conditions are required to determine a unique solution. In applied problems such conditions will usually arise naturally in the formulation of the particular problem. For example, in the

---

[1] Cf. S. Pincherle: Encyklopädie der mathematischen Wissenschaften, Vol. II, Part I, 2nd half, pp. 788−817. Leipzig 1904−1916. — Kamke, E.: Differential-gleichungen, Lösungsmethoden und Lösungen, Vol. I, pp. 630−636. Leipzig 1942.

theory of structures[1] problems occur in which the additional conditions necessary to determine a unique solution of difference equations of the form

$$\sum_{k=0}^{n} a_{r,k} y_{(x_0+(k+r)h)} = c_r, \qquad (r = 0, 1, \ldots, p)$$

(with given $a_{r,k}$, c_r, x_0, h) are provided by certain boundary conditions; one then has a system of linear equations for a finite number of values of y, which can be solved by one of the usual methods.

Another very simple functional equation occurs in spectroscopy. If the width of the slit in a spectrometer is s, the measured (known) energy distribution $E(x)$ is related to the "true" (required) energy distribution $J(x)$ by the equation[2]

$$\frac{dE(x)}{dx} = J_{\left(x+\frac{s}{2}\right)} - J_{\left(x-\frac{s}{2}\right)}.$$

This equation is solved with the aid of summation symbols[3].

A problem in kinetics leads to the difference equation[4]

$$y_{(x)} + y_{\left(x+\frac{2\pi}{3}\right)} + y_{\left(x+\frac{4\pi}{3}\right)} = h = \text{constant}.$$

The corresponding homogeneous equation, to which it can be reduced by putting $y_{(x)} = g_{(x)} + \frac{1}{3}h$, is satisfied by any trigonometric series of the form

$$g_{(x)} = \sum_{n=1}^{\infty} (a_n \cos n x + b_n \sin n x)$$

with $a_n = b_n = 0$ for all values of n which are divisible by 3. Thus there are infinitely many possible "guide" curves.

Another type of functional equation involves the "iterates" $\varphi_n(x)$ of a function $y = \varphi(x)$. These are defined successively for $n = 1, 2, \ldots$ by the iterative formula $\varphi_{n+1}(x) = \varphi(\varphi_n(x))$, $\varphi_1 = \varphi(x)$ [for example, $\varphi_2(x) = \ln \ln x$ if $\varphi(x) = \ln x$]. Such sequences of iterates are considered

---

[1] See, for instance, P. FUNK: Die linearen Differenzengleichungen und ihre Anwendung in der Theorie der Baukonstruktionen. Berlin 1920. — BLEICH, F., and E. MELAN: Die gewöhnlichen und partiellen Differenzengleichungen in der Baustatik. Berlin 1927. See also W. E. MILNE: Numerical calculus, 393 pp. Princeton 1949, in particular pp. 324—348.

[2] MEYER-EPPLER, W.: Z. Instrumentenkunde 60, 198 (1940).

[3] See N. E. NÖRLUND: Differenzenrechnung, Berlin 1924, or L. M. MILNE-THOMSON: Calculus of finite-differences, Ch. VIII, London 1933.

[4] FISCHER, H. J.: Kurven, in denen ein Drei- oder Vieleck so herumbewegt werden kann, daß seine Ecken die Kurve durchlaufen. Dtsch. Math. 1, 485—498 (1936).

in various contexts; for example, the repeated application of NEWTON's formula for the approximate determination of a zero of an algebraic or transcendental equation $f(x) = 0$ yields the iterates of the function $\varphi(x) = x - f(x)/f'(x)$. A question of interest is whether the sequence $\varphi_n(x)$ converges as $n \to \infty$, or, more precisely, for what values of x does it converge. Many investigations in function theory concern this iteration of functions, especially the case in which the functions are rational; these investigations are rather complicated in parts, and we cannot go into them here.

Functional-differential equations arise in the analysis of control processes in which time-lags are taken into account. Probably the simplest case is that of an oscillatory system of one degree of freedom which is acted upon by a force $P(t)$ whose magnitude at time t depends on the displacement of the system at a previous time $t - \tau$, where τ is a constant time-lag. If $x_{(t)}$ is the displacement of the system at time t, so that $P(t)$ is a function of $x_{(t-\tau)}$, the equation of the system might read

$$m\ddot{x}_{(t)} + k\dot{x}_{(t)} + c\,x_{(t)} = a + b\,x_{(t-\tau)}.$$

More general systems give rise to equations of the form

$$\sum_{r=0}^{n} \sum_{k=0}^{p} a_{r,k}\,y_{(t-\tau_k)}^{(r)} = r(t).$$

Several investigations[1] concern those solutions which, with their first $n - 1$ derivatives, grow no faster than some power of t as $|t| \to \infty$. The question of stability, i.e. what conditions on the coefficients yield systems for which all continuous solutions of the functional-differential equation remain bounded as $t \to +\infty$, is particularly important for control processes. Again one can first remove the inhomogeneous term from the equation. One can then obtain an indication of the stability by assuming a solution of the form $y(t) = e^{st}$; this leads to a transcendental equation for s:

$$\sum_{r=0}^{n} \sum_{k=0}^{p} a_{r,k}\,s^r e^{-s\tau_k} = 0.$$

If $s_1, s_2, \ldots$ are the roots of this equation, then $y(t) = \sum_{\sigma=1}^{\infty} c_\sigma e^{s_\sigma t}$, with constants c_σ such that the series converges, but otherwise arbitrary, is

---

[1] SCHMIDT, ERHARD: Über eine Klasse linearer funktionaler Differentialgleichungen. Math. Ann. 70, 499—524 (1911). — HILB, E.: Lineare funktionale Differentialgleichungen. Math. Ann. 78, 137—170 (1918), among others.

also a solution; and if any one of the roots s_σ has a positive real part, then the control system is unstable. The theory of nomograms is often made use of in stability investigations[1].

5.2. Examples of analytic, continuous and discontinuous solutions of functional equations

For many functional equations a geometrical interpretation of the equation can yield a graphical method for the construction of solutions.

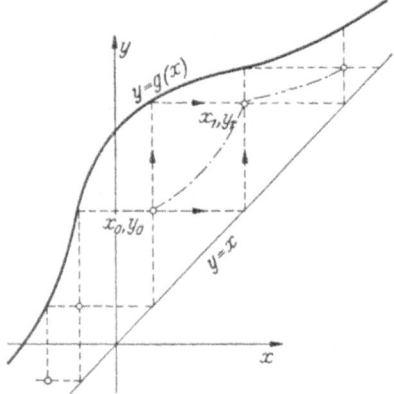

Example. Consider the functional equation

$$y\big(y(x)\big) = g(x). \qquad (5.1)$$

Thus we are required to find a function $y(x)$ whose second iterate $y_2(x)$ coincides with a given function $g(x)$.

Let the given function $g(x)$ be real and continuous, say, and assume for the present that

1. $g(x) > x$ for all x,

Fig. VI/9. Construction of solutions of the functional equation $y[y(x)] = g(x)$

2. $g(x)$ increases monotonically with x.

Let (x_0, y_0) be a point through which a curve $y(x)$ which satisfies (5.1) is to pass. Then a countably finite number of points on this curve can be computed successively by the formula

$$\left.\begin{array}{l} x_{n+1} = y(x_n) \\ y_{n+1} = g(x_n) \end{array}\right\} \quad (n = 0, 1, 2, \ldots);$$

alternatively these points may be located graphically by the self-evident construction shown in Fig. VI/9. This sequence of points can also be extended "backwards" $(n = -1, -2, \ldots)$. It can be seen from the construction that if the points (x_0, y_0) and (x_1, y_1) are joined by any curve C which is not crossed more than once by each parallel to the

---

[1] See R. C. OLDENBOURG and H. SARTORIUS: Dynamik selbsttätiger Regelungen, 2nd ed., 258 pp. Munich 1951. — ENGEL, F. V. A. in collaboration with R. C. OLDENBOURG: Mittelbare Regler und Regelanlagen. Berlin: VDI-Verlag 1944. — HAHN, W.: Bericht über Differential-Differenzengleichungen mit festen und veränderlichen Spannen. Jber. Dtsch. Math. Ver. **57**, 55—84 (1954). Another applied problem giving rise to a functional-differential equation is treated by C. MEISSNER (Zürich): Bestimmung des Profils einer Seilbahn, auf der unter Mitberücksichtigung des Gewichtes des Drahtseiles gleichförmige Bewegung möglich sein soll. Schweiz. Bauztg. **54**, No. 7, 96—98 (1909).

co-ordinate axes, then the construction can be repeated for all points of this curve; in this way we can complete a particular solution of (5.1) for each such curve C. This procedure can still be carried out if there are discontinuities in the curve C; hence we can also construct discontinuous solutions of (5.1) in the same way.

If $g(x)$ possesses a power series expansion:

$$g(x) = \sum_{n=0}^{\infty} g_n x^n,$$

we may expect there to be solutions possessing power series expansions:

$$y(x) = \sum_{n=0}^{\infty} a_n x^n.$$

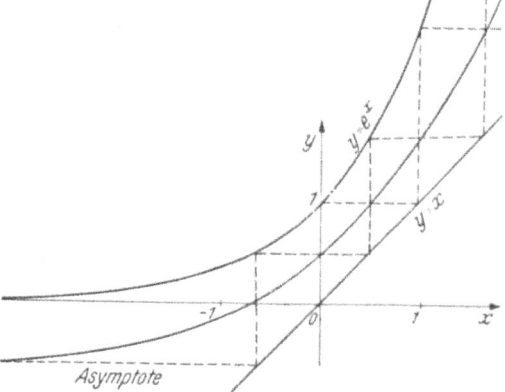

Fig. VI/10. An analytic solution of the equation $y[y(x)] = e^x$

We set out to find them by substituting the assumed form of solution into the functional equation and equating coefficients of powers of x. This yields an infinite system of non-linear equations for the unknowns a_ν:

$$
\left.
\begin{aligned}
g_0 &= a_0 & &+ a_0 a_1 + a_0^2 a_2 + a_0^3 a_3 & &+ \cdots \\
g_1 &= & &a_1^2 + 2 a_0 a_1 a_2 + 3 a_0^2 a_1 a_3 & &+ \cdots \\
g_2 &= 2 a_0 a_2^2 + a_1 a_2 + a_1^2 a_2 & &+ 3 a_0 a_3 (a_1^2 + a_0 a_2) & &+ \cdots \\
g_3 &= 2 a_0 a_2 a_3 + 2 a_1 a_2^2 + a_1 a_3 + a_1^3 a_3 & &+ 3 a_0^2 a_3^2 + 6 a_0 a_1 a_2 a_3 & &+ \cdots.
\end{aligned}
\right\} \quad (5.2)
$$

Normally an approximate solution of such a system of equations will be determined by solving a finite system obtained from the infinite system by truncation, i.e. by ignoring all but the first p (say) equations and putting $a_\nu = 0$ for $\nu \geq p$. In (5.2) the successive truncated systems are indicated by dotted lines.

We now consider some specific forms for $g(x)$.

1. $g(x) = e^x$. The above procedure for obtaining a power series solution[1] here yields

$$y \approx \tfrac{1}{2} + x$$

---

[1] The methods of function theory have been used to demonstrate the existence of an analytic solution for this case by H. KNESER: Reelle analytische Lösungen der Gleichung $\varphi(\varphi(x)) = e^x$ und verwandter Funktionalgleichungen. J. Reine Angew. Math. **187**, 56—67 (1949). Mathematicians' attention was directed to this equation by its occurrence in practical industrial problems.

as a first approximation ($p=2$) and

$$y \approx 0\cdot4979 + 0\cdot8781\,x + 0\cdot2618\,x^2$$

as a second approximation ($p=3$). These curves can be extended for large $|x|$ by means of the graphical construction (see Fig. VI/10); the construction shows that the solution has an asymptote parallel to the x axis.

2. $g(x) = 1 - x^2$. This function violates the two conditions that $g(x)$ shall increase monotonically with x and always be greater than x.

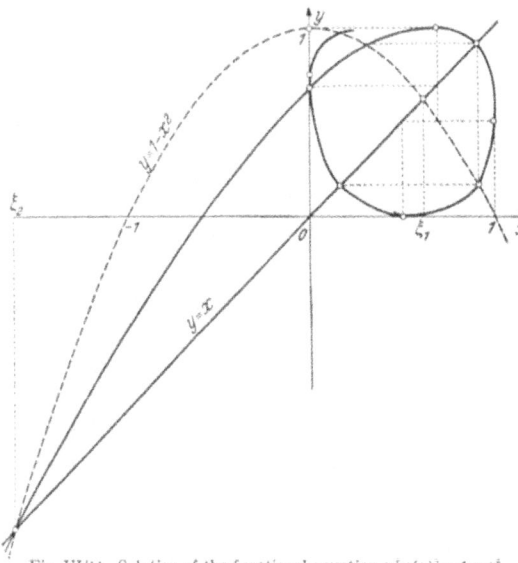

There are now two points where $g(x) = x$; their abscissae are $x=\xi_1$ and $x = \xi_2$, where $\xi_{1,2} = \tfrac{1}{2}(-1\pm\sqrt{5})$. No real differentiable solution can pass through the point $x = y = \xi_1 = \tfrac{1}{2}(-1+\sqrt{5})$, for differentiation of the functional equation yields

$$y'\big(y(x)\big) \cdot y'(x) = -2x,$$

from which it follows that $y'(\xi_1) = \sqrt{-2\xi_1}$ for $x = y = \xi_1$, and this value is imaginary. There is, however, an analytic solution through the

Fig. VI/11. Solution of the functional equation $y[y(x)] = 1 - x^2$

point $x = y = \xi_2 = -\tfrac{1}{2}(1+\sqrt{5})$, and for this solution we can assume a power series expansion; for the second approximation ($a_\nu = 0$ for $\nu > 2$) (5.2) reduces to three equations, which with $g_0 = 1$, $g_1 = 0$, $g_2 = -1$ lead to the quartic $a_1^2(a_1^2 + 2a_1) = 4$ for a_1 and thence to the two real approximations

$$y \approx \ \ \ 0\cdot648 + 1\cdot090\,x - 0\cdot842\,x^2,$$

$$y \approx -6\cdot244 - 2\cdot320\,x - 0\cdot186\,x^2.$$

Another way of obtaining a power series solution is to calculate the derivatives of $y(x)$ at the point $x = \xi_2$ from repeated differentiations of the equation and then write down the Taylor series for $y(x)$ at $x = \xi_2$. Thus we have

$$y''(y)\,[y'(x)]^2 + y'(y)\,y''(x) = -2, \quad \text{so that} \quad y''(\xi_2) = \frac{-2}{y'(\xi_2)\,(1+y'(\xi_2))};$$

similarly

$$y'''(\xi_2) = -3\frac{[y''(\xi_2)]^2}{1 + [y'(\xi_2)]^2} \quad \text{etc.,}$$

and we obtain

$$y = \xi_2 + 1{\cdot}799\,(x - \xi_2) - 0{\cdot}1986\,(x - \xi_2)^2 - 0{\cdot}0187\,(x - \xi_2)^3 + \cdots.$$

The curve corresponding to this solution is shown in Fig. VI/11, where the initial part calculated from the power series has been continued by means of the graphical construction. Once inside the square $0 \le (x, y) \le 1$, the curve cannot get out again and "inscribes" it an infinite number of times, clinging more and more closely to the sides of the square with each revolution.

5.3. Example of a functional-differential equation from mechanics

Consider the small oscillations of a mechanical system consisting of a homogeneous string of length l fixed at one end $x = l$ and attached at the other $x = 0$ to the centre of a transversely mounted spring as in Fig. VI/12.

The displacement of the string $u(x, t)$ at the point x at time t satisfies the wave equation

$$\frac{\partial^2 u}{\partial t^2} = C^2 \frac{\partial^2 u}{\partial x^2}, \qquad (5.3)$$

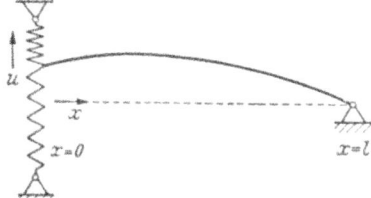

Fig. VI/12. A non-linear oscillatory system

where C is a given constant. A displacement $u(0, t) = u_0(t)$ at the point $x = 0$ produces an opposing force $H(u_0)$ in the spring, $H(u)$ being a given, in general non-linear, function of u; we will assume also that $H(u)$ is an odd function of u. The boundary conditions therefore read

$$\left.\begin{array}{c} \dfrac{\partial u(0, t)}{\partial x} = G\left(u(0, t)\right), \\[2mm] u(l, t) = 0, \end{array}\right\} \qquad (5.4)$$

where $G(u) = k H(u)$ is likewise a given, non-linear, odd function of u (k is a constant depending on the tension in the string). Our object is to investigate solutions which are periodic in time; the period T will depend on the maximum displacement. Let us look for a solution which passes through the position of equilibrium at time $t = 0$, say, i.e. a solution for which

$$u(x, 0) = 0 \quad \text{for} \quad 0 \le x \le l. \qquad (5.5)$$

Now the general solution of (5.3) can be expressed in the form

$$u = f(x + C t) - g(x - C t),$$

where f and g are arbitrary functions. The initial condition (5.5) then becomes

$$f(x) = g(x) \quad \text{for} \quad 0 \le x \le l,$$

and since the values of $f(x)$ for $x < 0$ and of $g(x)$ for $x > l$ do not affect the solution for $t > 0$, which is what interests us, we can put $f(x) = g(x)$ for all x. The boundary conditions (5.4) then reduce to

$$f(l + y) = f(l - y),\tag{5.6}$$
$$f'(y) - f'(-y) = G(f(y) - f(-y)),\tag{5.7}$$

where $y = C\,t$.

It remains to express the condition of periodicity in terms of f: we require that

$$f(x + \tau) = f(x - \tau)$$

for some real finite number τ; then $f(x)$ will have the period 2τ and $u(x, t)$, as a function of t, the period $T = 2\tau/C$.

Thus for given $G(u)$ we seek periodic solutions of the functional-differential equation (5.7) which are symmetric about the point $x = l$.

A simple way of obtaining a solution of this problem is to choose a value for the period 2τ and determine a solution of (5.7) with this period, say by the finite-

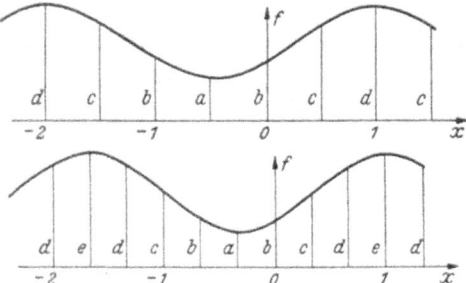

Fig. VI/13. Notation for the finite-difference solution of the functional equation

Fig. VI/14. Solution of the transcendental equations

difference method; naturally for given $G(u)$ there may be values of 2τ for which no solution exists.

We illustrate the procedure with $l = 1$, and choose first $2\tau = 3$. We need only consider a half-cycle, and if we divide this interval into three, i.e. use a finite-difference step $h = \frac{1}{3}\tau = \frac{1}{2}$, we have four unknown f values a, b, c, d as indicated in Fig. VI/13. The finite-difference equations read

$$\text{(for } x = \tfrac{1}{2}) \qquad d - b = G(c - a), \qquad \text{or} \qquad \beta = G(\alpha),$$
$$\text{(for } x = 1) \qquad c - a = G(d - b), \qquad \text{or} \qquad \alpha = G(\beta)$$

(no new equations are obtained for $x = \frac{3}{2}$ and $x = 2$); the α and β introduced here are corresponding values of $u(0, t)$:

$$\alpha = c - a = f\left(\frac{1}{2}\right) - f\left(-\frac{1}{2}\right) = u\left(0, \frac{T}{6}\right), \qquad \beta = d - b = f(1) - f(-1) = u\left(0, \frac{T}{3}\right).$$

For a normal spring $G(u)$ is monotonic, and the values of α and β are given by the intersections of the curve $v = G(u)$ with the straight line $v = u$ (Fig. VI/14).

Let us now choose another period, say $2\tau = \frac{8}{3}$, and this time use more pivotal points, say with $h = \frac{1}{4}\tau = \frac{1}{3}$. We now have five unknown f values a, b, c, d, e (as in Fig. VI/13) and three corresponding values of $u(0, t)$: $u\left(0, \dfrac{T}{8}\right) = f\left(\dfrac{1}{3}\right) -$

$f\left(-\dfrac{1}{3}\right) = c - a = \gamma,\ u\left(0, \dfrac{T}{4}\right) = d - b = \delta,\ u\left(0, \dfrac{3T}{8}\right) = e - c = \varepsilon.$ If we put

$G^* = 2hG = \frac{2}{3}G$, the difference equations read

$$\text{(for } x = \tfrac{1}{3}) \qquad \delta = G^*(\gamma),$$
$$\text{(for } x = \tfrac{2}{3}) \qquad \gamma + \varepsilon = G^*(\delta),$$
$$\text{(for } x = 1) \qquad \delta = G^*(\varepsilon)$$

(no new equations are obtained for $x = \frac{4}{3}, \frac{5}{3}, \dots$).

If $G(u)$ is monotonic, we must have $\gamma = \varepsilon$ and the equations simplify to the pair of equations

$$\delta = G^*(\gamma), \qquad 2\gamma = G^*(\delta).$$

These also may easily be solved graphically [see Fig. VI/14; in this sketch, which is only for illustration, we have not drawn a new curve for $G^*(u)$].

5.4. Miscellaneous exercises on Chapter VI

1. Consider a luminous, line "object" whose intensity of illumination is a function $z(\xi)$ of the distance ξ along the line, and let its image formed by an optical system (Fig. VI/15) be another line, say the x axis, illuminated with the intensity $y(x)$; further let all parts of the lines outside of the sections $|x| \leq 1$ and $|\xi| \leq 1$ be shielded by blinds. Then the intensity distributions $y(x)$ for $|x| \leq 1$ and $z(\xi)$ for $|\xi| \leq 1$ are related by an integral equation[1] of the form

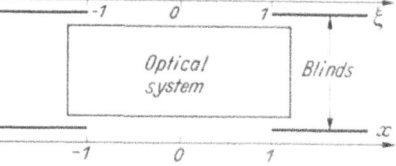

$$y(x) = \int_{-1}^{1} K(x, \xi)\, z(\xi)\, d\xi;$$

Fig. VI/15. Optical system with line object and image

the kernel $K(x, \xi)$ depends on the optical system used, but may be approximated by

$$K(x, \xi) = \begin{cases} 1 + \cos \pi (x - \xi) & \text{for} \quad |x - \xi| \leq 1, \\ 0 & \text{for} \quad |x - \xi| \geq 1. \end{cases}$$

For what intensity distributions are the object and image distributions similar, i.e. such that $z(x) = \lambda y(x)$? Calculate approximations for the first few eigenfunctions by the finite-sum method of § 1.2.

2. Apply the enclosure theorem (2.15), (2.16) of § 2.3 to the integral equation

$$y(x) = \lambda \int_{0}^{1} e^{x\xi}\, y(\xi)\, d\xi.$$

3. Determine an approximate solution of the non-linear integral equation

$$\int_{0}^{1} \frac{d\xi}{y(x) + y(\xi)} = 1 + x$$

by the finite-sum method of § 1.2. May the Ritz method of § 1.7 also be applied to this equation?

---

[1] FRANK, PH., and R. v. MISES: Die Differential- und Integralgleichungen der Mechanik und Physik, Vol. I, p. 473. Brunswick 1930.

4. Determine a real analytic solution of the functional equation

$$y_{(x + y(x))} = 1 - x^2$$

(where the argument is again written as a subscript so that it cannot be read as a factor) under the assumption that such a solution exists.

5. Use the finite-difference method to obtain approximate solutions of the eigenvalue problem presented by the functional-differential equation

$$- y''_{(x)} = \lambda y_{(1-x)}$$

with the boundary conditions

$$y(0) = y'(1) = 0.$$

6. Apply (a) the Ritz method and (b) the power series method to the problem of the last exercise.

7. Let us end by applying the two well-tried methods

(a) the finite-difference method

(b) the Ritz method,

which have been used repeatedly throughout this book, to the partial functional-differential eigenvalue problem

$$\nabla^2 u(x, y) + \lambda u(- x, - y) = 0$$

with the boundary conditions

$$u = 0 \quad \text{for} \quad x = 1 \quad \text{and for} \quad y = 1,$$

$$\frac{\partial u}{\partial v} = 0 \quad \text{for} \quad x = -1 \quad \text{and for} \quad y = - 1.$$

Calculate approximations for the first few eigenvalues.

5.5. Solutions

1. (a) Three pivotal points $x_j = j$ with $j = 0, \pm 1$. Let y_j be the approximate pivotal value of $y(x)$ at $x = x_j$. We must remember that there are discontinuities in the derivatives of the kernel and, as mentioned in § 1.2, must choose our quadrature formulae accordingly; if we evaluate the integral by the trapezium rule for $x = \pm 1$ and by Simpson's rule for $x = 0$, we obtain from the equations

$$y_{-1} = \Lambda y_{-1}, \quad y_0 = \Lambda \tfrac{1}{3} 8 y_0, \quad y_1 = \Lambda y_1$$

the three approximate values

$$\Lambda = \tfrac{3}{8}, 1, 1$$

for the eigenvalue λ.

(b) Five pivotal points $x_j = \tfrac{1}{2} j$ with $j = 0, \pm 1, \pm 2$. As usual with symmetric eigenvalue problems we can reduce the number of unknowns by treating the symmetric and antisymmetric solutions separately; thus if y_j is an approximation for $y(x_j)$, we postulate $y_j = y_{-j}$ for symmetric solutions and $y_j = - y_{-j}$ for antisymmetric. For the symmetric solutions we obtain the equations

$$y_2 = \frac{\Lambda}{6} (2 y_2 + 4 y_1),$$

$$y_1 = \frac{\Lambda}{8} \frac{3}{2} (y_2 + 6 y_1 + 3 y_0),$$

$$y_0 = \frac{\Lambda}{6} (8 y_1 + 4 y_0)$$

(it is a help in setting them up to record the values of the kernel $K(x, \xi)$ in an array as in Fig. VI/16); we have taken account of the points at which the kernel has discontinuous higher derivatives by using SIMPSON's rule for $x = 0, 1$ and the "three-eighths" rule for $x = \frac{1}{2}$.

For $x = 1/\Lambda$ we obtain the equation

$$- 24 x^3 + 51 x^2 - \frac{34}{3} x - 2 = 0,$$

which yields the approximate eigenvalues

$$\Lambda = 0 \cdot 541, \quad 2 \cdot 53, \quad - 8 \cdot 73.$$

By postulating antisymmetry we obtain similarly

$$x = \frac{1}{48} (35 \pm \sqrt{649}); \quad \Lambda = 0 \cdot 794, \quad 5 \cdot 04.$$

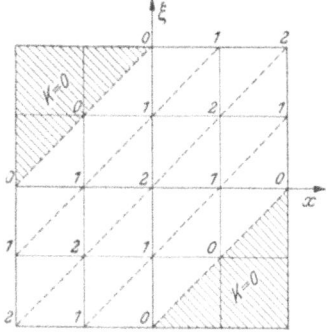

Fig. VI/16. Values of the kernel

The points given by the corresponding solutions of the homogeneous equations are plotted in Fig. VI/17 and joined by straight lines, so that the approximate eigenfunctions are represented by piecewise-linear functions; smooth approximations could be obtained by rounding off the corners.

2. An obvious choice for $F_0(x)$ is $e^{a x}$; then

$$F_1(x) = \int_0^1 e^{x \xi} e^{a \xi} d\xi = \frac{e^{a+x} - 1}{a + x}$$

and

$$\Phi(x) = \frac{F_0(x)}{F_1(x)} = \frac{e^{a x}(a + x)}{e^{a+x} - 1}.$$

Curves of $\Phi(x)$ against x are drawn in Fig. VI/18 for several values of a; the difference $\Phi_{max} - \Phi_{min}$ appears to be smallest for a value of a about $0 \cdot 59$; for this value we obtain the limits

$$\Phi_{min} = 0 \cdot 7338 \leq \lambda \leq 0 \cdot 7417 = \Phi_{max}.$$

The mean value, which can be in error by at most $0 \cdot 53\%$, gives $\lambda \approx 0 \cdot 7377$.

At the same time $F_1(x)$ with $a = 0 \cdot 59$ provides an approximation for the corresponding eigenfunction.

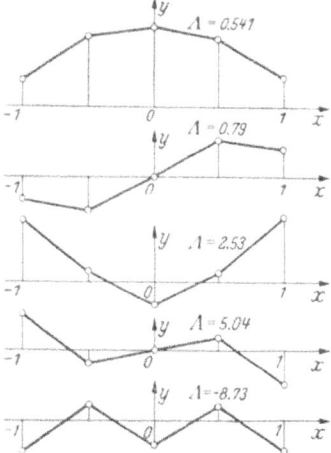

Fig. VI/17. Approximations obtained by the summation method for distributions of illumination which undergo pure magnification in an optical system

3. (a) First of all let us find a very crude approximation by taking only two pivotal points $x = 0$ and $x = 1$ and using the trapezium rule; then for y_0 and y_1, the approximate values of $y(0)$ and $y(1)$, we obtain the equations

$$\frac{1}{2 y_0} + \frac{1}{y_0 + y_1} = 2, \qquad \frac{1}{y_0 + y_1} + \frac{1}{2 y_1} = 4.$$

These yield a quadratic for the ratio $\eta = y_1/y_0$, from which we calculate the two values

$$\eta = \frac{y_1}{y_0} = -\frac{3}{4} \pm \frac{1}{4}\sqrt{17} = \begin{cases} 0 \cdot 2808 \\ -1 \cdot 7808; \end{cases}$$

we therefore obtain two solutions $(8y_0 = 1 \pm \sqrt{17},\ 16y_1 = 7 \mp \sqrt{17})$:

1st solution: $y_0 = 0 \cdot 640,$ $y_1 = 0 \cdot 180$

2nd solution: $y_0 = -0 \cdot 390,$ $y_1 = 0 \cdot 695.$

For the second solution, which changes sign, the integral equation is singular and the integral must be regarded as a Cauchy principal value (cf. § 3.2); con-

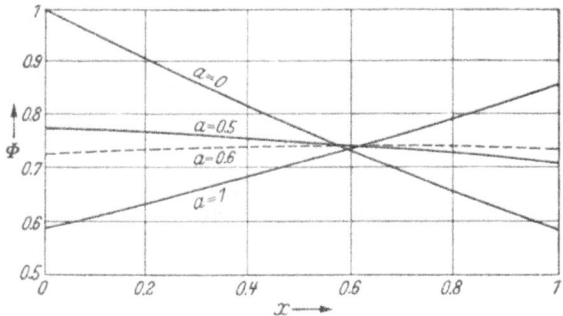

Fig. VI/18. A family of functions $\Phi(x)$ from which is to be chosen the one with smallest variation

sequently, indiscriminate use of the finite-sum method should not be carried any further in this case, and we will proceed to higher accuracy only for the solution with constant sign.

(b) Let $y_0,\ y_1,\ y_2$ be approximations for $y(0),\ y(\tfrac{1}{2}),\ y(1)$. Evaluating the integral by SIMPSON's rule we obtain

$$\frac{1}{2y_0} + \frac{4}{y_0 + y_1} + \frac{1}{y_0 + y_2} = 6,$$

$$\frac{1}{y_0 + y_1} + \frac{4}{2y_1} + \frac{1}{y_1 + y_2} = 9,$$

$$\frac{1}{y_0 + y_2} + \frac{4}{y_1 + y_2} + \frac{1}{2y_2} = 12.$$

If we write $a = y_0 + y_1,\ b = y_1,\ c = y_1 + y_2$, we have a system of non-linear equations for a, b, c for which an iterative solution is suitable; thus we calculate the $(n+1)$-th approximation $a_{n+1}, b_{n+1}, c_{n+1}$ from the n-th by the formulae

$$\left.\begin{aligned}
\frac{4}{a_{n+1}} &= 6 - \frac{1}{2(a_n - b_n)} - \frac{1}{a_n + c_n - 2b_n}, \\
\frac{3}{b_{n+1}} &= 9 - \frac{1}{a_{n+1}} - \frac{1}{c_n}, \\
\frac{4}{c_{n+1}} &= 12 - \frac{1}{a_{n+1} + c_n - 2b_{n+1}} - \frac{1}{2(c_n - b_{n+1})}
\end{aligned}\right\} \quad (n = 0, 1, 2, \ldots).$$

This iteration converges well, and yields the values

$$a = 0.990, \qquad b = 0.332, \qquad c = 0.505,$$

from which we obtain the approximate solution

$$y_0 = 0.658, \qquad y_1 = 0.332, \qquad y_2 = 0.173.$$

(c) In RITZ's method we have to find approximate solutions of the variational problem

$$J[u] = \tfrac{1}{2} \int_0^1 \int_0^1 \ln\left(u(x) + u(\xi)\right) dx\, d\xi - \int_0^1 (1+x)\, u(x)\, dx = \text{extremum}.$$

In the first approximation with $u = a$ we have $J = \tfrac{1}{2} \ln 2a - \tfrac{3}{2}a$, and $\partial J/\partial a = 0$ yields the value $a = \tfrac{1}{3}$; however, even for the second approximation with $u = a + bx$ the amount of calculation involved is already disproportionately large in comparison with the finite-sum method.

4. (a) Let us see first whether $y(x)$ can vanish at some point $x = \xi$; if, for the present, we consider only single-valued solutions, the functional equation at such a point would read $0 = 1 - \xi^2$, so that we must have $\xi = \pm 1$. Let us now determine the derivatives at $\xi = \pm 1$ of the solutions which vanish at these points. By repeated differentiation of the functional equation we obtain

$$y'_{(x+y(x))} [1 + y'(x)] = -2x,$$

$$y''_{(x+y(x))} [1 + y'(x)]^2 + y''(x)\, y'_{(x+y(x))} = -2;$$

now for $x = \xi = \pm 1$ and $y(\xi) = 0$ the first equation reduces to a quadratic for $y'(\xi)$:

$$[y'(\xi)]^2 + y'(\xi) + 2\xi = 0;$$

since this has real solutions only when $\xi = -1$ we need no longer consider the point $\xi = 1$. For $\xi = -1$ we have $y'(\xi) = 1$ or -2, and for each of these two values the higher derivatives at $\xi = -1$ are determined uniquely. To calculate the two solutions $y_1(x)$, $y_2(x)$ through the point $(-1, 0)$ we put $x + 1 = u$ and insert the power series

$$y_{(-1+u)} = \sum_{\nu=1}^{\infty} a_\nu u^\nu$$

into the functional equation, which then reads

$$y_{(-1+u+y(-1+u))} = 2u - u^2;$$

for $a_1 = -2$ and $a_1 = 1$ we obtain the respective expansions

$$y_{1(-1+u)} = -2u + u^2 - \frac{2}{3} u^3 + \frac{1}{3} u^4 - \cdots,$$

$$y_{2(-1+u)} = u - \frac{1}{5} u^2 - \frac{4}{225} u^3 - \frac{11}{3825} u^4 - \cdots.$$

(b) To pursue the solution $y_1(x)$ further, let us calculate sequences of points (x_j, y_j) from the equations

$$x_{j+1} = x_j + y_j, \qquad y_{j+1} = 1 - x_j^2 \qquad (j = 0, 1, 2, \ldots)$$

with $x_0 = 0$ and y_0 as a parameter; we choose several values of y_0 covering a range in which we expect to find the value associated with $y_1(x)$. Each of these sequences lies on a solution of the functional equation, and as can be seen in Fig. VI/19, which shows several points of the sequences for $y_0 = -1.4$, -1.38, -1.397, these curves

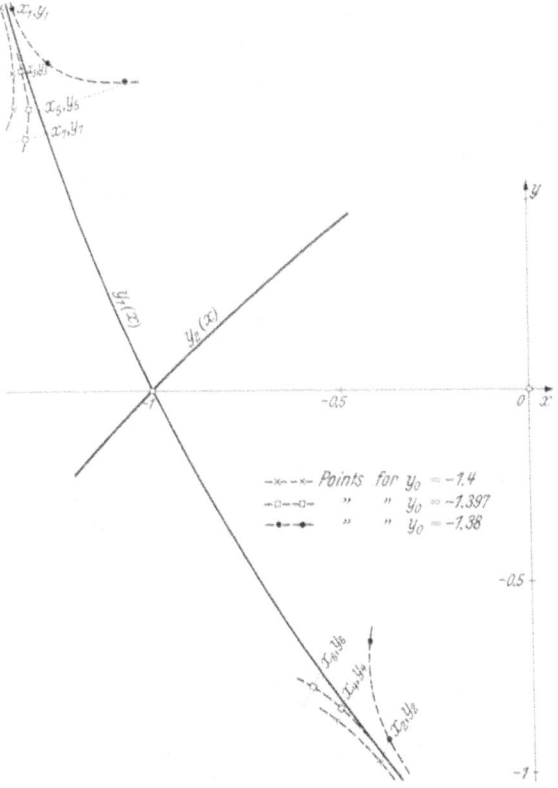

Fig. VI/19. Solution of the functional equation $y_{(x+y(x))} = 1 - x^2$

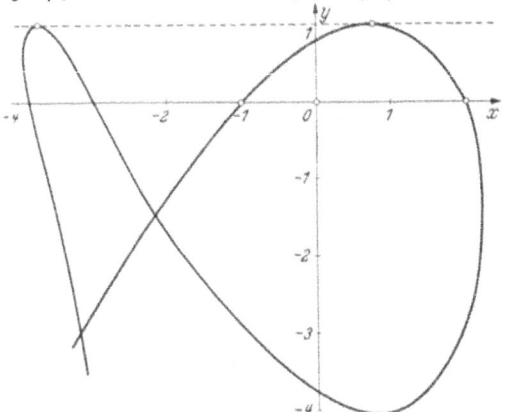

Fig. VI/20. A many-valued solution of the functional equation

diverge in varying degrees either side; by interpolation the initial value corresponding to the non-diverging curve is found to be $y_0 = -1{\cdot}396\,35$.

(c) By continuing the solution $y_2(x)$ by means of the functional equation we obtain a many-valued function; a part of it is reproduced in Fig. VI/20.

5. With the pivotal value notation a, b, c as in Fig. VI/21 and $\varLambda$ as an approximate value of λ the finite-difference equations read

$$9(b - 2a) + \varLambda b = 9(c - 2b + a) + \varLambda a = 9(2b - 2c) = 0,$$

and for non-trivial solutions we must have $\varLambda = 9(-1 \pm \sqrt{2})$. These values are expressed in decimals in Table VI/11, together with the values obtained for $h = \frac{1}{4}$ and $h = \frac{1}{8}$.

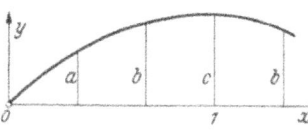

Fig. VI/21. Pivotal value notation for the functional-differential eigenvalue problem

Table VI/11. *Successive approximations for the eigenvalues*

Pivotal interval	$\varLambda_{-2}$	$\varLambda_{-1}$	$\varLambda_1$	$\varLambda_2$
$h = \frac{1}{3}$		$-21 \cdot 73$	$3 \cdot 728$	
$h = \frac{1}{4}$		$-22 \cdot 26$	$3 \cdot 634$	$50 \cdot 6$
$h = \frac{1}{8}$	$-87 \cdot 25$	$-22 \cdot 29$	$3 \cdot 591$	$56 \cdot 0$

The problem possesses infinitely many positive and infinitely many negative eigenvalues.

6. (a) The corresponding variational problem reads

$$J[u] = \int_0^1 [u'^2_{(x)} - \lambda u_{(x)} u_{(1-x)}] \, dx = \text{extremum}$$

with respect to the domain of functions u possessing continuous derivatives and satisfying the condition $u(0) = 0$. Thus if we put $u = y + \varepsilon\eta$, where $\eta(0) = 0$, in $J[u]$, the condition $(\partial J/\partial \varepsilon)_{\varepsilon=0}$, which is necessary if y is to minimize J, leads to

$$\int_0^1 \eta(x) \{ -y''_{(x)} - \lambda y_{(1-x)}\} \, dx + [\eta \, y']_0^1 = 0$$

on integration by parts, and the usual arguments show that y, which already satisfies $y(0) = 0$, must also satisfy the functional-differential equation and the boundary condition $y'(1) = 0$.

Let us use the two-parameter Ritz expression

$$u = c_1(2x - x^2) + c_2(3x - x^3),$$

which satisfies both boundary conditions [u need only satisfy the condition $u(0) = 0$, but here it is no more difficult to satisfy the other boundary condition at the same time].

The necessary conditions $\partial J/\partial c_1 = \partial J/\partial c_2 = 0$ yield two homogeneous linear equations for the c_j; their determinant

$$\begin{vmatrix} \dfrac{4}{3} - \varLambda \dfrac{11}{30} & \dfrac{5}{2} - \varLambda \dfrac{2}{3} \\[2ex] \dfrac{5}{2} - \varLambda \dfrac{2}{3} & \dfrac{24}{5} - \varLambda \dfrac{169}{140} \end{vmatrix}$$

vanishes for the values

$$\varLambda = \begin{cases} 3 \cdot 51984 \\ -23 \cdot 346. \end{cases}$$

A one-parameter expression ($c_2 = 0$) would have lead to the value $\varLambda = \dfrac{40}{11} \approx 3 \cdot 636$, while for a two-parameter expression satisfying the boundary condition $u(0) = 0$ only, i.e. $u = c_1 x + c_2 x^2$, we would have obtained the values

$$\varLambda = 4(-8 \pm \sqrt{79}) = \begin{cases} 3 \cdot 553 \\ -67 \cdot 55. \end{cases}$$

(b) At the point $x = 1$ we have $y'(1) = 0$, and from the functional equation $y''(1) = -\lambda y(0) = 0$. We therefore assume the power series solution

$$y_{(1-x)} = a_0 + a_3 x^3 + a_4 x^4 + a_5 x^5 \dots.$$

By inserting the series in the functional equation and equating coefficients we obtain an infinite system of equations for the a_j; with $\lambda = 20\mu$ they read

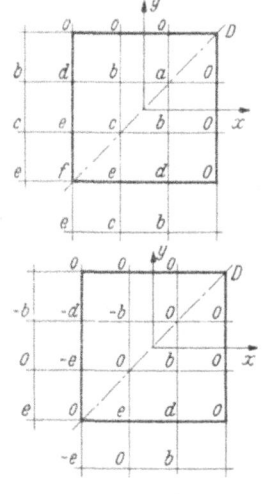

$$a_0 + a_3 + a_4 + a_5 + a_6 + \cdots = 0,$$
$$10\mu a_0 + 3a_3 + 6a_4 + 10a_5 + 15a_6 + \cdots = 0,$$
$$a_3 + 4a_4 + 10a_5 + 20a_6 + \cdots = 0,$$
$$a_4 + 5a_5 + 15a_6 + \cdots = 0,$$
$$-\mu a_3 \qquad + a_5 + 6a_6 + \cdots = 0,$$
$$\cdot \cdot \cdot \cdot \cdot \cdot \cdot \cdot \cdot \cdot \cdot \cdot \cdot \cdot$$

Apart from a missing line and the terms involving μ, the coefficients are binomial coefficients.

If we truncate the system by retaining only the first five equations and putting $a_j = 0$ for $j \geqq 7$, the zeros of the truncated determinant yield the approximate eigenvalues

$$\Lambda = \frac{1}{2}\left(-5 \pm \sqrt{145}\right) = \begin{cases} 3 \cdot 5208 \\ -8 \cdot 52. \end{cases}$$

7. (a) For the pivotal interval $h = \frac{2}{3}$ and with pivotal values $a, b, \dots, f$ as in Fig. VI/22 for solutions symmetrical about the diagonal D, the finite-difference equations read

$$-4a + 2b + \mu c \qquad\qquad = 0,$$
$$a + (\mu - 4)b + c + d \qquad = 0,$$
$$\mu a + 2b - 4c + 2e = 0,$$
$$2b - 4d + e = 0,$$
$$2c + d - 4e + f = 0,$$
$$4e - 4f = 0,$$

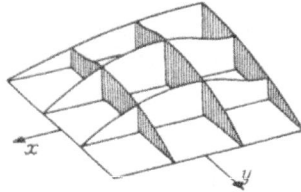

Fig. VI/22. An eigenvalue problem for a partial functional-differential equation

where $\mu = \Lambda h^2$ and Λ is an approximate value of λ.

For $\nu = \frac{1}{2}\mu$ we obtain the equation $11\nu^3 - 19\nu^2 - 41\nu + 26 = 0$, which yields

$$\nu = \begin{cases} -1 \cdot 580, \\ 0 \cdot 54100, \\ 2 \cdot 76 59, \end{cases} \quad \text{and hence} \quad \begin{cases} \Lambda_{-1} = -7 \cdot 11, \\ \Lambda_1 = 2 \cdot 434, \\ \Lambda_2 = 12 \cdot 45; \end{cases}$$

the eigenfunction corresponding to Λ_1 is depicted in Fig. VI/22. The value obtained for λ_1 by the simpler calculation with $h = 1$ is $\Lambda_1 = \frac{8}{3} = 2 \cdot 667$.

A corresponding calculation (with $h = \frac{2}{3}$) for the antisymmetric solutions (cf. Fig. VI/22) yields $\Lambda_{-2} = -7 \cdot 8$.

(b) A corresponding variational problem reads

$$J[\varphi] = \iint_Q [\varphi_x^2 + \varphi_y^2 - \lambda \varphi(x, y) \varphi(-x, -y)]\, dx\, dy = \text{extremum}$$

with $\varphi(1, y) = \varphi(x, 1) = 0$; Q is the square $|x| \leqq 1$, $|y| \leqq 1$.

This can be verified in the usual way by putting $\varphi = u + \varepsilon\eta$, where $\eta(1, y) = \eta(x, 1) = 0$, and calculating

$$\left(\frac{\partial J(u + \varepsilon\eta)}{\partial\varepsilon}\right)_{\varepsilon=0}$$

$$= \iint_Q [2(u_x\eta_x + u_y\eta_y) - \lambda\{u(x, y)\eta(-x, -y) + u(-x, -y)\eta(x, y)\}]\,dx\,dy.$$

To find the conditions on u necessary for this expression to be zero for any η we transform the first term by GREEN's formula (3.7) of Ch. I:

$$\iint_Q (u_x\eta_x + u_y\eta_y)\,dx\,dy = -\iint_Q \eta\,\nabla^2 u\,dx\,dy - \int_\Gamma \eta\,u_\nu\,ds,$$

and change the signs of x and y in the term in $\eta(-x, -y)$, obtaining the necessary condition

$$\iint_Q \eta\{\nabla^2 u(x, y) + \lambda u(-x, -y)\}\,dx\,dy + \int_\Gamma \eta\,u_\nu\,ds = 0.$$

By the usual arguments we deduce from the arbitrariness of η that the factor multiplying η in the double integral must vanish, i.e. u must satisfy the functional equation, and that on the part of the boundary where we have not required that $\eta = 0$ the normal derivative u_ν must also vanish, so that u must also satisfy all the given boundary conditions.

Let us now assume for φ the expression

$$\varphi = (1 - x)(1 - y)(\beta + x)(\beta + y)$$

(non-linear in the parameter β). Then J takes the form $J = f_1(\beta) - \Lambda f_2(\beta)$. In the usual way $\dfrac{\partial J}{\partial\beta} = f_1' - \Lambda f_2' = 0$ yields at the same time the value $\Lambda = \left(\dfrac{f_1}{f_2}\right)_{\text{extr.}}$ $\left[\text{from } \left(\dfrac{f_1}{f_2}\right)' = \dfrac{f_1'f_2 - f_1 f_2'}{f_2^2} = 0 \text{ it follows that } \dfrac{f_1}{f_2} = \dfrac{f_1'}{f_2'}\right].$ Here

$$\Lambda = 10\,\frac{(7 - 6\beta + 3\beta^2)(2 - 5\beta + 5\beta^2)}{(1 - 5\beta^2)^2},$$

and from a graph we find that the approximate position of the minimum is at the point $\beta \approx 2$ and the corresponding approximate value for λ_1 is $\Lambda = 2\cdot326$.

Appendix

Table I. *Approximate methods for ordinary differential equations of the first order*

$$y' = f(x, y)$$

Notation: y_s = approximation for $y(x_s)$; $f_s = f(x_s, y_s)$
(Explanations in the text of Chapter II)

	Name	Formulae
Formulae of Runge-Kutta type	1st order (polygon method)	$y_{r+1} = y_r + h f_r$
	2nd order (improved polygon method)	$f^*_{r+\frac{1}{2}} = f\left(x_{r+\frac{1}{2}}, y_r + \frac{1}{2} h f_r\right); \quad y_{r+1} = y_r + h f^*_{r+\frac{1}{2}}$
		$y^*_{r+1} = y_r + h f_r; \quad y_{r+1} = y_r + \frac{1}{2} h [f_r + f(x_{r+1}, y^*_{r+1})]$
	3rd order (HEUN)	$k_1 = h f_r; \quad k_2 = h f\left(x_{r+\frac{1}{2}}, y_r + \frac{1}{2} k_1\right);$ $k_3 = h f(x_{r+1}, y_r - k_1 + 2 k_2); \quad y_{r+1} = y_r + \frac{1}{6}(k_1 + 4 k_2 + k_3)$
		$k_1 = h f_r; \quad k_2 = h f\left(x_{r+\frac{1}{3}}, y_r + \frac{1}{3} k_1\right); \quad k_3 = h f\left(x_{r+\frac{2}{3}}, y_r + \frac{2}{3} k_2\right)$ $y_{r+1} = y_r + \frac{1}{4} k_1 + \frac{3}{4} k_3$
	4th order (KUTTA)	$k_1 = h f_r; \quad k_2 = h f\left(x_{r+\frac{1}{2}}, y_r + \frac{1}{2} k_1\right); \quad k_3 = h f\left(x_{r+\frac{1}{2}}, y_r + \frac{1}{2} k_2\right)$ $k_4 = h f(x_{r+1}, y_r + k_3); \quad y_{r+1} = y_r + \frac{1}{6}(k_1 + 2 k_2 + 2 k_3 + k_4)$
	DUFFING's formula	$y_{r+1} = y_r + \frac{1}{6} h \left[4 f_r + 2 f_{r+1} + h \left(\frac{d}{dx} f\right)_{x = x_r}\right]$
Formulae for finite-difference methods / Extrapolation	ADAMS	$y_{r+1} = y_r + h \left[f_r + \frac{1}{2} \nabla f_r + \frac{5}{12} \nabla^2 f_r + \frac{3}{8} \nabla^3 f_r + \frac{251}{720} \nabla^4 f_r + \right.$ $\left. + \frac{95}{288} \nabla^5 f_r + \frac{19087}{60480} \nabla^6 f_r + \cdots\right]$
	NYSTRÖM	$y_{r+1} = y_{r-1} + h \left[2 f_r + \frac{1}{3} \nabla^2 f_r + \frac{1}{3} \nabla^3 f_r + \frac{29}{90} \nabla^4 f_r + \frac{14}{45} \nabla^5 f_r + \cdots\right]$
Interpolation	ADAMS	$y_{r+1} = y_r + h \left[f_{r+1} - \frac{1}{2} \nabla f_{r+1} - \frac{1}{12} \nabla^2 f_{r+1} - \frac{1}{24} \nabla^3 f_{r+1} - \right.$ $\left. - \frac{19}{720} \nabla^4 f_{r+1} - \frac{3}{160} \nabla^5 f_{r+1} - \frac{863}{60480} \nabla^6 f_{r+1} - \cdots\right]$
	Central differences	$y_{r+1} = y_{r-1} + h \left[2 f_r + \frac{1}{3} \nabla^2 f_{r+1} - \frac{1}{90} \nabla^4 f_{r+2} + \frac{1}{756} \nabla^6 f_{r+3} - \cdots\right]$
	MILNE's formulae	$y^*_{r+1} = y_{r-3} + 4 h \left[f_{r-1} + \frac{2}{3} \nabla^2 f_r\right]; \quad f^*_{r+1} = f(x_{r+1}, y^*_{r+1})$ $y_{r+1} = y_{r-1} + h \left[2 f_r + \frac{1}{3} \nabla^2 f^*_{r+1}\right]$
	QUADE's formula	$y_{r+1} = \frac{8}{19}(y_r - y_{r-2}) + y_{r-3} + \frac{6 h}{19}(f_{r+1} + 4 f_r + 4 f_{r-2} + f_{r-3})$

Table II. *Approximate methods for ordinary differential equations of the second order*

$$y'' = f(x, y, y')$$

Notation: y_s, y_s' approximations for $y(x_s)$, $y'(x_s)$, resp.; $f_s = f(x_s, y_s, y_s')$. (When required, y_s' is to be calculated from the corresponding formula for a differential equation of the first order).

Name	Formulae
Runge-Kutta-Nyström	$k_1 = \dfrac{h^2}{2} f(x_r, y_r, y_r')$ $k_2 = \dfrac{h^2}{2} f\left(x_{r+\frac{1}{2}}, y_r + \dfrac{h}{2} y_r' + \dfrac{1}{4} k_1, y_r' + \dfrac{1}{h} k_1\right)$ $k_3 = \dfrac{h^2}{2} f\left(x_{r+\frac{1}{2}}, y_r + \dfrac{h}{2} y_r' + \dfrac{1}{4} k_1, y_r' + \dfrac{1}{h} k_2\right)$ $k_4 = \dfrac{h^2}{2} f\left(x_{r+1}, y_r + h y_r' + k_3, y_r' + \dfrac{2}{h} k_3\right)$ $y_{r+1} = y_r + h y_r' + \dfrac{1}{3} (k_1 + k_2 + k_3)$ $y_{r+1}' = y_r' + \dfrac{1}{3h} (k_1 + 2k_2 + 2k_3 + k_4)$
Adams extrapolation	$y_{r+1} = y_r + h y_r' + h^2\left(\dfrac{1}{2} f_r + \dfrac{1}{6} \nabla f_r + \dfrac{1}{8} \nabla^2 f_r + \dfrac{19}{180} \nabla^3 f_r + \dfrac{3}{32} \nabla^4 f_r + \cdots\right)$
Störmer-Nyström extrapolation	$y_{r+1} = 2y_r - y_{r-1} + h^2\left(f_r + \dfrac{1}{12} \nabla^2 f_r + \dfrac{1}{12} \nabla^3 f_r + \dfrac{19}{240} \nabla^4 f_r + \right.$ $\left. + \dfrac{3}{40} \nabla^5 f_r + \dfrac{863}{12096} \nabla^6 f_r + \cdots\right)$
Adams interpolation	$y_{r+1} = y_r + h y_r' + h^2\left(\dfrac{1}{2} f_{r+1} - \dfrac{1}{3} \nabla f_{r+1} - \dfrac{1}{24} \nabla^2 f_{r+1} - \right.$ $\left. - \dfrac{7}{360} \nabla^3 f_{r+1} - \dfrac{17}{1440} \nabla^4 f_{r+1} - \cdots\right)$
Central-difference method	$y_{r+1} = 2y_r - y_{r-1} + $ $+ h^2\left(f_r + \dfrac{1}{12} \nabla^2 f_{r+1} - \dfrac{1}{240} \nabla^4 f_{r+2} + \dfrac{31}{60480} \nabla^6 f_{r+3} - \cdots\right)$
Milne extrapolation[1]	$y_{r+1} = y_r + y_{r-2} - y_{r-3} + h^2\left(3f_{r-1} + \dfrac{5}{4} \nabla^2 f_r\right)$ $y_{r+1} = y_r + y_{r-4} - y_{r-5} + \dfrac{h^2}{48}(67f_r - 8f_{r-1} + 122f_{r-2} - 8f_{r-3} + 67f_{r-4})$
Milne interpolation	$y_{r+1} = y_r + y_{r-2} - y_{r-3} + $ $+ \dfrac{h^2}{240}(17f_{r+1} + 232f_r + 222f_{r-1} + 232f_{r-2} + 17f_{r-3})$

[1] Milne, W. E.: Amer. Math. Monthly **40**, 322—327 (1933). — Hartree, D. R.: Mem. Manchester **76**, 91—107 (1932).

Appendix

Table III. *Finite-difference expressions for ordinary differential equations*

		Formulae [Notation: $y_j = y(jh)$, $y_j' = y'(jh)$, etc.]	The next non-vanishing term of the Taylor expansion
Formulae for y'	symmetric	$y_0' = \frac{1}{2h}(-y_{-1} + y_1) +$	$-\frac{1}{6}h^2 y_0''' - \cdots$
		$y_0' = \frac{1}{12h}(y_{-2} - 8y_{-1} + 8y_1 - y_2) +$	$+\frac{1}{30}h^4 y_0^{V} + \cdots$
		$y_0' = \frac{1}{60h}(-y_{-3} + 9y_{-2} - 45y_{-1} + 45y_1 - 9y_2 + y_3) +$	$-\frac{1}{140}h^6 y_0^{VII} + \cdots$
		$y_{-1}' + 4y_0' + y_1' + \frac{3}{h}(y_{-1} - y_1) = 0 +$	$+\frac{1}{30}h^4 y_0^{V} + \cdots$
		$y_{-1}' + 3y_0' + y_1' + \frac{1}{12h}(y_{-2} + 28y_{-1} - 28y_1 - y_2) = 0 +$	$-\frac{1}{420}h^6 y_0^{VII} - \cdots$
		$y_{-2}' + 16y_{-1}' + 36y_0' + 16y_1' + y_2' + \frac{5}{6h}(5y_{-2} + 32y_{-1} - 32y_1 - 5y_2) = 0 +$	$+\frac{1}{630}h^8 y_0^{IX} + \cdots$
		$7y_{-2}' + 32y_{-1}' + 12y_0' + 32y_1' + 7y_2' + \frac{45}{2h}(y_{-2} - y_2) = 0 +$	$+\frac{4}{21}h^6 y_0^{VII} + \cdots$
		$y_{-2}' + 4y_{-1}' + 4y_1' + y_2' + \frac{1}{6h}(19y_{-2} - 8y_{-1} + 8y_1 - 19y_2) = 0 +$	$+\frac{1}{35}h^6 y_0^{VII} + \cdots$
	unsymmetric	$y_0' = \frac{1}{h}(-y_0 + y_1) +$	$-\frac{1}{2}h\, y_0'' - \cdots$
		$y_0' = \frac{1}{2h}(-3y_0 + 4y_1 - y_2) +$	$+\frac{1}{3}h^2 y_0''' + \cdots$
		$y_0' = \frac{1}{12h}(-3y_{-1} - 10y_0 + 18y_1 - 6y_2 + y_3) +$	$-\frac{1}{20}h^4 y_0^{V} + \cdots$
		$y_0' = \frac{1}{60h}(2y_{-2} - 24y_{-1} - 35y_0 + 80y_1 - 30y_2 + 8y_3 - y_4) +$	$+\frac{1}{105}h^6 y_0^{VII} + \cdots$
		$y_0' + y_1' + \frac{2}{h}(y_0 - y_1) = 0 +$	$+\frac{1}{6}h^2 y_0''' + \cdots$
		$y_{-1}' + 9y_0' + 9y_1' + y_2' + \frac{1}{3h}(11y_{-1} + 27y_0 - 27y_1 - 11y_2) = 0 +$	$+\frac{1}{140}h^6 y_0^{VII} + \cdots$
Formulae for y''	symmetric	$y_0'' = \frac{1}{h^2}(y_{-1} - 2y_0 + y_1) +$	$-\frac{1}{12}h^2 y_0^{IV} + \cdots$
		$y_0'' = \frac{1}{12h^2}(-y_{-2} + 16y_{-1} - 30y_0 + 16y_1 - y_2) +$	$+\frac{1}{90}h^4 y_0^{VI} + \cdots$
		$y_0'' = \frac{1}{180h^2}(2y_{-3} - 27y_{-2} + 270y_{-1} - 490y_0 + 270y_1 - 27y_2 + 2y_3) +$	$-\frac{1}{560}h^6 y_0^{VIII} + \cdots$
		$y_{-1}'' + 10y_0'' + y_1'' - \frac{12}{h^2}(y_{-1} - 2y_0 + y_1) = 0 +$	$+\frac{1}{20}h^4 y_0^{VI} + \cdots$
		$2y_{-1}'' + 11y_0'' + 2y_1'' - \frac{3}{4h^2}(y_{-2} + 16y_{-1} - 34y_0 + 16y_1 + y_2) = 0 +$	$-\frac{23}{5040}h^6 y_0^{VIII} + \cdots$

Table III (continued)

		Formulae [Notation: $y_j = y(jh)$, $y'_j = y'(jh)$, etc.]	The next non-vanishing term of the Taylor expansion
Formulae for y''	symmetric	$23y''_{-2} + 688y''_{-1} + 2358y''_0 + 688y''_1 + 23y''_2 -$ $- \frac{15}{h^2}(31y_{-2} + 128y_{-1} - 318y_0 + 128y_1 + 31y_2) = 0 +$	$+\frac{79}{1260}h^8 y_0^{X} + \cdots$
		$y''_{-1} - 8y''_0 + y''_1 + \frac{9}{h}(y'_{-1} - y'_1) +$ $\qquad + \frac{24}{h^2}(y_{-1} - 2y_0 + y_1) = 0 +$	$+\frac{1}{2520}h^6 y_0^{VIII} + \cdots$
		$y''_{-1} - y''_1 + \frac{1}{h}(7y'_{-1} + 16y'_0 + 7y'_1) +$ $\qquad + \frac{15}{h^2}(y_{-1} - y_1) = 0 +$	$-\frac{1}{315}h^5 y_0^{VII} + \cdots$
	unsymmetric	$y''_0 = \frac{1}{h^2}(2y_0 - 5y_1 + 4y_2 - y_3) +$	$+\frac{11}{12}h^2 y_0^{IV} + \cdots$
		$y''_0 = \frac{1}{12h^2}(11y_{-1} - 20y_0 + 6y_1 + 4y_2 - y_3) +$	$+\frac{1}{12}h^3 y_0^{V} + \cdots$
		$y''_0 = \frac{1}{180h^2}(-13y_{-2} + 228y_{-1} - 420y_0 + 200y_1 +$ $\qquad + 15y_2 - 12y_3 + 2y_4) +$	$-\frac{1}{90}h^5 y_0^{VII} + \cdots$
Formulae for y'''	symmetric	$y'''_0 = \frac{1}{2h^3}(-y_{-2} + 2y_{-1} - 2y_1 + y_2) +$	$-\frac{1}{4}h^2 y_0^{V} + \cdots$
		$y'''_0 = \frac{1}{8h^3}(y_{-3} - 8y_{-2} + 13y_{-1} - 13y_1 + 8y_2 - y_3) +$	$+\frac{7}{120}h^4 y_0^{VII} + \cdots$
		$y'''_{-1} + 2y'''_0 + y'''_1 + \frac{2}{h^3}(y_{-2} - 2y_{-1} + 2y_1 - y_2) = 0 +$	$-\frac{1}{60}h^4 y_0^{VII} + \cdots$
		$y'''_{-2} + 56y'''_{-1} + 126y'''_0 + 56y'''_1 + y'''_2 +$ $\qquad + \frac{120}{h^3}(y_{-2} - 2y_{-1} + 2y_1 - y_2) = 0 +$	$-\frac{1}{252}h^6 y_0^{IX} + \cdots$
	unsymmetric	$y'''_0 = \frac{1}{2h^3}(-3y_{-1} + 10y_0 - 12y_1 + 6y_2 - y_3) +$	$+\frac{1}{4}h^2 y_0^{V} + \cdots$
		$y'''_0 = \frac{1}{8h^3}(-y_{-2} - 8y_{-1} + 35y_0 - 48y_1 + 29y_2$ $\qquad - 8y_3 + y_4) +$	$-\frac{1}{15}h^4 y_0^{VII} + \cdots$
Formulae for y^{IV}	symmetric	$y_0^{IV} = \frac{1}{h^4}(y_{-2} - 4y_{-1} + 6y_0 - 4y_1 + y_2) +$	$-\frac{1}{6}h^2 y_0^{VI} + \cdots$
		$y_0^{IV} = \frac{1}{6h^4}(-y_{-3} + 12y_{-2} - 39y_{-1} +$ $\qquad + 56y_0 - 39y_1 + 12y_2 - y_3) +$	$+\frac{7}{240}h^4 y_0^{VIII} + \cdots$
		$y_{-1}^{IV} + 4y_0^{IV} + y_1^{IV} - \frac{6}{h^4}(y_{-2} - 4y_{-1} + 6y_0 - 4y_1 + y_2)$ $\qquad = 0 +$	$+\frac{1}{120}h^4 y_0^{VIII} + \cdots$
		$y_{-2}^{IV} - 124y_{-1}^{IV} - 474y_0^{IV} - 124y_1^{IV} + y_2^{IV} +$ $\qquad + \frac{720}{h^4}(y_{-2} - 4y_{-1} + 6y_0 - 4y_1 + y_2) = 0 +$	$+\frac{5}{21}h^6 y_0^{X} + \cdots$

Table IV. *Euler expressions for functions of one independent variable*

(To facilitate the setting up of variational problems corresponding to given ordinary differential equations.)

$F(x, y, y', y'', \ldots, y^{(n)})$	$\dfrac{\partial F}{\partial y} - \dfrac{d}{dx}\left(\dfrac{\partial F}{\partial y'}\right) + \dfrac{d^2}{dx^2}\left(\dfrac{\partial F}{\partial y''}\right) - \cdots + (-1)^n \dfrac{d^n}{dx^n}\left(\dfrac{\partial F}{\partial y^{(n)}}\right)$
$\frac{1}{2}h_n(x)\,[y^{(n)}]^2$	$(-1)^n\,[h_n(x)\,y^{(n)}]^{(n)}$
special case $\frac{1}{2}h_0(x)\,y^2$	$h_0(x)\,y$
„ „ $\frac{1}{2}h_1(x)\,y'^2$	$-[h_1(x)\,y']'$
„ „ $\frac{1}{2}h_2(x)\,y''^2$	$[h_2(x)\,y'']''$
$-h(x, y)\,y'^n$	$y'^{n-2}[n(n-1)\,h\,y'' + n\,h_x\,y' + (n-1)\,h_y\,y'^2]$
special case $h(x, y)$	$h_y(x, y)$
„ „ $h(x, y)\,y'$	$-h_x$
„ „ $f(y)\,y'$	0
„ „ $-h(x, y)\,y'^2$	$2h\,y'' + 2h_x\,y' + h_y\,y'^2$
„ „ $-f(y)\,y'^2$	$2f\,y'' + f'\,y'^2$
$h(x, y)\,y''^n$	$\begin{cases} y''^{n-3}[n(n-1)\,h\,y''\,y^{\mathrm{IV}} + n(n-1)\,y'''\{2h_x\,y'' + \\ \quad + 2h_y\,y'\,y'' + (n-2)\,h\} + y''^3(n+1)\,h_y + \\ \quad + y''^2 n\{h_{xx} + 2h_{xy}\,y' + h_{yy}\,y'^2\}] \end{cases}$
special case $h(x, y)\,y''$	$2h_y\,y'' + h_{xx} + 2h_{xy}\,y' + h_{yy}\,y'^2$
„ „ $h(x, y)\,y''^2$	$\begin{cases} 2h\,y^{\mathrm{IV}} + 4(h_x + h_y\,y')\,y''' + 3h_y\,y''^2 + \\ \quad + 2y''(h_{xx} + 2h_{xy}\,y' + h_{yy}\,y'^2) \end{cases}$
$(-1)^{n+1}f(x)\,[y^{(n-1)}y^{(n+1)} - y^{(n)2}]$	$4\,[f\,y^{(n)}]^{(n)} + [f''\,y^{(n-1)}]^{(n-1)}$
special case $f(x)\,[y\,y'' - y'^2]$	$4\,[f\,y']' + f''\,y$
$(-1)^{n+1}f(x)\,[y^{(n-1)}y^{(n+1)} + y^{(n)2}]$	$[f''\,y^{(n-1)}]^{(n-1)}$
special case $f(x)\,[y\,y'' + y'^2]$	$f''\,y$
„ „ $(a + bx)\times$ $\times [y^{(n-1)}y^{(n+1)} + y^{(n)2}]$	0
$h(y')$	$-h''(y')\,y''$
$h(y'')$	$h'''(y'')\,y^{\mathrm{IV}} + h^{\mathrm{IV}}(y'')\,y'''^2$
special case $\dfrac{1}{24}\,y''^4$	$y''\,y^{\mathrm{IV}} + y'''^2$
$h(x, y, y')\,y''$	$y''(2h_y + h_{xy'} + h_{yy'}\,y') + h_{xx} + 2h_{xy}\,y' + h_{yy}\,y'^2$
special case $p(x)\,q(y')\,y''$	$\dfrac{d}{dx}\,[p'(x)\,q(y')] = p'\,q'\,y'' + p''\,q$

Table V. *Euler expressions for functions of two independent variables*

(To facilitate the setting up of variational problems corresponding to given partial differential equations.)

$F(x, y, u, u_x, u_y, u_{xx}, u_{xy}, u_{yy})$	$F_u - \dfrac{\partial}{\partial x} F_{u_x} - \dfrac{\partial}{\partial y} F_{u_y} + \dfrac{\partial^2}{\partial x^2} F_{u_{xx}} + \dfrac{\partial^2}{\partial x \partial y} F_{u_{xy}} + \dfrac{\partial^2}{\partial y^2} F_{u_{yy}}$
$h(x, y) g(u)$	$h(x, y) g'(u)$
special case $h(x, y) u$	$h(x, y)$
,, ,, $\dfrac{1}{2} h(x, y) u^2$	$h(x, y) u$
$h(x, y) u_x$	$-h_x$
$-\dfrac{1}{2} h(x, y) u_x^2$	$\dfrac{\partial}{\partial x} (h u_x)$
$h(x, y) u u_x$	$-h_x u$
$-h(x, y) u_x u_y$	$\dfrac{\partial}{\partial x} (h u_y) + \dfrac{\partial}{\partial y} (h u_x) = 2 h u_{xy} + h_x u_y + h_y u_x$
special case $-\dfrac{1}{2} u_x u_y$	u_{xy}
$-\dfrac{1}{2} h(x, y) u_y^2$	$\dfrac{\partial}{\partial y} (h u_y)$
$-\dfrac{1}{2} (u_x^2 + u_y^2)$	$\nabla^2 u \equiv u_{xx} + u_{yy}$
$h(x, y) u u_{xx}$	$h u_{xx} + \dfrac{\partial^2}{\partial x^2} (h u) = 2 h u_{xx} + 2 h_x u_x + h_{xx} u$
$\dfrac{1}{2} h(x, y) u_{xx}^2$	$\dfrac{\partial^2}{\partial x^2} (h u_{xx})$
$\dfrac{1}{2} h(x, y) u_{xy}^2$	$\dfrac{\partial^2}{\partial x \partial y} (h u_{xy})$
$h(x, y) u_{xx} u_{yy}$	$\dfrac{\partial^2}{\partial x^2} (h u_{yy}) + \dfrac{\partial^2}{\partial y^2} (h u_{xx})$
u_{xy}^2	$\left.\vphantom{\begin{matrix}a\\b\end{matrix}}\right\} 2 u_{xxyy}$
$u_{xx} u_{yy}$	
$\dfrac{1}{2} (\nabla^2 u)^2 = \dfrac{1}{2} (u_{xx} + u_{yy})^2$	$\left.\vphantom{\begin{matrix}a\\b\\c\end{matrix}}\right\}$ $\nabla^4 u \equiv u_{xxxx} + 2 u_{xxyy} + u_{yyyy}$
$\dfrac{1}{2} (u_{xx}^2 + 2 u_{xy}^2 + u_{yy}^2)$	
$u_{xx} u_{yy} - u_{xy}^2$	0
$\displaystyle\int_0^u f(x, y, v) dv$	$f(x, y, u)$
$g(x, y, u) (u_x^2 + u_y^2)$	$-2 g \nabla^2 u - g_u (u_x^2 + u_y^2) - 2 g_x u_x - 2 g_y u_y$

Table VI. *Stencils for the differential operators ∇^2 and ∇^4 (for meshes formed from squares, equilateral triangles and cubes)*

Formula relates	Stencil	Stencil to be read as	Further terms of the Taylor expansion. Notation: $\zeta_{j,k} = \left(\dfrac{\partial^{j+k}u}{\partial x^j \partial y^k} + \dfrac{\partial^{j+k}u}{\partial x^k \partial y^j}\right)_{\substack{x=0\\y=0}}$
$h^2\nabla^2 u_{0,0}$ with values of u	$\begin{array}{ccc} & 1 & \\ 1 & -4 & 1 \\ & 1 & \end{array}$	$h^2\nabla^2 u_{0,0} = -4u_{0,0}+u_{1,0}+u_{0,1}+$ $+u_{-1,0}+u_{0,-1}+$	$-\dfrac{h^4}{12}\zeta_{0,4} - \dfrac{h^6}{360}\zeta_{0,6} - \cdots$
$12h^2\nabla^2 u_{0,0}$ with values of u	$\begin{array}{ccccc} & & -1 & & \\ & 16 & -60 & 16 & -1 \\ -1 & 16 & & & \\ & -1 & & & \end{array}$	$12h^2\nabla^2 u_{0,0} = -60u_{0,0}+16(u_{1,0}+\cdots)-$ $-(u_{2,0}+\cdots)+$	$\dfrac{2}{15}h^6\zeta_{0,6}+\dfrac{h^8}{84}\zeta_{0,8}+\cdots$
values of $\boxed{h^2\nabla^2 u}$ and values of u	$\begin{array}{ccc} -2 & -8\,(1) & -2 \\ -8\,(1) & 40\,(8) & -8\,(1) \\ -2 & -8\,(1) & -2 \end{array}$	$8h^2\nabla^2 u_{0,0}+h^2\nabla^2 u_{1,0}+h^2\nabla^2 u_{0,1}+$ $+h^2\nabla^2 u_{-1,0}+h^2\nabla^2 u_{0,-1}+$ $+40u_{0,0}-8(u_{1,0}+u_{0,1}+\cdots)-$ $-2(u_{1,1}+u_{1,-1}+\cdots)=0+$	$\dfrac{h^6}{60}(3\zeta_{0,6}-5\zeta_{2,4})+\cdots$
values of $\boxed{\dfrac{1}{12}h^2\nabla^2 u}$ and values of u	$\begin{array}{ccc} (1) & -4 & (1) \\ -4 & 20 & -4 \\ (1) & -4 & (1) \\ & & \\ (4) & (52) & (4) \\ (4) & & (4) \end{array}$	$\dfrac{1}{12}h^2[52\nabla^2 u_{0,0}+4(\nabla^2 u_{1,0}+\nabla^2 u_{0,1}+\cdots)+$ $+(\nabla^2 u_{1,1}+\nabla^2 u_{1,-1}+\cdots)]+$ $+20u_{0,0}-4(u_{1,0}+u_{0,1}+\cdots)-$ $-(u_{1,1}+u_{1,-1}+\cdots)=0+$	$\dfrac{h^6}{120}(3\zeta_{0,6}+5\zeta_{2,4})+\cdots$

values of $\dfrac{2}{3}h^2\nabla^2 u$ and values of u	 ``` -3 -16(1) -16(10) -16(1) -3 -16(10) 140(46) -16(10) -3 -16(1) -16(10) -16(1) -3 ``` 	$\dfrac{2}{3}h^2[46\nabla^2 u_{0,0} + 10(\nabla^2 u_{1,0} + \nabla^2 u_{0,1} + \cdot + \cdot) + (\nabla^2 u_{1,1} + \nabla^2 u_{1,-1} + \cdot + \cdot)] + 140 u_{0,0} - 16(u_{1,0} + u_{1,1} + \cdots) - 3(u_{2,0} + \cdot + \cdot) = 0 +$	$\dfrac{h^8}{1260}(-23\zeta_{0,8} + 42\zeta_{2,6}) + \cdots$
$\boxed{\dfrac{1}{10}h^4\nabla^4 u_{0,0}}$ with values of $\boxed{\dfrac{1}{15}h^2\nabla^2 u}$ and values of u	 ``` (1) -4 20 -4 (1) (1) 20 (82)(3) 20 (1) -4 20 -4 (1) (1) ``` 	$\dfrac{3}{10}h^4\nabla^4 u_{0,0} + \dfrac{1}{15}h^2[82\nabla^2 u_{0,0} + \nabla^2 u_{1,0} + \cdot + \cdot + \cdot + \cdot + \cdot + \cdot] + \nabla^2 u_{1,1} + \cdot + \cdot + \cdot + \cdot + \cdot + \cdot] + 20 u_{0,0} - 4(u_{1,0} + u_{0,1} + \cdot + \cdot) - (u_{1,1} + u_{1,-1} + \cdot + \cdot) = 0 +$	$\dfrac{h^8}{50400}(13\zeta_{0,8} + 168\zeta_{2,6} + 105\zeta_{4,4}) + \cdots$
$h^4\nabla^4 u_{0,0}$ with values of u	 ``` 1 2 -8 2 1 -8 20 -8 1 2 -8 2 1 ``` 	$h^4\nabla^4 u_{0,0} = 20 u_{0,0} - 8(u_{1,0} + u_{0,1} + \cdot + \cdot) + 2(u_{1,1} + u_{1,-1} + \cdot + \cdot) + (u_{2,0} + u_{0,2} + \cdot + \cdot) +$	$-\dfrac{h^6}{6}(\zeta_{0,6} + \zeta_{2,4}) - \cdots$
$6h^4\nabla^4 u_{0,0}$ with values of u	 ``` -1 -1 14 -1 -1 14 -77 20 -1 -1 20 -77 184 -77 20 -1 -1 20 -77 14 -1 14 -1 -1 ``` 	$6h^4\nabla^4 u_{0,0} = 184 u_{0,0} - 77(u_{1,0} + u_{0,1} + \cdot + \cdot) + 20(u_{1,1} + u_{1,-1} + \cdot + \cdot) + 14(u_{2,0} + u_{0,2} + \cdot + \cdot) - (u_{0,3} + u_{1,2} + u_{2,1} + \cdots) +$	$\dfrac{h^8}{120}(21\zeta_{0,8} + 16\zeta_{2,6} + 5\zeta_{4,4}) + \cdots$

Table VI continued

Formula relates	Stencil	Stencil to be read as	Further terms of the Taylor expansion. Notation: $u_{jk} = \left(\dfrac{\partial^{j+k} u}{\partial x^j \partial y^k}\right)_{\substack{x=0\\y=0}}$
values of $\boxed{\dfrac{1}{2} h^4 \nabla^4 u}$ and values of u	*(stencil diagram)* −1 −1 −1 2 10① 2 −1 10① −36② 10① −1 2 10① 2 −1 −1 −1 −1	$\dfrac{1}{2} h^4 [2\nabla^4 u_{0,0} + (\nabla^4 u_{1,0} + \nabla^4 u_{0,1} + \cdot + \cdot)] -$ $- 36 u_{0,0} + 10(u_{1,0} + u_{0,1} + \cdot + \cdot) +$ $+ 2(u_{1,1} + u_{1,-1} + \cdot + \cdot) -$ $- (u_{2,0} + u_{2,1} + u_{1,2} + \cdots) = 0 +$	$\dfrac{h^8}{240} (\zeta_{0,8} - 24\zeta_{2,6} - 15\zeta_{4,4}) + \cdots$
values of $\boxed{\dfrac{h^4}{20} \nabla^4 u}$ and values of u	*(stencil diagram)* 1 8 18① 8 1 8 −144 468② −144 8 18① 468② −332 468② 18① 8 −144 468② −144 8 1 8 18① 8 1	$\dfrac{h^4}{20} \Big[-332 \nabla^4 u_{0,0} - 72(\nabla^4 u_{1,0} + \nabla^4 u_{0,1} + \cdot) -$ $- 26(\nabla^4 u_{1,1} + \nabla^4 u_{1,-1} + \cdot + \cdot) +$ $+ (\nabla^4 u_{2,0} + \nabla^4 u_{0,2} + \cdot + \cdot)\Big] +$ $+ 468 u_{0,0} - 144(u_{1,0} + u_{0,1} + \cdot + \cdot) -$ $- 8(u_{1,1} + \cdot + \cdot + \cdot) + 18(u_{2,0} + \cdot + \cdot + \cdot) +$ $+ 8(u_{1,2} + u_{2,1} + \cdots) + (u_{2,2} + \cdots) = 0 +$	$\dfrac{h^{10}}{420} (5\zeta_{0,10} + 12\zeta_{2,8} + 21\zeta_{4,6}) + \cdots$
$\boxed{\dfrac{3}{2} h^2 \nabla^2 u_{0,0}}$ with values of u	*(stencil diagram)* 1 1 1 −6 1 1 1	$\dfrac{3}{2} h^2 \nabla^2 u_{0,0} = -6 u_{0,0} + (u_{1,0} + \cdots) +$	$-\dfrac{3 h^4}{32} (\nabla^4 u)_{0,0} - \dfrac{h^6}{3840} (11 u_{60} +$ $+ 15 u_{42} + 45 u_{24} + 9 u_{06}) + \cdots$

$9h^2\nabla^2 u_{0,0}$ with values of u	$9h^2\nabla^2 u_{0,0} = -48u_{0,0} + 9(u_{1,0}+\cdots) - \cdots +$	$\dfrac{3h^8}{160}(2u_{80}+15u_{42}+3u_{06})+\cdots$
values of $\boxed{\dfrac{h^2}{16}\nabla^2 u}$ and values of u	$\dfrac{h^2}{16}[18\nabla^2 u_{0,0}+(\nabla^2 u_{1,0}+\cdots)]+$ $+6u_{0,0}-(u_{1,0}+\cdots)=0+$	$\dfrac{h^6}{7680}(23u_{60}+105u_{42}+$ $+45u_{24}+27u_{06})+\cdots$
values of $\boxed{\dfrac{h^2}{8}\nabla^2 u}$ and values of u	$\dfrac{1}{8}h^2[-240\nabla^2 u_{0,0}-21(\nabla^2 u_{1,0}+\cdots)+\cdots]-$ $-168u_{0,0}+27(u_{1,0}+\cdots)+\cdots=0+$	$\dfrac{h^8}{71680}(307u_{80}+3892u_{62}+$ $+2730u_{44}-252u_{26}+603u_{08})+$ $+\cdots$
$\boxed{\dfrac{3}{32}h^4\nabla^4 u_{0,0}}$ and $\dfrac{3}{2}h^2\nabla^2 u_{0,0}$ with values of u	$\dfrac{3}{32}h^4\nabla^4 u_{0,0}+\dfrac{3}{2}h^2\nabla^2 u_{0,0}+$ $+6u_{0,0}-(u_{1,0}+\cdots)=0+$	$h^6\cdots$ (cf. last formula on p. 544)

Table VI continued

Formula relates	Stencil	Stencil to be read as	Further terms of the Taylor expansion. Notation: $u_{jk} = \left(\dfrac{\partial^{j+k} u}{\partial x^j \partial y^k}\right)_{\substack{x=0 \\ y=0}}$
$\dfrac{9}{16} h^4 \nabla^4 u_{0,0}$ with values of u	(hexagonal stencil: $1,\,-3,\,1$ / $-3,\,12,\,-3$ / $1,\,-3,\,1$)	$\dfrac{9}{16} h^4 \nabla^4 u_{0,0} = 12 u_{0,0} - 3(u_{1,0} + \cdots) + \cdots +$	$-\dfrac{h^6}{128}\,(7u_{60} + 39 u_{42} + 9 u_{24} + 9 u_{06}) + \cdots$
values of $\boxed{\dfrac{3}{16} h^4 \nabla^4 u}$ and values of u	(hexagonal stencil: $-1,\,-2,\,-1$ / $-2,\,10,\,10,\,-2$ / $-1,\,10,\,-42,\,10,\,-1$ / $-2,\,10,\,10,\,-2$ / $-1,\,-2,\,-1$)	$\dfrac{3}{16} h^4 [6 \nabla^4 u_{0,0} + (\nabla^4 u_{1,0} + \cdots)] - \\ -42 u_{0,0} + 10(u_{1,0} + \cdots) - \cdots = 0 +$	$\dfrac{h^8}{5120}\,(u_{80} + 76 u_{62} + 30 u_{44} - 36 u_{26} + 9 u_{08}) + \cdots$

remainder term of the 4th order

  ,, ,, ,,   6th ,,

  ,, ,, ,,   6th ,,

  ,, ,, ,,   8th ,,

Cubical mesh (notation $u_m, \Sigma u_s, \ldots, \Sigma u_s, \ldots, \nabla^2 u_m, \Sigma \nabla^2 u_s, \ldots$ as in Table X) [from J. ALBRECHT: Z. Angew. Math. Mech. 33, 48 (1953), where further formulae can be found]

$$\Sigma u_s - 6 u_m - h^2 \nabla^2 u_m =$$

$$\Sigma u_k + 2 \Sigma u_s - 24 \Sigma u_m - \tfrac{1}{2} h^2 [\Sigma \nabla^2 u_s + 6 \nabla^2 u_m] =$$

$$\Sigma u_k + 2 \Sigma u_s - 24 \Sigma u_m - \tfrac{1}{12} h^2 [\Sigma \nabla^2 u_k + 2 \Sigma \nabla^2 u_s + 48 \nabla^2 u_m] =$$

$$\Sigma u_e + 3 \Sigma u_k + 14 \Sigma u_s - 128 u_m - \tfrac{1}{3} h^2 [\Sigma \nabla^2 u_k - \Sigma \nabla^2 u_m + 84 \nabla^2 u_m] - \tfrac{3}{2} h^4 \nabla^4 u_m =$$

Table VII. Catalogue of examples treated

Abbreviations for the methods used: Bm = boundary-maximum theorem, Ch = method of characteristics, Cl = collocation, D = ordinary finite-difference method, Di = improved finite-difference method, Ds = finite-difference step-by-step method, Es = exact solution, Et = enclosure theorem, Fs = finite-sum method, G = Galerkin's method, It = iteration method, Ls = least-squares method, Mp = method based on the monotonic property, Pt = perturbation method, Ps = power series, Pl = polygon method, R = Ritz's method, RK = Runge-Kutta method, Se = Series expansion, T = Trefftz's method.

No.	Type	Equation(s)	Initial or boundary conditions	Possible physical applications and other remarks	Methods used and page numbers
Chapter II. Initial-value problems in ordinary differential equations					
1	Differential equation of the first order	$y' = y$	$y(0)=1$		Ls 138
2		$y' = x + y$	$y'(0)=0$		Ds 82
3		$y' = y - \dfrac{2x}{y}$	$y(0)=1$		Es 55, Pl 55, RK 73, Ps 80, Ds 92, It 115
4		$y' = x^2 + y^3$	$y(0)=-1$		Ps 137, Ds 137
5		$y' = -y - 2y^3 + \sin 2x$	$y'(0)=0$	Transient effects when an electro-magnet is switched on	Ds 91
6	System of first order equations	$\dot u_1 = u_1 u_3 - 0.6\,u_1$ $\dot u_2 = -u_1 u_3 - 0.3\,u_3$ $\dot u_3 = \dfrac{1}{3} u_1 u_2 - 0.2\,u_3$	$u_1(0)=u_2(0)=1$ $u_3(0)=0$	Motion of a body about a fixed point (Euler's equations)	RK 75
7	Diff. equ. of higher order	$y'' = -y^3$	$y(0)=0.2,\ y'(0)=0$	Oscillations of a mass attached to a spring	RK 137
8		$y''' = -xy$	$y(0)=0,\ y'(0)=1,\ y''(0)=0$		Ds 138
9		$y''' = -yy''$	$y(0)=y'(0)=0,\ y''(0)=1$	Boundary layer along a flat plate	RK 75, Ds 130, RK 138, Ps 130, 138
Chapter III. Boundary-value problems in ordinary differential equations					
1	Linear differential equation of the second order	$y''=0$	$y(0)=1,\ y(1)=y'(1)$	No solution	R 252
2		$y''+4y=2$	$y(\pm 1)=0$		R 252, Es 256
3		$y''+y+x=0$	$y(0)=y(1)=0$		R 220, Ls 220
4		$y'' - \dfrac{2}{x^2}\,y = -\dfrac{1}{x}$	$y(2)=y(3)=0$		Es 178, D 178
5		$y'' + (1+x^4)\,y + 1 = 0$	$y(\pm 1)=0$	Bending of a beam	D 143, 155, Di 163, 168, Cl 182, Pt 188, It 195, R 209, Ps 224
6		$y'' = \dfrac{1+x}{2+x}\,y$	$y(0)=1,\ y(\infty)=0$	Temperature distribution in an infinitely long rod	D 150, Cl 182
7		$(1+x^2)\,y'' + \left(\dfrac{3}{x} + 5x\right) y' + \dfrac{4}{3}\,y + 1 = 0$	$y(\pm 2)=0.6,\ y$ at $x=0$ regular	Stress distribution in a steam turbine rotor	D 251
8	Eigen-value problem	$y'' + \lambda(1+x^4)\,y = 0$	$y(\pm 1)=0$	Vibration of an inhomogeneous string	R 252, It 253
9		$y'' + \lambda(2-x^4)\,y = 0$	$y(0)=0,\ 2y(1)+y'(1)=0$		Et 252
10		$-[(1+x)y']' = \lambda(1+x)\,y$	$y'(0)=y(1)=0$	Longitudinal vibrations of a rod	D 147, Di 163,170, R 210, It 235, Et 237, R 248

Table VII (continued)

No.	Type	Equation	Initial or boundary conditions	Possible physical applications and other remarks	Methods used and page numbers
11	Non-linear differential equation of the second order	$y''=\dfrac{2x^2}{}$	$y(\pm1)=1$		Es 252, D 252
12		$y'y''=4x$	$y(\pm1)=0$		R 252
13		$y''=y^2$	$y(0)=0,\ y(2)=-2$	Only complex solutions	R 252
14		$y''=\dfrac{3}{2}y^2$	$y(0)=4,\ y(1)=1$		Es 145, D 145, 180, Ls 184, R 212, Ps 226
15		$y''+(1+e^y)=0$	$y(0)=0,\ y(1)=1$		D 159
16		$y''+6y+y^3+\dfrac{3}{2}\cos x=0$	$y(0)-y(2\pi)=y'(0)-y'(2\pi)=0$	Forced oscillations of a system with a non-linear restoring force	It 198
17		$y'=6xy^3$	$y(0)=y(1)=1$		Mp 201
18	Diff. equ. of the fourth order	$[(2-x^2)y'']'+40y=2-x^2$	$y''(\pm1)=y'''(\pm1)=0$	Transversely loaded, elastically embedded rail	D 152, R 219
19		$y^{IV}=\lambda(1+x)y$	$y(0)=y'(0)=y''(1)=y'''(1)=0$	Flexural vibrations of a rod with variable cross-section	R 253, Fs 472

Chapter IV. *Initial and initial-/boundary-value problems in partial differential equations*

No.	Type	Equation	Initial or boundary conditions	Possible physical applications and other remarks	Methods used and page numbers
1	Differential equation of the first order	$(0\cdot1\sqrt[3]{u}-0\cdot5)u_x+u_y=-0\cdot5$	$u(x,0)=2\dfrac{4-x}{5-x}$ for $0\le x\le4$	Motion of a glacier	Ch 309, Ps 310, D 311, It 316
2		$u_x-v_y=0$ $-4u_y+v_x=u$	$u(x,0)=\sin\dfrac{\pi}{2}x,\ v(x,0)=0$ for $0\le x\le2$ $u(0,y)=u(2,y)=0$ for $y\ge0$	Transient currents in an electric cable	Ch 324, Es 326, D 335
3	Parabolic type	$u_{xx}=c\,u_y$	$u(x,0)=0$ for $0\le x\le2$ $u(0,y)=\sin\dfrac{8\pi}{3c}y,\ u(2,y)=0$ for $y\ge0$	Heat conduction in a rod	D 267, 268, 334
4		$u_{xx}=2\,u_y$	$u(x,0)=0$ for $|x|\le1$ $u(\pm1,y)=1-e^{-\frac{5}{c}y}$ for $y\ge0$	Heat conduction in a rod	D 281
5		$u_{xx}-K\,u_y+\beta=0$	$u(0,y)=0,\ u(10,y)=100\,e^{-0\cdot1y}$ for $y\ge0$ $u(x,0)=10x$ for $0\le x\le10$	Heat conduction with constant heat generation	D 334
6		$u_{xx}=u_y+\dfrac{1}{2}u$	$u(x,0)=u(\pm1,y)=0$	Heat conduction with heat loss to the surroundings	Di 293, Mp 333
7		$u_{xx}+\dfrac{1}{x}u_x=u_y$	$\begin{cases}u(x,0)=1 & \text{for } |x|\le1\\ u+u_x=0 & \text{for } x=\pm1 \text{ and } y>0\end{cases}$		D 334
8			$\begin{cases}u\pm\dfrac{1}{2}u_x=1 & \text{for } x=\pm1 \text{ and } y>0\\ u(x,0)=0 & \text{for } |x|\le1\end{cases}$	Eddy current density in a metal cylinder	D 273, 282
9	Hyperbolic type	$u_{xx}=u_{yy}$	$u_y(x,0)=0,\ u(x,0)=\cos x$ for $|x|\le\dfrac{\pi}{2}$ $u\left(\pm\dfrac{\pi}{2},y\right)=-\sin y$ for $y\ge0$		Ps 261, Es 261
10		$u_{xx}=u_{yy}+\dfrac{1}{2}u_y$	$u(x,0)=0,\ u_y(x,0)=e^{-x}$ for $0\le x\le1$ $u(0,y)=u(1,y)=0$ for $y\ge0$	Transient current produced by an initially non-uniform charge distribution	D 335

#	Type	Equation	Boundary conditions	Application	Reference
11	Hyperb.	$u_{xx} + \dfrac{1}{x}u_x = u_{yy}$	$\begin{cases} u(x,0)=1-x^3,\ u_y(x,0)=0 & \text{for } |x|\le 1 \\ u(\pm1,y)=0 & \text{for } y\ge 0 \end{cases}$	Vibrations of a circular membrane	D 279
12	Parabol.	$u_{zz} + u_{yy} = u_t$	$u = \dfrac{\partial u}{\partial v}$ for $|x|=4$, $|y|=4$; $u=1$ for $t=0$	Cooling of a long square prism	D 335
13	Parabol.	$u_{zzzz} + K u_{yy} = 0$	$\begin{cases} u(x,0),\ u_y(x,0) \text{ prescribed for } 0\le x\le a, \\ u(0,y)=u_{zz}(0,y)=u(a,y)=u_{zz}(a,y)=0 \text{ for } y\ge 0 \end{cases}$	Flexural vibrations of a rod	D 303

Chapter V. *Boundary-value problems in partial differential equations*

#	Group	Equation	Boundary conditions	Application	Reference
1	Laplace's equation in two independent variables	$\nabla^2 u = u_{xx} + u_{yy} = 0$	$u=1$ on three sides of a square, $u=0$ on the fourth side	Temperature distribution in a plate	D 361 – 366
2			$u=x^2+y^2$ on all four sides of a square	Torsion of a square-sectioned girder	Bm 403
3			$\dfrac{\partial u}{\partial v}$ prescribed on all six sides of an L-shaped region	Fluid flow through a sharply bent channel	D 355, Di 380, 387
4			Two concentric squares: $u=1$ on inner boundary, $u=0$ on outer boundary	Transformer field	D 451
4			$u(x,\pm1)=0$, $u(0,y)=1-y^2$	Infinite fundamental region	R 440
5	Poisson's equation in two independent variables	$\nabla^2 u = u_{zz} + u_{yy} = -1$	$u=0$ for $|x|=1$ and for $|y|=1$	Torsion problem for a girder — Square section	Se 422, T 447, Ps 451, Se 451, D 451, Di 451
6			$u=0$ on the boundary	I section	D 451
7			$u=\dfrac{\partial u}{\partial v}$ for $|x|=1$ and for $|y|=1$	Temperature in a square-sectioned wire carrying a current	D 381, Di 381, 387, Bm 405, Cl 411, Gl 413, Ls 414, Ps 419, R 429
8		$\nabla^2 u = u_{zz} + u_{yy} = x^2 - 1$	$u=0$ for $|x|=1$ and for $|y|=\dfrac{1}{2}$	Distortion of a rectangular membrane by a given pressure distribution	Ls 417
9	Further elliptic differential equations	$u_{zz} + u_{yy} + \dfrac{3}{5-y}u_y = -1$	$u=0$ for $|x|=\dfrac{1}{2}$ and for $|y|=1$; $u=1$ for $z=1$, $u=-1$ for $z=-1$	Stresses in a helical spring	D 358, Di 383, 387, R 453
10		$\nabla^2 u = u_{rr} + \dfrac{1}{r}u_r + u_{zz} = 0$	$u=\dfrac{\partial u}{\partial v}$ for $r=1$	Temperature distribution in a cylinder	D 452
11		$\nabla^2 u = u_{xx} + u_{yy} + u_{zz} = 0$	$u_v=0$ over each face of cube, $u=\pm1$ at two opposite corners	Temperature distribution in a cube	D 368, R 437

Table VII (continued)

No.	Type	Equation	Initial or boundary conditions		Possible physical applications and other remarks	Methods used and page numbers
12	Eigenvalue problems with elliptic differential equations	$\nabla^2 u = u_{xx} + u_{yy} = -\lambda u$	$u=0$ on the boundary (of a regular hexagon)		Normal modes of vibration of a hexagonal membrane	D 392
13			$u=0$ along the sides of a square; special condition along the diagonal		Normal modes of vibration of a membrane containing an elastic thread	D 453, R 453
14			$u=0$ on the boundary			D 373
15		$\nabla^2 u = u_{xx}+u_{yy}+u_{zz} = -\lambda u$	$u_y=0$ on part of the boundary; $u=0$ on the remainder		Natural acoustic frequencies of a room	D 452
16	Non-lin. ell. d.e.	$\nabla^2 u = u_{xx}+u_{yy} = -\tfrac{1}{2}(1+e^u)$	$u=0$ for $|x|=1$ and for $|y|=1$		Temperature distribution in a plate	D 453, R 453, Cl 453
17	Biharmonic problems	$\nabla^4 u = \dfrac{p}{N}$	$u=u_x=0$ along the long side of a trapezium; $u=\nabla^2 u=0$ along the other three sides		Uniformly loaded trapezoidal plate with its long edge clamped and the others freely supported	D 391, 453
18		$\nabla^4 u = 1$	$u=u_x=0$ on the boundary (of the rectangle $|x|\leq 1,\ |y|\leq 2$)		Uniformly loaded rectangular plate clamped along its edges	R 434
19		$\nabla^4 u = x^2$	$u=u_z=0$ for $|x|=1$; $u=u_{yy}=0$ for $|y|=1$		Non-uniformly loaded square plate with the two pairs of opposite sides clamped and freely supported, resp.	D 359, Se 421, R 431
20		$\nabla^4 u = \lambda u$	$u=u_x=0$ on the boundary (of a regular hexagon)		Normal modes of vibration of a clamped hexagonal plate	D 395

Chapter VI. *Integral and functional equations*

No.	Type	Equation			Possible physical applications and other remarks	Methods used and page numbers
1	Inhomogeneous linear integral equations	$\pi y(x)+\displaystyle\int_0^{2\pi}\frac{2y(\xi)}{5-3\cos(x+\xi)}\,d\xi = \begin{cases}\sin x & \text{for } 0\leq x\leq\pi\\ 0 & \text{for } -\pi\leq x\leq 0\end{cases}$			First boundary-value problem of potential theory: steady temperature distribution in an elliptic cylinder	Fs 471
2		$y(x)+\displaystyle\int_{-1}^{1}K(x,\xi)\,y(\xi)\,d\xi = \frac{1}{1+x^2}$ with $K(x,\xi)=\begin{cases}1-(x-\xi)^2 & \text{for } |x-\xi|\leq 1\\ 0 & \text{for } |x-\xi|\geq 1\end{cases}$			Calculation of true intensity from readings affected by interference	R 487
3		$\displaystyle\int_0^x y(x-\xi)\left\{1+\frac{e^{-\xi}-1}{2\xi}\right\}d\xi = x$			Linear transmission system, sluggish measuring instrument, Volterra integral equation	Fs 514, It 516
4		$\dfrac{1}{2\pi}\displaystyle\oint_{-1}^{1}\frac{y(\xi)\,d\xi}{\xi-x} = |x|$			Singular integral equation	Se 509

		Equation	Conditions	Application	References
5	Eigen-value problem	$y(x) = \lambda \int_0^1 e^{x\xi} y(\xi)\, d\xi$			Et 527, It 527
6	Eigenvalue problems	$y(x) = \lambda \int_{-1}^{1} K(x,\xi) y(\xi)\, d\xi$ with $K(x,\xi) = \begin{cases} 1+\cos\pi(x-\xi) & \text{for } |x-\xi| \leq 1 \\ 0 & \text{for } |x-\xi| \geq 1 \end{cases}$		Distortion of intensity (of illumination) distribution by an optical system	Fs 527
7		$y(x) = \lambda \int_0^1 K(x,\xi) y(\xi)\, d\xi$ with $K(x,\xi) = \begin{cases} \dfrac{1}{6} x^2 (3\xi - x)(1+\xi) & \text{for } x \leq \xi \\ \dfrac{1}{6} \xi^2 (3x - \xi)(1+\xi) & \text{for } x \geq \xi \end{cases}$		Flexural vibrations of a beam of variable cross-section	R 253, Fs 472
8		$u(x,y) = \lambda \int_{-1}^{1}\int_{-1}^{1} |x\xi + y\eta|\, u(\xi,\eta)\, d\xi\, d\eta$			Fs 474, It 479
9	Non-linear integral equations	$\int_0^1 \dfrac{[y(x)+y(\xi)]^2}{1+x+\xi}\, d\xi = 1$			Fs 477, It 480, Se 497
10		$\int_0^1 \dfrac{d\xi}{y(x)+y(\xi)} = 1+x$			Fs 527
11		$\int_0^1 \sqrt{x+\xi}\,[y(\xi)]^2\, d\xi = y(x)$ $\int_0^1 \sqrt{x+y(\xi)}\, d\xi = y(x)$			It 481
12	Integro-differential equations	$\dfrac{1+x}{2} y'' + 10 \int_0^1 \dfrac{y(\xi)}{1+x+\xi}\, d\xi + 1 = 0$	$y(0) = y(1) = 0$		Se 496 Degenerate kernel 500
13		$(EJ y'')'' - H y'' + k \int_{-l}^{l} y(\xi)\, d\xi = m\omega^2 y$	$y(\pm l) = y''(\pm l) = 0$	Vertical oscillations of a suspension bridge	R 492
14	Functional equations	$y(y(x)) = e^x$	—		Ps 523
15		$y(y(x)) = 1 - x^2$	—		Ps 523
16		$y(x) - y(-x) = G[y(x) - y(-x)]$	—		D 526
17		$y(x+y(x)) = 1 - x^2$	—	Non-linear oscillations of a stretched string attached to a spring	Ps 528
18		$-y'_x(x) = \lambda y(1-x)$	$y(0) = y'(1) = 0$		D 528, R 528, Ps 528
19		$\nabla^2 u(x,y) + \lambda u(-x,-y) = 0$	$\begin{cases} u(1,y) = u(x,1) = 0 \\ u_x(-1,y) = u_y(x,-1) = 0 \end{cases}$		D 528, R 528

Table VIII. *Taylor expansion of a general finite expression involving the differential operators ∇^2 and ∇^4 for a square mesh. (All the formulae of Table VI for a square mesh may be derived from this Taylor expansion.)*

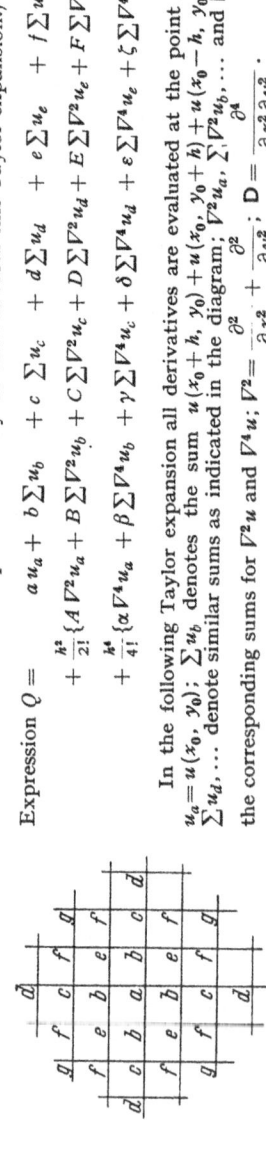

Expression $Q =$

$$a u_a + b \sum u_b + c \sum u_c + d \sum u_d + e \sum u_e + f \sum u_f + g \sum u_g$$
$$+ \frac{h^2}{2!} \{A \nabla^2 u_a + B \sum \nabla^2 u_b + C \sum \nabla^2 u_c + D \sum \nabla^2 u_d + E \sum \nabla^2 u_e + F \sum \nabla^2 u_f + G \sum \nabla^2 u_g\}$$
$$+ \frac{h^4}{4!} \{\alpha \nabla^4 u_a + \beta \sum \nabla^4 u_b + \gamma \sum \nabla^4 u_c + \delta \sum \nabla^4 u_d + \epsilon \sum \nabla^4 u_e + \zeta \sum \nabla^4 u_f + \eta \sum \nabla^4 u_g\}.$$

In the following Taylor expansion all derivatives are evaluated at the point $x = a$; $h =$ mesh width; $u_a = u(x_0, y_0)$; $\sum u_b$ denotes the sum $u(x_0 + h, y_0) + u(x_0 - h, y_0) + u(x_0, y_0 + h) + u(x_0, y_0 - h)$; $\sum u_c$, $\sum u_d, \ldots$ denote similar sums as indicated in the diagram; $\nabla^2 u_a, \sum \nabla^2 u_b, \ldots$ and $\nabla^4 u_a, \sum \nabla^4 u_b, \ldots$ denote the corresponding sums for $\nabla^2 u$ and $\nabla^4 u$; $\nabla^2 = \frac{\partial^2}{\partial x^2} + \frac{\partial^2}{\partial y^2}$; $\mathbf{D} = \frac{\partial^4}{\partial x^2 \partial y^2}$.

Then $Q =$

$\qquad u_a \left[4\{b + c + d + e + 2f + g\} + a\right]$

$+ \dfrac{h^2}{2!} \nabla^2 u_a \left[2\{b + 4c + 9d + 2e + 10f + 8g\} + 4\{B + C + D + E + 2F + G\} + A\right]$

$+ \dfrac{h^4}{4!} \nabla^4 u_a \left[2\{b + 16c + 81d + 2e + 34f + 32g\} + \dbinom{4}{2}2\{B + 4C + 9D + 2E + 10F + 8G\} + 4\{\beta + \gamma + \delta + \epsilon + 2\zeta + \eta\} + \alpha\right]$

$- \dfrac{h^4}{4!} \mathbf{D} u_a \left[4\{b + 16c + 81d + 4e - 14f - 64g\}\right]$

$+ \dfrac{h^6}{6!} \nabla^6 u_a \left[2\{b + 64c + 729d + 2e + 130f + 128g\} + \dbinom{6}{2}2\{B + 16C + 81D + 2E + 34F + 32G\} + \dbinom{6}{4}2\{\beta + 4\gamma + 9\delta + 2\epsilon + 10\zeta + 8\eta\}\right]$

$- \dfrac{h^6}{6!} \nabla^2 \mathbf{D} u_a \left[6\{b + 64c + 729d + 2e - 70f - 512g\} + \dbinom{6}{2}4\{B + 16C + 81D - 4E - 14F - 64G\}\right]$

$+ \dfrac{h^8}{8!} \nabla^8 u_a \left[2\{b + 256c + 6561d + 2e + 514f + 512g\} + \dbinom{8}{2}2\{B + 64C + 729D + 2E + 130F + 128G\} + \dbinom{8}{4}2\{\beta + 16\gamma + 81\delta + 2\epsilon + 34\zeta + 32\eta\}\right]$

$- \dfrac{h^8}{8!} \nabla^4 \mathbf{D} u_a \left[8\{b + 256c + 6561d - 12e - 438f - 3072g\} + \dbinom{8}{2}6\{B + 64C + 729D - 8E - 70F - 512G\} + \dbinom{8}{4}4\{\beta + 16\gamma + 81\delta - 4\epsilon - 14\zeta - 64\eta\}\right]$

$+ \dfrac{h^8}{8!} \mathbf{D}^2 u_a \left[4\{b + 256c + 6561d + 16 - 1054f + 4096g\}\right]$

$+$ terms of the 10th and higher orders.

Table IX. *Taylor expansion of a general finite expression involving the differential operators ∇^2 and ∇^4 for a triangular mesh.* (All formulae of Table VI for a triangular mesh may be derived from this Taylor expansion.) Explanation of notation as for Table VIII, together with

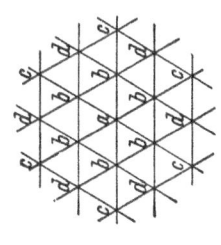

$$u_{kl} = \left(\frac{\partial^{k+l}u}{\partial x^k \partial y^l}\right)_{x_0,y_0}$$

and

$$\nabla_*^6 = u_{60} - 15u_{42} + 15u_{24} - u_{06},$$
$$\nabla_*^8 = u_{80} - 14u_{62} + 14u_{28} - u_{08},$$
$$\nabla_*^{10} = u_{10,0} - 13u_{82} - 14u_{64} + 14u_{46} + 13u_{28} - u_{0,10}.$$

Expression $Q =$

$$a u_a + b \Sigma u_b + c \Sigma u_c + d \Sigma u_d +$$
$$+ \frac{h^2}{2!}\{A\nabla^2 u_a + B\Sigma\nabla^2 u_b + C\Sigma\nabla^2 u_c + D\Sigma\nabla^2 u_d\} +$$
$$+ \frac{h^4}{4!}\{\alpha\nabla^4 u_a + \beta\,\Sigma\nabla^4 u_b + \gamma\,\Sigma\nabla^4 u_c + \delta\,\Sigma\nabla^4 u_d\}.$$

$Q = u_a[\; 6\;\{b + c + d\} + a]$

$+\dfrac{h^2}{2!}\nabla^2 u_a[\; 3\;\{b + 4c + 3d\} + 6\;\{B + C + D\} + A]$

$+\dfrac{h^4}{4!}\nabla^4 u_a\left[\;\dfrac{9}{4}\;\{b + 16c + 9d\} + \binom{4}{2}\,3\;\{B + 4C + 3D\} + 6\;\{\beta + \gamma + \delta\} + \alpha\right]$

$+\dfrac{h^6}{6!}\nabla^6 u_a\left[\;\dfrac{15}{8}\;\{b + 64c + 27d\} + \binom{6}{2}\,\dfrac{9}{4}\;\{B + 16C + 9D\} + \binom{6}{4}\,3\;\{\beta + 4\gamma + 3\delta\}\right]$

$+\dfrac{h^6}{6!}\nabla_*^6 u_a\left[\;\dfrac{3}{16}\;\{b + 64c - 27d\}\right]$

$+\dfrac{h^8}{8!}\nabla^8 u_a\left[\;\dfrac{105}{64}\;\{b + 256c + 81d\} + \binom{8}{2}\,\dfrac{15}{8}\;\{B + 64C + 27D\} + \binom{8}{4}\,\dfrac{9}{4}\;\{\beta + 16\gamma + 9\delta\}\right]$

$+\dfrac{h^8}{8!}\nabla_*^8 u_a\left[\;\dfrac{3}{8}\;\{b + 256c - 81d\} + \binom{8}{2}\,\dfrac{3}{16}\;\{B + 64C - 27D\}\right]$

$+\dfrac{h^{10}}{10!}\nabla^{10} u_a\left[\;\dfrac{189}{128}\;\{b + 1024c + 243d\} + \binom{10}{2}\,\dfrac{105}{64}\;\{B + 256C + 81D\} + \binom{10}{4}\,\dfrac{15}{8}\;\{\beta + 64\gamma + 27\delta\}\right]$

$+\dfrac{h^{10}}{10!}\nabla_*^{10} u_a\left[\;\dfrac{135}{256}\;\{b + 1024c - 243d\} + \binom{10}{2}\,\dfrac{3}{8}\;\{B + 256C - 81D\} + \binom{10}{4}\,\dfrac{3}{16}\;\{\beta + 64\gamma - 27\delta\}\right]$

$+$ terms of the 12th and higher orders.

Table X. *Taylor expansion of a general finite expression involving the differential operators* ∇^2 *and* ∇^4 *for a cubical mesh*
[from J. ALBRECHT: Z. Angew. Math. Mech. **33**, 41—48 (1953)]

In the following Taylor expansion all derivatives are evaluated at the point $m = (x_0, y_0, z_0)$; $h =$ mesh width; $u_m = u(x_0, y_0, z_0)$; $\sum u_s$, $\sum u_k$, $\sum u_e$ denote the sums over the u values at all mesh points whose distances from m are h, $\sqrt{2}\,h$, $\sqrt{3}\,h$, respectively; a corresponding notation is used for the values of $\nabla^2 u$ and $\nabla^4 u$; further

$$\nabla^{2n} = \left(\frac{\partial^2}{\partial x^2} + \frac{\partial^2}{\partial y^2} + \frac{\partial^2}{\partial z^2}\right)^n; \quad u_{jkl} = \frac{\partial^{j+k+l}u}{\partial x^j \partial y^k \partial z^l};$$

$$\nabla_3^{2k} u = u_{kk0} + u_{k0k} + u_{0kk} \quad \text{for } k > 0.$$

Expression $Q = m\,u_m + s\sum u_s \;+ k\sum u_k \;+ e\sum u_e$
$$\qquad + \frac{h^2}{2!}\{M\,\nabla^2 u_m + S\sum\nabla^2 u_s + K\sum\nabla^2 u_k + E\sum\nabla^2 u_e\}$$
$$\qquad + \frac{h^4}{4!}\{\mu\,\nabla^4 u_m + \sigma\sum\nabla^4 u_s + \varkappa\sum\nabla^4 u_k + \varepsilon\sum\nabla^4 u_e\}.$$

$Q =$

$\qquad u_m$ $[[\ 6s +\ 12k +\ 8e\ \}]$

$+\dfrac{h^2}{2!}\ \nabla^2 u_m$ $[\{\ 2s +\ 8k +\ 8e\} + \{\ \ 6S +\ 12K +\ 8E + M\}]$

$+\dfrac{h^4}{4!}\ \nabla^4 u_m$ $[\{\ 2s +\ 8k +\ 8e\} + \binom{4}{2}\{\ 2S +\ 8K +\ 8E\} + \{\ \ 6\sigma +\ 12\varkappa +\ 8\varepsilon + \mu\}]$

$+\dfrac{h^4}{4!}\ \nabla_3^4 u_m$ $[\{-\ 4s +\ 8k +\ 32e\}]$

$+\dfrac{h^6}{6!}\ \nabla^6 u_m$ $[\{\ 2s +\ 8k +\ 8e\} + \binom{6}{2}\{\ 2S +\ 8K +\ 8E\} + \binom{6}{4}\{\ 2\sigma +\ 8\varkappa +\ 8\varepsilon\}]$

$+\dfrac{h^6}{6!}\ \nabla^2\nabla_3^4 u_m$ $[\{-\ 6s +\ 36k +\ 96e\} + \binom{6}{2}\{-4S +\ 8K +\ 32E\}]$

$+\dfrac{h^6}{6!}\ u_{222m}$ $[\{\ 6s -\ 156k +\ 384e\}]$

$+\dfrac{h^8}{8!}\ \nabla^8 u_m$ $[\{\ 2s +\ 8k +\ 8e\} + \binom{8}{2}\{\ 2S +\ 8K +\ 8E\} + \binom{8}{4}\{\ 2\sigma +\ 8\varkappa +\ 8\varepsilon\}]$

$+\dfrac{h^8}{8!}\ \nabla^4\nabla_3^4 u_m$ $[\{-\ 8s +\ 80k +\ 192e\} + \binom{8}{2}\{-6S +\ 36K +\ 96E\} + \binom{8}{4}\{-4\sigma +\ 8\varkappa +\ 32\varepsilon\}]$

$+\dfrac{h^8}{8!}\ \nabla^2 u_{222m}$ $[\{\ 16s -\ 496k +\ 2304e\} + \binom{8}{2}\{\ 6S -\ 156K +\ 384E\}]$

$+\dfrac{h^8}{8!}\ \nabla_3^8 u_m$ $[\{\ 4s +\ 72k +\ 128e\}]$

+ terms of the 10th and higher orders.

Author Index

Subject Index

The manufacturer's authorised representative in the EU is Springer
Nature Customer Service Centre GmbH, Europaplatz 3, 69115 Heidelberg,
Germany. If you have any concerns regarding our products, please
contact ProductSafety@springernature.com

Printed and bound by CPI Group (UK) Ltd, Croydon, CR0 4YY
28/04/2026
02098508-0005